A MANUAL OF ACAROLOGY

A MANUAL OF ACAROLOGY

SECOND EDITION

G. W. Krantz
Department of Entomology
Oregon State University

1978
Oregon State University Book Stores, Inc.
Corvallis
Second Printing (emended 1986)

Published by
OREGON STATE UNIVERSITY BOOK STORES, INC.
Oregon State University
Corvallis, Oregon 97331

First published 1970
SECOND EDITION 1978
Printed in the United States of America

Library of Congress Catalog Card No. 78-56128
ISBN: 0-88246-064-1

PREFACE TO SECOND EDITION

When I first set about to revise **A Manual of Acarology** in 1974, my intention was merely to update the classification and bibliography of the 1970 edition, and perhaps to include a limited amount of new information in some of the introductory chapters. It soon became apparent, however, that an effective revision would entail a major reworking of the text and the addition of many new illustrations. The reader will note, therefore, that while the format of the **Manual** has remained essentially unchanged, the second edition is larger and more detailed than was its predecessor. The **Manual** remains a primarily systematic text, providing only basic information on acarine biology, morphology, behavior, rearing, collection, preparation, and preservation. References to more detailed works on these subjects are listed at the conclusion of appropriate chapters. Taxonomic diagnoses are presented for each suborder and superfamily, along with selected bibliographies for each major taxon. Feeding habits, importance, and distribution of member families are discussed under each superfamilial heading. Keys are included for currently recognized families of the Acari except in the suborder Oribatida, where only the superfamilial categories are distinguished. The Ixodida, or ticks, are keyed to genus.

The **Manual** contains over 800 illustrations and diagrams assembled in 163 plates and 47 text figures. The index comprises over 2000 entries, including citations to all supraspecific taxa noted in the text and figures.

It is a pleasure for me to acknowledge the assistance and encouragement of my colleagues during the preparation of the **Manual**. I am indebted to B.D. Ainscough, W.T. Atyeo, Y. Coineau and J.B. Kethley for providing essential unpublished information in their areas of specialty. Discussions with W.T. Atyeo, E.W. Baker, P.E. Hunter, E.E. Lindquist, J.H. Oliver, I.M. Smith, and P.-H Vercammen-Grandjean were likewise of great value to me. I wish also to thank the following people for their kindness in providing suggestions and corrections for inclusion in the second edition: A. Aeschlimann, C. Athias-Henriot, C. Bader, Y.S. Balashov, K.K. Bohnsack, J.M. Brennan, J.E.M.H. van Bronswijk, Y. Coineau, D.A. Crossley, H.A. Denmark, A. Edler, W.R. Enns, M.H. Farrier, N.J. Fashing, Z. Feider, C.H.W. Flechtmann, R.C. Funk, J.E. George, U. Gerson, E.A. Hicks, W. Hirschmann, R.K. Hodgson, R.W. Husband, K.H. Hyatt, K.E. Hyland, D.E. Johnston, H.L. Keegan, H.H. Keifer, D.C. Lee, F. Lukoschus, M. Luxton, B. McDaniel, J.C. Moser, W.W. Moss, M. Nadchatram, R.A. Norton, G.E. Oldfield, G. Rack, F.J. Radovsky, R.M. Reeves, K. Samšinák, H. Schubart, R. Schuster, R. Traub, J. Travé, A.E. Treat, D.M. Tuttle, J.P. Webb, G.W. Wharton, N.A. Wilson, J.P. Woodring, and C.E. Yunker. My sincere thanks also to the students of the 1977 classes of the Laboratory of Acarology, Ohio State University, who tested the keys included in this volume and offered valuable suggestions for their improvement.

Special thanks to my wife, Vida, who typed the original manuscript, collaborated on the preparation of the index, and proofread the final galley. Her many contributions were instrumental in bringing the **Manual** to completion.

G.W. Krantz
Corvallis, Oregon
November 1, 1977

TABLE OF CONTENTS

Page

Page

Page

"If it would take a cannon-ball 3 1/3 seconds to travel four miles, and 3 3/8 seconds to travel the next four, and 3 5/8 to travel the next four, and if its rate of progress continued to diminish in the same ratio, how long would it take to go fifteen hundred million miles?"

—"Arithmeticus", Virginia, Nevada

I don't know.

—Mark Twain

From "Answers to Correspondents" in **Sketches Old and New**,
reprinted by permission of Harper & Row

I. INTRODUCTION

Few animal groups illustrate the diversity in form, habitat and behavior seen in the subclass Acari—the mites and ticks. Unlike other assemblages in the class Arachnida, many acarine groups have evolved far beyond primitive predation. Some are exclusively phytophagous, while others have developed complex parasitic relationships with both vertebrate and invertebrate animals. Many are considered beneficial to man in that they prey on undesirable arthropods. Others aid in the breakdown of forest litter and implement nutrient recycling. However, a number of acarine groups are injurious to crops and to livestock, both because of their feeding activities and because of their capacity to carry and transmit a variety of disease organisms to their plant or animal hosts.

Acarines may be found in virtually any environment including severe desert and tundra situations, mountain tops, deep soil strata, subterranean caves, hot springs, and ocean floors. In short, mites have colonized almost every terrestrial, marine and fresh water habitat known to man.

The high degree of habitat diversity illustrated by the mites and ticks is no more remarkable than is their range of form, size, structure and behavior. On the basis of these variations, more than 30,000 species of Acari have been described to date. It has been estimated that up to a half million more await discovery. Such an estimate seems well within reason when one considers that a random examination of almost any organic substrate yields mite specimens representing undescribed species. Even a cursory examination of litter or soil samples from tropical areas such as equatorial Africa or Amazonia serves to illustrate the enormous species diversity in the Acari and gives an indication of how few species are actually described. Recent explorations into more exotic realms such as ultradeep soil strata, dermal and subdermal tissues of vertebrates, and the byssal ocean depths, have revealed entire mite faunas whose existence had previously been unsuspected. Thus, our concepts of acarine systematics are based on little more than a fragmentary understanding of the fauna.

The genesis of the discipline of acarology may be traced to 18th century Europe, but awareness of the Acari existed well before that time. A reference to "tick fever" was found on an Egyptian papyrus scroll dated 1550 B.C., and Homer mentioned the occurrence of ticks on Ulysses' dog in 850 B.C. Some 500 years later, Aristotle discussed a mite parasite of locusts (probably *Eutrombidium*) in *De Animalibus Historia Libri*. Other early references to the Acari appear in the writings of Hippocrates, Plutarch, Aristophanes and Pliny. Mites and ticks often were referred to as "lice", "beesties", or "little insects" during the Dark Ages and the Renaissance. Use of the terms "Akari" and "mite" originated about 1650.

In 1735, Linnaeus used the generic name *Acarus* in the first edition of the *Systema Naturae* for which he named the type, *A. siro*, in 1758. The tenth edition of the *Systema* included fewer than 30 mite species, all of which were relegated to the genus *Acarus*. During the 100 years that followed, major contributions to acarine systematics were made by Latreille, Leach, Dugès, De Geer, and C.L. Koch. Michael (1884) summarizes these and related pioneer works.

The emergence of acarology as a modern science occurred in Europe during the late 19th and early 20th centuries with the historic contributions of Kramer, Mégnin, Canestrini, Berlese, Michael, Reuter, Trägårdh, Oudemans, Vitzthum and Grandjean. Their work provided the basis for virtually all acarological research until the end of World War II.

Among those who participated in the post-war awakening of acarology were E.W. Baker and G.W. Wharton. Realizing the need for an up-to-date basic acarological text in English, Baker and Wharton published **An Introduction to Acarology** in 1952. The **Introduction** served as the standard systematics text in acarology until 1958, when Baker and colleagues published the **Guide to the Families of Mites**. The **Guide** included many families not treated in the **Introduction**, and contained several major changes in higher categories.

A number of general reference works in acarology have been published subsequent to the appearance of the **Guide to the Families of Mites** (Hughes 1959, Evans et al. 1961, Hirschmann 1966, Krantz 1970, van der Hammen 1972, Flechtmann 1976), as have many important research papers and books dealing with particular taxonomic groups, or with specific problems in mite biology, physiology, behavior, and systematics. Many of these works will be referred to in the sections to follow.

Useful References

Baker, E.W. and G.W. Wharton (1952). An Introduction to Acarology. MacMillan Co., New York: 465 pp. + xiii.

Baker, E.W., J.H. Camin, F. Cunliffe, T.A. Woolley and C.E. Yunker (1958). Guide to the Families of Mites. Institute of Acarology Contr. No. **3**:242 pp. + ix.

Berlese, A. (1882-1903). Acari, Myriapoda, et Scorpiones hucusque in Italia reperta; fasc. 1-101. Padova.

Berlese, A. (1899). Gli Acari agrarii. Riv. Pat. veg. **7**:312-344.

Brennan, J.M. and E.K. Jones (eds.) (1968). A Directory of Acarologists of the World, 9th edition. Rocky Mo. Lab., Hamilton, Montana: 131 pp.

Canestrini, G. (1891). Abbozzo del sistema Acarologico. Atti Istit. ven. **38**:699-725.

DeGeer, C. (1778). Mémoires pour servir a l'histoire des Insectes. Stockholm.

Dugès, A. (1839). Recherches sur l'ordre des Acariens. Ann. Sci. Nat. Zool. **1**(2):18-63.

Evans, G.O., J.G. Sheals and D. MacFarlane (1961). The Terrestrial Acari of the British Isles. Vol. 1. Introduction and Biology. British Museum (Natural History), London: 219 pp.

Flechtmann, C.H.W. (1976). Elementos de Acarologia. Livr. Nobel S.A., São Paulo: 344 pp.

Hammen, L. van der (1972). Mijten-Acarida. Algemene inleiding in de acarologie. Spinachtigen-Arachnidea IV. Wetensch. Meded. Konink. Nederlandse Natuurhist. Ver. 91:71 pp.

Hirschmann, W. (1966). Milben (Acari). Einführung in die Kleinlebewelt. Kosmos-Verlag, Stuttgart: 76 pp.

Hughes, T.E. (1959). Mites or the Acari. University of London, Athlone Press: 225 pp. + viii.

Koch, C.L. (1842). Uebersicht des Arachnidensystems. Nürnberg: Fasc. 3.

Kramer, P. (1877). Grundzüge zur Systematik der Milben. Arch. für Naturg. **2**:215-247.

Krantz, G.W. (1970). A Manual of Acarology, First Ed. Oregon State Univ. Bookstores, Corvallis: 335 pp.

Latreille, P.A. (1806-1809). Genera Crustaceorum et Insectorum. Paris: 4 vols.

Leach, W.E. (1815). A tabular view of the external characters of four classes of animals which Linné arranged under Insecta. Trans. Linn. Soc. London **11**(2):306-400.

Mégnin, P. (1876). Mémoire sur l'organisation et la distribution zoologique des Acariens de la famille des Gamasides. Paris.

Michael, A.D. (1884-1888). British Oribatidae, Vols. I-II. Ray Society, London: 657 pp. + xvii.

Oudemans, A.C. (1906). Nieuwe classificatie der Acari. Ent. Ber. **2**:43-46.

Oudemans, A.C. (1926). Kritisch Historisch Oversicht der Acarologie **1**:500 pp. + vii (850 V.C.-1758). E.J. Brill, Leiden.

Oudemans, A.C. (1929). Kritisch Historisch Oversicht der Acarologie **2**:1097 pp. + xvii (1759-1804). E.J. Brill, Leiden.

Oudemans, A.C. (1936-1937). Kritisch Historisch Oversicht der Acarologie **3**:3379 pp. + ci (1805-1850). E.J. Brill, Leiden.

Radford, C.D. (1950). Systematic check list of mite genera and type species. Internat. Union Biol. Sci. Ser. C (Ent.) **1**:232 pp.

Reuter, E. (1909). Zur Morphologie und Ontogenie der Acariden. Acta Soc. Sci. Fenn. **36**(4):1-288.

Sasa, M. (1965). Mites.. An introduction to classification, bionomics and control of Acarina. Univ. Tokyo Press:494 pp. (in Japanese).

Vitzthum, H.G. (1929). Acari. Die Tierwelt Mitteleuropas **3**(7):1-112.

Vitzthum, H.G. (1931). Acari. Kükenthals Handbuch der Zoologie **3**(2):1-160.

Vitzthum, H.G. (1940-42). Acarina. Bronn's Klassen und Ordnungen des Tierreichs 5(4), Book **5**:1-1011.

II. SYSTEMATIC POSITION OF THE ACARI

The Phylum Arthropoda includes a myriad of forms which share the characteristics of jointed legs and a chitinous exoskeleton. Within the Arthropoda is a group of animals which, unlike the insects or myriapods, have neither antennae nor mandibles. These are the Chelicerata, of which the Class Arachnida makes up the largest part.

The Arachnida comprises those chelicerates which may possess simple eyes and which are primarily terrestrial. It includes such diverse forms as scorpions, spiders, vinegaroons and the long-legged harvestmen so common to most temperate and tropical regions. Of the eleven extant divisions of Arachnida listed by Savory (1964), all but two are completely predaceous in habit, with representatives often displaying a variety of morphological characteristics well suited to a predatory existence. The two pairs of mouthparts—the *chelicerae* and *pedipalpi* (Fig. 2, p. 7) are adapted for grasping, piercing, sucking or tearing. Specialized pedipalpal or cheliceral venom glands are present in some of these orders, while others have terminal venomous stings (Plate 1, p. 6) or acid glands.

Although generally predaceous, members of the subclass Opiliones (Plate 1-5, p. 6) have been observed to feed on dead organic matter. They, along with the Acari, comprise the exceptions to the rule of total predation in the Arachnida.

The mites and ticks differ from most arachnids in that somatic segmentation generally is inconspicuous or absent.[1] Abdominal segmentation is a primary attribute in all of the other subclasses except the Araneae—the spiders. Thus the mites, ticks and spiders may easily be separated from other arachnids on this basis.

---

[1] Limited secondary development of somatic segmentation may be seen in certain of the Heterostigmae (Plate 74, p. 329). Evidence of primary segmentation occurs in the Suborder Opilioacarida (Plate 8, p. 108).

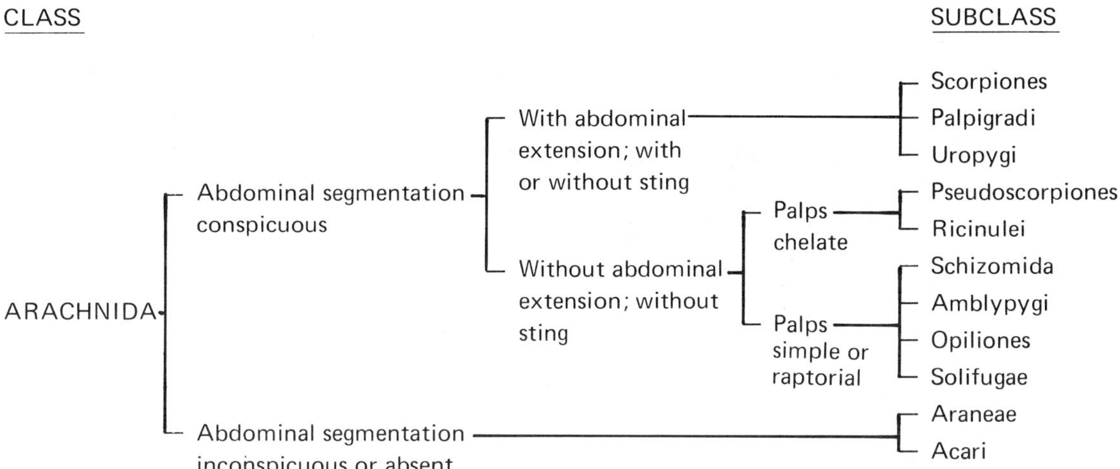

Fig. 1. Key to arachnid subclasses.

PLATE 1

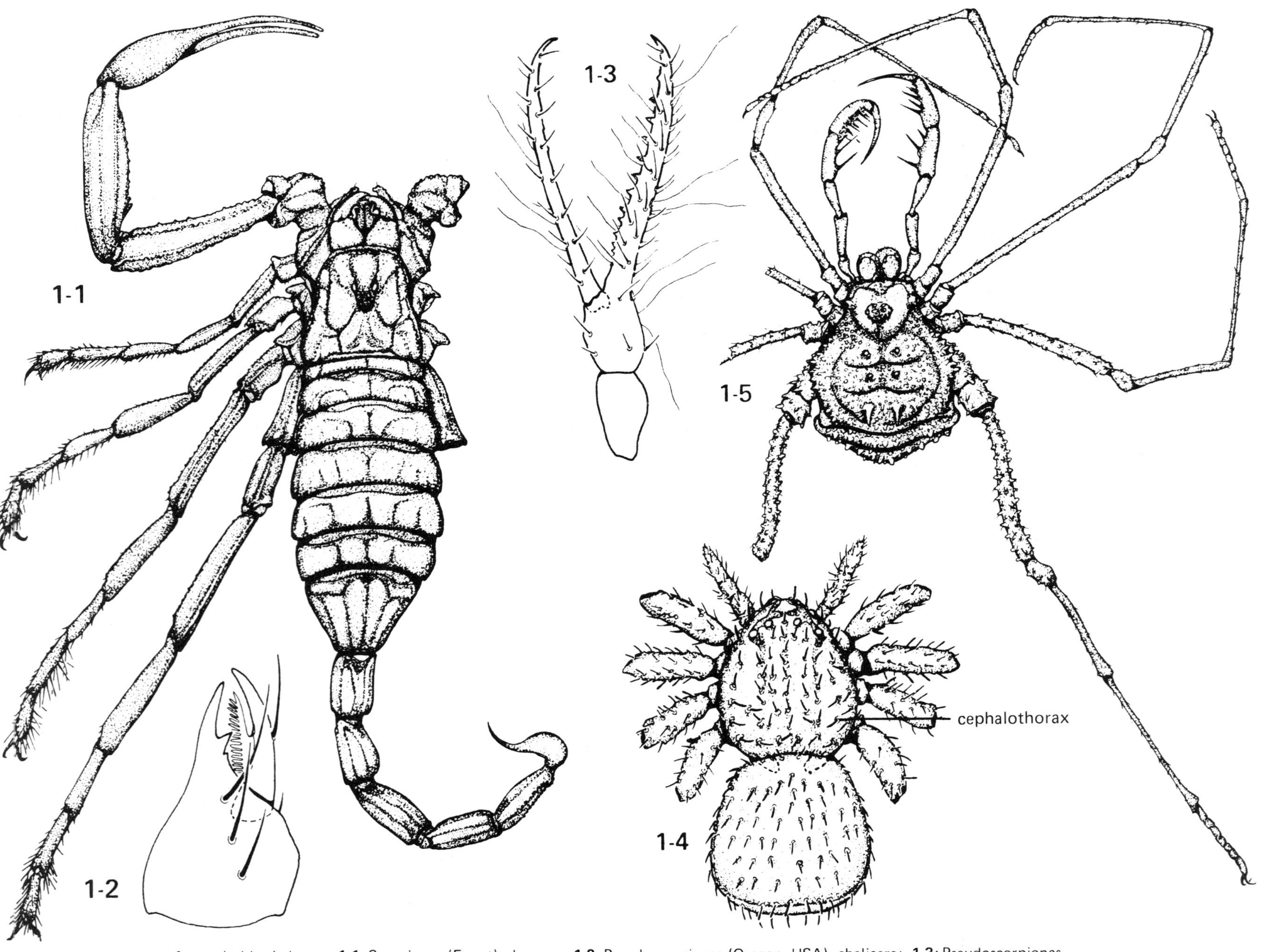

1; arachnid subclasses. **1-1;** Scorpiones (Egypt), dorsum: **1-2;** Pseudoscorpiones (Oregon, USA), chelicera: **1-3;** Pseudoscorpiones, palp: **1-4;** Araneae (Mexico), dorsum with leg extremities removed: **1-5;** Opiliones (Brazil), dorsum

Mites and ticks may be distinguished from spiders through the use of the following key:

1. Mouthparts inserted anteriorly on the *cephalothorax* (Plate 1-4, p. 6) which is composed of fused head and thoracic segments, and which is connected to the abdominal portion by a narrow *pedicel;* legs borne on cephalothorax . . . . . .Subclass ARANEAE

2. Mouthparts contained in a discrete anterior *gnathosoma* (Fig. 2); portion of the body on which the legs are inserted (the *podosoma*) broadly joined to the portion of the body behind the legs (the *opisthosoma*) to form the *idiosoma* . . . . . Subclass ACARI

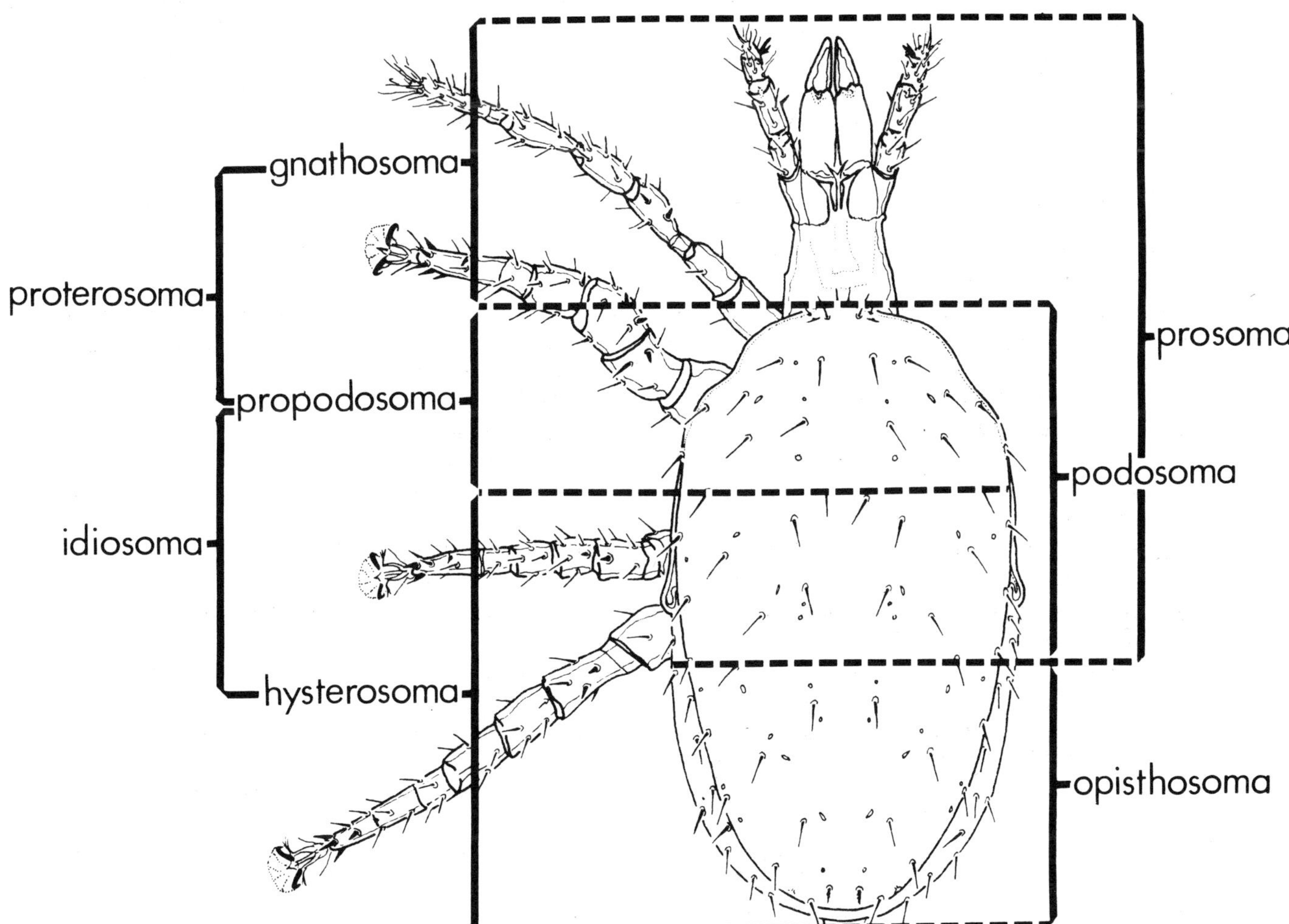

Fig. 2. Dorsum of *Macrocheles merdarius* Berlese (subclass Acari) illustrating major body divisions.

While lack of comprehensive fossil evidence does not permit a phylogenetic evaluation of the Arachnida in general, or of the Acari in particular, indications are that the arachnids were already well represented on land by the mid-Palaeozoic era (Main 1972) (the earliest known mite fossil, *Protacarus crani* Hirst, dates from this time). The fossil record also indicates that a major adaptive breakthrough occurred in the Acari during the late Mesozoic and early Cenozoic. Acarine radiation during that time reflected development of a wide range of non-predatory adaptations, which allowed considerable scope in niche exploitation and the development of a high degree of morphological diversity. Lindquist (1975) suggests that acarine radiation may have been stimulated by an "evolutionary synergism" between radiating biota during the late Mesozoic. This was a period of excessive development of angiosperm plants and a concomitant speciation "explosion" of the insects, a group with which the Acari have close ecological ties. A similar but less obvious reciprocal effect appears to have occurred between mites and evolving warm-blooded vertebrate animals.

Sharov (1966) feels that the Acari evolved from pedipalpid ancestral stock during the middle Devonian. Grandjean (1935), Zakhvatkin (1952), van der Hammen (1970, 1971), Athias-Henriot (1975) and others believe the Acari to be of diphyletic origin, with the Acariformes and Parasitiformes (see page 99) branching as separate entities from some primitive arachnid stock. Literature on phylogeny of the Acari has been reviewed by Woolley (1961).

Useful References

Athias-Henriot, C. (1975). The idiosomatic euneotaxy and epineotaxy in gamasids (Arachnida, Parasitiformes). Z. Zool. Syst. Evolut.-forsch. **13** (2):97-109.

Bekker, E.G. (1959). Concerning the Acarina as a natural grouping. Trud. Inst. Morf. zhiv. **27**:151-162.

Bekker, E.G. (1960). Systematics and comparative anatomy in solving the problem of the phylogeny of ticks and mites (Acarina). Rept. No. 1. Critical discussion of the views of acarologists-taxonomists on the polyphylogeny of the order Acarina. Vest. Mosk. Univ. Ser. 6:Biol. **15**(4):13-20.

Carpenter, F.M., J.W. Folsom, E.O. Essig, A.C. Kinsey, C.T. Brues, M.W. Boesel and H.E. Ewing (1937). Insects and arachnids from Canadian amber. Univ. Toronto Studies, Geol. Ser. **40**:7-62.

Dubinin, V.B. (1959a). Chelicerate animals (subphylum Chelicerophа W. Dubinin nom. n.) and their systematic position. Zool. Zh. **38**(8):1163-1189.

Dubinin, V.B. (1959b). Phylogenesis of chelicerate animals of the subphylum Chelicerophа W. Dub. and affinity of Chelicerata with pycnogonides. Trud. Inst. Morf. zhiv. **27**:134-150.

Grandjean, F. (1935). Observations sur les Acariens. Bull. Mus. nat. Hist. natur. Paris, sér. 2, **7**:201-208.

Hammen, L. van der (1964). The relation between phylogeny and post-embryonic ontogeny in actinotrichid mites. Proc. 1st Int. Congr. Acarology, Fort Collins. Acarologia **6**(fasc. h.s.):85-90.

Hammen, L. van der (1968). Introduction generale a la classification, la terminologie morphologique, l'ontogenese et l'evolution des Acariens. Acarologia **10**(3):401-412.

Hammen, L. van der (1970). La phylogenèse des Opilioacarides, et leurs affinites avec les autres acariens. Acarologia **12**(3):465-473.

Hammen, L. van der (1971). Classification and phylogeny of mites. Proc. 3rd Int. Congr. Acarology, Prague: 275-282.

Hammen, L. van der (1972). A revised classification of the mites (Arachnidea, Acarida) with diagnoses, a key, and notes on phylogeny. Zool. Med. **47**:273-292.

Hirst, S. (1923). On some arachnid remains from the Old Red Sandstone (Rhynie Chert Bed, Aberdeenshire). Ann. Mag. Nat. Hist. **12**(9):455-474.

Lindquist, E.E. (1975). Associations between mites and other arthropods in forest floor habitats. Can. Ent. **107**:425-437.

Main, B.Y. (1972). Subphylum V: Chelicerata. **In** Textbook of Zoology, Invertebrates. 7th Edition. A.J. Marshall and W.D. Williams, eds. Macmillan, London:411-480.

Petrunkevitch, A. (1949). A study of paleozoic Arachnida. Trans. Connecticut Acad. Art. Sci. **37**:69-315.

Savory, T. (1964). Arachnida. Academic Press, London: 291 pp. + viii.

Sharov, A.G. (1966). Basic Arthropodan Stock with Special Reference to Insects. Pergamon Press, Oxford: 271 pp. + xii.

Snodgrass, R.E. (1938). Evolution of the Annelida, Onychophora and Arthropoda. Smithsonian Misc. Coll. **97**(6):159 pp.

Southcott, R.V. and R.T. Lange (1971). Acarine and other microfossils from the Maslin Eocene, South Australia. Rec. S. Austral. Mus. **16**(7):1-21.

Tiegs, O.W. and S.M. Manton (1958). The evolution of the Arthropoda. Biol. Rev. **33**:255-337.

Woolley, T.A. (1961). A review of the phylogeny of mites. Ann. Rev. Ent. **6**:263-284.

Zakhvatkin, A.A. (1952). Division of the Acarina into orders and their position in the system of the Chelicerata. Mag. Parasitol. Moscow **14**:5-46.

III. MORPHOLOGY AND FUNCTION

A. EXTERNAL

The exoskeletal envelope of the typical acarine begins its development as undifferentiated tissue covered by a thin layer of *cuticulin* and separated from the underlying epidermis by an extremely thin, poorly defined granular *Schmidt* layer (Fig. 3). As development proceeds, surface portions of the undifferentiated layer often become sclerotized to varying degrees through orthoquinone tanning. These portions, the *epicuticle* and *exocuticle*, may then be distinguished from the underlying laminated layer, the *endocuticle*, as discrete shields or plates (Belozerov 1960). The laying down of new cuticle is preceded by secretion of numerous granules in the dermal layer, which finally coalesce into microfibers. The fibers are then consolidated to form laminations as the cuticle matures (Brody 1970).

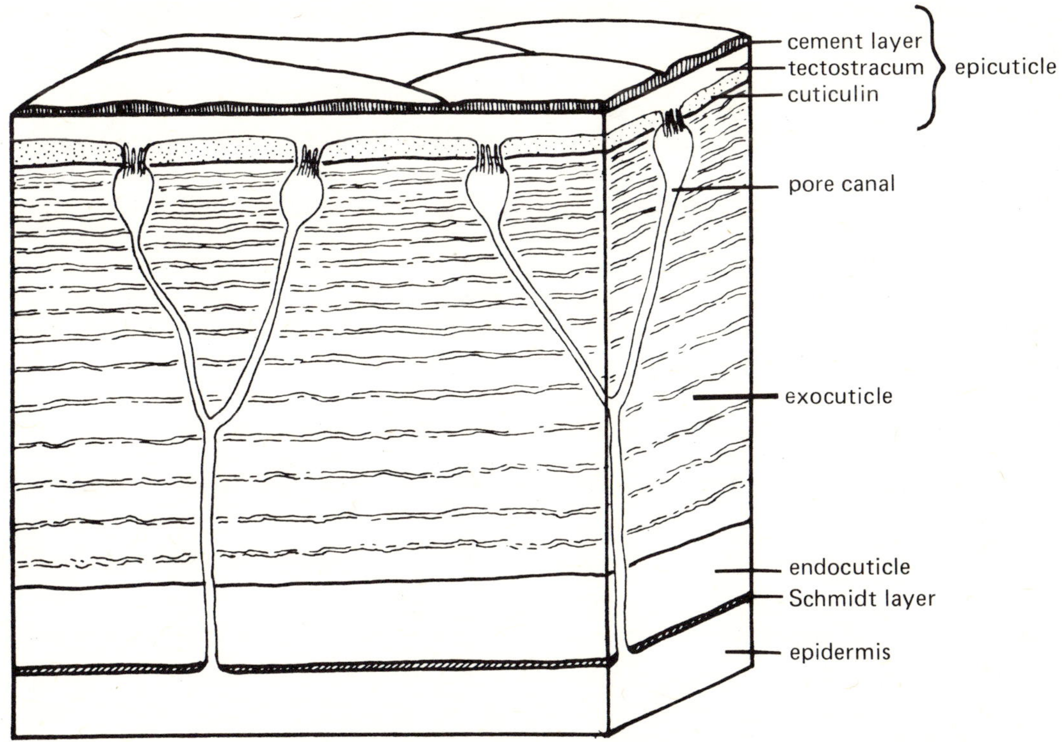

Fig. 3. Diagrammatic cross section of acarine cuticle.

The cuticulin surface layer may contain a profusion of micropores which are connected to *pore canals.* These canals originate in the epidermal cells underlying the Schmidt layer and pass through both the endo- and exocuticle (Wharton et al. 1968), branching terminally at the surface of the cuticulin layer. A possible function of the pore canals may be that of transporting an epidermal secretion to the cuticulin surface layer, where it forms a protective waxy coating referred to as the *tectostracum.* A thin overlying *cement layer* may also be laid down in this way. The tectostracal and cement layers may provide protection against excessive water loss from the body surface. Brody (1970) feels that the cement layer is of real significance in water balance maintenance in the immature instars of *Oppia coloradensis* Dolan (Oribatida). The underlying cuticle—or some portion of it—appears to be a major pathway for the uptake or sorption of water from unsaturated air (Wharton and Devine 1968, Devine and Wharton 1973).

In addition to micropores, various macropores occur both on the body and appendages of Acari. Like the micropores, some of these openings appear to have a secretory function while others are thought to be proprioceptors, mediating stimuli from within the animal itself. Athias-Henriot (1969a, 1969b) has developed a classification of cuticular pores in Gamasida (Parasitiformes) in which three categories are recognized:

1. *Poroidal*, which includes the lyriform pores or *lyrifissures* common both to the body and appendages (Plate 9-1, p. 112). Lyrifissures are considered to be proprioceptors.

2. *Setal*, which describes the insertions of tactile and chemosensory hairs.

3. *Glandular*, which includes a variety of cuticular openings consisting primitively of a gland cell, an excretory duct and an external opening. The sperm induction pore found in many Gamasida (page 24), the lateral dermal glands of Uropodina (Gamasida) (Woodring and Galbraith 1976), and the primary dorsal macropore system of the Parasitiformes in general, are examples of glandular openings. Cuticular glands are referred to as *crobylophores* by Athias-Henriot (1975).

The body of the typical acarine is composed of an anterior *gnathosoma* and a posterior *idiosoma*. They are separated by a *circumcapitular suture* which may or may not be obscured by secondary somatic development. The basic arachnid body plan is thought to comprise 13-21 somites, depending on stage of development and on individual interpretation. Coineau (1974) has presented a theoretical scheme of a hypothetical precursor of the actinedid family CAECULIDAE based in part on an earlier plan of Grandjean (1954). In this scheme, the archetypal acariform adult is considered to consist of 16 somites, plus a precheliceral segment (Plate 2-1, p. 13). The adult idiosoma comprises 14 segments, of which the first four are leg-bearing. An idiosomal segment is added terminally at each postlarval molt.

Primitive segmentation in the Parasitiformes is considerably more difficult to interpret than is that of the Acariformes (Evans et al. 1961). Zakhvatkin's (1952) scheme for the Gamasida, which includes 12 idiosomal segments plus a terminal telson, is one of several possible constructions.

Gnathosoma

The gnathosoma, or capitulum, resembles the head of the generalized arthropod only in that the mouthparts are appended to it. The brain lies in the idiosoma behind the gnathosoma rather than within it, and the ocelli (when present) are situated dorsally or dorsolaterally on the propodosoma. The gnathosoma, then, is little more than a tube through which food is carried to the esophagus. Despite its apparent simplicity, the gnathosoma represents a highly evolved and specialized body region when compared to the homologous parts of more primitive arachnids, particularly from the standpoint of mouthpart displacement (van der Hammen 1970a).

The roof of the gnathosomal tube is termed the *epistome* (tectum capituli of Evans and Till 1965), and the lateral walls are made up of the enlarged coxae of the *palpi* (Fig. 5, p. 14). The floor of the tube, the *subcapitulum* or basis capituli, represents a medial extension and coalescence of the ventral portions of the palpal coxae. The anterior coxal endites, or lateral lips, of the subcapitulum (Knülle 1959), along with associated anteroventral elements, comprise the *hypostome*. Lying above the buccal cavity are the paired *chelicerae* which generally are three-segmented (two-segmented in the suborder Ixodida), and may be retractible. The chelicerae, along with the palpi, are the primary organs of food acquisition.

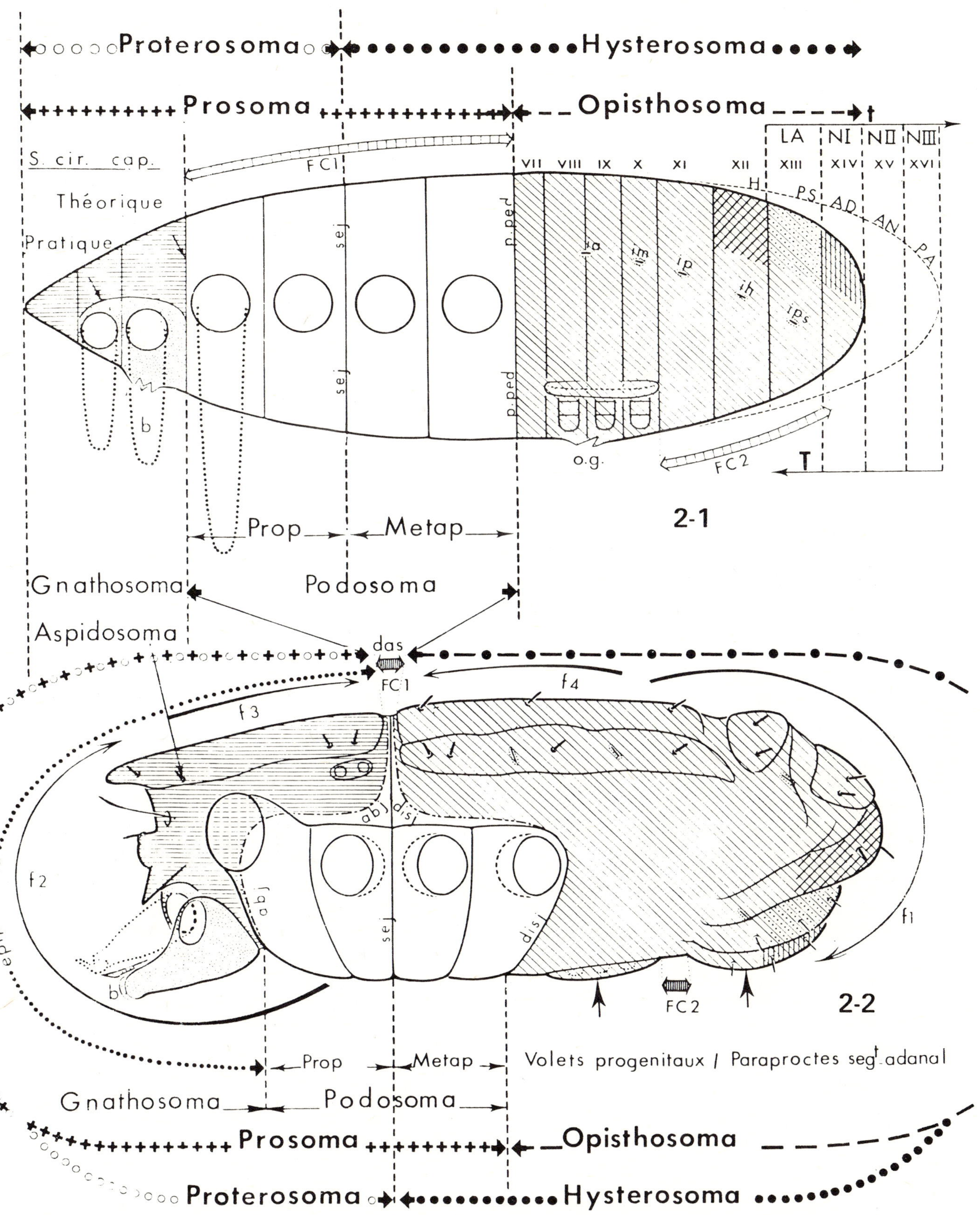

2; primitive segmentation and evolved modifications in the Acari (from Coineau 1974). **2-1**; primitive acarine condition: **2-2**; segmental fusions and migrations in a caeculid mite (CAECULIDAE, Actinedida)

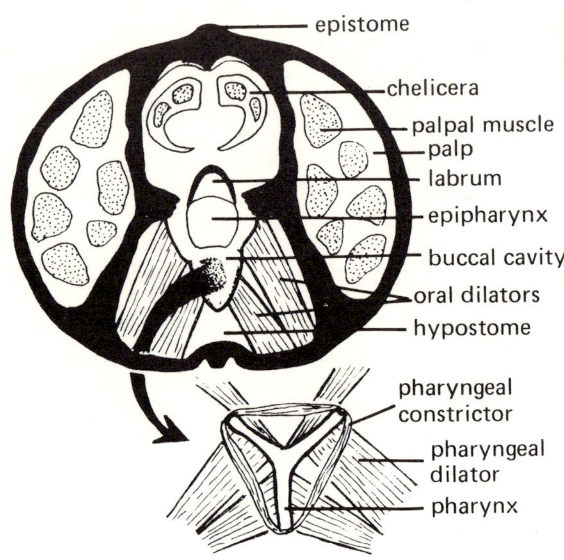

Fig. 4. Cross section of gnathosoma of *Ornithonyssus bacoti* (Hirst) at level of hypostomal setae 2-3. Pharynx (arrow) lies posterior to buccal cavity (after Hughes 1959).

Typically, the palpi are simple sensory appendages equipped with a variety of terminal chemosensory and thigmotactic hairs which aid the acarine in locating its food. However, the palpi may be modified as raptorial structures in certain predatory groups (Plate 3-1, p. 15), or as organs for anchoring parasites to their hosts. The simple palpi of predatory Gamasida (Plate 3-3) are utilized in manipulating the prey so as to bring it closer to the buccal area (Wernz and Krantz 1976).

The number of free palpal segments varies from one or two (most Acaridida, some Actinedida) to five (many Gamasida and Oribatida). Variations in palpal chaetotaxy in the Gamasida may be useful in determining relationships between higher taxa, and in separating immature instars of single species (Evans 1963b).

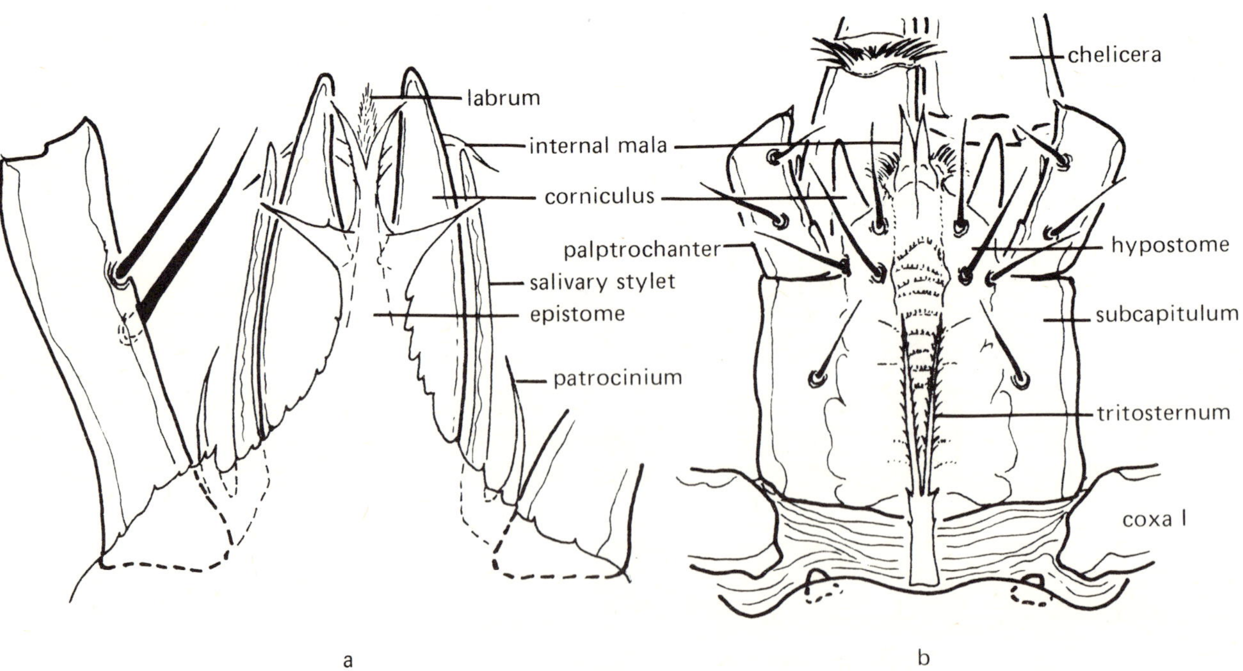

Fig. 5. Gnathosoma of gamasid type: **a**; *Macrocheles mycotrupetes* Krantz and Mellott, dorsal aspect, female (chelicerae removed): **b**; *Parasitus coleoptratorum* (L.), ventral aspect (deutonymph).

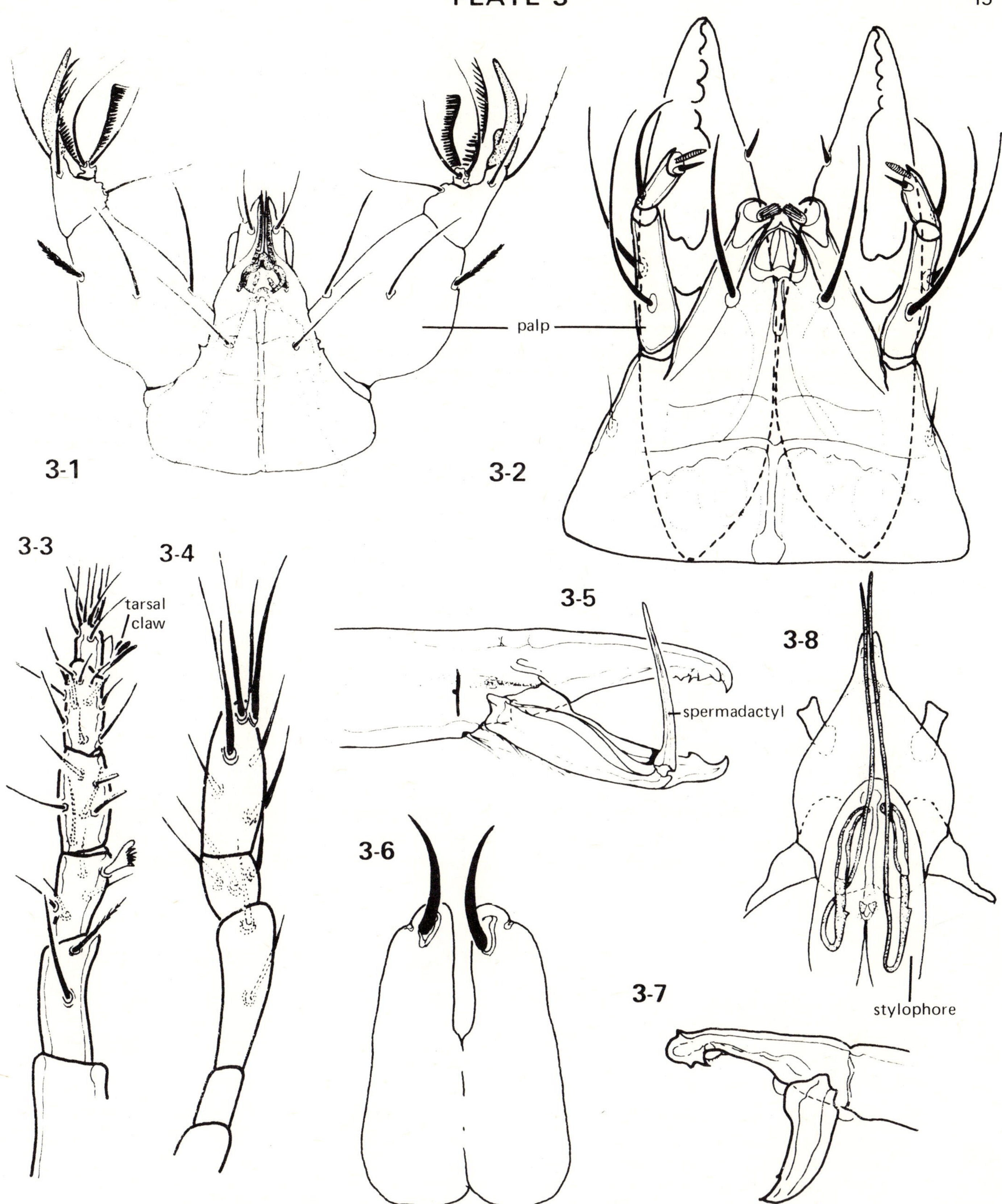

3-1; *Cheletomorpha lepidopterorum* (Shaw) (Actinedida, CHEYLETIDAE), dorsum of gnathosoma: **3-2**; *Glycyphagus* sp. (Acaridida, GLYCYPHAGIDAE), venter of gnathosoma: **3-3**; *Veigaia* sp. (Gamasida, VEIGAIIDAE), palp: **3-4**; *Rhagidia* sp. (Actinedida, RHAGIDIIDAE), palp: **3-5**; *Parholaspella spatulata* Krantz (Gamasida, PARHOLASPIDIDAE), chelicera of male: **3-6**; *Opsereynetes* sp. (Actinedida, EREYNETIDAE), chelicerae: **3-7**; *Phaulodinychus mitis* (Leonardi) (Gamasida, UROPODIDAE), chelicera of female: **3-8**; *Brevipalpus cardinalis* (Banks) (Actinedida, TENUIPALPIDAE), dorsum of gnathosoma

The chelicerae may differ considerably between taxa, but are generally adapted for piercing, sucking, or chewing. Primitively, the terminal third cheliceral segment is a movable digit which opposes the fixed distal portion of the second segment (Plate 3-8, p. 15). These opposed chelae may be edentate or variously toothed for grasping or grinding. Modifications in the basic cheliceral plan have occurred in many acarine groups with the acquisition of specialized feeding habits. For example, Karg (1961) has noted that, amongst free-living Gamasida, there is a correlation between cheliceral structure and the type of food taken; i.e. gamasids with short cheliceral digits and long teeth feed primarily on nematodes, while those with longer, slender digits and retrorse teeth feed on other mites or on Collembola. Cheliceral modification in phytophagous or parasitic acarines, or in highly specialized predatory groups, often involves fusion of the fixed digits into a *stylophore* and development of the movable digits into stylets or spines for piercing plant or animal tissues (Plate 3-7). In some acarine groups, the movable digit of the male is modified to effect sperm transfer to the female (Plate 3-5). Other cheliceral modifications are referred to in the sections to follow.

The buccal cavity lies below the chelicerae and opens into the pharynx. It is bordered ventrally by the hypostome and dorsally by the *labrum* (Fig. 4, p. 14), an extension of the dorsal pharyngeal wall. A discrete *epipharynx* may also be present, lying between the labrum and the pharynx. The pharynx itself serves as a suction pump for ingested food materials (Hughes 1959). It is controlled by several sets of constrictor and dilator muscles which, along with those muscles which control the movement of the palpi and chelicerae, virtually fill the gnathosomal cavity (Mitchell 1962). The labrum-epipharynx may function as a pre-pharyngeal valve in the Ixodida and in various Gamasida, closing off the buccal cavity when the pharyngeal muscles constrict and preventing loss of food (Hughes 1959, Sonenshine and Gregson 1970).

Salivary glands may be present in the anterior portion of the idiosoma, opening through paired ducts into the buccal cavity (the *salivarium* of Sonenshine and Gregson 1970), or through styli into the buccal area. These glands supply various enzymes which may not only aid in pre-oral digestion of food materials (Moss 1962, Roshdy 1972), but which may "cement" the acarine to its host (Chinery 1973). Other idiosomal glands open into the mouth area, including some which produce silk and others which may secrete a stringy exudate for securing prey (Alberti 1973). While it is not a gnathosomal appendage, the medioventral *tritosternum* (Fig. 5b, p. 14) of certain predatory Gamasida has been found to function in concert with ventral gnathosomal elements as a fluid transport mechanism, directing overflow prey fluids to the prebuccal region (Wernz and Krantz 1976).

Idiosoma

The acarine idiosoma assumes functions parallel to those of the abdomen, thorax, and portions of the head of insects. It may be protected by sclerotized shields or it may be soft and virtually without sclerotization. The great diversity in idiosomal shape and ornamentation is evident in the illustrations which accompany the keys to families.

Although the idiosoma is considered to be unsegmented, various grooves or sutures may be seen, especially in those groups where extensive scleritic fusion has not occurred. These sutures delimit discrete idiosomal regions which are recognizable throughout the Acari. Several systems of idiosomal nomenclature have been devised, all of which have been based primarily on those of Reuter (1909) and Oudemans (1911). Grandjean's (1969) scheme for the Oribatida is highly comprehensive and applies equally well to many other acarine groups as well. The Grandjean system is utilized where applicable in this text.

The idiosoma includes an anterior *propodosoma* and a posterior *hysterosoma* which may or may not be separated from each other by a *sejugal furrow* (Plate 2-1, p. 13; *sej*.). The anterior two pairs of legs are inserted ventrally on the propodosoma, while legs III-IV are located on the adjacent portion of the hysterosoma. The latter zone is the *metapodosoma* which, along with the propodosoma, constitutes the leg-bearing portion of the idiosoma, or the *podosoma*. A *postpedal furrow* (Plate 2-1, *p. ped*.) may separate the podosoma from the *opisthosoma*, which is that portion of the hysterosoma behind legs IV. The terms *opisthonotum* and *opisthogaster* are useful in describing dorsal or ventral shield attributes of the opisthosoma. The propodosoma, together with the gnathosoma, comprises the *proterosoma.*

Various groups in the Oribatida and Actinedida may undergo radical modification in body architecture, with various somatic regions developing to the point where they obscure primitive divisions. For example, movement of the mouthparts to a prognathous position from the primitive hypognathous state (Plate 2-1, p. 13) may involve a shift of dorsal gnathosomal entities so as to obscure the circumcapitular suture and the dorsum of the propodosoma. The resulting pronotal extension is the *aspidosoma*, and the suture or furrow which delineates it ventrally and anteroventrally is the *abjugal furrow* (Plate 2-2, *abj*). The gnathosoma and the aspidosomal extension are referred to as the *epiprosoma*. The opisthosomal tergites may expand anteriorly and posteriorly, obscuring the metapodosoma and postpedal furrow dorsally, and telescoping tergites terminally. Tergal extension to the sejugal furrow encloses the ventral and lateral portions of the podosoma in a ''box'' which is delineated anterodorsally by the abjugal furrow, and posterodorsally by the *disjugal furrow* (Plate 2-2, *disj*). The podosomal box and the anterodorsal epiprosoma constitute the *prosoma*, a homologue of the cephalothorax in Araneae. Terminal telescoping of opisthosomal tergites carries the anal opening from a terminal to a ventral position near the genital region. The term *stethosoma* is used by van der Hammen (1974) to describe the prosoma minus the gnathosoma (= podosoma, Fig. 2, p. 7).

Shields or platelets commonly cover portions of the idiosoma (Plate 4, p. 18), with the degree of sclerotization usually increasing with ontogenetic development. Shields not only serve as sites of muscle attachment, but provide a degree of protection from desiccation and predation. An anterior dorsal shield often occurs in the Acari, covering only the prodorsal region or extending to cover the entire propodosoma. A separate hysterosomal shield, or series of hysterosomal sclerites comprising *mesonotal scutella* and a terminal *pygidial shield,* is widespread in most groups. Fusion of propodosomal and hysterosomal shield elements occurs commonly in the Oribatida, less frequently in the Gamasida, and occasionally in other acarine suborders. Most acariform taxa have two or more independent dorsal shields, which allows greater flexibility and ease of movement. The fused idiosomal shield found in many Oribatida not only forms a covering for the dorsum of the hysterosoma, but characteristically extends laterally to form a shell-like *notogaster* (Plate 148-2, p. 473). The proterosoma of Oribatida may also be covered by a single shield which completely obscures the gnathosoma dorsally.

Ventrally, the idiosoma may carry a variety of shields or be virtually unsclerotized, but the genital-anal openings and leg articulations rarely are without some type of surrounding sclerite. Shield expansion and fusion is widespread in the Oribatida, with the fused ventral shield or shields abutting the dorsal notogaster. Expanded ventral shields form characteristic patterns in adult and immature Gamasida.

The primary external structures found on the idiosoma are *locomotory, respiratory, copulatory, sensory,* and *secretory* in function.

PLATE 4

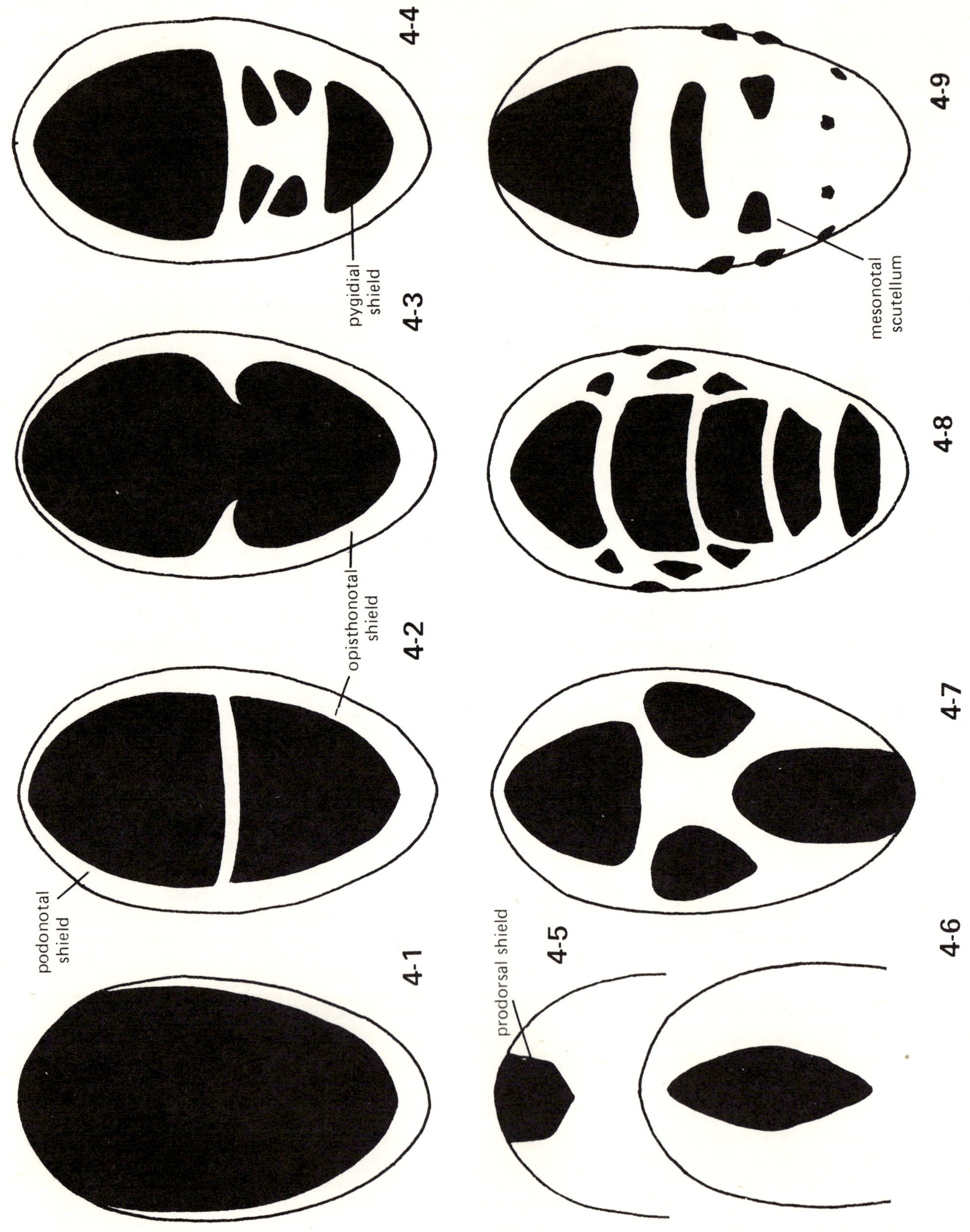

4-1 to 4-9; common dorsal shield configurations in the Acari. **4-1,2,3,4,6**; Gamasida: **4-2,7,8**; Actinedida: **4-5**; Acaridida: **4-9**; Oribatida

Locomotion

With few exceptions, adult and nymphal Acari possess four pairs of jointed legs, while the larva has three pairs. Legs IV appear with the first nymphal instar.

The legs are divided into seven primary segments (Fig. 7, p. 20). Beginning with the most proximal, these are the *coxa, trochanter, femur, genu* (patella of some authors), *tibia, tarsus,* and *pretarsus.* Secondary division of the trochanter may occur (Plate 8-1, p. 108), and tarsal and/or femoral sutures may give the impression of even further primary division. These are secondary articulations, however, which lack the tendons common to primary leg segments (van der Hammen 1970b). Loss of segments through fusion may occur in some groups (Hughes 1959).

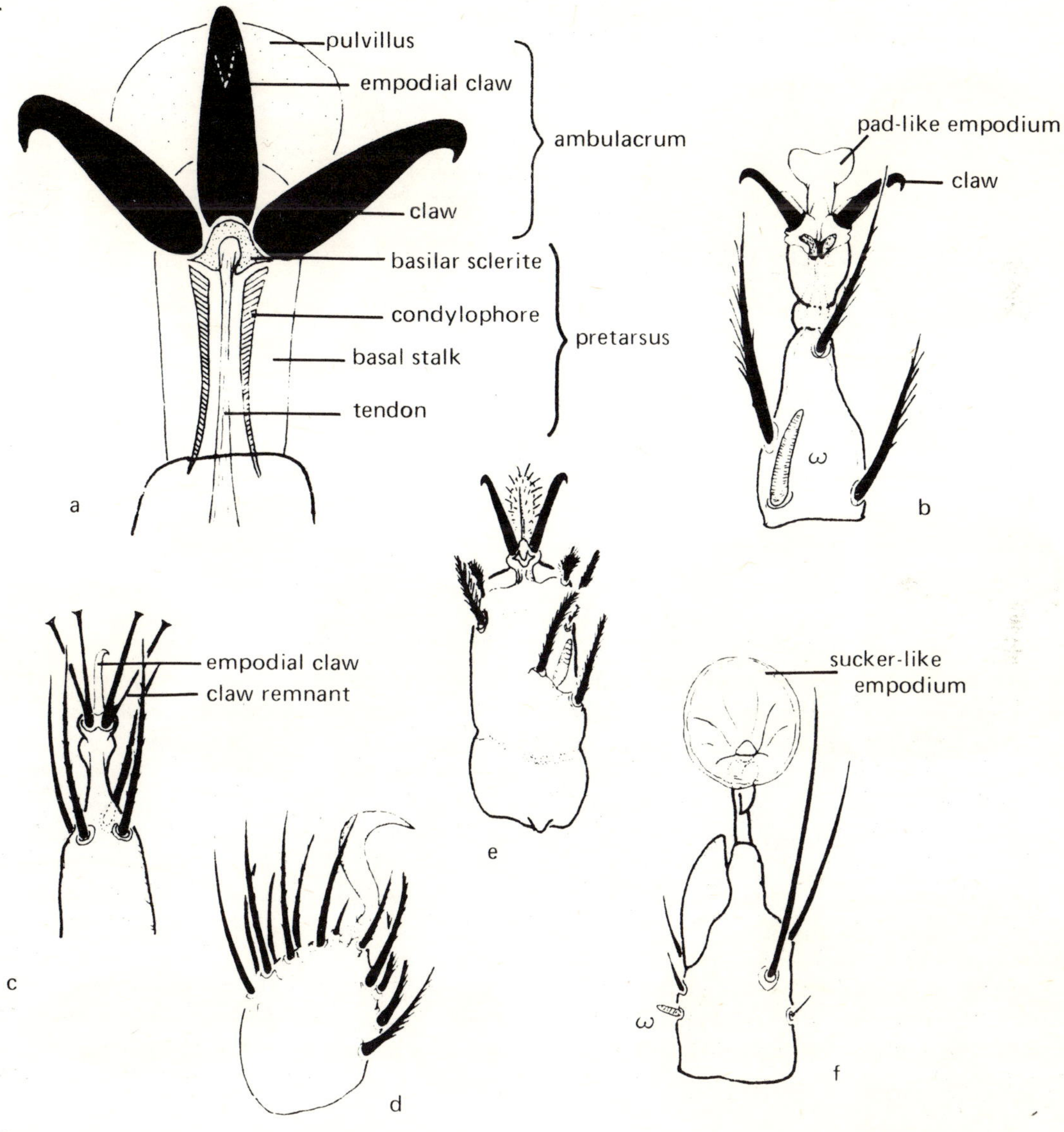

Fig. 6. Pretarsi and ambulacra of some representative acarine types: **a**; hypothetical: **b**; *Pygmephorus* sp. (Actinedia, PYGMEPHORIDAE), tarsus II: **c**; *Oligonychus* sp. (Actinedida, TETRANYCHIDAE), tarsus III: **d**; Oribatida (OPPIOIDEA), tarsus II: **e**; *Ricardoella limacum* (Schrank) (Actinedida, EREYNETIDAE), tarsus II: **f**; *Heterocheylus* sp. (Actinedida, HETEROCHEYLIDAE), tarsus II (see also Fig. 38, p. 373).

The pretarsus may articulate with a terminal *ambulacrum* or apotele (Fig. 6a, p. 19) which usually consists of a *basilar sclerite,* paired claws, and/or a median *empodium.* A membranous *pulvillus* may also be present. The basilar sclerite is motivated by levator and depressor tendons operating in conjunction with adjacent paired sclerotized *condylophores.* The empodium frequently persists in the absence of true claws as a claw-like pretarsal extension (Fig. 6d). It is often absent in acariform mites, so that only a pair of claws remain (Plate 58-3, p. 313). In some groups only one true claw is lost and the empodium assumes its position, structure and function (Plate 95-7, p. 350).

Leg segment musculature in the Acari is reduced, and leg extension is brought about by hydrostatic pressure. Flexion is achieved by individual flexor muscles in each segment (Fig. 7). Coxal protractors and retractors provide backward and forward leg movement.

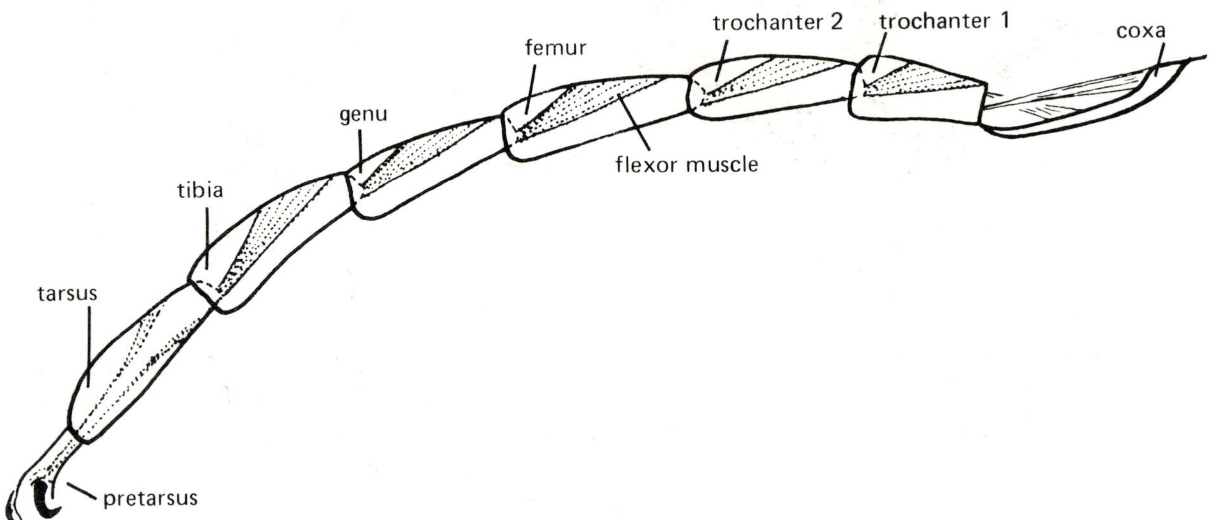

Fig. 7. Typical acarine leg (diagrammatic) showing segmental musculature.

While the legs are primarily ambulatory, they may be modified to serve other functions. Legs I are especially subject to modification in that they make initial contact with the food substrate. Consequently, legs I serve as sensory structures in a number of acarine groups, assuming an elongate antenniform appearance (Plate 63-2, p. 318). They also may be equipped with spine-like internal apophyses to aid in capturing and holding prey (Plate 91-1, p. 346) and in certain parasitic families, legs I are modified for clasping hairs or for adhering to the skin of host animals (Plate 82-1, p. 337). Feider (1971) observed that certain nasal mites of the gamasid family RHINONYSSIDAE (*Rhinonyssus, Mesonyssus*) use the claws of legs I as surrogate chelicerae, employing them to pierce and tear the nasal mucous membranes of their bird hosts. The chelicerae in these groups are weakly developed membranous structures which appear to function only as part of the pathway for liquid food uptake. Legs I-IV often function as grasping or adhering organs in parasitic or phoretic acarines (Plate 33-2, p. 191). Legs II and/or IV of male Gamasida commonly have spur-like apophyses for grasping the female during mating (Plate 13-2, p. 171).

The legs of Acari may be smooth or variously ornamented and usually possess a number of tactile and sensory hairs (Fig. 14, p. 28) which follow fixed patterns of insertion in a given taxon, both in position and in number. The setal distribution of any given leg segment may be reduced to a formula (Fig. 26, p. 117) which often is useful in establishing systematic relationships (Evans 1963a).

Respiration

Exchange of carbon dioxide and oxygen in the Acari is accomplished in ways which are so diverse as to rule out any theory of single line evolution of respiratory systems. The presence or absence of spiracular openings and their relative position provides a major diagnostic feature for identifying the acarine suborders. Thus the presence of a pair of spiracles, or *stigmata*, on the lateromedian aspect of the idiosoma characterizes the members of the Gamasida (Fig. 8c), and two pairs of lateromedian stigmata (occasionally three) distinguish the Holothyrida (Fig. 8b). Stigmatal openings in the Actinedida may be behind or between the cheliceral bases (Figs. 9c, 10a,b, pp. 22-23), or on the lateral humeral angles of the propodosoma (Fig. 9b). Stigmata are obscure or absent in the Oribatida. Tracheal ducts of many adult Oribatida open ventrolaterally between legs II-III and in the acetabular cavities of legs I-III (Fig. 10c, p. 23), while other secondary tracheal systems open at the surface of the idiosoma through localized porose areas and at the bases of special paired dorsal sensory hairs. Discrete stigmata are located in the region of coxae IV in the Ixodida (Fig. 9a, p. 22) and on the dorsum of the idiosoma in the Opilioacarida (Fig. 8a). In those mites which have no apparent stigmata or tracheal system (the Acaridida, immature and "inferior" Oribatida, and certain families in other suborders), gaseous exchange is thought to occur through the integument. Lack of discrete external respiratory atria does not always connote lack of a respiratory system. Genital tracheae occur commonly in the Acaridida and Actinedida, and are apparently associated with the papillae which border the genital opening. Larvae of many Actinedida and Acaridida, and all of the Oribatida, possess paired Claparède organs or *urstigmata* (Grandjean 1946) ventrally on coxae I or II, or between them (Fig. 11, p. 23). These variously shaped structures are considered homologous to the genital papillae (Knülle 1959, van der Hammen 1969, Vercammen-Grandjean 1975). Based on examination of representative genera in the Actinedida, Vercammen-Grandjean has concluded that both the urstigmata and genital papillae function as respiratory organs.

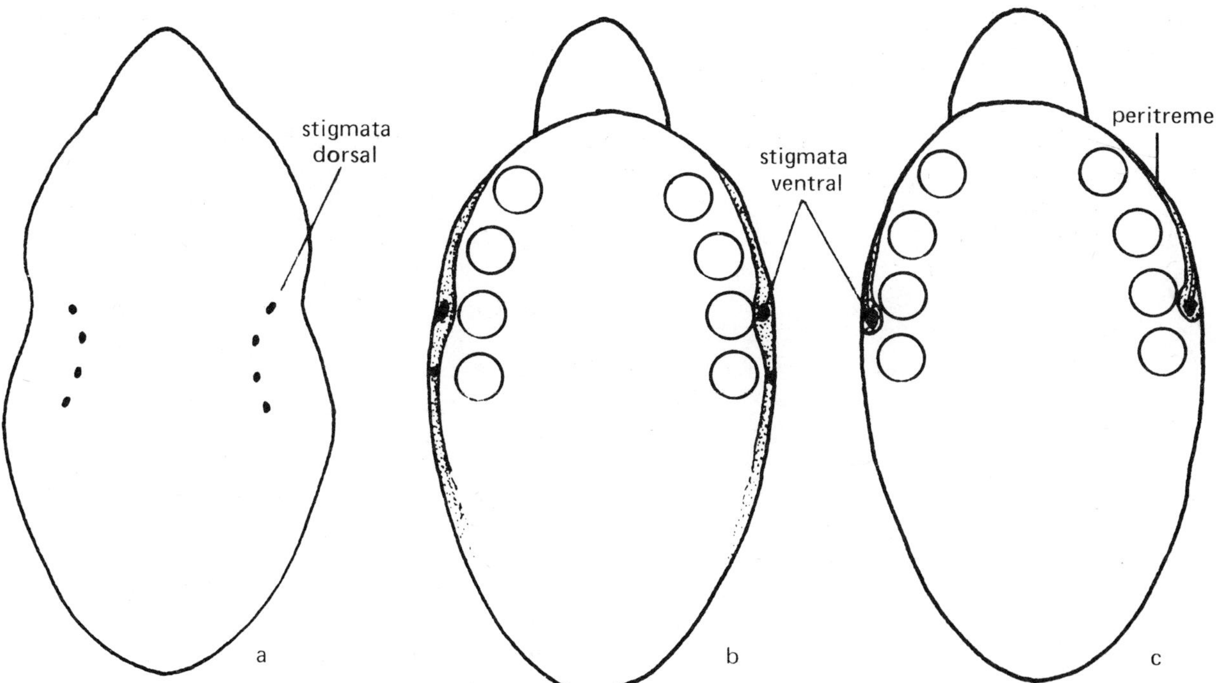

Fig. 8. External respiratory and associated structures in the Opilioacarida **(a)**, Holothyrida **(b)**, and Gamasida **(c)**.

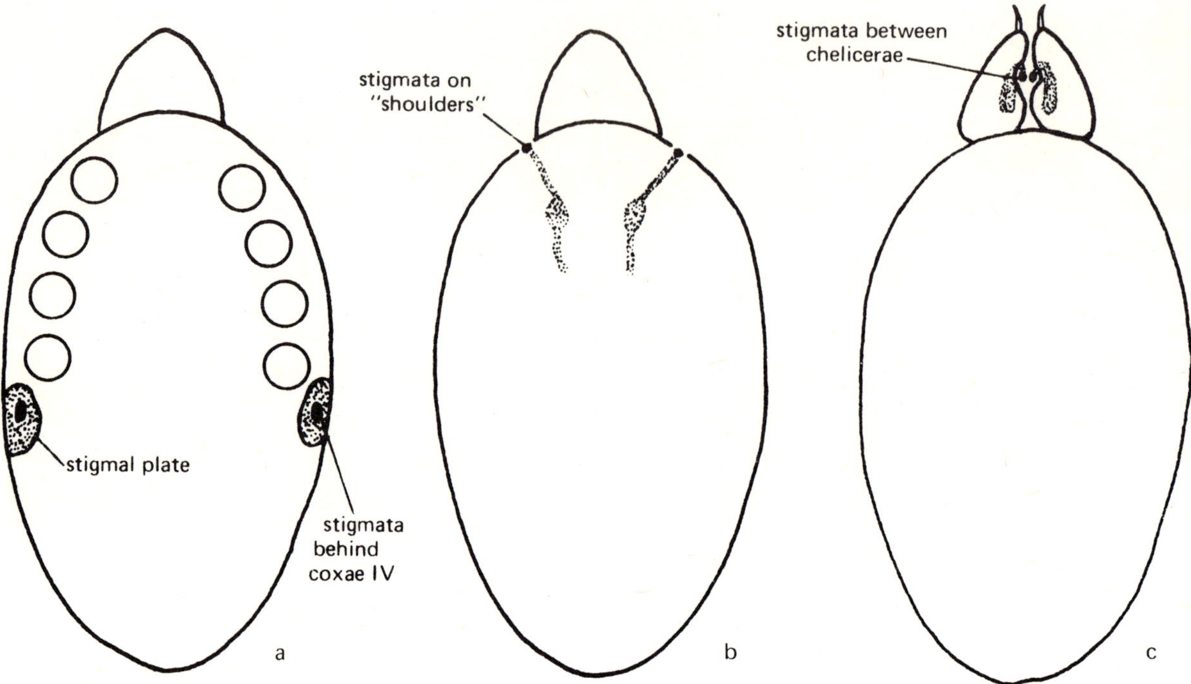

Fig. 9. External respiratory and associated structures in the Ixodida **(a)**, Actinedida-Heterostigmae **(b)**, and Actinedida-Parasitengonae **(c)**.

Most stigmate acarines possess tracheal trunks or air chambers which connect the stigmata to a ramifying tracheal system that serves the various organs. Exceptions are found in the TROMBICULIDAE and in certain water mites (Actinedida, Hydrachnidia) where tracheae either do not occur, or where they form a subcuticular network with independent tracheal strands serving particular organs (Mitchell 1972).

The stigmata in the Parasitiformes generally are associated with discrete sclerotized processes which either take the form of a simple encircling plate (Fig. 9a) or a groove of varying length and complexity which is directed anteriorly—or both anteriorly and posteriorly—from the stigma. The groove or *peritreme* of the Gamasida (Fig. 8c, p. 21) appears to function as an extension of the stigma, ensuring that blockage of the stigmata will not necessarily interfere with respiration (Radovsky 1969). The peritreme is lined with cuticular processes reminiscent of the filtering devices found in the spiracular chambers of insects. Peritremes are adapted for plastron respiration in various gamasids found in aquatic or marine habitats (Hinton 1971, Krantz 1974).

The stigmata in Ixodida each are surrounded by a *stigmal plate* (Fig. 9a) which may contain circles of fine pores or *aeropyles*. Woolley (1972) describes aeropyles as air gates which allow air to pass into the atrial and subatrial chambers and thence into the tracheae. According to Roshdy and Hefnawy (1973), however, aeropyles are unrelated to gas exchange. Aeropyles are found only in hard ticks (IXODIDAE); soft ticks (ARGASIDAE) have a simple ostium set in a hinged flap which opens directly into the atrial cavity (Hinton 1967, Sonenshine and Gregson (1970).

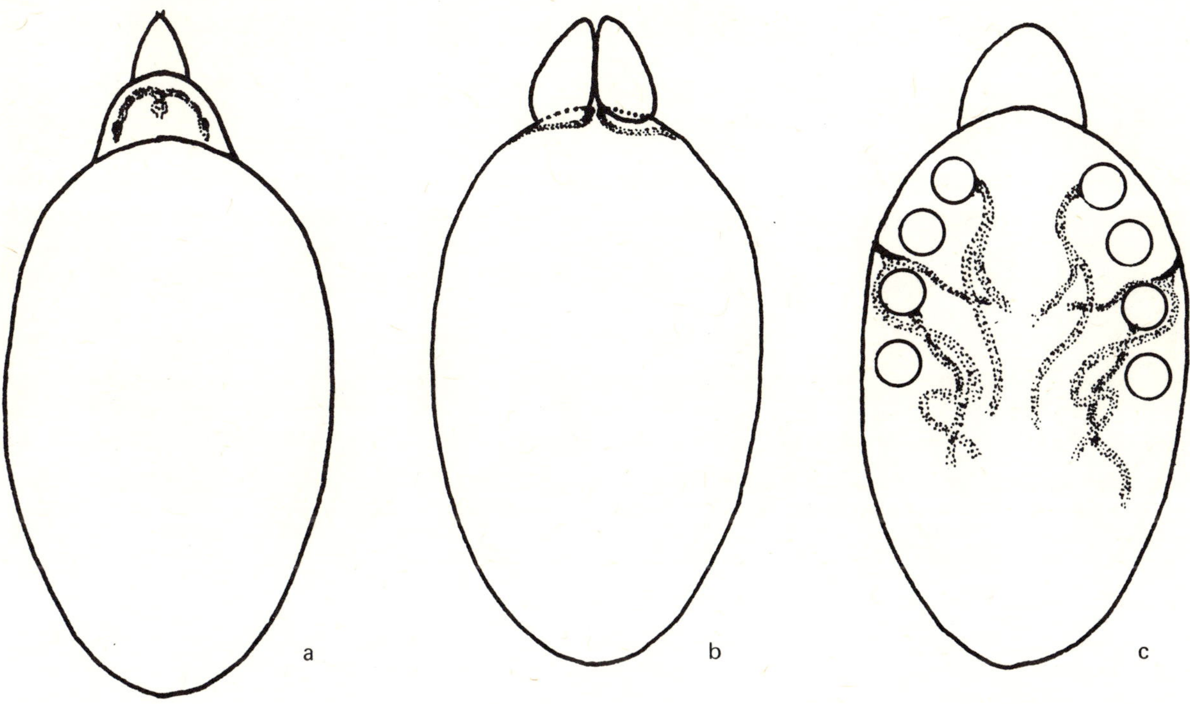

Fig. 10. External respiratory and associated structures in the Actinedida-Eleutherengonina (**a** and **b**) and the Oribatida (**c**).

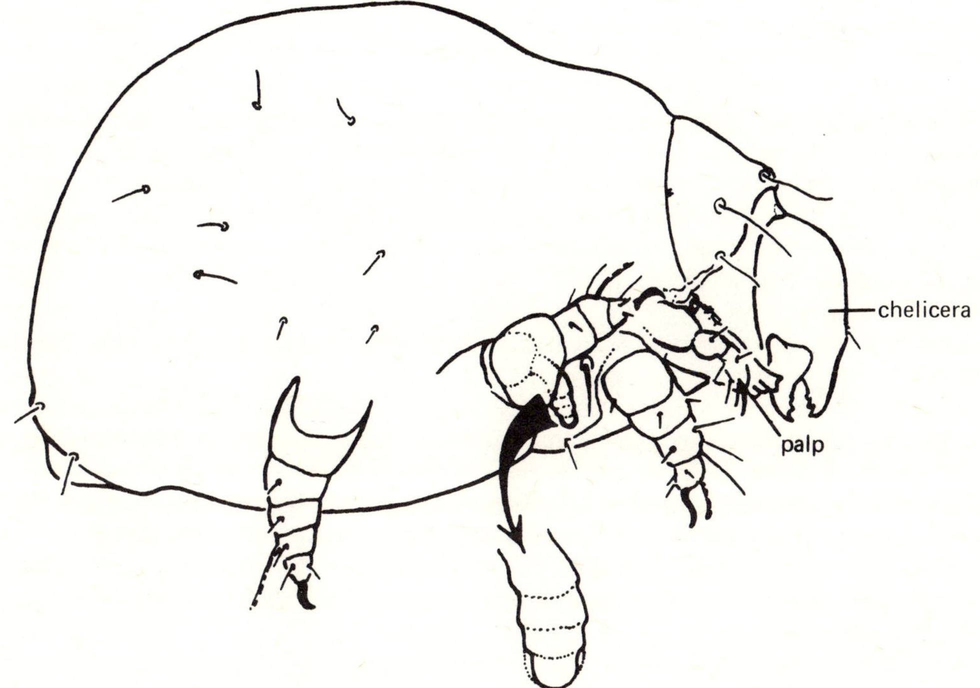

Fig. 11. Larva of *Phthiracarus* sp. (Oribatida, PHTHIRACAROIDEA), with detail of the urstigma (Claparède organ).

Copulation and Non-Copulatory Sperm Transfer

Sperm transfer methods in the Acari illustrate a diversity which rivals that of the respiratory systems described above. In those mite groups where a ventral or dorsal male intromittent organ, or *aedeagus*, is present (Plate 5, p. 25), transfer of sperm may be made directly to the female genital opening (as in various Actinedida) or to a special extragenital structure located terminally or posterodorsally on the female idiosoma (as in the Acaridida). The structure generally takes the form of a simple *bursa copulatrix* (Plate 5-1). However, in some acaridid families (e.g. the CRYPTUROPTIDAE and ATOPOMELIDAE), it is represented by a greatly elongate tubular *sperm duct* (Plate 140-5, p. 438). Sperm placed in the bursa or sperm duct moves internally to a seminal receptacle and thence to paired *ovaries* (Fig. 15e, p. 33) where fertilization of the ova occurs.

Indirect or non-copulatory sperm transfer occurs throughout the Acari, but the actual method of transfer varies widely. In every case, a sperm packet or droplet is produced by the male and presented to the female by means other than an aedeagus. Perhaps the simplest transfer method is illustrated in the gamasid cohort Uropodina, where a sperm packet is transferred directly from the male to the female genital aperture. In *Leiodinychus krameri* (Can.), a complex pre-mating ritual is followed by a period of venter-to-venter contact during which the sperm packet is affixed to the anterior edge of the epigynial shield. Assimilation of the contents of the packet occurs soon afterwards (Radinovsky 1965).

Many members of the suborder Gamasida utilize the chelicerae in sperm transfer. Sperm may be placed directly into the female genital opening by the male chela, or transferred instead to female extragenital openings which are analogous to the bursa copulatrix in the Acaridida. Transfer to the female genital opening (*tocospermy* of Athias-Henriot 1970) occurs in the PARASITIDAE (Micherdzinski 1969) and has been observed in *Cercoleipus coelonotus* Kinn, a member of the family CELAENOPSIDAE (Kinn 1971).

Sperm transfer to paired female extragenital openings is widespread in the Gamasida. These openings, or *sperm induction pores*, are found on the posterodorsal faces of coxae III or IV, trochanters III, femora III, the endopodal shields, the metapodal shields and, rarely, the sternal shield (Lee 1970, Athias-Henriot 1970). Sperm transfer to the sperm induction pores is correlated with the presence of a distinctive sperm transfer organ, or *spermadactyl*, on the movable digit of the male chelicera (Plate 3-5, p. 15). Sperm passed to the chelicera from the male genital opening (Plate 14-5, p. 172) moves into the hollow spermadactyl and through its narrowed tip into the sperm induction pore. The term *podospermy* has been applied to this phenomenon by Athias-Henriot (1970).

The means of sperm transfer from the male genital opening to the spermadactyl in most of the podospermous and tocospermous Gamasida is poorly understood. Lee (1974) has observed that the sperm droplet produced by male OLOGAMASIDAE is picked up between the chelicerae following a downward and backward flexion of the gnathosoma. Contents of the droplet are then transferred to the female sperm induction pore by the tips of the paired spermadactyli. The sequence of events leading to sperm transfer in the MACROCHELIDAE, however, follows a completely different pattern (Krantz and Wernz, 1979). The sperm droplet passes directly to either chelicera by means of a membranous, telescoping genital tube which extrudes over the tritosternum to the tip of the hypostome. The chelicerae are extruded so that the arthrodial brush of one of the chelicerae "captures" the droplet, the contents of which are then transferred to the open median groove of the opposing chela. Sperm passes from the groove into the contiguous closed canal of the spermadactyl and thence to the female sperm induction pore. Only one spermadactyl is utilized in a given transfer episode.

PLATE 5

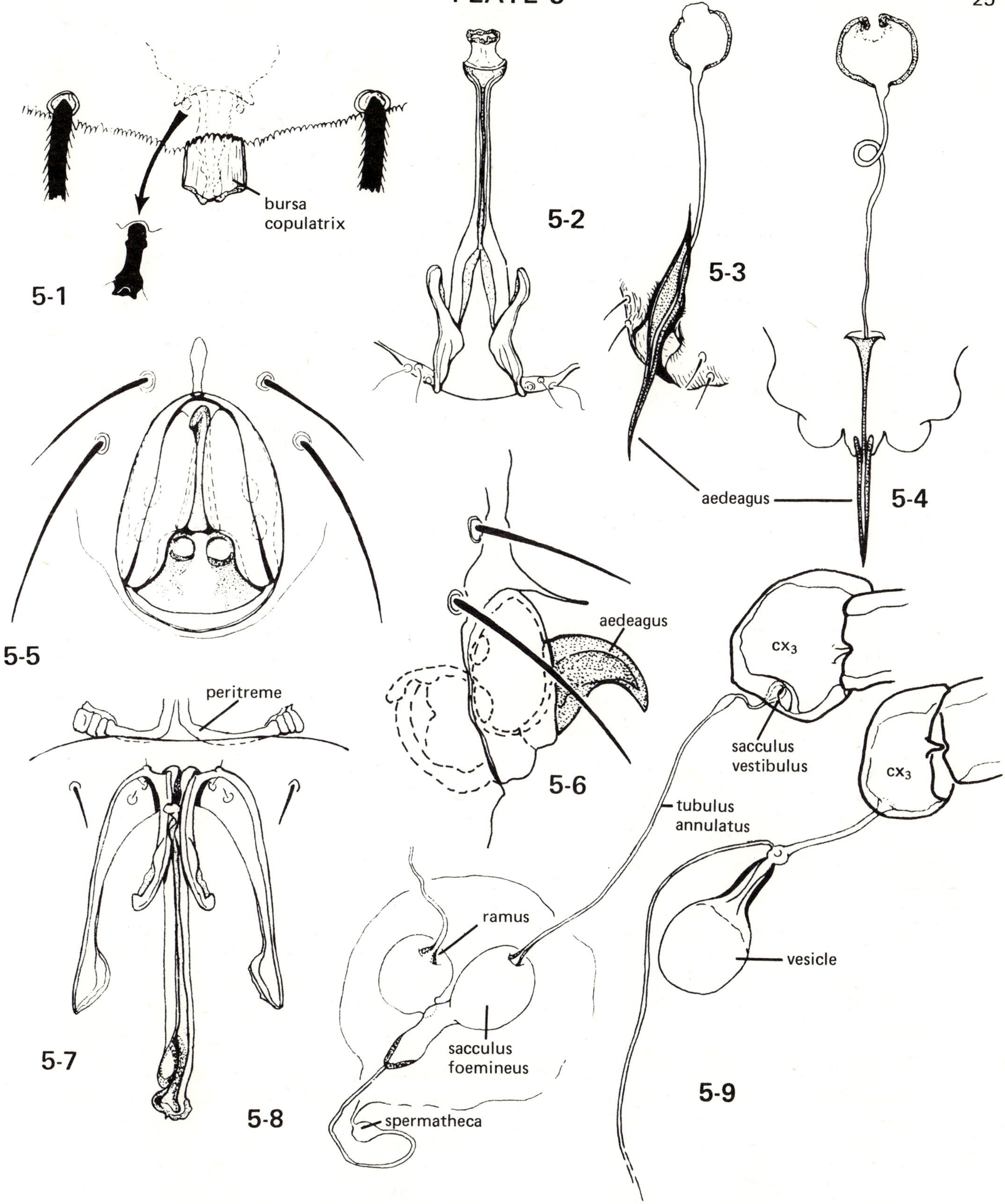

5-1; *Glycyphagus domesticus* (DeGeer) (Acaridida, GLYCYPHAGIDAE), posteroventral aspect of female

5-2 to 5-7; examples of male acariform genital structure. **5-2;** ? genus (Actinedida, EUPALOPSELLIDAE): **5-3;** *Eotetranychus carpini carpini* (Oudemans) (Actinedida, TETRANYCHIDAE): **5-4;** *Brevipalpus* sp. (Actinedida, TENUIPALPIDAE): **5-5 and 5-6;** *Glycyphagus destructor* (Schrank) (Acaridida, GLYCYPHAGIDAE): **5-7;** *Harpyrhynchus* sp. (Actinedida, HARPYRHYNCHIDAE)

5-8 and 5-9; sacculus foemineus and related structures in Gamasida. **5-8;** type "A" sacculus complex (*Macrocheles merdarius* (Berlese), MACROCHELIDAE): **5-9;** type "B" sacculus complex (*Amblyseius stramenti* Karg, PHYTOSEIIDAE)

Transfer of sperm in ticks (Ixodida) is of the tocospermous type, although the male hypostome and palpi often are inserted in the female genital opening along with the sperm-bearing chelicerae (Oliver et al. 1974).

The method of sperm transfer in certain Actinedida and Oribatida differs in a peculiar way from those previously discussed. The sperm packet is deposited by the males at the tip of a stalk which is extruded through a median "penis" as a fluid thread (Woodring 1970, Alberti 1974), and which hardens upon contact with the air (Mitchell 1958, Schuster and Schuster 1966, Theis and Schuster 1974). The stalk, or *spermatophore* (Fig. 12) supports the sperm packet so as to make it more easily available to nearby females. The packet is picked up by the genitalia of the female, which then abandons the bare spermatophore (Lipovsky et al. 1957, Woodring and Cook 1962). Sperm apparently is pressed from the packet by the female of *Aculus cornutus* (Banks) (Actinedida, ERIOPHYIDAE) so that it remains essentially intact following sperm uptake (Oldfield et al. 1970 and 1972).

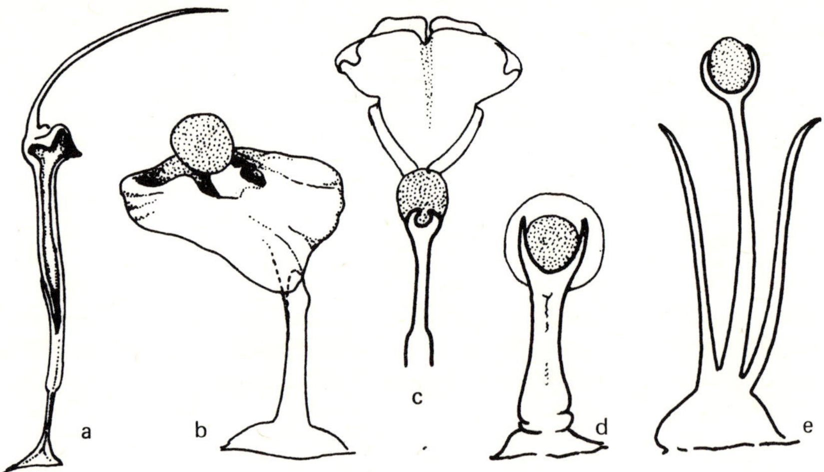

Fig. 12. Representative spermatophore types in the Acariformes (stippled portions are the sperm packets): **a**; *Neomolgus littoralis* (L.) (Actinedida, BDELLIDAE) (after Alberti 1974): **b**; *Phyllocoptruta oleivora* (Ashmead) (Actinedida, ERIOPHYIDAE) (after Jeppson et al. 1975): **c**; *Abrolophus passerinii* (Berlese) (Actinedida, ERYTHRAEIDAE) (after Witte 1975): **d**; *Pergalumna omniphagous* (Oribatida, GALUMNOIDEA) (after Woodring 1970): **e**; Actinedida, TROMBICULIDAE.

Deposition of the spermatophore by male Oribatida may be preceded by a ritualistic series of movements (Schuster 1962) which attracts females to the deposition site. Apparently, a sex attractant chemical substance or *pheromone* is deposited by males of *Erythraeus regalis* (C.L. Koch) (Actinedida, ERYTHRAEIDAE) coincident with the circular dance which precedes spermatophore placement. The females have no difficulty finding the spermatophore following its deposition (Witte 1975).

Spermatophore morphology is highly distinctive within families and genera, and often characterizes individual species as well. Spermatophores may vary from elongate, often intricately sculptured forms as in the BDELLIDAE (Fig. 12a), to miniscule leaf-like structures as in the actinedid superfamily Eriophyoidea (Fig. 12b). Spermatophores are known not only for terrestrial acarines, but have also been observed in aquatic habitats. Males of several water mite species (Actinedida, Hydrachnidia) produce spermatophores (Mitchell 1958, Efford 1966, Lanciani 1972), although other water mites are known to engage in copulative mating. Kirchner (1967) observed spermatophore deposition by a marine mite (Actinedida, HALARIDAE) in the laboratory.

The mating response in male Acari may be influenced by the production of pheromones by the female. Cone et al. (1971a and 1971b) and Penman and Cone (1974) reported that male *Tetranychus urticae* Koch (Actinedida, TETRANYCHIDAE) are attracted to quiescent pre-imaginal female nymphs (deutonymphs) and suggested the presence of a sex pheromone. Regev and Cone (1975) found that extracts of the female nymphs contain farnesol, a sesquiterpene alcohol. Males were strongly attracted to a farnesol isomer. Later tests revealed that pharate female *T. urticae* produce another sex attractant, nerolidol (Regev and Cone 1976).

A pheromonal response by males of the ixodid species *Amblyomma americanum* (L.), *A. maculatum* Koch and *Dermacentor variabilis* (Say) has been demonstrated by Berger et al. (1971). The pheromonal substance in *A. americanum* was identified by Berger (1972) as 2,6-dichlorophenol. Pheromonal response has since been verified in *D. variabilis,* and recorded in *D. andersoni* (Stiles) by Sonenshine et al. (1974 and 1976). A similar response has been observed in *Ixodes ricinis* L. by Graf (1976).

Sensory Structures

The idiosoma of the typical acarine is well equipped with a variety of cuticular sensory receptors, almost all of which are setal. These are primarily tactile in function, although certain setae are adapted for chemoreception. Grandjean (1935a) found that tactile and chemosensory hairs of certain acarine suborders contain a layer of optically active iodophilic material, *actinopiline* (= actinochitin), which exhibits birefringence in polarized light. Actinopiline may occur as a solid core or as a layer of material surrounding a protoplasmic extension of the basal setal nerve cell. Acarine suborders which possess optically active setae, and appendages derived from these setae (i.e. the Acaridida, Oribatida and Actinedida (the Acariformes)), are classified together as the Actinotrichida (van der Hammen 1972). The Opilioacarida, Holothyrida, Gamasida and Ixodida (the Parasitiformes) lack actinopiline, and are referred to as the Anactinotrichida (Fig. 25, p. 106).

Tactile setae are characteristically hollow both in the Actinotrichida and the Anactidotrichida, and form distinctive patterns on the idiosoma which often are useful in systematic studies. These setae are commonly simple and spinose, but may be variously expanded or ornamented (Fig. 14, p. 28). Specialized tactile setae often are found on the legs and idiosoma of Actinotrichida. Unlike other tactile hairs, *trichobothria* have a solid core of actinopiline (Fig. 14d). Leg trichobothria are simple and greatly elongate, but may assume a variety of forms on the idiosoma. Only one pair of propodosomal trichobothria, or *sensilla,* occurs in the Oribatida, and these are inserted in distinct cavities called *bothridia* (Fig. 13b, p. 28). One or two pairs of simple, pectinate or clavate propodosomal trichobothria may be found in the Actinedida (Fig. 13a), and certain members of the actinedid family EREYNETIDAE have a pair on the posterodorsal aspect of the hysterosoma (Plate 67-3, p. 322).

While the function of the leg trichobothria is clearly tactile, that of the idiosomal trichobothria may be somewhat more complex. Pauli (1958) determined that the sensilla are vibro- and anemoreceptors in the oribatid family BELBIDAE. Beklemishev (1969) classifies idiosomal sensilla of other Actinotrichida as balancing organs.

Chemosensory setae of the Actinotrichida are of several basic types, but most are optically active and are found on the terminal segments of legs I and II and on the palpi. *Eupathidia* (= acanthoides and pseudacanthoides of Grandjean 1946) are spinose structures which have an actinopiline sheath surrounding a core of protoplasm. They are found on the tarsi of legs and palpi. The *famulus* is similar to the eupathidium internally, but is small and often expanded distally (Fig. 14c, p. 28). Famuli generally are confined to tarsi I. The

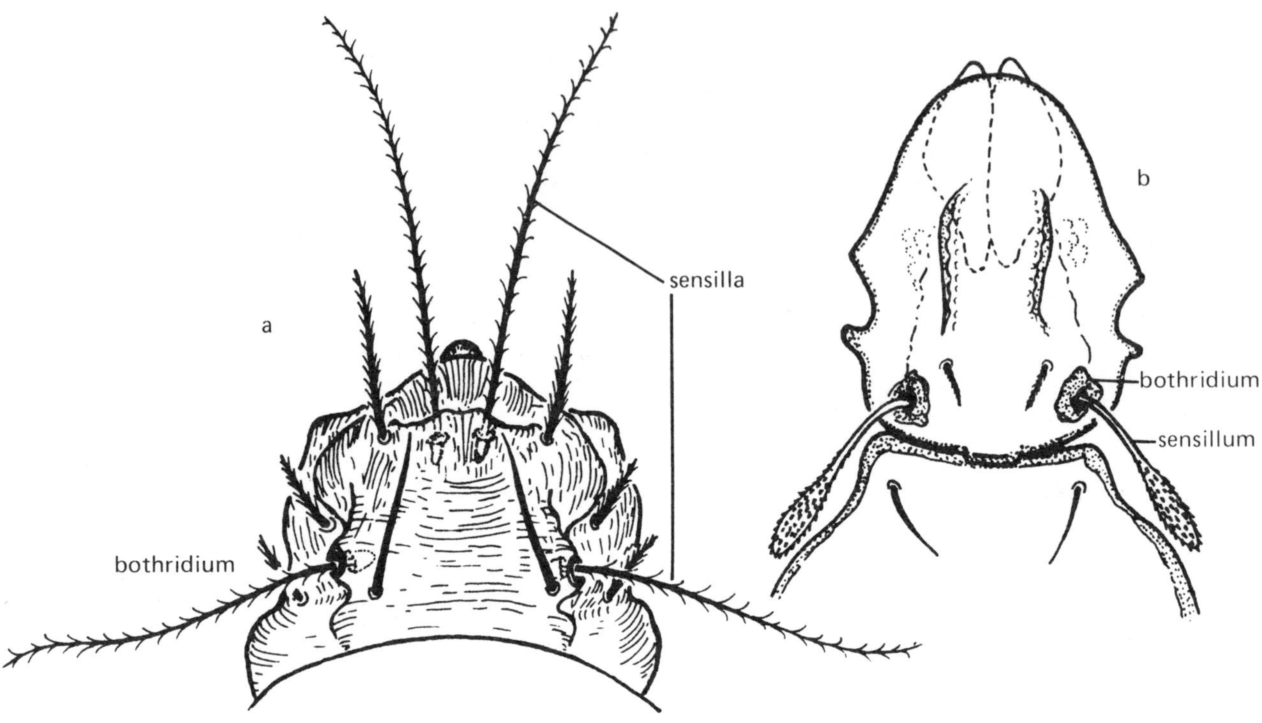

Fig. 13. Position of prodorsal bothridial sensilla (trichobothria) of: **a**; *Nanorchestes* sp. (Actinedida, NANORCHESTIDAE) and **b**; *?Phthiracarus* sp. (Oribatida, PHTHIRACAROIDEA).

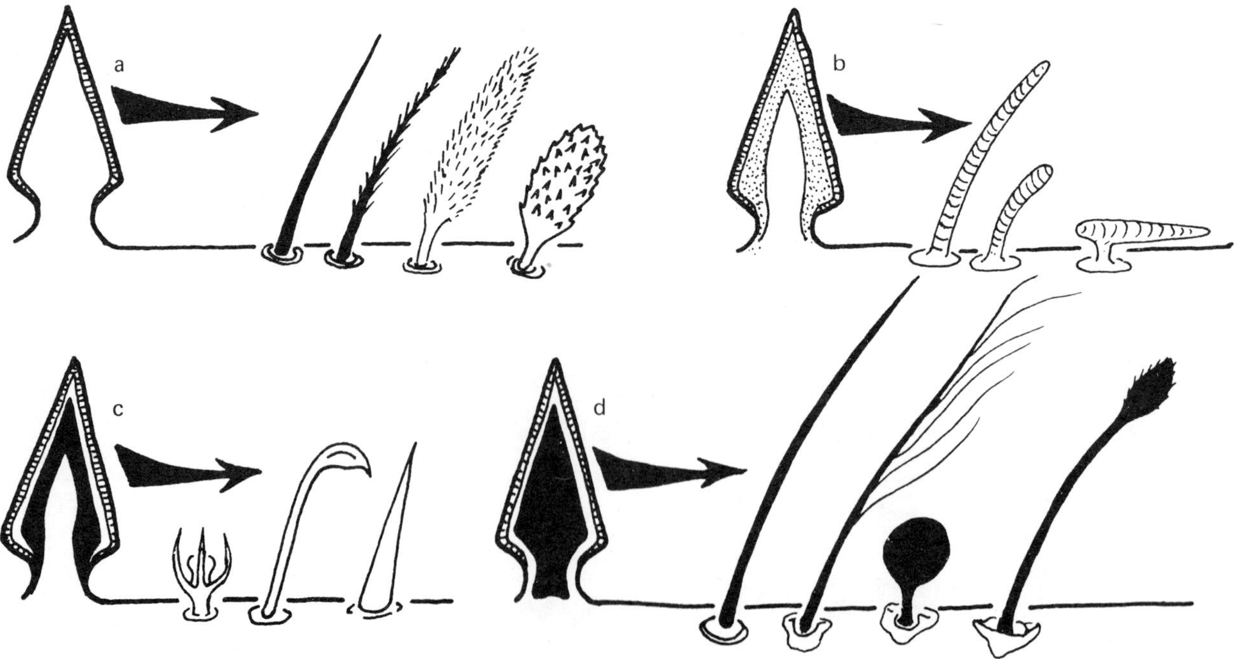

Fig. 14. Acarine setal types: **a**; tactile setae (hollow): **b**; solenidia (protoplasmic core): **c**; famuli (two types, left) and eupathidium (right) with a sheath of actinopiline (drawn in black) around the core: **d**; trichobothria, with core of actinopiline.

chemosensory *solenidia* contain no sheath or core of actinopiline, although a protoplasmic layer is present. The walls of most solenidia are transversely striated, and their insertions on the legs or palpi are broad and flat. Solenidia may be long and setiform, or short and rounded terminally. They may be erect or closely appressed to the segment on which they are inserted (Fig. 14b).

Optically inactive setae of several types form sensory fields on the appendages of Gamasida (Haarlov 1943), and of Holothyrida and Opilioacarida (van der Hammen 1968a) (Anactinotrichida). The sensory field on tarsus I of the Ixodida is contained in paired dorsal depressions referred to collectively as *Haller's organ* (Fig. 30, p. 211). The setae of Haller's organ may be contact chemoreceptors or long range olfactory receptors, apparently depending on how the tick attacks its host (Balashov 1972).

In addition to setal receptors, ocelli or other photosensitive organs occur in many acarine groups. One or two pairs of simple ocelli may be present laterally on the propodosoma of Opilioacarida, Actinedida, Ixodida and—exceptionally—the Acaridida and Oribatida. An unpaired anteromedian propodosomal ocellus or pigmented light-sensitive zone is visible in certain Actinedida (HALACARIDAE, for example). However, the median eye is more commonly carried inferiorly on a *naso*, an anteromedian propodosomal projection which occurs in a number of actinedid and oribatid families (Grandjean 1943, van der Hammen 1969, Coineau 1970). The ocelli serve as photoreceptors (Naegele et al. 1965), although response to that portion of the far red spectrum near infrared may be mediated elsewhere (McEnroe 1971). The median lenticule (Grandjean 1951) on the anterior border of the notogaster of *Hydrozetes* species (Oribatida) appears to function as a photosensitive structure (Newell 1945), as do the paired anteromedian pigment spots on the propodosoma of certain Acaridida (Plate 122-1, p. 420).

Camin (1953) discovered photosensitive spots on the pulvillar membrane of tarsi I of the snake mite, *Ophionyssus natricis* (Gerv.) (Gamasida, MACRONYSSIDAE). Although they lack obvious photosensitive centers, other gamasids react strongly to changes in light intensity. Since olfactory stimuli are mediated by certain setae of tarsi I in various Gamasida (Farish and Axtell 1966, Jalil and Rodriguez 1970, Kinn 1971), it is not unreasonable to suspect a photosensitive role for these or other tarsal setae of legs I, or of the palpi.

Secretion

In addition to the various cuticular pores referred to earlier, acarines possess complex secretory systems which may be expressed externally in various ways. Coelomoducts or *coxal glands* of mites and other chelicerates are excretory structures derived from onychophoran nephridia, and consist of a coelomic sac and a winding canal. They tend to develop in different body segments in different chelicerate groups, but usually open on the coxa or femur as a simple pore (Beklemishev 1969). The coxal glands of *Peripatus*, some millipedes, apterygote insects, decapod crustaceans, and all Arachnida are homologous structures.

Coxal glands apparently serve an osmoregulatory function in the Acari (Balashov 1972, Woodring 1973). In soft ticks (Ixodida, ARGASIDAE), coxal organs contain elaborate canal systems which function in water balance (Obenchain and Oliver 1974) and in ion concentration regulation (Araman and Said 1972).

The coxal gland in many groups of Actinotrichida is the most posterior portion of a system of propodosomal glands which empty into a lateral *podocephalic canal* (Fig. 16, p. 34). The paired podocephalic canals may be internal tubes in the higher Actinedida, but often take the form of external grooves in other acarines. They are short and distinct in the Acaridida (Plate 116-5, p. 414), but are long, sinuous and often obscure in Oribatida, running from coxa I to the pedipalpal bases. Podocephalic canals are considered to be secretory in function (Grandjean 1938, 1944, 1968, 1970, 1971; van der Hammen 1968; Coineau 1974). The paired opisthonotal glands of Acaridida (Plate 116-2, p. 414) also are secretory structures which may produce alarm pheromones (citral or neryl formate) in certain ACARIDAE, CARPOGLYPHIDAE and PYROGLYPHIDAE (Kuwahara 1976).

B. INTERNAL

Internally, the idiosoma is a surprisingly complex series of organ systems bathed in a colorless plasma called *haemolymph*. Tick haemolymph contains three primary types of cells: minute (5-7.6μ) proleucocytes with large nuclei, oval basophilous haemocytes (10-20μ) containing glycogen, and amoeboid eosinophilous haemocytes (12-25μ) (Teravsky 1957). Other transitional cells also have been observed in acarine haemolymph, as have amino acids (Boctor 1972), lipids (Hajjar 1972, Woodring and Galbraith 1976), and glucose (Aboul-Nasr and Bassal 1971).

The haemolymph moves freely throughout the haemocoel, circulating primarily as a result of bodily movement. However, a flat ostiate dorsomedian heart may aid circulation in some acarines. Circulation of the haemolymph to the legs and other extremities is facilitated by contraction of the powerful dorsoventral idiosomal muscles, an action which also brings about leg extension and cheliceral extrusion. Thus, although the postcoxal leg segments are equipped only with flexor muscles (Fig. 7, p. 20), hydrostatic pressure provides the countermovement.

The major internal organ complexes of the idiosoma are the *digestive, reproductive, glandular* and *nervous* systems.

Digestion

While the post-oral digestive system varies somewhat between acarine suborders, certain features are more or less constant (Plate 6-1,2, p. 31). The buccal cavity opens internally into the heavily muscled pharynx, as described under the section on the gnathosoma. The stomodaeal pharynx opens posteriorly into an *esophagus* which in turn passes through a *brain* comprised of fused supra- and subesophageal ganglia. The esophagus leads into a midgut or *ventriculus* which is considerably larger in actinotrichidous than in anactinotrichidous Acari (Woodring and Galbraith 1976). Two or more diverticula, or *gastric caeca* (Blauvelt 1945, Hughes 1952), may be present. The caeca provide additional surface area for digestive processes to take place. Two or three long thin caecal pairs are found commonly in the Gamasida (Plate 6-2), while members of the suborder Ixodida (Plate 6-3) may have five or more primary pairs in addition to several smaller diverticula (Balashov 1961, Sonenshine and Gregson 1970). The Acaridida and Oribatida typically have a single pair of short thick caeca, and the Opilioacarida and Holothyrida apparently have none (Hughes 1959).

Digestion is intracellular, at least in the anterior portion of the ventriculus. Here the gut wall is lined with vacuolated epithelial cells which absorb soluble food, and which eventually are sloughed into the lumen. Brody et al. (1972) have noted that extracellular digestion may occur in the posterior portion of the midgut of *Dermatophagoides farinae* Hughes (Acaridida, PYROGLYPHIDAE). This process allows assimilation of large particles of solid food which are ingested by *D. farinae*. Protection of the delicate epithelial layer from the amorphous, often jagged, food bolus is provided by a *peritrophic membrane* which envelops the bolus as it moves

PLATE 6

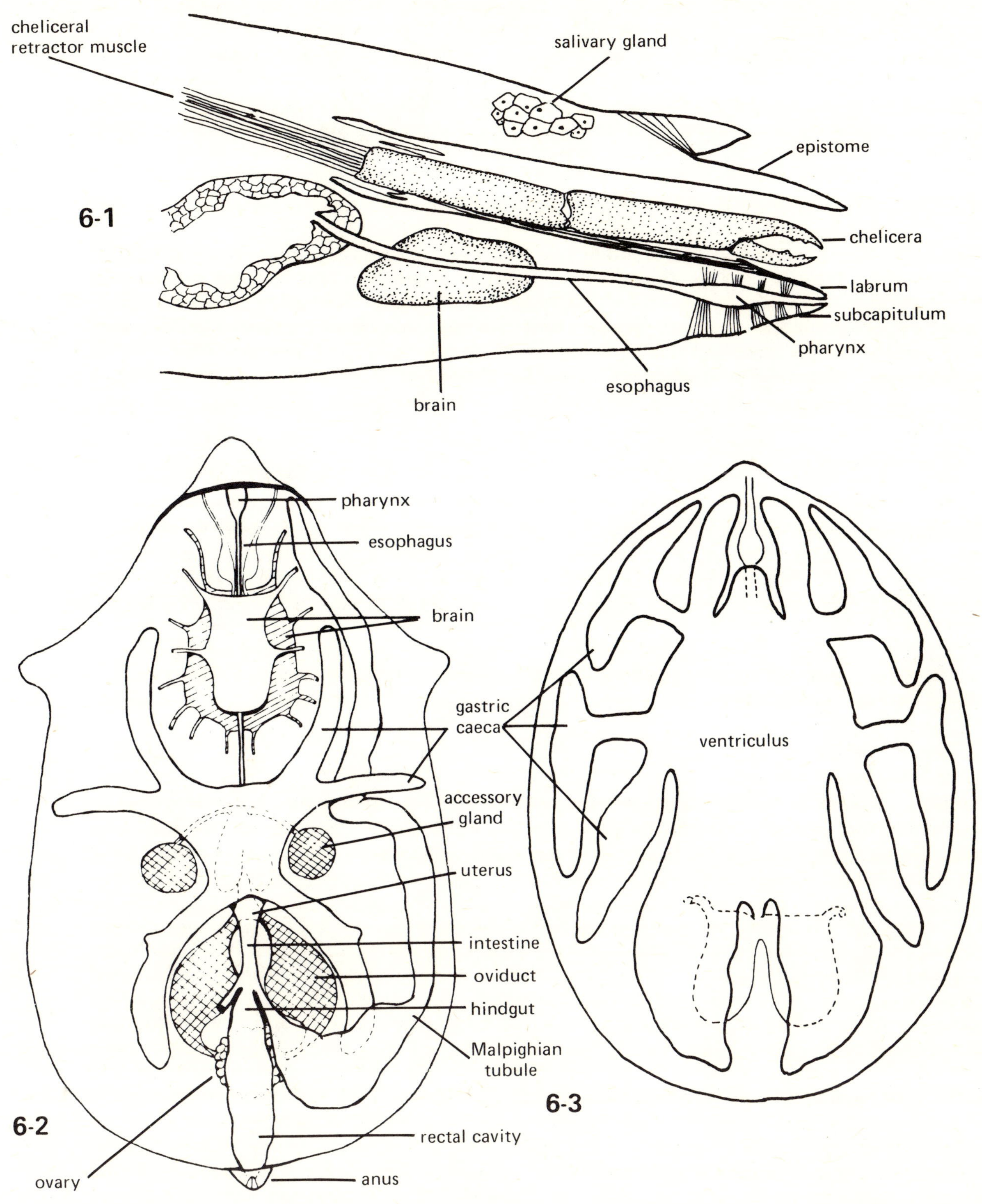

6-1 and 6-2; *Caminella peraphora* Krantz and Ainscough (Gamasida, DITHINOZERCONIDAE) (after Ainscough 1960). **6-1;** midsagittal section of gnathosoma and propodosoma: **6-2;** dorsal internal aspect of female

6-3; *Ornithodoros kelleyi* Kohls (Ixodida, ARGASIDAE), dorsal internal aspect illustrating gastric caeca (after Sonenshine and Gregson 1970)

to the posterior midgut. The membrane remains in place around the digesting bolus and resulting fecal pellet which finally passes to the *hindgut* (Wharton and Brody 1972). It seems likely that other acarines which ingest solid food particles also produce a peritrophic membrane.

A short *intestine* joins the ventriculus to the hindgut, on which one or two pairs of *Malpighian tubules* may be inserted (Plate 6-2). As in other arthropods, the tubules (when present) collect and excrete guanine in the form of birefringent spherules which form a paste-like substance (Hughes 1959). The hindgut leads to a proctodaeal *rectum* which opens externally at the *anus*. The fate of solid byproducts of digestion in the ventriculus is bizarre in adults of some of the higher Actinedida (Parasitengonae). Here the indigestible residues accumulate in the gut cells, which break away and move into the posterodorsal gut lobes (Mitchell 1964). When a lobe has become filled with gut cells it loses its connection with the gut proper and, through a process known as *schizeckenosy* (Mitchell and Nadchatram 1969), passes through a horizontal split of the posterodorsal cuticle. The split soon heals, leaving a scar which indicates that schizeckenosy has occurred. In contrast, waste products may be sequestered in the fat body or parenchyma in the Acaridida (Woodring and Galbraith 1976).

Reproductive System

The reproductive system of both male and female acarines consists of a series of paired, fragmented, or fused elements. The *testes* may be an unpaired organ as in many of the lower Gamasida (Fig. 15a, p. 33) and most Ixodida (Roshdy 1966). It is a paired structure in the gamasid family UROPODIDAE (Woodring and Galbraith 1976) (Fig. 15b), certain Ixodida (Douglas 1943, Wagner-Jevseenko 1958) and in the acaridid ACARIDAE (Rohde and Oemick 1967), but a multiple organ in some of the higher Actinedida (Fig. 15c). The testes manufacture sperm cells which are passed through the paired or fused *vas deferens* to the *ejaculatory duct*. Various accessory glands usually occur between the duct and the vas deferens. It is probable that one of their functions is that of a *seminal vesicle.*

The female reproductive system has a single, paired or cluster *ovary* (Figs. 15d-g) which joins to the *oviduct,* or oviducts. An unpaired *uterus* is found both in the Gamasida and Actinedida (Figs. 15d and 15f). The uterus opens into a *vagina* which may be median or posterior on the venter of the idiosoma. A separate uterus is not apparent in some of the Acaridida. Where a bursa copulatrix occurs, it opens into a *seminal receptacle* which is connected to the ovary (Fig. 15e).

In those gamasid mites where extragenital insemination occurs (see page 24), each of the paired external orifices or sperm induction pores leads internally to a vestibule which opens into a tube (Athias-Henriot 1968). The vestibule is the *sacculus vestibulus* (Krantz and Mellott 1968) and the tube into which it empties is the *tubulus annulatus.* The organ complex which receives sperm from the tubulus assumes two major forms in the Gamasida (Plate 5-8 and 5-9, p. 25). Sperm deposited in either sacculus vestibulus of the "type A" complex passes through the tubulus and contiguous *ramus sacculi* to a medial *sacculus foemineus* (Michael 1892), and then into an appended *spermatheca* (cornu sacculi of Michael). The spermatheca opens via a minute lumen (Pound and Oliver 1976) into the ovary, which may have lateral paired ovulogenous glands of unknown function (Young 1968). In the "type B" complex (Dosse 1958), the tubuli lead into rami and thence into paired vesicles which appear homologous with the unpaired sacculus of the "type A" complex. In addition to their connections with the vesicles, the rami open into paired ducts which lead to the unpaired ovary. The vesicles themselves end blindly, and discrete spermathecae are absent or obscure.

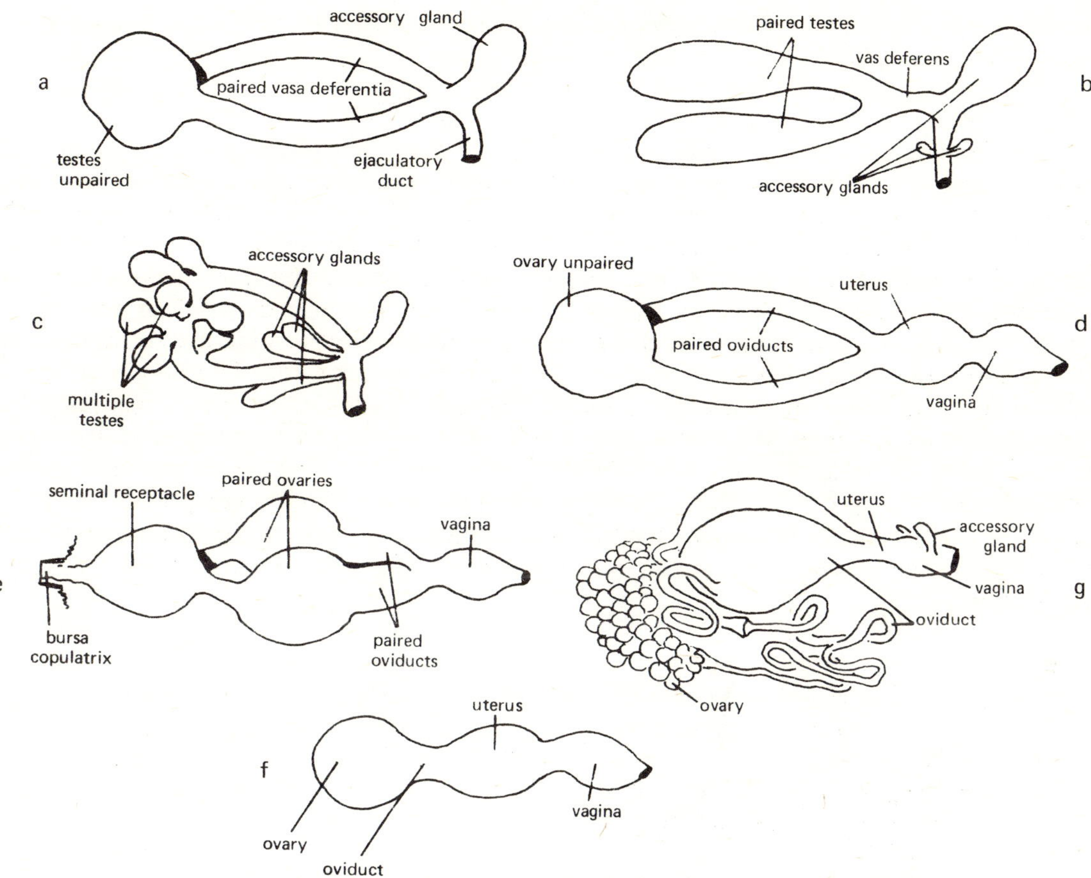

Fig. 15. Representative acarine reproductive systems (after Hughes 1959 and Sonenshine 1970): **a**; Gamasida-PARASITIDAE, male: **b**; Gamasida-UROPODIDAE, male: **c**; Actinedida-ERYTHRAEIDAE, male: **d**; generalized female system, Gamasida and Actinedida: **e**; Acaridida-ACARIDAE, female: **f**; Actinedida-ERIOPHYIDAE, female: **g**; Ixodida-ARGASIDAE, female.

Based on his studies of *Haemogamasus reidi* Ewing (Gamasida, LAELAPIDAE), Young (1968) feels that the sacculus and rami may represent the fourth or fifth pair of coelomoducts, and function in osmoregulation and water storage in addition to serving as sperm receptors.

Glands

Two basic idiosomal gland patterns have been identified in the Acari. The simpler of these is found in the Parasitiformes (page 103) and consists of paired acinous salivary glands in the dorsal region of the propodosoma. The gland components feed into paired common ducts which empty in the buccal area. In Gamasida, the ducts open into elongate stylets (*siphunculi* of van der Hammen 1964) which empty in the region of the hypostome. Their products may assist in pre-oral digestion. A pair of globose glands of uncertain function open into gastric caeca III in certain Gamasida (Woodring and Galbraith 1976).

The acinous glands of Ixodida feed into common ducts which empty at the level of the capitulum. These glands appear to serve multiple functions. A cement-like substance is produced by the salivary glands of many hard ticks (IXODIDAE) which serves to secure the tick to its host (Chinery 1965). Roshdy (1966) identified two types of alveoli in salivary gland tissues of a soft tick, *Argas (Persicargas) persicus* (Oken) (ARGASIDAE), one of which appears to secrete an anticoagulant.

A more complex gland pattern is found in the Acariformes (page 103). Paired secretory ducts, or podocephalic canals, receive the products of the coxal glands and one to three additional gland pairs. In the Oribatida, certain of these glands may have an endocrine function associated with molting, while others appear to function as salivary glands (Woodring 1973). Alberti (1973) and Alberti and Storch (1973, 1974) studied the podocephalic gland complex in the BDELLIDAE and TETRANYCHIDAE (Actinedida) and found that a major function of the podocephalic system is the production of silk. They have postulated that podocephalic glands 3 (Fig. 16, gl_3), in concert with the independent infracapitular glands, are involved in silk production in the BDELLIDAE. Silk production in the TETRANYCHIDAE is considered to be the function of large paired unicellular glands which open at the palpal tips. Silk is produced as a thread with the aid of a hollow terminal palptarsal hair. Neither glands nor hollow hairs are present in *Bryobia rubrioculus* (Scheuten) and *B. praetiosa* Koch, tetranychids which do not spin silk. Mills (1973) considered the entire podocephalic complex of *Tetranychus urticae* Koch to be involved in silk production, with the tubular coxal glands (= nephridia) of the female secreting a pheromone additive which attracts nearby males to the female webbing.

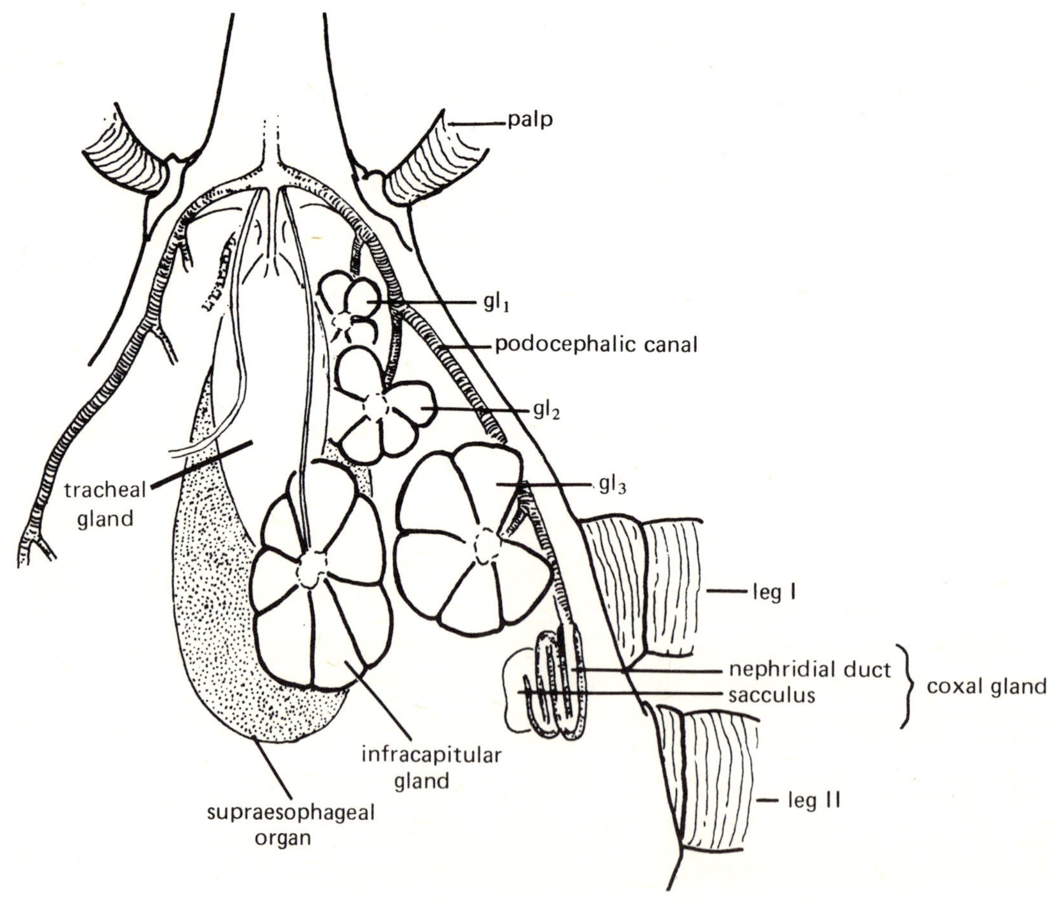

Fig. 16. Idiosomal gland complex of the bdellid type (after Grandjean 1938, Alberti 1973, and Alberti and Storch 1973).

Moss (1962) identified additional idiosomal glands in *Allothrombium* (Actinedida, TROMBIDIIDAE) including a salivary gland independent of the podocephalic system, and two venom glands associated with the chelicerae and with the hypostome.

Nervous System

Lying around the esophagus is a well developed central nervous system consisting of supra- and subesophageal ganglia from which a series of nerves radiate (Plate 6-2, p. 31). The legs, digestive system, musculature and genitalia appear to be innervated by nerves arising on the subesophageal portion of the brain (Winkler 1888, Ainscough 1960, Moss 1962, Sonenshine 1970, Woodring and Galbraith 1976). The mouthparts receive impulses from the dorsal ganglion, although Vijayambika and John (1975, 1976a and 1976b) found that the chelicerae of *Lardoglyphus konoi* (Sasa and Asanuma) (Acaridida, ACARIDAE) are innervated by subesophageal nerves. The optic nerves also are derived from the supraesophageal ganglion. Tsvileneva (1965) feels that only motor and association neurons are found in the brain of ixodid ticks (Ixodida), and that sensory neuron bodies lie in a peripheral sensory neuropile. The structure and function of the acarine peripheral nervous system as a receiver of stimuli from surface receptors has not been studied.

Useful References

Aboul-Nasr, A.E. and T.T.M. Bassal (1971). Biochemical and physiological studies of certain ticks (Ixodoidea). The sugar content and concentration in *Argas* and *Hyalomma* biological fluids. J. Med. Ent. **8**(5):521-524.

Ainscough, B.D. (1960). The internal morphology of *Caminella peraphora* Krantz and Ainscough, with descriptions of the immature stages. Oregon State University Thesis LD 4330:53 pp.

Alberti, G. (1973). Ernährungsbiologie und Spinnvermögen der Schnabelmilben (Bdellidae, Trombidiformes). Z. Morph. Tiere **76**:285-338.

Alberti, G. (1974). Fortpflanzungsverhalten und Fortpflanzungsorgane der Schnabelmilben (Acarina:Bdellidae, Trombidiformes). Z. Morph. Tiere **78**:111-157.

Alberti, G. and V. Storch (1973). Zur Feinstruktur der "Munddrüsen" von Schnabelmilben (Bdellidae, Trombidiformes). Z. wiss. Zool., Leipzig **186**(1/2):149-160.

Alberti, G. and V. Storch (1974). Über Bau und Funktion der Prosoma-Drüsen von Spinnmilben (Tetranychidae, Trombidiformes). Z. Morph. Tiere **79**:133-153.

Araman, S.F. and A. Said (1972). Biochemical and physiological studies of certain ticks (Ixodoidea). The ionic regulatory role of the coxal organs of *Argas (Persicargas) persicus* (Oken) and *A. (P.) arboreus* Kaiser, Hoogstraal and Kohls. J. Parasitol. **58**(2):348-353.

Arthur, D.R. (1956). The morphology of the British Prostriata with particular reference to *Ixodes hexagonus* Leach. III. Parasitol. **46**:261-307.

Athias-Henriot, C. (1968). L'appareil d'insémination laelapoïde (Acariens anactinotriches:Laelapoidea, ♀♀) Possibilité d'emploi à des fins taxonomiques. Bull. Sci. Bourgogne **25**:229-274.

Athias-Henriot, C. (1969a). Les organes cuticulaires sensoriels et glandulaires des gamasides. Poroïdotaxie et adénotaxie. Bull. Soc. Zool. France **94**(3):485-492.

Athias-Henriot, C. (1969b). Notes sur la morphologie externe des gamasides (Acariens anactidotriches). Acarologia **11**(4):609-629.

Athias-Henriot, C. (1970). Observations sur la morphologie externe des gamasides. Mises au point terminologiques. Acarologia **12**(1):25-27.

Athias-Henriot, C. (1975). The idiosomatic euneotaxy and epineotaxy in gamasids (Arachnida, Parasitiformes). Z. Zool. Syst-Evolut.-forsch. **13**(2):97-109.

Balashov, Y.S. (1959). The excretion processes and activity of Malpighian tubes of the ticks. Parazit. Sborn Zool. Inst., Akad. Nauk SSSR. **18**:120-128.

Balashov, Y.S. (1961). The structure of digestive organs and the blood digestion in Argasidae. Parazit. Sbornik. Zool. Inst. Akad. Nauk. SSSR **20**:185-225.

Balashov, Y.S. (1972). Bloodsucking ticks (Ixodoidea)-vectors of diseases of man and animals. Misc. Publ. Ent. Soc. Amer. **8**(5):161-376.

Beklemishev, W.N. (1969). Principles of Comparative Anatomy of Invertebrates. Vol. 2. Organology. Univ. Chicago Press: 529 p. + vii.

Belozerov, V.N. (1960). Structure of the integument of gamasid mites (Parasitiformes, Gamasoidea). Ent. oboz. **39**(4):850-859.

Berger, R.S. (1972). 2,6-dichlorophenol, sex pheromone of the lone star tick. Science 177:704-705.

Berger, R.S., J.C. Dukes and Y.S. Chow (1971). Demonstration of a sex pheromone in three species of hard ticks. J. Med. Ent. **8**(1):84-86.

Blauvelt, W.E. (1945). The internal morphology of the common red spider *(Tetranychus telarius* Linn.). Cornell Univ. Agr. Exp. Sta. Memoir 270:35 pp. + 11 plates.

Boctor, F.N. (1972). Biochemical and physiological studies of certain ticks (Ixodoidea). Free amino acids in female *Argas (Persicargas) arboreus* Kaiser, Hoogstraal and Kohls (Argasidae) analyzed by gas-liquid chromatography. J. Med. Ent. **9**(3):201-204.

Brody, A.R. (1970). Observations on the fine structure of the developing cuticle of a soil mite *Oppia coloradensis* (Acarina:Cryptostigmata). Acarologia **12**(2):421-431.

Brody, A.R., J.C. McGrath and G.W. Wharton(1972). *Dermatophagoides farinae:* the digestive system. J. New York Ent. Soc. **80**(3):152-177.

Camin, J.H. (1953). Observations on the life history and sensory behavior of the snake mite, *Ophionyssus natricis* (Gervais) (Acarina:Macronyssidae). Chicago Acad. Sci. Special Publ. 10:75 pp.

Chinery, W.A. (1965). Studies on the various glands of the tick *Haemaphysalis spinigera* Neum. 1897. Part III. The salivary glands. Acta Trop. **22**:321-349.

Chinery, W.A. (1973). The nature and origin of the "cement" substance at the site of attachment and feeding of adult *Haemaphysalis spinigera* (Ixodidae). J. Med. Ent. **10**(4):355-362.

Clifford, C.M. and G. Anastos. (1960). The use of chaetotaxy in the identification of larval ticks. J. Parasitol. **46**(5):567-578.

Coineau, Y. (1964). Contribution à l'etude des Caeculidae. Première série: Developpement postlarvaire de *Allocaeculus catalanus* Franz 1954. Deuxième partie: La chètotaxie des pattes. Acarologia **6**(1):47-72.

Coineau, Y. (1970). A propos de l'oeil antérieur du naso des Caeculidae. Acarologia **12**(1):109-118.

Coineau, Y. (1974). Éléments pour une monographie morphologique, écologique et biologique des Caeculidae (Acariens). Mém. Mus. nat. d'Hist. natur. (N.S.) A, 81:299 pp. + vi.

Cone, W.W., L.M. McDonough, J.C. Maitlen and S. Burdajewicz (1971a). Pheromone studies of the twospotted spider mite. 1. Evidence of a sex pheromone. J. Econ. Ent. **64**(2):358-361.

Cone, W.W., S. Predki and E.C. Klostermeyer (1971b). Pheromone studies of the twospotted spider mite. 2. Behavioral response of males to quiescent deutonymphs. J. Econ. Ent. **64**(2):379-382.

Costa, M. (1966). Notes on macrochelids associated with manure and coprid beetles in Israel. 1. *Macrocheles robustulus* (Berlese 1904), development and biology. Acarologia **8**(4):532-548.

Cutler, B. and A.G. Richards (1972). Sclerotization and the localization of brown and black colors in chelicerates (Arthropoda). Zool. Jahr. Anat. **89**:404-421.

Devine, T.L. and G.W. Wharton (1973). Kinetics of water exchange between a mite, *Laelaps echidnina,* and the surrounding air. J. Insect Physiol. **19**:243-254.

Douglas, J.R. (1943). The internal anatomy of *Dermacentor andersoni* Stiles. Univ. Calif. Publ. Ent. **7**:207-272.

Dosse, G. (1958). Die Spermathecae, ein zusatzliches Bestimmungsmerkmal bie Raubmilben (Acar., Phytoseiidae). Pflanz. Ber. **20**(1-2):1-11.

Efford, I.E. (1966). Observations on the life history of three stream-dwelling watermites. Acarologia **8**(1):86-93.

Ehara, S. (1960). Comparative studies on the internal anatomy of three Japanese trombidiform acarinids. J. Fac. Sci. Hokkaido Univ., ser. **VI** (Zool)**14**:410-434.

Evans, G.O. (1963a). Observations on the chaetotaxy of the legs in the free-living Gamasina (Acari:Mesostigmata). Bull. Brit. Mus. (Nat. Hist.) Zool. **10**(5):277-303.

Evans, G.O. (1963b). Some observations on the chaetotaxy of the pedipalps in the Mesostigmata (Acari). Ann. Mag. Nat. Hist **6**(13):513-527.

Evans, G.O., J.G. Sheals and D. MacFarlane (1961). The Terrestrial Acari of the British Isles. Vol. 1. Introduction and Biology. British Museum (Natural History), London: 219 pp.

Evans, G.O. and W.M. Till (1965). Studies on the British Dermanyssidae (Acari:Mesostigmata). Part I. External morphology. Bull. Brit. Mus. (Nat. Hist.) Zool. **13**(8):249-294.

Fain, A. (1963). La spermatheque et ses canaux adducteurs chez les acariens mesostigmatiques. Acarologia **5**(4):463-479.

Fain, A. and R. Domrow (1974). The subgenus *Cytostethum* Domrow (Acari:Atopomelidae): multiple speciation in the marsupial *Potorous tridactylus* (Kerr.). Austral. J. Zool. **22**:549-572.

Farish, D.J. and R.C. Axtell (1966). Sensory functions of the palps and first tarsi of *Macrocheles muscaedomesticae* (Acarina:Macrochelidae), a predator of the house fly. Ann. Ent. Soc. Amer. **59**(1):165-170.

Gorirossi, F.E. (1956). The gnathosoma of *Megalolaelaps ornata* (Acarina Mesostigmata Gamasides). Amer. Midland Nat. **55**(2):357-362.

Graf, J.-F. (1975). Écologie et éthologie d'*Ixodes ricinus* L. en Suisse (Ixodoidea; Ixodidae). Clinquième note: mise en evidence d'une phéromone sexuelle chez *Ixodes ricinus.* Acarologia **17**(3):436-441.

Grandjean, F. (1934). Les organes respiratoires secondaires des Oribates (Acariens). Ann. Soc. ent. France **103**:109-146.

Grandjean, F. (1935a). Les poiles et les organes sensitifs portées par les pattes et le palpe chez les Oribates. Bull. soc. zool. France **60**(1):6-39.

Grandjean, F. (1935b). Observations sur les Acariens (1ère sér.). Bull. Mus. nat. Hist. natur. Paris (2) **7**:119-126.

Grandjean, F. (1936). Un acarien synthétique: *Opilioacarus segmentatus* With. Bull. Soc. Afr. Nord **27**:413-444.

Grandjean, F. (1938). Observations sur les Bdelles (Acariens). Ann. Soc. ent. France **107**:1-24.

Grandjean, F. (1943). Quelques genres d'Acariens appartenant au groupe des Endeostigmata (2e série). Ann. Sci. nat., Zool. (11)**4**:85-135, **5**:1-59.

Grandjean, F. (1944). Les "taenidies" des Acariens. C.R. Séanc. Soc. Phys. Hist. nat. Genève **61**:142-146.

Grandjean, F. (1946). Au sujet de l'organe de Claparède des eupathidies multiples et des taenidies mandibulaires chez les Acariens actinochitineux. Arch. Sci. phys. natur. Genève **28**(5):63-87.

Grandjean, F. (1951). Comparaison du genre *Limnozetes* au genre *Hydrozetes* (Oribates). Bull. Mus. nat. Hist. natur. (2e série) **23**(2):200-207.

Grandjean, F. (1956). Observations sur les Oribates (34e série). Bull. Mus. nat. Hist. natur. **28**(2):205-212.

Grandjean, F. (1964). La solenidiotaxie des Oribates. Acarologia **6**(3):529-556.

Grandjean, F. (1968). Nouvelles observations sur les Oribates (6e série). I. Tecta acetabulaires, fossé podocéphalique, canal podocéphalique, épine supracoxale du segment des pattes I, trachées. Acarologia **10**(2):357-391.

Grandjean, F. (1969). Stases. Actinopiline. Rappel de ma classification des Acariens en 3 groupes majeurs. Terminologie en soma. Acarologia **11**(4):796-827.

Grandjean, F. (1970). Nouvelles observations sur les Oribates (7e série). Acarologia **12**(2):432-460.

Grandjean, F. (1971). Nouvelles observations sur les Oribates (8e série). Acarologia **12**(4):849-876.

Hajjar, N.P. (1972). Biochemical and physiological studies of certain ticks (Ixodoidea). Phospholipid and sterol patterns in biological fluids of nymphal and adult *Hyalomma (H.) dromedarii* Koch and *H. (H.) anatolicum excavatum* Koch (Ixodidae). J. Med. Ent. **9**(4):281-285.

Hammen, L. van der (1961). Description of *Holothyrus grandjeani* nov. spec., and notes on the classification of the mites. Nova Guinea n.s. **10**(9):173-194.

Hammen, L. van der (1963). The addition of segments during the postembryonic ontogenesis of the Actinotrichida (Acarida) and its importance for the recognition of the primary subdivision of the body and the original segmentation. Acarologia **5**(3):443-454.

Hammen, L. van der (1964). The morphology of *Glyptholaspis confusa* (Foá, 1900) (Acarida, Gamasina). Zool. Verhandl. **71**:56 pp.

Hammen, L. van der (1966). Studies on Opilioacarida I. Description of *Opilioacarus texanus* (Chamberlain and Mulaik) and revised classification of the genera. Zool. Verhandl. 86:80 pp.

Hammen, L. van der (1968a). Stray notes on *Acarida* (Arachnida) 1. Zool. Meded. **42**(35):261-280.

Hammen, L. van der (1968b). The gnathosoma of *Hermannia convexa* (C.L. Koch) (Acarida:Oribatina) and comparative remarks on its morphology in other mites. Zool. Verhandl. 94:45 pp.

Hammen, L. van der (1969). Notes on the morphology of *Alycus roseus* C.L. Koch. Zool. Meded. **43**(15): 177-202.

Hammen, L. van der (1970a). Remarques générales sur la structure fondementale du gnathosoma. Acarologia **12**(1):16-22.

Hammen, L. van der (1970b). La segmentation des appendices chez les Acariens. Acarologia **12**(1):11-15.

Hammen, L. van der (1974). La terminologie acarologique. La Banque des Mots **8**:205-218.

Hazan, A., U. Gerson and A.S. Tahori (1974). Spider mite webbing I. The production of webbing under various environmental conditions. Acarologia **16**(1):68-84.

Hazan, A., U. Gerson and A.S. Tahori (1975a). Spider mite webbing II. The effect of webbing on egg hatchability. Acarologia **17**(1):270-273.

Hazan, A., A.S. Tahori and U. Gerson (1975b). Spider mite webbing III. Solubilization and amino acid composition of the silk protein. Comp. Biochem. Physiol. **51B**:457-462.

Hinton, H.E. (1967). The structure of the spiracles of the cattle tick, *Boophilus microplus.* Austral. J. Zool. **15**:941-945.

Hinton, H.E. (1971). Plastron respiration in the mite, *Platyseius italicus.* J. Ins. Physiol. **17**:1185-1199.

Hughes, T.E. (1949). The functional morphology of the mouthparts of *Liponyssus bacoti.* Ann. Trop. Med. Parasitol. **43**:349-360.

Hughes, T.E. (1952). The morphology of the gut of *Bdellonyssus bacoti* (Hirst 1913, Fonesca 1941). Ann. Trop. Med. Parasitol. **46**:54-60.

Hughes, T.E. (1958). The respiratory system of the mite *Cheyletus eruditus* (Schrank, 1781). Proc. Zool. Soc. London **130**(2):231-239.

Hughes, T.E. (1959). The gnathosoma and the appendages. **In** Mites, or the Acari. Univ. London, Athlone Press: 135-156.

Jalil, M. and J.G. Rodriguez (1970). Biology of and odor perception by *Fuscuropoda vegetans* (Acarina: Uropodidae), a predator of the house fly. Ann. Ent. Soc. Amer. **63**(4):935-938.

Karg, W. (1961). Ökologische Untersuchungen von edaphischen Gamasiden (Acarina, Parasitiformes). Pedobiologia **1**(1,2):53-74, 77-98.

Kinn, D.N. (1971). The life cycle and behavior of *Cercoleipus coelonotus* (Acarina:Mesostigmata) including a survey of phoretic mite associates of California Scolytidae. Univ. Calif. Publ. Ent. 65:66 pp.

Kirchner, W.P. (1967). Spermatophoren bei Halacariden (Acarina). Naturw. **54**(13):345-346.

Knülle, W. (1959). Morphologische und Entwicklungsgeschictliche untersuchungen zum phylogenetischen System der Acari: Acariformes Zachv. II. Acaridiae:Acaridae. Mitt. Zool. Mus. Berlin **35**(2):347-417.

Krantz, G.W. (1974). *Phaulodinychus mitis* (Leonardi 1899) (Acari:Uropodidae), an intertidal mite exhibiting plastron respiration. Acarologia **16**(1):11-20.

Krantz, G.W. and J.L. Mellott (1968). Two new species of *Macrocheles* (Acarina:Macrochelidae) with notes on their host-specific relationships with geotrupine beetles (Scarabaeidae:Geotrupinae). J. Kansas Ent. Soc. **41**(1):48-56.

Krantz, G.W. and J.G. Wernz (1979). Sperm transfer in Glyptholaspis americana. Recent Advances in Acarology **2**:441-446.

Kuwahara, Y. (1976). Alarm pheromones produced by several grain mites. Proc. Symposium "Insect Pheromones and their Applications." Agr. Forestry Fish. Res. Council, Tokyo: 65-76.

Lanciani, C.A. (1972). Mating behavior of water mites of the genus *Eylais.* Acarologia **14**(4):631-637.

Lee, D.C. (1970). The Rhodacaridae (Acari:Mesostigmata); classification, external morphology and distribution of genera. Rec. S. Austral. Mus. **16**(3):1-219.

Lee, D.C. (1974). Rhodacaridae (Acari:Mesostigmata) from near Adelaide, Australia III. Behaviour and development. Acarologia **16**(1):21-44.

Lipovsky, L.J., G.W. Byers and E.H. Kardos (1957). Spermatophores—the mode of insemination of chiggers (Acarina:Trombiculidae). J. Parasitol. **43**:256-262.

McEnroe, W.D. (1971). The red photoresponse of the spider mite *Tetranychus urticae* (Acarina:Tetranychidae). Acarologia **13**(1):113-118.

Manton, S.M. (1958). Hydrostatic pressure and leg extension in arthropods, with special reference to arachnids. Ann. Mag. Nat. Hist. ser. 13, **1**:161-182.

Michael, A.D. (1892). On the variations in the internal anatomy of the Gamasinae, especially in that of the genital organs, and their mode of coition. Trans. Linn. Soc. London, ser. 2, **5**:281-324.

Michael, A.D. (1896). The internal anatomy of *Bdella.* Linn. Soc. London (Zool.) **6**(7):477-528.

Mills, L.R. (1973). Morphology of glands and ducts in the two-spotted spider-mite, *Tetranychus urticae* Koch, 1836. Acarologia **15**(2):218-236.

Mitchell, R.D. (1958). Sperm transfer in the water-mite *Hydryphantes ruber* Geer. Amer. Midl. Nat. **60**(1): 156-158.

Mitchell, R.D. (1962). The structure and evolution of water mite mouthparts. J. Morphol. **110**(1):41-59.

Mitchell, R.D. (1964). The anatomy of an adult chigger mite *Blankaartia acuscutellaris* (Walch). J. Morphol. **114**(3):373-391.

Mitchell, R.D. (1972). The tracheae of water mites. J. Morphol. **136**(6):327-335.

Mitchell, R.D. and M. Nadchatram (1969). Schizeckenosy: the substitute for defecation in chigger mites. J. Nat. Hist. **3**:121-124.

Moss, W.W. (1962). Studies on the morphology of the trombidiid mite *Allothrombium lerouxi* Moss (Acari). Acarologia **4**(3):313-345.

Naegele, J.A., W.D. McEnroe and A.B. Soans (1965). Spectral sensitivity and orientation in the two-spotted spider mite, *Tetranychus urticae* K. from 350 mμ to 700 mμ. J. Ins. Physiol. **12**(9):1187-1195.

Newell, I.M. (1945). *Hydrozetes* Berlese (Acari, Oribatoidea); the occurrence of the genus in North America, and the phenomenon of levitation. Trans. Connecticut Acad. Arts and Sci. **36**:253-275.

Obenchain, F.D. and J.H. Oliver Jr. (1974). Undescribed glandular tissues from *Amblyomma tuberculatum* (Arachnida:Parasitiformes:Ixodidae). Bull. ASB **21**(2):72.

Oldfield, G.N., R.F. Hobza and N.S. Wilson (1970). Discovery and characterization of spermatophores in the Eriophyoidea (Acari). Ann. Ent. Soc. Amer. **63**(2):520-526.

Oldfield, G.N., I.M. Newell and D.K. Reed (1972). Insemination of protogynes of *Aculus cornutus* from spermatophores and description of the sperm cell. Ann. Ent. Soc. Amer. **65**(5):1080-1084.

Oliver, J.H. Jr., Z. Al-Ahmadi and R.L. Osburn (1974). Reproduction in ticks (Acari:Ixodoidea). 3. Copulation in *Dermacentor occidentalis* Marx and *Haemaphysalis leporispalustris* (Packard) (Ixodidae). J. Parasitol. **60**(3):499-506.

Oudemans, A.C. (1911). Acarologische Aanteekeningen XXXVIII. Namen voor Lichaamsafdeelingen. Ent. Ber. **3**:183-184.

Pauli, F. (1948). Zur Biologie einiger Belbiden und zur Funktion über pseudostigmatischen Organe. Zool. Jahr. **84**:275-328.

Popp, E. (1967). Die Begattung bei den Vogelmilben *Pterodectes* Robin (Analgesoidea, Acari). Z. Morph. Ökol. Tiere **59**:1-32.

Pound, J.M. and J.H. Oliver, Jr. (1976). Reproductive morphological spermatogenesis in *Dermanyssus gallinae* (DeGeer) (Acari:Dermanyssidae). J. Morphol. **150**(4):825-842.

Radinovsky, S. (1965). The biology and ecology of granary mites of the Pacific Northwest. IV. Various aspects of the reproductive behavior of *Leiodinychus krameri* (Acarina:Uropodidae). Ann. Ent. Soc. Amer. **58**(3):267-272.

Radovsky, F.J. (1969). Adaptive radiation in the parasitic Mesostigmata. Acarologia **11**(3):450-483.

Regev, S. and W.W. Cone (1975). Evidence of farnesol as a male sex attractant of the twospotted spider mite, *Tetranychus urticae* Koch (Acarina:Tetranychidae). Environ. Ent. **4**(2):307-311.

Regev, S. and W.W. Cone (1976). Analysis of pharate female twospotted spider mites for nerolidol and geraniol: evaluation for sex attraction of males. Environ. Ent.**5**(1):133-138.

Reuter, E. (1909). Zur Morphologie und Ontogenie der Acariden mit besonderer Berücksichtigung von *Pediculopsis graminum*. Acta Soc. Sci. Fenn. **36**(4):1-287.

Rohde, C.J. and D.A. Oemick (1967). Anatomy of the digestive and reproductive systems in an acarid mite (Sarcoptiformes). Acarologia **9**(3):608-616.

Roshdy, M.A. (1966). Comparative internal morphology of subgenera of *Argas* ticks (Ixodoidea, Argasidae). 4. Subgenus *Ogadenus: Argas brumpti* Neumann, (1907). J. Parasitol. **52**(4):776-782.

Roshdy, M.A. (1972). The subgenus *Persicargas* (Ixodoidea, Argasidae, *Argas*). 15. Histology and histochemistry of the salivary glands of *A. (P.) persicus* (Oken). J. Med. Ent. **9**(2):143-148.

Roshdy, M.A., R.F. Foelix and R.C. Axtell (1972). The subgenus *Persicargas* (Ixodoidea:Argasidae:*Argas*). 16. Fine structure of Haller's organ and associated tarsal setae of adult *A. (P.) arboreus* Kaiser, Hoogstraal and Kohls. J. Parasitol. **58**(4):805-816.

Roshdy, M.A. and T. Hefnawy (1973). The functional morphology of *Haemaphysalis* spiracles (Ixodoidea: Ixodidae). Z. Parasitenk. 42:10 pp.

Schulz, P. (1942). Über die Hautsinnesorgane der Zecken, besonders über eine bisher unbekannte Art von Arthropodensinnesorgane, die Krobylophoren. Zeitschr. Morph. Ökol. Tiere **38**:379-419.

Schuster, R. (1962). Nachweis eines Paarhungszeremoniells bei den Hornmilben (Oribatei, Acari). Naturw. **49**(21):502-503.

Schuster, R. (1972). Spinnvermögen der Tydeiden (Milben). Naturw. **59**(6):275.

Schuster, R. and I.J. Schuster (1966). Über das Fortpflanzungsverhalten von Anystiden-Männchen (Acari, Trombidiformes). Naturw. **53**(6):162-163.

Schuster, R. and I.J. Schuster (1969). Gesteilte Spermatophoren bei Labidostomiden (Acari, Trombidiformes). Naturw. **56**(3):145-146.

Schuster, I.J. and R. Schuster (1970). Indirekte Spermaübertragung bei Tydeidae (Acari, Trombidiformes). Naturw. **57**(5):256-257.

Sonenshine, D.E. and J.D. Gregson (1970). A contribution to the internal anatomy and histology of the bat tick *Ornithodoros kelleyi* Cooley and Kohls, 1941 I. The alimentary system, with notes on the food channel in *Ornithodoros denmarki* Kohls, Sonenshine and Clifford, 1965. J. Med. Ent. **7**(1):46-64.

Sonenshine, D.E., R.M.Silverstein, E.C. Layton and P.J. Homsher (1974). Evidence for the existence of a sex pheromone in 2 species of ixodid ticks (Metastigmata:Ixodidae). J. Med. Ent. **11**(3):307-315.

Sonenshine, D.E., R.M. Silverstein, E. Plummer, J.R. West and Br.T. McCullough (1976). 2,6-dichlorophenol, the sex pheromone of the Rocky Mountain wood tick, *Dermacentor andersoni* Stiles and the American dog tick, *Dermacentor variabilis* (Say). J. Chem. Ecol. **2**(2):201-209.

Sternlicht, M. and D.A. Griffiths (1974). The emission and form of spermatophores and the fine structure of adult *Eriophyes sheldoni* Ewing (Acarina, Eriophyoidea). Bull. Ent. Res. **63**:561-565.

Summers, F.M. and R.L. Witt (1971). The gnathosoma of *Cheyletus cacahuamilpensis* Baker (Acarina:Cheyletidae). Proc. Ent. Soc. Wash. **73**(2):158-168.

Summers, F.M., R.H. Gonzalez-R. and R.L. Witt (1973). The mouthparts of *Bryobia rubrioculus* (Sch.) (Acarina:Tetranychidae). Proc. Ent. Soc. Wash. **75**(1):96-111.

Teravsky, I.K. (1957). On the formed elements of the haemolymph of ticks of the family Argasidae. Zool. Zh. **36**(10):1448-1454 (Translation 256, NAMRU-3, Cairo).

Theis, G. and R. Schuster (1974). Gesteilte Tröpfchenspermatophoren bei Calyptostomiden (Acari, Trombidiformes). Mitt. Naturw. Ver. Steiermark **104**:183-185.

Tsvileneva, V.A. (1965). Architectonics of nervous elements in the brain of ixodid ticks (Acarina, Ixodidae). Ent. Obozr. **44**:241-257.

Vannier, G. (1976). Principaux modes d'étude de la balance hydrique chez les acariens. Acarologia **18**(1):3-19.

Vercammen-Grandjean, P.-H. (1975). Les organes des Claparède et les papilles génitales de certains acariens sont-ils des organes respiratoires? Acarologia **17**(4):624-630.

Vijayambika, V. and P.A. John (1975). Internal morphology and histology of the fish mite *Lardoglyphus konoi* (Sasa and Asanuma) (Acarina:Acaridae). 3. Nervous system. Acarologia **17**(1):114-119.

Vijayambika, V. and P.A. John (1976a). Internal morphology and histology of the post-embryonic stages of the fish mite *Lardoglyphus konoi* (Sasa and Asanuma). Acarina:Acaridae. 2. Protonymph. Acarologia **18**(1):133-137.

Vijayambika, V. and P.A. John (1976b). Internal morphology and histology of the post-embryonic stages of the fish mite *Lardoglyphus konoi* (Sasa and Asanuma). Acarina:Acaridae. 3. Deutonymph. Acarologia **18**(1):138-142.

Wagner-Jevseenko, O. (1958). Fortpflanzung bei *Ornithodoros moubata* und genitale ubertragung von *Borrelia duttoni.* Acta Trop. **15**(2):117-168.

Wallace, M.M.H. and J.A. Mahon (1972). The taxonomy and biology of Australian Bdellidae (Acari) 1. Subfamilies Bdellinae, Spinibdellinae and Cytinae. Acarologia **14**(4):544-580.

Wernz, J.G. and G.W. Krantz (1976). Studies on the function of the tritosternum in selected Gamasida (Acari). Can. J. Zool. **54**:202-213.

Wharton, G.W. and A.R. Brody (1972). The peritrophic membrane of the mite *Dermatophagoides farinae:* Acariformes. J. Parasitol. **58**(4):801-804.

Wharton, G.W. and T.L. Devine (1968). Exchange of water between a mite, *Laelaps echidnina,* and the surrounding air under equilibrium conditions. J. Ins. Physiol. **14**:1303-1318.

Wharton, G.W., W. Parrish and D.E. Johnston (1968). Observations on the fine structure of the cuticle of the spiny rat mite, *Laelaps echidnina* (Acari-Mesostigmata). Acarologia **10**(2):206-214.

Winkler, W. (1888). Anatomie der Gamasiden. Arbeit. Zool. Inst. Univ. Wien **7**(3):1-38.

Witte, H. (1975). Funktionsanatomie der Genitalorgane und Fortpflanzungsverhalten bei dem Männchen der Erythraeidae (Acari, Trombidiformes). Z. Morph. Tiere **80**:137-180.

Woodring, J.P. (1970). Comparative morphology, homologies, and functions of the male system in oribatid mites (Arachnida:Acari). J. Morphol. **132**(4):425-451.

Woodring, J.P. (1973). Comparative morphology, functions, and homologies of the coxal glands in orabatid mites (Arachnida:Acari). J. Morphol. **139**(4):407-429.

Woodring, J.P. and C.A. Galbraith (1976). The anatomy of the adult uropodid *Fuscuropoda agitans* (Arachnida:Acari), with comparative observations on other Acari. J. Morphol. **150**(1):19-58.

Woolley, T.A. (1972). Scanning electron microscopy of the respiratory apparatus of ticks. Trans. Amer. Microscop. Soc. **91**(3):348-363.

Wright, K.A. and I.M. Newell (1964). Some observations on the fine structure of the midgut of the mite *Anystis* sp. Ann. Ent. Soc. Amer. **57**(6):684-693.

Young, J.H. (1968). The morphology of *Haemogamasus ambulans* II. Reproductive system. J. Kansas Ent. Soc. **41**(4):532-543.

Zakhvatkin, A.A. (1952). Division of the Acarina into orders and their position in the system of the Chelicerata. Mag. Parasitol. Moscow **14**:5-46.

Zolotarev, E.K. (1962). The leg and its terminology in parasitiform ticks. Zool. Zh. **41**(11):1739-1741. (Translation 306, NAMRU-3, Cairo).

IV. REPRODUCTION AND EMBRYOGENESIS

Although reproduction in the Acari generally follows the classic pattern of fertilization and production of male and female progeny, facultative parthenogenesis is known to occur throughout the subclass (Oliver 1971). *Arrhenotoky*, the production of haploid males from unfertilized eggs, has been observed in the Gamasida and the Actinedida (Boudreaux 1946, Filipponi 1957, Hansell et al. 1964, Summers and Witt 1972, Cicolani et al. 1975). Production of females from unfertilized eggs, or *thelytoky,* is common amongst certain Gamasida Filipponi et al. 1971) and Actinedida. Thelytoky has also been observed in ixodid ticks (Belikova 1966, Hoogstraal et al. 1968, Oliver 1971, 1974) and in some Oribatida (Grandjean 1947, Woodring and Cook 1962). *Amphoterotoky,* in which unfertilized eggs give rise to both male and female offspring, has been reported or suggested only in the Acaridida (Hughes and Jackson 1958, Heinemann and Hughes 1969).

Embryonic development in the Acari is, at best, poorly known. Citations to literature on acarine embryology are few, and nearly half of them are dated before 1920 (van der Hammen 1972). Based on available information, it appears that total cleavage of the primordial cytoplasmic mass does not occur, although exceptions have been noted in the PYGMEPHORIDAE (Reuter 1909) and TETRANYCHIDAE (Dittrich 1965). The nucleus divides within the cytoplasm and migrates to the surface, or migration occurs prior to cleavage. A period of great cytoplasmic growth follows. The nuclei continue to divide and eventually form an envelope called the *blastoderm,* within which the yolk is deposited. A few blastoderm nuclei enter the yolk area where, as *vitellophage cells,* they liquefy the yolk and make it readily available to the developing blastoderm. Aeschlimann (1958) observed that the vitellophage cells of *Ornithodoros moubata* (Murray) (Ixodida, ARGASIDAE) persist following yolk liquefaction and take part in the formation of the digestive system. According to Aeschlimann, both the nuclei and chromosomes dissolve during oogenesis in *O. moubata,* only to reform again following oviposition.

Yolk liquefaction signals the development of the *germinal band* which differentiates ventrad of the primordial ventriculus. The germinal area eventually gives rise to both the gnathosomal and idiosomal appendages, the latter usually consisting of only three pairs of limb buds. Four pairs have been observed in early developmental stages in some species (Sicher 1891, Supino 1895) but the fourth pair is resorbed at the time of pedipalpal differentiation. Aeschlimann (1958) found that legs IV withdraw toward the end of embryogenesis and reappear after the larval molt.

Balashov (1972) recognizes five stages in oocyte development in the Ixodida, beginning with a phase in which slight growth occurs, passing through periods of "great cytoplasmic growth" and yolk deposition, and ending with the passage of the mature oocyte into the lumen or to the oviducts after detaching from the ovary walls.

Beament (1951) has studied egg formation in *Panonychus ulmi* (Koch) (Actinedida, TETRANYCHIDAE), and has determined that the eggshell is picked up by the mature embryo in the ovary prior to passage into the ovipositor. According to Witte (1975), the eggshell of ERYTHRAEIDAE (Actinedida) is a product of the oocyte itself.

Acarine chromosomes generally are few in number. Diploid chromosome counts as high as 36 are recorded in the Acari (Oliver 1967), but counts of 2-4 have been observed most frequently (Hansell et al. 1964, Oliver and Nelson 1967, Regev 1974). Few animal or plant groups share this characteristic.

Useful References

Aeschlimann, A. (1958). Développement embryonnaire d'*Ornithodoros moubata* (Murray) et transmission transovarienne de *Borrelia duttoni.* Acta Trop. **15**(1):15-64.

Balashov, Y.S. (1972). Bloodsucking ticks (Ixodoidea)-vectors of diseases of man and animals. Misc. Publ. Ent. Soc. Amer. **8**(5):161-376.

Beament, J.W.L. (1951). The structure and formation of the egg of the fruit tree spider mite, *Metatetranychus ulmi* Koch. Ann. Appl. Biol. **38**:1-24.

Belikova, N.P. (1966). Observations on development of the bisexual and parthenogenetic population of *Haemaphysalis neumanni* D. in Primor'ye region. Tezisy Dokl. Yubil. Konf., Vladivostok:3 pp. (Translation 295, NAMRU-3, Cairo).

Boudreaux, H.G. (1956). Revision of the two-spotted spider mite (Acarina, Tetranychidae) complex, *Tetranychus telarius* (Linnaeus). Ann. Ent. Soc. Amer. **49**(1):43-48.

Cicolani, B., L. Bullini and S.G. Montalenti (1975). Richerche sulla biologia reproduttiva e sull'ecologia dei Macrochelidi. I. Analisi del rapporto sessi in una popolazione di laboratorio di *Macrocheles matrius* (Acarina:Mesostigmata). Acad. Naz. Lincei 8, **59**(5):481-492.

Dittrich, V. (1965). Embryonic development of tetranychids. Boll. Zool. Agr. Bachic Ser. (2)7:101-104.

Filipponi, A. (1957). Arrenotochia in *Macrocheles subbadius* (Acarina, Mesostigmata). Est. Rend. Ist. Sup. Sanitá **20**:1037-1044.

Filipponi, A., G. Petrelli and S. Passariello (1971). Contributi sperimentali di laboratorio sulla autoecologia e demoecologia di *Macrocheles penicilliger* (Berl.) (Acari:Mesostigmata). Boll. Zool. **38**(1):1-33.

Grandjean, F. (1947). Observations sur les Oribates (18e Série). Bull. Mus. nat. Hist. natur. Paris **19**(2):165-172.

Hammen, L. van der (1972). Reflexions sur la valeur des donnes embryologiques pour la morphologie. Acarologia **14**(4):520-523.

Hansell, R.I.C., M. Mollison and W.L. Putman (1964). A cytological demonstration of arrhenotoky in three mites of the family Phytoseiidae. Chromosoma **15**:562-567.

Heinemann, R.L. and R.D. Hughes (1969). The cytological basis for reproductive variability in the Anoetidae (Sarcoptiformes:Acari). Chromosoma **28**(3):346-356.

Helle, W. and H.R. Bolland (1967). Karyotypes and sex-determination in spider mites (Tetranychidae). Genetica **38**:43-53.

Hoogstraal, H.F., H.S. Roberts, G.M. Kohls and V.J. Tipton (1968). Review of *Haemaphysalis (Kaiseriana) longicornis* Neumann (resurrected) of Australia, New Zealand, New Caledonia, Fiji, Japan, Korea and northeastern China and U.S.S.R.. and its pathenogenetic and bisexual populations (Ixodoidea: Ixodidae). J. Parasitol. **58**:950-959.

Hughes, R.D. and C.G. Jackson (1958). A review of the Anoetidae (Acari). Virginia J. Sci. **9**, n.s. (1):5-198.

Hughes, T.E. (1950). The embryonic development of the mite *Tyroglyphus farinae* Linnaeus 1758. Proc. Zool. Soc. London **119**(4):873-886.

Oliver, J.H. Jr. (1967). Cytogenetics of acarines. **In** Genetics of Insect Vectors of Disease. J. Wright and R. Pal, eds. Elsevier Publ. Co., Amsterdam:417-439.

Oliver, J.H. Jr. (1971). Parthenogenesis in mites and ticks (Arachnida:Acari). Amer. Zool. **11**(2):283-299.

Oliver, J.H. Jr. (1974). Symposium on reproduction of arthropods of medical and veterinary importance. IV. Reproduction in ticks (Ixodoidea). J. Med. Ent. **11**(1):26-34.

Oliver, J.H. Jr. and C.S. Herrin (1974). Morphometrics of sexual dimorphism in an arrhenotokous mite, *Ornithonyssus bacoti* (Acari:Mesostigmata). J. Exp. Zool. **189**(3):291-301.

Oliver, J.H. Jr. and B.C. Nelson (1967). Mite chromosomes: an exceptionally small number. Nature **214** (5090):809.

Regev. S.(1974). Cytological and radioassay evidence of haploid parthenogenesis in *Cheyletus malaccensis* (Acarina:Cheyletidae). Genetica **45**:125-132.

Reuter, E. (1909). Zur Morphologie und Ontogenie der Acariden mit besonderer Berucksichtigung von *Pediculopsis graminum.* Acta Soc. Sci. Fenn. **36**(4):1-288.

Sicher, E. (1891). Contributione alla embriologica degli acari. Atti Soc. Veneto—Trent., Padua 1, **12**.

Summers, F.M. and R.L. Witt (1972). Nesting behavior of *Cheyletus eruditus* (Acarina:Cheyletidae). Pan-Pac. Ent. **48**(4):261-269.

Supino, F. (1895). Embryologia degli Acari. Atti Soc. Veneto—Trent., Padua 2, **2**(1):242-261.

Witte, H. (1975). Funktionsanatomie des weiblichen Genitaltraktes und Oogenese bei Erythraeiden (Acari, Trombidiformes). Zool. Beitr. N.F. **21**(2):247-277.

Woodring, J.P. and E.F. Cook (1962). The biology of *Ceratozetes cisalpinus* Berlese, *Scheloribates laevigatus* Koch, and *Oppia neerlandica* Oudemans (Oribatei), with a description of all stages. Acarologia **4**(1): 101-137.

V. OVIPOSITION AND LIFE STAGES

While embryonic or postembryonic ovoviviparity has been observed in the Acari (Strandtmann and Wharton 1958, Filipponi and Francaviglia 1963, Cross 1965, Mitchell 1968, Egan and Hunter 1975), typical oviposition occurs in the majority of acarines which have been studied. Oviposition is accomplished in several ways. Typically, the oval, ovoid, elongate or flattened eggs are passed through the genital valves and dropped either singly or in clusters. The eggs are highly elastic and resilient at the time of oviposition, which helps prevent egg damage during passage through the comparatively small genital aperture of the female. Acarine eggs may be smooth or sculptured and many are variously ornamented with striations, reticulations, or waxy extrusions. Members of the genera *Bdella* and *Spinibdella* (Actinedida, BDELLIDAE) may spin a protective silken cocoon around each egg (Wallace and Mahon 1972, Alberti 1973), while eggs of the bdellid genera *Cyta* and *Bdellodes* often have a covering of fine hairlike projections or spinose capitate protrusions, presenting a pincushion appearance (Wallace and Mahon, op. cit., Wallace and Mahon 1976). A waterproof waxy coating or thin lipid layer may be applied to each egg prior to deposition (Lees and Beament 1948, Beament 1951).

Mites which feed on plants or on plant parts stored in bulk tend to oviposit freely on their plentiful food substrate, often countering losses through heavy predation or exposure with multivoltinism and/or high fecundity (Solomon 1945, Huffaker et al. 1969, Hussey et al. 1969). Where the food source is transient, and contact with it is fortuitous, high fecundity is a common survival strategy. Ticks (Ixodida) may deposit several thousand eggs at a time (Drummond et al. 1971) so as to ensure at least a minimal chance of survival to the adult stage. Acarines which must locate a limited food source in a broad habitat such as soil or water generally secrete their eggs where loss is minimized. To this end, many Oribatida and Actinedida possess an extrusible *ovipositor* (Fig. 17, p. 48) which holds the egg while simultaneously probing for a suitable niche in which to place it. Females of *Pergamasus quisquiliarum* Can. (Gamasida, PARASITIDAE) attach their eggs to the rootlets of plants where hatching larvae may more quickly contact the symphylans on which they prey (Berry 1973). Parasitic mites often are larviparous, but those which oviposit generally choose a particular host tissue as an oviposition site. Others place their eggs in protected situations where access of hatching larvae to the host will be virtually assured (Camin 1953, Yunker 1973). *Dicrocheles phalaenodectes* (Treat) (LAELAPIDAE), a gamasid parasite of noctuid moths, prepares the oviposition site by kneading the tissue upon which the egg is to be deposited (Treat 1958, 1975).

Members of the genus *Cheyletus* (Actinedida, CHEYLETIDAE) often brood their clusters of eggs in concavities or caverns in the substrate (Summers and Witt 1972). Females of *C. eruditus* (Schrank), a grain-inhabiting predatory mite, have been observed to drive other cheyletids or larger arthropods from the "nest" area. Hatching larvae, however, are eaten by the mother if they tarry too long after eclosion. Many spider mites (Actinedida, TETRANYCHIDAE) deposit their eggs in silken webbing on their plant hosts, providing them a degree of protection from mite and insect predators. Hazan et al. (1975) found that removal of eggs of *Tetranychus cinnabarinus* (Boisd.) from the surrounding webbing reduces egg hatchability under conditions of extremely low or high humidity. Development of large numbers of eggs simultaneously is made possible in some Actinedida through *physogastry*. Here the idiosoma may swell to many times its normal size so as to accommodate developing embryos. Members of the Heterostigmae commonly exhibit extreme physogastry (Plate 73-7, p. 328), with development of the progeny proceeding within the chorion until the adult stage is reached (Herfs 1926, Moser et al. 1971). Species of the pygmephorid genus *Trochometridium*, on the other hand, oviposit their eggs but hatching does not occur until the adult stage is reached (Cross 1965).

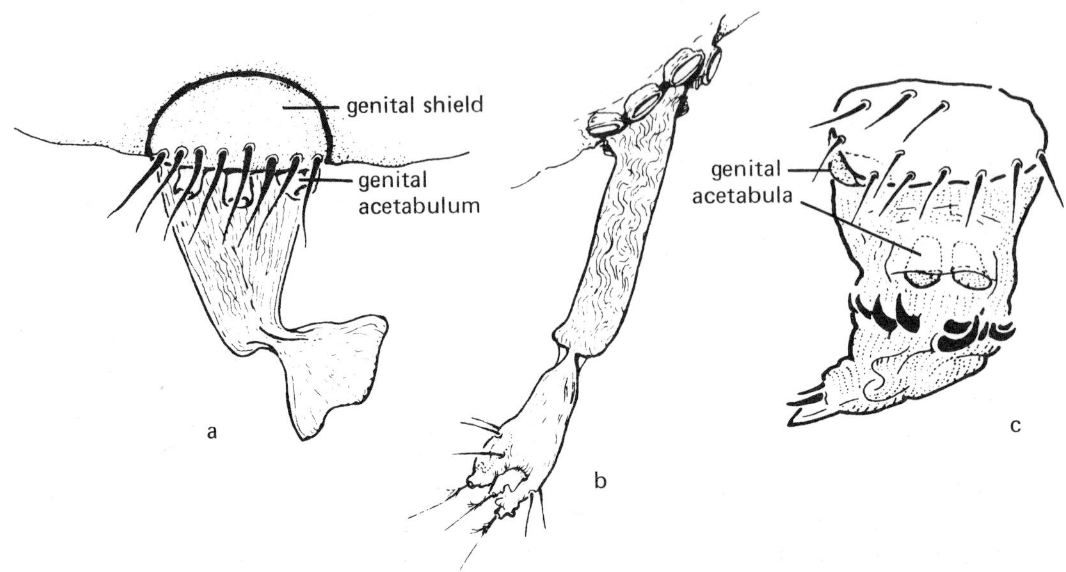

Fig. 17. Ovipositor types: **a**; *Nanhermannia* sp. (Oribatida, NANHERMANNIOIDEA): **b**; *Nanorchestes* sp. (Actinedida, NANORCHESTIDAE): **c**; *Acaronychus tragardhi* Grandjean (Oribatida, ARCHEO-NOTHROIDEA) (after Grandjean 1954).

While oviparity is the general rule in both the Acaridida and Orbatida, a peculiar type of brooding has been noted in species of both suborders. It involves the viviparous development of progeny within the dead body of the mother and their feeding on her tissues. Lipa and Chmielewski (1966) refer to this phenomenon as *aparity*.

Mites may be uni- or multivoltine and some require more than a year to complete their development. Multiple generations are common in phytophagous, coprophagous, nidicolous and parasitic groups where the food source is essentially inexhaustible for extended periods of time (Hussey et al. 1969, Yunker 1973, Jeppson et al. 1975). Predatory acarines which feed in these habitats also tend to be multivoltine (Chant 1959, Axtell 1969). Univoltine acarines occur commonly in soil, litter, aquatic and marine habitats (Hartenstein 1962, Newell 1971, Alberti 1973). Univoltinism also is suspected or known among certain ectoparasites of insects (Anderson 1968, Treat 1975).

Development from egg to adult may require as little as 4-5 days in the Acari, but more often extends to several weeks or months. The common itch mite, *Sarcoptes scabiei* (De Geer) (Acaridida, SARCOPTIDAE), may complete an entire life cycle in ten days (Yunker 1973), while species of the family MACROCHELIDAE (Gamasida) require as little as 60 hours (Axtell 1969). Cycles of five months to one year are reported for various Oribatida (Haarlov 1960, Hartenstein 1962, Shereef 1972) and Actinedida (Vistorin-Theis 1975). Late-hatching larvae of *Dermacentor andersoni* Stiles (Ixodida, IXODIDAE) often overwinter as nymphs and may require nearly two years to complete their life cycle (Hunter and Bishopp 1911). *Ixodes uriae* White may have a life cycle of 4-5 years in cooler climates (Eveleigh and Threlfall 1975).

Temperature, humidity, and food availability often have a significant effect on development time in a given population. Mite longevity also is highly variable. Species of *Tyrophagus* (Acaridida, ACARIDAE) may have a life span of less than one month, while ticks (Ixodida) may survive for several years (Balashov 1972).

Acari may pass through six (occasionally more) instars after eclosion. These are the *prelarva, larva, protonymph, deutonymph, tritonymph,* and *adult.*

Prelarva

Typically, the prelarva is a quiescent, non-feeding primitive form which occurs not only in certain acarine groups, but in other arachnids as well. Grandjean (1938) theorized that the ancestral prelarva was an active stage which routinely appeared prior to the larval molt. The prelarva is considered an inhibited rather than a regressive form since its occurrence does not necessarily signify regressive tendencies in subsequent stages (Coineau 1974a, 1974b).

In some cases, the prelarva appears to be little more than a featureless sac without legs or mouthparts (Fig. 18c). Extreme inhibition of this type is referred to as *calyptostasis,* and is a feature of prelarvae of higher Oribatida (Lions 1967, 1973) and certain of the higher Actinedida (Grandjean 1962). In other actinedid groups, the inactive prelarva may possess three pairs of legs, mouthparts and setae (Robaux 1971, Coineau 1974a and 1976b), although development of the chelicerae and of leg segmentation may be regressive. Such incomplete calyptostasis is referred to by Grandjean (1957) as *elattostasis* (Fig. 18e). An extreme example of elattostasis is found in the actinedid family ADAMYSTIDAE, where the prelarva is not only highly developed, but active as well (Coineau 1976a).

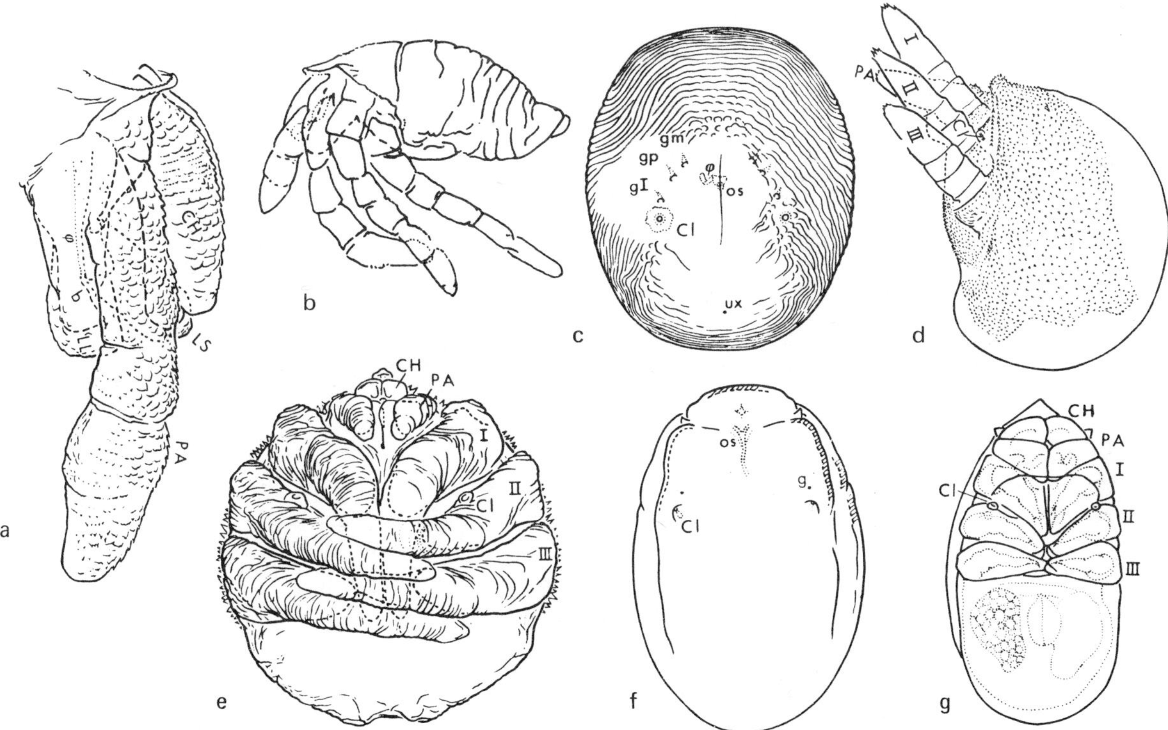

Fig. 18 Acarine prelarvae (from Coineau 1973 and 1974, and van der Hammen 1976): **a;** *Phalangiacarus* brosseti Coineau and van der Hammen, 1979. (Opilioacarida, OPILIOACARIDAE), gnathosoma: **b;** *P. brosseti,* lateral aspect: **c;** *Balaustium florale* (Grandjean) (Actinedida, ERYTHRAEIDAE), ventral aspect: **d;** *Anystis* sp. (Actinedida, ANYSTIDAE), lateral aspect: **e;** *Microcaeculus steineri delamarei* Coineau (Actinedida, CAECULIDAE), ventral aspect: **f;** *Pilogalumna allifera* (Oudemans) (Oribatida, GALUMNOIDEA), ventral aspect: **g;** *Trhypochthonius tectorum* (Berlese) (Oribatida, NOTHROIDEA), ventral aspect. (PA = palp, CH = chelicera, Cl = urstigma).

Acariform prelarvae share certain characteristics including *urstigmata* (Figs. 18c-g, *Cl*) and a non-functioning stomodeum. Other features such as pretarsi, an anal opening, and distinct coxal glands may or may not occur in a given group.

An extremely primitive prelarval stage has been discovered in *Phalangiacarus brosseti* Coineau and van der Hammen (OPILIOACARIDAE) from Gabon (Coineau 1973, van der Hammen 1976). Legs and mouthparts are well developed and apparently functional (Figs. 18a-b).

The prelarva molts to the larval stage, occasionally with the aid of an anterior "opening mechanism" (Ehrnsberger 1974). Passage from the prelarval to the larval stage may require as little as 1½ days (Coineau 1976a) or as much as 14-15 days (Robaux 1971) in Actinedida.

Larva

The typical acarine larva is a hexapod form with little or no sclerotization and without indication of external genitalia (Fig. 11, p. 23). Idiosomal sclerotization, when present, often is confined to the region of the podosoma. Ventral shields are absent or obscure in the Acaridida, and greatly restricted in other suborders (Plate 10a, p. 116). In some groups (e.g., the Gamasida), the larva may be a weak, sluggish non-feeding instar while in others it is a voracious predator (as in the actinedid family CHEYLETIDAE) or an aggressive parasite (as in the actinedid family TROMBICULIDAE). Lack of development of many key structures often makes identification of larvae difficult, except perhaps to suborder or superfamily. Exceptions are seen in the Actinedida and Ixodida, where key adult structures often are apparent throughout ontogeny. Members of the family TROMBICULIDAE are routinely identified through larval rather than adult characteristics (Brennan and Jones 1959, Vercammen-Grandjean and Langston 1971a and b).

The larva usually completes its development with little or no change in initial form, other than that occasioned by engorgement (Plate 99-3, p. 354). An exception is found in the larvae of sea snake parasites of the genus *Vatacarus* (TROMBICULIDAE). Members of the genus exhibit *neosomy,* a phenomenon entailing a significant enlargement in overall size, and formation of new external structures. These changes result, at least in part, from the secretion of new cuticle during larval stasis (Audy et al. 1972). Adventitious ontogenetic abbreviation resulting in the suppression of a free larval instar may occur in some groups under certain environmental conditions. Mating of sexually mature males and females of *Siteroptes cerealium* Kirchner (Actinedida, PYGMEPHORIDAE) results in the production of larvae which subsequently molt into males and females. During the summer, however, females generally mate with male nymphs and produce no free larval forms. Females and male nymphs appear following physogastric development within the fertilized female (Rack 1972).

While most acarine larvae have six legs, Coineau (1973) noted distinct vestiges of legs IV in the larva of an African opilioacarid (Opilioacarida). Although vestiges of legs IV have been observed in embryological studies (Aeschlimann 1958), their appearance in the larval instar is unusual.

Nymphs

With the exception of the hard ticks (Ixodida, IXODIDAE) and certain of the higher Actinedida, a single free nymphal stage is uncommon in the Acari. Two or three nymphal instars usually appear between the larval and adult stases, with as many as eight occurring in some ARGASIDAE (Ixodida). Nymphs are generally octopod, and undergo progressive differentiation of shields along with setal augmentation of the idiosoma and appendages at each nymphal molt (Plate 10, p. 116). These interstadial changes, as well as others which apply to specific taxa, often provide useful characters for taxonomic diagnoses. Primordial genital structures appear in the nymphal stadia of some taxa (Acariformes) but not in others (Parasitiformes).

A protonymph, deutonymph and tritonymph occur in the Oribatida, and all may occur in certain Actinedida and Acaridida. Only two nymphal stadia occur in most Gamasida (the proto- and deutonymph) while in some Actinedida (Heterostigmae) there may be no free nymphal stage at all. Only one active nymphal instar, the deutonymph, occurs in the Parasitengonae and certain Anystoidea (Actinedida). Development of all nymphal stadia within the larval skin occurs rarely in the TROMBICULIDAE (Audy et al. 1972).

Protonymph

The first nymph or protonymph usually is a free active instar which may or may not feed. When it occurs, the protonymph ordinarily is adapted to a substrate which is similar or identical to that of later instars. Radovsky et al. (1971) have described an unusual instance of independent protonymphal adaptation in three species of *Radfordiella* (Gamasida, MACRONYSSIDAE). All three feed in the mouth tissues of phyllostomatid bats, but only as protonymphs. The protonymph and the tritonymph of the family PTERYGOSOMATIDAE, as well as those of most Parasitengonae (Actinedida), are *pharate* forms which develop within the skin of the preceding instar (Johnston and Wacker 1967, Newell 1971). Thus, while only one active nymphal instar occurs in these groups, there are two quiescent, sequestered calyptostases, one between the larva and active deutonymphal instar, and the other between the deutonymph and adult.

Deutonymph

The second nymphal stage or deutonymph assumes the general nonsexual characteristics of the adult, differing from it only in size and in sclerotization pattern. Deutonymphs of Acaridida, however, are exceptional in that they are completely unlike the preceding and succeeding nymphal instars both in morphology and behavior. This heteromorphic nymph, or *hypopus* (Fig. 19, p. 52), occurs only sporadically within the Acaridida and may or may not appear in a given generation (*facultative hypopody*). Hypopodes are highly resistant to environmental stresses and commonly have ventral suckers or claspers with which they secure attachment to passing animals (Michael 1901, Fain 1971). The relationship generally is a phoretic one, since hypopodes lack functional mouthparts (a cheliceral anlage has been identified in sections of *Sancassania boharti* (Cross) by Woodring and Carter (1974)). Inert hypopodes lack suckers or claspers (Fig. 40, p. 377) and often rely on air currents for transportation. The hypopodes of certain GLYCYPHAGIDAE also lack distinctive organs of attachment (Fig. 19a), and are found in the subdermal tissues of a variety of animal hosts (Fain 1965, 1967, 1969, Fain et al. 1973, Lukoschus et al. 1972).

Deutonymphs of parasitic Gamasida often are non-feeding (Evans and Till 1965), as are the phoretic deutonymphs of the gamasid cohort Uropodina. Two deutonymphal morphotypes may occur in the Uropodina, only one of which is phoretic (Athias 1975).

Development of male parasitic mites may be accelerated by the omission of the deutonymphal instar (Evans 1963, Treat 1975). In these cases, the adult may possess features typical of earlier stages *(neoteny).*

Tritonymph

Where it occurs, the tritonymph usually is an active instar. It is, however, a pharate calyptostatic stage in some Actinedida. In those acaridid species where hypopody occurs, the second homeomorphic nymph is considered by some authors to be the tritonymph, since it is the third post-larval preimaginal instar. As mentioned earlier, the tritonymph is found only in a few acarine groups. Thus, the final molt to the adult usually occurs at the conclusion of the second nymphal stage.

Molting during adult stasis is rare, but has been observed in some groups. Michener (1946), Imamura (1952) and Furumizo and Wharton (1975) have reported cases of post-imaginal molting in species of TROMBIDIIDAE and ARRENURIDAE (Actinedida), and in PYROGLYPHIDAE (Acaridida).

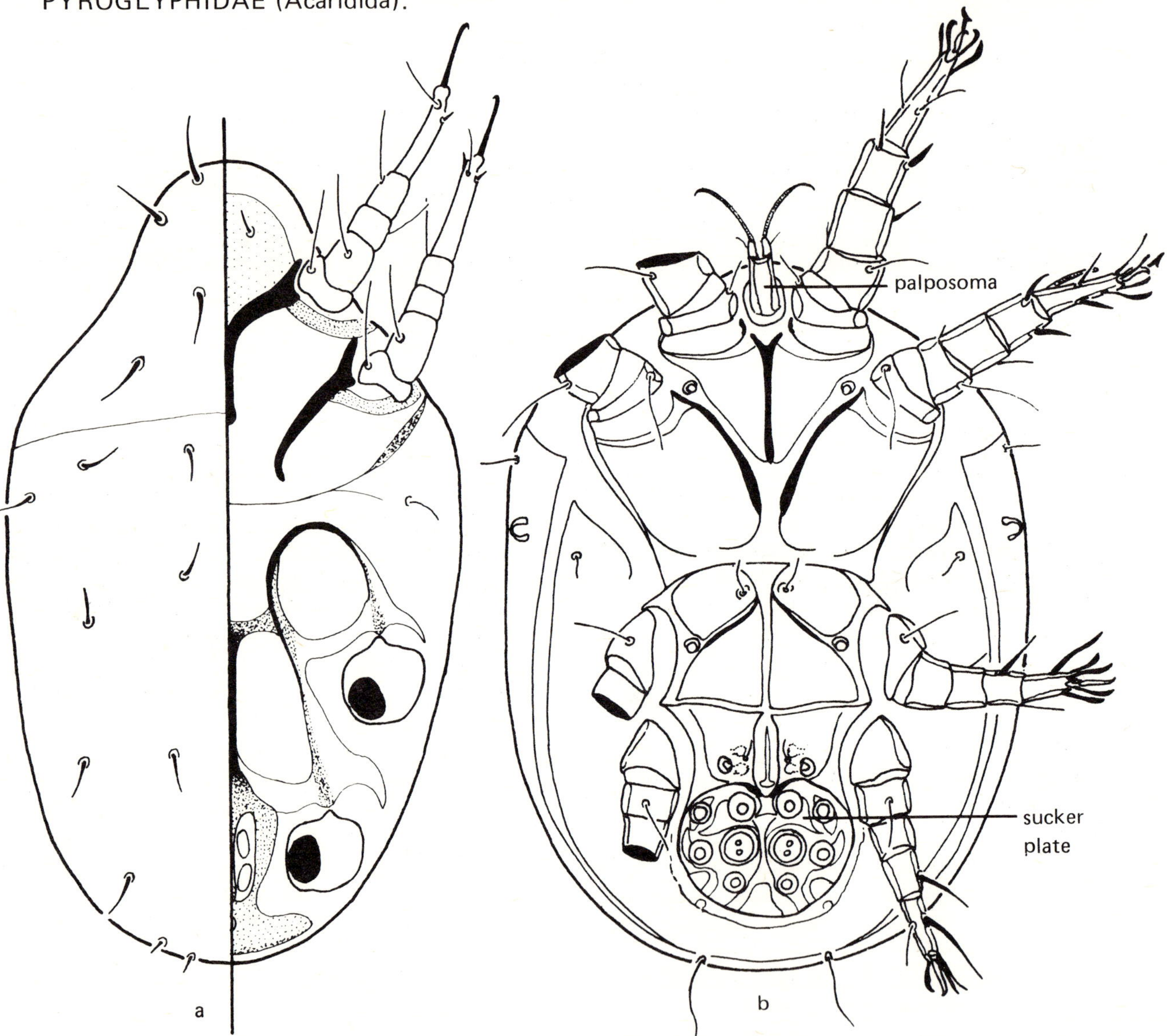

Fig. 19. Acaridid hypopodes: **a**; *Rodentopus sciuri* Fain (GLYCYPHAGIDAE) (adapted from Fain 1965): **b**; *Sancassania* sp. (ACARIDAE).

Useful References

Aeschlimann, A. (1958). Développement embryonnaire *d'Ornithodoros moubata* (Murray) et transmission transovarienne de *Borrelia duttoni.* Acta Trop. **15**(1):15-64.

Alberti, G. (1973). Ernährungsbiologie und spinnvermögen der Schnabelmilben. Z. Morph. Tiere **76**:285-338.

Anderson, R.C. (1968). Ecological observations on three species of *Pimeliaphilus* parasites of Triatominae in the United States (Acarina:Pterygosomidae) (Hemiptera:Reduviidae). J. Med. Ent. **5**(4):459-464.

Athias, F. (1975). Observations morphologiques sur *Polyaspis patavinus* Berlese 1881 (Acariens:Uropodides). I. Morphologie de l'idiosoma au cours du développement postembryonnaire. Acarologia **17**(3):410-435.

Audy, J.R., F.J. Radovsky and P.-H. Vercammen-Grandjean (1972). Neosomy: radical intrastadial metamorphosis associated with arthropod symbioses. J. Med. Ent. **9**(6):487-494.

Axtell, R.C. (1969). Macrochelidae (Acarina:Mesostigmata) as biological control agents for synanthropic flies. Proc. 2nd Int. Congr. Acarology, Nottingham:401-416.

Balashov, Y.S. (1972). Bloodsucking ticks (Ixodoidea)-vectors of diseases of man and animals. Misc. Publ. Ent. Soc. Amer. **8**(5):161-376.

Beament, J.W.L. (1951). The structure and formation of the egg of the fruit tree spider mite, *Metatetranychus ulmi* Koch. Ann. Appl. Biol. **38**:1-24.

Berry, R.E. (1973). Biology of the predaceous mite, *Pergamasus quisquiliarum* on the garden symphylan, *Scutigerella immaculata,* in the laboratory. Ann. Ent. Soc. Amer. **66**(6):1354-1356.

Brennan, J.M. and E.K. Jones (1959). Keys to the chiggers of North America with synonymic notes and descriptions of two new genera (Acarina:Trombiculidae). Ann. Ent. Soc. Amer. **52**(1):7-16.

Camin, J.H. (1953). Observations on the life history and sensory behavior of the snake mite, *Ophionyssus natricis* (Gervais) (Acarina:Macronyssidae). Chicago Acad. Sci. Spec. Publ. 10:75 pp.

Chant, D.A. (1959). Phytoseiid mites (Acarina:Phytoseiidae). Part I. Bionomics of seven species in southeastern England. Can. Ent. **91** (suppl. 12):5-44.

Coineau, Y. (1973). A propos de quelques caractères particulierement primitifs de la prélarve et de la larve d'un Opilioacaridae du Gabon (Acariens). C.R. Acad. Sci. Paris **276** (Série D):1181-1184.

Coineau, Y. (1974a). Éléments pour une monographie morphologique, écologique et biologique des Caeculidae (Acariens). Mem. Mus. Nat. d'Hist. nat. (N.S.)A, **81**:299 pp + vi.

Coineau, Y. (1974b). Les Acariens. **In** Introduction a l'Étude des Microarthropodes du Sol et de ses Annexes. Documents Ecol. DOIN, éds:57-83.

Coineau, Y. (1976a). Les Adamystidae, une etonnante famille d'Acarines prostigmates primitifs. Proc. 4th Int. Congr. Acarology, Saalfelden (in press).

Coineau, Y. (1976b). La première prélarve connue du genre *Eupodes strandtmanni* n. sp. Acarologia **18**(1):56-64.

Coineau, Y. and L. van der Hammen (1979). The postembryonic development of Opilioacarida, with notes on new taxa and on a general model for the evolution. Proc. IV Congress of Acarology (Saalfelden): 437-441.

Cooley, R.A. (1932). The Rocky Mountain wood tick. Montana State Coll. Bull. 268:58 pp.

Costa, M. (1966). The biology and development of *Hypoaspis (Penumolaelaps) hyatti* (Acari: Mesostigmata). J. Zool. **148**:191-200.

Cross, E.A. (1965). The generic relationships of the family Pyemotidae (Acarina: Trombidiformes). Univ. Kansas Sci. Bull. **45**(2):29-275.

Dosse, G. and I. Schneider (1957). Biologie und Lebensweise von *Czenspinskia lordi* Nesbitt (Acar., Sarcoptiformes). Zeit. ang. Ent. **44**:403-418.

Drummond, R.O., T.M. Whetstone and W.J. Gladney (1971). Oviposition of the lone star tick. Ann. Ent. Soc. Amer. **64**(1):191-194.

Egan, M.E. and P.E. Hunter (1975). Redescription of a cockroach mite, *Proctolaelaps nauphoetae,* with notes on its biology. Ann. Ent. Soc. Amer. **68**(2):361-364.

Ehrnsberger, R. (1974). Prälarval- und Larvalentwicklung bei Rhagidiien (Acarina:Prostigmata). Osnabrücker Naturw. Mitt. **3**:85-117.

Evans, G.O. (1963). Observations on the classification of the family Otopheidomenidae (Acari:Mesostigmata) with description of two new species. Ann. Mag. Nat. Hist. ser 13, **5**:609-620.

Evans, G.O. and W.M. Till (1965). Studies on the British Dermanyssidae (Acari:Mesostigmata). Pt. I. External morphology. Bull. Brit. Mus. (Nat. Hist.) Zool. **13**(8):249-294.

Eveleigh, E.S. and W. Threlfall (1975). The biology of *Ixodes (Ceratixodes) uriae* White, 1852 in Newfoundland. Acarologia **16**(4):621-635.

Fain, A. (1965). Un nouveau type d'hypope, parasite cuticole de rongeurs africains (Acari:Sarcoptiformes). Zeit. f. Parasit. **26**:82-90.

Fain, A. (1967). Les hypopes parasites des tissus cellulaires des oiseaux (Hypodectidae:Sarcoptiformes). Bull. Inst. roy. Sci. nat. Belg. **43**(4):1-139.

Fain, A. (1969). Les deutonymphs hypopiales vivant en association phoretique sur les mammiferes (Acarina: Sarcoptiformes). Bull. Inst. roy. Sci. nat. Belg. **45**(33):1-262.

Fain, A. (1971). Évolution de certains groupes d'hypopes en fonction du parasitisme (Acarina:Sarcoptiformes). Acarologia **13**(1):171-175.

Fain, A., F.S. Lukoschus, J.M.W. Louppen and E. Méndez (1973). *Echimyopus dasypus*, n.sp., a hypopus from *Dasypus novemcinctus* in Panama (Glycyphagidae, Echimyopinae:Sarcoptiformes). J. Med. Ent. **10**(6): 552-555.

Filipponi, A. and G. Francaviglia (1963). Oviparitá e larviparitá in *Macrocheles peniculatus* Berl. (Acari: Mesostigmata) regolate da fattori ecologici. Parassitol. **24**(2):81-104.

Furumizo, R.T. and G.W. Wharton (1975). A case of postimaginal molt in the American house dust mite *Dermatophagoides farinae* Hughes, 1961 (Acari:Pyroglyphidae). Acarologia **17**(4):730-733.

Grandjean, F. (1938). Observations sur les Acariens (4e série). Bull. Mus. nat. Hist. natur. 2e sér. **10**(1):64-71.

Grandjean F. (1957). L'evolution selon l'âge. Arch. Sci. Phys. nat. Genève **10**(4):477-526.

Grandjean, F. (1962). Prélarves des Oribates. Acarologia **4**(3):423-439.

Griffiths, D.A. (1966). Nutrition as a factor influencing hypopus formation in the *Acarus siro* species complex (Acarina, Acaridae). J. Stored Prod. Res.**1**:325-340.

Haarlov, N. (1960). Microarthropods from Danish soils. Ecology, phenology. Oikos, Supp. 3:176 pp.

Hammen, L. van der (1976). Glossaire de la terminologie acarologique. Vol. II. Opilioacarida. W. Junk B.V., The Hague:137 pp. + plates.

Hartenstein, R. (1962). Soil Oribatei V. Investigations on *Platynothrus peltifer* (Acarina:Camisiidae). Ann. Ent. Soc. Amer. **55**(6):709-713.

Hazan, A., V. Gerson and A.S. Tahori (1975). Spider mite webbing II. The effect of webbing on egg hatchability. Acarologia **17**(1):270-273.

Herfs, A. (1926). Okologische Untersuchungen an *Pediculoides ventricosus* (Newp.) Berlese. Zoologia. Gesamtgeb. Zool. **74**:3-68.

Huffaker, C.B., M. van de Vrie and J.A. McMurtry (1969). The ecology of tetranychid mites. Ann. Rev. Ent. **14**:125-173.

Hunter, W.D. and F.C. Bishopp (1911). The Rocky Mountain spotted fever tick. USDA Bull. 105:47 pp.

Hussey, N.W., W.H. Read and J.J. Hesling (1969). Order Acarina: Mites. **In** The Pests of Protected Cultivation. Amer. Elsevier Publ. Co., New York:190-228.

Imamura, T. (1952). Notes on the moulting of the adult of the water mite, *Arrenurus uchidai* n.sp. Zool. Soc. Japan **25**(4):447-451.

Jeppson, L.R., H.H. Keifer and E.W. Baker (1975). Mites Injurious to Economic Plants. Univ. Calif. Press, Berkeley:614 pp. + 74 plates + xix.

Lees, A.D. and J.W.L. Beament (1948). An egg waxing organ in ticks. Quart. Jour. Micr. Sci. London **98**:291-332.

Lions, J.-C. (1967). La prélarve de *Rhysotritia ardua* (C.L. Koch) 1836 (Acarien, Oribate). Acarologia **9**(1):273-283.

Lions, J.-C. (1973). Quelques prélarves nouvelles d'Oribates. Acarologia **15**(2):356-370.

Lipa, J.J. and W. Chmielewski (1966). Aparity observed in the development of *Caloglyphus* mite (Acarina: Acaridae). Ekol. Polska-Ser. A, **14**(37):741-748.

Lukoschus, F.S., A. Fain and F.M. Driessen (1972). Life cycle of *Apodemopus apodemi* (Fain, 1965) (Glycyphagidae:Sarcoptiformes). Tijdschr. Ent. **115**(8):325-339.

Michael, A.D. (1901). British Tyroglyphidae. Ray Society, London **1**:291 pp. + vii.

Michener, C.D. (1946). The taxonomy and bionomics of some Panamanian trombidiid mites (Acarina). Ann. Ent. Soc. Amer. **39**(3):349-380.

Mitchell, C.J. (1968). Biological studies on *Laelaps myonyssognathus* G. and N. (Acarina:Laelapidae). J. Med. Ent. **5**(1):99-107.

Moser, J.C., E.A. Cross and L.M. Roton (1971). Biology of *Pyemotes parviscolyti* (Acarina:Pyemotidae). Entomophaga **16**(4):367-379.

Newell, I.M. (1971). The protonymph of *Pimeliaphilus* (Pterygosomatidae) and its significance relative to the calyptostases in the Parasitengona. Proc. 3rd Int. Congr. Acarology, Prague:789-795.

Radovsky, F.J., J.K. Jones Jr. and C.J. Phillips (1971). Three new species of *Radfordiella* (Acarina:Macronyssidae) parasitic in the mouth of phyllostomatid bats. J. Med. Ent. **8**(6):737-746.

Rack, G. (1972). Pyemotiden an Gramineen in schwedischen landwirtschaftlichen Betrieben. Ein Beitrag zur Entwicklung von *Siteroptes graminum* (Reuter 1900) (Acarina, Pyemotidae). Zool. Anz. Leipzig **188**(3/4):157-174.

Robaux, P. (1971). Recherches sur le développement et la biologie des Acariens Thrombidiidae. PhD Thesis, Faculty of Science, Univ. Paris. C.N.R.S. Reg. No. 5616.

Shereef, G.M. (1972). Observations on oribatid mites in laboratory cultures. Acarologia **14**(2):281-291.

Solomon, M.E. (1945). Tyroglyphid mites in stored products. Methods for the study of population density. Ann. Appl. Biol. **32**:71-75.

Strandtmann, R.W. and G.W. Wharton (1958). A Manual of Mesostigmatid Mites Parasitic on Vertebrates. Inst. Acarology Contrib. 4:330 pp. + xi + plates.

Summers, F.M. and R.L. Witt (1972). Nesting behavior of *Cheyletus eruditus* (Acarina:Cheyletidae). Pan-Pac. Ent. **48**(4):261-269.

Travé, J. (1976). Les prélarves d'Acariens. Mise au point et données récentes. Rev. Écol. Biol. Sol **13**(1): 161-171.

Treat, A.E. (1958). Social organization in the moth ear mite *(Myrmonyssus phalaenodectes).* Proc. 10th Int. Cong. Ent. **2**:475-480.

Treat, A.E. (1975). Mites of Moths and Butterflies. Cornell Univ. Press, Ithaca, N.Y.: 362 pp.

Vercammen-Grandjean, P. -H. and R.L. Langston (1971a). Revision of the *Leptotrombidium* generic complex based on palpal setation combined with other morphological characters. J. Med. Ent. **8**(4):445-449.

Vercammen-Grandjean, P. -H. and R.L. Langston (1971b). The chigger mites of the world. Vol. 8 *Guntherana* complex: Section A Genus *Guntherana.* G.W. Hooper Foundation, San Francisco: 153 pp. + 62 pl. + vii.

Vistorin-Theis, G. (1975). Entwicklungszyklus der Calyptostomiden (Acari, Trombidiformes). Acarologia **17**(4):683-692.

Wallace, M.M.H. and J.A. Mahon (1972). The taxonomy and biology of Australian Bdellidae (Acari). I. Subfamilies Bdellinae, Spinibdellinae and Cytinae. Acarologia **14**(4):544-580.

Wallace, M.M.H. and J.A. Mahon (1976). The taxonomy and biology of Australian Bdellidae (Acari). II. Subfamily Odontoscirinae. Acarologia **18**(1):65-123.

Woodring, J.P. and S.C. Carter (1974). Internal and external morphology of the deutonymph of *Caloglyphus boharti* (Arachnida:Acari). J. Morphol. **144**(3):275-295.

Yunker, C.E. (1973). Mites. **In** Parasites of Laboratory Animals. R.J. Flynn, ed. Iowa State Univ. Press, Ames: 425-492.

VI. HABITS AND HABITATS

The remarkable diversity in morphology of the Acari is more than equalled by the variety of behavioral characteristics seen in the subclass. Specializations in habitat often are paralleled by specializations in structure. It is therefore essential that both habits and habitats of acarines be examined if an understanding of their classification is to be achieved.

The Acari may be grouped under two major habital headings: the *free-living* forms and the *parasitic* forms. Species which may be beneficial or injurious to man occur in both groups.

Free-Living Forms

The free-living acarines comprise a vast complex represented in all of the mite suborders except the Ixodida. The complex includes predaceous mites of infinite variety, mites which feed on plants or their derivatives, and others which utilize various organic substrates as food. Free-living mites may be roughly categorized on the basis of habitat, although some families are represented in more than one category.

A. Predaceous Mites

1. Ground species. These mites are common in the upper layers of soil and in moss, humus, and animal waste products where they feed on small arthropods or their eggs, on nematodes, and occasionally on each other. Predaceous ground species are commonly long-legged, fast-moving mites with strong chelate-dentate chelicerae or raptorial palps (Plate 3, p. 15) for capturing or macerating their prey. Idiosomal shields are generally well developed, although lack of tanning in some species may render sclerites indistinguishable from the surrounding integument. Typical predaceous ground species may be found in many families of the Gamasida and Actinedida including the PARASITIDAE, (Plate 13, p. 171), SEJIDAE (Plate 12, p. 170), MACROCHELIDAE (Plate 27, p. 185), ASCIDAE (Plate 20, p. 178), RHAGIDIIDAE (Plate 64, p. 319), LABIDOSTOMMATIDAE (Plate 62, p. 317) and CHEYLETIDAE (Plate 81, p. 336).

Many predaceous ground species are considered beneficial to man since they feed on harmful arthropods (Hurlbutt 1958, Axtell 1963 and 1969, Berry 1973). Members of the family MACROCHELIDAE have been utilized successfully in housefly control in manure. For example, *Macrocheles muscaedomesticae* (Scop.) achieved a 99% kill of housefly eggs in poultry manure (Rodriguez et al. 1970), substantially reducing the housefly population. *Fuscuropoda vegetans* (DeGeer) (Gamasida, UROPODIDAE) is a common predator of early instar fly larvae (O'Donnell and Axtell 1965), as is *Parasitus coleoptratorum* (L.) (PARASITIDAE) (Wernz and Krantz 1976).

At least one moss-inhabiting species of Oribatida is known to supplement its normally saprophytic diet with live nematodes (Rockett and Woodring 1966). Other records of predation in the Oribatida are cited by Luxton (1972).

2. Aerial species. Like the terricolous species, predaceous aerial mites often are long-legged and rapid in movement, preying primarily on phytophagous mites or their eggs (Jeppson et al. 1975). Idiosomal shields may be extensive although they are often weakly defined. Actinedid aerial species are frequently brightly colored in shades of red, yellow or green. Aerial forms occur in several gamasid and actinedid families including the PHYTOSEIIDAE (Plate 21, p. 179), BDELLIDAE (Plate 68, p. 323), STIGMAEIDAE (Plate 79, p. 334), and ANYSTIDAE

(Plate 92, p. 347). Numerous species of the slow-moving predaceous genus *Asca* (ASCIDAE, Plate 20, p. 178) have been collected on plants in the tropics (Moutia 1958, DeLeon 1967a, 1967b) but are equally common in dead leaves on or under plants.

Aerial species of the family PHYTOSEIIDAE are of considerable importance as predators in pest management programs against tetranychid and eriophyoid mites (Actinedida), particularly on orchard and vine crops (Chant 1961, Hoyt 1969, Flaherty and Huffaker 1970, Westigard 1971, Hoyt and Caltagirone 1971, McMurtry and Scriven 1971). Many of these same species may feed and develop successfully on other foods including insects or insect eggs, nematodes, pollen or honeydew (Huffaker and Kennett 1956, Chant 1960, McMurtry and Scriven 1964, Kamburov 1971).

3. Storage species. Predators associated with food storages usually are small ($<700\mu$) weakly sclerotized species which may either move rapidly over and through their substrate in search of food, or remain more or less immobile until the prospective prey comes within reach. Members of the gamasid families ASCIDAE (Plate 20, p. 178), PARASITIDAE (Plate 13, p. 171) and LAELAPIDAE (Plate 36, p. 194) are especially common food storage predators (Hughes 1961), as are species of the actinedid family CHEYLETIDAE (Plate 81, p. 336). Cheyletids are considered by some observers to be effective control agents of ACARIDAE (Plate 116, p. 414) in grain storage situations (Pulpan and Verner 1966).

Some gamasid predators in storage facilities are facultatively parasitic on rodents (Evans and Till 1966), feeding on arthropods more or less incidentally. However, while *Haemogamasus pontiger* (Berlese), a facultative parasite of rodents and other vertebrates, feeds alternatively on acaridid grain mites, it has also been observed to complete its entire development on a diet of wheat germ (Hughes 1961).

4. Littoral-intertidal-marine species. The littoral-intertidal habitat has been exploited by many predaceous species of Gamasida and Actinedida. These mites feed on arthropods and other invertebrates which are attracted by accumulations of organic material in the tidal region. Halbert (1920) divided the intertidal zone into bands which are dominated by particular algae or lichens. Evans et al. (1961) recognized these divisions and designated two additional zones—the tidal debris and estuarine—with each zone characterized by its own assemblage of mite predators and prey. The degree of habital specificity in the major littoral-intertidal zones is substantial, according to estimates by Costa (1974).

Members of the families PARASITIDAE (Plate 13, p. 171), MACROCHELIDAE (Plate 27, p. 185), OLOGAMASIDAE (Plate 18, p. 176), HALOLAELAPIDAE (Plate 19, p. 177) and LAELAPIDAE (Plate 36, p. 194) are common gamasid predators of the littoral-intertidal zone (Halbert 1920, André 1934, Schuster 1957, Evans et al. 1961, Haq 1965, Luxton 1968, Lee 1970, Costa 1974). Littoral-intertidal predaceous Actinedida include members of the families BDELLIDAE (Plate 68, p. 323), RHAGIDIIDAE (Plate 64, p. 319), ERYTHRAEIDAE (Plate 94, p. 349), and a number of intertidal species of the marine family HALACARIDAE (Plate 69, p. 324).

Aside from a few intertidal species of water mites (Actinedida, Hydrachnidia) which have adapted to a marine existence, only the HALACARIDAE have succeeded in colonizing the oceans and seas (André 1934, Newell 1947 and 1971). Many halacarids are predatory, preying on other invertebrates in a variety of intertidal, subtidal and byssal habitats. Unlike many aquatic mites (see below), halacarids do not swim but rather crawl on their diverse substrates. The crawling habit may be a selective response to the hazards faced by small swimming animals in habitats constantly swept by tidal currents.

5. Aquatic species. The actinedid phalanx Hydrachnidia contains the majority of predaceous mite species found in aquatic situations (Cook 1974). A common but not constant feature in this highly diverse group is the presence of long "swimming hairs" on the legs (Plate 109-2, p. 364). Many species are brightly colored in red, orange, green or blue tones (Soar and Williamson 1925). Eyes usually are present and the palpi often are modified for grasping. Predaceous adults and nymphs feed on other mites and on small crustaceans, isopods and insects. Larval forms commonly are collected on a variety of insects (Sparing 1959). Some Hydrachnidia have adapted to hot springs habitats, swimming at water temperatures up to $45^{o}C$ (Mitchell 1974). The genera *Thermacarus* and *Partnuniella* are examples of obligate thermophiles which require water temperatures of more than $32^{o}C$ (Mitchell 1960).

While the actinedid family HALACARIDAE is primarily marine in habitat, some species have invaded fresh water situations. Aquatic species representing several genera are found in cave pools, driven wells, lakes and subterranean continental waters (Imamura 1968 and 1970, Petrova 1971 and 1972). The presence of halacarids in these habitats may be the result of secondary invasions of fresh waters from marine intertidal habitats.

Few gamasid predators have adapted to aquatic habitats (Willmann 1942, Hinton 1971), but predaceous nymphs and adults of the actinedid supercohort Parasitengonae (Plates 94 and 96, pp. 349 and 351) are frequently encountered in wet or submerged gravel or soil bordering waterfalls, streams and lakes (Newell 1957, Hughes 1959).

B. Phytophagous Mites

1. Ground species. Few groups of mites are adapted to feeding on live plant tissues in soil. Phytophagous ground species feed on root tissue, corms or bulbs, and often are responsible for economic injury to ornamentals and vegetables. Members of this group are mostly opaque white or translucent, slow moving forms with short legs and with little or no distinctive idiosomal sclerotization. Some have chelate-dentate chelicerae for grinding and macerating plant tissues (Acaridida, ACARIDAE, Plate 116, p. 414), while others have stylettiform chelae for piercing plant cells (Actinedida, TARSONEMIDAE, Plate 74, p. 329) (Hussey et al. 1969). At least one oribatid mite (PERLOHMANIIDAE) has been implicated as a phytophagous soil species (Evans et al. 1961).

2. Aerial species. Aerial phytophagous species are slow-moving or sedentary mites which are weakly sclerotized. The majority of species are red, yellow, or green in color while some may appear white or translucent. They feed by inserting stylet-like chelicerae into the cells of the plant host and sucking up the contents. Included in this group are some of our most important arthropod pests of plants. The spider mites (Actinedida, TETRANYCHIDAE, Plate 86, p. 341), and the three families of the actinedid superfamily Eriophyoidea (ERIOPHYIDAE, SIERRAPHYTOPTIDAE and RHYNCAPHYTOPTIDAE, Plates 89 and 90, pp. 344-345) are of particular concern to agriculture (Jeppson et al. 1975). Aside from their feeding injury, at least seven eriophyoid species have been found to transmit plant viruses (Oldfield 1970, Slykhuis 1972). Other important aerial phytophagous species occur in the actinedid families TARSONEMIDAE (Plate 74, p. 329) and TENUIPALPIDAE (Plate 88, p. 343).

3. Storage species. Stored grains and other stored products often are infested by various kinds of mites which feed on the product itself (Hughes 1961, Cusack et al. 1975) or on fungi which infest it (Smith 1964, Sinha and Harasymek 1974). Storage species are white or brownish-white in color, and are commonly slow-moving and sac-like. The chelicerae are blunt and toothed, and are useful for scraping and gouging the food material. Graminivorous mites feed on the germ tissue of the grain and may move into the surrounding endosperm as well.

Dried fruit, linseed oil, stored tubers and bulbs also are subject to injury. Common storage species are found in the acaridid families ACARIDAE (Plate 116, p. 414) and GLYCYPHAGIDAE (Plate 118, p. 416). The grain mite, *Acarus siro,* an important acarid contaminant of processed grain products throughout the world, was chosen by Linnaeus in 1758 as the type for the first formally named mite genus.

C. Mycophagous Mites

Aside from a tendency to be slow-moving or sedentary, the mycophagous mites do not lend themselves to categorizing. Species of all suborders except the Ixodida have been found to feed on fungi in habitats ranging from tree buds to stored grain. Members of the gamasid family UROPODIDAE (Plate 41, p. 199) consume fungi in ground and storage situations as do some of the weakly sclerotized species of the acaridid families ACARIDAE and GLYCYPHAGIDAE. Several species of the aerial predaceous gamasid family PHYTOSEIIDAE (Plate 21, p. 179) have been observed to complete their development on fungi (Jeppson et al. 1975). *Calvolia lordi* (Nesbitt) (Acaridida, SAPROGLYPHIDAE), a common inhabitant of filbert trees, feeds avidly on *Cladosporium* isolated from filbert bud tissues.

Commercial mushroom houses often are infested by mycophagous mites. Species of the genus *Pygmephorus* (Actinedida, PYGMEPHORIDAE, Plate 73, p. 328) may occur on commercial mushrooms in great numbers (Gurney and Hussey 1967, Wicht 1970), creating subsequent contamination problems in soup processing plants. Members of the actinedid genera *Linopodes* and *Tarsonemus* (EUPODIDAE and TARSONEMIDAE), and of the acaridid genera *Tyrophagus* and *Sancassania* (ACARIDAE), also are important fungivorous pests in mushroom houses (Davis 1944, Hussey et al. 1969).

Many representatives of the Oribatida feed on fungi in woody plant tissues (Woolley 1960, Luxton 1972), and in soil or humus (Cancela da Fonseca and Kiffer 1969, Kiffer and da Fonseca 1971). Luxton refers to these forms as microphytophages (see section D, below). Soil fungi also are utilized by other acarine groups including the TARSONEMIDAE (Plate 74, p. 329) and EUPODIDAE (Plate 63, p. 318).

Mycophagous mites often illustrate selective mycophagy (Hartenstein 1962b), and may even prefer different fungi as juveniles than as adults (Luxton 1972). The vertical distribution of these mites in forest soils appears to be based, at least indirectly, on the location of preferred food substrates. Fungivorous acaridid species of *Acarus, Sancassania, Suidasia* and *Tyrophagus* (ACARIDAE), and of *Glycyphagus* (GLYCYPHAGIDAE), also show varying degrees of selectivity in choosing fungi normally occurring on seeds (Sinha 1968).

D. Saprophagous Mites

Saprophagy occurs in most mite suborders either as a way of life or as a facultative or sporadic phenomenon. Saprophagous mites occur in highly diverse habitats, but the great majority are found in soil and litter where they feed on dead and dying plant or animal tissues. Most of the known saprophagous species are members of the suborder Oribatida.

Decomposition of organic materials by saprophagous Oribatida has been demonstrated to be a vital factor in forest litter breakdown and consequent nutrient recycling (Crossley and Witkamp 1963, Berthet 1964 and 1971, Edwards et al. 1970, Wallwork 1970, Fujikawa 1970 and 1972, Lebrun 1971, McBrayer et al. 1976). Some forest soil oribatids feed on dead Collembola, earthworms and fallen leaves, reducing them to a state in which microorganisms may decompose them. Other species feed on fungi which in turn are primary decomposers. Ingestion of cellulose by oribatid primary decomposers has been reported (Hartenstein 1962), and Luxton (1972) has proven that these mites are able to digest pectin as well.

Based on earlier work by Schuster (1956), Luxton (1972) recognizes three major feeding categories amongst the Oribatida observed in soil in a Danish beech forest. The *macrophytophages* feed on decaying higher plant materials, while *microphytophages* consume fungi, yeasts, bacteria and algae. *Panphytophages* combine the attributes of both the first and second categories. Saprophagous macrophytophages and panphytophages are often dependent on fungi to soften and decay plant tissues so that the mites may consume them.

Examples of oribatid saprophages are legion, since the Oribatida probably comprises the largest assemblage of species found in forest soil and litter habitats (Hartenstein 1961). Woody twigs and vascular elements of leaves may be consumed by members of the superfamily PHTHIRACAROIDEA (Plate 153, p. 478) (Wallwork 1958 and 1967), while decaying leaf tissues are fed upon by species of NOTHROIDEA (Plate 155, p. 480), PHTHIRACAROIDEA, ORIBATELLOIDEA (Plate 160, p. 485), LIACAROIDEA (Plate 157, p. 482), CERATOZETOIDEA (Plate 162, p. 487), and many others (Luxton 1972). *Rostrozetes flavus* Woodring (ORIBATULOIDEA) shows a distinct preference for decomposing outer root sheaths (Woodring 1965).

While the litter and humus layers are attractive substrates for macro- and microphytophages, the underlying mineral subsoil should not be neglected in sampling for these forms. Price (1973) found that more than 50% of many ground-inhabiting groups in the pine forest soils of California occur in the mineral subsoil during both the wet and dry seasons. Some subterranean acarines may, in fact, be found at depths exceeding 200 cm (Price 1976).

Amongst non-oribatids, *Sancassania berlesei* (Michael) (ACARIDAE) may be classed as a phytophage, but it is also found as a necrophage on dead soil insects. Another acaridid acarid, *Rhizoglyphus echinopus* F. and R., may be a macrophytophage, feeding on bulb tissues damaged by other invading organisms or by improper handling (Hussey et al. 1969).

House dust mites (Acaridida, PYROGLYPHIDAE, Plate 130, p. 428) are saprophages which feed on sloughed human skin scales and hair, and are often a significant component of circulating house dust (van Bronswijk and Sinha 1971). Other pyroglyphids are found in the nests or on the skin of birds and mammals, and in stored products. Species of the genus *Lardoglyphus* (ACARIDAE) feed on animal hides, as well as on fertilizer offal (Hughes 1959).

E. Other Microphytophages

In addition to mycophagous mites, Luxton's (1972) microphytophage category includes species which feed on bacteria *(bacteriophages)* and those which consume algae *(phycophages).* An additional category, *lichenophages,* is included here.

Comparatively few acarines feed exclusively on bacteria, although some may be ingested by mycophages along with fungi or yeasts. *Gustavia microcephala* (Nic.) (Oribatida, LIACAROIDEA, Plate 157, p. 482) prefers bacteria to fungi (Luxton 1972), as do immatures of the oribatid *Belba corynopus* (Hermann). Members of the acaridid family ANOETIDAE (Plate 126, p. 424) have highly modified mouthparts which allow them to filter microorganisms from liquefied substrates (Hughes and Jackson 1958). Bacteria probably constitute a significant portion of their diet.

Few terrestrial mites feed on algae alone. Some species of the family TARSONEMIDAE (Actinedida) are known to depend on algae as a primary food source (Evans et al. 1961), a propensity shared by certain species of Oribatida (Sengbusch and Sengbusch 1970). Marine habitats probably support more phycophages than do terrestrial habitats, since algae are the predominant plant group in most marine situations. Marine mites of the subfamily Rhombognathinae (Actinedida, HALACARIDAE, Plate 69, p. 324) are phycophagous (Newell 1947), feeding primarily in the intertidal zone. Members of the acaridid family HYADESIIDAE (Plate 125, p. 423) also are intertidal phycophages (André 1931), but are found also in brackish and fresh water (Fain 1974, Fain and Johnston 1975). Luxton (1966) found that *Hygroribates schneideri* (Oud.) (Oribatida, AMERONOTHROIDEA), an intertidal oribatid, readily consumes filamentous green algae, and that *Ameronothrus* sp. feeds on algal sporelings in and around barnacle growths.

Gerson (1973) reviewed lichen-arthropod associations, and noted that lichenophagous mites occur both in the Actinedida and Oribatida. The oribatid *Camisia segnis* (Hermann) (NOTHROIDEA, Plate 155, p. 480) feeds on and overwinters in lichens (Grandjean 1948), and *Scapheremaeus petrophagus* (Banks) (CYMBAEREMAEOIDEA, Plate 160, p. 485) is lichenophagous on water-splashed rocks near waterfalls (Hughes 1961). *Dometorina plantivaga* (Berlese) (ORIBATULOIDEA, Plate 163, p. 488) not only feeds on lichens, but burrows into the thalli (Grandjean 1950). Certain members of the actinedid family TYDEIDAE (Plate 66, p. 321) also feed on lichens (Gerson 1968).

F. Coprophagous and Necrophagous Mites

Dung and carrion offer attractive niches for a variety of acarines, including many which prey on primary dung or carrion feeders. Foremost in this group are the predaceous ground forms mentioned earlier. However, some mites actually feed on dung or carrion itself. Nymphs of *Euphthiracarus* and *Steganacarus* (Oribatida, EUPHTHIRACAROIDEA and PHTHIRACAROIDEA, Plate 153, p. 478) are associated with bark beetles and may require beetle feces as food for continued development (Woolley 1961). A similar relationship has been implied by Wallwork (1958) for *Galumna formicarius* (Berlese) (Oribatida, GALUMNOIDEA, Plate 163, p. 488) and for species of *Oppia* (OPPIOIDEA, Plate 159, p. 484). Necrophagous mites generally are repelled by larger animals on the soil surface (Luxton 1972) and play only a minor role in their decomposition (Bornemissza 1957). Dead insects and other invertebrates, however, are attractive to various Oribatida (Wallwork 1958 and 1967, Luxton op. cit.), and to some soil-dwelling Acaridida. An example of the latter group is *Sancassania berlesei* (Michael) (ACARIDAE), which feeds on dead soil insects. Some gamasid predators are facultatively necrophagous or corprophagous (Weis-Fogh 1948).

G. Phoretic Mites

Farish and Axtell (1971) define phoresy as "a phenomenon in which one animal actively seeks out and attaches to the outer surface of another animal for a limited time during which the attached animal (termed the phoretic) ceases both feeding and ontogenesis, such attachment presumably resulting in dispersal from areas unsuited for further development, either of the individual or of its progeny." Deutonymphs and adults of many species of Gamasida, Actinedida and Acaridida are phoretic on both invertebrates and vertebrates. Woolley (1969) reported a phoretic association between *Licnocepheus reticulatus* Woolley, an oribatid mite, and an elaterid beetle, but phoresy among the Oribatida is unusual. Phoresy crosses a number of the habital lines discussed earlier in this chapter, and so is perhaps best considered as a phenomenon rather than as a life style.

Farish and Axtell recognize four categories of phoresy which might be referred to as the "parasitid", "macrochelid", "uropodid" and "acarid" types. Both the parasitid and macrochelid types involve casual attachment to the carrier without the aid of specialized devices, devices which are developed by both the uropodid and acarid types. Because a number of known phoretic relationships fit into a given category only with difficulty, a more general approach is taken in the discussion to follow.

Deutonymphs and adults of several gamasid families have established phoretic relationships with a variety of arthropods. Included here are the PARASITIDAE (Plate 13, p. 171), OLOGAMASIDAE (Plate 18, p. 176), ASCIDAE (Plate 20, p. 178), AMEROSEIIDAE (Plate 23, p. 181), PHYTOSEIIDAE (Plate 21, p. 179), DIGAMASELLIDAE (Plate 17, p. 175), LAELAPIDAE (Plate 36, p. 194), EVIPHIDIDAE (Plate 26, p. 184), PACHYLAELAPIDAE (Plate 28, p. 186), MACROCHELIDAE (Plate 27, p. 185), and several families of the gamasid supercohort Trigynaspides (page 146) (Hughes 1961, Krantz 1967, Costa 1969, Kinn 1968 and 1971, Binns 1972 and 1973, Treat 1975). Some of these groups prey on various life stages of their insect carriers, while others are nematophagous in the host's substrate. Phoretic deutonymphs of the gamasid cohort Uropodina may attach to their arthropod associates by means of an *anal pedicel,* a liquid strand drawn from the anus which hardens upon contact with the air (Fig. 20, p. 64). Other uropodines (some POLYASPIDIDAE, for example) grasp the integument or setae of the insect carrier with claws and chelicerae. Phoretic adults of the genus *Dinogamasus* (LAELAPIDAE) may be found in a special abdominal pouch, or *acarinarium,* of their carpenter bee hosts (Le Veque 1930).

The associations of many species of Trigynaspides and Uropodina (Gamasida) with their hosts appear to be more intimate than simple phoresy. *Antennophorus grandis* Berlese (ANTENNOPHORIDAE, Plate 49, p. 207) usually attaches to the head of its ant host and strokes the mouth of the ant with its legs to induce disgorgement of a droplet, which is eaten by the mite (Evans et al. 1961). A similar relationship has been noted between *Echinomegistus wheeleri* (Wasmann) (PARAMEGISTIDAE, Plate 47, p. 205) and its carabid beetle hosts (Nickel and Elzinga 1970a). *Micromegistus bakeri* Trägårdh (PARANTENNULIDAE, Plate 50, p. 208) also appears to feed on secretions of its carabid associates (Nickel and Elzinga 1970b). *Urodiscella philoctena* (UROPODIDAE) clings to the forelegs of its ant host and feeds on substances which the ant scrapes from its body and legs.

Phoretic adults of many species of Heterostigmae (Actinedida) have been collected from bark beetles and other insects (Schaarschmidt 1958, Krczal 1959, Karafiat 1959, Lindquist and Bedard 1961, Lindquist 1969a and 1969b, Mahunka 1974a and b, 1975a), while others are associated with mammals (Mahunka 1973, 1974c, and 1975b). Members of the genus *Pyemotes* found in bark beetle galleries are phoretic only on the adult beetles, and may feed on any or all of the immature stages (Cross and Moser 1971, Moser et al. 1971). Many species of the pygmephoroid families SCUTACARIDAE (Plate 72, p. 327) and PYGMEPHORIDAE (Plate 73, p. 328) have been found in association with a variety of insects, especially bees and wasps (Mahunka 1974d, Delfinado et al. 1976). The relationships apparently are phoretic, although little information is available on the biology of these groups.

The heteromorphic deutonymph or hypopus of the Acaridida is a phoretic form (see Chapter V) which may be adapted for transfer via arthropods or vertebrates. Hypopodes of the families SAPROGLYPHIDAE (Plate 122, p. 420), ACARIDAE (Plate 115, p. 372), ANOETIDAE (Plate 125, p. 423) and CHAETODACTYLIDAE (Plate 121, p. 419) attach to their invertebrate hosts by means of a ventral sucker plate. Hypopodes of certain groups of GLYCYPHAGIDAE, however, have ventral claspers (Plate 120, p. 418) by which they attach to the hairs of mammals (Fain 1971). Others lack developed suckers or claspers and attach to the base of the hair follicle, or secrete themselves in the endofollicular spaces (Fain 1969b).

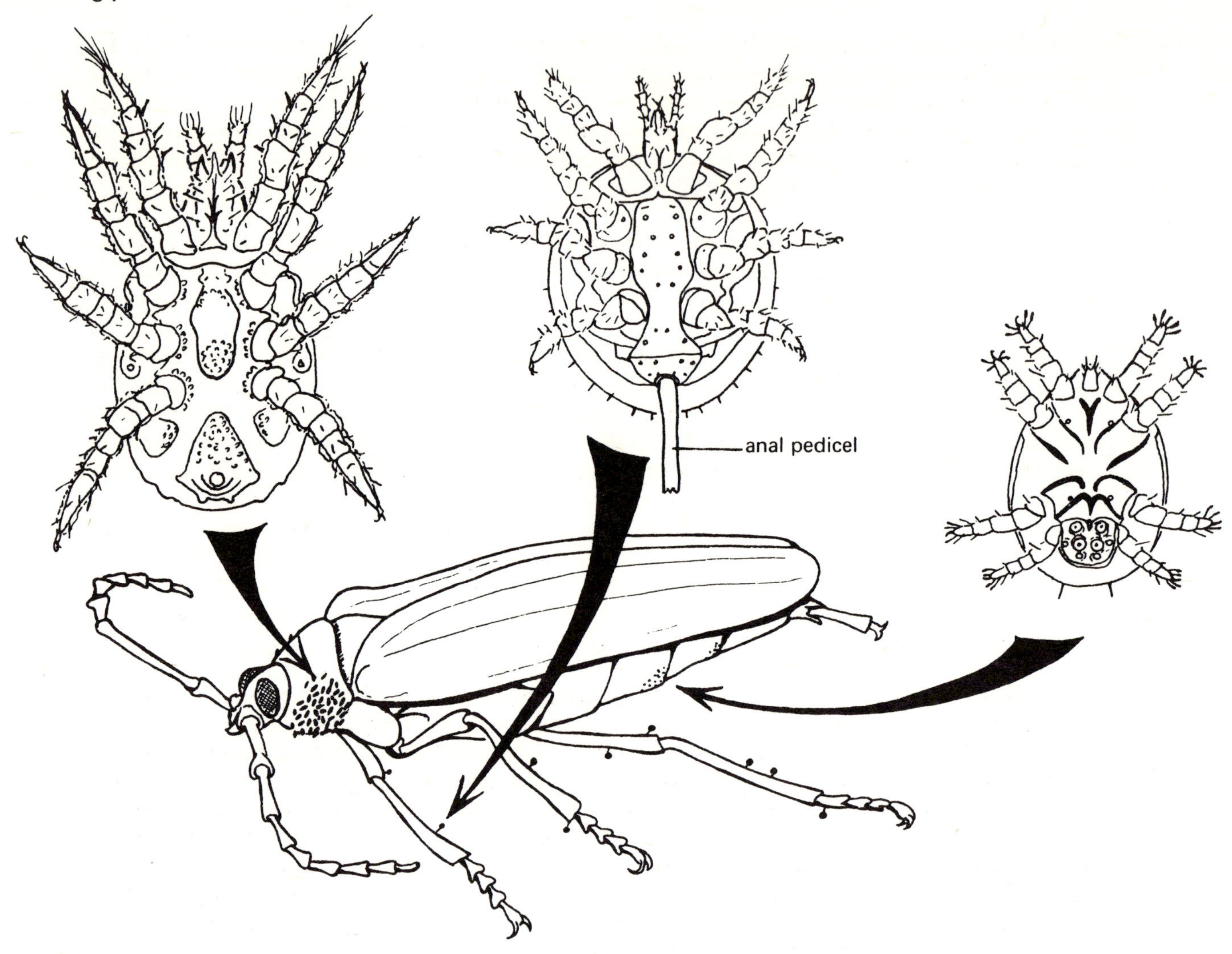

Fig. 20. Phoretic deutonymphs of (from left) POLYASPIDIDAE, UROPODIDAE and ACARIDAE on a prionid beetle. Arrows denote typical attachment sites.

Phoretic specificity has been reported or suspected in various Gamasida and Actinedida (Krantz 1967, Krantz and Mellott 1972, Costa 1967 and 1969, Moser et al. 1971, Kinn 1971, Binns 1972). The positive or negative response of a mite species to a given insect carrier may be due to pheromonal secretions produced by the insect. Site specificity on the host has been observed in some phoretic acarine groups including the MACROCHELIDAE (Krantz and Mellott op. cit.) and the various phoretic associates of the alkali bee, *Nomia melanderi* (Cross and Bohart 1969). Attachment patterns of four phoretic mite species on adult *N. melanderi* apparently are influenced by the age and sex of the bee carrier.

Parasitic Forms

Acarine parasites of animals occur in all but one principal suborder (the Oribatida), and many are of major importance to man. Pathogenic organisms of many types are transmitted by various mites or ticks to man or to his domesticated animals. Some of these pathogens, such as the organisms causing scrub typhus (Audy 1968) and Rocky Mountain spotted fever (Philip 1963), have altered the course of human migration, of economic development, and of military operations. Aside from transmission of pathogens, acarines may be vectors for internal parasites such as tapeworms and filarial worms. Members of both the Oribatida and Acaridida may serve as intermediate hosts for a variety of cestodes including the sheep tapeworm (Allred 1954). The simple process of feeding by mites may result in damage to their animal hosts through exsanguination, irritation, or by providing sites for invasion by secondary disease organisms.

The parasitic acarines may be divided into two groups on the basis of their feeding sites. A few examples of each group are cited below.

A. Ectoparasitic Mites

1. Vertebrate ectoparasites. Almost every animal group has its complex of external acarine parasites. Many show varying degrees of host specificity (Nutting 1968), as well as some remarkable morphological and biological adaptations to a parasitic existence (Fain 1969a, Radovsky 1969).

Development of vertebrate parasitism in the Acari proceeded, in all likelihood, through forms which made initial contact with prospective hosts through adaptation to nidicoly. Adaptation to the nest habitat has led to great structural and behavioral diversity amongst parasitic Acari. In the nidicolous parasitic Gamasida, for example, concentration of the food supply and shelter from external environmental forces has brought about an increase in feeding capacity of many species and concomitant reduction in protective and confining shields (Radovsky 1969). Furthermore, concentration of the parasite population allows males to find females quickly, obviating the need for the male to feed. As a result, male chelicerae have tended to become highly specialized for sperm transfer, often to the point where feeding is impossible (Plate 37-3, p. 195). Dimorphism in other nidicolous parasitic suborders also may be pronounced. Development of obligate ectoparasitism and of endoparasitism has led to additional changes including, in some cases, the repression of various preimaginal instars and the virtual disappearance of males.

Zemskaya (1968 and 1971) has observed that a transition from nest parasitism to constant host parasitism is occurring in the genus *Dermanyssus* (DERMANYSSIDAE, Plate 33, p. 191). Nest species such as *D. gallinae* (DeGeer) and *D. hirudinus* (Hermann) are long-legged, ovate forms which feed heavily on their hosts and oviposit up to 20 eggs in the nest material. *D. grochovskae* Zemskaya and *D. quintus* Vitzthum, however, have relatively shorter thicker legs and flat bodies with modified setae, and are more frequently collected on the host rather than in the nest. They ingest considerably less blood at each feeding than do *D. gallinae* and *D. hirudinis* and, soon after digestion, they oviposit a few eggs on the feathers of the host. Moss (1978) discusses transitional host parasitism in *Dermanyssus* in his review of the genus.

Ectoparasitic mites and ticks are recovered from animals such as bats, armadillos, birds, marsupials, reptiles and primates—including man. They include acarines which feed on blood, lymph, sebaceous secretions or digested tissues of their hosts by puncturing the skin, or by invading wounds or surface pores.

The chicken mite, *Dermanyssus gallinae* (DeGeer) and the northern fowl mite, *Ornithonyssus sylviarum* C. & F. (Gamasida, DERMANYSSIDAE and MACRONYSSIDAE, Plates 33 and 34, pp. 191 and 192), the mange and scab mites of the acaridid families SARCOPTIDAE (Plate 145, p. 443) and PSOROPTIDAE (Plate 128, p. 426), the ticks (Ixodida, IXODIDAE and ARGASIDAE (Plates 53-55, pp. 221-223), and the many species of chiggers (larval forms of the actinedid family TROMBICULIDAE, Plate 99, p. 354), are only a few common examples of the hundreds of ectoparasitic acarines found on vertebrates (Wharton and Fuller 1952, Baker et al. 1956, Strandtmann and Wharton 1958, Evans and Till 1966, Hoogstraal 1973, Yunker 1973). Their feeding sites and methods of attack generally reflect the type of food material ingested. For example, *Demodex folliculorum* (Simon) and *D. brevis* Akbulatova, the actinedid follicle mites of man (DEMODICIDAE, Plate 84, p. 339) (Spickett 1961, Desch and Nutting 1972) feed on sebaceous materials in hair follicles of the eyebrows and foreheads of virtually the entire world population. The primarily ectoparasitic trombiculid chiggers secrete a digestive enzyme at the point of attack and ingest the dissolved host tissues (Fig. 36, p. 280). Sarcoptid mange mites burrow into the skin of their host, while psoroptid scab mites attack the surface skin, causing a weeping lesion which finally hardens into a protective scab (Yunker 1973).

The pelage of a variety of mammals is infested by fur mites of the acaridid superfamilies Listrophoroidea and Psoroptoidea (pages 386, 394), some of which may cause mange in their hosts (Baker et al. 1956, Fain et al. 1970, Yunker 1973). In like manner, the skin of mammals is attacked by members of the actinedid families PSORERGATIDAE (Plate 83, p. 338) and MYOBIIDAE (Plate 82, p. 337), and that of birds by the acaridid family EPIDERMOPTIDAE (Plate 133, p. 431). Other acaridid mites are found on the feathers of birds (e.g., the ANALGIDAE, ALLOPTIDAE, DERMOGLYPHIDAE, PROCTOPHYLLODIDAE (Plates 131, 133-135, pp. 429, 431-433), and the many smaller families of Analgoidea (Radford 1953, Dubinin 1957, Atyeo and Braasch 1966, Lukoschus et al. 1974)), while actinedid mites of the family SYRINGOPHILIDAE (Plate 83, p. 338) are parasitic within the feather quills (Kethley 1970). The pit mites of snakes (Actinedida, OPHIOPTIDAE, Plate 82, p. 337) form cutaneous depressions at the base of scales in which mite development takes place (Fain 1964c). Although it lacks functional mouthparts, the hypopodial stage of the glycyphagid subfamily Hypodectinae apparently is capable of moving into and deriving nourishment from the subdermal tissues of birds and (more rarely) mammals in which they are found (Fain 1967, 1969a and 1969b, Pence 1972).

Gamasid mites of the families DERMANYSSIDAE (Plate 33, p. 191), MACRONYSSIDAE (Plate 34, p. 192), LAELAPIDAE (Plates 36-37) and related groups feed primarily on the blood and tissue secretions of their vertebrate hosts (Radovsky 1969, Yunker 1973), while the ticks (Ixodida) are essentially haematophagous.

Viruses, rickettsias, bacteria, spirochaetes, protozoans and helminths have been isolated from vertebrate ectoparasites. Many of these organisms cause virulent or debilitating diseases in man and animals throughout the world (Smith et al. 1944, Arthur 1962, Philip 1963, Audy 1968).

2. Invertebrate ectoparasites. Comparatively few mites have established a truly parasitic association with invertebrates. Species of the actinedid families TROMBIDIIDAE (Plate 97, p. 352), ERYTHRAEIDAE (Plate 94, p. 349), SMARIDIDAE (Plate 95, p. 350) and JOHNSTONIANIDAE (Plates 95-96) are parasitic on insects, but only in the larval stage (Severin 1944, Southcott 1961). Nymphs and adults of these mites are predatory. A similar situation exists in the phalanx Hydrachnidia where the larval stage (Fig. 21, p. 67) may parasitize various aquatic and semiaquatic insects (Mitchell 1961, Prasad and Cook 1972, Mullen 1974). Larvae of TROMBIDIIDAE and ERYTHRAEIDAE parasitize arachnids as well as insects. Members of the erythraeid genus *Leptus* are fairly common ectoparasites of opilionids and occur less frequently on scorpions. *Isothrombium oparbellae* André is an ectoparasite of a solfugid (André 1949). Nymphs and adults of some hygrobatoid Hydrachnidia are found in mollusks and sponges (Mitchell 1955).

Species of the actinedid families PODAPOLIPIDAE (Plate 75, p. 330), PTERYGOSOMATIDAE (Plate 93, p. 348), and various members of the superfamily Pyemotoidea (Plates 70-72, pp. 325-327) may be parasitic on their invertebrate hosts throughout their lives (Hirst 1921, Newell and Ryckman 1966). *Ricardoella limacum* (Schrank) (Actinedida, EREYNETIDAE, Plate 67, p. 322) feeds on slugs and snails, and occasionally causes their death under conditions of high mite population. Baker (1970) has shown that *R. limacum* is haematophagous on its gastropod hosts.

Butterflies and moths are the hosts for members of the parasitic gamasid genera *Dicrocheles* (LAELAPIDAE, Plate 36, p. 194), *Otopheidomenis, Noctuiseius* and *Prasadiseius* (OTOPHEIDOMENIDAE, Plate 22, p. 180) (Treat 1975). Species of the otopheidomenid genera *Treatia* and *Hemipteroseius* parasitize hemipterans (Evans 1963). Hemipterous insects also are utilized by actinedid mites of the genus *Coreitarsonemus* (TARSONEMIDAE) which live in the tissues of the odor gland vestibules of their plant bug hosts (Fain 1970a).

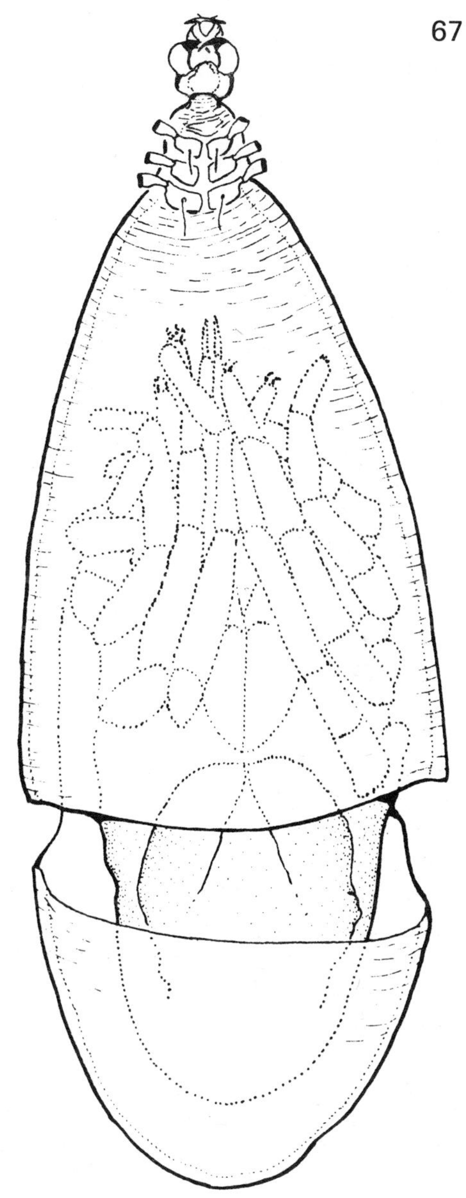

Fig. 21. Water mite calyptostase, with a portion of developing nymph exposed by break in larval integument.

A cockroach, *Nauphoeta cinerea,* serves as host for the gamasid *Proctolaelaps nauphoetae* (Womersley) (ASCIDAE, Plate 20, p. 178) (Egan and Hunter 1975). At least two other gamasid parasites have been implicated as economic pests of honeybees. *Tropilaelaps clareae* Delfinado and Baker (LAELAPIDAE) feeds on honeybee larvae, and has been observed to cause up to 50% mortality in severely infested colonies (Atwal and Goyal 1971). "Varrosis", or injury to bees caused by *Varroa jacobsoni* (Oud.) (VARROIDAE, Plate 35, p. 193) is considered a serious threat to apiculture in parts of Europe and southern Asia (Kulikov 1965, Haragasim and Samšiňák 1972, Akratanakul and Burgett 1975). Species of another gamasid genus, *Ljunghia* (LAELAPIDAE), are parasites of mygalomorph spiders (Domrow 1975).

A type of hyperparasitism exists between species of the genus *Myialges* (Acaridida, EPIDERMOPTIDAE, Plate 133, p. 431) and their insect associates. *Myialges (Promyialges) macdonaldi* is a skin parasite of the passeriform bird, *Parus caeruleus,* but females attach to hippoboscid flies to oviposit. Emerging larvae then disperse from the fly to the bird host (Evans et al. 1963). Other *Myialges* species are found on a variety of hippoboscid species, and on menoponid lice (Fain 1965). Members of another acaridid family, the EWINGIIDAE (Plate 126, p. 424) parasitize pagurid crabs, attaching in the gills of the host. Species of the ewingiid genera *Ewingia, Hoogstraalacarus* and *Askinasia* are crab parasites in Africa and the neotropics (Yunker 1970).

B. Endoparasitic Mites

1. Vertebrate endoparasites. The majority of vertebrate endoparasites show a reduction in sclerotization when compared to related external forms. Many have reduced mouthparts and legs, and none has functional eyes. Most of the vertebrate endoparasites are associated with the respiratory systems of their hosts. For example, the gamasid family RHINONYSSIDAE (Plate 31, p. 189) is largely confined to the nasal passages of birds, although *Sternostoma* may invade tracheal tissues also. The nasal and/or lung habitats of birds have also been exploited by the speleognathine EREYNETIDAE (Plate 67, p. 322), the acaridid family TURBINOPTIDAE (Plate 135, p. 433) (Domrow 1969, Fain 1970b), and certain groups of chiggers (Actinedida, TROMBICULIDAE) (Yunker and Jones 1961). Some members of the gamasid family HALARACHNIDAE (Plate 30, p. 188) occur in the nasal passages of seals and walruses (Strandtmann and Wharton 1958) while others may be embedded in the sinuses, bronchiae, tracheae or lungs of various mammal hosts (Hull 1970, Kim and Bang 1970, Yunker 1973).

The gamasid family MACRONYSSIDAE (Plate 34, p. 192) is generally considered an ectoparasitic group, but some species may be found in the mouth tissues of bats (Radovsky et al. 1971) or in the nasal cavities of lizards (Yunker and Radovsky 1966). The ENTONYSSIDAE (Plate 30, p. 188) is exclusively endoparasitic in the lungs and tracheae of reptiles (Fain 1961).

The nasal fossae of bats are infested by the acaridid family GASTRONYSSIDAE (Plate 146, p. 444) (Fain1964a). A similar habitat is colonized by the LEMURNYSSIDAE in lemurs (Fain 1964b). *Cytonyssus* species (Acaridida, CYTODITIDAE, Plate 146, p. 444) are found in the nasal passages of birds, while *Cytodites* invade the lungs and air sacs (Pence 1975). Members of the acaridid family PNEUMOCOPTIDAE (Plate 145, p. 443) often are encountered in the lungs of rodents (Baker et al. 1956).

Endoparasitic mites may occasionally infest portions of their vertebrate hosts other than the respiratory system. A species of the acaridid genus *Laminosioptes* (family LAMINOSIOPTIDAE, Plate 145, p. 443) is subcutaneous in domestic fowl. Dead mites serve as loci for the formation of subcutaneous cysts which may be felt through the skin of the bird host, particularly in the pectoral area. *Cytodites nudus* (Vizioli) (family CYTODITIDAE, Plate 146, p. 444) is found primarily in the air sacs of chickens. However it also occurs in the body cavity or alimentary canal of its host where its presence may produce a spectrum of symptoms which occasionally result in death (Yunker 1973). *Gastronyssus bakeri* Fain (Acaridida, GASTRONYSSIDAE) usually is encountered in the nasal tissues of its chiropteran hosts, but has also been recovered from the intestines of African fruit bats (Fain 1955).

Accidental ingestion of live mites by vertebrates may result in a condition known as *acariasis,* in which the mites survive and reproduce in the alimentary tract. Acariasis occurs in cattle which are fed mite-infested grain (families ACARIDAE and GLYCYPHAGIDAE). Symptoms include vomiting and diarrhea (Hinman and Kampmeier 1934).

2. Invertebrate endoparasites. Few mite groups have adapted to an endoparasitic habitat in invertebrates, and these are confined to the Gamasida and Actinedida. The actinedid genus *Acarapis* (family TARSONEMIDAE, Plate 74, p. 329) includes at least three species which parasitize honeybees. One of these, *Acarapis woodi* (Rennie), invades the prothoracic tracheae and air sacs of the bee host and is considered by some to be a serious problem in commercial honeybee hives throughout much of Europe and Asia (Hirst 1921, Hirschfelder and Sachs 1952). The tracheae of hymenopterans and of orthopterans may be invaded by members of the genus *Locustacarus* (Actinedida, PODAPOLIPIDAE) (Husband and Sinha 1970).

Otopheidomenis species (Gamasida, OTOPHEIDOMENIDAE) are frequently found in the thoraco-abdominal cleft or wing bases of their moth hosts but, unlike the moth ear mites of the genus *Dicrocheles* (LAELAPIDAE) which invade the tympanic and countertympanic cavities to feed and lay eggs (Treat 1975), lepidopterophilous otopheidomenids utilize the tympanic recesses or corresponding thoraco-abdominal areas of their noctuid and sphingid hosts only to oviposit.

The monotypic genus *Enterohalacarus* (Actinedida, HALACARIDAE) is recorded as an internal parasite of deep sea urchins, at least in the larval and nymphal instars (Viets 1938). Other halacarids *(Halixodes* and *Astacopsiphagus)* are gill parasites of chitons and of freshwater decapods (Brucker and Trouessart 1900, Viets 1931).

Useful References

Akratanakul, P. and D.M. Burgett (1975). *Varroa jacobsoni:* a prospective pest of honeybees in many parts of the world. Bee World **56**(3):119-121.

Allred, D.M. (1954). Mites as intermediate hosts of tapeworms. Proc. Utah Acad. Sci., Arts and Letters **31**:44-51.

André, M. (1931). Sur le genre *Hyadesia* Mégnin, 1889 (Sarcoptides Hydrophiles). Bull. Mus. nat. Hist. natur. Paris, 2ᵉ série, **3**:496-506.

André, M. (1934). Acariens terrestres adaptés à la vie marine. C.R. 67ᵉ Congr. Soc. Savantes, Paris: 134-156.

André, M. (1949). Nouvelle forme larvaire de Thrombidion (*Isothrombium oparbellae* n.g., n.sp.) parasite d'un Solifuge. Bull. Mus. nat. Hist. natur. Paris, 2ᵉ série, **21**:354-357.

Aoki, J. (1967). Microhabitats of oribatid mites on a forest floor. Nat. Sci. Mus. Tokyo **10**(2):133-138.

Arthur, D.R. (1962). Ticks and Disease. Pergamon Press, Oxford: 445 pp.

Atwal, A.S. and N.P. Goyal (1971). Infestation of honeybee colonies with *Tropilaelaps,* and its control. J. Apic. Res. **10**(3):137-142.

Atyeo, W.T. and N.L. Braasch (1966). The feather mite genus *Proctophyllodes* (Sarcoptiformes:Proctophyllodidae). Bull. Univ. Nebraska State Mus. **5**:1-354.

Audy, J.R. (1968). Red Mites and Typhus. Univ. London, Athlone Press: 191 pp. + x.

Axtell, R.C. (1963). Effect of Macrochelidae (Acarina:Mesostigmata) on house fly production from dairy cattle manure. J. Econ. Ent. **56**(3):317-321.

Axtell, R.C. (1969). Macrochelidae (Acarina:Mesostigmata) as biological control agents for synanthropic flies. WHO/VBC 119:17 pp.

Baker, E.W., T.M. Evans, D.J. Gould, W.B. Hull and H.L. Keegan (1956). A Manual of Parasitic Mites of Medical or Economic Importance. Natl. Pest Control Assoc. Tech. Publ.: 170 pp.

Baker, R.A. (1970). The food of *Ricardoella limacum* (Schrank)-Acari-Trombidiformes and its relationship with pulmonate molluscs. J. nat. Hist. **4**:521-530.

Berry, R.E. (1973). Biology of the predaceous mite, *Pergamasus quisquiliarum* on the garden symphylan, *Scutigerella immaculata,* in the laboratory. Ann. Ent. Soc. Amer. **66**(6):1354-1356.

Berthet, P. (1964). L'activité des Oribatides (Acari:Oribatei) d'une chenâie. Mém. Inst. roy. Sci. natur. Belg. **152**:1-152.

Berthet, P. (1971). Mites. **In** Methods of Study in Quantitative Soil Ecology: Population, Production and Energy Flow. Int. Biol. Prog. Handbook 18:186-208.

Binns, E.S. (1972). *Arctoseius cetratus* (Sellnick) (Acarina:Ascidae) phoretic on mushroom sciarid flies. Acarologia **14**(3):350-356.

Binns, E.S. (1973). *Digamasellus fallax* Leitner (Mesostigmata:Digamasellidae) phoretic on mushroom sciarid flies. Acarologia **15**(1):10-17.

Bornemissza, G.F. (1957). An analysis of arthropod succession in carrion and the effect of its decomposition on the soil fauna. Aust. J. Zool. **5**:1-12.

Bronswijk, J.E.M.H. van and R.N. Sinha (1971). Pyroglyphid mites (Acari) and house dust allergy. J. Allergy **47**(1):31-52.

Brucker, E.A. and E.L. Trouessart (1900). Descriptions d'espéces nouvelles d'Halacaridae (par le Dr. Trouessart) et description d'un genre nouveau (par Brucker et Trouessart). Bull. Etud. sci. Angers **29**: 209-234.

Cancela da Fonseca, J.P. and E. Kiffer (1969). A propros des rapports entre certaines espèces d'acariens oribates et certaines espèces de champignons existant dans un meme sol. C.R. Acad. Sci. **269**:386-389.

Chant, D.A. (1960). Phytoseiid mites (Acarina:Phytoseiidae). Part I Bionomics of seven species in southeastern England. Can. Ent. **91**(suppl. 12):5-44.

Chant, D.A. (1961). An experiment in biological control of *Tetranychus telarius* (L) (Acarina:Tetranychidae) in a greenhouse using the predaceous mite *Phytoseiulus persimilis* Athias-Henriot (Phytoseiidae). Can. Ent. **93**(6):437-443.

Chant, D.A. and C.A. Fleschner (1960). Some observations on the ecology of phytoseiid mites (Acarina: Phytoseiidae) in California. Entomophaga **5**:131-139.

Cook, D.R. (1974). Water mite genera and subgenera. Mem. Amer. Ent. Inst. 21:860 pp. + vii.

Costa, M. (1967). Notes on macrochelids associated with manure and coprid beetles in Israel. II. Three new species of the *Macrocheles pisentii* complex. Acarologia **9**(2):304-329.

Costa, M. (1969). The association between mesostigmatid mites and coprid beetles. Acarologia **11**(3):411-428.

Costa, M. (1974). Mesostigmatic mites (Acari:Mesostigmata) from the Mediterranean shores of Israel I. The genus *Hypoaspis* Canestrini, 1884. Israel J. Ent. **9**:219-228.

Cross, E.A. and G.E. Bohart (1969). Phoretic behavior of four species of alkali bee mites as influenced by season and host sex. J. Kansas Ent. Soc. **42**:195-219.

Cross, E.A. and J.C. Moser (1971). Taxonomy and biology of some Pyemotidae (Acarina:Tarsonemoidea) inhabiting bark beetle galleries in North American conifers. Acarologia **13**(1):47-64.

Crossley, D.A. and M. Witkamp (1963). Forest soil mites and mineral cycling. Proc. 1st Int. Congr. Acarology, Fort Collins. Acarologia **6**(fasc. h.s.):137-145.

Cusack, P.D., G.O. Evans and P.A. Brennan (1975). A survey of the mites of stored grain and grain products in the Republic of Ireland. Sci. Proc. Roy. Dublin Soc. Ser. B, **3**(20):273-329.

Davis, A.C. (1944). The mushroom mite (*Tyrophagus lintneri* (Osborn)) as a pest of cultivated mushrooms. USDA Tech. Bull. 879:26 pp.

De Leon, D. (1967a). Some mites of the Caribbean area. Part I. Acarina on plants in Trinidad, West Indies. Allen Press, Lawrence, Kansas: 1-46.

De Leon, D. (1967b). Some mites of the Caribbean area. Part II. The genus *Asca* in the Caribbean area. Allen Press, Lawrence, Kansas: 47-66.

Delfinado, M.D., E.W. Baker and M.J. Abbatiello (1976). Terrestrial mites of New York - III. The family Scutacaridae (Acarina). J. New York Ent. Soc. **84**(2):106-145.

Domrow, R. (1969). The nasal mites of Queensland birds (Acari:Dermanyssidae, Ereynetidae and Epidermoptidae). Linn. Soc. N.S.W. **93**(3):297-426.

Domrow, R. (1975). *Ljunghia* Oudemans (Acari:Dermanyssidae), a genus parasitic on mygalomorph spiders. Rec. S. Austral. Mus. **17**(4):31-39.

Desch, C. and W.B. Nutting (1972). *Demodex folliculorum* (Simon) and *D. brevis* Akbulatova of man: redescription and reevaluation. J. Parasitol. **58**(1):169-177.

Dubinin, W.B. (1951). Feather mites (Analgesoidea). Part I. Introduction to their study. Fauna USSR **6**(5): 1-363.

Edwards, C.A., D.E. Reichle and D.A. Crossley Jr. (1970). The role of soil invertebrates in turnover of organic matter and nutrients. **In** Analysis of Temperate Forest Ecosystems. D.E. Reichle, ed. Springer-Verlag, Berlin: 147-172.

Egan, M.E. and P.E. Hunter (1975). Redescription of a cockroach mite, *Proctolaelaps nauphoetae,* with notes on its biology. Ann. Ent. Soc. Amer. **68**(2):361-364.

Evans, G.O. (1955). Identification of terrestrial mites. **In** Soil Zoology. D.K. McE. Kevan, ed. Academic Press, London: 55-61.

Evans, G.O. (1963). Observations on the classification of the family Otopheidomenidae (Acari:Mesostigmata) with descriptions of two new species. Ann. Mag. Nat. Hist. **5**(13):609-620.

Evans, G.O., A. Fain and J. Bafort (1963). Découverte du cycle évolutif du genre *Myialges* avec description d'une espèce nouvelle (Myialgesidae:Sarcoptiformes). Bull. Ann. Soc. roy. Ent. Belg. **99**(34):486-500.

Evans, G.O., J.G. Sheals and D. Macfarlane (1961). Free-living Acari. **In** The Terrestrial Acari of the British Isles. Vol. I Introduction and biology. British Museum (Natural History), London: 89-106.

Evans, G.O. and W.M. Till (1966). Studies on the British Dermanyssidae (Acari:Mesostigmata). Part II. Classification. Bull. Brit. Mus. (Nat. Hist.) Zool. **14**(5):109-370.

Fain, A. (1955). Un acarien remarquable vivant dans l'estomac d'une chauve-souris:*Gastronyssus bakeri* n.g., n.sp. Ann. Soc. Belg. Méd. Trop. **35**(6):681-688.

Fain, A. (1961). Les acariens parasites endopulmonaires des serpents (Entonyssidae:Mesostigmata). Inst. roy. Sci. nat. Belg. **37**(6):1-135.

Fain, A. (1964a). Chaetotaxie et classification des Gastronyssidae avec description d'un nouveau genre parasite nasicole d'un ecureuil sudafricain (Acarina:Sarcoptiformes). Rev. Zool. Bot. Afr. **70**(1-2):40-52.

Fain, A. (1964b). Les Lemurnyssidae parasites nasicoles des Lorisidae africains et des Cebidae sud-américaines. Description d'une espèce nouvelle (Acarina:Sarcoptiformes). Ann. Soc. Belg. Méd. Trop. **44**(3):453-458.

Fain, A. (1964c). Les Ophioptidae acariens parasites des ecailles des serpents (Trombidiformes). Inst. roy. Soc. nat. Belg. **40**(15):1-57.

Fain, A. (1965). A review of the family Epidermotidae Trouessart parasitic on the skin of birds (Acarina: Sarcoptiformes). Part 1. Konink. Acad. Wetensch., Lett. Schone Kunsten Belg. (Wetensch.) **27**(84):176 pp. + ix.

Fain, A. (1967). Les hypopes parasites des tissues cellularies des oiseaux (Hypodectidae:Sarcoptiformes). Bull. Inst. roy. Sci. nat. Belg. **43**(4):1-139.

Fain, A. (1969a). Adaptation to parasitism in mites. Acarologia **11**(3):429-449.

Fain, A. (1969b). Les deutonymphs hypopiales vivant en association phoretique sur les mammiferes (Acarina: Sarcoptiformes). Bull. Inst. roy. Sci. nat. Belg. **45**(33):1-262.

Fain, A. (1970a). *Coreitarsonemus* un nouveau genre d'Acariens parasitant la glande oderiférante des Hémiptères Coreidae (Tarsonemidae:Trombidiformes). Rev. Zool. Bot. Afr. **82**(3-4):315-334.

Fain, A. (1970b). Notes sur les spéléognathines parasites nasicoles des mammifères (Ereynetidae:Trombidiformes). Acarologia **12**(3):509-521.

Fain, A. (1971). Évolution de certains groupes d'hypopes en fonction du parasitisme (Acarina:Sarcoptiformes). Acarologia **13**(1):171-175.

Fain, A. (1974). Acariens récoltés par le Dr. J. Trave aux iles subantarctiques. I. Familles Saproglyphidae et Hyadesiidae (Astigmates). Acarologia **16**(4):684-708.

Fain, A. and D.E. Johnston (1975). A new algophagin mite, *Algophagopsis pneumatica* gen.n., sp.n. living in a river (Astigmata:Hyadesiidae). Bull. Ann. Soc. roy. Ent. belge 111:66-70.

Fain, A., A.J. Munting and F. Lukoschus (1970). Les Myocoptidae parasites des rongeurs en Hollande et en Belgique. Acta Zool. Path. Antverp. 50:67-172.

Farish, D.J. and R.C. Axtell (1971). Phoresy redefined and examined in *Macrocheles muscaedomesticae* (Acarina:Macrochelidae). Acarologia **13**(1):16-29.

Flaherty, D.L. and C.B. Huffaker (1970). Biological control of Pacific mites and Willamette mites in San Joaquin Valley, I. Role of *Metaseiulus occidentalis.* II. Influence of dispersion patterns of *M. occidentalis.* Hilgardia **40**(10):267-330.

Fujikawa, T. (1970). Relation between oribatid fauna and some environments of Napporo National Forest in Hokkaido (Acarina:Cryptostigmata). II. Oribatid fauna in soils under four different vegetations. Appl. Ent. Zool. **5**(2):69-83.

Fujikawa, T. (1972). Preliminary survey on the relationship between orbatid mites and the decomposition of fresh leaves. Appl. Ent. Zool. **7**(4):181-189.

Gerson, U. (1968). Five tydeid mites from Israel (Acarina:Prostigmata). Israel J. Zool. **17**:191-198.

Gerson, U. (1973). Lichen-arthropod associations. Lichenologist **5**:434-443.

Grandjean, F. (1948). Sur l'elevage de certains oribates en vue d'obtenir des clones. Bull. Mus. nat. Hist. natur. Paris **20**:450-457.

Grandjean F. (1950). Sur deux espèces du genre *Dometorina* N.G. et les moeurs de *D. plantivaga* (Berl.) (Acariens, Oribates). Bull.Soc. Zool. France **75**:224-242.

Gurney, B. and N.W. Hussey (1967). *Pygmephorus* species (Acarina:Pyemotidae) associated with cultivated mushrooms. Acarologia **9**(2):353-358.

Halbert, J.N. (1920). The Acarina of the seashore. Proc. Roy. Irish Acad. **35**(B,7):106-152.

Haq, J. (1965). Records of some interstitial mites from Nobska Beach together with a description of a new genus and species *Psammonsella nobskae,* of the family Rhodacaridae (Acarina Mesostigmata). Acarologia **7**(3):411-419.

Haragsim, O. and K. Samšiňák (1972). L'acarien *Varroa jacobsoni* Oudemans decouvert en Europe. Bull. Apic. **15**(2):1-14.

Hartenstein, R. (1961). On the distribution of forest soil microarthropods and their fit to contagious distribution functions. Ecol. **42**:190-194.

Hartenstein, R. (1962). Soil Oribatei I. Feeding specificity among forest soil Oribatei (Acarina). Ann. Ent. Soc. Amer. **55**(2):202-206.

Hinman, E.H. and R.H. Kampmeier (1934). Intestinal acariasis due to *Tyrophagus longior* Gervais. Amer. J. Trop Med. **14**:355-362.

Hinton, H.E. (1971). Plastron respiration in the mite *Platyseius italicus.* J. Ins. Physiol. **17**:1185-1199.

Hirschfelder, H. and H. Sachs (1952). Recent research on the acarine mite. Bee World **33**(12):201-209.

Hirst, S. (1921). On the mite (*Acarapis woodi,* Rennie) associated with the Isle of Wight Bee Disease. Ann. Mag. Nat. Hist. ser. 9, **7**:509-519.

Hoogstraal, H. (1973). Ticks. **In** Parasites of Laboratory Animals. R.J. Flynn, ed. Iowa State Univ. Press, Ames:398-424.

Hoyt, S.C. (1969). Integrated chemical control of spider mites on pears. J. Econ. Ent. **62**(1):74-86.

Hoyt, S.C. and L.E. Caltagirone (1971). The developing program of integrated control of pests of apples in Washington and peaches in California. **In** Biological Control, C.B. Huffaker, ed. Plenum Press, New York: 395-421.

Huffaker, C.B. and C.E. Kennett (1956). Experimental studies in predation: predation and cyclamen-mite populations on strawberries in California. Hilgardia **26**(4):191-222.

Hughes, A.M. (1976). The Mites of Stored Food and Houses. Second Edition. Min. Agr. Fish. Food, London, Tech. Bull. **9**:400 pp.

Hughes, R.D. and C.G. Jackson (1958). A review of the Anoetidae (Acari). Virginia J. Sci. **9** N.S.(1):5-198.

Hughes, T.E. (1959). The free-living Acari. **In** Mites, or the Acari. Athlone Press, London:1-22.

Hull, W.B. (1970). Respiratory mite parasites in nonhuman primates. Lab. Animal Care **20**(2):402-406.

Hurlbutt, H.W. (1958). A study of soil-inhabiting mites from Connecticut apple orchards. J. Econ. Ent. **51**(6):767-772.

Husband, R.W. and R.N. Sinha (1970). A revision of the genus *Locustacarus* with a key to genera of the family Podapolipidae (Acarina). Ann. Ent. Soc. Amer. **63**(4):1152-1162.

Hussey, N.W., W.H. Read and J.J. Hesling (1969). Order Acarina: Mites. **In** The Pests of Protected Cultivation. Amer. Elsevier Publ. Co., New York: 190-228.

Imamura, T. (1968). Results of the speleological survey in South Korea 1966. IX. Halacaridae (Acari) found in a limestone cave of South Korea. Bull. Nation. Sci. Mus. Tokyo **11**(3):281-284.

Imamura, T. (1970). Subterranean water mites (Limnohalacarinae and Hydrachnellae) of the Tsushima Islands. Bull. Nation. Sci. Mus. Tokyo **13**(2):249-262.

Jeppson, L.R., H.H. Keifer and E.W. Baker (1975). Mites Injurious to Economic Plants. Univ. California Press, Berkeley: 614 pp. + 74 pl. + xix.

Kamburov, S.S. (1971). Feeding, development, and reproduction of *Amblyseius largoensis* on various food substances. J. Econ. Ent. **64**(3):643-648.

Karafiat, H. (1959). Systematik und Ökologie der Scutacariden. **In** Beiträge zur Systematik und Ökologie mitteleuropäischer Acarina. Bd. 1, Abs. IV: 627-712.

Kethley, J.B. (1970). A revision of the family Syringophilidae (Prostigmata:Acarina). Contrib. Amer. Ent. Inst. **5**(6):1-76.

Kiffer, E. and J.P. Cancela da Fonseca (1971). A simultaneous study of microarthropod populations and fungus populations of cellulose traps from a podzolized soil. Zentralbl. Bakteriol., Parasit., Infektionshrank. Hyg. **126**(5):510-520.

Kim, C.S. and B.G. Bang (1970). Nasal mites parasitic in the nasal and upper skull tissues in the baboon (*Papio* sp.). Science **169**:372-373.

King, L.A.L. (1914). Notes on the habits and characteristics of some littoral mites of Newport. Proc. Roy. Physiol. Soc. **19**:129-141.

Kinn, D.N. (1968). A new species of *Pleuronectocelaeno* (Acarina:Celaenopsidae) associated with bark beetles in North and Central America. Acarologia **10**(2):191-205.

Kinn, D.N. (1971). The life cycle and behavior of *Cercoleipus coelonotus* (Acarina:Mesostigmata) including a survey of phoretic mite associates of California Scolytidae. Univ. Calif. Publ. Ent. **65**:66 pp.

Krantz, G.W. (1967). I. A review of the genus *Holocelaeno* Berlese 1910. II. A review of the genus *Holostaspella* Berlese 1903 (Acarina:Macrochelidae). Acarologia **9**, fasc. suppl.: 146 pp.

Krantz, G.W. and J.L. Mellott (1972). Studies on phoretic specificity in *Macrocheles mycotrupetes* and *M. peltotrupetes* Krantz and Mellott (Acari:Macrochelidae), associates of geotrupine Scarabaeidae. Acarologia **14**(3):317-344.

Krczal, H. (1959). Systematik und Ökologie der Pyemotiden. **In** Beiträge zur Systematik und Ökologie mitteleuropäischer Acarina. Bd.1, Abs. III: 385-625.

Kulikov, N.S. (1965). Varrosis of bees. Pchelovodstvo **11**:15-16.

Lebrun, P. (1971). Écologie et biocenotique de quelques peuplements d'arthropodes édaphiques. PhD Thesis, Univ. Louvain.

Lee, D.C. (1970). The Rhodacaridae (Acari:Mesostigmata); classification, external morphology and distribution of genera. Rec. S. Austral. Mus. **16**(3):1-219.

Le Veque, N. (1930). Mites of the genus *Dinogamasus (Dolaea)* found in the abdominal pouch of African bees known as *Mesotrichia* or *Koptorthosoma* (Xylocopidae). Amer. Mus. Nov. **434**:1-19.

Lindquist, E.E. (1969a). New species of *Tarsonemus* (Acarina:Tarsonemidae) associated with bark beetles. Can. Ent. **101**(12):1291-1314.

Lindquist, E.E. (1969b). Review of holarctic tarsonemid mites (Acarina:Prostigmata) parasitizing eggs of ipine bark beetles. Mem. Ent. Soc. Canada 60: 111 pp.

Lindquist, E.E. and W.D. Bedard (1961). Biology and taxonomy of mites of the genus *Tarsonemoides* (Acarina:Tarsonemidae) parasitizing eggs of bark beetles of the genus *Ips*. Can. Ent. **83**:982-999.

Lukoschus, F., J.M.W. Louppen and T.C. Maa (1974). *Psorergates squamipes*, n.sp. (Acari:Psorergatidae), a skin mite from *Anourosorex squamipes* (Insectivora) in Taiwan. Pac. Insects **16**:51-56.

Luxton, M. (1968). A new genus and species of littoral mite (Acari:Mesostigmata) from New Zealand. N.Z.J. Marine Freshwater Res. **2**(3):497-504.

Luxton, M. (1972). Studies on the oribatid mites of a Danish beech wood soil. Pedobiol. **12**:434-463.

Luxton, M. (1975). Studies on the oribatid mites of a Danish beech wood soil. II. Biomass, colorimetry, and respirometry. Pedobiol. **15**:161-200.

Mahunka, S. (1973). *Pygmephorus* species (Acari, Tarsonemida) from North American small mammals. Parasit. Hung. **6**:247-259.

Mahunka, S. (1974a). Auf Insekten lebende Milben (Acari:Acarida, Tarsonemida) aus Afrika. III. Acta Zool. Acad. Sci. Hung. **20**(1-2):137-154.

Mahunka, S. (1974b). Auf Insekten lebende Milben (Acari:Acarida, Tarsonemida) aus Afrika. IV. Acta Zool. Acad. Sci. Hung. **20**(3-4):367-402.

Mahunka, S. (1974c). New data to the knowledge of *Pygmephorus*-species (Acari:Tarsonemida) living on small mammals in America. Parasit. Hung. **7**:197-200.

Mahunka, S. (1974d). Beiträge zur Kenntnis der an Hymenopteren lebenden Milben (Acari), I. Ann. Hist. nat. Mus. Nat. Hung. **66**:389-394.

Mahunka, S. (1975a). Auf Insekten lebende Milben (Acari:Acarida und Tarsonemida) aus Afrika. V. Acta. Zool. Acad. Sci. Hung. **21**(1-2):39-72.

Mahunka, S. (1975b). Further data to the knowledge of Tarsonemida (Acari) living on small mammals in North America. Parasit. Hung. **8**:85-94.

McBrayer, J.F., J.M. Ferris, L.J. Metz, C.S. Gist, B.W. Carnaby, Y. Kitazawa, T. Kitazawa, J.G. Wernz, G.W. Krantz and H. Jensen (1977). Decomposer invertebrate populations in U.S. forest biomes. Pedobiol. **17**:89-96.

McMurtry, J.A. and G.T. Scriven (1964). Biology of the predaceous mite *Typhlodromus rickeri* (Acarina: Phytoseiidae). Ann. Ent. Soc. Amer. **57**(3):362-367.

McMurtry, J.A. and G.T. Scriven (1971). Predation by *Amblyseius limonicus* on *Oligonychus punicae* (Acarina): effects of initial predator-prey ratios and prey distribution. Ann. Ent. Soc. Amer. **64**(1):219-224.

Mitchell, R.D. (1955). Anatomy, life history, and evolution of the mites parasitizing fresh-water mussels. Univ. Mich. Misc. Publ. Zool. **89**:28 pp.

Mitchell, R.D. (1960). The evolution of thermophilous water mites. Evol. **14**(3):361-377.

Mitchell, R.D. (1961). Behavior of the larvae of *Arrenurus fissicornis* Marshall, a water mite parasitic on dragonflies. Animal Behav. **9**(3-4):220-224.

Mitchell, R.D. (1974). The evolution of thermophily in hot springs. Quart. Rev. Biol. **49**:229-242.

Morse, R.B. (1974). An annotated bibliography on *Varroa jacobsoni* and *Tropilaelaps clareae.* Cornell Univ. Dept. Ent. (mimeo): 7 pp.

Moser, J.C., E.A. Cross and L.M. Roton (1971). Biology of *Pyemotes parviscolyti* (Acarina:Pyemotidae). Entomophaga **16**(4):367-379.

Moss, W.W. (1978). The mite genus *Dermanyssus:* a survey, with description of *Dermanyssus trochilinis* n.sp., and a revised key to the species (Acari:Mesostigmata:Dermanyssidae). J. Med. Ent. **14**(6):627-640.

Moutia, L.A. (1958). Contribution to the study of some phytophagous Acarina and their predators in Mauritius. Bull. Ent. Res. **49**:59-75.

Mullen, G.R. (1974). Acarine parasites of mosquitoes. II. Illustrated larval key to the families and genera of mites reportedly parasitic on mosquitoes. Mosquito News **34**(2):183-195.

Newell, I.M. (1947). A systematic and ecological study of the Halacaridae of eastern North America. Bull. Bingham Ocean. Coll. **10**(3):1-232.

Newell, I.M. (1957). Studies on the Johnstonianidae (Acari, Parasitengona). Pac. Sci. **11**:396-466.

Newell, I.M. (1971). Halacaridae (Acari) collected during Cruise 17 of the R/V Anton Bruun, in the southeastern Pacific Ocean. Anton Bruun Rept. 8:3-58.

Newell, I.M. and R.E. Ryckman (1966). Species of *Pimeliaphilus* (Acari:Pterygosomidae) attacking insects, with particular reference to the species parasitizing Triatominae (Hemiptera:Reduviidae). Hilgardia **37**(12):403-436.

Nickel, P.A. and R.J. Elzinga (1970a). New host records and biological notes for *Echinomegistus wheeleri* (Wasmann). J. Kansas Ent. Soc. **43**(1):32-34.

Nickel, P.A. and R.J. Elzinga (1970b). Biology of *Micromegistus bakeri* Trägårdh, 1948 (Acarina:Parantennulidae) with descriptions of immatures and redescription of adults. Acarologia **12**(2):234-243.

Nutting, W.B. (1968). Host specificity in parasitic acarines. Acarologia **10**(2):165-180.

O'Donnell, A.E. and R.C. Axtell (1965). Predation by *Fuscuropoda vegetans* (Acarina:Uropodidae) on the house fly *(Musca domestica).* Ann. Ent. Soc. Amer. **58**:403-404.

Oldfield, G.N. (1970). Mite transmission of plant viruses. Ann. Rev. Ent. **15**:343-380.

Pence, D.B. (1972). The hypopi (Acarina:Sarcoptiformes:Hypoderidae) from the subcutaneous tissues of birds in Louisiana. J. Med. Ent. **9**(5):435-438.

Pence, D.B. (1975). Keys, species and host list, and bibliography for nasal mites of North American birds (Acarina:Rhinonyssinae, Turbinoptinae, Speleognathinae and Cytoditidae). Spec. Publ. Mus. Texas Tech. Univ. 8:148 pp.

Petrova, A. (1971). On halacarid migration into freshwaters. Proc. 3rd Int. Congr. Acarology, Prague: 169-171.

Petrova, A. (1972). Sur la présence d'*Halacarellus subterraneus* Schulz, 1933 et *Halacarellus phreaticus* n.sp. (Halacaridae, Acari) en Bulgarie. Acarologia **13**(2):367-373.

Philip, C.B. (1963). Ticks as purveyors of animal ailments: a review of pertinent data and recent contributions. Advances in Acarology **1**:285-325.

Prasad, V. and D.R. Cook (1972). The taxonomy of water mite larvae. Mem. Amer. Ent. Inst. 18:326 pp. + ii.

Price, D.W. (1973). Abundance and vertical distribution of microarthropods in the surface layers of a California pine forest soil. Hilgardia **42**(4):121-147.

Price, D.W. (1976). Vertical distribution of pomerantziid mites (Acarina:Pomerantziidae). Proc. Ent. Soc. Wash. **78**(3):309-313.

Pulpan, J. and P.H. Verner (1965). Control of tyroglyphoid mites in stored grain by the predatory mite *Cheyletus eruditus* (Schrank). Can. J. Zool. **43**:417-432.

Radford, C.D. (1953). The mites living on or in the feathers of birds. Parasitol. **43**(3-4):199-230.

Radovsky, F.J. (1969). Adaptive radiation in the parasitic Mesostigmata. Acarologia **11**(3):450-483.

Radovsky, F.J., J.K. Jones Jr. and C.J. Phillips (1971). Three new species of *Radfordiella* (Acarina:Macronyssidae) parasitic in the mouth of phyllostomid bats. J. Med. Ent. **8**(6):737-746.

Rapp, A. (1959). Zur Biologie und Ethologie der Kafermilbe *Parasitus coleoptratorum* L. (Ein Beitrag zum Phoresie - Problem). Zool. Jahr. Syst. **86**(4-5):303-366.

Rockett, C.L. and J.P. Woodring (1966). Oribatid mites as predators of soil nematodes. Ann. Ent. Soc. Amer. **59**(4):669-671.

Rodriguez, J.G., P. Singh and B. Taylor (1970). Manure mites and their role in fly control. J. Med. Ent. **7**(3):335-341.

Schaarschmidt, L. (1959). Systematik und Ökologie der Scutacariden. **In** Beiträge zur Systematik und Ökologie mitteleuropäischer Acarina. Bd. 1, Abs. V: 713-823.

Schuster, R. (1956). Der Anteil der Oribatiden an den Zergetzungsvorgängen in Boden. Z. Morph. Ökol. Tiere **45**:1-33.

Schuster, R. (1957). Die terrestrische Kleinarthropenfauna in den *Tenarea*-Troittoirs des Westmediterranean Litorals. Kieler Meeresforsch. **13**(2):244-262.

Sengbusch, H.G. and C.H. Sengbusch (1970). Post-embryonic development of *Oppia nitens.* J. New York Ent. Soc. **78**(4):207-214.

Severin, H.C. (1944). The grasshopper mite *Eutrombidium trigonum* (Hermann) an important enemy of grasshoppers. S. Dakota Agr. Ext. Sta. Tech. Bull.3:36 pp.

Sinha, R.N. (1964). Ecological relationships of stored products mites and seed-borne fungi. Proc. 1st. Int. Congr. Acarology, Fort Collins. Acarologia **6**(h.s.):372-389.

Sinha, R.N. (1968). Adaptive significance of mycophagy in stored-product Arthropoda. Evol. **22**(4):785-798.

Sinha, R.N. and L. Harasymek (1974). Survival and reproduction of stored-product mites and beetles on fungal and bacterial diets. Environ. Ent. **3**(2):243-246.

Slykhuis, J.T. (1972). Transmission of plant viruses by eriophyid mites. **In** Principles and Techniques in Plant Virology. Kado and Agarwal, eds. Van Nostrand-Reinhold Co., Princeton, New Jersey: 204-225.

Smith, M.G., R.J. Blattner and F.M. Heys (1944). The isolation of the St. Louis encephalitis virus from chicken mites *(Dermanyssus gallinae)* in nature. Science **100**:362-363.

Soar, C.D. and W. Williamson (1925). The British Hydracarina, Vol. 1. Ray Society, London: 214 pp. + x.

Southcott, R.V. (1961). Notes on the genus *Caeculisoma* (Acarina:Erythraeidae) with comments on the biology of the Erythraeoidea. Trans. Roy. Soc. S. Australia **84**:163-178.

Sparing, I. (1959). Die Larven der Hydrachnellae, ihre parasitische Entwicklung und ihre Systematik. Parasit. Schrift. 10:168 pp.

Strandtmann, R.W. and G.W. Wharton (1958). A manual of mesostigmatid mites parasitic on vertebrates. Inst. Acarology, Univ. Maryland, Contrib. **4**:330 pp.

Treat, A.E. (1975). Mites of Moths and Butterflies. Cornell Univ. Press, Ithaca, N.Y.: 362 pp.

Viets, K. (1931). Über die an Krebskiemen parasitierende Süsswassermilbe *Astacocroton* Haswell, 1922. Zool. Anz. **97**(3,4):85-93.

Viets, K. (1938). Eine merkwürdige, neue, in tiefsee—echiniden schmarotzende Halacariden-gattung und-Art (Acari). Z. Parasitenk. **10**(2):210-216.

Vukasović, P. (1947). Contribution to the study of *Pediculoides ventricosus* Newport (Acarina, Arachnoidea). Acta Med. Jugoslav. **1**(1-2):76-128.

Wallwork, J.A. (1958). Notes on the feeding behavior of some forest soil Acarina. Oikos **9**(2):260-271.

Wallwork, J.A. (1959). The distribution and dynamics of some forest soil mites. Ecol. **40**(4):557-563.

Wallwork, J.A. (1967). Acari. **In** Soil Biology. Academic Press, New York: 363-395.

Wallwork, J.A. (1970). Ecology of Soil Animals. McGraw-Hill, London: 283 pp.

Weis–Fogh, T. (1948). Ecological investigations on mites and collemboles in the soil. Nat. Jutland **1**:135-270.

Wernz, J.G. and G.W. Krantz (1976). Studies on the function of the tritosternum in selected Gamasida (Acari). Can. J. Zool. **54**:202-213.

Westigard, P.H. (1971). Integrated control of spider mites on pears. J. Econ. Ent. **64**(2):496-501.

Wharton, G.W. and H.S. Fuller (1952). A manual of the chiggers: the biology, classification, distribution, and importance to man of the larvae of the family Trombiculidae (Acarina). Mem. Ent. Soc. Wash. **4**:185 pp.

Wicht, M.C. Jr. (1970). Three new species of pyemotid mites associated with commercial mushrooms. Acarologia **12**(2):262-268.

Willmann, C. (1942). Milben aus deutschen Mineralquellen. Zool. Anz. **139**:242-244.

Woodring, J.P. (1963). The nutrition and biology of saprophytic Sarcoptiformes. Advances in Acarology **1**:89-111.

Woodring, J.P. (1965). The biology of five new species of oribatids from Louisiana. Acarologia **7**(3):564-576.

Woolley, T.A. (1961). Some interesting aspects of oribatid ecology (Acarina). Ann. Ent. Soc. Amer. **53**(2): 251-253.

Woolley, T.A. (1969). A new and phoretic oribatid mite (Acarina:Cryptostigmata:Licnodamaeidae). Proc. Ent. Soc. Wash. **71**(4):476-481.

Yunker, C.E. (1970). New genera and species of Ewingidae (Acari:Sarcoptiformes) from pagurids (Crustacea), with notes on *Ewingia coenobitae* Pearse, 1929. Rev. Zool. Bot. Afr. **81**(3-4):237-254.

Yunker, C.E. (1973). Mites. **In** Parasites of Laboratory Animals. R.J. Flynn, ed. Iowa State Univ. Press, Ames: 425-492.

Yunker, C.E. and E.K. Jones (1961). Endoparasitic chiggers. I. Chiroptera, a new host order for intranasal chiggers, with descriptions of two new genera and species (Acarina:Trombiculidae). J. Parasitol. **47**(6): 995-1000.

Yunker, C.E. and F.J. Radovsky (1966). The dermanyssid mites of Panama (Acarina:Dermanyssidae). Ectoparasites of Panama: 83-103.

Zemskaya, A.A. (1968). Sparrow mite *Dermanyssus passerinus* Berlese and Trouessart. Med. Parazitol. i. Parazitar. Bolzeni **37**(3):313-319.

Zemskaya, A.A. (1971). Mites of the family Dermanyssidae Kolenati, 1859, of the U.S.S.R. fauna. Med. Parazitol. i. Parazitar. Bolzeni **40**(6):709-717.

VII. COLLECTION, REARING, AND PREPARATION FOR STUDY

Collection

Successful collection of mites and ticks depends to a great extent on proper selection of collection sites. Yields of predaceous ground forms, for example, generally will be sparse in exposed dry humus or in sandy soil. Except for relatively few species, most ground and aerial forms prefer a substrate which is not exposed to direct sunlight or wind, and one which provides enough moisture for the mite to maintain a satisfactory water balance (Vannier 1976). Phytophagous species may be less dependent on a protected substrate since they have a constant source of moisture from the juices of their plant hosts. Ticks and other vertebrate ectoparasites often may be collected in nesting areas or on runway or watering routes of their hosts, but only rarely through indiscriminant search.

The small size of most acarines often makes collection of individual specimens from a given substrate impractical. Thus it has become customary, under most conditions, to collect samples of the habitat for subsequent separation of the acarines hidden within.

A. Terrestrial Free-Living Mites

Predaceous, fungivorous and saprophagous ground and soil forms may be extracted in substantial numbers through the use of a *modified Tullgren apparatus* (Plate 7-1, p. 78) (Haarlov 1947, MacFadyen 1953), based on the Berlese funnel. Habitat samples are placed on a screen within the funnel, and an incandescent light is appended over it. The heat from the light bulb desiccates the sample, which presumably forces the arthropods in the sample to burrow deeper into the substrate. Nef (1971) observed that the movement of Oribatida is not the result of progressive downward desiccation, but rather a geopositive response triggered at substrate humidities of less than 20-25%. In any event, the mites eventually reach the screen, fall through, and are collected in a screw-top jar or vial suspended below the funnel neck.

The wattage of light bulbs utilized in Berlese-Tullgren funnel separation depends in large part on the size and water content of the samples and on the distance of the bulb from the sample surface. Although high wattage lights may be more effective than cooler bulbs in effecting separation of large damp samples, they often kill high percentages of lightly sclerotized, slow-moving acarines before the mites can work down through the sample and into the collecting jar (Fujikawa 1970). On the other hand, wattages of less than 40 may greatly prolong the separation period and, in large damp samples, may even enhance continued feeding and reproduction by mites in the sample substrate. Ideally, Berlese-Tullgren samples should be no more than 12 liters in volume, and should be separated using a frosted light bulb rated at no more than 75 watts. Coineau (1974b) suggests that no heat be applied to the samples during the first day on the funnel and that, if possible, heat should be progressively increased during the sample run. Samples should be allowed to remain on the funnels for at least 4 days, although a longer period is desirable for moist substrates.

Collection of mites from Berlese-Tullgren funnels usually is made into a preservative such as 70% alcohol. Prolonged preservation in alcohol, however, tends to harden internal tissues of small arthropods and may render them unsatisfactory for whole mounts. Therefore, when prolonged preservation in alcohol is anticipated (more than 3 months), mites should be collected in or transferred to another preservative. A mixture of glycerine (50 parts), water (40 parts) and glacial acetic acid is recommended for this purpose. *Oudeman's fluid* (Hughes 1961) substitutes alcohol for water, and is mixed as follows:

PLATE 7

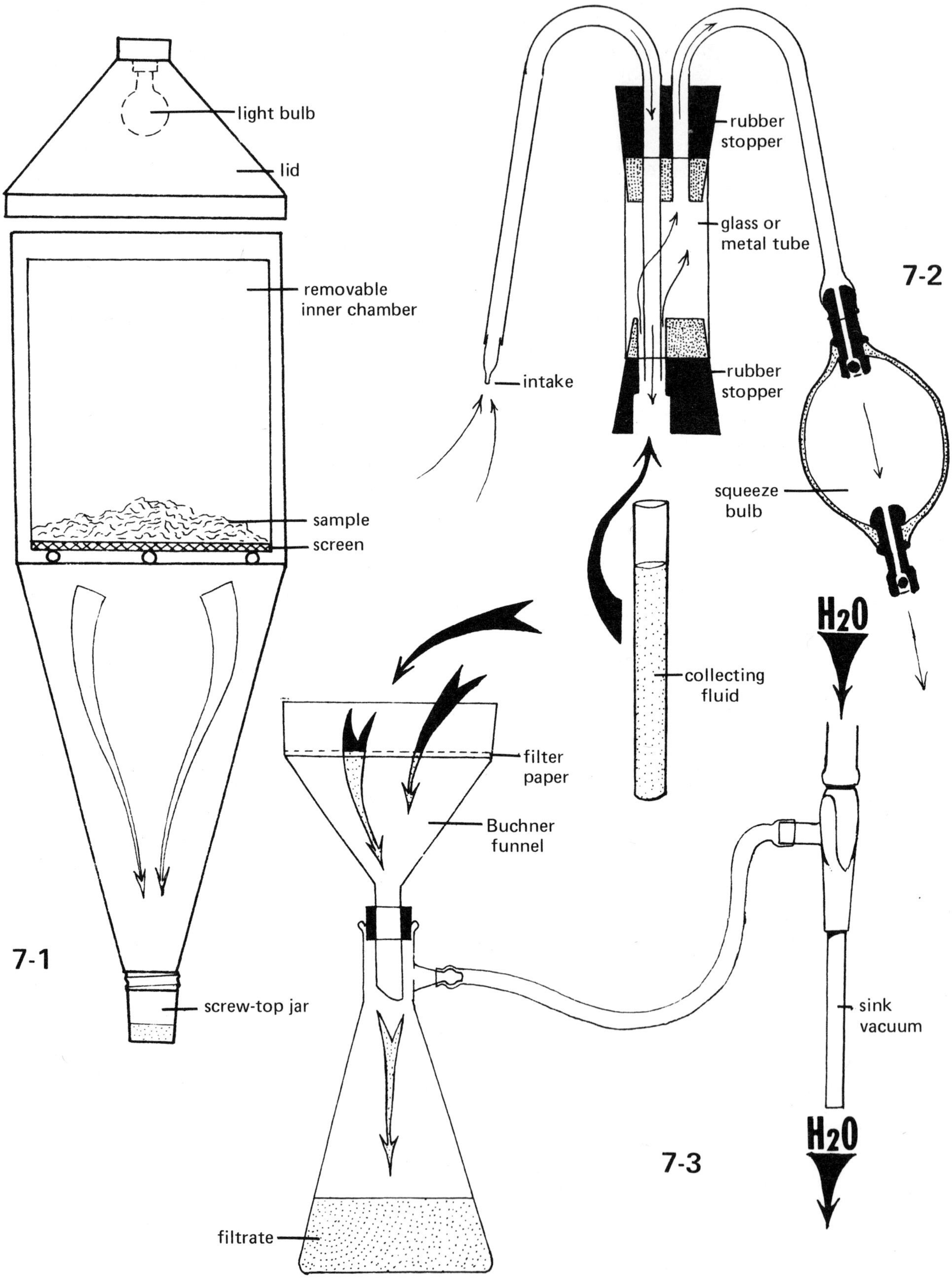

7-1; a modified Tullgren apparatus: **7-2**; a Singer aspirator (after Singer 1964): **7-3**; a Buchner funnel apparatus for vacuum filtration

Glycerine. . . . . 5 parts
Alcohol (70%) . . . . . . .87 parts
Glacial acetic acid . . . . 8 parts

Collection of live mites is best accomplished by placing moistened paper towel strips in the collection jar rather than a preservative. The towel strips provide both moisture and hiding places, reducing mortality caused by desiccation and by predation.

While the Berlese-Tullgren funnel technique is an extremely useful qualitative tool for faunal census in a variety of substrates, it is of questionable value in quantitative measurements. Macfadyen (1953) devised a modification of the Berlese-Tullgren apparatus for soil arthropods which has an extraction efficiency significantly greater than could be achieved with a conventional funnel. Macfadyen's modification entails development of wide humidity and temperature gradients within small samples. The temperature gradient between the surface of each sample and its base is increased by circulating cold water or cold air around the bottom of the sample in an open system (Macfadyen 1962, Southwood 1966). Merchant and Crossley (1970) devised a simplified high gradient extractor in an operating household refrigerator, with heat supplied to the top of each compact funnel by a 7.5 watt bulb. It is of interest to note that the magnitude of the temperature gradient may not be as important as maintenance of temperature stability during extraction (Auerbach and Crossley 1960).

Collection of soil and litter samples for quantitative research necessitates use of a method which delivers non-compacted samples of standard size. A variety of *core samplers* are available for this purpose, most of which resemble the apparatus described by Murphy (1962).

Extraction of mites from soil or stored products substrates may be accomplished by flotation. Here, extraneous particles are separated from the acarines and other arthropods by precipitation in a variety of liquids so that the arthropods may be decanted or picked from the surface film (Edwards and Fletcher 1971). Effective flotation liquids generally have specific gravities of 0.7-1.6 (Gridelet-de-Saint-Georges 1975).

Extraction of soil mites onto a film of grease from a mixture of the soil sample and water is described by Aucamp (1969). The method is based on the principle of differential wetting. Since most soil-dwelling Acari possess a water-repellent cuticle, they tend to stick to a suitable grease in a soil-water mixture.

A history of development of soil arthropod extraction, and a comparison of efficacy of various modifications, is provided in a comprehensive review of the subject by Edwards and Fletcher (1971).

Large numbers of acaridid stored products mites (*Rhizoglyphus* and *Tyrophagus* (ACARIDAE)) may be separated from laboratory cultures by precipitating migration from the food substrate through overcrowding. Mites are collected as they leave the culture, eliminating the problem of adhering debris (Woodring 1968, Stepien and Rodriguez 1973). The overcrowding principle could prove useful as a collecting technique for other mass laboratory separations where debris is a consideration.

Phytophagous and predaceous aerial species may be extracted from plant materials with a Berlese-Tullgren apparatus as described above, but other methods are available. Handpicking with the aid of a hand lens and a fine brush is practiced on a variety of habitats, including plants. Bulb aspirators of the Singer type (Plate 7-2, p. 78) are recommended in these situations, since they collect directly from the habitat into the preservative, eliminating handling of the specimens (Singer 1964). Samples of mite-infested foliage may be beaten on a mesh screen covering a large tray, and the dislodged mites picked up with a fine brush or aspirator and placed into preservative. This method is particularly useful in the field, where large bulky samples of foliage are an inconvenience. A power-driven brushing apparatus is available for the collection and assessment of plant mite populations on infested foliage (Chant and Muir 1955). Johnson et al. (1957) utilized a portable electric blower for collecting plant mites from rough grassland and from low-growing plants.

Mite galls or other mite-induced abnormalities may be removed intact from plants for later dissection and mite recovery (Jeppson et al. 1975). Mite-infested leaves or twigs should be placed directly into vials of alcohol or transported back to the laboratory in plastic or paper bags. Bags may be refrigerated at 10-15°C. for several days before mites suffer significant mortality.

Collection and counting of eriophyoid bud mites (Actinedida) presents a special problem since the mites often occur in enormous numbers within a single bud and cannot be extracted efficiently by conventional methods. Sternlicht (1966, 1969) devised a centrifugation method for extraction of citrus bud mites, *Eriophyes sheldoni* Ewing, in which the buds are soaked and then centrifuged for 20 minutes at 300 rpm. However, the technique apparently is no more effective than direct counting methods.

Flotation has been used successfully in separation of house dust mites (Acaridida, PYROGLYPHIDAE) from their substrate (Shamiyeh et al. 1971). House dust is collected from a vacuum sweeper, mixed with a detergent and a saturated NaCl solution, and exposed to ultrasonic treatment prior to low speed centrifugation. Crude separation also is possible through sieving or by Berlese-Tullgren extraction.

Vertical sampling of stored products is simplified by use of a grain spear sampler of the type used by Brett (1969). A simple grain sieve suffices to separate and concentrate arthropods from stored seed samples in the field. Laboratory extraction of the concentrate may then be carried out in a Berlese-Tullgren apparatus.

B. Aquatic and Marine Mites

Aquatic mites (Actinedida, phalanx Hydrachnidia) are collected in a variety of ways, depending on whether the mites are free-swimming, benthic or parasitic. Birge net or bottom dredge samples should be examined in a white porcelain tray in which the mites may easily be seen and from which they may be removed with an eye dropper. Benthic mites may be collected in an inverted funnel bottom trap such as those devised by Pieczynski (1961) and Conroy (1971). A small silk stocking net, or a tea strainer on a long stick, is invaluable in collecting individual free-swimming Hydrachnidia from ponds or sluggish streams. Extraction of water mites from bottom mud may be accomplished by placing the sample in a concentrated magnesium sulfate solution and decanting the floating animals after 10 minutes has elapsed (Efford 1965).

The larvae of Hydrachnidia often are parasitic on aquatic or semiaquatic insects or on mollusks. Hemipterans, beetles, dragon flies, may flies, damsel flies, and mosquitoes frequently are found to carry unattached or feeding larvae, or sac-like opaque white or bright orange quiescent pharate protonymphs (calyptostases) attached to wings or bodies. The gill and excurrent siphon tissues of mollusks may harbor pharate protonymphal and tritonymphal water mite calyptostases in larval or deutonymphal skins (Mitchell 1955). Larvae of actinedid families of Trombidia found in aquatic and semi-aquatic habitats commonly parasitize insects in these habitats (Mullen 1974a), and may be recovered with careful examination.

Unlike many of the aquatic mites, the marine mites (Actinedida, HALACARIDAE) are non-swimming, mostly non-parasitic forms which cannot be collected by conventional means. Those halacarids not found in bottom sediments or interstitially in tidal sands cling tightly to their plant or animal substrates, and are extremely difficult to see. Their recovery depends on successful collection of mite-infested substrates and on application of specific extraction techniques.

Littoral and intertidal marine samples may usually be collected by hand, while subtidal and benthic substrates are harvested by means of dredges, trawls, or grabs. Newell (1971) described several types of deep-sea samplers which have been used with varying degrees of success for collecting marine mites.

Extraction of marine mites from bottom sediments is simplified by initial reduction of sediment volume through differential sedimentation and screening, followed by silt removal with a 105 μ screen (Newell 1971). Sediment samples in the 105 μ-1 mm range are then examined for mites.

Mite extraction from algae, coral, or colonial invertebrates is best accomplished by separation of the live mites from their habitat. The recovered substrate is broken or torn apart (if possible) and placed in a 12 liter plastic bucket filled with fresh sea water, to which 15 cc of chloroform or one of the halogenated ether compounds is added. This anesthetizes the mites, causing them to relinquish their hold on the substrate. Pouring the mixture from one bucket to another several times hastens anesthetization. After 30 minutes have elapsed, the substrate material is washed vigorously and removed from the treated sea water. The water may then be filtered through a muslin bag, with more sea water being added as necessary for retrieving all of the precipitate. The bag is then labeled, tied shut, and placed in 95% alcohol for subsequent study (Newell, personal communication). Newell (1971) found that staining of marine sample precipitates with a solution of Rose Bengal (1.5 ml of 5% dye/100 ml of preservative) imparts a bright red color to otherwise inconspicuous halacarids and makes their recovery much simpler.

C. Parasitic Mites

Collection of mite parasites of vertebrates is accomplished both through collection of the host habitat (nests, runway litter, etc.) and through examination of the host itself. Nest and litter material lends itself well to Berlese-Tullgren funnel separation.

Chiggers (larval TROMBICULIDAE) have been collected successfully in light traps (Cockings 1948, Steffey and Wingo 1975), and postlarval stages may be separated from soil substrates through water flotation. Collection of unengorged trombiculid larvae from ground holes was achieved by Kundin et al. (1966) by insertion of a flat strip of black formica into the holes. The mites apparently were stimulated to drop onto the strips from the roof of the chamber, where they are most frequently situated. Foreign objects of almost any kind attract unattached trombiculid larvae. Porcelain or plastic discs, or squares of oilcloth dropped on the ground in an infested area, will almost always have chiggers crawling on it within a few minutes (Smith and Gouck 1944, Williams 1946, Wharton and Fuller 1952).

Since carbon dioxide is attractive to ticks (Miles 1968, Garcia 1969, Hair et al. 1972) and other parasitic Acari, subliming dry ice can be used to attract aggregates of these species in woodlots and other areas of concentration. Collection is simplified with vacuum equipment such as that described by Hair et al. (op. cit.). Unattached ticks may also be collected in runways by dragging a flannel cloth over suspected infestation sites *(flagging)*. Ticks attach to the cloth as it passes over them.

Individual parasitic mites often are discovered in bird or rodent nests, or in recently abandoned nesting material. The use of a bulb-operated Singer aspirator (Singer 1964) for collecting nest Acari is recommended over the oral vacuum aspirator, since it eliminates the possibility of accidentally inhaling potentially dangerous disease vectors or acariasis-producing species.

Collection of host animals is advisable or necessary for many external and for all internal parasites. Live collection of rodents and other small mammals is effected with a variety of baited live traps, while birds and bats may be captured in mist nets. Since many acarine ectoparasites are not obligate and detach from the host animal after feeding, steps should be taken to prevent their escape in the field. Animals should be transferred to individual heavy cloth bags for transport to the laboratory. Small mammals should be bagged while still in their traps, if trap supplies are adequate. After removal of the animal for study, the transport bag may be inverted and examined microscopically for mites and ticks. For collection of live ticks from infested small animal hosts under field conditions, Kaiser and Hoogstraal (1968) recommend confinement of the animal in a cylindrical metal cage which is screened at both ends (16 sq/inch), and placed in a cloth bag.

Collection of ectoparasites from the living host animal in the laboratory may be accomplished by anesthetizing the host with an intramuscular injection of a 10% solution of phencyclidine hydrochloride (1 mg/kg). Such a procedure eliminates the necessity of sacrificing the host (Vercammen-Grandjean 1971). Furman et al. (1974) found that *Pneumonyssus simicola* Banks (Gamasida, HALARACHNIDAE), a lung mite of monkeys, is effectively recovered from anesthetized macaques by means of tracheo-bronchial swab irrigation.

If infested animals are maintained for several days in screen cages suspended over water, engorged ectoparasites may be collected from the water surface as they drop from the host.

Removal of mites from dead rodents and birds is carried out by immersing the animal in water-detergent mixture which is shaken vigorously (Lipovsky 1951). This detaches most of the ectoparasites, which are then decanted into another container for collection with brush, eye dropper or forceps. The liquid may also be poured into a Buchner funnel (Plate 7-3, p. 78) in which filter paper has been inserted, and drawn through the filter paper with a sink vacuum. The paper may then be examined for mites (Watson and Amerson 1967). Mite parasites of furred animals also are recovered through vigorous fine-combing and brushing over a suitable pan or dish. Examination of particular sites on the host is advisable, since parasites often are site-specific. For example, chiggers on small mammals are most commonly encountered in the ears, on the muzzle or chin, in the axillary and inguinal areas, and on the midline of the belly (Nadchatram 1968). Removal of mites from within feather quills (family SYRINGOPHILIDAE) or from lung tissue (families ENTONYSSIDAE, RHINONYSSIDAE and others) must be done through examination and dissection of the infested structure or organ. Nasal mites are recovered by flushing the nasal cavities with a stream of water under high pressure (Yunker 1961). Splitting the bill between the nares often simplifies recovery.

Invertebrate ectoparasites are found on their hosts in a wide range of locations. Antennal bases, wing axillaries, coxal cavities, eyes, spiracular plates and abdominal conjunctiva of insects and other arthropods shelter a variety of mites representing over a dozen families. These may be removed with a fine brush, a pin, or a pair of forceps. Recovery of mite parasites and phoretic associates of insects from museum collections is often difficult in that the mites are dry and brittle, and often are secreted in deep recesses or cavities of the insect host. Breakage may be kept to a minimum by wetting with a small droplet of alcohol, carefully removing mites with a small brush or pin, and placing them in 70% alcohol for temporary storage prior to restoration. Where mites are secreted under the wings or in the wing axillaries of dried museum specimens, it is advisable to relax the host prior to examination so that the wings may be lifted without damaging the specimen. Husband (1974) relaxes mite-infested Coleoptera in distilled water at 60°C. for one hour. Excess water is removed with tissue paper, the elytra are lifted, and the wings are moved aside for visual inspection of abdominal tergites and wing axillaries. The exposed areas are then forcefully sprayed with 70% alcohol over a small petri dish, and the contents held for later examination. The beetle host is dried again, and the wings and elytra are returned to their original position.

Internal parasites of invertebrates (families TARSONEMIDAE, PODAPOLIPIDAE and others) may be recovered only through careful dissection of the host. Husband (1967) suggests that tracheal podapolipid parasites of bombine bees may be recovered by examining the excised air sacs which are revealed by making an incision between sternites I-II and carefully pulling them apart. The exposed crop is removed, after which the air sacs are punctured and inspected internally.

Rearing

A number of methods for laboratory rearing of free-living Acari have been developed over the past century. Mites are cultured primarily in various types of glass or plexiglass cells or dishes, developed from a prototype consisting of a glass ring cemented to a microscope slide and closed at the top by a coverslip (Michael 1884, Jacot 1937). The earlier types utilized a moistened filter paper substrate on which a suitable food material was placed (Michael 1901, Grandjean 1950, Camin 1953, Sengbusch 1954, Pauly 1956, Evans et al. 1961, Hartenstein 1962). Other variations of the basic cell were used by Garman (1917), Davis (1944), Robertson (1944), Rivard (1958), Hughes (1961), and Woodring (1963).

While the filter paper substrate provides a means of maintaining a moist environment in the cell, relative humidity is difficult to control since evaporation makes necessary the addition of water to the filter paper at frequent intervals. Also, the paper provides an ideal substrate for the growth of fungi which may either foul the culture or confuse feeding experiment results (Evans et al. 1961). Substitution of a plaster-charcoal or plaster-soil substrate in rearing cells has proven highly successful in the rearing of a variety of free-living Acari on both natural and artificial diets (Lipovsky 1953, Rohde 1956, Radinovsky and Krantz 1961, Woodring 1963, Arlian and Woolley 1970). Humidity is more easily controlled in cells with a plaster base, primarily because plaster holds more water for longer periods of time than does a thin filter paper disc. Contamination problems also are reduced in that surface contaminants may often be eliminated without disturbing the cell as a whole.

Predatory instars of water mites (Actinedida, Hydrachnidia) may be reared from captured engorged larvae in 2-dram vials partially filled with water and stoppered with cotton to retard evaporation. A strip of filter paper in the vial provides a site for attachment during molting. Ostracods are added at regular intervals to serve as prey (Mullen 1974b).

Rearing of plant mites in mass culture requires little more than healthy vigorous host plants maintained under optimal growing conditions. Difficulties often arise, however, in that mite contamination of nearby clean plants is likely to occur. Thus, the rearing of spider mites (Actinedida, TETRANYCHIDAE) on a limited number of plants in the greenhouse typically leads to a general infestation of the entire facility. Inasmuch as observation of individual mites on infested plant parts *in situ* also presents obvious problems, plant mites usually are cultured on isolated plant parts in small cages which are constructed so that the plant material is kept moist and fresh for an extended period. Cages of the Huffaker type (Huffaker 1948, Munger and Gilmore 1963, Tashiro 1967, Beavers and Oldfield 1970) employ a series of wood or acrylic plates, layers of blotting paper or cheesecloth, and a water source. Tashiro (op. cit.) designed a cage in which a leaf is placed on a layer of cheesecloth set on a base (Fig. 22,d) through which a cotton wick protrudes. A 9 mm thick acrylic plate with a 29 mm hole in its center (b) is placed over the leaf, with the opening centered over a rubber band (c) equal in inside diameter to the opening itself. A top plate (a) with a 160 mesh screen over the central opening provides a lid for the cage. Water is supplied to the leaf via the cotton wick (e) in the cage base. The Tashiro cage is secure enough to confine the smallest of phytophagous mites. In addition, isolated leaves held in this fashion may retain their viability for periods up to 4 weeks. Cages of other types have been affixed to isolated or *in situ* fruit for the study of citrus mites, with varying results (Swirski and Amitai 1957, Munger and Gilmore 1963).

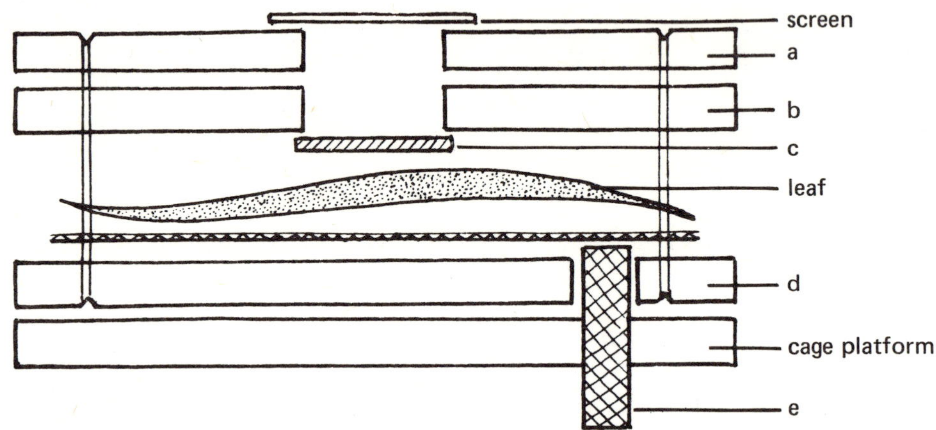

Fig. 22. Exploded diagram of Tashiro rearing cage (adapted from Tashiro 1967).

Plant mites have been reared successfully on synthetic media through a leaf-like membrane composed of a film of collodion formed on water by the use of a collodion-amylacetate mixture (Rodriguez 1969). The resulting membrane is semi-permeable and, like a leaf, transpires. *Tetranychus urticae* Koch, a spider mite (Actinedida, TETRANYCHIDAE) has been maintained on a medium consisting of a number of amino acids, sugars, vitamins A and E, and RNA (Ekka et al. 1971).

The culturing of bud or gall mites (Actinedida, Eriophyoidea) does not lend itself to the cage approach, since the mites tend to sequester themselves in the plant tissues. Microscopic study of excised or teased tissues is recommended in these instances.

Many ectoparasitic mites may be mass-reared successfully on confined host animals in the laboratory. Birds, mammals or reptiles, along with their mite parasites, are placed in mite-tight containers partially filled with clean hay, straw or wood shavings and held in this fashion for the requisite period of time. Such a method is relatively simple and quite successful, but the container soon becomes fouled and must be cleaned to prevent buildup of secondary organisms (Strandtmann and Wharton 1958). While the sanitation problem is alleviated by reduced feeding of the host animal (Cross and Wharton 1954), or by shortening the period of host presence in the container (Chamberlain and Sikes 1950), it is virtually eliminated by allowing the host to void through the bottom of the container into a sand substrate (Bertram et al. 1946).

Jones (1951) reared larvae of *Trombicula autumnalis* Shaw (TROMBICULIDAE) on young mice held on wire gauze over a petri dish lined with moist cotton wool, and covered with black filter paper. Engorged larvae fell through the gauze onto the filter paper. They were then placed in rearing cells in which the emerging nymphs were fed on a mixture of yeast, molasses, agar, and chicken feces. Lipovsky (1951) substituted Collembola and their eggs as the food source. Like other free-living mites, the non-parasitic stadia of TROMBICULIDAE are reared satisfactorily on a charcoal-plaster or soil-plaster substrate. Nadchatram (1968) suggests rearing trombiculids in clay pots with a soil substrate. The pots are suspended over a water trough, to which they are joined by filter paper strips to prevent desiccation.

Rearing of ticks and other ectoparasitic Acari *in situ* generally entails confinement of the parasites to a particular site on the host animal. Cages and bags of many types have been used for this purpose, although grooming or rubbing by the host often dislodges them (Evans et al. 1961). Watts et al. (1972) succeeded in feeding ticks on the ears of laboratory rabbits without dislodgement by placing a wide polyethylene collar around the neck of the host. The collar prevented ear grooming and possible removal of the ear bags in which the ticks were confined.

Feeding of individual mites is best accomplished either by placing them in small vials or plastic cups attached directly to the host animal (Camin 1953, Cross and Wharton 1954), by allowing mites to feed on heparinized blood through a bolting silk or animal-derived membrane (Cross 1954, Clark 1958, Tarshis 1958), or by force-feeding immobilized individuals by immersing the hypostome into the food substrate (Burgdorfer 1957, Furman et al. 1961).

Preparation for Study

Various microscopic techniques have been developed over the past 50 years which make possible the critical study of fine structure of properly prepared acarine specimens. Those techniques which are most widely used are based either on optical or scanning electron principles. Optical phase contrast and interference systems are more widely used at present but they are limited in that the wavelength of light restricts effective magnification to approximately 1000X (Griffiths et al. 1971). Moreover, the depth of field in optical systems is minimal, and specimens must be flattened or dissected for optimal resolution. The scanning electron microscope or SEM eliminates these problems, making it possible to obtain usable magnification of 25,000X and greatly increasing depth of field.

Preparation of preserved material for use in optical and scanning systems differs considerably. Each is discussed in the sections to follow.

Optical

With the advent of phase contrast and interference microscopy, it has become necessary to prepare specimens in such a way that a high degree of transparency is achieved. Since the degree of clearing which occurs in most mounting media is minimal, maceration or removal of opaque tissues from all but smaller, weakly sclerotized mites should be attempted before the mite specimen is placed on the slide. This is accomplished either by 1) the use of a clearing or digesting agent, or 2) dissection.

1. Clearing agents. Various chemicals are effective in macerating internal tissues of preserved mites with little or no damage to the exoskeleton. One of the most popular for general use is *lactophenol,* which is prepared with the following ingredients added in sequence:

Lactic acid. . . . . .50 parts
Phenol crystals . . . . .25 parts
Distilled water. . . . . .25 parts

Specimens placed in lactophenol at room temperature may be left for a week or more without harm to exoskeletal structure. Lactophenol is an acid corrosive and does not tend to soften integument as do the basic corrosives (potassium hydroxide, for example). Acceleration of maceration occurs at higher temperatures. Larger specimens are punctured to allow easy entrance of the lactophenol into the body cavity. Blood-engorged mites or mites containing large amounts of pigment should be punctured and gently squeezed so as to remove these substances. Additional material may be squeezed out after immersion in lactophenol for 24-48 hours. Immersion of dried or brittle mite specimens in lactophenol for 48 hours at room temperature restores most treated specimens to a condition resembling freshly collected material. It is therefore extremely useful in preparing collections from dried insects, preserved bird or mammal skins, or from pressed plant materials in herbarium collections.

Other acid corrosive materials which may be recommended for general use include pure lactic acid, and *Andre's fluid* (1/3 glacial acetic acid, 1/3 chloral hydrate, 1/3 water, all by weight). Like lactophenol, *Vitzthum's fluid* utilizes phenol as a clearing agent, but the lactic acid component is replaced with chloral hydrate:

Chloral hydrate . . . . 10 parts
Phenol . . . . 9 parts
Distilled water . . . . 1 part

Nesbitt's fluid is a powerful clearing agent which is recommended only for old, alcohol-stored specimens which do not yield to clearing by other methods. Nesbitt's fluid is prepared as follows:

Chloral hydrate . . . . 40 grams
Distilled water . . . . 25 ml
Concentrated hydrochloric acid . . . . 2.5 ml

Several preparation fluids have been developed for the minute plant mites of the superfamily Eriophyoidea, which often are collected dry from preserved plant parts (Keifer 1952, Keifer 1954). One of the simpler of these fluids is recommended prior to mounting in aqueous media. *Kono's fluid* (Jeppson et al. 1975) is prepared with the following ingredients, and is warmed during use:

Chloral hydrate . . . . 100 grams
Glycerine . . . . 10 grams
Concentrated hydrochloric acid . . . . 1 ml
Distilled water . . . . 50 ml

It should be mentioned that, although prolonged treatment with lactophenol and other comparatively slow-acting acid corrosives will not harm well-sclerotized specimens, submersion of lightly sclerotized specimens for more than 48 hours tends to weaken leg conjunctiva and shields. This breakdown makes dissection difficult, since the techniques involved (see next section) require considerable manipulation and handling. Dorsal shields are particularly prone to breakage during dissection of over-treated specimens. Another possible problem is that the cleared specimens may become so transparent that they are virtually impossible to find in the preparation dish. The problem may be avoided by adding lignin pink dye to the clearing agent (Evans and Browning 1955). Chlorazol (Michrome 336), a differential integumental strain (Coineau 1974b) might also be used for this purpose.

Enzymatic digestion of internal tissues of HALACARIDAE and other strongly pigmented mites may be accomplished through the use of trypsin in a toluene atmosphere, or with pepsin.

As is the case with lactophenol, pigmented mites treated with trypsin or pepsin solutions must first be punctured so as to allow easy entrance of the digesting agent into the body cavity (Newell 1947).

2. **Dissection.** Heavily sclerotized mites, particularly those of the suborders Gamasida and Oribatida, often are difficult to study despite the use of clearing procedures outlined above. The thickness of the dorsal and ventral shields prevents easy observation of surface structures through the microscope and makes the use of phase contrast or interference techniques impossible. Also, the growing use of microphotography in acarology demands clarity of structure often beyond that obtainable through standard clearing procedures. A satisfactory solution to this problem is to carefully separate the dorsum and venter with appropriate dissection tools and, in cases where the shape or structure of the epistome or chelicerae is of importance, to remove the chelicerae as well. The following tools are recommended:

Watchmaker's microdissection forceps—straight shaft

Minuten pin, finely sharpened on carborundum stone, inserted in a wood matchstick base and fixed in place with paraffin wax

Microscalpel, made of razor blade fragment embedded in glass or wood handle

Specimens to be dissected are placed in lactophenol for 48-72 hours in order to soften the integument. They are then washed in several changes of water and transferred to 20-40% ethyl alcohol in a spot plate or small watch glass for dissection under a stereoscopic microscope.

The specimen is held gently but firmly with the fine tips of the forceps while the minuten pin is inserted into the soft integument laterad of the dorsal shield or shields. If the integument has been sufficiently softened, it can be torn by the pin without harm to the shields. The pin should be pulled through the integument from the vertex to a point behind coxae IV. The forceps may then be released for the second part of the operation. Since most of the delicate setae usually are found on the dorsum, the dissection should be completed with the forceps holding only the venter of the mite. The minuten pin is inserted into the integument on the intact side where the tearing operation is repeated. The gnathosoma thus remains with the venter.

Once the dorsum and venter have been separated, the pin may be used to hold the dorsal portion in place while the internal tissues are carefully picked out with the forceps. The venter is cleared in a similar manner although the muscle tissues associated with the sternal and coxal areas may be more difficult to extract. The chelicerae may be removed from the gnathosoma by first inserting the minuten pin into the open proximal end of each chelicera and forcing it to protrude from the epignathal foramen. The forceps may then be used to pull the extruded chelicerae from the foramen. Where the proximal cheliceral segment is elongate and intact, the chelicera may be removed by drawing the proximal segment back through the gnathosoma with the forceps and removing the entire structure from the body cavity.

Separation of dorsum and venter often is difficult in weakly sclerotized acarines with heavy integument; i.e., Ixodida. Here it is advisable to use a microscalpel for the initial incision and for subsequent separation of dorsum and venter.

Examination and Mounting Techniques

The small size of most acarines requires that they be observed under a compound or a dissecting microscope at substantial magnifications. Specimens to be studied must therefore be placed temporarily or permanently on microscope slides or other suitable platforms in an appropriate fluid. The techniques and materials utilized in this operation vary considerably with each researcher and with the type of acarine being observed. For the larger, heavily sclerotized Oribatida, and for other large acarines, many workers favor using a temporary mount consisting of a cavity slide, a coverslip, and lactic acid as the mounting medium. Mites placed in a drop of lactic acid near the side wall of the slide cavity may be manipulated into any position by movement of the coverslip which partially covers the cavity. Additional movement is achieved by means of a pin inserted into the exposed side of the cavity (Grandjean 1949, Coineau 1974a). Vertical illumination is advisable for study of temporary cavity slide mounts.

An alternate method for examination of Oribatida in a temporary preparation is the carbon block technique developed by Grandjean (1949). The specimen is placed on the surface of a carbon block moistened by 70% alcohol which is delivered from a reservoir by a capillary tube. The specimen is clearly observed under incident light since the moist carbon block provides a black background for it. The specimen is moved back into the alcohol near the capillary if excessive drying occurs during examination.

Clarity of structure in large, heavily tanned specimens to be studied in temporary mounts may be improved by clearing prior to examination. Balogh (1972) suggests that heavily sclerotized Oribatida be transferred from a 75-85% alcohol preservative into equal parts of 90% ethyl alcohol and lactic acid in a small open vial placed in a dust-proof cabinet. The alcohol soon evaporates, leaving the mite in lactic acid. Mites become transparent within one-three weeks and are well suited for cavity slide study. Clearing of mites in lactic acid is greatly accelerated by the application of heat. Cavity slide preparations may be satisfactorily cleared by gentle warming (Evans et al. 1961).

In recent years, most of the permanent or semi-permanent mounting media used by acarologists have been aqueous; i.e., they contain, and are soluble in, water. Aqueous mounting media have replaced oil soluble resins such as balsam or dammar primarily because specimens may be mounted in aqueous media without preliminary fixing operations as required for resin mounts. A serious disadvantage in using aqueous media is that they are hygroscopic, taking up water from the atmosphere. An aqueous mount thus is subject to displacement of medium by atmospheric moisture, and to eventual crystallization and breakdown. To remedy this situation, it has become customary to ring the completed slide preparation with a non-soluble protectant, making the preparation more or less permanent.

Most aqueous mountants now in use are modifications of Berlese's gum chloral fluid, and contain water, chloral hydrate, gum arabic and glycerine in various concentrations. *Hoyer's medium,* one of the more widely used variations, is prepared by mixing the following ingredients in sequence:

Distilled water. . . . . 50 ml
Gum arabic (amorphic) . . . . 30 grams
Chloral hydrate . . . . .200 grams
Glycerine. . . . . 20 ml

Singer's (1967) formula for Hoyer's medium calls for 50 grams of gum arabic, 30 ml of glycerine and only 125 grams of chloral hydrate. In either case, solid ingredients should be completely dissolved before addition of succeeding reagents. The resulting material should be filtered through clean cheesecloth in order to remove bits of wood or other impurities from the gum arabic. Powdered gum arabic should not be substituted for crystalline gum arabic, since the minute particles are difficult or impossible to wet.

According to Jeppson et al. (1975), crystallization of Hoyer's may be alleviated to some degree by replacing 60% of the gum arabic component with sorbitol, a stable sugar.

Another popular modification of Berlese's mountant is *de Faure's* medium:

Distilled water	50 ml
Gum arabic	30 grams
Chloral hydrate	50 grams
Glycerine	20 ml

A methyl cellulose aqueous medium devised by Clark and Morishita (1950) has proven suitable for many mites (Evans et al. 1961), although it is somewhat more difficult to prepare than the Berlese modifications. The following ingredients are used:

Methocellulose	5 grams
Carbowax 4000	2 grams
Diethylene glycol	1 ml
95% ethyl alcohol	25 ml
Lactic acid	100 ml
Distilled water	75 ml

The methocellulose and alcohol are mixed, added to the remaining ingredients and filtered through glass wool. The medium is heated at 40-45°C for 3-5 days, or until the desired consistency is obtained. Viscosity may be reduced by adding small amounts of 95% alcohol or water, or by gentle heating (Evans et al. 1961). The Clark-Morishita medium, or *C-M,* may soon enjoy increased popularity in those countries where chloral hydrate has become a regulated drug.

Special permanent mountants have been developed for marine mites (HALACARIDAE) by Newell (1947), and for eriophyoid mites by Keifer (Jeppson et al. 1975). However, mites in both of these groups may be satisfactorily observed in aqueous media.

The following mounting procedures are recommended:

1. Remove lactophenol-treated specimens from lactophenol and rinse in 3-4 changes of water in a porcelain spot plate. Continue to wash until cloudy interface of lactophenol and water disappears.

2. Place a drop of Hoyer's medium or other suitable aqueous mountant in the center of a clean 1 x 3 microscope slide.

3. Lift specimen from spot plate with a fine wire loop affixed to a matchstick base, or gently grasp it with watchmaker's forceps.

4. If forceps are used, slowly separate tines so that the mite adheres to one or the other tine, and the water trapped between the tines dissipates.

5. Touch wire loop, or forceps tine on which mite has adhered, to the droplet of mountant on the slide.

6. Carefully press the mite to the bottom of the droplet and arrange it on a vertical axis with a minuten pin probe. If the specimen is on the surface of the droplet when the coverslip is applied, it will roll to the edge of the coverslip.

7. Using a clean pair of forceps, pick up a coverslip at its rim, apply the opposite edge to the rim of the Hoyer's droplet, and allow the coverslip to fall into place. Last second orientation of the specimen may be accomplished under the dissecting microscope with gentle pressure of a probe on the coverslip surface. A 12 mm, 0-thickness coverslip or a 15 mm, 1-thickness coverslip, is recommended except for specimens exceeding 3 mm. Larger, thicker coverslips should be used in this case.

8. Turn slide so that the gnathosoma of the mite is directed posteriorly. Mark the right side of the slide with an identifying number or letter (a diamond pen is recommended, since wax markings often become obliterated in succeeding steps).

9. Place slide in an oven at 45°C for 48 hours—one week. Temperatures above 55°C cause the medium to bubble at the edge of the coverslip, while those between 46-54°C may bring about excessive contraction of the mountant.

10. Heat-treated slides should be held for one week at room temperature to allow the extra-thin coverslip to return to its normal flat state. Heating causes drying and contraction of the medium at the edges, bringing about "bowing" or "bubbling" of the coverslip. Decrease of heat often results in pulling away of the medium from the straightening coverslip, or in cracking. After one week, edges should be filled if necessary, and cleaned with a razor blade.

11. A ring of Glyptal, a waterproofing paint for electrical circuits (General Electric Co.) is recommended for sealing the coverslip to the slide surface (Travis 1968). Application of Glyptal or any other non-soluble sealant is most easily done with a No. 1 camel's hair brush on a slide turntable. A second coat applied after the first ring has dried assures a more impervious seal.

12. A locality label should be placed where the slide originally was marked. The label should include date, host or substrate and collector as well as the locality. Additional detailed information on habitat or other aspects of the collection should be written in a data book with the information referenced by an identification number both on the slide and in the book. An identification label should be affixed on the left side of the slide. The mounting medium used in making the slide should be noted at the bottom of the identification label (Fig. 23).

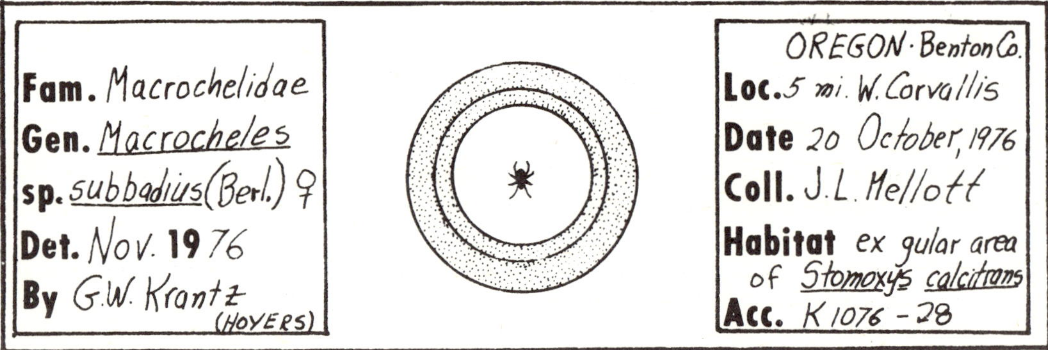

Fig. 23. Completed microslide preparation.

Dissected specimens should be cleaned as in step 1. The dorsum, venter and chelicerae should be placed in the Hoyer's medium as in steps 3 and 4, with the dorsum and venter arranged side by side and the chelicerae occupying a position anterior to the venter.

Properly mounted and ringed slides may be expected to last indefinitely. When, for reasons of inadequate cleaning or poor ringing, deterioration of the preparation occurs, the specimen is easily retrieved by scoring the ring with a steel needle and soaking off the coverslip in water. The freed specimen may be remounted directly to a new slide.

More assured permanency of slide preparations may be possible by using oil soluble resins, or by employing a double coverslip technique, using glycerine jelly or other aqueous media as the mountant between the coverslips, and balsam as the permanent "bed" for the preparation (Mitchell and Cook 1952, Travis 1968).

Serial sections often are desirable for embryological or critical internal and external morphological studies. Useful sectioning and staining techniques are outlined by Rohde (1964) and by Evans and Loots (1972). As pointed out by the latter authors, certain difficulties may be experienced during fixation and sectioning. The hardness of the surface sclerites of most acarines and the failure of the fixative to properly infiltrate the cuticle often cause displacement and shearing of the embedded specimen. Evans and Loots have solved these problems to a large extent by using *Henning's fixative,* which is prepared as follows:

Nitric acid . . . . 16 ml
5% aqueous chromic acid . . . . 16 ml
Saturated mercuric chloride in
60% alcohol. . . . . 24 ml
Saturated aqueous picric acid . . . . 12 ml
Absolute alcohol . . . . 42 ml

Fixation for 12-96 hours is recommended, depending on the degree of sclerotization of the specimen. The specimen is then washed in iodine 70% alcohol. (2.5 grams sublimated iodine in 1000 ml of 70% alcohol.).

Subsequent embedding and sectioning procedures are outlined here from Evans and Loots (op. cit.):

1. Attach small (<400 μ) weakly sclerotized specimens to previously fixed 1 cm^3 liver block, in a small groove at side of block. Attach with concentrated albumen in 90% alcohol, and then dehydrate in ethyl alcohol as follows:

   a. 80% for 2 hours
   b. 96% for ½ hour
   c. 100% for 2 hours

2. Leave in 1% Necolodine, or celloidin in methyl benzoate, until transparent.

3. Remove and wash several times in benzene. Add molten Paraplast-Plus tissue embedding medium (MP 56-57°C) to last wash. Mix and pour off most of solution and add more wax 2-3 times until benzene is removed.

4. Place specimen in pure wax in a watchglass at 60°C. Transfer and place specimen into molten wax in second greased watchglass, cool in cold water, and release block.

5. Section at 6-10 μ with freshly honed knife.

6. Mount on dry slide previously cleaned with acid alcohol. Apply a thin film of concentrated albumen to slide surface, add 2-3 drops of water and place on hot plate at 45°C, orienting ribbons to desired position. Allow to dry, under cover, for 12 hours.

Evans and Loots recommend *Mayer's Haemalum* for staining prepared sections. Mayer's Haemalum is comprised of the following ingredients:

Haematoxylin . . . . .1 gram
Sodium iodate. . . . . .0.2 gram
Potassium alum . . . . .50 grams
Distilled water. . . . . .1000 ml

Shake frequently until blue-violet in color and add:

Chloral hydrate . . . . .50 grams
Citric acid crystals. . . . . 1 gram

The following mounting procedure is followed by Evans and Loots:

1. Remove wax in xylol.

2. Hydrate through alcohols (96, 90, 50, 35%) to water.

3. Stain sections in Mayer's Haemalum until nuclei turn bright red (4-6 minutes).

4. Wash thoroughly in tap water until blue (10 minutes).

5. Pass through alcohols to 90%.

6. Counterstain in saturated solution of eosin in 90% alcohol (2-5 minutes). Differentiate in 90% aocohol.

7. Pass through absolute alcohol to xylol.

8. Mount in balsam or euparal.

Storage

Unmounted specimens to be held for future study should be preserved in a fluid other than simple solutions of alcohol. Progressive hardening of tissues in alcohol-preserved material greatly complicates slide preparation and often renders specimens unusable for phase or interference observation. *Koenike's fluid* is suggested as a permanent or semipermanent storage fluid in that it maintains tissues and appendages in a softened or flexible state and alleviates problems of breakage during mounting or dissection. Koenike's fluid is prepared as follows:

Glacial acetic acid . . . . .10 parts
Glycerine. . . . . .50 parts
Distilled water. . . . . .40 parts

Although it contains alcohol, many workers favor *Oudeman's fluid* as a long-term preservative, as well as a collecting fluid. The formula for Oudeman's fluid is given earlier in this section.

Unmounted specimens should be stored in rubber-stoppered vials and held in one pint museum jars. Collecting data for each vial should be typed or written in India ink on high quality paper and placed in the collection vial. Museum jars should be labeled in turn, indicating the taxa or general habitats of the material stored therein.

Completed microslides should be stored in slide drawers or boxes so that the slides lie flat. Slides should not be placed in permanent storage prior to affixing a complete locality label to each slide preparation.

Scanning Electron Microscopy

Unlike optical systems, the SEM operates on a high energy electron beam principle similar to that used in transmission electron microscopes. However, the beam is deflected so that it moves in zig-zag fashion over the surface of the specimen. Current passing over the scanning coils of a cathode-ray tube reproduces a magnified image on the tube's surface. A secondary low energy electron transmission is then set off by the high energy beam, modulating the intensity of the cathode tube beam and resulting in the formation of an image of the scanned surface. The image may then be transferred to a permanent micrograph for study.

Griffiths et al. (1971) and Evans and Loots (1972) outline preparation procedures for SEM. Specimens for study are examined dry and require little preliminary treatment. Best results are obtained with material freshly killed in liquid air or nitrogen, in 5% potassium hydroxide, or in boiling water. Good preparations have been obtained with fresh freeze-dried specimens, and even with air-dried material preserved in alcohol. Prior to freeze drying, Loots and Evans recommend that study specimens be immersed in 5% potassium hydroxide for 4-6 hours, washed in distilled water and transferred to an aqueous solution of picric acid. Following fixation, specimens are washed again and placed on a fine mesh sieve (400 sq/cm) with a drop of distilled water. After excess water has been removed by resting the sieve on blotting paper, the sieve is placed in a container floating on liquid air or nitrogen for 5-10 minutes. Specimens are then fixed to a metal stub with nail varnish, gum chloral, epoxy resin or adhesive tape prior to coating them with a thin conducting layer of gold or gold-palladium under vacuum.

Mites with an essentially impermeable cuticle containing so little water that it insulates rather than conducting a charge (alcohol-preserved material, for example) should be treated with an anti-static agent. Brody and Wharton (1971) recommend a mixture of 96.6% glycerol, 0.05% potassium chloride and 3.35% distilled water, all by weight. Anti-static agents eliminate a surface glow which is not corrected by gold plating.

SEM techniques have produced excellent results with lightly sclerotized mites (Griffiths et al. 1971), although highly satisfactory micrographs of more heavily sclerotized Oribatida, Ixodida and Gamasida are also attainable with scanning methods (Woolley 1970, 1971 and 1972, Evans and Loots 1972, Evans and Loots 1975).

Humphreys et al. (1971) succeeded in viewing and photographing live eriophyoid gall mites on an exposed leaf gall surface in an SEM apparatus. Although the vacuum treatment tended to immobilize the mites, normal activity was resumed after 5-10 minutes.

Useful References

Arlian, L.G. and T.A. Woolley (1970). Observations on the biology of *Liacarus cidarus* (Acari:Cryptostigmata, Liacaridae). J. Kansas Ent. Soc. **43**(3):297-301.

Aucamp, J.L. (1969). The grease film extractor and its application in soil acarology. Proc. 2nd Int. Congr. Acarology, Nottingham: 553-556.

Auerbach, S.I. and D.A. Crossley Jr. (1960). A sampling device for soil microarthropods. Acarologia **2**(3): 279-287.

Balogh, J. (1972). The Oribatid Genera of the World. Akadémiai Kaido, Budapest: 188 pp. + 71 plates.

Beavers, J.B. and G.N. Oldfield (1970). Portable platforms for watering leaves in acrylic cages containing small leaf-feeding arthropods. J. Econ. Ent. **63**(1):312-313.

Bertram, D.S., K. Unsworth and R.M. Gordon (1946). The biology and maintenance of *Liponyssus bacoti* Hirst, 1913, and an investigation into its role as a vector of *Litomosoides carinii* to cotton rats and white rats, together with some observations of the infection in the white rats. Ann. Trop. Med. Parasitol. **40**:228-254.

Brett, G.A. (1969). Distribution of mites and moisture in long stored flour, as shown by a sampler of new design. Proc. 2nd Int. Congr. Acarology, Nottingham: 235-240.

Brody, A.R. and G.W. Wharton (1971). The use of glycerol-KCl in scanning microscopy of Acari. Ann. Ent. Soc. Amer. **64**(2):528-530.

Burgdorfer, W. (1957). Artificial feeding of ixodid ticks for studies on the transmission of disease agents. J. Infect. Dis. **100**:212-214.

Camin, J.H. (1953). Observations on the life history and sensory behavior of the snake mite, *Ophionyssus natricis* (Gervais) (Acarina:Macronyssidae). Chicago Acad. Sci. Spec. Publ. 10:75 pp.

Chamberlain, R.W. and R.K. Sikes (1950). Laboratory rearing methods for three common species of bird mites. J. Parasitol. **36**:461-465.

Chant, D.A. and R.C. Muir (1955). A comparison of the imprint and brushing machine methods for estimating the numbers of fruit tree red spider mite, *Metatetranychus ulmi* (Koch), on apple leaves. Rept. E. Malling Res. Sta. for 1954: 141-145.

Clark, E.W. and F. Morishita (1950). C-M medium: A mounting medium for small insects, mites and other whole mounts. Science **112**:789-790.

Clark, G.M. (1958). *Hepatozoon griseisciuri* n. sp.; a new species of *Hepatozoon* from the grey squirrel (*Sciurus carolinensis* Gmelin, 1788), with studies on the life cycle. J. Parasitol. **44**(1):52-63.

Clifford, C.M. and D.T. Lewers (1960). A rapid method for clearing and mounting the genitalia of female ixodid ticks. J. Parasitol. **46**(6):802.

Cockings, K.L. (1948). Successful methods of trapping *Trombicula* (Acarina) with notes on rearing *T. deliensis* Walch. Bull. Ent. Res. **39**:281-296.

Coineau, Y. (1974a). Introduction a l'Étude des Microarthropodes du Sol et de ses Annexes. Documents Ecol. DOIN, éds.: 118 pp.

Coineau, Y. (1974b). Nouvelles techniques pour l'étude de la morphologie des formations chitineuses des Acariens. Acarologia **16**(1):4-10.

Conroy, J.C. (1971). A new method for trapping water-mites in the benthos of a lake. Proc. 3rd Int. Congr. Acarology, Prague: 151-157.

Cross, H.F. (1954). Feeding tests with blood sucking mites on heparinized blood. J. Econ. Ent. **47**(6):1155.

Cross, H.F. and G.W. Wharton (1954). Techniques for testing the attachment and feeding rate of mites on living hosts. J. Econ. Ent. **47**(6):1153-1154.

Davis, A.C. (1944). The mushroom mite (*Tyrophagus lintneri* (Osborn)) as a pest of cultivated mushrooms. USDA Tech. Bull. 879:26 pp.

Edwards, C.A. and K.E. Fletcher (1971). A comparison of extraction methods for terrestrial arthropods. **In** Methods of Study in Quantitative Soil Ecology: Population, Production and Energy Flow. IBP Handbook No. 18:150-185.

Efford, I.E. (1965). The ecology of the watermite *Feltria romijni* Besseling. J. Anim. Ecol. **34**:233-251.

Ekka, I., J.G. Rodriguez and D.L. Davis (1971). Influence of dietary improvement on oviposition and egg viability of the mite, *Tetranychus urticae.* J. Ins. Physiol. **17**:1393-1399.

Evans, G.O. and E. Browning (1955). Techniques for the preparation of mites for study. Ann. Mag. Nat. Hist. **8**(12):631-635.

Evans, G.O. and G.C. Loots (1972). Scanning electron microscopy in the study of the gnathosoma of the Acari. Wetensk. Bydr. Potchefstroomse Univ. B (Natuur.)49:13 pp.

Evans, G.O. and G.C. Loots (1975). Scanning electron microscope study of the structure of the hypostome of *Phityogamasus, Laelaps* and *Ornithonyssus* (Acari:Mesostigmata). J. Zool., London **176**:425-436.

Evans, G.O., J.G. Sheals and D. Macfarlane (1961). Techniques (Chapter 3). **In** The Terrestrial Acari of the British Isles, Vol. I. Introduction and Biology. British Museum (Natural History), London: 61-88.

Fujikawa, T. (1970). Notes on the efficiency of a modified Tullgren apparatus for extracting oribatid mites. Appl. Ent. Zool. **5**(1):42-44.

Furman, D.P., H. Bonasch, R. Springsteen, D. Stiller and D.A. Rahlmann (1974). Studies on the biology of the lung mite, *Pneumonyssus simicola* Banks (Acarina:Halarachnidae) and diagnosis of infestation in macaques. Lab. Anim. Sci. **24**(4):622-629.

Furman, D.P., S.F. Quan, L. Kartman and F.M. Prince (1961). Experimental attempts to transmit *Pasteurella pestis* with the mite *Haemogamasus liponyssoides hesperus* Radovsky (Acarina:Haemogamasidae). Amer. J. Trop. Med. Hyg. **10**(4):551-555.

Garcia, R. (1969). Reaction of the winter tick, *Dermacentor albipictus* (Packard) to CO_2. J. Med. Ent. **6**(3): 286.

Garman, P. (1917). *Tarsonemus pallidus* Banks, a pest of geraniums. Maryland Agr. Exp. Sta. Bull. 208:327-342.

Grandjean, F. (1949). Observation et conservation des trés petits arthropodes. Bull. Mus. nat. Hist. natur. Paris (2)**21**:363-370.

Grandjean, F. (1950). Observations éthologiques sur *Camisia segnis* et *Platynothrus peltifer.* Bull. Mus. nat. Hist. natur. Paris (2)**22**:224-231.

Gridelet-de-Saint-Georges, D. (1975). Techniques d'extraction applicables a l'étude écologique des acariens des poussières de maison. Comparaison qualitative et quantitative de divers types de poussières. Acarologia **17**(4):693-708.

Griffiths, D.A., D. Macfarlane and J.G. Sheals (1971). The scanning electron microscope in acarology. Ann. Zool. Écol. anim. **3**(1):49-55.

Haarlov, N. (1947). A new modification of the Tullgren apparatus. J. Animal Ecol. **16**(2):115-121.

Hair, J.A., A.L. Hoch, R.W. Barker and P.J. Semtner (1972). A method of collecting nymphal and adult lone star ticks, *Amblyomma americanum* (L.) (Acarina:Ixodidae), from woodlots. J. Med. Ent. **9**(2):153-155.

Hartenstein, R. (1962). Soil Oribatei V. Investigations on *Platynothrus peltifer* (Acarina:Camisiidae). Ann. Ent. Soc. Amer. **55**(6):709-713.

Huffaker, C.B. (1948). An improved cage for work with small insects. J. Econ. Ent. **41**:648-649.

Hughes, A.M. (1961). Preparation and storage of mites. **In** The Mites of Stored Food. Min. Agr. Fish., London: 281-283.

Humphreys, W.J., P.E. Hunter and H.E. Barké (1970). Live organisms viewed moving in the scanning electron microscope. Proc. Cambridge Stereoscan Colloquium: 177-187.

Husband, R.W. (1969). Technique for inspecting bees for internal mites. Proc. 2nd Int. Congr. Acarology, Nottingham: 581.

Husband, R.W. (1974). *Ovacarus peellei,* a new species of mite (Acarina:Podapolipidae) associated with the carabid *Pasimachus elongatus.* Great Lakes Ent. **7**(1):1-7.

Jacot, A.P. (1937). Culture of non-predaceous, non-parasitic mites (Oribatoidea and Tyroglyphoidea). **In** Culture Methods for Invertebrate Animals. Comstock Publ. Co., Ithaca, N.Y.: 245-246.

Jeppson, L.R., H.H. Keifer and E.W. Baker (1975). Mites Injurious to Economic Plants. Univ. California Press, Berkeley:614 pp. + 74 plates + xix.

Johnson, C.G., T.R.E. Southwood and H.M. Entwistle (1957). A new method of extracting arthropods and molluscs from grassland and herbage with a suction apparatus. Bull. Ent. Res. **48**:211-218.

Jones, B.M. (1951). The growth of the harvest mite, *Trombicula autumnalis* Shaw. Parasitol. **41**:229-248.

Kaiser, M.N. and H. Hoogstraal (1968). Simple field and laboratory method for recovering living ticks (Ixodoidea) from hosts. J. Parasitol. **54**(1):188-189.

Keifer, H.H. (1952). The eriophyid mites of California (Acarina:Eriophyidae). Bull. California Insect Survey **2**:1-123.

Keifer, H.H. (1954). Eriophyid studies XXII. Bull. California Dept. Agr., Sacramento: 121-127.

Kempson, D., M. Lloyd and R. Ghelardi (1963). A new extractor for woodland litter. Pedobiol. **3**:1-21.

Kundin, W.D., M. Nadchatram, R.W. Upham Jr. and G. Rapmund (1966). Recovery of unengorged larval trombiculid mites (Acarina) from ground holes. Nature 5054:1213.

Lipovsky, L.J. (1951). A washing method of ectoparasite recovery with particular reference to chiggers (Acarina—Trombiculidae). J. Kansas Ent. Soc. **24**:151-156.

Lipovsky, L.J. (1953). Improved technique for rearing chigger mites. Ent. News **64**:4-7.

Macfadyen, A. (1953). Notes on methods for the extraction of small soil arthropods. J. Anim. Ecol. **22**:65-77.

Macfadyen, A. (1962). Control of humidity in three funnel-type extractors for soil arthropods. **In** Progress in Soil Zoology. Butterworth, London: 158-168.

Merchant, V.A. and D.A. Crossley Jr. (1970). An inexpensive, high-efficiency Tullgren extractor for soil microarthropods. J. Georgia Ent. Soc. **5**(2):83-87.

Michael, A.D. (1884). Development and immature stages. **In** British Oribatidae, Vol. I. Ray Society, London: 65-88.

Michael, A.D. (1901). Development and immature stages. **In** British Tyroglyphidae, Vol. I. Ray Society, London: 126-183.

Miles, V.I. (1968). A carbon dioxide bait trap for collecting ticks and fleas from animal burrows. J. Med. Ent. **5**:491-495.

Mitchell, R.D. (1955). Anatomy, life history, and evolution of the mites parasitizing fresh-water mussels. Univ. Michigan Misc. Publ. Zool. 89:28 pp.

Mitchell, R.D. and D.R. Cook (1952). The preservation and mounting of water mites. Turtox News **39**(9): 4 pp.

Mullen, G.R. (1974a). Acarine parasites of mosquitoes. II. Illustrated larval key to the families and genera of mites reportedly parasitic on mosquitoes. Mosquito News **34**(2):183-195.

Mullen, G.R. (1974b). Acarine parasites of mosquitoes. III. Collection, preservation and rearing techniques used to study water mites (Acarina:Hydrachnellae) parasitic on mosquitoes. Proc. 61st Ann. Meeting, New Jersey Mosquito Extermination Assoc., Atlantic City: 117-122.

Munger, F. and J.E. Gilmore (1963). Equipment and techniques used in rearing and testing the citrus red mite. **In** Advances in Acarology, Vol. I. J.A. Naegele, ed. Comstock Publ. Co., Ithaca, N.Y.: 157-168.

Murphy, P.W. (1962). Sample preparation for funnel extraction and routine methods for handling the catch. **In** Progress in Soil Zoology. P.W. Murchy, ed. Butterworth, London: 189-198.

Nadchatram, M. (1968). A technique for rearing trombiculid mites (Acarina) developed in a tropical laboratory. J. Med. Ent. **5**(4):465-469.

Nef, L. (1971). Influence de l'humidité sur le géotactisme des Oribates (Acarina) dans l'extracteur de Berlese-Tullgren. Pedobiol. **11**:433-445.

Newell, I.M. (1947). A systematic and ecological study of the Halacaridae of eastern North America. Bull. Bingham Ocean. Coll. **10**(3):1-232.

Newell, I.M. (1971). Halacaridae (Acari) collected during Cruise 17 of the R/V ANTON BRUUN, in the southeastern Pacific Ocean. Anton Bruun Rept. 8:3-58.

Pauly, F. (1956). Zur Biologie einiger Belbiden und zur Funktion ihrer pséudostigmatischen Organe. Zool. Jahrb. (Syst.) **84**:275-328.

Pieczynski, E. (1961). The trap method of capturing water mites (Hydracarina). Ekol. Pol. B, 7:111-115.

Radinovsky, S. and G.W. Krantz (1961). The biology and ecology of granary mites of the Pacific Northwest II. Techniques for laboratory observation and rearing. Ann. Ent. Soc. Amer. **54**(4):512-518.

Rivard, I. (1958). A technique for rearing tyroglyphid mites in mould cultures. Can. Ent. **90**(3):146-147.

Robertson, P.L. (1944). A technique for biological studies of cheese mites. Bull. Ent. Res. **35**:251-255.

Rodriguez, J.G. (1964). Nutritional studies in the Acarina. Proc. 1st Int. Congr. Acarology, Fort Collins. Acarologia **6**(fasc. h.s.):324-337.

Rodriguez, J.G. (1969). Dietectics and nutrition of *Tetranychus urticae* Koch. Proc. 2nd Int. Congr. Acarology, Nottingham: 469-475.

Rodriguez, J.G. and A.M. Laskeen (1971). Axenic culture of *Tyrophagus putrescentiae* in a chemically defined diet and determination of essential amino acids. J. Ins. Physiol. **17**(6):979-985.

Rohde, C.J. (1956). A modification in the plaster-charcoal technique for the rearing of mites and other small arthropods. Ecol. **37**(4):843-844.

Rohde, C.J. (1964). Some techniques in the preparation of stained whole mounts and serial sections of mite embryos and adults. Proc. 1st Inst. Congr. Acarology, Fort Collins. Acarologia **6**(fasc. h.s.):208-214.

Sengbusch, H.G. (1954). Studies on the life history of three oribatoid mites with observations on other species (Acarina, Oribatei). Ann. Ent. Soc. Amer. **47**(4):646-667.

Sengbusch, H.G. (1956). A modified Tullgren funnel for the collection of small invertebrates (Mesobiota) in soil. Turtox News **34**(11):226-228.

Shamiyeh, N.B., S.E. Bennett, R.P. Hornsby and N.L. Woodiel (1971). Isolation of mites from house dust. J. Econ. Ent. **64**(1):53-55.

Singer, G. (1964). A simple aspirator for collecting small arthropods directly into alcohol. Ann. Ent. Soc. Amer. **57**(6):796-798.

Singer, G. (1967). A comparison between different mounting techniques commonly employed in acarology. Acarologia **9**(3):475-484.

Smith, C.N. and H.K. Gouck (1944). DDT, sulfur and other insecticides for the control of chiggers. J. Econ. Ent. **37**(1):131-132.

Southwood, T.R.E. (1966). Ecological Methods. Methuen Publ., London: 391 pp.

Steffey, K.L. and C.W. Wingo (1975). Methods for collecting the overwintering and larval forms of *Neoschöngastia americana.* J. Econ. Ent. **68**(4):471-472.

Stepien, Z. and J.G. Rodriguez (1973). Collecting large quantities of acarid mites. Ann. Ent. Soc. Amer. **66**(2):478-480.

Sternlicht, M. (1966). Trials in the control of the citrus bud mite, *Aceria sheldoni* (Ewing) in Israel. Israel J. Agr. Res. **16**(3):115-124.

Sternlicht, M. (1969). A study of fluctuations in the citrus bud mite population. Ann. Zool. Écol. anim. **1**(2):127-147.

Strandtmann, R.W. and G.W. Wharton (1958). A manual of mesostigmatid mites parasitic on vertebrates. Inst. Acarology, Univ. Maryland, Contrib. No. 4:330 pp.

Swirski, E. and S. Amitai (1957). Techniques for breeding the citrus rust mite (*Phyllocoptruta oleivora* Ashm., Acarina, Eriophyidae). Bull. Res. Council Israel **6**(8):251-252.

Tarshis, I.B. (1958). A preliminary study on feeding *Ornithodoros savigni* (Audouin) on human blood through animal derived membranes (Acarina:Argasidae). Ann. Ent. Soc. Amer. **51**(3):294-299.

Tashiro, H. (1967). Self-watering acrylic cages for confining insects and mites on detached leaves. J. Econ. Ent. **60**(2):354-356.

Travis, B.V. (1968). Glyptal—a useful slide ringing compound. J. Med. Ent. **5**(1):24.

Vannier, G. (1976). Principaux modes d'étude de la balance hydrique chez les acariens. Acarologia **18**(1):3-19.

Vercammen-Grandjean, P.-H. (1971). Of techniques and ortho-iconography. Proc. 3rd Int. Congr. Acarology, Prague: 321-328.

Wallace, M.M.H. (1972). A portable power-operated apparatus for collecting epigaeic Collembola and Acari. J. Austral. Ent. Soc. **11**:261-263.

Watson, G.E. and A.B. Amerson Jr. (1967). Instructions for collecting bird parasites. Smithsonian Inst. Mus. Nat. Hist. Information Leaflet 477: 11 pp.

Watts, B.P. Jr., J.M. Pound and J.H. Oliver Jr. (1972). An adjustable plastic collar for feeding ticks on ears of rabbits. J. Parasitol. **58**(6):1105.

Wharton, G.W. and H.S. Fuller (1952). A manual of the chiggers: The biology, classification, distribution, and importance to man of the larvae of the family Trombiculidae (Acarina). Mem. Ent. Soc. Wash. **4**:185 pp.

Williams, R.W. (1946). A contribution to our knowledge of the common North American chigger *Eutrombicula alfreddugesi* (Oudemans) with a description of a rapid collecting method. Amer. J. Trop. Med. **26**:243-250.

Woodring, J.P. (1963). The nutrition and biology of saprophytic Sarcoptiformes. **In** Advances in Acarology, Vol. I. J.A. Naegele, ed. Comstock Publ. Co., Ithaca, N.Y.: 89-111.

Woodring, J.P. (1968). An automatic collecting device for tyroglyphid (Acaridae) mites. Ann. Ent. Soc. Amer. **61**(4):1030-1031.

Woolley, T.A. (1970). Some observations on external anatomy of oribatid mites by the scanning electron microscope. Bioscience **20**(23):1253-1257.

Woolley, T.A. (1971). Scanning electron microscopy of oribatid mites. Proc. 3rd Int. Congr. Acarology, Prague: 55-56 + plates.

Woolley, T.A. (1972). Scanning electron microscopy of the respiratory apparatus of ticks. Trans. Amer. Microscop. Soc. **91**(3):348-363.

Yunker, C.E. (1961). A sampling technique for intranasal chiggers (Trombiculidae). J. Parasitol. **47**(5):720.

VIII. CLASSIFICATION

The fragmentary state of knowledge existing in the field of acarological systematics makes any attempt at a higher classification of the Acari a difficult task. Discoveries of new mite species and genera occur by the hundreds every year, and many of these finds invalidate existing family diagnoses. Acarology is, in fact, in a state of systematic turmoil similar to that experienced in the field of entomology a century ago. A so-called "natural" classification for the Acari is impossible at our present level of understanding, nor may a phylogenetic system ever be realized in this fossil-poor group. The perplexing problem of natural categories in the Acari, however, has engendered some innovative and useful approaches to acarological classification in recent years. These include 1) application of numerical taxonomic methods, 2) development of formulized leg, palpal and idiosomal chaetotaxy and poroidotaxy, 3) chromosomal mapping, and 4) recognition and utilization of behavioral and ecological data as tools in systematic research.

The classification and keys to families presented below is an attempt to bring our knowledge of the subclass up to date. It should be recognized, however, that the scheme presented is only one of a number of possible treatments. Suffixes for supercohort and cohort names have been standardized throughout the subclass ("-ides", "-ina"). Where they are used, subcohort names have been given the ending "-ae". Phalanx names cited in the dendogram for the Actinedida (page 228) end in "-idia".

Key references to earlier literature on higher classification are listed in Chapter 1 (page 2).

CLASSIFICATION OF THE HIGHER CATEGORIES OF THE SUBCLASS ACARI[1]

I. ORDER PARASITIFORMES[2]
  - A. Suborder Opilioacarida[3]
    - 1. Superfamily Opilioacaroidea
  - B. Suborder Holothyrida[4]
    - 1. Superfamily Holothyroidea
  - C. Suborder Gamasida[5]
    - Supercohort Monogynaspides
      - a. Cohort Sejina[6]
        - 1. Superfamily Sejoidea

---

[1]The subordinal categories used here are taken from van der Hammen 1972.

[2]Superorder ANACTINOTRICHIDA of van der Hammen 1972.

[3]Suborder ONYCHOPALPIDA, Superfamily Opilioacaroidea of Baker et al. 1958; Order NOTOSTIGMATA of Evans et al. 1961; Order OPILIOACARIFORMES of Johnston 1968; Suborder NOTOSTIGMATA of Krantz 1970.

[4]Suborder ONYCHOPALPIDA, Superfamily Holothyroidea of Baker et al. 1958; Order TETRASTIGMATA of Evans et al. 1961; Suborder HOLOTHYRINA of Johnston 1968; Suborder TETRASTIGMATA of Krantz 1970.

[5]Suborder MESOSTIGMATA of Baker and Wharton 1952, Johnston 1968, and Krantz 1970; Order MESOSTIGMATA of Evans et al. 1961.

[6]Epicriina *pars* of Athias-Henriot 1972.

b. Cohort Gamasina[1]
1. Superfamily Parasitoidea
2. Superfamily Rhodacaroidea
3. Superfamily Ascoidea
4. Superfamily Phytoseioidea
5. Superfamily Eviphidoidea
6. Superfamily Heterozerconoidea
7. Superfamily Dermanyssoidea

c. Cohort Uropodina[2]
1. Superfamily Thinozerconoidea
2. Superfamily Polyaspidoidea
3. Superfamily Uropodoidea
4. Superfamily Diarthrophalloidea

Supercohort Trigynaspides[3]

a. Cohort Cercomegistina
1. Superfamily Cercomegistoidea

b. Cohort Antennophorina
1. Superfamily Antennophoroidea
2. Superfamily Aenictequoidea
3. Superfamily Celaenopsoidea
4. Superfamily Megisthanoidea
5. Superfamily Fedrizzioidea
6. Superfamily Parantennuloidea

D. Suborder Ixodida[4]
1. Superfamily Ixodoidea

II. ORDER ACARIFORMES[5]

A. Suborder Actinedida[6]

Supercohort Endeostigmatides

a. Cohort Pachygnathina
1. Superfamily Pachygnathoidea[7]

b. Cohort Adamystina
1. Superfamily Adamystoidea

---

[1]Derived, in part, from Karg (1965, 1971, 1973). Karg has divided the Gamasina into five superfamilies—the Eugamasoidea, Ascaoidea, Phytoseioidea, Eviphidoidea and Dermanyssoidea (Dermanyssidae of Evans and Till 1965)—based on relationships of character states, or "gangmerkmals", in adults and immatures of "sister" groups. Karg has utilized shield size and setation, and form of the chelicerae, hypostome, sacculus and epistome in deriving his higher classification. While Karg's system may be based on valid assumptions (Hennig 1966), lack of imaginal characters limits its usefulness in a general treatment. Nevertheless, I have incorporated a slightly revised version of his scheme here and in the treatment of the Gamasina in the following section.

[2]Derived, in part, from Evans (1972).

[3]Derived from Kethley (1977).

[4]Suborder IXODIDES of Baker et al. 1958 and Johnston 1968; Order METASTIGMATA of Evans et al. 1961; Suborder METASTIGMATA of Krantz 1970.

[5]Superorder ACTINOTRICHIDA of van der Hammen 1972.

[6]Suborder TROMBIDIFORMES, Supercohors PROSTIGMATA, HETEROSTIGMATA, and PARASITENINI of Baker et al. 1958; Order PROSTIGMATA of Evans et al. 1961; Suborder PROSTIGMATA of Krantz 1970. The present classification of the ACTINEDIDA is based, in part, on Lindquist (1976).

[7]Theron (1974) has alluded to the suppression of the family name PACHYGNATHIDAE Lavoipierre 1946 in favor of the name ALYCIDAE Berlese 1905. Inasmuch as the synonomy between *Pachygnathus* Dugès 1834 and *Alycus* Koch 1841 established by Grandjean (1936) is not entirely convincing (van der Hammen 1969), I choose to retain the names Pachygnathoidea and Pachygnathina as derivatives of the older generic name, *Pachygnathus.*

Supercohort Promatides

a. Cohort Labidostommatina

1. Superfamily Labidostommatoidea

b. Cohort Eupodina

1. Superfamily Eupodoidea
2. Superfamily Tydeoidea
3. Superfamily Bdelloidea
4. Superfamily Halacaroidea

c. Cohort Eleutherengonina

Subcohort Heterostigmae

1. Superfamily Pyemotoidea
2. Superfamily Pygmephoroidea
3. Superfamily Tarsonemoidea
4. Superfamily Tarsocheyloidea
5. Superfamily Heterocheyloidea

Subcohort Raphignathae

1. Superfamily Cheyletoidea
2. Superfamily Raphignathoidea
3. Superfamily Tetranychoidea
4. Superfamily Eriophyoidea

Subcohort Anystae

1. Superfamily Caeculoidea
2. Superfamily Anystoidea

Subcohort Parasitengonae

Phalanx Trombidia[1]

1. Superfamily Calyptostomatoidea
2. Superfamily Erythraeoidea
3. Superfamily Trombidioidea

Phalanx Hydrachnidia[2]

1. Superfamily Hydrovolzioidea
2. Superfamily Hydrachnoidea
3. Superfamily Eylaoidea
4. Superfamily Hydryphantoidea
5. Superfamily Lebertioidea
6. Superfamily Hygrobatoidea
7. Superfamily Arrenuroidea

B. Suborder Acaridida[3]

Supercohort Acaridides

1. Superfamily Acaroidea
2. Superfamily Anoetoidea
3. Superfamily Canestrinioidea
4. Superfamily Ewingioidea

Supercohort Psoroptides

1. Superfamily Pterolichoidea[4]
2. Superfamily Freyanoidea[4]
3. Superfamily Analgoidea
4. Superfamily Psoroptoidea

---

[1]Supercohort PARASITENINI of Baker et al. 1958; Supercohort PARASITENGONA of Krantz 1970; Suborder PARASITENGONA of Prasad and Cook (1972).

[2]Supercohort (?) HYDRACHNELLAE of Baker and Wharton 1952, and of Krantz 1970; HYDRACARINA of Cook 1974.

[3]Suborder SARCOPTIFORMES, Supercohort ACARIDIAE of Baker et al. 1958; Order ASTIGMATA of Evans et al. 1961; Suborder ASTIGMATA of Krantz 1970.

[4]Analgoidea *pars* of Krantz 1970. Derived from Gaud and Atyeo 1977.

5. Superfamily Listrophoroidea
6. Superfamily Sarcoptoidea
7. Superfamily Cytoditoidea

C. Suborder Oribatida[1]

Supercohort Macropylides[2]

a. Cohort Bifemoratina
   1. Superfamily Archeonothroidea
   2. Superfamily Ctenacaroidea
   3. Superfamily Palaeacaroidea
b. Cohort Ptyctimina
   1. Superfamily Prothoplophoroidea
   2. Superfamily Mesoplophoroidea
   3. Superfamily Phthiracaroidea
   4. Superfamily Euphthiracaroidea
c. Cohort Arthronotina
   1. Superfamily Parhypochthonoidea
   2. Superfamily Hypochthonoidea
   3. Superfamily Brachychthonoidea
   4. Superfamily Phyllochthonoidea
   5. Superfamily Heterochthonoidea
   6. Superfamily Cosmochthonoidea
d. Cohort Holonotina
   1. Superfamily Lohmannioidea
   2. Superfamily Nothroidea
   3. Superfamily Eulohmannioidea
   4. Superfamily Epilohmannioidea
   5. Superfamily Perlohmannioidea
   6. Superfamily Collohmannioidea

Supercohort Brachypylides[3]

a. Cohort Apterogasterina[4]

   Subcohort Polytrichae
   1. Superfamily Nanhermannioidea
   2. Superfamily Hermannioidea
   3. Superfamily Hermannielloidea
   4. Superfamily Liodoidea
   5. Superfamily Gymnodamaeoidea

   Subcohort Oligotrichae
   1. Superfamily Cepheoidea
   2. Superfamily Carabodoidea
   3. Superfamily Polypterozetoidea
   4. Superfamily Zetorchestoidea
   5. Superfamily Eremaeoidea
   6. Superfamily Eremuloidea
   7. Superfamily Damaeoidea
   8. Superfamily Oppioidea
   9. Superfamily Hydrozetoidea

---

[1]Suborder SARCOPTIFORMES, Supercohort ORIBATEI of Baker et al. 1958; Order CRYPTOSTIGMATA of Evans et al. 1961; Suborder CRYPTOSTIGMATA of Krantz 1970. Classification of the ORIBATIDA is derived from Balogh (1972).

[2]Supercohors PALAEACARI and ORIBATEI INFERIORES of Krantz 1970.

[3]CIRCUMDEHISCENTIAE of Grandjean 1954; Supercohort ORIBATEI SUPERIORES of Krantz 1970.

[4]Cohort PYCNONOTICINA of Krantz 1970; APTEROGASTERINA-GYMNONOTA of Balogh 1972.

10. Superfamily Ameronothroidea
11. Superfamily Cymbaeremaeoidea
12. Superfamily Otocepheoidea
13. Superfamily Liacaroidea

b. Cohort Pterogasterina[1]

1. Superfamily Passalozetoidea
2. Superfamily Pelopoidea
3. Superfamily Galumnoidea
4. Superfamily Microzetoidea
5. Superfamily Oribatelloidea
6. Superfamily Oribatuloidea
7. Superfamily Ceratozetoidea

---

[1]Cohort PORONOTICINA of Krantz 1970; PTEROGASTERINA-PORONOTA of Balogh 1972.

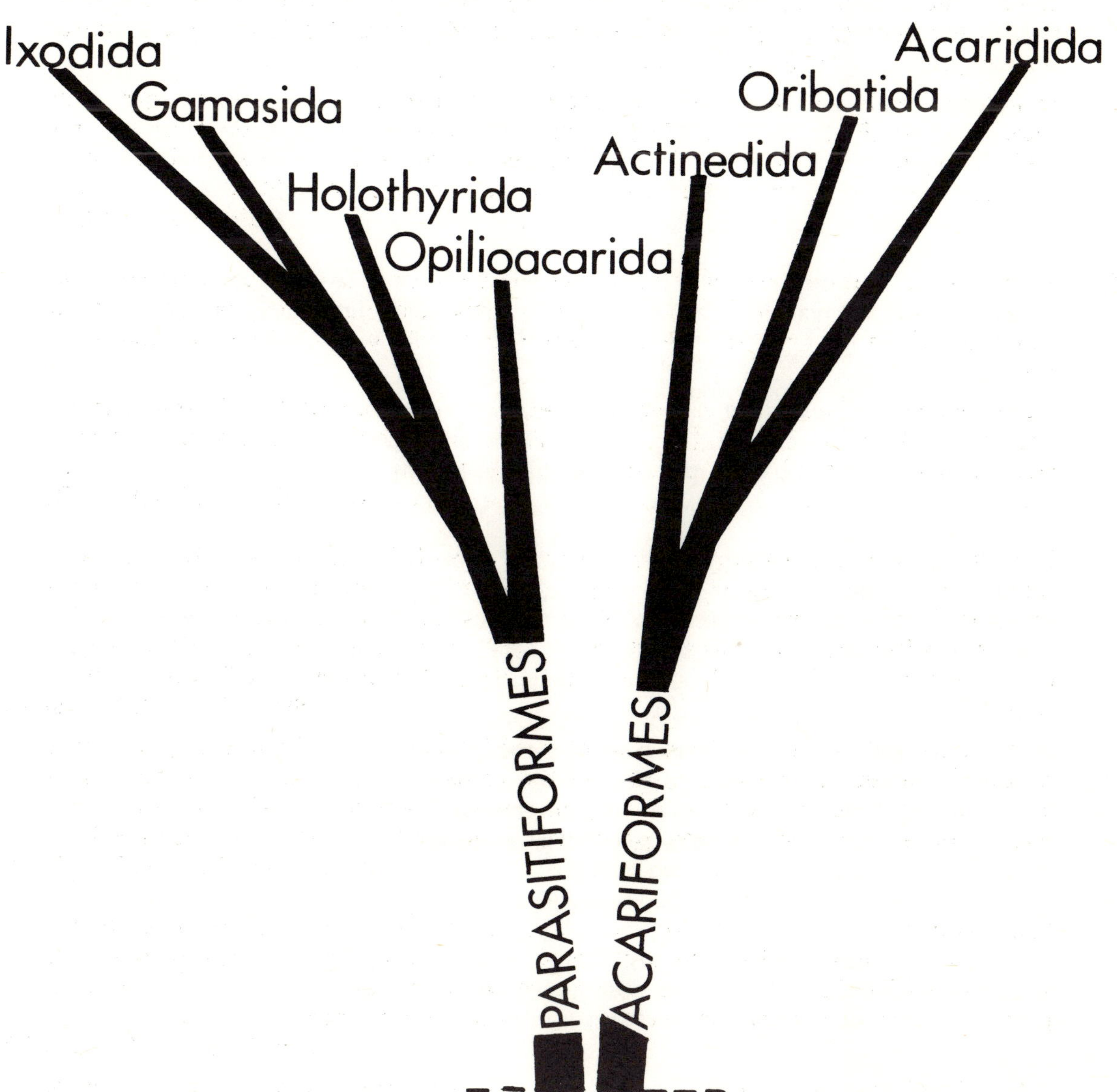

Fig. 24. Dendogram illustrating possible relationships within the subclass Acari (derived, in part, from van der Hammen 1972).

KEY TO THE ORDERS AND SUBORDERS OF ACARI

1. With one to four pairs of dorsolateral or ventrolateral stigmata posterior to coxae II (Figs. 8 and 9a, pp. 21-22); specialized propodosomal sensory organs and podocephalic canals absent; coxae free, distinct . . . . .ORDER PARASITIFORMES ... 2

— Without visible stigmata posterior to coxae II; propodosomal sensory organs, when present, in the form of simple sensilla or modified specialized structures in distinctive insertions (Fig. 13, p. 28); with a pair of podocephalic canals often visible (Fig. 16, p. 34); coxae often fused into ventral body wall, forming coxisternal regions delimited by epimera (Plate 115, p. 372) (number of legs occasionally reduced) . . . . .ORDER ACARIFORMES ... 5

2. With a terminal, subterminal or basal simple or tined apotelic claw on the palpal tarsus (Plate 3-3, p. 15; Plates 8-2 and 9-3, pp. 108, 112); hypostome serving only as part of the floor of the gnathosoma (Fig. 5b, p. 14); tarsus I only rarely with dorsal sensory pit. 3

— Pedipalpal tarsus without claws; hypostome modified into a piercing organ with retrorse teeth (Plate 54-7, p. 222); dorsum of tarsus I with a distinct sensory pit Haller's organ, (Fig. 30, p. 211); stigmata behind coxae IV or laterad above coxal intervals III-IV, each surrounded by a stigmal plate (Plate 55-2, p. 223) . . . . SUBORDER IXODIDA

3. Hysterosoma without primary segmentation, with one or two pairs of stigmata between coxal intervals II-III and III-IV, commonly ventrolateral in position (Figs. 8b and c, p. 21); palpal tarsal apotele never terminal, rutella absent; with or without peritremes 4

— Hysterosoma with weak but distinct sutures, with four pairs of dorsolateral stigmata posterior to level of coxae III (Fig. 8a, p. 21), tarsus of palp with one or two terminal claws, rutella present (Plate 8, p. 108); peritremes absent . . . . SUBORDER OPILIOACARIDA

4. Hypostome with a maximum of three pairs of setae (Fig. 5b, p. 14); with a 2- or 3-tined apotele near the inner basal angle of the palpal tarsus (rarely absent in endoparasitic forms); tritosternum (Fig. 5b, p. 14) usually present and commonly with one or two laciniae; anal valves nude or at most with a pair of setae; peritremes generally present (Fig. 8, p. 21); epistome (Fig. 5a, p. 14) present, roofing the gnathosoma . . . . SUBORDER GAMASIDA

— Hypostome with more than three pairs of setae (Plate 9-2, p. 112); apotele sometimes divided, inserted basally or medially on the palpal tarsus (Plate 9-3); tritosternum absent (present but reduced in *Allothyrus* van der Hammen); anal valves each with two or more pairs of small setae; peritremes present (Plate 9-1), epistome indistinct, reduced . . . . SUBORDER HOLOTHYRIDA

5. Palpi with only two segments (Plate 3-2, p. 15), discrete stigmatal openings absent; pretarsus with an empodial claw and fleshy pulvillus (Fig. 38a, p. 373), or pretarsus sucker-like (Fig. 38b), true claws absent, pretarsi III-IV often modified or absent in parasitic forms; without specialized sensory organs on propodosoma . . . . .SUBORDER ACARIDIDA

— Palpi sometimes minute but typically with 3-5 recognizable segments (Plate 3-4, p. 15), stigmatal openings present or absent; pretarsi various but generally with true claws on at least some legs; propodosomal sensory organs often present 6

6. Chelicerae typically stylettiform (Plate 3-8, p. 15) or hooklike (Plate 3-6), rarely chelate; palpi simple or modified into a thumb-claw process (Plate 3-1), rutella present or absent; stigmata, when present, open at or between the bases of the chelicerae (Fig. 9c, p. 22), at the base of the gnathosoma (Figs. 10a and 10b, p. 23) or on the humeral angles of the propodosoma (Fig. 9b); propodosomal sensory organs, when present, often elongate (Plate 59-8, p. 314) or short and capitate (Plate 74-2, p. 329); empodial processes commonly rayed or padlike (Plates 66-7, 74-3, pp. 321, 329), often with tenent hairs (Plate 87-5, p. 342), occasionally claw- or sucker-like and rarely absent. Generally weakly sclerotized species . . . . .SUBORDER ACTINEDIDA

— Chelicerae typically chelate-dentate (Fig. 11, p. 23), sometimes attenuate (Plate 158-1, p. 483); palpi simple, never with thumb-claw process, rutella present; discrete stigmata absent, tracheal system (when present) opens in the acetabular cavities of legs I and II (Fig. 10c, p. 23), on the legs themselves, or through the prodorsal bothridia (Fig. 13b, p. 28), bothridial sensilla generally present, often assuming distinctive forms (Plates 152-1 and 160-1, pp. 477 and 485), empodial processes clawlike or absent. Generally well-sclerotized species. . . . . .SUBORDER ORIBATIDA

Useful References

Athias-Henriot, C. (1972). Gamasides Chiliens (Arachnides). II. Revision de la famille Ichythyostomatogasteridae Sellnick, 1953 (= Uropodellidae Camin 1955). Arq. Zool., São Paulo **22**(3):113-191.

Baker, E.W., J.H. Camin, F. Cunliffe, T.A. Woolley and C.E. Yunker (1958). Guide to the Families of Mites. Institute of Acarology Contr. No. 3:242 pp. + ix.

Balogh, J. (1972). The Oribatid Genera of the World. Akad. Kaidó, Budapest: 188 pp. + 71 plates.

Cook, D.R. (1974). Water mite genera and subgenera. Mem. Amer. Ent. Inst. 21:860 pp.

Evans, G.O. (1972). Leg chaetotaxy and the classification of the Uropodina (Acari:Mesostigmata). J. Zool., London **167**:193-206.

Evans, G.O., J.G. Sheals and D. MacFarlane (1961). The Terrestrial Acari of the British Isles. Vol. I. Introduction and Biology. British Museum (Natural History), London: 219 pp.

Evans, G.O. and W.M. Till (1966). Studies on the British Dermanyssidae (Acari:Mesostigmata). Part II. Classification. Bull. Brit. Mus. (Nat. Hist.) Zool. **14**(5):109-370.

Grandjean, F. (1936). Le genre *Pachygnathus* Dugès (*Alycus* Koch) (Acariens) 1re partie. Bull. Mus. nat. Hist. natur. Paris **8**(2):398-405.

Grandjean, F. (1954). Essai de classification des Oribates (Acariens). Bull. Soc. Zool. France **78**(5-6):421-446.

Hammen, L. van der (1969). Notes on the morphology of *Alycus roseus* C.L. Koch. Zool. Med. **43**(15):177-202.

Hammen, L. van der (1972). A revised classification of the mites (Arachnidea, Acarida) with diagnoses, a key, and notes on phylogeny. Zool. Med. **47**(22):273-292.

Hennig, W. (1966). Phylogenetic Systematics. Univ. Illinois Press, Urbana: 263 pp.

Johnston, D.E. (1968). An Atlas of Acari I. The Families of Parasitiformes and Opilioacariformes. Acarology Lab. Publ. 172. Acarology Publ., Columbus: 110 pp. + x.

Karg, W. (1965). Larvalsystematische und phylogenetische Untersuchung sowie Revision des Systems der Gamasina Leach, 1915 (Acarina, Parasitiformes). Mitt. Zool. Mus. Berlin **41**(2):193-340.

Karg, W. (1971). Die freilebenden Gamasina (Gamasides), Raubmilben. Die Tierwelt Deutschlands **59**:475 pp. Fischer Verlag, Jena.

Karg, W. (1973). Begründung der Überfamilie Eviphidoidea (Acarina, Gamasina) und Darstellung der verfolgten Arbeitsweise in Form eines heuristischen Programms. Zool. Anz. **190**(5/6):386-400.

Kethley, J.B. (1977). A review of the higher categories of Trigynaspida (Acari:Parasitiformes). Int. J. Acarology **3**(2):129-149.

Krantz, G.W. (1970). A Manual of Acarology. Oregon State Univ. Bookstores, Corvallis: 335 pp.

Lindquist, E.E. (1976). Transfer of the Tarsocheylidae to the Heterostigmata, and reassignment of the Tarsonemina and Heterostigmata to lower hierarchic status in the Prostigmata (Acari). Can. Ent. **108**:23-48.

Prasad, V. and D.R. Cook (1972). The taxonomy of water mite larvae. Mem. Amer. Ent. Inst. 18:326 pp. + ii.

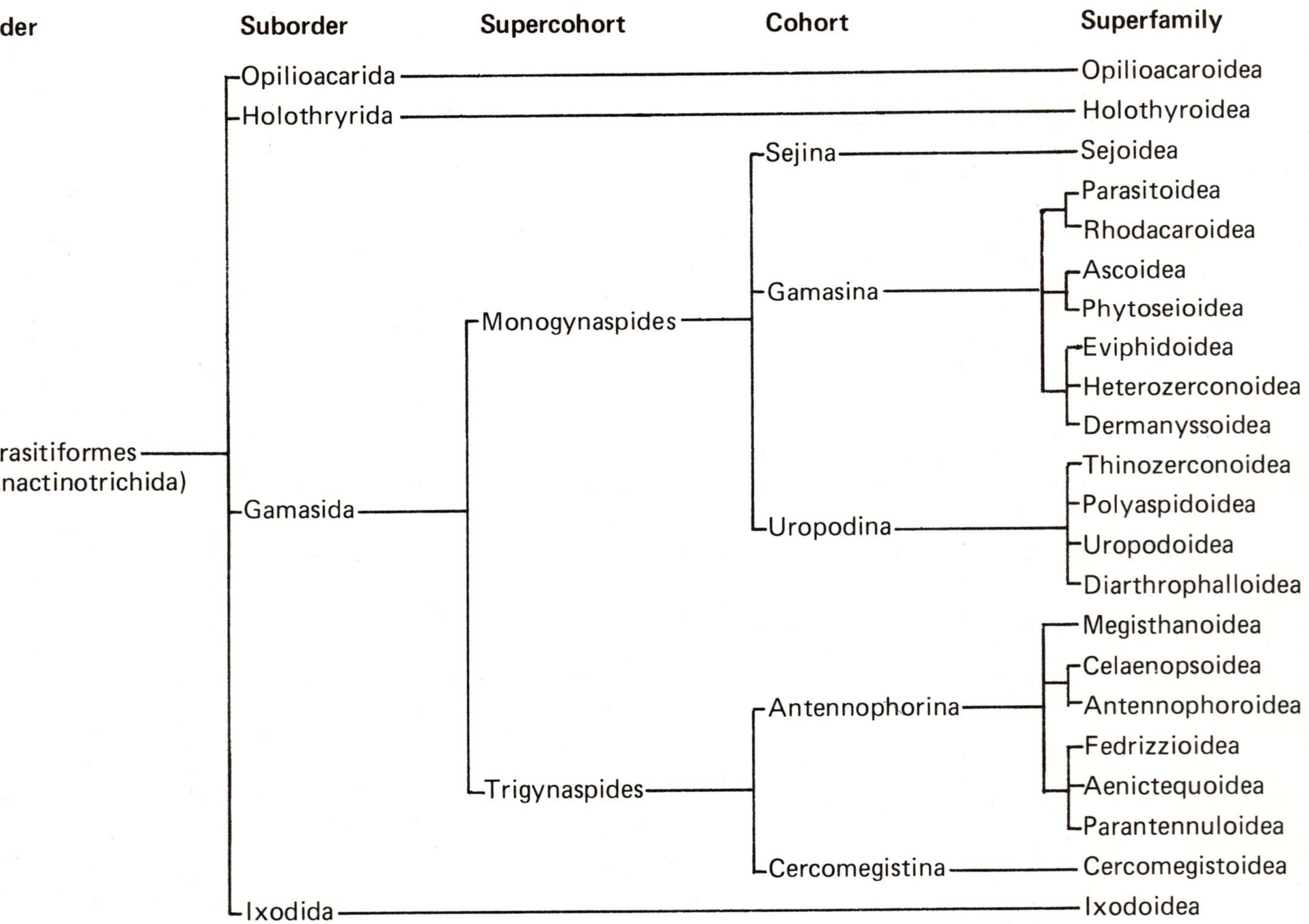

Fig. 25. Dendrogram illustrating possible relationships within the order Parasitiformes of the subclass Acari.

SUBORDER OPILIOACARIDA

The Opilioacarida are large (1000 μ+) elongate leathery mites which superficially resemble members of the arachnid group Opiliones. They are considered to be the most primitive of the Parasitiformes since they illustrate the greatest number of ancestral characters (van der Hammen 1972). Important identifying features of the suborder include:

1. A series of transverse sutures which are demarcated by round or elliptical muscle attachment sites, and which divide the opisthosoma into 12 postgenital "segments".

2. Four pairs of small dorsolateral stigmatal openings on the opisthosoma (Plate 8-1, p. 108). Peritremes are absent. Two or three pairs of ocelli are present on the propodosoma.

3. A terminal palptarsal claw which permits the palp to be used as a grasping structure (Plate 8-2) (the more advanced parasitiform mites have a subterminal claw).

4. One or two pairs of strong sclerotized *rutella* on the anterolateral angles of the hypostome. Another pair of paralabial hypertrophied setae, *With's organs*, are inserted mediad from the rutella (Plate 8-4) (may be absent in *Adenacarus*). Always with more than seven pairs of normally produced setae on the venter of the adult gnathosoma. An epistome is present, but indistinct.

5. A large number of *lyrifissures* (>200) may be found on the opisthosoma (Plate 8-1).

6. Three or four dorsal setae on the fixed cheliceral digit, and usually an additional seta on the dorsodistal portion of the basal cheliceral segment (Plate 8-3).

7. A divided tritosternal base (Plate 8-4), rather than the fused base found in more advanced parasitiform mites.

8. Divided trochanters III-IV, plus secondary articulations of femora, tarsi, and tibiae I.

9. Transverse genital aperture uncovered, between coxae III-IV. Anal valves without setae.

Opilioacarids prefer dark protected semi-arid habitats, and generally secrete themselves under stones or other suitable retreats during the day. While they are often thought of as predators, opilioacarids may in fact prefer pollen or fungi as food. Pollen grains or fungal spores have been recovered from the ventricular contents of dissected specimens of *Opilioacarus texanus* (Chamberlin and Mulaik) (van der Hammen 1966). Opilioacarids are known from the southwestern United States, Puerto Rico, South America, central Asia, Africa, and southern Europe.

The Opilioacarida are considered to be unique in the Acari, but their affinities with the Parasitiformes prompts their placement in that order (van der Hammen 1972). However, opilioacarids do possess rutella, a trait otherwise restricted to certain acariform suborders (see page 105). The recent discovery of a prelarval stage in an African species of OPILIOACARIDAE (Coineau 1973, Coineau and van der Hammen 1979) further confuses the affinities of the Opilioacarida, since prelarvae formerly were known only in the Acariformes.

PLATE 8

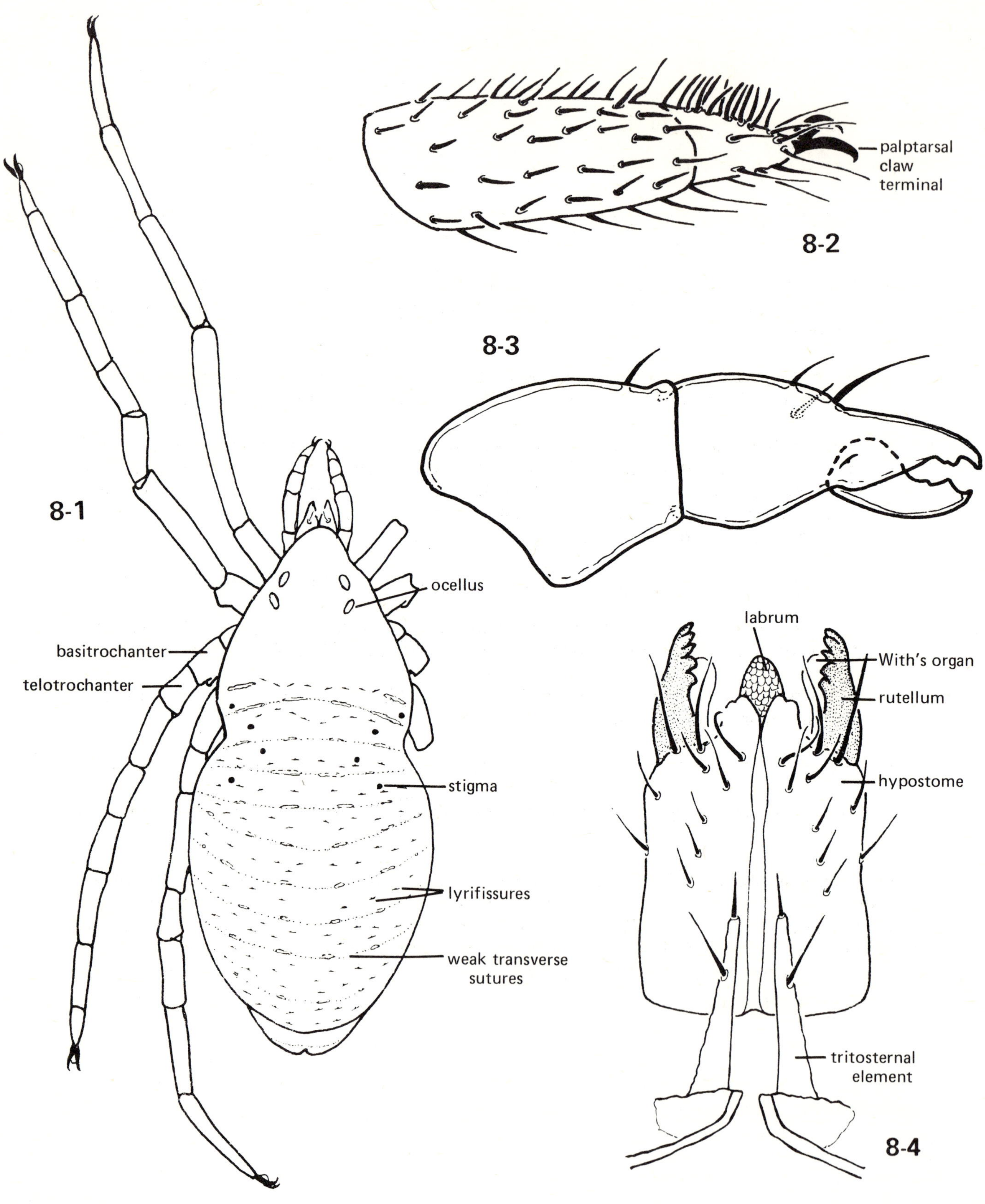

8-1 to 8-4; family OPILIOACARIDAE, *Opilioacarus* sp. **8-1**; dorsum of female: **8-2**; palpal tibia and tarsus: **8-3**; chelicera (antiaxial): **8-4**; venter of gnathosoma and divided tritosternal elements

The four described genera of Opilioacarida (*Opilioacarus, Paracarus, Adenacarus* and *Panchaetes*) are relegated to a single family, the OPILIOACARIDAE.

Useful References

Chamberlin, R.V. and S. Mulaik (1942). On a new family in the Notostigmata. Proc. Biol. Soc. Wash. **55**: 125-132.

Coineau, Y. (1973). A propos de quelques caractères particulierment primitifs de la prélarve et de la larve d'un Opilioacaridae du Gabon (Acariens). C.R. Acad. Sci. Paris **276**(Série D): 1181-1184.

Coineau, Y. and L. van der Hammen (1979). The postembryonic development of Opilioacarida, with notes on new taxa and on a general model for the evolution. Proc. IV Congress of Acarology (Saalfelden): 437-441.

Grandjean, F. (1936). Un acarien synthétique: *Opilioacarus segmentatus* With. Bull. Soc. Hist. natur. Afr. Nord. **27**:413-444.

Hammen, L. van der (1966). Studies on Opilioacarida (Arachnida). 1. Description of *Opilioacarus texanus* (Chamberlin and Mulaik) and revised classification of the genera. Zool. Verh. 86:80 pp.

Hammen, L. van der (1968). Studies on Opilioacarida (Arachnida) II. Redescription of *Paracarus hexophthalmus* (Redikorzev). Zool. Meded. **43**(5):57-76.

Hammen, L. van der (1969). Studies on Opilioacarida (Arachnida) III. *Opilioacarus platensis* Silvestri, and *Adenacarus arabicus* (With). Zool. Meded. **44**(8):113-131.

Hammen, L. van der (1970). La phylogenèse des Opilioacarides, et leurs affinités avec les autres acariens. Acarologia **12**(3):465-473.

Hammen, L. van der (1972). A revised classification of the mites (Arachnidea, Acarida) with diagnoses, a key, and notes on phylogeny. Zool. Meded. **47**(22):273-292.

Hammen, L. van der (1976). Glossaire de la terminologie acarologique (Glossary of acarological terminology). Vol. II. Opilioacarida. W. Junk B.V., The Hague: 137 pp. + plates.

Naudo, M.H. (1963). Acariens Notostigmata de l'Angola. Publ. Cult. Co. Diamante Angola, Lisboa **63**:13-24.

With, C.J. (1904). The Notostigmata a new suborder of Acari. Vidensk. Medd. Natur. Foren. Kobenh.: 137-192.

SUBORDER HOLOTHYRIDA

Unlike the Opilioacarida, members of the suborder Holothyrida are strongly armored species without indications of opisthosomal segmentation. They are large (2-7000 μ) and oval or rounded in shape. Useful identifying features include:

1. A pair of stigmata and peritremes laterad from coxae I-III, and a second more posterior pair of stigmatal pores on the lateral margins of the dorsal shield, connected internally to a system of air sacs (Plate 9-1, p. 112). *Allothyrus australasiae* (Womersley) apparently has three pairs of stigmata (van der Hammen 1972). Ocelli are absent.

2. A 2- or 3-tined subterminal palptarsal claw (Plate 9-4).

3. A strongly raduliform median labrum, and a pair of horn-like or furcate *corniculi* (Plate 9-2). Rutella and With's organs are absent. With more than four, but usually less than seven, pairs of normally produced setae on the venter of the gnathosoma. An epistome is present.

4. Few lyrifissures (typically 3-4 pairs) on the dorsum (Plate 9-2); with additional pairs laterally.

5. Usually a single seta at the base of the fixed digit, corresponding to the most distal seta on the fixed digit of Opilioacarida (Plate 9-5). *Allothyrus* may have more than one cheliceral seta.

6. A transverse genital opening covered, in the female, by a complex of four shields hinged along their external margins (Plate 9-6), and in the male by two setate genital valves. Anal valves each with more than two setae (Plate 9-7).

The Holothyrida apparently are predaceous and may be found under stones or in decaying vegetation where their prey occurs (Hughes 1959). They are reported from Australia, New Zealand, New Guinea, Ceylon, Seychelles and Mauritius.

At least one species, *Holothyrus coccinella* Gervais, has been responsible for several cases of human poisoning (Hirst 1917) involving severe inflammation of the mucous membranes and pharynx following handling of the mites. Southcott (1976) described the reactions of a person who accidentally put his fingers in his mouth after touching one specimen. There was an initial "extraordinarily pungent, galvanic sensation or taste" which spread rapidly to the throat, followed by excessive salivation.

The Holothyrida comprises two families, the HOLOTHYRIDAE and the ALLOTHYRIDAE (van der Hammen 1972), which are separated as follows:

1. With two pairs of lateral stigmata; fixed cheliceral digit with only one seta dorsally; tritosternum absent . . . .(Plate 9, p. 112) Family HOLOTHYRIDAE

— With three pairs of lateral stigmata; fixed cheliceral digit with more than one dorsal seta; with a weakly produced tritosternum consisting of paired unfused elements . . . . Family ALLOTHYRIDAE

PLATE 9

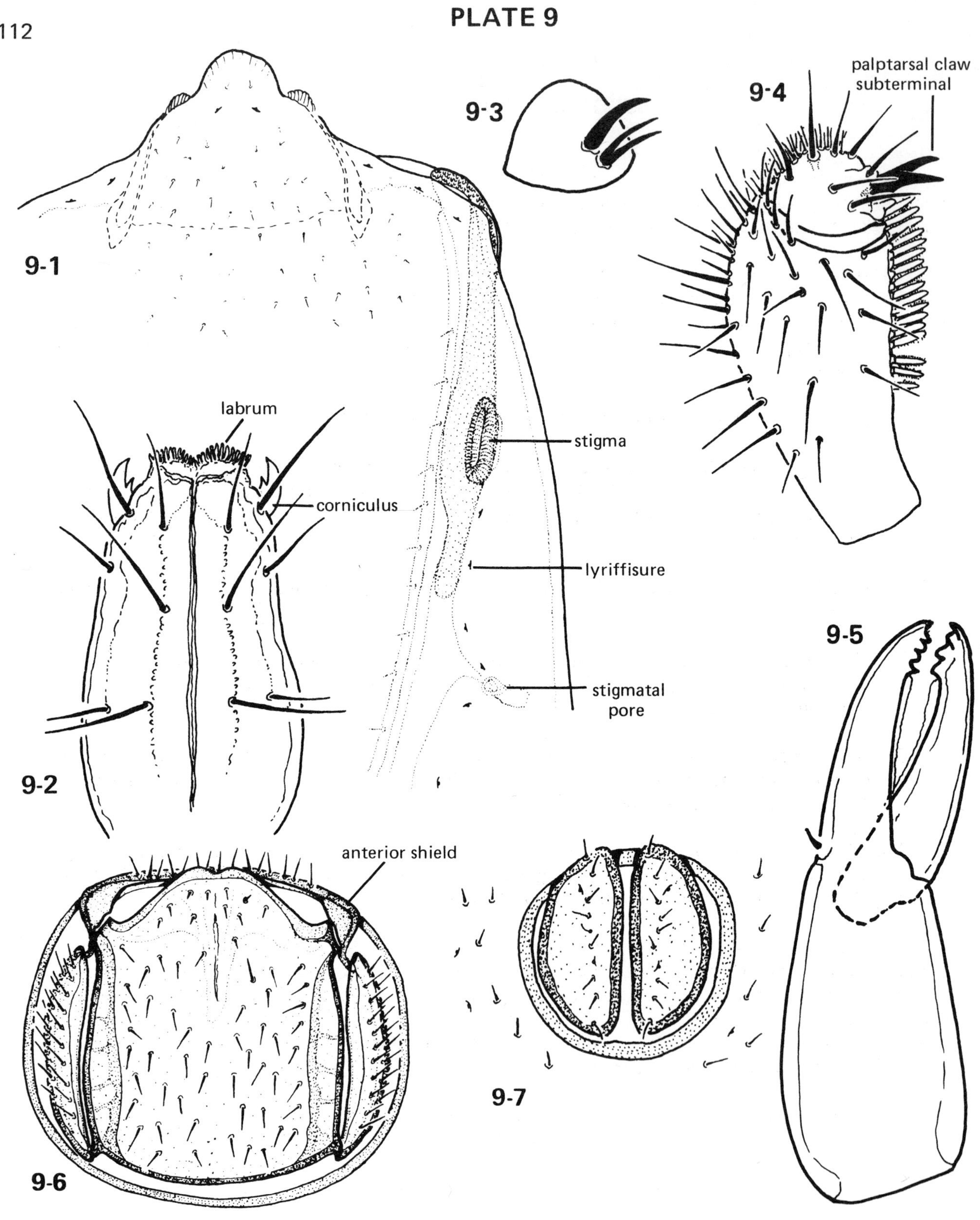

9-1 to 9-7; family HOLOTHYRIDAE, *Holothyrus* spp. **9-1;** anterodorsal aspect: **9-2;** hypostome: **9-3;** palpal tarsus, with divided subterminal claw (after Camin et al. 1958): **9-4;** palpal tibia and tarsus, illustrating entire palptarsal claw: **9-5;** *Holothyrus grandjeani* van der Hammen, chelicera (antiaxial) (after van der Hammen 1961): **9-6 and 9-7;** *Holothyrus* sp. (New Guinea). **9-6;** genital valves of female: **9-7;** anal valves of female

Useful References

Hammen, L. van der (1961). Description of *Holothyrus grandjeani* nov. spec., and notes on the classification of the mites. Nova Guinea n.s. **10**(9):173-194.

Hammen, L. van der (1965). Further notes on the Holothyrina (Acarida) I. Supplementary description of *Holothyrus coccinella* Gervais. Zool. Meded. **40**(28):253-276.

Hammen, L. van der (1972). A revised classification of the mites (Arachnidea, Acarida) with diagnoses, a key, and notes on phylogeny. Zool. Meded. **47**(22):273-292.

Hirst, S. (1917). Species of Arachnida and Myriopoda (Scorpions, spiders, mites, ticks and centipedes) injurious to man. Brit. Mus. (Nat. Hist.) Econ. Ser. 6:60 pp.

Hughes, T.E. (1959). The free-living Acari. **In** Mites, or the Acari. Athlone Press, London: 1-22.

Southcott, R.V. (1976). Arachnidism and allied syndromes in the Australian region. Rec. Adelaide Children's Hosp. **1**(1):97-186.

SUBORDER GAMASIDA

The Gamasida is a large group of acarines which have adapted to highly diverse habitats. The majority of gamasids are free-living predators (Karg 1971) but many species are external or internal parasites of mammals, birds, reptiles, or invertebrates (Strandtmann and Wharton 1958, Yunker 1973, Treat 1975).

Gamasids range in size from 200 to over 2000 μ and usually possess a number of distinctive sclerotized shields on the dorsum and venter. These shields generally illustrate characteristic incremental development, fusion, or fragmentation beginning with the larval stage and ending with the imago (Plate 10, p. 116). Ontogenetic development in the Gamasida also is marked by the appearance and elongation of the peritreme in the postlarval stages and, except where neoteny occurs, an increase in numbers of setae and pores on the idiosoma and on the appendages. The palpi of gamasids provide a useful means for determining particular instars in that certain segments acquire additional setae at the protonymphal and deutonymphal molts (Evans 1964). The following chart lists setal numbers on palpal segments in each instar of a typical gamasid. Other features of ontogenetic development are illustrated in Plate 10.

	Number of setae on				
	Trochanter	Femur	Genu	Tibia	Tarsus
Larva	0	4	5	12	11
Protonymph	1	4	5	12	15
Deutonymph, Adult	2	5	6	14	15

There is pronounced sexual dimorphism in gamasids. Females generally have a distinctive *epigynial* complex (Figs. 28 and 29; pp. 122, 147) and, in many cases, concomitant fragmentation of the surrounding ventral armature. Males have an antero- or medioventral genital opening (Plates 14-5 and 19-3; pp. 172, 177) usually set in a holoventral or sternoventral shield complex. In addition, cheliceral dimorphism occurs in gamasid groups where the male chelicerae are modified for sperm transfer to the female (see Chapter III).

Gamasids share a number of identifying characteristics including:

1. A pair of lateroventral or laterodorsal stigmatal openings at the level of coxae II-IV, usually associated with elongate peritremes in postlarval stages (Fig. 27a, p. 120). Peritremes may be reduced or absent in certain families. Ocelli are absent.

2. A basal palptarsal claw with two or three tines (Fig. 27a). The claw may show considerable variation in some parasitic species (Pence and Casto 1976), and may be absent in certain obligate parasites.

3. A pair of horn-like corniculi at the terminus of the hypostome (Fig. 5b, p. 14), usually well sclerotized and occasionally dentate or furcate. With's organs and rutella absent. Three pairs of setae inserted on the hypostome may form a triangle (Fig. 5b) or a straight line. Two pairs are present in certain parasitic forms.

4. A pair of anteromedial propodosomal lyrifissures usually discernible (Fig. 2, p. 7). Other characteristic pores and gland openings also are found dorsally in many families.

PLATE 10

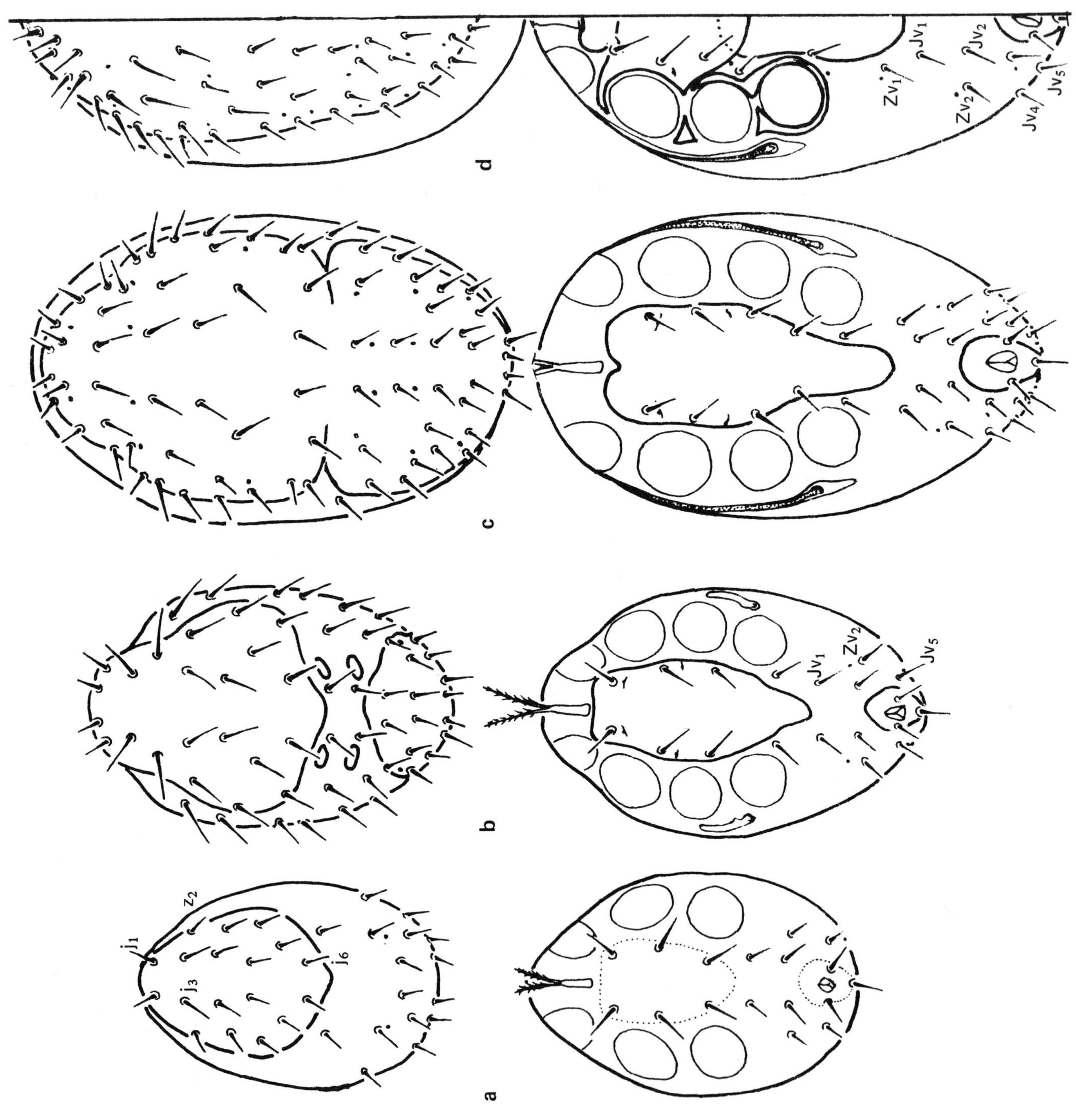

10; ontogenetic development of *Hypoaspis aculeifer* (Canestrini) (LAELAPIDAE), showing dorsal (top) and ventral aspects of each instar: a, larva; b, protonymph; c, deutonymph; d, adult female

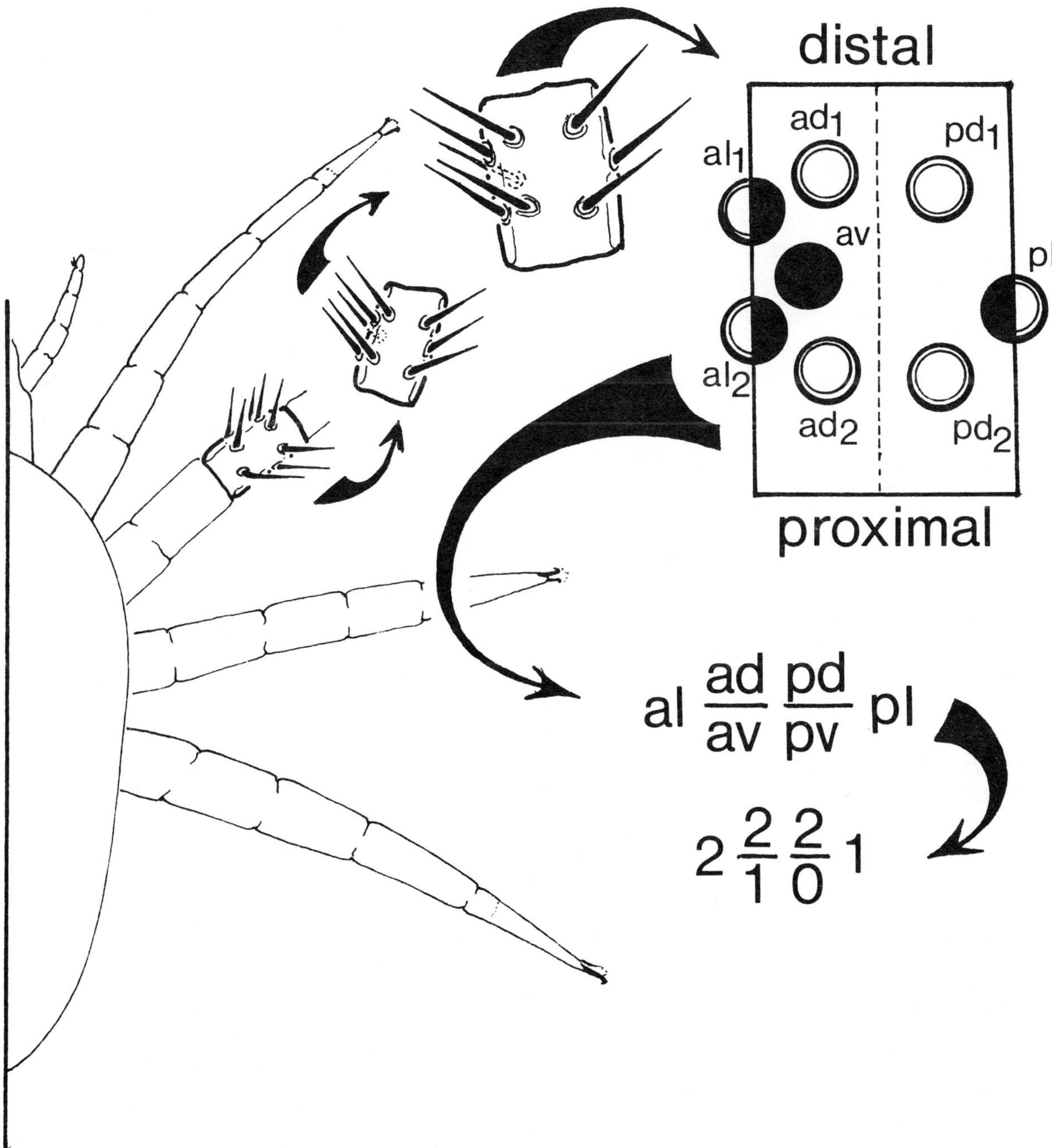

Fig. 26. Leg segment chaetotaxy in the Gamasida. Genu II has been separated from leg II and oriented so that its leading edge (the anterolateral face, or *al*) corresponds to the vertical left-hand margin in the segment diagram. Genu II is viewed in dorsal aspect, and its top margin corresponds to the distal extremities of the segment. Dorsal setal insertions are indicated by open circles, and ventral insertions by solid circles.

5. An undivided tritosternum (sternapophysis of van der Hammen 1966) with one, two, or three laciniae which may be secondarily anastomosed. The laciniae or the entire tritosternum may be absent in certain parasitic families.

6. An anteriorly projecting epistome (tectum capituli of Evans and Till 1965) which may be simple or highly ornamented (Fig. 5a, p. 14) (the epistome is absent in *Uropodella,* a genus of the cohort Sejina). The labrum may be well developed, but is not raduliform.

7. At most a single seta dorsally at the base of the fixed cheliceral digit, often short and broad. An antiaxial and a paraxial lyrifissure may be present on the fixed digit, along with a *pilus dentilis* (cheliseta of van der Hammen 1964) (Plate 16-7, p. 174).

8. A transverse genital aperture in the intercoxal region of the female covered by one, three or four shields. The male genital aperture is covered by two valves. Anal valves usually without setae, rarely with a single pair.

Members of the Gamasida are found throughout the world in association with soil, litter, plants, and animals. The suborder has been divided here into two supercohorts, based primarily on genital shield configuration. Approximately 65 families are included in the key to the suborder (page 157), and these are grouped in five cohorts and 19 superfamilies.

Chaetotactic patterns on particular leg segments in the Gamasida provide valuable taxonomic criteria for establishing relationships within the suborder (Evans 1963 and 1972, Evans and Till 1965). A system of setal nomenclature devised by Evans permits characterization of a given segment by means of a simple formula. Setae inserted on the various faces of the segment are recorded in the following form:

$$\text{anterolaterals (al)} \; \frac{\text{anterodorsals (ad)}}{\text{anteroventrals (av)}} \; \frac{\text{posterodorsals (pd)}}{\text{posteroventrals (pv)}} \; \text{posterolaterals (pl)}$$

Thus, a leg segment with two anterolateral and two posterolateral setae, three anterodorsals and three posterodorsals, one anteroventral and no posteroventrals, would be characterized as follows:

$$\text{al} \; \frac{\text{ad}}{\text{av}} \; \frac{\text{pd}}{\text{pv}} \; \text{pl} = 2 \, \frac{3}{1} \, \frac{3}{0} \, 2$$

A graphic example of chaetotactic formula derivation is given in Fig. 26, p. 117.

Useful References

Athias-Henriot, C. (1969). Notes sur la morphologie externe des gamasides (Acariens anactinotriches). Acarologia **11**(4):609-629.

Athias-Henriot, C. (1971). Un progrès dans la connaisance de la composition metamerique des gamasides: leur sigillotaxie idiosomale (Arachnides). Bull. Soc. zool. France **96**:73-85.

Athias-Henriot, C. (1974). The idiosomatic euneotaxy and epineotaxy in gamasids (Arachnida, Parasitiformes). Z. Zool. Sept. Evol.-forsch. **13**(2):97-109.

Camin, J.H. and F. Gorirossi (1955). A revision of the suborder Mesostigmata (Acarina), based on new interpretations of comparative morphological data. Chicago Acad. Sci. Spec. Publ. 11:70 pp.

Evans, G.O. (1957). An introduction to the British Mesostigmata (Acarina) with keys to the families and genera. Linn. Soc. J. Zool. **43**:203-259.

Evans, G.O. (1963). Observations on the chaetotaxy of the legs in the free-living Gamasina (Acari:Mesostigmata). Bull. Brit. Mus. (Nat. Hist.) Zool. **10**(5):277-303.

Evans, G.O. (1964). Some observations on the chaetotaxy of the pedipalps in the Mesostigmata (Acari). Ann. Mag. Nat. Hist. **6**(13):513-527.

Evans, G.O (1972). Leg chaetotaxy and the classification of the Uropodina (Acari:Mesostigmata). J. Zool., London **167**:193-206.

Evans, G.O. and W.M. Till (1965). Studies on the British Dermanyssidae (Acari:Mesostigmata). Part I. External morphology. Bull. Brit. Mus. (Nat. Hist.) Zool. **13**(8):249-294.

Hammen, L. van der (1964). The morphology of *Glyptholaspis confusa* (Foà, 1900) (Acarida, Gamasina). Zool. Verh. 71:56 pp.

Hammen, L. van der (1966). Studies on Opilioacarida (Arachnida) 1. Description of *Opilioacarus texanus* (Chamberlin and Mulaik) and revised classification of the genera. Zool. Verh. 86:80 pp.

Hirschmann, W. (1957-1976). Gangsystematik der Parasitiformes. Folge 1-22, Teilen 1-232. Acarologie. Schriftenreihe für vergleichende Milbenkunde. Hirschmann-Verlag Inh., Fürth/Bayern.

Johnston, D.E. (1968). An Atlas of Acari I. The Families of Parasitiformes and Opilioacariformes. Acarology Lab. Publ. 172. Acarology Publ., Columbus: 110 pp. + x.

Karg, W. (1965a). Larvalsystematische und phylogenetische Untersuchung sowie Revision des Systems der Gamasina Leach, 1915 (Acarina, Parasitiformes). Mitt. Zool. Mus. Berlin **41**(2):193-340.

Karg, W. (1965b). Neue Erkenntnisse zum System der Gamasina (Acarina, Parasitiformes) durch Larvalsystematische Untersuchungen. Zesz. Problem. Post. Nauk Roln. **65**:89-114.

Karg, W. (1965c). Die Anwendung Systematisch-Phylogenetischer Arbeitsmethoden bei einer Bearbeitung der Gamasina (Acarina, Parasitiformes). Zesz. Problem. Post. Nauk Roln. **65**:115-138.

Karg, W. (1965d). Entwicklungsgeschichtliche Betrachtung zur Ökologie der Gamasina (Acarina, Parasitiformes). Zesz. Problem. Post. Nauk Roln. **65**:139-155.

Pence, D.B. and S.D. Casto (1976). Studies on the variation of morphology of the *Ptilonyssus "sairae"* complex (Acarina:Rhinonyssidae) from North American birds. J. Med. Ent. **13**(1):71-95.

Schweizer, K. (1961). Die Landmilben der Schweiz (Mittelland, Jura und Alpen). Parasitiformes Reuter. Mem. Soc. Helvet. Sci. nat. **84**:207 pp. + vii.

Strandtmann, R.W. and G.W. Wharton (1958). Manual of mesostigmatid mites parasitic on vertebrates. Inst. Acarology, Univ. Maryland, Contr. 4:330 pp. + vii + 69 plates.

Trägårdh, I. (1946). Outlines of a new classification of the Mesostigmata (Acarina) based on comparative data. Kungl. Fysiogr. Sällska. Handl. **57**:1-37.

Treat, A.E. (1975). Mites of Moths and Butterflies. Cornell Univ. Press, Ithaca, N.Y. 362 pp.

Yunker, C.E. (1973). Mites. **In** Parasites of Laboratory Animals. R.J. Flynn, ed. Iowa State Univ. Press, Ames: 425-492.

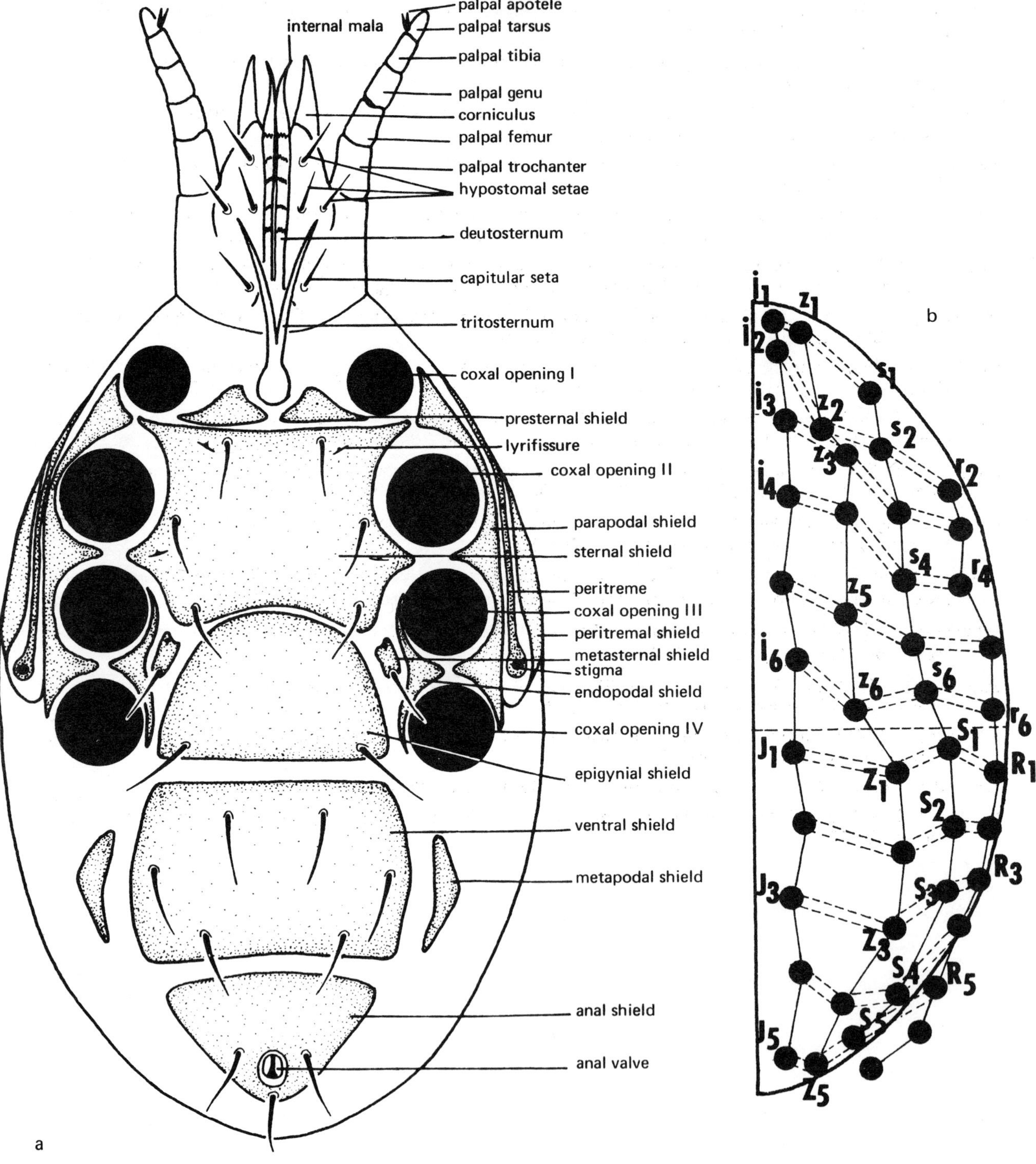

Fig. 27. Venter (a) and dorsum (b) of generalized gamasid mite. Dorsal chaetotactic system is that of Lindquist and Evans (1965); dotted and solid lines denote assumed vertical and horizontal relationships of setal bases so connected.

Supercohort Monogynaspides

The Monogynaspides are a large and extremely diverse group of free-living, phoretic and parasitic gamasids which are widely distributed throughout the world. The supercohort includes the majority of gamasid species, and these are assigned to 12 superfamilies (Fig. 25, p. 106).

Monogynaspide females usually are distinguished from other Gamasida by the presence of a single sclerotized genital flap, the *epigynial shield* (Fig. 28, p. 122). The shield opens outwardly at its anterior margin, and may be hinged or fixed posteriorly in the ventral integument or to ventral shield elements. Reduction of the epigynial shield occurs in several families of the gamasid cohort Gamasina and, in certain parasitic gamasine taxa, it may be completely absent (Plate 30-4, p. 188). Conversely, the shield often is expanded through coalescence with adjacent ventral, metapodal and/or anal elements (Fig. 28g, p. 122). The epigynial shield or epigynial portion of the shield complex may carry several setae (i.e., the Sejina), but more often has only a single pair (i.e., many Gamasina), or none (i.e., most Uropodina). The ventral shields of gamasid males differ greatly from those of the female, as has been noted for the suborder (page 115). Male spermadactyli often are present in the Gamasina, but do not appear in other gamasid cohorts. One or two arthrodial brushes or a setal coronet may be found at the base of the movable cheliceral digit of adults and immatures (Plate 13-5, p. 171), but ventromedial or ventrodistal excrescences do not occur.

Adult and deutonymphal Monogynaspides generally have five ventral setae on tarsi II-IV ($3\ \frac{3}{2}\ \frac{1}{1}\ \frac{3}{2}\ 3$), although certain of these setae may be lost in some parasitic gamasine groups. Tarsi II-IV of the protonymph lack a medioventral hair ($3\ \frac{3}{2}\ \frac{1}{0}\ \frac{3}{2}\ 3$).

The Monogynaspides comprise three cohorts—the Sejina, Gamasina and Uropodina.

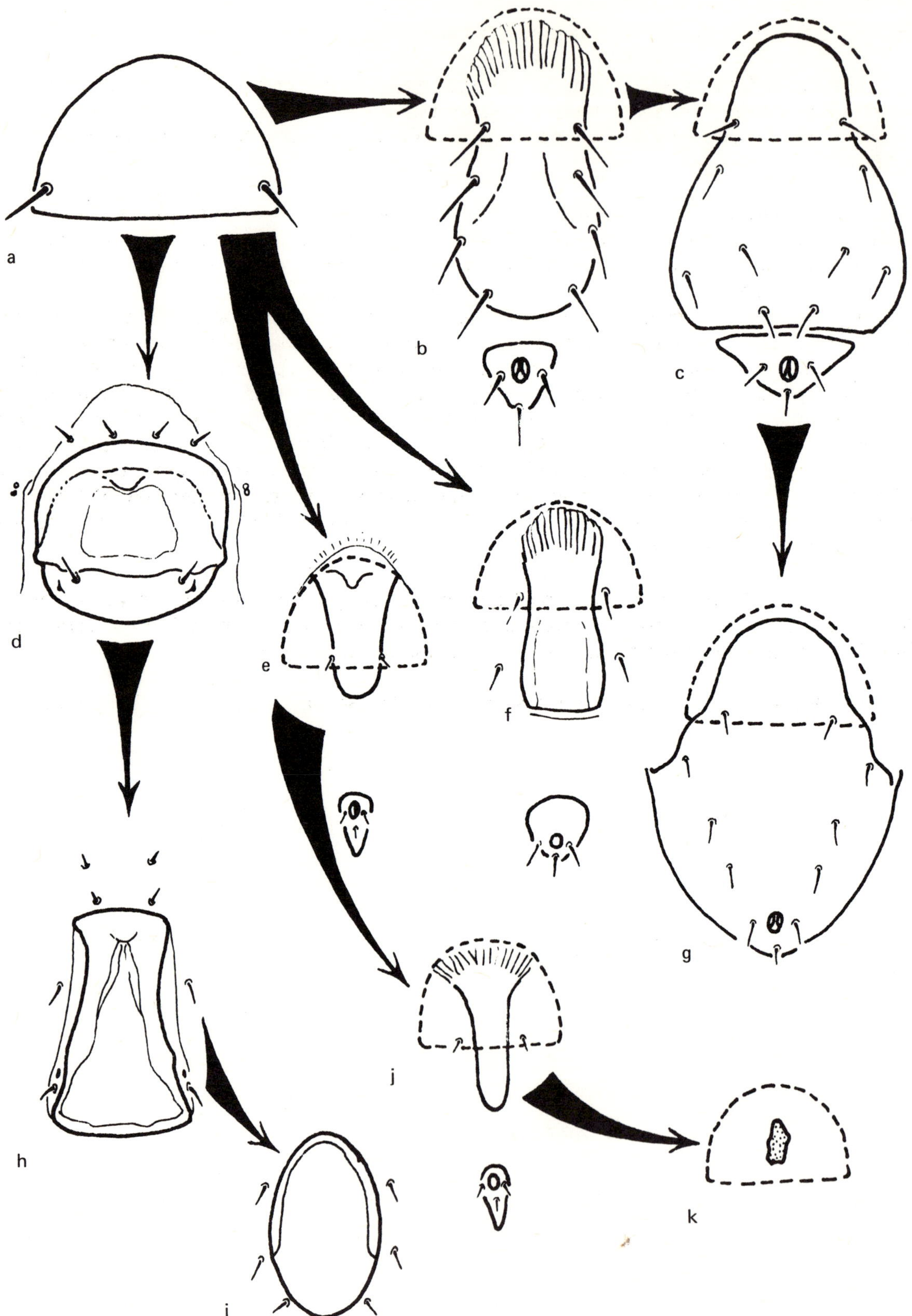

Fig. 28. Representative epigynial and genitiventral shield types in the supercohort Monogynaspides. Arrows indicate possible evolutionary trends in expansion or reduction of shield size from a simple truncate type (**a**). Figs. **b, c, e, g**; LAELAPIDAE: **d, h**; POLYASPIDIDAE: **f**; ASCIDAE: **i**; DIARTHROPHALLIDAE: **j**; MACRONYSSIDAE: **k**; RHINONYSSIDAE.

Cohort Sejina

Members of the Sejina are distinctive in that adults and deutonymphs have seven setae on femur IV, a feature shared in the Gamasida only by certain uropodine groups (Evans 1972). Other distinguishing characters are listed in the following diagnosis of the single representative superfamily, the Sejoidea.

Superfamily Sejoidea
(Plates 11-12, pp. 169-170)

DIAGNOSIS: Female with a single epigynial shield on which one, three, or several pairs of setae are inserted; sternal shield divided or fragmented behind sternal setae II; with one or several dorsal shields plus marginal platelets. Tarsus I with or without claws, palptarsal claw 2-tined. Male genital aperture within genital shield, male chelicerae unmodified for sperm transfer.

The Sejoidea includes three recognized families, of which the SEJIDAE (Plate 12, p. 170) is the best known. Sejids are common in forest litter and humus in the tropics, but some are found also in temperate North America and Europe. One species, *Zuluacarus termitophilus* Trägårdh, is a scavenger in termite nests. A sejid was discovered on a rat in Puerto Rico (Fox 1947) although an ectoparasitic relationship was not established. Only three species of MICROGYNIIDAE (Plate 11, p. 169) are described, one from the northwestern United States and the other two from Scandinavia. An undescribed species has recently been recorded from litter in central Oregon, USA. Like the sejids, microgyniids occur in humus and litter, and under the bark of rotting tree stumps. Their feeding habits are unknown. Sejoid mouthparts do not appear to be adapted for predation.

The ICHTHYOSTOMATOGASTERIDAE (= Uropodellidae Camin 1955) (Plate 11, p. 169) are forest litter inhabitants which are thought to feed on fungi or particulate organic debris. Most ichthyostomatogasterids have been collected in northern Chile where the family is thought to have had its origins (Athias-Henriot 1972). Species of *Asternolaelaps* are known also from Argentina, Uruguay, Europe and Australia. *Uropodella* species have been found in Brazil, Argentina, Paraguay and the United States. Radovsky (1969) mentions the occurrence of an undescribed species of *Asternolaelaps* on a rodent, *Ctenomys* sp., in Uruguay and a *Uropodella* species on a mouse, *Peromyscus* sp. He notes that sejines have been collected on rodents and bats with sufficient regularity to suggest phoretic or nidicolous associations between certain of these mites and small vertebrates.

Useful References

Athias-Henriot, C. (1960). Contribution aux Mesostigmates d'Algerie (Parasitiformes:Liroaspidae, Veigaiidae). Acarologia **2**(2):159-174. [SEJIDAE]

Athias-Henriot, C. (1972). Gamasides Chiliens (Arachnides). II. Revision de la famille Ichthyostomatogasteridae Sellnick, 1953 (= Uropodellidae Camin 1955). Arq. Zool., São Paulo **22**(3):113-191.

Balogh, J. (1938). Systematische Studien über eine neue Milbengattung: *Willmannia* gen. nov. Zool. Anz. **123**:259-265. [SEJIDAE]

Camin, J.H. (1955). Uropodellidae, a new family of mesostigmatid mites based on *Uropodella laciniata* Berlese, 1888 (Acarina:Liroaspina). Bull. Chicago Acad. Sci. **10**(5):65-81. [ICHTHYOSTOMATOGASTERIDAE]

Camin, J.H. and F.E. Gorirossi (1955). A revision of the suborder Mesostigmata based on new interpretations of comparative morphological data. Chicago Acad. Sci. Spec. Publ. 11:70 pp.

Evans, G.O. (1972). Leg chaetotaxy and the classification of the Uropodina (Acari:Mesostigmata). J. Zool., London **167**:193-206.

Fox, I. (1947). Seven new mites from rats in Puerto Rico. Ann. Ent. Soc. Amer. **40**:598-603. [SEJIDAE]

Krantz, G.W. (1961). A re-evaluation of the Microgynioidea, with a description of a new species of *Microgynium* (Acarina:Mesostigmata). Acarologia **3**(1):1-10.

Radovsky, F.J. (1969). Adaptive radiation in the parasitic Mesostigmata. Acarologia **11**(3):450-483.

Sellnick, M. (1953). *Ichthyostomatogaster nyleni,* eine neue Acaride aus Schweden. Ent. Tidsk. **74**(1-2):24-37. [ICHTHYOSTOMATOGASTERIDAE]

Trägårdh, I. (1942). Microgyniina, a new group of Mesostigmata. Ent. Tidsk. **63**(3-4):120-133. [MICROGYNIIDAE]

Cohort Gamasina

The Gamasina is a diverse group of seven superfamilies which share the character of six setae on femur IV in deutonymphs and adults. Gamasines have a single epigynial shield (Fig. 27, p. 120) with no more than one pair of epigynial setae in the intercoxal region. These setae may be inserted in the marginal integument bordering the shield or, occasionally, they may be absent. The *sternal shield* (Fig. 27) usually is entire and may be fused with the metasternal elements. The dorsum generally is covered by one or two shields, although shield fragmentation or abbreviation may occur. Claws may or may not be present on tarsus I. The male genital aperture is either at the anterior margin of the sternal shield or within it. A spermadactyl may be present on the internal or ventral face of the movable cheliceral digit.

Superfamily Parasitoidea[1]
(Plates 13-15, pp. 171-173)

DIAGNOSIS: *Sternal shield of female with three pairs of setae, metasternal setae on separate shields or free in the integument, with a ventrianal or anal shield. Tibia I with two pv setae* ($2\,\frac{3}{2}\,\frac{3}{2}\,2$)*, chaetotaxy of genu IV is* $2\,\frac{2}{1}\,\frac{3}{1}\,1$*, tibia III with eight or nine setae* ($2\,\frac{1}{1}\,\frac{2}{1}\,1(2)$)*, palptarsal claw 3-tined. Male genital aperture anterior or medial; when present, male spermadactyl ventrally situated on the movable cheliceral digit, free throughout its length or coalesced with digit; females usually tocospermous (podospermous in VEIGAIIDAE). Larva without pygidial shield; protonymphal pygidial shield laterally reduced, generally with 4-5 pairs of setae (without setae* S_4-S_5*), shield sometimes absent; dorsal shield of deutonymph divided or deeply incised medially.*

The Parasitoidea includes three families of free-living predators found in soil, humus or moss, and in various types of organic debris which support microarthropod and nematode populations (Karg 1971a). The PARASITIDAE (Plate 13, p. 171) is a large cosmopolitan family which is common in manure and vegetable compost, in moss, and in animal nests (Bhattacharyya 1963, Micherdzinski 1969). Kinn (1971) cites records of a *Eugamasus* species feeding on nematodes and mites in bark beetle galleries, and of species of *Parasitus* preying on the bark beetles themselves. *P. coleoptratorum* (L.) is a common predator of housefly larvae in manure, and *Eugamasus butleri* Hughes is an important predator in stored products. *Pergamasus crassipes* (L.) often occurs in or adjacent to the littoral zone in rocky coastal areas of England (Evans et al. 1961).

Parasitid deutonymphs often are phoretic on insects (Rapp 1959), utilizing them for transport from one feeding locus to another. Deutonymphs of *Parasitus fucorum* (De Geer) and *P. bombophilus* Vitzthum are phoretic on bumblebees, while the other life stages are found free in the bee nest (Karg 1971a). Other species of *Parasitus* are common phoretics of carabid and silphid beetles. Deutonymphs of *Pergamasus* frequently are encountered in mammal nests (Evans et al. 1961) where they prey on other nidicolous arthropods.

---

[1]Includes the Parasitinae plus Veigaiainae (EUGAMASIDAE pars) of Karg (1971a). The ARCTACARIDAE is included in the Parasitoidea, although its affinities are not entirely clear.

The VEIGAIIDAE (Plate 14, p. 172) favor organic humus and litter habitats and are especially plentiful in forest soils (Metz and Farrier 1969, Karg 1971a). *Cyrthydrolaelaps* species are found intertidally, preying on the microfauna of beach wrack.

Arctacarus rostratus Evans is a free-living representative of the seldom seen family ARCTACARIDAE (Evans 1955c). An undescribed species occurs in forest litter in western Oregon (Plate 15, p. 173).

Superfamily Rhodacaroidea[1]
(Plates 16-18, pp. 174-176)

DIAGNOSIS: *Sternal shield generally with four pairs of setae (metasternal and sternal elements fused), with a ventrianal shield posteriorly; often with two pairs of scleritic nodules (scleronoduli) on the propodosoma. Tibia I with five or six dorsal and one or two pv setae* ($2\,\frac{3}{2}\,\frac{2(3)}{1}\,2$ *or* $2\,\frac{3}{2}\,\frac{3}{2}\,2$), *genu IV often lacking one al and with one, or no, pv seta* ($1\,\frac{2}{1}\,\frac{2}{0}\,1$ *or* $2\,\frac{2}{1}\,\frac{3}{1}\,1$); *tibia III with seven or eight setae* ($1\,\frac{1}{1}\,\frac{2}{1}\,1$ *or* $2\,\frac{1}{1}\,\frac{2}{1}\,1$). *Palptarsal claw 2- or 3-tined. Male genital aperture anterior; spermadactyl fixed at the base of the movable digit, free or fused distally to the digit. Female podospermous; sperm induction pores on coxae, trochanters or femora III, or on the endopodal, metapodal or (rarely) the sternal shield; with a "type A" sacculus. Larva with or without pygidial shield; protonymphal pygidial shield typically with 7-8 pairs of setae, setae* S_4-S_5 *may be inserted marginally on the shield or in the adjacent integument; dorsal shield of deutonymph divided.*

Rhodacaroids are primarily free-living soil forms which are especially common in the northern and southern temperate and subantarctic zones (Hirschmann 1960, Hunter 1970, Lee 1974, Lee and Hunter 1974). The RHODACARIDAE (Plate 16, p. 174) includes small, delicate species frequently encountered in deeper soil layers. Rhodacarids prey on nematodes and, probably, on collembolans. Species of *Rhodacarus* and *Rhodacarellus* often are found in soil core samples in both the Old and New World (Evans et al. 1961, Price 1973).

Members of the family DIGAMASELLIDAE (Plate 17, p. 175) are widespread in surface and subsurface soils and in organic compost where they feed on collembolans, nematodes, arthropod eggs, and possibly on fungi (Lindquist (1975). *Digamasellus punctum* (Berlese), one of the two described species of the type genus, is phoretic on geotrupine dung beetles. Although most of the approximately 100 species of *Dendrolaelaps* are soil or litter forms (Coineau 1974, Lindquist op. cit.), many are found in the galleries of bark beetles (Hirschmann 1960, McGraw and Farrier 1969, Kinn 1971). *Dendrolaelaps* species feed readily on nematodes in the beetle galleries, although *D. quadrisetus* (Berlese) is thought to feed also on fungal and fecal material in the subcortical habitat. Deutonymphs of the nematophage *D. fallax* (Leitner) are phoretic on sciarid flies (Binns 1973). While other recognized digamasellid genera occupy diverse habitats, *Longoseius* species are exclusively subcortical forms found in association with bark beetles (Lindquist, op. cit.). Their feeding habits are unknown.

[1]Comprises the RHODACARIDAE and the Gamasellinae (EUGAMASIDAE pars) of Karg (1971a). The family RHODACARIDAE is considered here in the strict sense of Johnston (1968) (Rhodacarinae of Lee (1970)). The OLOGAMASIDAE includes all of Lee's non-rhodacarine rhodacarids (the Gamasellinae of Hirschmann (1962) and of Karg (1971a)).

The OLOGAMASIDAE (Plate 18, p. 176) is a large, widely distributed group of predators which are encountered in soil, humus and compost. *Cyrtolaelaps* and *Euryparasitus* species commonly occur in nests of small mammals (Karg 1971a), while *Hydrogamasus, Psammonsella* and certain other ologamasid genera have adapted to intertidal marine habitats (Haq 1965, Lee 1970). *H. antarcticus* Trägårdh has been collected from tussock grass and from wet moss in Antarctica (Hunter 1967).

Prey specificity has been observed in a number of ologamasid species kept in laboratory culture. Lee (1974) noted that *Gamasellus concinnus, Euepicrius filamentosus* and *Athiasella dentata* fed preferentially on collembolans, while *Gamasiphis fornicatus, Acugamasus semipunctatus* and *Gamasiphoides propinquus* favored mites as prey.

Superfamily Ascoidea[1]
(Plates 19-20, pp. 177-178)

DIAGNOSIS: *Sternal shield of female with three pairs of setae, metasternal setae on separate shields or free in the integument, with a ventrianal or anal shield. Chaetotaxy of tibia I and genu IV highly variable (Evans 1963a, Lindquist and Evans 1965); tibia III with eight or nine setae. Palptarsal claw 2-tined. Male genital aperture generally anterior (medial in ZERCONIDAE); spermadactyl, when present, free terminally. Female podospermous or tocospermous. Larva with or without pygidial shield; protonymphal pygidial shield present, with 7-12 pairs of setae (setae S_4-S_5 on shield); dorsal shield of deutonymph entire or divided.*

Ascoids are widely distributed both in tropical and temperate habitats and feed on nematodes, insects, and on mites in soil or humus in both damp and dry niches. Some are thought to feed on fungi at least incidentally (Muma 1975). The ZERCONIDAE (Plate 19, p. 177) comprises a large assemblage of well-sclerotized slow-moving species which are routinely recovered from litter, humus and moss samples in a variety of habitats. Blaszak (1975) recognizes ten zerconid genera based on relative peritremal length. Members of the four genera of HALOLAELAPIDAE (Plate 19, p. 177) favor wet organic substrates, dung, and coastal-intertidal habitats (Karg 1969, 1971a). Several species of *Halolaelaps* and *Halodarcia* have been found on marine algae and beach wrack in the intertidal zone. *Halolaelaps octoclavatus* (Vitzthum) is recorded from an histerid beetle, although phoretic relationships with insects must be considered exceptional for this family.

The ASCIDAE (Plate 20, p. 178) is a large and diverse group which has adapted to a broad spectrum of terrestrial and subaquatic habits. Some are more or less restricted to coastal and intertidal regions, and others inhabit vertebrate nests. Certain species are common predators of mites and insects in stored products. For example, *Blattisocius tarsalis* (Berlese) feeds on immature stadia of acarid mites and of grain moths and beetles in food storages (Lindquist and Evans 1965). The same species, along with others of the genus *Blattisocius,* attach phoretically to moths and feed on their eggs and larvae (Treat 1975).

---

[1]Sensu Karg (1971a), but with the following changes:

a. The ASCIDAE as considered in this treatment includes Karg's subfamily Blattisocinae (PHYTOSEIIDAE), the genera *Neojordensia, Aceoseius* (=*Lasioseius* pars of Lindquist and Evans 1965), and *Lasioseius* of Karg's family PODOCINIDAE.

b. The genus *Antennoseius* is here considered a taxon in the ASCIDAE rather than in the HALOLAELAPIDAE.

While most ascids are predatory, a few species appear to have adapted to polleniphagy, or even to parasitism. Species of the genus *Rhinoseius* are phoretic on hummingbirds, utilizing them for transportation from flower to flower where the mites feed either on other arthropods or on pollen (Baker and Yunker 1964, Dusbábek and Cerný 1970). A similar relationship may exist between the ascid *Proctolaelaps vandenbergi* (Ryke) and its honey bird carrier (Lindquist and Evans 1965). *P. bombophilus* (Westerboer) and *P. longisetosus* (Westerboer) probably feed on pollen in the bumblebee nests where they are found. *P. nauphoetae* (Womersley) feeds on the blood of its cockroach host, and thus may be considered a parasite (Egan and Hunter 1975).

A casual relationship exists between lepidopterans and species of *Proctolaelaps* and *Lasioseius.* The mites utilize the well-protected interpalpal region of the moth for shelter during phoretic transfer (Treat 1975). Other ascids are found in the spiracular atria of orthopterans, coleopterans and polydesmoid millipedes, although parasitic relationships have not been demonstrated (Lindquist 1962, Chant 1963).

Superfamily Phytoseioidea
(Plates 21-25, pp. 179-183)

DIAGNOSIS: Sternal shield of female with 0-3 pairs of setae, sternals III often on separate platelets, with a ventrianal or anal shield. Tibia I with four or five dorsal setae ($2\ \frac{2}{1}\ \frac{2}{1}\ 2$ *or* $2\ \frac{3}{2}\ \frac{2}{1}\ 2$), *genu IV with one or two al setae, pv setae absent; tibia III with seven or eight setae (rarely nine). Palptarsal claw usually 2-tined. Male genital aperture generally anterior (medial in EPICRIIDAE); spermadactyl, when present, free distally. Female podospermous or tocospermous, podospermous forms typically with a "type B" sacculus. Larva without pygidial shield; pygidial shield of protonymph generally broad, with fewer than seven pairs of setae* (S_4-S_5 *on shield), sometimes reduced (OTOPHEIDOMENIDAE) or absent (some Ameroseius); dorsal shield of deutonymph generally entire, often incised laterally.*

The Phytoseioidea includes predatory, parasitic and polleniphagous species found in both ground and aerial habitats. Members of the family PHYTOSEIIDAE (Plate 21, p. 179) often occur on vegetable crops, perennial shrubs and trees where they prey on phytophagous mites. Several phytoseiid species are considered to be important predators of spider mites (McMurtry et al. 1970), and have been used successfully in integrated control programs (Hoyt 1969, Westigard 1971, Hoyt and Burts 1974). Jeppson et al. (1975) list 28 phytoseiid species which prey on spider mites (TETRANYCHIDAE) on economic crops. Some of these predators have specific preferences for certain tetranychid species. Preferences often are based on the webbing habits of the prospective prey, or on the location of the spider mites on the plant substrate. *Phytoseiulus persimilis* Athias-Henriot, an important greenhouse predator, is solely dependent on spider mites as a food source. Some members of the genera *Typhlodromus* and *Amblyseius,* however, may feed alternately on eriophyoid or tenuipalpid mites, or on small insects and their eggs. Some phytoseiids apparently prefer eriophyoids as a primary food source.

Phytoseiid species may supplement their arthropod diets with pollen, fungi, honeydew, or plant nectar. *Typhlodromus pyri* Scheuten, T. *rhenanus* (Oudemans) and *Amblyseius finlandicus* (Oud.) are known to extract plant juices from hosts on which prey species are scarce (Jeppson et al. 1975). Treat (1975) records *Amblyseius potentillae* (Garman) and two species of *Typhlodromus* as moth associates.

The family OTOPHEIDOMENIDAE[1] (Plate 22, p. 180) is represented by a distinctive group of parasitic mites found in the thoraco-abdominal and wing axillary areas of noctuid and sphingid moths, and in the tegular region of hemipterans, in both the Old and New World (Treat 1975). Eggs of *Otopheidomenis* and *Noctuiseius* generally are oviposited bilaterally in the tympanic recesses of the moth hosts, but feeding appears to occur in the tegular region. Aside from a single observation of a live female of *Otopheidomenis zalelestes* Treat, all records of otopheidomenid-moth associations are based on preserved material (Treat op. cit.). Species of *Hemipteroseius* and *Treatia* may be found in all stages under the hemielytra of their hemipteran hosts (Evans 1963c).

Mites of the family AMEROSEIIDAE (Plate 23, p. 181) occupy a wide range of terrestrial substrates including moss, rotting straw and compost, forest humus and manure (Karg 1971a). *Kleemannia plumosa* (Oudemans) is found in stored products and in nests of small mammals, ants, and bees. Other ameroseiids have established phoretic associations with insects. *K. plumigera* Oudemans, collected in Britain from grain siftings, may be the same species described by Treat (1975) from the tympanic membranes of a pyralid moth. Members of the ameroseiid genus *Neocypholaelaps* occur on butterflies and bees in the Old World tropics, apparently feeding on pollen of flowers visited by their insect carriers. The mites have also been collected from flowers normally favored by the insects (Evans 1963b, Treat 1975). Adult *N. indica* (Evans) may attach to the petite workers of *Apis indica*, the Indian honey bee, in great numbers. Kshirsagar (1968) found an average of 47 mites on each bee, corresponding to approximately ½ the body weight of the bee carrier.

Most of the described species of *Ameroseius* inhabit terrestrial substrates but some have evolved what appears to be a phoretic association with birds. *A. wilsoni* Allred was taken from the nasal cavities of a psittacid bird in New Guinea, while *A. senarius* Allred was recovered from a columbiform bird (Allred 1970).

Many ameroseiids may feed preferentially on fungi (Rack 1963) or pollen. Fungivores, however, may also be predaceous (Karg 1971a).

The free-living members of the family PODOCINIDAE (Plate 24, p. 182) inhabit humus and litter habitats throughout the tropics, with the possible exception of Australia (Evans and Hyatt 1957). *Podocinum pacificum* Berlese, however, is known from Europe, North America, central Asia and Japan (Evans and Hyatt op. cit., Ishikawa 1970). *P. sagax* (Berlese), a widely distributed tropical species, has been collected in the Boboli Gardens of Florence, Italy, where it may have been accidentally introduced with ornamental plants. The EPICRIIDAE (Plate 25, p. 183) are found in moss, humus, and in the decaying bark of trees (Evans 1957, Athias-Henriot 1961, Karg 1971b). Their feeding habits have not been studied.

Superfamily Eviphidoidea
(Plates 26-28, pp. 184-186)

DIAGNOSIS: *Sternal shield of female with three pairs of setae, metasternal setae on shields which are free or amalgamated with adjacent endopodal elements, or occasionally free in integument; with a ventrianal or anal shield. Tibia I with one pv seta and one or two al setae* ($1\ \frac{3}{2}\ \frac{2}{1}\ 2$ *or* $2\ \frac{3}{2}\ \frac{2}{1}\ 2$), *tibia III with seven setae* ($1\ \frac{1}{1}\ \frac{2}{1}\ 1$); *genu IV usually with seven setae* ($1\ \frac{2}{1}\ \frac{2}{0}\ 1$), *pl seta occasionally absent* ($1\ \frac{2}{1}\ \frac{2}{0}\ 0$). *Palptarsal claw 2- or 3-tined. Male genital aperture anterior,*

---

[1] Otopheidomeninae, family PHYTOSEIIDAE, of Chant (1965).

spermadactyl often arising medially on the movable digit on the dorsal or ventral aspect, free distally. Female podospermous, with a "type A" sacculus. Larva with or without pygidial shield; pygidial shield of protonymph broad, with 10-11 pairs of setae (S_4-S_5 inserted on shield); dorsal shield of deutonymph divided or entire.

While they are basically humus and litter inhabitants, many eviphidoids have adapted to transient habitats such as dung, carrion, or open coast beach wrack, and have developed phoretic associations with insects or other invertebrates as a means of escape from deteriorating substrates. The PARHOLASPIDIDAE (Plate 26, p. 184) is exceptional in that none of the species which make up the family are known to be phoretic.

The genera *Eviphis* and *Evimirus* of the family EVIPHIDIDAE (Plate 26, p. 184) include some free-living, non-phoretic forms encountered in humus, moss, compost, and other relatively stable terrestrial substrates. Representatives of the remaining six genera are predators in temporary habitats and are phoretic on arthropods which exploit these same habitats (Evans 1969, Potter and Johnston 1976). Deutonymphs and/or adults of *Copriphis, Alliphis,* and *Scarabaspis* often are found on scarabaeid dung beetles, while amphipods serve as phoretic carriers for members of the genus *Thinoseius* in intertidal habitats (Karg 1963 and 1971a, Evans 1969).

Like the EVIPHIDIDAE, certain genera of the family MACROCHELIDAE (Plate 27, p. 185) are totally independent of phoretic associations with arthropods. *Geholaspis* and *Macrholaspis* species are free-living predators commonly observed in forest litter and humus habitats shared by parholaspidid mites. Members of the genera *Holocelaeno, Macrocheles* and *Neopodocinum* exhibit varying degrees of phoretic specificity to their scarabaeid beetle or muscid fly carriers (Costa 1967, Jalil and Rodriguez 1970, Krantz and Mellott 1972). Generally speaking, only female macrochelids are phoretic, attaching to the gular or coxal regions of the host insect. Silkworm larvae are reported to carry females of *Macrocheles bombycis* Kishida and *M. muscaedomesticae* (Scopoli), a unbiquitous inhabitant of dung which is commonly phoretic on flies (Treat 1975). Various species of *Macrocheles* have also been found in the nests of birds, mammals, bees or ants, in the fur of mammals, and occasionally on birds (Evans and Browning 1956, Bregetova and Koroleva 1960, Karg 1971a). Macrochelids are predaceous on nematodes, fly larvae, and other microfauna occurring in a variety of organic substrates.

Members of the PACHYLAELAPIDAE (Plate 28, p. 186) share the habits of the MACROCHELIDAE, preying on microfauna in organic substrates and often moving from place to place by means of phoretic attachment to arthropods, primarily scarabaeid beetles (Evans and Hyatt 1956, Hirschmann and Krauss 1965, Karg 1971a). Like the macrochelids, great numbers of pachylaelapid species have been collected in the Old and New World tropics, with many as yet undescribed. *Pachylaelaps* is also well represented in the temperate regions of Europe and North America.

Superfamily Heterozerconoidea
(Plate 28, p. 186)

DIAGNOSIS: Sternal shield absent, with lateral endopodal fragments which may carry some sternal setae, sternal pores II-III absent; dorsal shield entire, well sclerotized. Tibia I with four dorsal setae, with one or two pl setae ($2\ \frac{2}{1}\ \frac{2}{1}\ 2$, $2\ \frac{2}{1}\ \frac{2}{1}\ 1$, *or* $2\ \frac{2}{2}\ \frac{2}{1}\ 1$), *tibia III with eight or nine setae* ($2\ \frac{1}{1}\ \frac{2}{1}\ 2$ or $2\ \frac{1}{1}\ \frac{2}{1}\ 1$); *genu IV with one pv seta and with one, or no, pl seta* ($2\,\frac{2}{1}\,\frac{3}{1}\,1$, $2\,\frac{2}{1}\,\frac{2}{1}\,1$, *or* $2\,\frac{2}{1}\,\frac{2}{1}\,0$)*; femora III-IV without pl seta. Palptarsal claw 2-tined. With a pair of large discoid adhesive organs ventrolaterally on the idiosoma, secondarily lost in some species. Male spermadactyl often greatly elongated and convoluted, arising either from the movable or the fixed cheliceral digit.*

The HETEROZERCONIDAE (Plate 28-3 to 5) comprise a bizarre group of large, poorly known acarines which are commensals or parasites of millipedes and, rarely, of snakes. Scolopendrine centipedes are hosts for the few described species of DISCOZERCONIDAE (Plate 28-2). Members of both families utilize large paired adhesive organs to secure themselves to their hosts (Berlese 1910, Finnegan 1931, Baker and Wharton 1952). A smooth, heavily sclerotized domed dorsal shield offers maximum protection from injury while on the exposed attachment sites of their diplopod and chilopod hosts.

While the HETEROZERCONIDAE and DISCOZERCONIDAE have been relegated here to a single superfamily, uncertainties exist as to the systematic relationships between the two assemblages. Furthermore, the presence of a spermadactyl on the fixed cheliceral digit of male heterozerconids rather than on the movable digit as in other podospermous Gamasina, raises the question of whether the HETEROZERCONIDAE should be given separate cohortal rank. Dr. John B. Kethley of the Field Museum, Chicago, currently is reviewing the status of these remarkable acarines, basing his studies on recently acquired material representing over 30 new genera and 100 new species. Pending his findings, the two families are treated here as members of a single superfamily, the Heterozerconoidea, of the Cohort Gamasina.

Superfamily Dermanyssoidea[1]
(Plates 29-37, pp. 187-195)

DIAGNOSIS: Sternal shield of female normally developed, reduced or absent, sternal setae or setal remnants generally present; metasternal setae present or absent. Dorsal shield entire, divided, or variously fragmented, sometimes greatly reduced. Chaetotaxy of legs highly variable except in free-living forms; tibia I typically with six dorsal setae (four or five in certain obligate parasites); genu IV nearly always with two al setae, and with one, or no, pv setae; palptarsal claw 2- (rarely 3-) tined. Male genital aperture anterior; spermadactyl arises externally on the movable digit, free distally. Female podospermous. Larva without pygidial shield; protonymphal pygidial shield broad, narrowed, or represented by a series of mesonotal scutella, entire shields with 4-8 pairs of setae (S_4*-*S_5

---

[1] A number of attempts have been made in past years to arrive at a satisfactory higher classification for this large and complex group of parasitic gamasids. Most of these schemes have been based on works by Berlese (1892, 1913), Vitzthum (1943), Baker and Wharton (1952) and Evans (1957). Evans and Till (1966) included Vitzthum's DERMANYSSIDAE and LAELAPIDAE in a single family, and added four additional subfamilial categories to the assemblage. The result was an extremely broad taxon including free-living forms, nidicoles, vertebrate and invertebrate paraphages, facultative and obligatory ectoparasites, and endoparasites grouped in 15 subfamilies. Based on family name priorities amongst those groups included, Evans and Till applied Kolenati's (1859) family name DERMANYSSIDAE, relegating the LAELAPIDAE to subfamily status.

Radovsky (1967) conducted a critical morphological and behavioral review of dermanyssids ectoparasitic on bats, and concluded that Evans and Till's subfamilies Macronyssinae and Dermanyssinae were valid familial entities. Furthermore, Radovsky (1969) chose to use the family name LAELAPIDAE in preference to DERMANYSSIDAE *s. lat.*, based on the long established usage of the former name. In the present treatment, the family DERMANYSSIDAE is considered in the narrow sense; i.e., in the sense of Evans and Till's Dermanyssinae. The family LAELAPIDAE is retained essentially as envisioned by Radovsky. Rules of priority are followed in assigning the superfamilial name Dermanyssoidea to the 15 families included in this assemblage.

either inserted on the shield or in the adjacent integument where the shield is reduced); dorsal shield of deutonymph entire and generally incised, or retaining a separate small pygidial element posteriorly.

The Dermanyssoidea is a large and diverse assemblage of free-living and parasitic forms which, despite their morphological and behavioral heterogeneity, appear to be derivative groups of a basic free-living stock exemplified by the subfamily Hypoaspidinae (LAELAPIDAE) (Radovsky 1969).

The LAELAPIDAE (Plates 36-37, pp. 194-195) is here considered to be comprised of nine subfamilies which are free-living or parasitic. They include forms which show extensive sclerotization and hypertrichy of the opisthosoma (e.g., *Ololaelaps* or *Haemogamasus*), hypotrichy of the dorsal shield *(Ondatralaelaps, Oryctolaelaps),* or reduction of dorsal sclerotization *and* setae *(Hirstionyssus).* Hypoaspidine laelapids often are collected in litter or soil substrates, but species of *Hypoaspis, Ololaelaps, Gaeolaelaps, Laelaspis, Gymnolaelaps, Ayersacarus,* and many other hypoaspidine genera are routinely encountered in the nests of mammals or arthropods, or on insects (Hunter 1964, Evans and Till 1966, Karg 1971a). Certain species of *Hypoaspis* are found under the elytra of a variety of beetles, while *Coleolaelaps* species are confined to the subelytral spaces of melolonthine scarabs of the genera *Anoxia* and *Polyphylla* (Costa and Hunter 1970). Similarly, members of the hypoaspidine genus *Pneumolaelaps* are restricted to the nests and bodies of bumble bees (Evans and Till 1966, Hunter and Husband 1973). The Haemogamasinae are facultative parasites of rodents and (occasionally) of birds, and share the nidicolous habits of many hypoaspidines. However, most haemogamasines are polyphagous, feeding on dead arthropods, nematodes, or vegetable matter. Some also penetrate intact skin of young nestlings and feed (*Haemogamasus reidi* Ewing is an example). Other species of *Haemogamasus,* along with members of the haemogamasine genus *Brevisterna,* are obligatory haemotophages (Evans and Till 1966).

The Laelapinae includes over 15 genera which are facultative parasites of mammals (Strandtmann and Wharton 1938, Radovsky 1967). Species of *Androlaelaps,* a comparatively unspecialized laelapine genus, often are nidicolous or free-living in organic debris. Nidicolous forms may penetrate intact skin of young rodents or birds and feed on blood. *Laelaps* species appear to feed on secretions of their mammal hosts, (a trait shared by the Myonyssinae), and occasionally imbibe blood. Many laelapine genera favor hosts with common habitats. For example, a number of *Laelaps* species appear to prefer myomorph rodents and, moreover, myomorphs which occur in moist habitats. *Mysolaelaps* and other laelapine genera show a similar preference for moist situations. On the other hand, *Steptolaelaps* is found on hosts which favor arid regions in the Nearctic and Neotropical realms (heteromyid rodents). Members of the Hirstionyssinae appear to be obligatory haemotophages (Herrin 1970). Other laelapine subfamilies include arthropod paraphages (Iphiopsinae and Melittiphinae) (Evans 1955a, Evans and Till 1966, Delfinado and Baker 1974) and a peculiar parasite of the mountain beaver, *Aplodontia rufa* (Alphalaelapinae) (Tipton 1960).

Most of the remaining 14 dermanyssoid families comprise vertebrate parasites, and many are specific to particular animal groups. For example, the families SPINTURNICIDAE, SPELAEORHYNCHIDAE[1] and, to a large extent, the MACRONYSSIDAE (Plates 29, 33, 34, pp. 187, 191, 192), are found on bats (Rudnick 1960, Fain et al. 1967, Radovsky 1967, Hoffman and Barrera 1970, Herrin and Tipton 1975), the HALARACHNIDAE,

[1]Radovsky (1969) notes that these peculiar mites illustrate a stronger similarity to the Sejina than to the Gamasina. The overly heavy chelicerae, ventrally placed camerostome, spinose denticulations, slender legs with weak pretarsi, and several other features suggest to Radovsky a possible phylogenetic relationship with sejines of the family ICHTHYOSTOMATOGASTERIDAE (page 123).

DASYPONYSSIDAE and MANITHERIONYSSIDAE (Plates 30, 32, pp. 188, 190) are confined to mammals (Newell 1947, Radovsky and Yunker 1971, Yunker 1973), the RHINONYSSIDAE (Plate 31, p. 189) to birds (Pence 1975), and the ENTONYSSIDAE, OMENTOLAELAPIDAE, and IXODORHYNCHIDAE (Plates 29, 30, 35, pp. 187-8, 193) to reptiles (Fain 1961a and b, Fain 1962). Members of the families DERMANYSSIDAE and MACRONYSSIDAE (Plates 33 and 34) parasitize mammals or birds, but some macronyssids also infest reptiles (Evans and Till 1966, Saunders 1975). *Hystrichonyssus turneri* Keegan, Yunker and Baker (HYSTRICHONYSSIDAE, Plate 34, p. 192) is an ectoparasite of a Malayan porcupine and a tree snake. Dermanyssids, spinturnicids, hystrichonyssids, most macronyssids, and (probably) ixodorhynchids are obligate haematophages. The VARROIDAE (Plate 35, p. 193) parasitize honeybees, causing serious bee losses through feeding injury *(varrosis)* in European and Asian apiaries (Delfinado and Baker 1974, Akratanakul and Burgett 1975).

Primary injury to host animals by dermanyssoid parasites may be occasioned by the piercing of the host's skin (as occurs with *Dermanyssus* (DERMANYSSIDAE), *Haemogamasus* (LAELAPIDAE), *Raillietia* (HALARACHNIDAE) and *Ornithonyssus* (MACRONYSSIDAE), among others), and through the occurrence of allergic reactions in sensitive host animals (Yunker 1973, Southcott 1976). Secondary injury to broken skin may be caused by *Laelaps* species and other facultative laelapids. Macronyssid fowl mites *(Ornithonyssus)* and snake mites *(Ophionyssus)* often exsanguinate their hosts to the point of anemia or death (Baker et al. 1956, Yunker 1973). Respiratory parasites of the families RHINONYSSIDAE, ENTONYSSIDAE and HALARACHNIDAE may precipitate lung congestion, rhinitis or sinusitis. The common lung mite of monkeys, *Pneumonyssus simicola* Banks, causes few overt symptoms but the lung parenchyma of autopsied infested animals have characteristic spotty lesions (Yunker op. cit.) similar to those caused by tuberculosis.

Secondary host injury by dermanyssoid parasites also occurs as the result of transmission of disease organisms to host animals. Species of DERMANYSSIDAE, LAELAPIDAE and MACRONYSSIDAE are proven transmittors of diseases in birds, reptiles and mammals—including man.

Useful References

Akratanakul, P. and D.M. Burgett (1975). *Varroa jacobsoni:* a prospective pest of honeybees in many parts of the world. Bee World **56**(3):119-121. [VARROIDAE]

Allred, D.M. (1970). New ameroseiid mites from birds of New Guinea. J. Med. Ent. **7**(1):99-102.

Athias-Henriot, C. (1961). Mesostigmates (Urop. excl.) edaphiques Méditerranéens (Acaromorpha, Anactinotrichida). Prem. Sér. Acarologia **3**(4):381-509.

Athias-Henriot, C. (1968). Observations sur les *Pergamasus* IV. Un essai de coordination de la taxonomie et de la chorologie du sous-genre *Pergamasus* s.s. (Acariens Anactinotriches, Parasitidae). Acarologia **10**(2):181-190.

Athias-Henriot, C. (1972). Gamasides Chiliens (Arachnides). II. Revision de la famille Ichthyostomatogasteridae Sellnick, 1953 (= Uropodellidae Camin 1955). Arq. Zool., São Paulo **22**(3):113-191.

Baker, E.W., T.M. Evans, D.J. Gould, W.B. Hull and H.L. Keegan (1956). A Manual of Parasitic MItes of Medical or Economic Importance. Natl. Pest Control Assoc. Tech. Publ.: 170 pp.

Baker, E.W., J.H. Camin, F. Cunliffe, T.A. Woolley and C.E. Yunker (1958). Guide to the Families of Mites. Institute of Acarology Contr. No. 3:242 pp. + ix.

Baker, E.W. and G.W. Wharton (1952). An Introduction to Acarology. Macmillan Co., N.Y.: 465 pp. + xiii.

Baker, E.W. and C.E. Yunker (1964). New blattisociid mites (Acarina:Mesostigmata) recovered from neotropical flowers and hummingbird's nares. Ann. Ent. Soc. Amer. **57**(1):103-126. [ASCIDAE]

Berlese, A. (1892). Acari, Myriapoda et Scorpiones hucusque in Italia reperta. Ordo Mesostigmata (Gamasidae): 1-143.

Berlese, A. (1910). Brevi diagnosi di generi e specie nuovi di Acari. Redia **6**:346-388. [HETEROZERCONIDAE]

Berlese, A. (1913). Acarotheca Italica, Firenze (1,2):1-221.

Bhattacharyya, S.K. (1963). A revision of the genus *Pergamasus* Berlese s. lat. (Acari, Mesostigmata). Bull. Brit. Mus. (Nat. Hist.) Zool. **11**:133-242. [PARASITIDAE]

Binns, E.S. (1973). *Digamasellus fallax* Leitner (Mesostigmata:Digamasellidae) phoretic on mushroom sciarid flies. Acarologia **15**(1):10-17.

Blaszak, C. (1975). A revision of the family Zerconidae (Acari, Mesostigmata) (Systematic studies on family Zerconidae-I). Acarologia **17**(4):553-569.

Bregetova, N.G. (1961). The veigaiaid mites (Gamasoidea, Veigaiaidae) in the USSR. Parazit. sborn. Zool. Inst. Akad. Nauk SSSR **20**:10-107.

Bregetova, N.G. (1967). Ontogenesis in the gamasid mites as a criterion for the erection of their natural system. Akad Nauk SSSR, Parasit. **1**(6):465-479.

Bregetova, N.G. and E.V. Koroleva (1960). The Macrochelidae Vitzthum, 1930 in the USSR. Parazit. sborn. Zool. Inst. Akad. Nauk. SSSR **19**:32-154.

Chant, D.A. (1959). Phytoseiid mites (Acarina:Phytoseiidae). Part I: Bionomics of seven species in southeastern England. Part II. A taxonomic review of the family Phytoseiidae, with descriptions of 38 new species. Can. Ent. suppl. 12:166 pp.

Chant, D.A. (1963). The subfamily Blattisocinae Garman (= Aceosejinae Evans) (Acarina:Blattisociidae Garman) (=Aceosejidae Baker and Wharton) in North America, with descriptions of new species. Can. J. Zool. **41**:243-305. [ASCIDAE]

Chant, D.A. (1965). Generic concepts in the family Phytoseiidae (Acarina:Mesostigmata). Can. Ent. **97**(4): 351-374. [OTOPHEIDOMENIDAE]

Coineau, Y. (1974). Introduction a l'Étude des Microarthropodes du Sol et des ses Annexes. Documents Ecol. DOIN, éds.: 118 pp.

Costa, M. (1967). Notes on macrochelids associated with manure and coprid beetles in Israel. II. Three new species of the *Macrocheles pisentii* complex. Acarologia **9**(2):304-329.

Costa, M. and P.E. Hunter (1970). The genus *Coleolaelaps* Berlese, 1914 (Acarina:Mesostigmata). Redia **52**:323-360. [LAELAPIDAE]

Delfinado, M.D. and E.W. Baker (1974). A new record for the bee mite *Mellitiphis* in New Zealand. Bee World **55**(4):148-149. [LAELAPIDAE]

Domrow, R. (1965). New laelapid nasal mites from Australian birds. Acarologia **7**(3):430-460. [RHINONYSSIDAE]

Domrow, R. (1975). *Ljunghia* Oudemans (Acari:Dermanyssidae), a genus parasitic on mygalomorph spiders. Rec. S. Austral. Mus. **17**(4):31-39. [LAELAPIDAE]

Dove, W.E. and B. Shelmire (1932). Some observations on tropical rat mites and endemic typhus. J. Parasitol. **18**:159-168. [MACRONYSSIDAE]

Dusbábek, F. and V. Černý (1970). The nasal mites of Cuban birds. I. Ascidae, Ereynetidae, Trombiculidae (Acarina). Acarologia **12**(1):269-281.

Egan, M.E. and P.E. Hunter (1975). Redescription of a cockroach mite, *Proctolaelaps nouphoetae,* with notes on its biology. Ann. Ent. Soc. Amer. **68**(2):361-364. [ASCIDAE]

Evans, G.O. (1955a). A review of the laelaptid paraphages of the Myriapoda with descriptions of three new species (Acarina:Laelaptidae). Parasitol. **45**:352-368.

Evans, G.O. (1955b). A revision of the family Epicriidae (Acarina:Mesostigmata). Bull. Brit. Mus. (Nat. Hist.) Zool. **3**(4):171-200.

Evans, G.O. (1955c). A collection of mesostigmatid mites from Alaska. Bull. Brit. Mus. (Nat. Hist.) Zool. **2**:287-306. [ARCTACARIDAE]

Evans, G.O. (1957). An introduction to the British Mesostigmata (Acarina) with keys to families and genera. J. Linn. Soc.-Zool. **43**(291):203-259.

Evans, G.O. (1958). A revision of the British Aceosejinae (Acarina:Mesostigmata). Proc. Zool. Soc. London **131**(1):177-229. [ASCIDAE]

Evans, G.O. (1963a). Observations on the chaetotaxy of the legs in the free-living Gamasina (Acari:Mesostigmata). Bull. Brit. Mus. (Nat. Hist.) Zool. **10**(5):277-303.

Evans, G.O. (1963b). The genus *Neocypholaelaps* Vitzthum (Acari:Mesostigmata). Ann. Mag. Nat. Hist. 6, **13**:209-230. [AMEROSEIIDAE]

Evans, G.O. (1963c). Observations on the classification of the family Otopheidomenidae (Acari:Mesostigmata) with description of two new species. Ann. Mag. Nat. Hist. 5, **13**:609-620.

Evans, G.O. (1969). A new mite of the genus *Thinoseius* Halbt. (Gamasina:Eviphididae) from the Chatham Islands New Zealand. Acarologia **11**(3):505-514.

Evans, G.O. and E. Browning (1956). British mites of the subfamily Macrochelinae Trägårdh (Mesostigmata-Macrochelidae). Bull. Brit. Mus. (Nat. Hist.) Zool. **4**:1-54.

Evans, G.O. and K.H. Hyatt (1956). British mites of the genus *Pachylaelaps* Berlese (Gamasina-Pachylaelaptidae). Ent. monthly Mag. **92**:118-129.

Evans, G.O. and K.H. Hyatt (1958). The genera *Podocinum* Berl. and *Podocinella* gen. nov. (Acarina:Mesostigmata). Ann. Mag. Nat. Hist. 12, **10**:913-932. [PODOCINIDAE]

Evans, G.O. and K.H. Hyatt (1960). A revision of the Platyseiinae (Mesostigmata:Aceosejidae). Bull. Brit. Mus. (Nat. Hist.) Zool. **6**(2):25-101. [ASCIDAE]

Evans, G.O. and K.H. Hyatt (1963). Mites of the genus *Macrocheles* Latr. (Mesostigmata) associated with coprid beetles in the collections of the British Museum (Natural History). Bull. Brit. Mus. (Nat. Hist.) Zool. **9**(9):327-401. [MACROCHELIDAE]

Evans, G.O., J.G. Sheals and D. Macfarlane (1961). The Terrestrial Acari of the British Isles. Vol. I. Introduction and Biology. British Museum (Natural History), London: 219 pp.

Evans, G.O. and W.M. Till (1965). Studies on the British Dermanyssidae (Acari:Mesostigmata). Part I. External morphology. Bull. Brit. Mus. (Nat. Hist.) Zool. **13**(8):249-294. [DERMANYSSOIDEA]

Evans, G.O. and W.M. Till (1966). Studies on the British Dermanyssidae (Acari:Mesostigmata). Part II. Classification. Bull. Brit. Mus. (Nat. Hist.) Zool. **14**(5):109-370. [DERMANYSSOIDEA]

Fain, A. (1961a). Une nouvelle famille d'Acariens. Rev. Zool. Bot. Afr. **66**(3-4):283-296. [OMENTOLAELAPIDAE]

Fain, A. (1961b). Les acariens parasites endopulmonaires des serpentes. Bull. Inst. roy. Sci. natur. Belg. **37**(6):1-135. [ENTONYSSIDAE]

Fain, A. (1962). Les acariens mesostigmatiques ectoparasites des serpentes. Bull. Inst. roy. Sci. natur. Belg. **38**(18):1-149. [IXODORHYNCHIDAE]

Fain, A. (1967). Les acariens parasites nasicoles des oiseaux de Trinidad Indes Occidentales. I. Rhinonyssidae:Mesostigmates. Bull. Inst. roy. Sci. natur. Belg. **43**:1-44.

Fain, A., G. Anastos, J. Camin and D. Johnston (1967). Notes on the genus *Spelaeorhynchus*. Description of *S. praecursor* Neumann and of two new species. Acarologia **9**(3):535-556. [SPELAEORHYNCHIDAE]

Farrier, M.H. (1957). A revision of the Veigaiidae (Acarina). N. Carolina Agr. Exp. Sta. Tech. Bull. 124: 103 pp.

Filipponi, A. (1962). Metodi sperimentali nella sistematica degli acari macrochelidi (Acarina, Mesostigmata, Macrochelidae). Parassit. **4**(2-3):113-146.

Finnegan, S. (1931). On a new species of mite of the family Heterozerconidae parasitic on a snake. Proc. Zool. Soc. London 1931: 1349-1357.

Furman, D.P. and A.W. Smith (1973). In vitro development of two species of *Orthohalarachne* (Acarina: Halarachnidae) and adaptations of the life cycle for endoparasitism in mammals. J. Med. Ent. **10**(4): 415-416.

Hammen, L. van der (1964). The morphology of *Glyptholaspis confusa* (Foà, 1900) (Acarida, Gamasina). Zool. Verh. 71:56 pp. [MACROCHELIDAE]

Haq, S. (1965). Records of some interstitial mites from Nobska Beach together with a description of a new genus and species *Psammonsella nobskae,* of the family Rhodacaridae (Acarina Mesostigmata). Acarologia **7**(3):411-419. [OLOGAMASIDAE]

Herrin, C.S. (1970). A systematic revision of the genus *Hirstionyssus* (Acari:Mesostigmata) of the nearctic region. J. Med. Ent. **7**(4):391-437. [LAELAPIDAE]

Herrin, C.S. and V.J. Tipton (1975). Spinturnicid mites of Venezuela (Acarina:Spinturnicidae). Brigham Young Univ. Sci. Bull. Biol. Ser. **20** 1(2):1-72.

Herrin, C.S. and V.J. Tipton (1976). A systematic revision of the genus *Laelaps* s. str. (Acari:Mesostigmata) of the Ethiopian region. Great Basin Natur. **36**(2):113-205.

Herrin, C.S. and C.E. Yunker (1975). Systematics of neotropical *Hirstionyssus* mites with special emphasis on Venezuela (Acarina:Mesostigmata). Brigham Young Univ. Sci. Bull. Biol. Ser. **20**, 3(2):93-127. [LAELAPIDAE]

Hirschmann, W. (1960). Die Gattung *Dendrolaelaps* Halbert 1915. Gangsyt. Parasitiformes. Schrift. für vergleich. Milben. **3**(3):27 pp. + plates. [DIGAMASELLIDAE]

Hirshmann, W. (1962). Gamasiden. Gangsyst. Parasitiformes. Schrift. für vergleich. Milben. **5**(5):80 pp. + plates.

Hirschmann, W. and W. Krauss (1965). Gamasiden Bestimmungstafeln von 55 *Pachylaelaps*-Arten. Gangsyst. Parasitiformes. Shrift. für vergleich. Milben. **7**(8):32 pp.

Hoffmann, A. and I.B. de Barrera (1970). Acaros de la familia Spelaeorhynchidae. Rev. lat.-amer. Microbiol. **12**:145-149.

Hoyt, S.C. (1969). Integrated chemical control of insects and biological control of mites on apple in Washington. J. Econ. Ent. **62**(1):74-86.

Hoyt, S.C. and E.C. Burts (1974). Integrated control of fruit pests. Ann. Rev. Ent. **19**:231-252. [PHYTOSEIIDAE]

Hull, W.B. (1970). Respiratory mite parasites in nonhuman primates. Lab. Animal Care **20**(2):402-406.

Hunter, P.E. (1964). Laelaptid mites from Auckland and Macquarie Islands. (Acarina:Laelaptidae). Pac. Insects Monogr. **7**(suppl.):630-641.

Hunter, P.E. (1967). Rhodacaridae and Parasitidae mites (Acarina:Mesostigmata) collected by the British Antarctic Survey, 1961-64. Brit. Antarct. Surv. Bull. 13:31-39.

Hunter, P.E. and R.W. Husband (1973). *Pneumolaelaps* (Acarina:Laelapidae) mites from North America and Greenland. Florida Ent. **56**(2):77-91.

Hurlbutt, H. (1967). Digamasellid mites associated with bark beetles in North America. Acarologia **9**(3):497-534. [DIGAMASELLIDAE]

Ishikawa, K. (1970). Studies on the mesostigmatid mites in Japan. III. Family Podocinidae Berlese. Annot. Zool. Japon. **43**(2):112-122.

Jeppson, L.R., H.H. Keifer and E.W. Baker (1975). Biological enemies of mites. **In** Mites Injurious to Economic Plants. Univ. Calif. Press, Berkeley: 75-90.

Johnston, D.E. (1968). An Atlas of Acari I. The Families of Parasitiformes and Opilioacariformes. Acarology Lab. Publ. 172. Acarology Publ., Columbus: 110 pp. + x.

Karg, W. (1963). Systematische Untersuchung der Eviphididae Berlese 1913 (Acarina, Parasitiformes) mit einer neuen Art aus Ackerboden. Zool. Anz. **170**(7-8):269-281.

Karg, W. (1965). Larvalsystematische und phylogenetische Untersuchung sowie Revision des Systems der Gamasina Leach, 1915 (Acarina, Parasitiformes). Mitt. Zool. Mus. Berlin **41**(2):193-340.

Karg, W. (1969). Untersuchungen zur Kenntnis der Ascaoidea Karg, 1965 (Acarina, Parasitiformes) mit der Beschreibung von acht neuen Arten. Zool. Anz. **182**(5/6):393-406. [HALOLAELAPIDAE]

Karg, W. (1971a). Die freilebenden Gamasina (Gamasides), Raubmilben. Die Tierwelt Deutschlands **59**: 475 pp. Fischer Verlag, Jena.

Karg, W. (1971b). Zur Kenntnis der Epicriidae Berlese, 1885 (Acarina, Anactinochaeta). Zool. Anz. **186**(1/2):105-114.

Karg, W. (1973). Begründung der Überfamilie Eviphidoidea (Acarina, Gamasina) und Darstellung der verfalgten Arbeitsweise in Form eines heuristischen Programms. Zool. Anz. **190**(5/6):386-400.

Keegan, H.L., C.E. Yunker and E.W. Baker (1960). Malaysian Parasites. XLVI. *Hystrichonyssus turneri* n. sp., n.g., representing a new subfamily of Dermanyssidae (Acarina) from a Malayan porcupine. Inst. Med. Res. Fed. Malaya **29**:205-208. [HYSTRICHONYSSIDAE]

Kinn, D.N. (1971). The life cycle and behavior of *Cercoleipus coelonotus* (Acarina:Mesostigmata) including a survey of phoretic mite associates of California Scolytidae. Univ. Calif. Publ. Ent. **65**:66 pp.

Krantz, G.W. (1960). A re-evaluation of the Parholaspinae Evans 1956 (Mesostigmata:Macrochelidae). Acarologia **2**(4):393-433. [PARHOLASPIDIDAE]

Krantz, G.W. (1962). A review of the genera of the family Macrochelidae Vitzthum 1930 (Acarina:Macrochelidae). Acarologia **4**(2):143-173.

Krantz, G.W. and J.L. Mellott (1968). Two new species of *Macrocheles* (Acarina:Macrochelidae) from Florida, with notes on their host-specific relationships with geotrupine beetles (Scarabaeidae:Geotrupinae). J. Kansas Ent. Soc. **41**(1):48-56.

Krantz, G.W. and J.L. Mellott (1972). Studies on phoretic specificity in *Macrocheles mycotrupetes* and *M. peltotrupetes* Krantż and Mellott (Acari:Macrochelidae), associates of geotrupine Scarabaeidae. Acarologia **14**(3):317-344.

Kshirsagar, K.K. (1968). Further observations on phoretic bee mites *Neocypholaelaps indica* (Evans). Indian Bee J. **30**(2):68-69. [AMEROSEIIDAE]

Lee, D.C. (1970). The Rhodacaridae (Acari:Mesostigmata); classification, external morphology and distribution of genera. Rec. S. Austral. Mus. **16**(3):1-219. [RHODACARIDAE, OLOGAMASIDAE]

Lee, D.C. (1974). Rhodacaridae (Acari:Mesostigmata) from near Adelaide, Australia. III. - Behaviour and development. Acarologia **16**(1):21-44. [OLOGAMASIDAE]

Lee, D.C. and P.E. Hunter (1974). Arthropoda of the subantarctic islands of New Zeland 6. Rhodacaridae (Acari:Mesostigmata). New Zealand J. Zool. **1**(3):295-328. [OLOGAMASIDAE]

Lindquist, E.E. (1962). *Mucroseius monochami,* a new genus and species of mite (Acarina:Blattisocidae) symbiotic with sawyer beetles. Can. Ent. **94**:972-980. [ASCIDAE]

Lindquist, E.E. (1975). *Digamasellus* Berlese, 1905, and *Dendrolaelaps* Halbert, 1915, with descriptions of new taxa of Digamasellidae (Acarina:Mesostigmata). Can. Ent. **107**(1):1-43.

Lindquist, E.E. and G.O. Evans (1965). Taxonomic concepts in the Ascidae, with a modified setal nomenclature for the idiosoma of the Gamasina (Acarina:Mesostigmata). Mem. Ent. Soc. Canada **47**:64 pp.

Loots, G.C. (1969). Notes on *Rhodacarus* Oudemans and its related genera with descriptions of new species from the Ethiopian region. Publ. Cult. Companhia Diamantes Angola, Lisboa 81:45-82. [RHODACARIDAE]

McGraw, J.R. and M.H. Farrier (1969). Mites of the superfamily Parasitoidea (Acarina:Mesostigmata) associated with *Dendroctonus* and *Ips* (Coleoptera:Scolytidae). N. Carolina Agr. Exp. Sta. Tech. Bull. 192: 162 pp.

McMurtry, J.S., C.B. Huffaker and M. van de Vrie (1970). Ecology of tetranychid mites and their natural enemies. Pt. I. Tetranychid enemies: their biological characters and the impact of spray practices. Hilgardia **40**(11):331-390.

Merwe, G. van der (1968). A taxonomic study of the family Phytoseiidae (Acari) in South Africa with contributions to the biology of two species. S. Afr. Dept. Agr. Tech. Serv., Ent. Mem. 18:198 pp.

Metz, L.J. and M.H. Farrier (1969). Acarina associated with decomposing forest litter in the North Carolina piedmont. Proc. 2nd Int. Congr. Acarology, Nottingham: 43-52.

Micherdzinski, W. (1969). Die Familie Parasitidae Oudemans 1901 (Acarina, Mesostigmata). Zak. Zool. Syst. Pol. Acad. Nauk, Krakow: 690 pp.

Muma, M.H. (1975). Mites associated with citrus in Florida. Bull. Univ. Fla. Agr. Exp. Sta. 640A:92 pp. + iv.

Newell, I.M. (1947). Studies on the morphology and systematics of the family Halarachnidae Oudemans 1906 (Acari, Parasitoidea). Bull. Bingham Oceanogr. Coll. **10**(4):235-266.

Pence, D.B. (1975). Keys, species and host list, and bibliography for nasal mites of North American birds (Acarina:Rhinonyssinae, Turbinoptinae, Speleognathinae and Cytoditidae). Spec. Publ. Mus. Texas Tech. Univ. 8:148 pp.

Petrova-Nikitina, A.D. (1969). Analysis of the family Parholaspidae Evans, 1956 (Parasitiformes, Gamasoidea). Proc. 2nd Int. Congr. Acarology, Nottingham: 187-190.

Phillips, C.J., J.K. Jones Jr. and F.J. Radovsky (1969). Macronyssid mites in oral mucosa of long-nosed bats: occurrence and associated pathology. Science **165**:1368-1369.

Potter, D.A. and D.E. Johnston (1976). *Canestriniphis megalodacne* n.g., n. sp. (Acari:Eviphididae) from a pleasing fungus beetle, *Megalodacne heros.* Ann. Ent. Soc. Amer. **69**(3):494-496.

Price, D.W. (1973). Abundance and vertical distribution of microarthropods in the surface layers of a California pine forest soil. Hilgardia **42**(4):121-147.

Rack, G.S. (1963). *Kleemannia* (Acarina, Ameroseiidae) ein neuer Wohnunglästling. Ent. Mitt. Zool. Hamburg **2**(44):1-7.

Radovsky, F.J. (1967). The Macronyssidae and Laelapidae (Acarina-Mesostigmata) parasitic on bats. Univ. Calif. Publ. Ent. **46**:288 pp.

Radovsky, F.J. (1969). Adaptive radiation in the parasitic Mesostigmata. Acarologia **11**(3):450-483.

Radovsky, F.J. and C.E. Yunker (1971). *Xenarthronyssus furmani,* n. gen., n. sp. (Acarina:Dasyponyssidae), parasites of armadillos, with two subspecies. J. Med. Ent. **8**(2):135-142. [DASYPONYSSIDAE, MANITHERIONYSSIDAE]

Rapp, A. (1959). Zur Biologie und Ethologie der Kafermilbe *Parasitus coleoptratorum* L. 1758 (Ein Beitrag zum Phoresie-Problem). Zool. Jahr. Syst. **86**(4-5):303-366. [PARASITIDAE]

Redington, B.C. (1970). Studies on the morphology and taxonomy of *Haemogamasus reidi* Ewing, 1925 (Acari:Mesostigmata). Acarologia **12**(4):643-667. [LAELAPIDAE]

Rudnick, A. (1960). A revision of the mites of the family Spinturnicidae (Acarina). Univ. Calif. Publ. Ent. **17**(2):157-284.

Ryke, P.A.J. (1962). The subfamily Rhodacarinae with notes on a new subfamily Ologamasinae (Acarina: Rhodacaridae). Ent. Ber. **8**(22,1):155-162. [RHODACARIDAE, OLOGAMASIDAE]

Saunders, R.C. (1975). Venezuela Macronyssidae (Acarina:Mesostigmata). Brigham Young Univ. Sci. Bull. Biol. Ser. **20**, 2(2):75-90.

Sellnick, M. (1958). Die Familie Zerconidae Berlese. Acta Zool. **3**(3-4):313-368.

Southcott, R.V. (1976). Arachnidism and allied syndromes in the Australian region. Rec. Adelaide Children's Hosp. **1**(1):97-186.

Stammer, H.J. (ed.) (1963). Beiträge zur Systematik und Ökologie Mitteleuropäischer Acarina. Zool. Inst. Freidrich - Alexander - Univ. Bd. II. Mesostigmata **1**:804 pp. + vii.

Strandtmann, R.W. and G.W. Wharton (1958). Manual of mesostigmatid mites parasitic on vertebrates. Inst. Acarology, Univ. Maryland, Contr. **4**:330 pp. + vii + 69 plates.

Till, W.M. (1963). Ethiopian mites of the genus *Androlaelaps* Berlese s. lat. (Acari:Mesostigmata). Bull. Brit. Mus. (Nat. Hist.) Zool. **10**(1):104 pp. [LAELAPIDAE]

Tipton, V.J. (1960). The genus *Laelaps* with a review of the Laelaptinae and a new subfamily Alphalaelaptinae (Acarina:Laelaptidae). Univ. California Publ. Ent. **16**(6):233-356 + 25 plates.

Treat, A.E. (1975). Mites of Moths and Butterflies. Cornell Univ. Press, Ithaca, N.Y.: 362 pp.

Vitzthum, H.G. (1931). Resultats scientifiques du voyage aux Indes Orientales Neerlandaises de LL.AA. Belgique. Acarinen. Mem. Mus. Hist. natur. Belg. (H.S.) **3**(5):1-55. [PACHYLAELAPIDAE]

Vitzthum, H.G. (1940-43). Acarina. **In** Bronn's Klassen und Ordnungen des Tierreiches **5**(4), Buch 5:1-1011. Leipzig.

Westigard, P.H. (1971). Integrated control of spider mites on pears. J. Econ. Ent. **64**(2):496-501. [PHYTOSEIIDAE]

Womersley, H. (1956). On some new Acarina - Mesostigmata from Australia, New Zealand and New Guinea. Linn. Soc. J. Zool. **42**(288):505-599.

Yunker, C.E. (1973). Mites. **In** Parasites of Laboratory Animals. R.J. Flynn, ed. Iowa State Univ. Press, Ames: 425-492.

Zemskaya, A.A. and A.A. Pchelkina (1967). Gamasid mites and Q fever. **In** Problemy parazitologii. A.P. Markevich, ed.: 612. (Translation 296, NAMRU-3, Cairo).

Zumpt, F. (1961). The arthropod parasites of vertebrates in Africa south of the Sahara (Ethiopian region). Vol. 1. (Chelicerata). S. Afr. Inst. Med. Res. **9**(1):1-457.

Cohort Uropodina[1]

The Uropodina is a large and diverse group of mites which differ from the Gamasina in shield morphology and, to some degree, in chaetotaxy. Females generally have no setae on the epigynial shield, nor is the shield commonly fused posteriorly with ventral shield elements as occurs in many gamasine families (Plate 40, p. 198). The male genital aperture is well removed from the anterior border of the sternal shield and, since sperm transfer is tocospermous throughout the cohort, the spermadactyl is absent. Extensive fusion of ventral shields occurs in both sexes. Dorsal sclerotization generally consists of an entire dorsal shield or a large dorsal shield bordered posteriorly by a pygidial shield. Further fragmentation occasionally occurs. A marginal shield or shields, or marginal platelets, usually are present. Femur IV bears six or seven setae, and claws may be present or absent on tarsus I. The gnathosoma in both sexes is contained in an anteroventral idiosomal depression, the *camerostome.* The chelicerae typically are elongate and extrusible, with small chelate digits.

The Uropodina are here considered to comprise four superfamilies and seven families, based primarily on chaetotactic and shield characters.

Superfamily Thinozerconoidea[2]
(Plates 38-39, pp. 196-197)

DIAGNOSIS: Dorsal shield of adults divided into large anterior and small posterior pygidial shields, or entire; with a separate anterior marginal sclerite covering the region of the vertex, lateral shields or platelets undeveloped. Tritosternal base broad or narrow, separated from widely separated coxal elements. Femur I with two anterolateral and five dorsal setae ($2\,\frac{5}{3}\,2$ *or* $2\,\frac{5}{4}\,1$), *femur II with 11 setae; genua I-IV with five or six dorsal setae* ($2\,\frac{3}{1}\,\frac{3}{1}\,2$, $2\,\frac{3}{1}\,\frac{2}{1}\,2$, $2\,\frac{3}{1}\,\frac{2}{0}\,2$, *or* $2\,\frac{3}{1}\,\frac{3}{0}\,1$). *Tibia I with 11-13 setae, including three or four ventrals* ($2\,\frac{2}{2}\,\frac{3}{2}\,2$ *or* $2\,\frac{2}{1}\,\frac{2}{2}\,2$). *Palpgenu with six setae. Larva and protonymph with four dorsal setae on tibia I.*

The Thinozerconoidea includes two families of poorly known holarctic mites from seashore or aquatic habitats.[3] *Thinozercon michaeli* Halbert (THINOZERCONIDAE, Plate 39-5 to 7) is an intertidal form which has been encountered in a number of localities in Ireland (Evans 1957). *Protodinychus punctatus* Evans (PROTODINYCHIDAE, Plate 39-1 to 4) was taken from flood water debris in England (Evans op.cit.). Deutonymphs of a second species have been taken in North America from beetles found in the beaver lodge habitat of the adult mites (Johnston 1961). Feeding habits of the Thinozerconoidea are unknown.

Superfamily Polyaspidoidea[4]
(Plates 38, 40, pp. 196, 198)

DIAGNOSIS: Dorsum of adults with a large anterior and small pygidial shield; marginal shields present, often fragmented into smaller platelets, separate vertex shield absent. Tritosternal base broader than long, coxae I widely separated. Femur I

---

[1]Includes the cohorts Uropodina and Trachyuropodina (supercohort Atrichopygidiina) of Hirschmann (1975).

[2]Polyaspidoidea (pars) of Evans (1972); UROPODIDAE, Polyaspidini (pars) of Hirschmann and Zirngiebl-Nicol (1964).

[3]An undescribed thinozerconid has been reported from the Subantarctic realm by B.D. Ainscough (personal communication).

[4]UROPODIDAE, Polyaspidini (pars) of Hirschmann and Zirngiebl-Nicol (1974).

with one anterolateral and four dorsal setae $(1\ \frac{4}{3}\ 1)$, *femur II with nine setae; genua I-IV with four dorsal setae, genua II-III with two anterolateral and two posterolateral setae. Tibia I with eight, nine, or ten setae, including two ventrals* $(2\ \frac{2}{1}\ \frac{2}{1}\ 2,\ 2\ \frac{2}{1}\ \frac{2}{1}\ 1,$ *or* $1\ \frac{2}{1}\ \frac{2}{1}\ 1)$. *Palpgenu with five setae. Larva and protonymph with four dorsal setae on tibia I.*

Two families comprise the Polyaspidoidea—the POLYASPIDIDAE (Plate 40-4 to 8) and the DITHINOZERCONIDAE (Plate 40-1 to 3). Representatives of both families are found in forest and treehole litter, moss, under tree bark and on insects throughout most of the world. Polyaspidid deutonymphs often are found as phoretics on insects, but some have been collected in ant nests as free-living residents. One of the six described species of *Polyaspis—P. patavinus* Berlese—has been shown to have two morphologically distinct deutonymphal types, only one of which is phoretic (Athias 1975). Other species of *Polyaspis* apparently show similar deutonymphal heteromorphy.

Many species of the polyaspidid genus *Trachytes* are known from forest humus and litter, primarily in moist temperate habitats (Pecina 1970, Hutu 1973, Zirngiebl-Nicol 1973), as are species of the genera *Calotrachytes* and *Polyaspinus.* Similar niches have been colonized in the Neotropical realm by members of the genera *Tetrasejaspis* and *Baloghjkaszabia* (Hirschmann 1973a, b, c and e). None is known to be phoretic.

The DITHINOZERCONIDAE are widespread in both the Old and New Worlds (Hirschmann and Hutu 1974). Deutonymphs of various species of *Dithinozercon* and *Apionoseius,* like those of many polyaspidids, are phoretic on or associated with insects. *Apionoseius* species have been found on trogine and geotrupine scarabs, and *A. acuminatus* (C.L. Koch) was collected from honeybee hives in Europe (Evans et al. 1961). *Caminella peraphora* Krantz and Ainscough is virtually aquatic in habitat, being found only in wet moss anchored in a mountain stream in western Oregon (Krantz and Ainscough 1960).

Adult polyaspidoid mites are thought to feed on organic detritus or fungus. However, *Caminella peraphora* feeds readily on nematodes in laboratory culture, often digging deep into the wet filter paper substrate to retrieve them.

Superfamily Uropodoidea
(Plates 38, 41-44, pp. 196, 199-202)

DIAGNOSIS: Dorsal shield of adults entire, or with a separate pygidial shield; with a marginal shield free or fused anteriorly to dorsal shield, often disappearing posteriorly; marginal platelets may be present posteriorly. Tritosternal base generally longer than broad, arising between and partially covered by the converging internal angles of coxae I; fovae pedales or scabellum usually present. Femur I with one anterolateral and four dorsal setae $(1\ \frac{4}{3}\ 1$ *or* $1\ \frac{4}{2}\ 1)$, *femur II with nine setae; genua I-IV usually with four dorsal setae, genua II-III with one anterolateral and one posterolateral seta. Tibia I typically with seven setae, including two ventrals* $(1\ \frac{1}{1}\ \frac{2}{1}\ 1)$. *Larva and protonymph with three dorsal setae on tibia I.*

The Uropodoidea, or "higher uropodines" (Evans 1972) is a large cosmopolitan group consisting of fungivores (Radinovsky and Krantz 1961), insect associates (Elzinga and Rettenmeyer 1966, Treat 1975), nidicoles (Berlese 1904, Hughes 1959) and occasional predators (Reid 1957). Uropodoids often are referred to as "tortoise mites" due to the ability of most species to withdraw their appendages into ventrolateral concavities or *fovae pedales,* or behind a protective lamellar ridge or *scabellum* (Plate 43-1, p. 201). Most species are associated with insects during one or more instars.

Deutonymphs of the family UROPODIDAE[1] (Plates 41-43) often attach phoretically to the cuticle of insects and other animals by means of an anal pedicel (Fig. 20, p. 64) leaving the carrier only after molting to the adult stage. Many uropodids occur commonly in forest litter and detritus (Hirschmann 1972c), and others have been collected in soil, moss, rotting wood and the nests or galleries of insects (Hirschmann 1972d and e). Generalized uropodid nest and gallery dwellers (*Uroobovella* and *Trichouropoda,* for example) are considered to be scavengers which probably feed on organic detritus and fungi. The mycetophagous habit apparently is shared by uropodid species which inhabit stored products or animal wastes (*Leiodinychus, Trematura* and *Fuscuropoda* spp.). *Fuscuropoda agitans* (Banks) may occur in great numbers in commercial fishworm beds, competing with the worms for food and creating a source of concern for producers (Stone and Ogles 1953).

Members of the genus *Phaulodinychus* are routinely encountered in litter, but many occur intertidally (Halbert 1915, Willmann 1957, Hirschmann 1972c). *P. mitis* (Leonardi) is regularly submerged by tidal flooding on the northern Adriatic coast, but is capable of plastron respiration during these periods (Krantz 1974). The uropodid genera *Deraiophorus* and *Eutrachytes* are primarily tropical inhabitants of soil, humus and litter (Hirschmann 1973e), while *Prodinychus, Dinychus* and *Discourella* generally occupy similar niches in temperate realms (Hirschmann and Hutu 1974).

The TRACHYUROPODIDAE[2] (Plate 44, p. 202) is comprised exclusively of species which are myrmecophilous at least in the adult stage. Unlike the uropodids, deutonymphs do not attach to their arthropod carriers by means of a secreted anal pedicel. The family is heterogenous, including both smooth and punctate forms of moderate size (*Trichocylliba*[3] and *Oplitis,* for example) and a variety of heavily sclerotized, often quite large ($> 1000\ \mu$) uropodoids which may be ornamented in the adult stage with a series of dorsal and/or ventral pits, depressions, or ridges. These are often bordered by hypertrophied setate excresecences

---

[1]Hirschmann and Zirngiebl-Nicol (1964, summarized by Hirschmann (1971)) divide the UROPODIDAE into three subfamilies—the Uropodinae, Oplitinae and Uroactiniinae—basing their intrafamilial divisions solely on cheliceral characters. The Uropodinae are characterized by having a short rounded seta ("Kolbenpilus") on the fixed cheliceral digit. The subfamily includes, among other taxa, the families TRACHYTIDAE, DINYCHIDAE and EUTRACHYTIDAE of earlier authors. The Oplitinae have a duplex seta ("Doppelpilus") on the fixed digit and include the POLYASPIDIDAE and PROTODINYCHIDAE, in addition to *Oplitis* and *Trachyuropoda.* The Uroactiniinae are distinguished by their terminal or subterminal brush-like cheliceral seta ("Pinselpilus"), and include the family DIARTHROPHALLIDAE. Seven uropodid tribes are differentiated by Hirschmann and Zirngiebl-Nicol on the basis of cheliceral dentition and the presence or absence of a sclerotized node on the levator tendon near the base of the movable digit ("Rollplatte"). The resulting dissolution of the "lower" and "higher" Uropodina (Camin and Gorirossi 1955, Evans 1957, Johnston 1961) is refuted by Evans (1972) who examined leg chaetotactic characters of a variety of uropodines, and concluded that retention of the "lower" (Protodinychoidea/Polyaspidoidea) and "higher" (Uropodoidea) categories (both sensu Evans 1964) is entirely justified. In addition, he determined that Hirschmann and Zirngiebl-Nicol's tribe Diarthrophallini (Uroactiniinae) could not logically be accommodated in the Uropodidae, nor in the superfamily Uropodoidea. He suggested a separate superfamilial, or possibly a cohortal, division for this group. The former alternative is followed in this treatment.

[2]Subfamily Oplitinae, Tribe Trachyuropodini of Hirschmann and Zirngiebl-Nicol (1964), cohort Trachyuropodina pars of Hirschmann (1975). Evans (1972) points out that the subfamily name Trachyuropodinae Berlese has priority over Hirschmann and Zirngiebl-Nicol's Oplitinae.

[3]Hirschmann (1973d) recently revised the genus *Trichocylliba* to include *Circocylliba* (CIRCOCYLLIBIDAE of previous authors), *Planodiscus* (PLANODISCIDAE of previous authors), *Coxequesoma* and *Antennequesoma* (COXEQUESOMIDAE of previous authors). *T. ablesi* Hirschmann is considered a transitional species between typical *Trichocylliba* species and the abovementioned groups in that *T. ablesi* shares the characters of pronounced metapodal separation and a distinct V-shaped suture demarcating the anal region. While I accept Hirschmann's conclusion that a close relationship exists between *Trichocylliba* and the aforementioned genera, I have followed Ainscough (personal communication) in restricting *Trichocylliba* to include holarctic myrmecophiles exemplified by *T. ablesi.*

which entirely obscure the associated integumental structure.[1] Various Neotropical species illustrate remarkable adaptations for attachment to ants, and many are site specific on their hosts. For example, *Oplitis philoctena* (Trouessart) and other species of the genus attach to the tibial spur of legs I of their ant hosts (Hunter and Farrier 1975, 1976), apparently feeding on debris which the ants discard as they groom themselves. *Circocylliba camerata* Sellnick and other species of the genus described by Elzinga and Rettenmeyer (1974) are found only on the head, mandibles, thorax or gaster of ecitoniine army ants. *Planodiscus squamatim* Sellnick, *P. foreli* Elzinga and Rettenmeyer, *P. hamatus* E. & R. and *P. burchelli* E. & R., on the other hand, are confined to the legs of their army ant associates (Elzinga and Rettenmeyer 1970). *Coxequesoma collegianorum* Sellnick, *Antennequesoma reichenspergeri* S. and *A. lujai* S. are capsulate forms which cover the terminal antennal segments of their ant carriers, presenting a smooth hard exterior to the potential hazards of the ant nest environment (Sellnick 1926).[2]

Trachyuropodids are primarily tropical forms, although the genera *Trachyuropoda, Leonardiella, Oplitis* and *Urodiscella* are cosmopolitan (Hirschmann and Hutu 1974).

Superfamily Diarthrophalloidea[3]
(Plate 38, p. 196)

DIAGNOSIS: Dorsal shield of adults entire or divided, marginal shields absent. Tritosternal base various, flanked by a pair of setae, separated from widely divergent coxae I. Adanal setae long, often more than ½ length of body. Leg chaetotaxy hypotrichous; femur I with one anterolateral and three dorsal setae $(1\ \frac{3}{2}\ 1)$, *femur II with eight setae; genua I-IV with only two dorsals* $(1\ \frac{1}{0}\ \frac{1}{0}\ 1)$. *Tibia I with five setae, anterodorsals absent* $(1\ \frac{0}{1}\ \frac{1}{1}\ 1)$. *Palpgenu with four setae.*

The Diarthrophalloidea is represented by the family DIARTHROPHALLIDAE, an assemblage of relatively few described genera and species of tiny, weakly sclerotized hypotrichous mites which are intimately associated with beetles of the family Passalidae (Trägårdh 1946). Diarthrophallids have been described from New Guinea, Australia, and various localities in the Neotropical realm (Womersley 1961, Hunter and Glover 1968a, b, c). One species, *Diarthrophallus quercus* (Pearse and Wharton), is associated with the American passalid *Popilius disjunctus* in the eastern United States. A family revision currently in progress indicates that the world diarthrophallid fauna is considerably more extensive than was previously estimated (R.O. Schuster, personal communication). Furthermore, there appears to be a unique diarthrophallid species associated with each species of beetle in a number of passalid genera. The occurrence of all life stages of diarthrophallid mites in the subelytral niche suggests a mite-beetle relationship more intimate than phorsey.

---

[1]Based on their general facies, Hirschmann (1976a) includes a number of species which have no idiosomal pits or depressions. Their dorsal ornamentation, however, suggests an emerging pattern similar to that of other *Trachyuropoda* species with conservative ridge development.

[2]The genera *Circocylliba, Planodiscus, Coxequesoma, Antennequesoma* and *Trichocylliba* are included in the TRACHYUROPODIAE primarily on ecological grounds. Lack of information on immature stages of these genera, however, makes their familial placement somewhat arbitrary.

[3]UROPODIDAE, Diarthrophallini of Hirschmann and Zirngiebl-Nicol (1964).

Useful References

Athias, F. (1975). Observations morphologiques sur *Polyaspis patavinus* Berlese 1881 (Acariens:Uropodides). I. Morphologie de l'idiosoma au cours du developpement postembryonnaire. Acarologia **17**(3):410-435.

Berlese, A. (1904). Acari mirmecofili. Redia **1**:299-474 + plates.

Camin, J.H. (1953a). Metagynellidae, a new family of uropodine mite with the description of *Metagynella parvula*, a new species from tree holes. Bull. Chicago Acad. Sci. **9**(18):391-409.

Camin, J.H. (1953b). A revision of the cohort Trachytina Trägårdh, 1938, with the description of *Dyscritaspis whartoni*, a new genus and species of polyaspid mites from treeholes. Bull. Chicago Acad. Sci. **9**(17): 335-385.

Camin, J.H. and F. Gorirossi (1955). A revision of the suborder Mesostigmata (Acarina), based on new interpretations of comparative morphological data. Chicago Acad. Sci. Spec. Publ. 11:70 pp.

Elzinga, R.J. and C.W. Rettenmeyer (1966). A neotype and new species of *Planodiscus* (Acarina:Uropodina) found on doryline ants. Acarologia **8**(2):191-199.

Elzinga, R.J. and C.W. Rettenmeyer (1970). Five new species of *Planodiscus* (Acarina:Uropodina) found on doryline ants. Acarologia **12**(1):59-70.

Elzinga, R.J. and C.W. Rettenmeyer (1974). Some new species of *Circocylliba* (Acarina:Uropodina) found on army ants. Acarologia **16**(4):595-611.

Evans, G.O. (1957). An introduction to the British Mesostigmata (Acarina) with keys to families and genera. Linn Soc. J. Zool. **43**(291):203-259.

Evans, G.O. (1964). Some observations on the chaetotaxy of the pedipalps in the Mesostigmata (Acari). Ann. Mag. Nat. Hist. **6**(13):513-527.

Evans, G.O. (1972). Leg chaetotaxy and the classification of the Uropodina (Acari:Mesostigmata). J. Zool., London **167**:193-206.

Evans, G.O., J.G. Sheals and D. Macfarlane (1961). Mites associated with insects and other invertebrates. **In** Terrestrial Acari of the British Isles, Vol. I. Introduction and Biology. British Museum (Natural History), London: 166-185.

Halbert, J.N. (1915). Clare Island Survey, Part 39. Acarinida: ii - Terrestrial and marine Acarina. Proc. Royal Irish Acad. **31**:45-136. [THINOZERCONIDAE]

Hirschmann, W. (1961). Die Gattung *Trichouropoda* Berlese 1916 nov. comb., die Cheliceren und das System der Uropodiden. Gangsyst. Parasitiformes. Schrift. für vergleich. Milben. **4**(4):1-41 + plates. [UROPODIDAE]

Hirschmann, W. (1971). "Gangsystematik" of the Parasitiformes and the family Uropodidae Berlese. Proc. 3rd Int. Congr. Acarology, Prague: 287-292.

Hirschmann, W. (1972a). Adulten-Gruppen und Ruckenflächenbestimmungstabelle von 34 *Discourella*-Arten (Uropodini, Uropodinae). Gangsyst. Parasitiformes. Schrift. für vergleich. Milben. **18**(114):26-29. [UROPODIDAE]

Hirschmann, W. (1972b). Teilgänge und Stadien von 22 neuen *Discourella*-Arten (Uropodini, Uropodinae). Gangsyst. Parasitiformes. Schrift. für vergleich. Milben. **18**(115):29-41. [UROPODIDAE]

Hirschmann, W. (1972c). Teilgänge, Stadien von 21 neuen *Uropoda* (*Phaulodinychus*)-Arten (Uropodini, Uropodinae). Gangsyst. Parasitiformes. Schrift. für vergleich. Milben. **18**(123):79-92. [UROPODIDAE]

Hirschmann, W. (1972d). Gang, Teilgäng, Stadien von 7 neuen *Nenteria*-Arten (Trichouropodini, Uropodinae). Gangsyst. Parasitiformes. Schrift. für vergleich. Milben. **18**(106):6-9. [UROPODIDAE]

Hirschmann, W. (1972e). Teilgänge, Stadien von 8 neuen *Trichouropoda*-Arten (Trichouropodini, Uropodinae). Gangsyst. Parasitiformes. Schrift. für vergleich. Milben. **18**(108):11-15. [UROPODIDAE]

Hirschmann, W. (1973a). Gang, Teilgäng, Stadien von 8 neuen *Tetrasejaspis*-Arten (Dinychini, Uropodinae). Gangsyst. Parasitiformes. Schrift. für vergleich. Milben. **19**(159):91-100. [POLYASPIDIDAE]

Hirschmann, W. (1973b). Teilgänge, Stadien von 3 neuen *Baloghjkaszabia*-Arten (Uropodini, Uropodinae). Gangsyst. Parasitiformes. Schrift. für vergleich. Milben. **19**(162):105-107. [POLYASPIDIDAE]

Hirschmann, W. (1973c). Stadien von 5 neuen *Kaszabjbaloghia*-Arten (Uropodini, Uropodinae). Gangsyst. Parasitiformes. Schrift. für vergleich. Milben. **19**(163):107-110. [POLYASPIDIDAE]

Hirschmann, W. (1973d). Adulten-Bastimmungstabelle von 19 *Trichocylliba*-Arten und Operculum-Bestimmungstabelle von 18 *Trichocylliba*-Weibchen (Dinychini, Uropodinae). Gangsyst. Parasitiformes. Schrift. für vergleich. Milben. **19**(172):124-127. [UROPODIDAE]

Hirschmann, W. (1973e). Gänge, Teilgänge, Stadien von 22 neuen *Deraiophorus*-Arten (Dinychini, Uropodinae). Gangsyst. Parasitiformes. Schrift. für vergleich. Milben. **19**(151):60-81. [POLYASPIDIDAE]

Hirschmann, W. (1975). Larvalsystematische, Gleiderung des Suborder Mesostigmata (Teilgang: Larve, Protonymphe, Deutonymphe). Novae Supercohortes Trichopygidiina Hirschmann 1975, Atrichopygidiina Hirschmann 1975, Nova Cohors Trachyuropodina Hirschmann 1975. Teilgangsyst. Parasitiformes. Schrift. für vergleich. Milben. **21**(1):93-100.

Hirschmann, W. (1976a). Adulten-Gruppen und Bestimmungstabelle von 81 *Trachyuropoda*-Arten (Trachyuropodini, Oplitinae). Gangsyst. Parasitiformes. Schrift. für vergleich. Milben. **22**(215):4-15. [TRACHYUROPODIDAE]

Hirschmann, W. (1976b). 3 neue *Trachyuropoda*-Arten der *excavata*-Gruppe (Trachyuropodini, Oplitinae). Gangsyst. Parasitiformes. Schrift. für vergleich. Milben. **22**(218):21-23. [TRACHYUROPODIDAE]

Hirschmann, W. (1976c). 1 neue *Trachyuropoda*-Art der *troguloides*-Gruppe (Trachyuropodini, Oplitinae). Gangsyst. Parasitiformes. Schrift. für vergleich. Milben. **22**(221):25-26. [TRACHYUROPODIDAE]

Hirschmann, W. and M. Hutu (1974). Uropodiden-Forschung und die Uropodiden der Erde, geordnet nach dem Gangsystem und nach den Ländern in zoogeographischen Reichen und Unterreichen. Gangsyst. Parasitiformes. Schrift. für vergleich. Milben. **20**(187):6-36.

Hirschmann, W. and I. Zirngiebl-Nicol (1964). Uropodiden-Das Gangsystem der Familie Uropodidae (Berlese 1892) Hirschmann u. Zirngiebl-Nicol nov. comb. Bestimmungstabellen Kurzdiagnosen Operculum-Bestimmungstabellen. Gangsyst. Parasitiformes. Schrift. für vergleich. Milben. **6**(7):2-18.

Hirschmann, W. and I. Zirngiebl-Nicol (1965). Uropodiden-Bestimmungstabellen von 300 Uropodiden-Arten (Larven, Protonymphen, Deutonymphen, Weibchen, Männchen). Gangsyst. Parasitiformes. Schrift. für vergleich. Milben. **8**(9):2-9, 12-13, 19-22.

Hirschmann, W. and I. Zirngiebl-Nicol (1972). Teilgänge, Stadien von 19 neuen *Urobovella*-Arten (Dinychini, Uropodinae). Gangsyst. Parasitiformes. Schrift. für vergleich. Milben. **18**(127):110-121. [UROPODIDAE]

Hunter, J.E. III and M.H. Farrier (1975). Mites of the genus *Oplitis* Berlese (Acarina:Uropodidae) associated with ants (Hymenoptera:Formicidae) in the southeastern United States. Part I. Acarologia **17**(4):595-623.

Hunter, J.E. III and M.H. Farrier (1976). Mites of the genus *Oplitis* Berlese (Acarina:Uropodidae) associated with ants (Hymenoptera:Formicidae) in the southeastern United States. Part II. Acarologia **18**(1):20-50.

Hunter, P.E. and Sandra Glover (1968a). The genus *Passalobia* Lombardini, 1926, with description of a new species (Acarina:Diarthrophallidae). Proc. Ent. Soc. Wash. **70**(1):38-42.

Hunter, P.E. and Sandra Glover (1968b). The genus *Brachytremella* Tragardh, 1946, with descriptions of three new species (Acarina:Diarthrophallidae). Proc. Ent. Soc. Wash. **70**(2):114-125.

Hunter, P.E. and Sandra Glover (1968c). The genus *Diarthrophallus* Tragardh 1946 (Acarina:Diarthrophallidae). Proc. Ent. Soc. Wash. **70**(3):193-197.

Hughes, T.E. (1959). Associations with other animals. **In** Mites or the Acari. Athlone Press, London: 23-42.

Hutu, M. (1972). Aktuelle Kenntnisse über die weltweite Verbreitung der Uropodiden (Acari, Parasitiformes). Gangsyst. Parasitiformes. Schrift. für vergleich. Milben. **18**(125):95-106.

Hutu, M. (1973). Zur Kenntnis der Uropodiden-Fauna Rumäniens. Neue Uropodiden-Arten der Gattungen *Trachytes* Michael 1894, *Dinychus* Kramer 1886 and *Trachyuropoda* (Berlese 1888) Hirschmann u. Zirngiebl-Nicol 1961 nov. comb. Gangsyst. Parasitiformes. Schrift. für vergleich. Milben. **19**(145):45-51. [POLYASPIDIDAE]

Johnston, D.E. (1961). A review of the lower uropodoid mites (former Thinozerconoidea, Protodinychoidea, and Trachytoidea) with notes on the classification of the Uropodina. Acarologia **3**(4):522-545.

Krantz, G.W. and B.D. Ainscough (1960). *Caminella peraphora,* a new genus and species of mite from Oregon (Acarina:Trachytidae). Ann. Ent. Soc. Amer. **53**(1):27-34. [DITHINOZERCONIDAE]

Pecina, P. (1970). Czechoslavak uropodid mites of the genus *Trachytes* Michael, 1894 (Acari, Mesostigmata). Acta Univ. Carolinae - Biol. 1969: 39-59. [POLYASPIDIDAE]

Radinovsky, S. and G.W. Krantz (1961). The biology and ecology of granary mites of the Pacific Northwest. II. Techniques for laboratory observation and rearing. Ann. Ent. Soc. Amer. **54**(4):512-518. [UROPODIDAE]

Reid, R.W. (1957). The bark beetle complex associated with lodgepole pine slash in Alberta. Part III. Notes on the biology of several predators with special reference to *Enoclerus sphegeus* Fab. (Coleoptera: Cleridae) and two species of mites. Can. Ent. **89**(2):111-120. [UROPODIDAE]

Sellnick, M. (1926). Alguns novos acaros (Uropodidae) myrmecophilos e termitophilos. Arch. Mus. Rio de Janeiro **26**:29-56. [TRACHYUROPODIDAE]

Stone, P. and G.D. Ogles (1953). *Uropoda agitans*, a mite pest in commercial fishworm beds. J. Econ. Ent. **46**:711. [UROPODIDAE]

Trägårdh, I. (1941). Further contributions towards the comparative morphology of the Mesostigmata. III. On the Polyaspididae Berl. Zool. Bidrag Fran Uppsala Bd. **20**:345-357.

Trägårdh, I. (1943). Zur Kenntnis der Prodinychidae (Acarina). Ark. Zool. **34A**(21):1-29. [UROPODIDAE]

Trägårdh, I. (1944). Zur Systematik der Uropodiden. Ent. Tidskr. **65**:173-186.

Trägårdh, I. (1946). Diarthrophallina, a new group of Mesostigmata, found on passalid beetles. Ent. Medd. **24**:369-394. [DIARTHROPHALLIDAE]

Treat, A.E. (1975). Some vagrants. **In** Mites of Moths and Butterflies. Cornell Univ. Press, Ithaca: 63-88.

Willmann, C. (1957). Revision einiger Milbengattungen und -arten von den Küsten der Nord- und Ostsee. Abh. naturw. Ver. Bremen **35**(1):162-188. [UROPODIDAE]

Womersley, H. (1961). On the family Diarthrophallidae (Acarina - Mesostigmata-Monogynaspida) with particular reference to the genus *Passalobia* Lombardini 1926. Trans. Royal Soc. S. Australia **84**:27-44.

Zirngiebl-Nicol, I. (1973). Wiederbeschreibung von 7 bekannten *Trachytes*-Arten (Uropodini, Uropodinae). Gangsyst. Parasitiformes. Schrift. für vergleich. Milben. **19**(135):10-20. [POLYASPIDIDAE]

Supercohort Trigynaspides

The Trigynaspides are distinctive, often remarkably ornamented gamasid mites which are largely tropical or subtropical associates of arthropods and, occasionally, of lizards and snakes (Trägårdh 1943 and 1950, Camin and Gorirossi 1955, Strandtmann and Wharton 1958, Womersley 1959, Kinn 1970 and 1971, Kethley 1974, Funk 1974). They usually are large (1000-2000 μ) and heavily sclerotized, and may easily be seen on their hosts. Some species apparently feed on the secretions of their hosts (see Chapter VI, page 63) while others are truly phoretic.

Trigynaspides are distinguished from other gamasids primarily by ventral shield and cheliceral characters. The primary genital arrangement in female trigynaspides comprises an epigynial complex consisting of three elements: a median *mesogynial* and paired lateral *latigynial* shields (Fig. 29a, p. 147). These may be coalesced or fused in distinctive ways, often providing a basis for familial separations within superfamilial taxa. Various sternal elements may also coalesce with the epigynial complex. For example, the posterior metasternal elements may lose connection with the sternal shield proper (as in many Gamasina) and form closely flanking shields anterior to the genital flap (i.e., the EUZERCONIDAE, Plate 51-5, p. 209). Where the fused or paired independent metasternal elements have only metasternal pores and no setae, the shield or shields are referred to as the *sternogynial(s)* (Figs. 29e-f, p. 147). The paired or fused sternogynial elements may expand so as to virtually cover the epigynial region (e.g., the MEGISTHANIDAE (Plate 47-3, p. 205) and the FEDRIZZIIDAE (Plate 46-2, p. 204)).

Where the metasternal elements remain a part of the sternal shield, paired posterior extensions of the sternal shield occasionally are observed beneath the mesogynial and latigynial elements. The *sternovaginal sclerites* have neither setae nor pores (Plate 49-2, p. 207), and may be seen clearly only in dissected or manipulated specimens. Sternovaginal sclerites may be homologous with the *vaginal sclerites* seen in celaenopsoid Trigynaspides (Plate 50-5, p. 208). The anterior portion of the sternal shield may undergo characteristic separation from the shield proper, forming paired *jugular shields* (Plate 48-1, p. 206) or a single fused *tetartosternum* (Plate 46-2, p. 204). A poststernal sclerite bearing neither setae nor pores (*pseudosternogynium,* Plate 47-1, p. 205) may be present.

The movable cheliceral digit of trigynaspide mites is ornamented with filamentous, brush-like or dendritic escrescences which arise medially or terminally on the ventral and (occasionally) the dorsal aspect of the digit (Plate 51-4, p. 209). Females are tocospermous, and male spermadactyli are absent.

Adult and nymphal Trigynaspides usually have an extra pair of ventral setae on tarsus IV ($4\ \frac{3}{3}\ \frac{1}{1}\ \frac{3}{3}\ 3$), a character which they share only with the gamasid cohort Sejina. Setae av_3 and pv_3 are inserted anterior to the suture separating basi- and telotarsus IV, and may sometimes be borne on a separate intercalary sclerite (Plate 47-6, p. 205). Deutonymphs and adults carry a fourth *al* seta (al_x) and a medioventral hair on tarsi II-IV, a setal character combination which is peculiar to the Trigynaspides (Evans 1965).[1]

The Trigynaspides are relegated to two cohorts, the Cercomegistina and the Antennophorina.

---

[1]Setae av_3, pv_3 and al_x are absent on tarsus IV of *Neotenogynium malkini* Kethley (Celaenopsoidea, NEOTENOGYNIIDAE). Their absence indicates neotenic suppression in the post-larval stages (Kethley 1974).

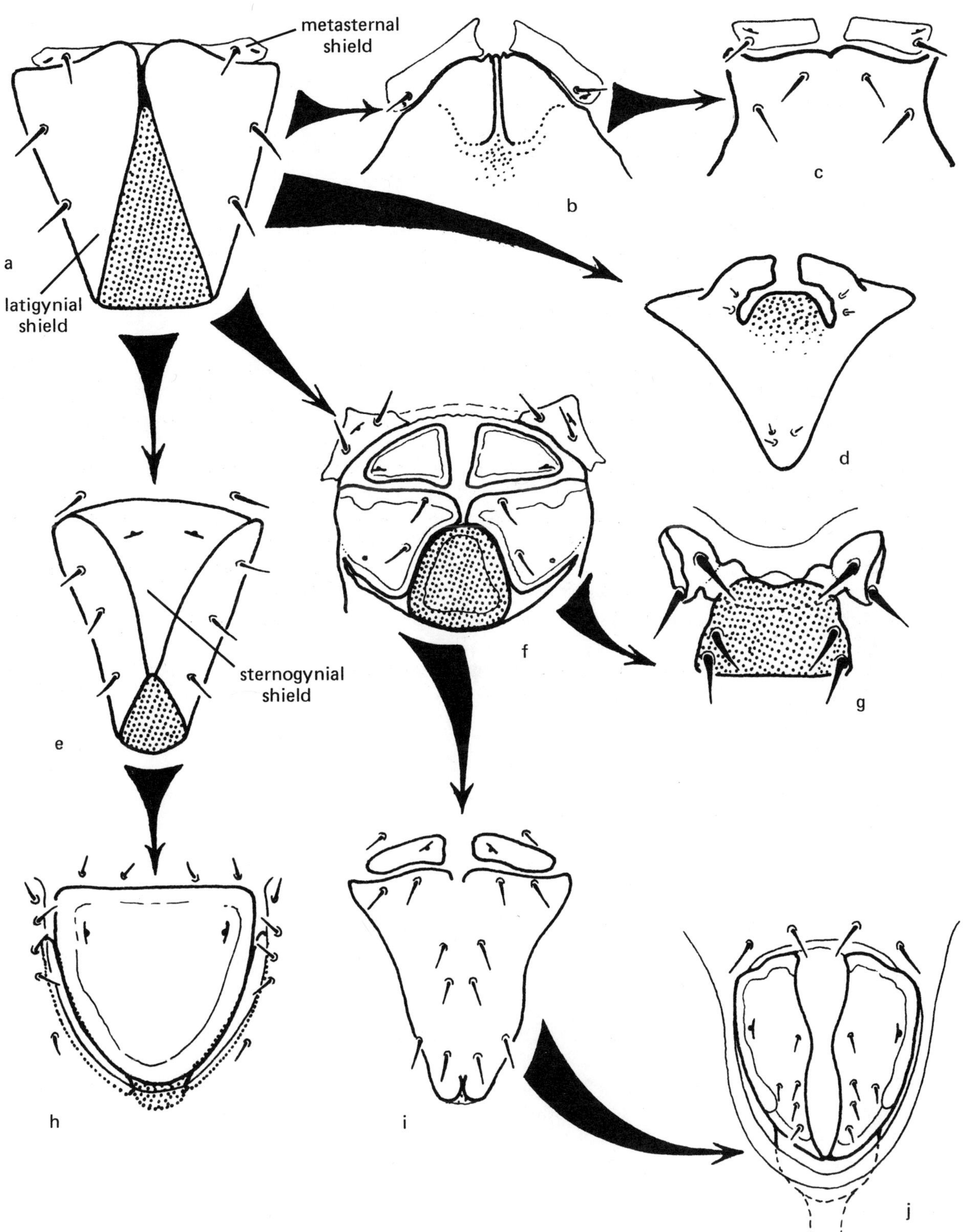

Fig. 29. Representative genital shield types and associated structures in the supercohort Trigynaspides. Arrows indicate possible evolutionary trends in expansion, fusion, or loss of various components, with the diplogyniid genital apparatus (**a**) considered as the basic type. Fig. **b**; EUZERCONIDAE: **c**; CELAENOPSIDAE: **d**; NEOTENOGYNIIDAE: **e**; CERCOMEGISTIDAE: **f**; PARAMEGISTIDAE: **g**; PARANTENNULIDAE: **h**; FEDRIZZIIDAE: **i**; HOPLOMEGISTIDAE: **j**; MEGISTHANIDAE.

Useful References

Camin, J.H. and F.E. Gorirossi (1955). A revision of the suborder Mesostigmata based on new interpretations of comparative morphological data. Chicago Acad. Sci. Spec. Publ. 11:70 pp.

Evans, G.O. (1965). The ontogenetic development of the chaetotaxy of the tarsi of legs II-IV in the Antennophorina (Acari:Mesostigmata). Ann. Mag. Nat. Hist. 6, **13**:513-527.

Funk, R.C. (1974). Megalocelaenopsidae, a new family of Celaenopsoidea (Acari:Mesostigmata). Acarologia **16**(3):382-393.

Kethley, J.B. (1974). Developmental chaetotaxy of a paedormorphic celaenopsoid, *Neotenogynium malkini* n.g., n.sp. (Acari:Parasitiformes:Neotenogyniidae, n.fam.) associated with millipedes. Ann. Ent. Soc. Amer. **67**(4):571-579.

Kethley, J.B. (1977). A review of the higher categories of Trigynaspida (Acari:Parasitiformes). Int. J. Acarology **3**(2):129-149.

Kinn, D.N. (1970). A new genus of Celaenopsidae from California with a key to the genera (Acari:Mesostigmata). Pan-Pac. Ent. **46**(2):91-95.

Kinn, D.N. (1971). The life cycle and behavior of *Cercoleipus coelonotus* (Acarina:Mesostigmata) including a survey of phoretic mite associates of California Scolytidae. Univ. Calif. Publ. Ent. **65**:66 pp.

Strandtmann, R.W. and G.W. Wharton (1958). Manual of mesostigmatid mites parasitic on vertebrates. Inst. Acarology, Univ. Maryland, Contr. 4:330 pp. + vii + 69 plates.

Trägårdh, I. (1943). Further contributions toward the comparative morphology of the Mesostigmata (Acarina). The Antennophoridae and the Megisthanidae. Ark. Zool. **34A**(20):1-10.

Trägårdh, I. (1946). Outlines of a new classification of the Mesostigmata (Acarina) based on comparative morphological data. Kungl. Fisiograf. Sällskapets Handl. N.F. **57**(4):1-37.

Trägårdh, I. (1950). Studies on the Celaenopsidae, Diplogyniidae and Schizogyniidae (Acarina). Ark. Zool. 2, **1**(23):361-451.

Womersley, H. (1959). Some Acarina from Australia and New Guinea paraphagic upon millipedes and cockroaches and on beetles of the family Passalidae. Pt. 2. Fedrizziidae. Trans. Roy. Soc. S. Austral. **82**:11-54.

Cohort Cercomegistina

A single superfamily, the Cercomegistoidea, accommodates the four recognized families of Cercomegistina. All possess an epistome with anterior projections but without a median keel, and fused or partially fused tritosternal laciniae. Cercomegistines may have one, two or more dorsal shields which often are hypertrichous. Claws are present or absent on legs I.

Superfamily Cercomegistoidea
(Plate 45, p. 203)

DIAGNOSIS: *With the characters of the cohort; mesogynial shield reduced or well developed, usually at least partially overlapped by strong, often elongate latigynial shields; latigynials and mesogynial occasionally fused, indistinguishable. Sternal setae I free in integument, on jugular shields, on a tetartosternum, or inserted on entire sternal shield. Palpal tibia and tarsus not fused; chelicerae robust, heavily dentate, movable digit with one or more pilose excrescences.*

Members of the CERCOMEGISTIDAE (Plate 45-1 to 5) occupy a variety of phoretic and free-living habitats in both temperate and tropical realms throughout the world. *Cercomegistus simplicitor* Vitzthum has been collected from ferns in the southwest Pacific (Vitzthum 1935), while other cercomegistid species *(Cercomegistus* and *Cercoleipus)* occur in the galleries of bark beetles in more temperate climates (Kinn 1967, 1970a). *Holocercomegistus agelenophilus* Evans was found in association with social spiders in Africa (Evans 1958), and species of *Vitzthumegistus* (Kethley 1977) are encountered on pagurid crabs (André 1937). Kinn (1971) has found that *Cercoleipus coelonotus* Kinn is phoretic in the adult stage on the bark beetles *Ips confusus* and *I. montanus,* preferring these species to other bark beetles. The phoretic response in *C. coelonotus* is triggered by decrease in phloem moisture and increase in gallery temperature. Unattached mites feed on nematodes but will also prey on *Digamasellus quadrisetus* Berlese, an important egg and larval predator of *Ips confusus.*

Members of the remaining families of Cercomegistoidea are free-living species. *Asternoseius ciliatus* Berlese (ASTERNOSEIIDAE, Plate 45-6) was taken from moss in Italy (Berlese 1910), while *Davacarus gressitti* Hunter was collected from bird nests and tussock grass in the Antarctic realm (Hunter 1970).[1]

Useful References

André, M. (1937). Description de trois espèces d'Acariens (Gamasoidea) pagurophiles. Bull. Soc. Zool. France **62**:45-68. [CERCOMEGISTIDAE]

Athias-Henriot, M. (1959). Redescription du stade adulte de *Seiodes ursinus* Berlese, 1887 (Parasitiformes, Antennophorina). Zool. Anz. **163**(1/2):11-25. [SEIODIDAE]

Berlese, A. (1910). Acari nuovi, Manipuli V, VI. Redia **6**:199-120. [ASTERNOSEIIDAE]

Berlese, A. (1914). Acari nuovi, Manipulus IX. Redia **10**:113-150. [CERCOMEGISTIDAE]

Evans, G.O. (1958). Some mesostigmatid mites from a nest of social spiders in Uganda. Ann. Mag. Nat. Hist. 13, **1**:580-590. [CERCOMEGISTIDAE]

Hunter, P.E. (1970). Acarina: Mesostigmata: free-living mites of South Georgia and Heard Island. Pac. Insects Monogr. 23:43-70. [DAVACARIDAE]

---

[1] Kethley (1977) has designated the monotypic genus *Davacarus* as the sole representative of a new family DAVACARIDAE (Plate 45-7,8). Similarly, Kethley has erected the family SEIODIDAE (Plate 46-1) for *Seiodes ursinus* Berlese (Athias-Henriot 1959). Key characters for these families may be found on page 165).

Kethley, J.B. (1977). A review of the higher categories of Trigynaspida (Acari:Parasitiformes). Int. J. Acarology **3**(2):129-149.

Kinn, D.N. (1967). A new species of *Cercomegistus* (Acari:Mesostigmata) from California. Acarologia **9**(3): 488-496.

Kinn, D.N. (1970). A new genus and species of Cercomegistidae (Acarina:Mesostigmata) from North America. Acarologia **12**(2):244-252.

Kinn, D.N. (1971). The life cycle and behavior of *Cercoleipus coelonotus* (Acarina:Mesostigmata) including a survey of phoretic mite associates of California Scolytidae. Univ. Calif. Publ. Ent. **65**:66 pp. [CERCOMEGISTIDAE]

Kinn, D.N. (1972). A new species of *Holocercomegistus,* including some observations on the chaetotaxy of the pedipalpal and ambulatory appendages of the Cercomegistidae (Acarina). Acarologia **13**(2):258-265.

Valle, A. (1954). Biogeografia dell'Isola di Zannone (Archipelago Pontino), Acarina. Rend. Acad. Nazionale **40**(4/5):142-148. [ASTERNOSEIIDAE]

Vitzthum, H.G. (1935). Terrestrische Acarinen von dem Marquesas. Bernice P. Bishop Mus. Bull. 142:64-99. [CERCOMEGISTIDAE]

Vitzthum, H.G. (1937). Acari in der Kiemenhöhle von *Birgus latro* (Crustacea, Macrura). Zeitschr. Parasit. **9**(5):638-647. [CERCOMEGISTIDAE]

Cohort Antennophorina

Antennophorines comprise the majority of described trigynaspide gamasids. They are assigned to six superfamilies which share the adult characters of a triangular keeled epistome (Plate 51-6, p. 209), free unfused tritosternal laciniae and, in most cases, a single dorsal shield. Claws are absent on legs I.

Superfamily Antennophoroidea
(Plate 49, p. 207)

DIAGNOSIS: *Mesogynial shield reduced, with anterior pedicellate ventrianal element dividing the narrow latigynial shields; latigynials sclerotized along anterior and mesal margins, with more than two pairs of setae; ventrianal shield ampulliform, strongly setate. Sternal setae 1 free in integument, on separate anterior shield or shields, or incorporated into entire sternal shield. Palpal tibia and tarsus distinct, unfused; chelicerae with minute teeth, movable digit with filamentous excrescences.*

A single monogeneric family, the ANTENNOPHORIDAE, is assigned to the Antennophoroidea. *Antennophorus* species are broad, nearly circular mites which are associated with ants in the Nearctic and Neotropical realms, and in Europe (Berlese 1904). *A. uhlmanni* Haller and *A. pubescens* Wasmann commonly attach to the abdomen of their ant hosts (Banks 1915), but others favor the gular region of the head. *A. foreli* Wasmann and *A. grandis* Berlese position themselves facing forward under the ant's head and stroke its mouthparts with their antenniform legs I. The ant responds by disgorging a droplet of fluid which the mite imbibes.

Superfamily Aenictequoidea
(Plates 48-49, pp. 206-207)

DIAGNOSIS: *Mesogynial shield triangular, well developed, separate from ventrianal shield and lying between or beneath well developed latigynial shields, latigynials with 2-10 pairs of setae. Sternal setae I on separate jugulars or on entire sternal shield. Palpal tibia and tarsus insensibly fused; chelicerae edentate.*

Aenictequoids are large, heavily sclerotized mites associated with ants. They are assigned to two families—the monogeneric AENICTEQUIDAE (Plate 48, p. 206) and the PTOCHACARIDAE (Plate 49, p. 207).[1] *Aenicteques chapmani* Jacot, the only nominate species of AENICTEQUIDAE, was described from the ant *Aenictus martini* in the Philippines (Jacot 1939). *Ptochacarus banksi* and *P. silvestrii* Womersley (PTOCHACARIDAE) are ant associates in the Australian and Oriental realms (Womersley 1958).

---

[1] Kethley (personal communication) plans to include an additional family category in the Aenictequoidea, the PHYSALOZERCONIDAE (ANTENNOPHORIDAE pars of Baker and Wharton 1952). I have included this taxon in the key to the Gamasida, pending its formal description by Kethley at a later date. A second undescribed family recognized by Kethley—the MESSORACARIDAE (*Messoracarus mirandus* Silvestri and *M. australis* (Banks))—is provisionally included in the Aenictequoidea, although Kethley indicates that the erection of a new superfamily to contain these and related unassigned species might be more appropriate.

Superfamily Celaenopsoidea
(Plates 50-52, pp. 208-210)

DIAGNOSIS: *Mesogynial shield amalgamated with latigynial and/or ventral elements, or distinct; latigynial shields various, sometimes undifferentiated; vaginal sclerites well developed. Ventral sclerotization various, often with a single extensive shield which encompasses coxal insertions II-IV and the anal aperture. Sternal setae I on entire sternal shield or on a tetartosternum, dorsum of adult with one or three shields. Palpal tibia and tarsus distinct, unfused; chelicerae strong, dentate, movable digit with dendritic or brush-like excrescences.*

The Celaenopsoidea includes six families, most species of which are associates of insects or myriapods. Members of the DIPLOGYNIIDAE (Plate 51, p. 209) are common on passalid and histerid beetles in the New World and in Asia, and on a variety of scarabaeine beetles in Africa. *Passalacarus sylvestris* Pearse and Wharton occurs commonly on the North American passalid *Popilius disjunctus.* Adults of *Ophiocelaeno sellnicki* Johnston and Fain are found on snakes (Johnston and Fain 1964). Other collections include species from under bark (Trägårdh 1950), from litter and treeholes in north central United States, with dampwood termites in Oregon (Krantz 1958), and with a palm weevil in Nicaragua (Hicks 1958). The feeding habits of diplogyniids are not known.

Several species of CELAENOPSIDAE are known to occur with bark beetles in Europe, North and Central America, and Africa (Vitzthum 1926, Trägårdh 1950, Ryke 1957, Kinn 1968), and one has been collected with ants in Tasmania (Trägårdh 1951). Galleries of the bark beetle *Scolytus laevis* shelter two celaenopsids, *Celaenopsis cuspidata* (Kramer) and *Pleuronectocelaeno austriaca* Vitzthum. Kinn (1968) described *P. drymoecetes* from galleries of *Ips* and *Dendroctonus* bark beetles in the New World.

While there are a number of celaenopsid species known from tropical and subtropical realms, the CELAENOPSIDAE is primarily a holarctic group. According to Trägårdh, it may be a relict assemblage which has flourished in temperate climates only by adapting to the galleries of its bark beetle associates.

Species of the family EUZERCONIDAE (Plate 51, p. 209) are common associates of passalid beetles in the New and Old World tropics, but some have invaded the temperate Nearctic realm as well. Observations on *Euzercon latus* (Banks), an associate of the passalid *Popilius disjunctus* in the eastern United States, indicate that immature *E. latus* are scavengers or fungivores in the decaying wood and litter frequented by the beetles (Hunter and Davis 1965). Only adults are found on the beetle host. Hunter and Davis observed adults to feed on the eggs of Collembola and on inactive collembolans, but not on actively moving forms.

Species of SCHIZOGYNIIDAE (Plate 50, p. 208) occur with passalid beetles in South Africa (Trägårdh 1950, Ryke 1957) and India (Sellnick 1954). *Choriarchus regina* Kinn is found in the galleries of bark beetles in western United States (Kinn 1966, 1971). One schizogyniid genus, *Indogynium,* is associated with snakes.

Neotenogynium malkini Kethley, the only described member of the family NEOTENOGYNIIDAE (Plate 50, p. 208), occurs in association with spirostreptid millipedes in the neotropics (Kethley 1974). Unlike other celaenopsoids which are found on their host only in the adult stage, all of the postembryonic instars of *N. malkini* occur on the millipede carrier. Host information on the MEGALOCELAENOPSIDAE (Plate 52, p. 210) is lacking, although a close association with beetles inhabiting bananas is implied for *Megalocelaenopsis oudemansi* Funk, the sole representative of the Neotropical type genus (Funk 1974).

Superfamily Megisthanoidea
(Plates 47-48, pp. 205-206)

DIAGNOSIS: *Mesogynial and latigynial shields represented by narrow fused elements bordering expanded sternogynial shield; sternogynial entire or divided, bearing metasternal pores.*[1] *Typically with a ventral shield separated posteriorly from a short, broad ventrianal shield.*[2] *Sternal setae I on jugular shields or fleshy projections flanking tritosternal base, metasternal and sternal elements fused. Palpal tibia and tarsus separate; chelicerae distinctly toothed, movable digit with dendritic excrescences.*

Two families are assigned to the Megisthanoidea—the MEGISTHANIDAE (Plate 47, p. 205) and the HOPLOMEGISTIDAE (Plate 48). Megisthanids are large ($> 1000\ \mu$) mites associated primarily with passalid beetles, possibly feeding on glandular secretions of their hosts. *Megisthanus floridanus* Banks is a common associate of the passalid *Popilius disjunctus* in the eastern United States. Other species of *Megisthanus* are found clinging to passalid and dung beetles throughout much of the world, including the tropical and Australian realms (Stoll 1886, Womersley 1937, Hunter and Costa 1970).

Nematodes and fungi comprise the primary diet of *M. floridanus* although immatures and adults have been observed to feed on Collembola, collembolan eggs, and immature mites and insects in the laboratory (Butler and Hunter 1968). Larval and nymphal *M. floridanus* construct cells of frass in which they conceal themselves, possibly to avoid desiccation.

Members of the HOPLOMEGISTIDAE also utilize passalid beetles, and are common in the neotropics (Stoll 1886, Turk 1948, Butler and Hunter 1966). Their feeding habits are unknown.

Superfamily Fedrizzioidea
(Plates 46-47, pp. 204-205)

DIAGNOSIS: *Mesogynial and latigynial shields often well developed; mesogynial may be reduced, or mesogynial-latigynial complex may be absent or nearly so, complex sometimes represented by a narrow band bordering an expanded sternogynial shield; sternogynial shield or bare poststernal sclerite always present, sternogynial entire or divided. Ventral and anal shield components variable, either fused or separate. Sternal setae 1 on separate jugular shields or on a tetartosternum, rarely on larger sternal shield fragment. Palpal tibia and tarsus fused or separate; movable digit of chelicera with filamentous excrescences.*

Members of the four families of Fedrizzioidea are tropical in distribution and are encountered on insects, myriapods and snakes. Species of FEDRIZZIIDAE (Plate 46, p. 204) and KLINCKOSTROEMIIDAE (Plate 46, p. 204) are associates of passalid beetles, apparently taking nourishment from them in the form of glandular secretions. Representatives of the FEDRIZZIIDAE are reported from passalid species in Central and South America and the southwest Pacific. Some species have not been found in direct contact with a passalid host (*Neofedrizzia vitzthumi* (Oudemans) and *N. camini* Womersley, for example), but rather in a locale which suggests the presence of passalids (rotting logs, fungi etc.) (Womersley 1959a). The three recognized genera of KLINCKOWSTROEMIIDAE apparently are restricted to the Neotropical realm (Womersley 1959b, Hunter and Butler 1966), where they associate with passalid beetles.

---

[1] The sternogynial shield of *Megisthanus jacobsoni* Warburton also carries the metasternal setae, but this is an exceptional condition.

[2] The ventral and anal elements are fused in *M. jacobsoni.*

The PARAMEGISTIDAE (Plate 47, p. 205) is comprised of species associated with insects and myriapods, primarily in the Old World tropics. Members of the genus *Ophiomegistus* are encountered on lizards and snakes in the southwest Pacific region (Voss 1967), and appear to be obligately parasitic on their reptilian hosts. *Paramegistus* and *Neomegistus* species are found on myriapods (Trägårdh 1906, 1907), while *Antennomegistus* and *Echinomegistus* species are especially common on carabid beetles. Although early literature on *E. wheeleri* (Wasmann) suggested a primary association with North American ants, Nickel and Elzinga (1970a) found adult male and female *E. wheeleri* on two carabid species in Kansas. The mites were observed to feed on organic particles adhering to their beetle hosts.

Promegistus armstrongi Womersley is the sole representative of a new family, the PROMEGISTIDAE (Plate 47, p. 205) (Kethley 1977). *P. armstrongi* was recovered from a passalid beetle *(Mastochilus)* and a carabid *(Pamborus)* in Australia (Womersley 1958).

Superfamily Parantennuloidea
(Plate 50, p. 208)

DIAGNOSIS: Mesogynial shield, when present, always separate from ventrianal shield, latigynial shields generally well developed. Sternal shield variously fragmented, poorly defined, sternal pores usually lost. Anus on separate shield which bears 2-6 setae. Palpal tibia and tarsus unfused; chelicerae edentate, digits reduced, movable digit with filamentous excrescences. Generally weakly sclerotized forms.

A single family, the PARANTENNULIDAE, currently comprises the superfamily, although Kethley (personal communication) has material which represents two new family categories.

The three recognized genera of PARANTENNULIDAE *(Parantennulus, Diplopodophilus* and *Micromegistus)* are associated with myriapods and with carabid beetles in both the Old and New Worlds. *Micromegistus bakeri* Trägårdh has been recovered from the venter of three carabid species in Kansas, USA (Nickel and Elzinga 1970b). It is interesting to note that all stages of *M. bakeri* are found on their carabid hosts. Females are viviparous and give birth while attached to the beetle. *M. bakeri* may feed on dermal secretions of the host or on organic debris adhering on or near the host's mouthparts. Nickel and Elzinga observed a larva and a protonymph of *M. bakeri* feeding on mealworm remains. Predation was never observed.

Useful References

Baker, E.W. and G.W. Wharton (1952). An Introduction to Acarology. Macmillan Co., New York: 465 pp. + xiii.

Banks, N. (1915). The Acarina or mites. USDA Rept. 108:86 pp.

Berlese, A. (1904). Acari mirmecofili. Redia **1**:299-474 + plates. [ANTENNOPHORIDAE]

Butler, L. and P.E. Hunter (1966). A new species of *Hoplomegistus* from Costa Rica (Acarina:Hoplomegistidae). Acarologia **8**(4):525-531.

Butler, L. and P.E. Hunter (1968). Redescription of *Megisthanus floridanus* with observations on its biology (Acarina:Megisthanidae). Florida Ent. **51**:187-197.

Camin, J.H. and F.E. Gorirossi (1955). A revision of the suborder Mesostigmata based on new interpretations of comparative morphological data. Chicago Acad. Sci. Spec. Publ. 11:70 pp.

Donisthorpe, H. St. J.K. (1927). The Guests of British Ants. London: 244 pp. + 16 plates.

Funk, R.C. (1964). An investigation of the Euzerconidae (Mesostigmata:Celaenopsoidea) based on the procedures of numerical taxonomy. Acarologia **6**(fasc. h.s.): 127-132.

Funk, R.C. (1974). Megalocelaenopsidae, a new family of Celaenopsoidea (Acari:Mesostigmata). Acarologia **16**(3):382-393.

Hicks, E.A. (1958). A new genus and species of diplogyniid from Nicaragua (Order Acarina, Family Diplogyniidae). Iowa State Coll. J. Sci. **33**(2):103-110.

Hunter, P.E. (1970). Acarina: Mesostigmata: free-living mites of South Georgia and Heard Island. Pac. Insects Monogr. **23**:43-70. [PARANTENNULIDAE]

Hunter, P.E. and L. Butler (1966). New *Klinckowstroemia* mites from Costa Rican passalid beetles (Acarina: Klinckowstroemiidae). J. Georgia Ent. Soc. **1**(4):24-30.

Hunter, P.E. and M. Costa (1970). Two new African species of *Megisthanus* Thorell (Mesostigmata:Megisthanidae). Florida Ent. **53**(4):233-240.

Hunter, P.E. and R. Davis (1965). Mites associated with the passalus beetle. III. Life stages and observations on the biology of *Euzercon latus* (Banks) (Acarina:Euzerconidae). Acarologia **7**(1):30-42.

Jacot, A.P. (1939). A new antennophorid mite, rider of the Philippine ant, *Aenictus martini.* Philippine J. Sci. **69**(4):433-434. [AENICTEQUIDAE]

Johnston, D.E. and A. Fain (1964). *Ophiocelaeno sellnicki*, a new genus and species of Diplogyniidae associated with snakes (Acari - Mesostigmata). Bull. Soc. roy. Ent. Belg. **100**(6):79-91.

Kethley, J.B. (1974). Developmental chaetotaxy of a paedomorphic celaenopsoid, *Neotenogynium malkini* n.g., n.sp. (Acari:Parasitiformes:Neotenogyniidae, n.fam.) associated with millipedes. Ann. Ent. Soc. Amer. **67**(4):571-579.

Kethley, J.B. (1977). A review of the higher categories of Trigynaspida (Acari:Parasitiformes). Int. J. Acarology **3**(2):129-149.

Kinn, D.N. (1966). A new genus and species of Schizogyniidae from North America with a key to the genera. Acarologia **8**(4):576-586.

Kinn, D.N. (1968). A new species of *Pleuronectocelaeno* (Acarina:Celaenopsidae) associated with bark beetles in North and Central America. Acarologia **10**(2):191-205.

Krantz, G.W. (1958). *Lobogyniella tragardhi,* a new genus and species of diplogyniid mite associated with dampwood termites in Oregon (Acarina:Diplogyniidae). Proc. Ent. Soc. Wash. **60**(3):127-131.

Nickel, P.A. and R.J. Elzinga (1970a). New host records and biological notes for *Echinomegistus wheeleri* (Wasmann) (Acarina:Paramegistidae). J. Kansas Ent. Soc. **43**(1):32-34.

Nickel, P.A. and R.J. Elzinga (1970b). Biology of *Micromegistus bakeri* Trägårdh, 1948 (Acarina:Parantennulidae) with descriptions of immatures and redescription of adults. Acarologia **12**(2):234-243.

Ryke, P.A.J. (1957). *Mixogynium proteae,* a new genus and species of Celaenopsoidea (Mesostigmata: Acarina) from South Africa. Ann. Mag. Nat. Hist. **10**:579-584. [SCHIZOGYNIIDAE]

Samšinák, K. (1957). Die mitteleuropäischen Arten der Familie Diplogyniidae (Acari). Acta Soc. Ent. Cechoslov. **54**(1):1-6.

Sellnick, M. (1954). *Indogynium lindbergi* nov. gen. nov. spec., eine neue Acaride aus Indien. Ent. Tidskr. **75**:285-291. [SCHIZOGYNIIDAE]

Stoll, O. (1886). Arachnida Acaridea. Biol. Centrali-Americana: 55 pp. + xvi + plates. [MEGISTHANIDAE, HOPLOMEGISTIDAE]

Trägårdh, I. (1906). Neue Acariden aus Natal und Zululand. Zool. Anz. **30**:870-877. [PARAMEGISTIDAE]

Trägårdh, I. (1907). Description of two myriopodophilous genera of Antennophoridae with notes on their development and biology. Ark. Zool. **3**(28):1-33 + plate. [PARAMEGISTIDAE]

Trägårdh, I. (1937). Zur Systematik der Mesostigmata. Ark. Zool. **29B**(11):1-8. [HOPLOMEGISTIDAE]

Trägårdh, I. (1941). Contributions towards the comparative morphology and phylogeny of the Mesostigmata. IV. On the Celaenopsidae and Euzerconidae. Ent. Tidskr. **62**:169-176.

Trägårdh, I. (1943). Further contributions towards the comparative morphology of the Mesostigmata (Acarina). The Antennophoridae and the Megisthanidae. Ark. Zool. **34A**(20):1-10.

Trägårdh, I. (1950). Studies on the Celaenopsidae, Diplogyniidae and Schizogyniidae (Acarina). Ark. Zool. ser. 2, **1**(25):361-451.

Trägårdh, I. (1951). *Brachycelaenopsis,* a new genus of Celaenopsidae (Acarina) from Tasmania. Ent. Tidskr. **72**(1-2):60-64.

Turk, F.A. (1948). Insecticolous Acari from Trinidad, B.W.I. Proc. Zool. Soc. London **118**(1):82-125. [KLINCKOWSTROEMIIDAE]

Vitzthum, H.G. (1926). Acari als Commensalen von Ipiden. Zool. Jahr. (Syst.) **52**:407-503. [CELAENOPSIDAE]

Voss, W.J. (1967). Three trigynaspid mites from Philippine reptiles (Acarina:Paramegistidae). J. Med. Ent. **3**(3):261-268.

Womersley, H. (1937). Australian Acari of the genus *Megisthanus* Thorell. Trans. Roy. Soc. S. Austral. **61**: 175-180.

Womersley, H. (1958). Some new or little known Mesostigmata (Acarina) from Australia, New Zealand and Malaya. Trans. Roy. Soc. S. Austral. **81**:115-130. [PTOCHACARIDAE, PROMEGISTIDAE]

Womersley, H. (1959a). Some Acarina from Australia and New Guinea paraphagic upon millipedes and cockroaches and on beetles of the family Passalidae. Pt. 2. Fedrizziidae. Trans. Roy. Soc. S. Austral. **82**:11-54.

Womersley, H. (1959b). *Klinckowstroemiella helleri* (Ouds., 1929) nov. comb. for *Fedrizzia helleri* Ouds., 1929 (Acarina-Klinckowstroemiidae). Zool. Meded. **34**(19):281-288.

SUBORDER GAMASIDA
(Plates 11-52, pp. 169-210)

KEY TO THE FAMILIES

1. One primary genital cover (epigynial shield) well developed and functional or, if reduced and non-functional, usually represented by remnant flanked by or bearing a pair of genital setae (Fig. 28, p. 122). Tarsi II-IV of deutonymph and adult typically with 18 setae $(3\,\frac{3}{2}\,\frac{1}{1}\,\frac{3}{2}\,3)$ (if additional ventral setae are present then without accessory al seta (al_x)), protonymph without medioventral tarsal hair $(3\,\frac{3}{2}\,\frac{1}{0}\,\frac{3}{2}\,3)$ (various tarsal setae may be reduced or lacking in certain obligate parasites); tarsus I with or without claws. Movable digit of chelicera without medially or terminally inserted ventral excrescences, but arthrodial brushes or setal coronets may occur at the base of the digit. Males often with sperm transfer organ, or spermadactyl (Plate 3-5, p. 15), originating on the movable digit (the family HETEROZERCONIDAE is exceptional in that the spermadactyl arises on the fixed digit), commonly with spurs, spines or apophyses on legs II-IV which are used to grasp female during sperm transfer; male hypostomal setae and processes generally similar to those of female (Fig. 5b, p. 14) . . . . Supercohort MONOGYNASPIDES ... 2

— Three primary genital covers (two latigynials and one mesogynial shield) functional, or variously coalesced or reduced (Fig. 29, p. 147); when reduced, often with sternogynial elements or a fused sternogynial shield covering genital aperture (Plate 46-2, p. 204). Tarsi II-III of deutonymph and adult with 19 setae $(4\,\frac{3}{2}\,\frac{1}{1}\,\frac{3}{2}\,3)$, av_3 and pv_3 added to normal complement on tarsus IV $(4\,\frac{3}{3}\,\frac{1}{1}\,\frac{3}{3}\,3)$; protonymph with av_3 and pv_3 on tarsus IV, but lacking one anterolateral and the medioventral seta $(3\,\frac{3}{3}\,\frac{1}{0}\,\frac{3}{3}\,3)$; tarsus I usually without claws. Movable digit of chelicera with medial or terminal dendritic, brush-like or filamentous excrescences (Plates 46-4 and 47-4, pp. 204 and 205), occasionally with additional excrescences at base of digit. Males without cheliceral spermadactyli or distinctive spurs or spines on legs; often with hypertrophied or extra hypostomal setae and/or processes (Plate 52-3, p. 210) . . . . Supercohort TRIGYNASPIDES ... 45

2. Epigynial shield well developed or reduced, occasionally lost in some endoparasitic groups, often extending posteriorly beyond the podosomal region as a fused genitiventral or genitiventrianal entity (Fig. 28, p. 122); with one pair of setae in the podosomal region. Sternal shield usually entire, occasionally reduced or fragmented; typically with one or two dorsal shields, separate marginal shields or platelets absent. Femur IV of deutonymph and adult typically with six setae, hypostomal setae 2-3, when present, generally inserted at approximately the same level (Fig. 5b, p. 14). Male genital aperture at anterior edge of sternal shield or within it; spermadactyli or other sperm-holding devices present on the chelicerae of those forms with an anterior genital opening, males often with leg armature . . . . Cohort GAMASINA ... 6

— Epigynial shield well developed, with or without genital setae, shield confined to the podosomal region, rarely displaced posteriorly or fused with ventral elements. Sternal shield may be entire, divided behind sternal setae 2, or variously fragmented. With one to several dorsal shields; marginal shields or platelets generally present. Femur IV of deutonymph and adult with seven or eight setae (if only six setae are present, then hypostomals 3 are always behind hypostomals 2). Male genital aperture within sternal region, never at anterior shield margin, spermadactyli absent; spines or apophyses for grasping female may be present on legs II 3

3. Sternal shield fragmented or simply divided behind sternals 2; epigynial shield with one, three, or several pairs of setae in the podosomal region. Femur IV of adults and deutonymphs with seven setae; hypostomals 2 and 3 inserted at more or less the same level. . . . . Cohort SEJINA, Superfamily SEJOIDEA ... 4

— Sternal shield entire, rarely fragmented (*Thinozercon,* Plate 39, p. 197), often fused with adjacent podal and peritremal shields; epigynial shield truncate posteriorly, rarely fused with ventral elements (*Thinozercon*), occasionally displaced behind podosomal region (*Metagynella,* Plate 42, p. 200), without genital setae (*Protodinychus,* Plate 39-1, p. 197, and certain species of POLYASPIDIDAE, Plate 40, p. 198, are exceptions). Femur IV of adults with six, seven or eight setae, hypostomals 3 always well behind hypostomals 2, describing a nearly straight line with hypostomals 1 (Plate 41-1, p. 199) .Cohort UROPODINA ... 39

4. With one pair of epigynial setae; male genital aperture behind coxae II. . . . . (Plate 11, p. 169) Family MICROGYNIIDAE

— With more than one pair of epigynial setae; male genital aperture between coxae II . . . 5

5. Epigynial shield subquadrangular, extended anteriorly so as to overlap posterior sternal shield elements, ornamented with at least 20 unpaired setae; with one to four major dorsal shields. . . . . (Plate 11, p. 169) Family ICHTHYOSTOMATOGASTERIDAE

— Epigynial shield more or less reniform, not overlapping posterior sternal shield elements, with three pairs of laterally inserted setae; with two to seven dorsal shields. . . . . (Plate 12, p. 170) Family SEJIDAE

6. Sternal and metasternal shields of female fused, with four pairs of setae,[1] sternals 1 may be inserted in a weakly defined anteromarginal extension. Epigynial shield usually rounded anteriorly, separated posteriorly from ventrianal shield. . . . . Superfamily RHODACAROIDEA ... 14

— Sternal shield free or marginally fused with sternal-metasternal elements, with 3 or fewer pairs of setae,[2] shield may be reduced or absent in parasitic forms; metasternal shields typically free, occasionally absent. Epigynial shield rounded, pointed, or truncate anteriorly, fused posteriorly to ventral or ventrianal elements or freely articulated, sometimes absent 7

7. With the following combination of characters: non-parasitic forms with 5 dorsal setae on tibia I $(1\frac{3}{2}\frac{2}{1}2$ or $2\frac{3}{2}\frac{2}{1}2)$[3] and 7 setae on tibia III $(1\frac{1}{1}\frac{2}{1}1)$[4]. Free-living, phoretic and nidicolous species . . . . Superfamily EVIPHIDOIDEA ... 22

— Parasitic, phoretic, or free-living forms without the above combination of chaetotactic characters; tibia I typically with 4 or 6 dorsal setae[5] and with 8 or more setae on tibia III. 8

8. Sternal shield fragmented and reduced, represented by lateral remnants which are contiguous with endopodal shields, sternal pores 2 and 3 absent. Venter of opisthosoma with a pair of distinctive sucker-like adhesive organs (secondarily lost in some species). Males often with spermadactyli arising on the fixed cheliceral digits . . . . .Superfamily HETEROZERCONOIDEA ... 25

---

[1]The metasternal setae (sternals 4) may be separated from the sternal shield in certain genera of the family OLOGAMASIDAE (*Psammonsella* Haq. for example). These forms are distinguished by their distinctive tubuli annulati (as in Plate 18-1, p. 176), combined with presence of four ventral setae on tibia I $(2\frac{3}{2}\frac{3}{2}2)$.

[2]Complete sternal-metasternal fusion occurs in some PACHYLAELAPIDAE (e.g. *Pachylaelaps, Olopachys,* and *Sphaerolaelaps*). In these cases, the sternal shield has 4 pairs of setae and extends posteriorly to the level of coxae IV (Plate 28-1, p. 186).

[3]If 4 dorsal setae, then only 1 al seta is present $(1\frac{2}{1}\frac{2}{1}1)$, and the papal apotele is 3-tined.

[4]If tibia III has 8 setae (the PARHOLASPIDIDAE are exceptional in this respect - $1\frac{2}{1}\frac{2}{1}1$),then genu IV also has 8 setae $(2\frac{2}{1}\frac{2}{0}1)$.

[5]If 5 dorsal setae, then femur I usually has less than 13 setae.

— Sternal shield may be reduced but not as described above, sternal pores 2 and 3 present (with few exceptions amongst the obligate parasites); without opisthosomal adhesive organs. Male spermadactyl, when present, always arising on the movable digit. 9

9. Sternal shield well developed and entire, with three pairs of setae and two or three pairs of pores; peritremes present and extending at least to level of coxae II; dorsal shield entire or divided, occasionally reduced posteriorly; palptarsal claw always 3-tined. With the following chaetotactic combination: with a posteroventral seta on genu III $(2\ \frac{2}{1}\ \frac{2}{1}\ 1$ or $2\ \frac{2}{1}\ \frac{2}{1}\ 2)$ and two on tibia I $(2\ \frac{3}{2}\ \frac{3}{2}\ 2)$; tibia IV with three posterodorsal setae $(2\ \frac{1}{1}\ \frac{3}{1}\ 2)$. Free-living predaceous ground species . . . . Superfamily PARASITOIDEA ... 12

— Sternal shield variously developed or reduced, peritremes sometimes absent or greatly reduced in length, palptarsal claw 2- or 3-tined, absent in some parasitic forms; without the above chaetotactic combination. Free-living and parasitic species 10

10. Epigynial shield truncate or weakly convex posteriorly and widely separated from anal shield, sometimes abutting a broad ventrianal shield (Plate 20-6, p. 178) (if epigynial shield is rounded posteriorly, then anal shield is usually not triangular in shape). Free-living, phoretic or parasitic on arthropods, or phoretically associated with birds 11

— Epigynial shield rounded or pointed posteriorly, usually widely separated from triangular anal shield, sometimes absent;[1] epigynial shield occasionally expanded so as to nearly abut anal shield, assuming a flattened or invaginated aspect posteriorly (Fig. 28c, p. 122). Predators, nidicoles, insect associates, and ecto- and endoparasites of vertebrates . . . . Superfamily DERMANYSSOIDEA ... 26

11. Deutonymphs and adults with less than 20 pairs of dorsal shield setae (if more than 20 pairs are present, then the peritremes are absent or tibia IV has two anterodorsal setae $(2\ \frac{2}{1}\ \frac{2}{1}\ 1)$). Free-living, or phoretic or parasitic on insects. . . . . Superfamily PHYTOSEIOIDEA ... 18

— Deutonymphs and adults with more than 20 pairs of dorsal shield setae. Free-living or phoretically associated with arthropods or birds. . . . . . Superfamily ASCOIDEA ... 16

12. Epigynial shield triangular and flanked by large discrete metasternal shields. Male spermadactyli represented by sperm-holding device which is coalesced distally with movable digit; males with highly developed spurs and apophyses on legs II, femoral spur often as long or longer than femur. Free-living; deutonymphs often phoretic on insects . . . . (Plate 13, p. 171) Family PARASITIDAE

— Epigynial and metasternal shields not as above. Male spermadactyli free, not coalesced distally with movable digit; apophyses may be present on legs II, rarely developed as described above. Free-living. 13

13. Distal extensions of internal malae of hypostome broadly fringed, often moustache-like; palpal claw with a hyaline, membranous excrescence arising at the apotelic base. Epigynial and ventral shields usually separated by incomplete suture; with claws on tarsus I. Male genital opening at anterior margin of sternal shield. . . . . (Plate 14, p. 172) Family VEIGAIIDAE

---

[1] Fusion of genitiventral and anal elements occurs in the laelapid genus *Ololaelaps* (Plate 37-1, p. 195).

— Internal malae terminate in unmodified, lightly fringed elements; palpal claw without adjacent hyaline element. Epigynial shield rounded posteriorly, without adjacent ventral shield; tarsus I terminates in a series of short hairs, claws present or absent. Male genital aperture well within sternal region . (Plate 15, p. 173) Family ARCTACARIDAE[1]

14. Dorsal shield divided; adults usually with distinctive sclerotic nodules *(scleronoduli)* in region of anterior dorsal shield setae i_4 and i_5 (Plate 17-5, p. 175) (absent in *Digamasellus* and *Longoseius*); anterior portion of sternal shield weakly defined, usually carrying sternal setae 1. Spermadactyli of male often recurved basally. 15

— Dorsal shield divided or entire, without scleronoduli; anterior portion of sternal shield undifferentiated from more posterior portions. Spermadactyli of male not recurved basally. Free-living predators . . . .(Plate 18, p. 176) Family OLOGAMASIDAE

15. Palptarsal claw 2-tined; usually with seven setae on genu IV $(1\ \frac{2}{1}\ \frac{2}{0}\ 1)$ and seven on tibia IV $(1\ \frac{1}{1}\ \frac{2}{1}\ 1)$. Male with sternal setae 5 ("epigynial" setae) on separate platelets. Predators and insect associates . . . .(Plate 17, p. 175) Family DIGAMASELLIDAE

— Palptarsal claw 3-tined; with ten setae on genu and tibia IV $(2\ \frac{2}{1}\ \frac{3}{1}\ 1$ and $2\ \frac{1}{1}\ \frac{3}{1}\ 2)$. Male with sternal setae 5 on sternitigenital shield. Free-living soil predators. . . . . (Plate 16, p. 174) Family RHODACARIDAE

16. Dorsal shield divided medially, with a transverse row of four posteromarginal fossae usually clearly defined on opisthonotal shield. Peritremes reduced, never extending anteriorly beyond coxae III. Genu IV with ten setae $(2\ \frac{2}{1}\ \frac{3}{1}\ 1)$.. With a ventrianal shield; male genital aperture well within sternal region. Free-living . . . . (Plate 19, p. 177) Family ZERCONIDAE

— Dorsal shield divided or entire, without posteromarginal fossae; peritremes usually extend beyond coxae III. Genu IV typically with nine setae $(2\ \frac{2}{1}\ \frac{3}{0}\ 1)$ (only seven setae on genu IV of certain groups). With an anal or ventrianal shield; male genital aperture at anterior margin of sternal shield 17

17. Without intercoxal exopodal development, sternal shield lacking intercoxal endopodal extensions (*Halodarcia incideta* Karg is an exception), free endopodalia present or absent. Tarsi II-IV often with a pair of elongate, membranous, distally acuminate dorsodistal setae which extend well beyond tarsal claws; tibia IV typically with eight setae $(2\ \frac{1}{1}\ \frac{2}{1}\ 1)$. Free-living . . . .(Plate 19, p. 177) Family HALOLAELAPIDAE

— Endopodal and exopodal elements generally distinct (exopodal elements reduced to fragments in the subfamily Arctoseiinae), endopodal elements usually present and fused with sternal shield (exceptions in the Arctoseiinae). Elongate dorsodistal tarsal setae present or absent, if present then accompanied by a median elongate or broadly rounded pulvillar lobe (Plate 20-7, p. 178); tibia IV typically with ten setae $(2\ \frac{1}{1}\ \frac{3}{1}\ 2)$ (only seven setae on tibia IV of the Arctoseiinae $(1\ \frac{1}{1}\ \frac{2}{1}\ 1)$, and nine on tibia IV of *Protogamasellus* $(2\ \frac{1}{1}\ \frac{3}{1}\ 1)$). Free-living or phoretically associated with insects or birds . . . . (Plate 20, p. 178) Family ASCIDAE

18. Stigmata located dorsolaterally in a distinctive tuberculate-reticulate dorsal shield, peritremes absent. Sternal setae 1 on presternal shield or paired jugularia. Male with genital opening within sternal shield, spermadactyli absent. Tarsus I with some dorsal setae minutely clubbed, claws absent. Free-living . (Plate 25, p. 183) Family EPICRIIDAE

---

[1]The position of the male genital aperture in the ARCTACARIDAE is similar to that in the families ZERCONIDAE (Ascoidea) and EPICRIIDAE (Phytoseioidea). The coincident occurrence of this unusual character in the three families has led to their clustering in some treatments (e.g. Baker et al. 1958).

— Stigmata ventrolateral, peritremes present; dorsal shield variously ornamented. Sternal setae 1 on sternal shield or free in integument. Male with genital opening at or near anterior margin of sternal shield element, spermadactyli present. Tarsus I without clubbed setae, with or without claws 19

19. Legs I greatly elongated; genual, tibial and tarsal segments attenuate, subequal, tarsus I terminating in one or two whiplike setae. With a pair of large, distinctive dorsal pores between setae J_4 and Z_4. Free-living. . . . (Plate 24, p. 182) Family PODOCINIDAE

— Legs I not as above, claws usually present on tarsus I. Dorsal pores present but without a distinct pair between the J and Z series 20

20. Sternal shield typically with two pairs of setae (sternals 1-2), sternals 3 occasionally on sternal shield or on adjacent platelets. Corniculi may be divided distally or otherwise modified in the distal portion; posterior row of deutosternal denticles extending laterally beyond insertions of capitular setae (Plate 23-2, p. 181). Chaetotaxy of femur II is $2\ \frac{4}{3}\ 1$, tibia IV is $2\ \frac{2}{1}\ \frac{2}{1}\ 1$. Free-living or phoretically associated with insects or birds . . . . (Plate 23, p. 181) Family AMEROSEIIDAE

— Sternal shield with 0-3 pairs of setae; corniculi not modified distally, posterior row of deutosternal denticles more or less confined to deutosternal groove. Chaetotaxies of femur II and tibia IV not as above, tibia IV with only one anterolateral seta 21

21. Fixed cheliceral digit absent or reduced to less than 1/4 the length of the movable digit; tritosternum commonly absent or reduced to a basal remnant. Anal opening terminal (occasionally posteroventral), in an anal shield. Parasites of insects . . . . (Plate 22, p. 180) Family OTOPHEIDOMENIDAE

— Fixed cheliceral digit normally developed, subequal in length to movable digit; tritosternum present. Anal opening subterminal, in a ventrianal shield. Free-living aerial predators. . . . . (Plate 21, p. 179) Family PHYTOSEIIDAE

22. Tibia and genu I each with one anterolateral seta ($1\ \frac{3}{2}\ \frac{2}{1}\ 2$ or $1\ \frac{2}{1}\ \frac{2}{1}\ 1$). Free-living or associated with insects or amphipods . . . . (Plate 26, p. 184) Family EVIPHIDIDAE

— Tibia and genu I each with two anterolateral setae 23

23. Chelicerae with one or two plumose or filamentous arthrodial processes, or with an arthrodial brush and an adjacent coronet of setae. Typically with a ventrianal shield (if anal shield is present, it is widely separated from genitiventral elements) 24

— Arthrodial process, when apparent, rarely forming a distinct brush. Typically with an anal shield closely bordered by a genitiventral shield, or with a fused genitiventrianal shield, occasionally with a ventrianal shield; peritremal, parapodal and metapodal elements usually fused, extending posteriorly beyond coxae IV. Free-living or associated with insects . . . .(Plate 28, p. 186) Family PACHYLAELAPIDAE

24. Peritremes generally looped proximally, joining the stigmata posteriorly, tarsus I usually without claws; *posterior paradactyli* of pretarsi II-IV membranous, distally divided or deeply serrate when observed obliquely. With a pair of strong accessory sclerites beneath the lateral margins of the epigynial shield. Free-living or associated with insects . . . . (Plate 27, p. 185) Family MACROCHELIDAE

— Peritremes normal, joining the stigmata anteriorly, tarsus I with or without claws; posterior paradactyli of pretarsi II-IV setate, undivided distally. Lateral accessory sclerites weak or absent. Free-living . . . . (Plate 26, p. 184) Family PARHOLASPIDIDAE

25. Genua III and IV each with three posterodorsal setae ($2\ \frac{2}{1}\ \frac{3}{1}\ 1$); male spermadactyl arising on fixed cheliceral digit. Associated with millipedes and snakes . . . . (Plate 28, p. 186) Family HETEROZERCONIDAE

— Genua III and IV each with two posterodorsal setae ($2\ \frac{2}{1}\ \frac{2}{1}\ 1$ or $2\ \frac{2}{1}\ \frac{2}{1}\ 0$); male spermadactyl arising on movable cheliceral digit. Associated with scolopendrine centipedes. . . . . (Plate 28, p. 186) Family DISCOZERCONIDAE

26. Sternal shield over six times wider than long at its widest point; epigynial membrane broad, convoluted, epigynial setae flanking narrow scleritic epigynial remnant; anal shield separate. Opisthosoma considerably broader than long, with a fringe of spatulate setae. Parasites of snakes . . . . (Plate 29, p. 187) Family OMENTOLAELAPIDAE

— Without the above combination of characters 27

27. Chelicerae massive, hooked; with a broad, heavy epistome overlying gnathosoma. Parasites of Neotropical bats . . . . (Plate 29, p. 187) Family SPELAEORHYNCHIDAE

— Epistome and chelicerae not as above 28

28. Peritremes usually absent or greatly reduced. Respiratory tract parasites of mammals, reptiles or birds 29

— Peritremes rarely absent, occasionally reduced. Free-living or ectoparasites of vertebrates or insects. 31

29. Epigynial shield absent or rudimentary (distinct in *Zumptiella bakeri*); sternal shield generally present. Stigmata ventral or lateroventral. Respiratory tract parasites of mammals. . . . . (Plate 30, p. 188) Family HALARACHNIDAE

— Epigynial shield distinct, sometimes reduced; sternal shield present or absent. Stigmata lateral or dorsal. Parasites of snakes or birds 30

30. Sternal and epigynial shields well developed but often weakly sclerotized, sternal and epigynial setae usually minute or absent. Stigmata lateral. Respiratory tract parasites of snakes . . . . (Plate 30, p. 188) Family ENTONYSSIDAE

— Sternal shield reduced or absent, but with distinct—or distinctive—sternal setae; epigynial shield well developed or reduced. Stigmata dorsal. Respiratory tract parasites of birds . . . . (Plate 31, p. 189) Family RHINONYSSIDAE

31. Legs I extremely stout, with heavy sessile claws; other legs slender, with long pretarsi and small claws; coxae widely separated. Sternal shield absent or barely visible as an interruption in the ventral integumentary striae; posteroventral portion of idiosoma ornamented with distinctive spur-like or broad flattened setae. Parasites of armadillos . . . . (Plate 32, p. 190) Family DASYPONYSSIDAE

— Legs I-IV of comparable thickness, or coxae I-IV contiguous; sternal shield well developed or reduced, but distinctly tanned and easily visible; without spur-like or flattened setae as above. 32

32. Tritosternum absent or represented by tritosternal base remnant (if tritosternal base is well developed, then the peritremes are reduced and extend only to the level of the anterior edge of coxae III). Sternal setae inserted at margins of reduced shield or in integument bordering it; epigynial shield reduced, with or without setae. Parasites of bats . . . . (Plate 33, p. 191) Family SPINTURNICIDAE

— Tritosternum well developed, with laciniae 33

33. Sternal shield subrectangular, reduced laterally, carrying only sternal setae 1 and associated pores, sternals 2 and 3 in adjacent integument. Anal shield narrowly extended posterior to postanal seta; posterior margin of idiosoma with two pairs of long flagellate setae. Parasites of edentates . . . . (Plate 32, p. 190) Family MANITHERIONYSSIDAE

— Sternal and anal shields variously developed, often reduced or expanded but not as above 34

34. Chelicerae of female whip-like, stylettiform, digits minute; corniculi membranous, indistinct 35

— Chelicerae normally produced, elongate or attenuate but not stylettiform; corniculi strongly or weakly sclerotized, well developed or obscure 36

35. Second cheliceral segment of female elongate, far exceeding basal segment in length; male chelicerae normally produced. Idiosoma broadly rounded posteriorly. Parasites of mammals and birds . . . . (Plate 33, p. 191) Family DERMANYSSIDAE

— Second cheliceral segment normally developed, considerably shorter than greatly elongated basal segment. Idiosoma strongly narrowed posteriorly. Parasites of porcupines and snakes . . . . (Plate 34, p. 192) Family HYSTRICHONYSSIDAE

36. Chelicerae elongate, edentate; corniculi membranous, usually lobate; palptrochanter often with a raised medioventral keel. With a large anterior nonsetigerous spur on coxa II (rarely minute or absent), other coxae without spurs but occasionally with small ridges (e.g. *Chirocetes*); chaetotaxy of genu IV diverse but commonly with two ventral setae. Parasites of mammals, birds and reptiles . . . . (Plate 34, p. 192) Family MACRONYSSIDAE

— Chelicerae various, dentate or edentate; corniculi strongly sclerotized or membranous, horn-like, barbed or lobate; without raised medioventral keel on palptrochanter. Generally either with more than one large nonsetigerous coxal spur (e.g. *Hirstionyssus*), or coxal spurs absent, (a single anterior spur is present on coxa II of *Scutanolaelaps* (IXODORHYNCHIDAE)). Chaetotaxy of genu IV diverse, but commonly with one ventral seta 37

37. Corniculi attenuate-acuate, often barbed;[1] with spur-like setae on coxae II, I-II, or I-III. Palptarsal claw present but greatly reduced, generally with a single tine. External parasites of snakes . . . . (Plate 35, p. 193) Family IXODORHYNCHIDAE

— Corniculi not as above, spur-like setae present or absent. Palptarsal claw present, 2-3 tined 38

38. Fixed cheliceral digit absent; with only two pairs of hypostomal setae. Peritremes of female short, looped medially or terminally, confined to level of coxae III, or coxae III-IV. Male may be without peritremes. Parasites of bees . . . . (Plate 35, p. 193) Family VARROIDAE

[1] Cornicular barbs are absent in the genera *Hemilaelaps, Scutanolaelaps, Strandtibbettsia* and *Asiatolaelaps*.

— Chelicerae dentate or edentate, fixed digit present; with three pairs of hypostomal setae in nymphs and adults. Peritremes variously produced, typically well developed and elongate, occasionally absent. A heterogenous group comprising free-living forms, and facultative and obligate parasites. . . . . (Plates 36-37, pp. 194-195) Family LAELAPIDAE

39. Coxae I widely separated; with a separate anteromarginal shield dorsally. Palpgenu of adults with six setae; tibia of leg I with three or four ventral setae ($2\,\frac{2}{1}\,\frac{2}{2}\,2$ or $2\,\frac{2}{2}\,\frac{3}{2}\,2$), genua each with at least five dorsal setae. . . . Superfamily THINOZERCONOIDEA ... 42

— Coxae I widely separated or nearly contiguous; anteromarginal shields absent or fused anteriorly with dorsal shield, or extending posteriorly beyond the level of the podosoma. Palpgenu of adults never with more than five setae; tibia of leg I with no more than two ventral setae, genua with no more than four dorsal setae 40

40. Palpgenu with five setae; tibia of leg I with four dorsal setae ($2\,\frac{2}{1}\,\frac{2}{1}\,2$ or $1\,\frac{2}{1}\,\frac{2}{1}\,1$), rarely with three (e.g. *Caminella* and *Iphidinychus*, $1\,\frac{1}{1}\,\frac{2}{1}\,1$): genua II-III with two anterolateral setae . . . . Superfamily POLYASPIDOIDEA ... 43

— Palpgenu with four or five setae; tibia of leg I with fewer than four dorsal setae, genua II-III with one anterolateral seta 41

41. Coxae I nearly contiguous, partially overlapping or covering tritosternal base; peritreme elongate but often strongly convoluted to follow lateral ridge contours. Palpgenu with five setae; tibiae of legs I-IV with three dorsal setae ($1\,\frac{1}{1}\,\frac{2}{1}\,1$); femur IV with six setae ($1\,\frac{1}{0}\,\frac{2}{1}\,1$) . . . . .Superfamily UROPODOIDEA ... 44

— Coxae I widely separated, not overlapping tritosternal base (if coxae I contiguous, then tritosternal base is completely anterior to coxal confluence). Palpgenu with four setae; tibiae I-IV with only one dorsal seta ($1\,\frac{0}{1}\,\frac{1}{1}\,1$); femur IV with eight setae ($1\,\frac{2}{1}\,\frac{2}{1}\,1$) . . . . Superfamily DIARTHROPHALLOIDEA, Family DIARTHROPHALLIDAE

42. Sternal setae 1 on jugular shields, remainder of sternal region unsclerotized, with a pair of presternal setae flanking the tritosternal base. Epigynial shield linguiform, non-setate, fused posteriorly to ventral elements. Femur I with four ventral setae ($2\,\frac{5}{4}\,1$), genu I with six dorsal setae ($2\,\frac{3}{1}\,\frac{3}{1}\,2$) . . . .(Plate 39, p. 197) Family THINOZERCONIDAE

— Sternal setae 1 inserted in fused sternal-podal-peritremal shield which may be contiguous with ventral shield; epigynial shield free posteriorly, with a pair of setae. Femur I with three ventral setae ($2\,\frac{5}{3}\,2$), genu I with five dorsal setae ($2\,\frac{3}{1}\,\frac{2}{1}\,2$). . . . .(Plate 39, p. 197) Family PROTODINYCHIDAE

43. Genu IV often with only seven setae ($1\,\frac{2}{1}\,\frac{2}{0}\,1$). Podonotal and pygidial shields of deutonymph insensibly fused to form a single plate . . . .(Plate 40, p. 198) Family DITHINOZERCONIDAE

— Genu IV with more than seven setae. Podonotal and pygidial shields of deutonymph usually separate (fused in non-phoretic deutonymphs of *Polyaspis s.l.*). . . . .(Plate 40, p. 198) Family POLYASPIDIDAE

44. Podonotal and mesonotal shields of protonymph separate. Variously ornamented mites encountered in a wide range of organic substrates, including ant nests; deutonymphs may attach phoretically to ants or other arthropods by means of an anal pedicel. . . . . (Plates 41-43, pp. 199-201) Family UROPODIDAE

— Podonotal and mesonotal shields of protonymph insensibly fused. Heavily sclerotized or highly modified, often large (> 1000 μ), exclusively myrmecophilous mites; adults may attach to ants by means of pretarsi and/or mouthparts; deutonymph does not secrete an anal pedicel. . . . . .(Plate 44, p. 202) Family TRACHYUROPODIDAE

45. With two or more dorsal shields (if one, then a line of fusion exists between anterior and posterior shields), usually with a pelage of setae. Epistome with anterior projections or serrations, without median keel; tritosternal laciniae fused, occasionally separated terminally . . . . .Cohort CERCOMEGISTINA, Superfamily CERCOMEGISTOIDEA ... 46

— With a single dorsal shield (if more than one shield, then vaginal sclerites are well developed). Epistome triangular, ususally smooth and with a median keel; tritosternum usually with two distinct laciniae, occasionally fused in the proximal portion. . . . . Cohort ANTENNOPHORINA ... 49

46. With four dorsal shields, metapodal shields free. Mesogynial shield well developed, elongate-subrectangular. . . . .(Plate 45, p. 203) Family DAVACARIDAE

— With two subequal dorsal shields, metapodal shields contiguous with exopodal and peritremal elements. Mesogynial shield reduced or lost, or triangular in shape 47

47. Sternal shield fragmented, sternal setae 1 inserted in integument posterior to paired presternal shields. Mesogynial shield triangular, more than ½ the length of elongate latigynial shields . . . . (Plate 45, p. 203) Family ASTERNOSEIIDAE

— Sternal setae 1 inserted on entire sternal shield, or on anterior jugularia or tetartosternum. Mesogynial shield reduced in size and partially fused with latigynial elements, or entirely lost 48

48. Sternal setae 1 on separate jugularia or tetartosternum; latigynial shields distinct, elongate and with membranous mesal margins, abutting or partially overlapping small triangular mesogynial shield. Ventrianal shield free posteromarginally, not fused with dorsal shield . . . .(Plate 45, p. 203) Family CERCOMEGISTIDAE

— Sternal setae 1 on an entire sternal shield; latigynial and mesogynial shields insensibly fused. Ventrianal shield fused posteromarginally with dorsal shield . . . . (Plate 46, p. 204) Family SEIODIDAE

49. Sternogynial shield(s) present, without sternal setae[1] but bearing sternal pores 3 (rarely with a poststernal sclerite bearing neither setae nor pores *(pseudosternogynium)* in place of a sternogynial shield). 50

— Without sternogynial shield or pseudosternogynium, sternal pores 3 on sternal or metasternal shields, or absent 55

50. Chelicerae with filamentous excrescences; anal shield contiguous with or insensibly fused to ventral elements. Often with fovae pedales to accommodate folded legs; palpal tibia and tarsus may be fused . . . . .Superfamily FEDRIZZIOIDEA ... 51

— Chelicerae with dendritic excrescences; anal opening in a narrow ventrianal shield which usually is separated from more anterior ventral elements by at least a narrow band of integument (*Megisthanus jacobsoni* Warburton is an exception). Fovae pedales never present, palpal tibia and tarsus normally articulated . . . . Superfamily MEGISTHANOIDEA ... 54

[1]Small secondary setae are found on the sternogynial shields of *Megisthanus remilleti* Hunter and Costa, *M. berlesei* H. and C., *M. jacobsoni* Warburton, and certain species of *Klinckowstroemia.*

51. Sternogynial shield entire, inversely triangular; tetartosternum, fovae pedales present. Turtle-like mites associated with passalid beetles. 52

— With divided sternogynial elements or with a pseudosternogynium; sternals 1 in jugularia, in a tetartosternum, or in an entire sternal shield. Fovae pedales absent. Associated with arthropods or with reptiles 53

52. Latigynial and mesogynial shields well developed; vaginal sclerites present but reduced. Male genital aperture ellipsoid, located between coxae III . . . . (Plate 46, p. 204) Family KLINCKOWSTROEMIIDAE

— Latigynial and mesogynial shields lost or represented only by a narrow band bordering the posterior edge of the enlarged sternogynial shield; vaginal sclerites absent or obscure. Male genital aperture circular or only slightly wider than long, located between coxae II-III . . . . (Plate 46, p. 204) Family FEDRIZZIIDAE

53. Sternal setae 1 on jugularia or (rarely) on a tetartosternum; with divided sternogynials. Palpal tibia and tarsus fused. Associated with insects, millipedes and reptiles. . . . . (Plate 47, p. 205) Family PARAMEGISTIDAE

— Sternal setae 1 on sternal shield; with a narrow, horizontally oriented pseudosternogynium bordering the latigynial and mesogynial shields, metasternal shields fused to endopodal elements; mesogynial shield fused insensibly with ventrianal shield. Palpal tibia and tarsus normally articulated. Chaetotaxy of tibiae I-III unique to supercohort ($2\,\frac{2}{1}\,\frac{2}{2}\,1$, $2\,\frac{2}{1}\,\frac{2}{1}\,1$, and $2\,\frac{2}{1}\,\frac{2}{1}\,0$). Associated with passalid and carabid beetles . . . . (Plate 47, p. 205) Family PROMEGISTIDAE

54. Sternal setae 1 on fleshy projections flanking tritosternum, or on narrow divided or entire presternal platelets; latigynial and mesogynial shields lost, represented by a ridge bordering the sternogynial elements posteriorly, sternogynial shield divided or entire, completely covering the genital field; vaginal sclerites reduced, without heads but often with thickened arms. Associated with passalid beetles . . . . (Plate 47, p. 205) Family MEGISTHANIDAE

— Sternal setae 1 on well developed jugularia; latigynial shields fused, forming an independent entity between the divided sternogynials anteriorly and a ventral shield posteriorly. Vaginal sclerites well developed, with broadened heads anteriorly. Associated with insects . . . . (Plate 48, p. 206) Family HOPLOMEGISTIDAE

55. Palpal tibia and tarsus insensibly fused, chelicerae edentate. Sternal shield with paired posterior extensions devoid of pores and setae, and usually covered by latigynial and mesogynial shields (*sternovaginal sclerites,* Plate 49-2, p. 207). Latigynial shields well developed, separate from and partially or completely overlapping mesogynial shield. Associated with ants . . . . .Superfamily AENICTEQUOIDEA ... 56

— Palpal tibia and tarsus normally articulated, chelicerae robust and strongly dentate, or tapered and with or without minute teeth. Sternovaginal sclerites absent; latigynial shields various, but not overlapping mesogynial shield 59

56. Latigynial shields elongate, rectangular or subrectangular, with closely appressed parallel mesal margins; latigynials extend posteriorly well beyond coxae IV, abutting a ventrianal shield which does not exceed 1/3 the total body length; mesogynial shield obscure or lost. (Plate 48, p. 206) Family MESSORACARIDAE

— Latigynial shields subtriangular in outline, mesal margins convex, overlapping margins of triangular mesogynial shield; latigynials and mesogynial extending only slightly (if at all) beyond coxae IV, abutting a ventrianal shield at least ½ the length of the idiosoma 57

57. Latigynial shields each with two pairs of setae; with a distinctive peritreme-like groove arising on the external aspect of coxa IV, branching anteriorly and posteriorly to the level of coxa II and to the metapodal region, separate from the true peritreme. Chaetotaxy of tibia IV $(2\ \frac{2}{1}\ \frac{2}{1}\ 2)$ unique to supercohort . . . . (Plate 48, p. 206) Family AENICTEQUIDAE

— Latigynial shields each with ten or more setae; peritreme-like grooves absent 58

58. Sternal setae 1 inserted in well developed jugular shields; latigynial shields each with ten setae. Chaetotaxy of genu IV $(2\ \frac{3}{1}\ \frac{3}{1}\ 0)$ unique to supercohort . . . . (Plate 49, p. 207) Family PTOCHACARIDAE

— Sternal setae 1 inserted in entire sternal shield; latigynial shields each with more than 30 setae. . . . . (Plate 49, p. 207) Family PHYSALOZERCONIDAE

59. Sternal setae free in soft integument, on jugularia, or on a fused tetartosternum; vaginal sclerites reduced, without heads or bow-shaped basal pieces. Chelicerae tapered, often edentate or with numerous minute teeth, movable digit with filamentous excrescences. Associated with insects. 60

— Sternal setae 1 on a sternal shield or a tetartosternum; vaginal sclerites well developed, usually with heads. Chelicerae robust, with large proximal tooth on movable digit; with dendritic or brush-like cheliceral excrescences. Associated with arthropods and snakes . . . . .Superfamily CELAENOPSOIDEA ... 61

60. With a ventrianal shield bearing more than eight pairs of setae; mesogynial and ventral elements fused. Chelicerae with minute teeth. Associated with insects . . . . Superfamily ANTENNOPHOROIDEA, (Plate 49, p. 207) Family ANTENNOPHORIDAE

— With a terminal anal shield bearing one-six pairs of setae; mesogynial shield, if present, always separate from ventral elements. Chelicerae edentate or with minute teeth, digits reduced. Associated with insects and myriapods. . . . . Superfamily PARANTENNULOIDEA, (Plate 50, p. 208) Family PARANTENNULIDAE

61. Sternal setae 1 on tetartosternum, sternal shield strap-like, bearing metasternal setae. With three dorsal shields; peritremes greatly reduced. Associated with millipedes . . . . (Plate 50, p. 208) Family NEOTENOGYNIIDAE

— Sternal seate 1 on an entire sternal shield; with a single dorsal shield 62

62. Latigynial shields well developed, free mesally over their entire length, free posteriorly or fused with ventral elements. Mesogynial shield free and distinct or fused with ventral shield. Associated with insects or snakes. 63

— Latigynial shields fused medially with each other and posteriorly with ventral and mesogynial elements, at most with a notch or short separation at anterior margin of insensibly fused latigynial-mesogynial-ventral complex 64

63. Latigynial shields elongate, often extending posteriorly beyond coxae IV, shields fused to ventral-mesogynial complex. . . . . .(Plate 50, p. 208) Family SCHIZOGYNIIDAE

— Latigynial shields rarely elongate, usually extending no further than the posterior margins of coxae III and rarely to the middle of coxae IV, shields free posteriorly, hinged to ventral shield; mesogynial shield distinct or fused to ventral elements. . . . .(Plate 51, p. 209) Family DIPLOGYNIIDAE

64. With an anterior mesal separation of the fused latigynial shield elements. Associated with passalid beetles . . . . (Plate 51, p. 209) Family EUZERCONIDAE

— Latigynial shields entirely fused with each other and with mesogynial-ventral elements except for a shallow notch anteromedially 65

65. Endopodal, metapodal and ventral elements fused as a single shield, anal shield free or fused to ventral shield complex. Small mites with dorsal shield $< 800\ \mu$ in length. Associated with insects . . . . (Plate 52, p. 210) Family CELAENOPSIDAE

— Fused endopodal-metapodal shields separated from ventrianal shield. Large mites with dorsal shield $> 1000\ \mu$ in length . . . . (Plate 52, p. 210) Family MEGALOCELAENOPSIDAE

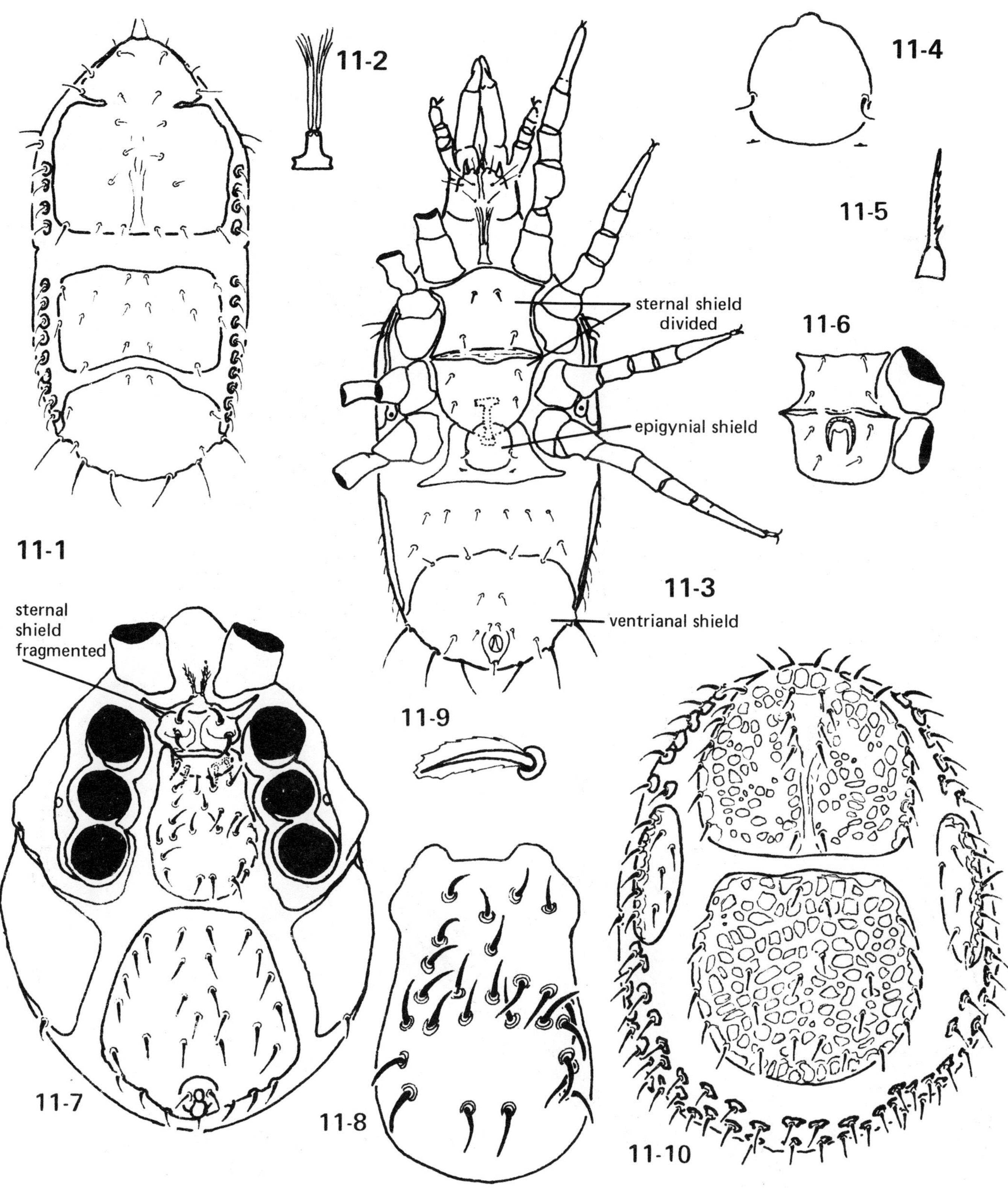

11-1 to 11-6; family MICROGYNIIDAE. **11-1**; *Microgynium incisum* Krantz (Oregon, USA), dorsum of female: **11-2**; *M. incisum*, tritosternum: **11-3**; *M. incisum*, venter of female: **11-4**; *M. incisum*, epigynial shield: **11-5**; *Microsejus truncicola* Trägårdh (Sweden), tritosternum: **11-6**; *M. incisum*, sternitigenital region of male

11-7 to 11-10; family ICHTHYOSTOMATOGASTERIDAE, *Uropodella* sp. (Ohio, USA). **11-7**; venter of female: **11-8**; epigynial shield: **11-9**; dorsal seta: **11-10**; dorsum of female

PLATE 12

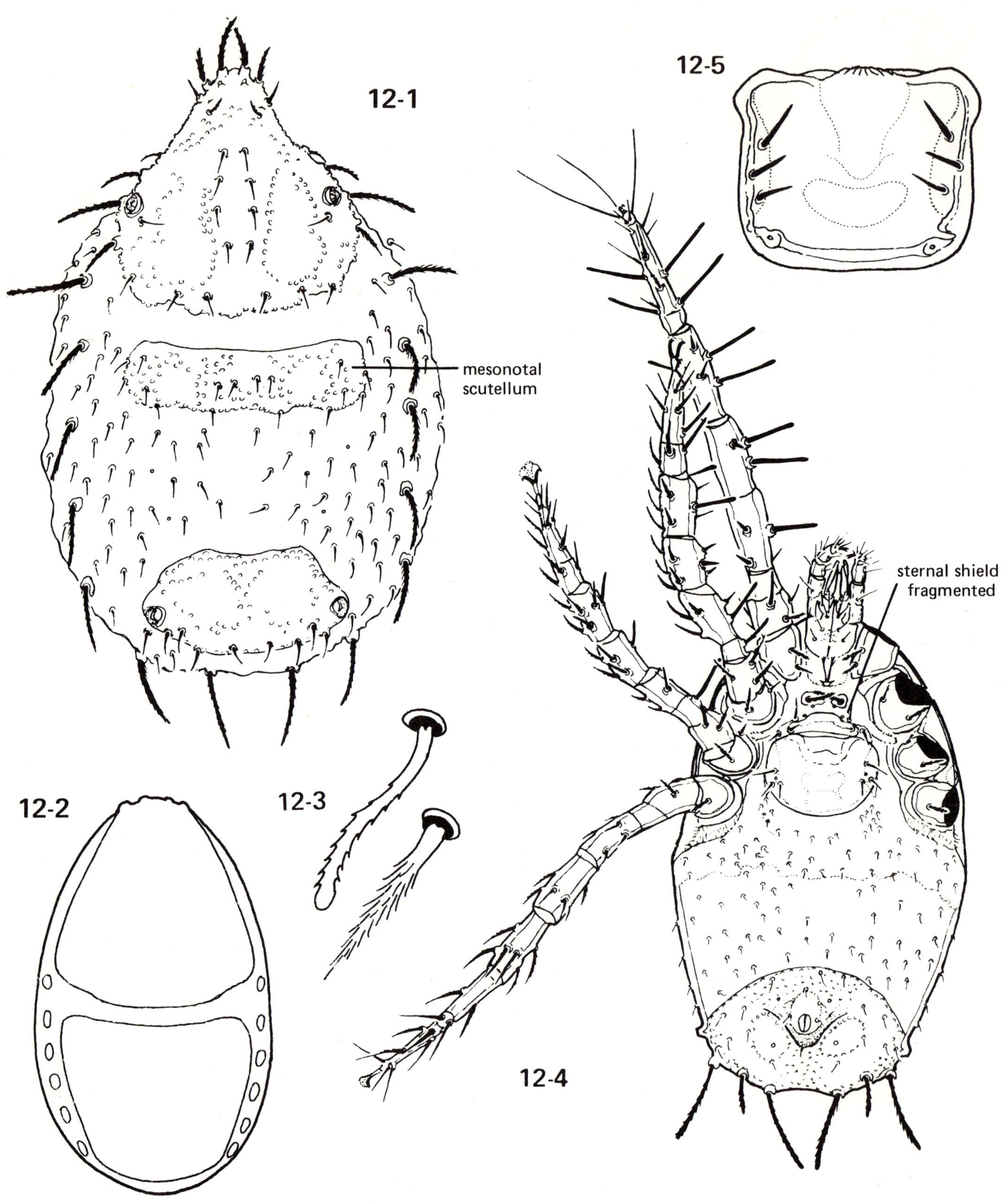

12-1 to 12-5; family SEJIDAE, *Sejus* sp. (Oregon, USA). **12-1;** dorsum of female: **12-2;** dorsum of male: **12-3;** types of dorsal setae: **12-4;** venter of female: **12-5;** epigynial shield

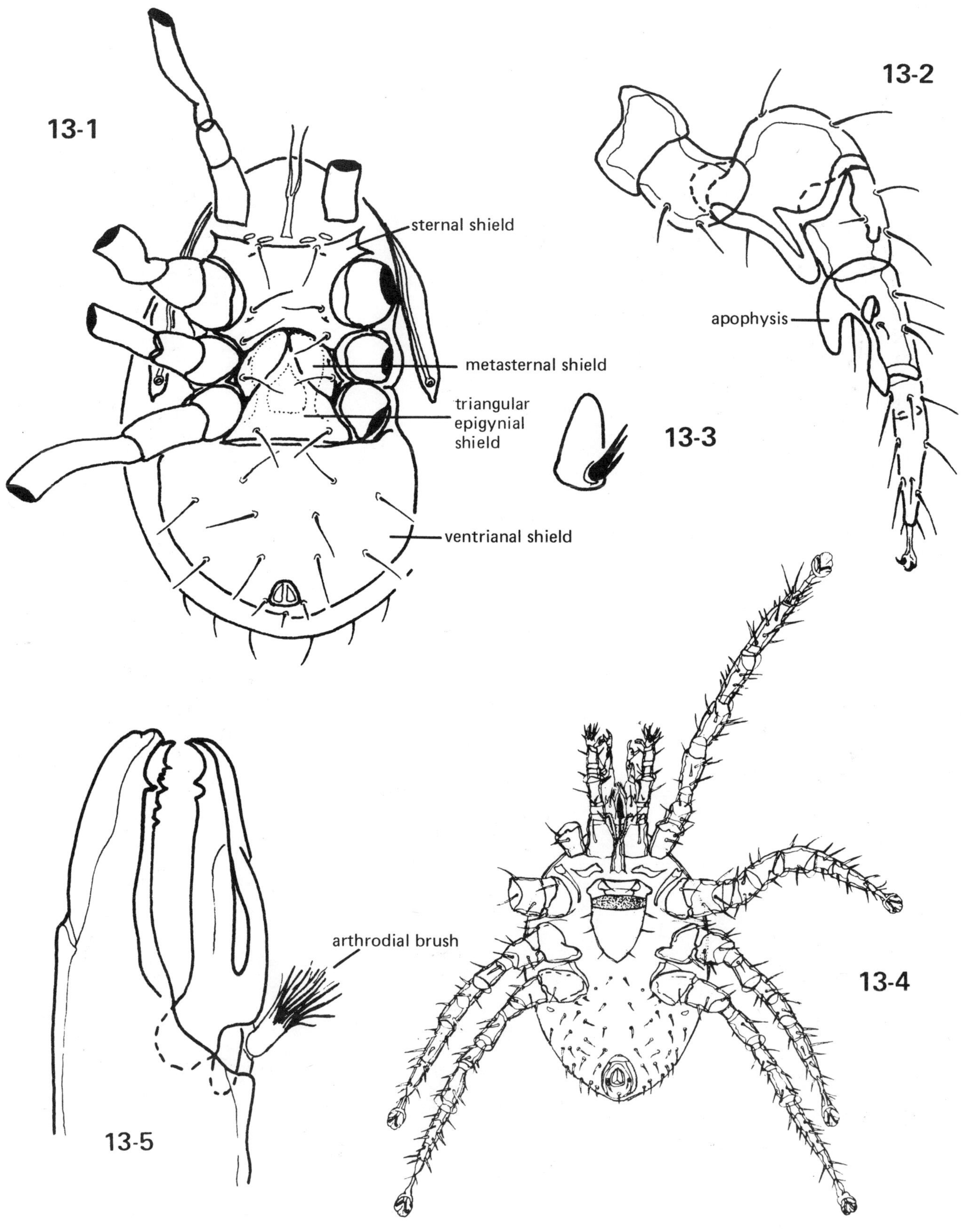

13-1 to 13-5; family PARASITIDAE. **13-1**; *Parasitus* sp. (Oregon, USA), venter of female: **13-2**; leg II of male parasitid (Oregon, USA): **13-3**; palptarsal claw: **13-4**; *Poecilochirus necrophori* Vitzthum (Oregon, USA), venter of deutonymph: **13-5**; *Parasitus diversus* Halbert, male chelicera (after Karg 1971).

PLATE 14

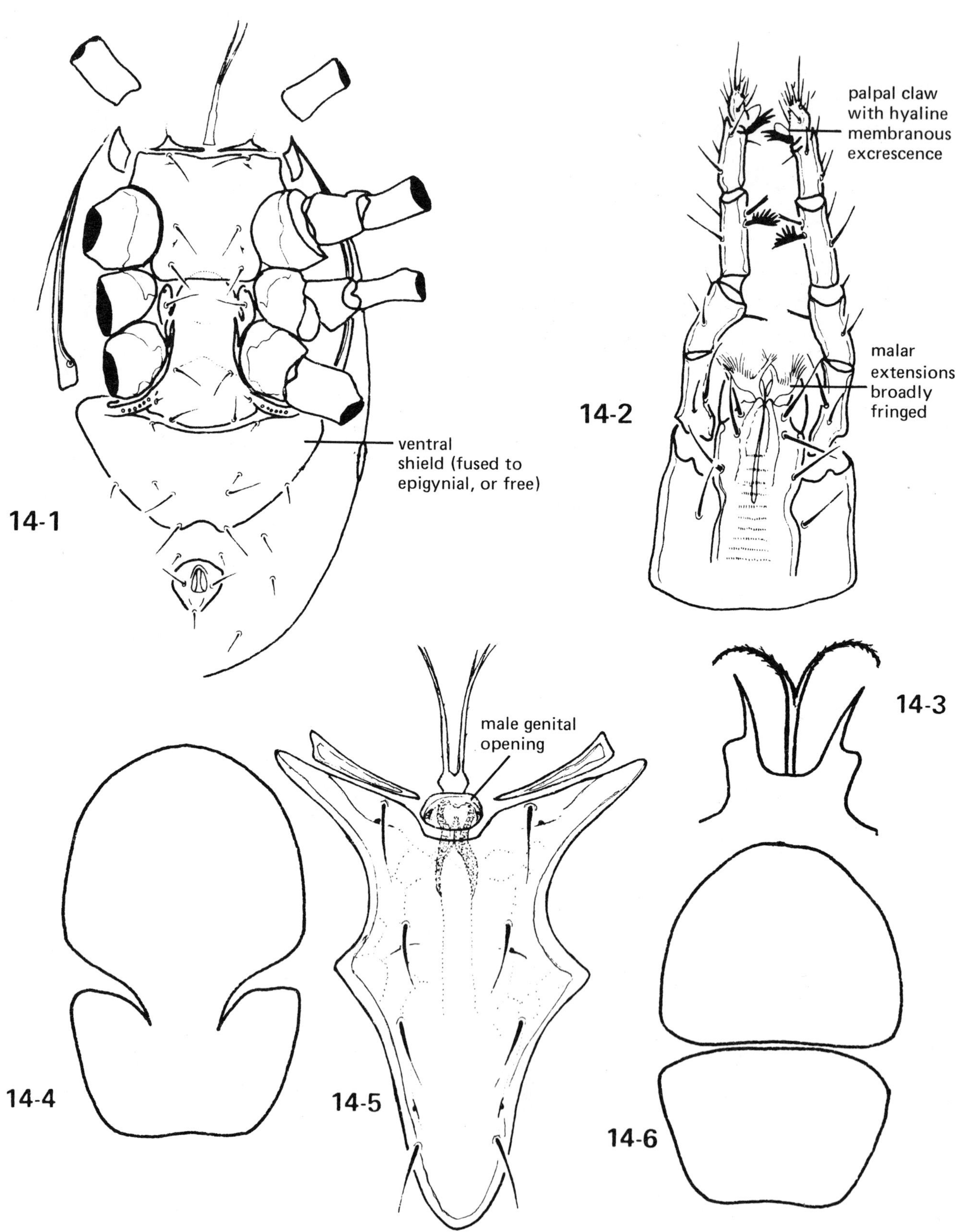

14-1 to 14-6; family VEIGAIIDAE. **14-1**; *Veigaia* sp. (Oregon, USA), venter of female: **14-2**; venter of female gnathosoma: **14-3**; epistome of female: **14-4 and 14-6**; dorsal shield types: **14-5**; *Veigaia* sp. (Washington, USA), male sternitigenital shield

PLATE 15

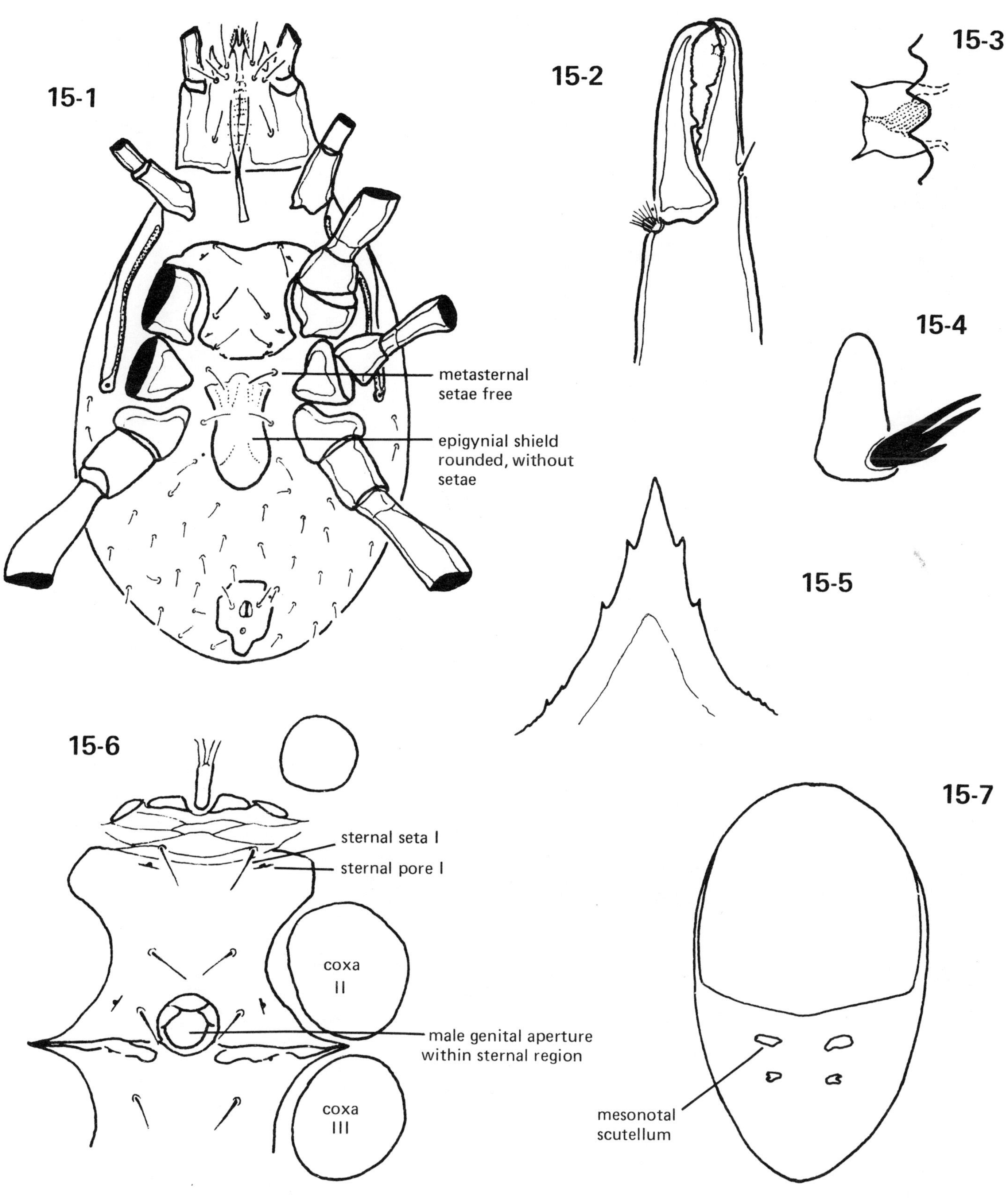

15-1 to 15-7; family ARCTACARIDAE, *Arctacarus* sp. (Oregon, USA). **15-1**; venter of female: **15-2**; chelicera of female: **15-3**; pilus dentilis of female chelicera: **15-4**; palptarsal claw: **15-5**; epistome of female: **15-6**; sternitigenital region of male: **15-7**; dorsum of female

PLATE 16

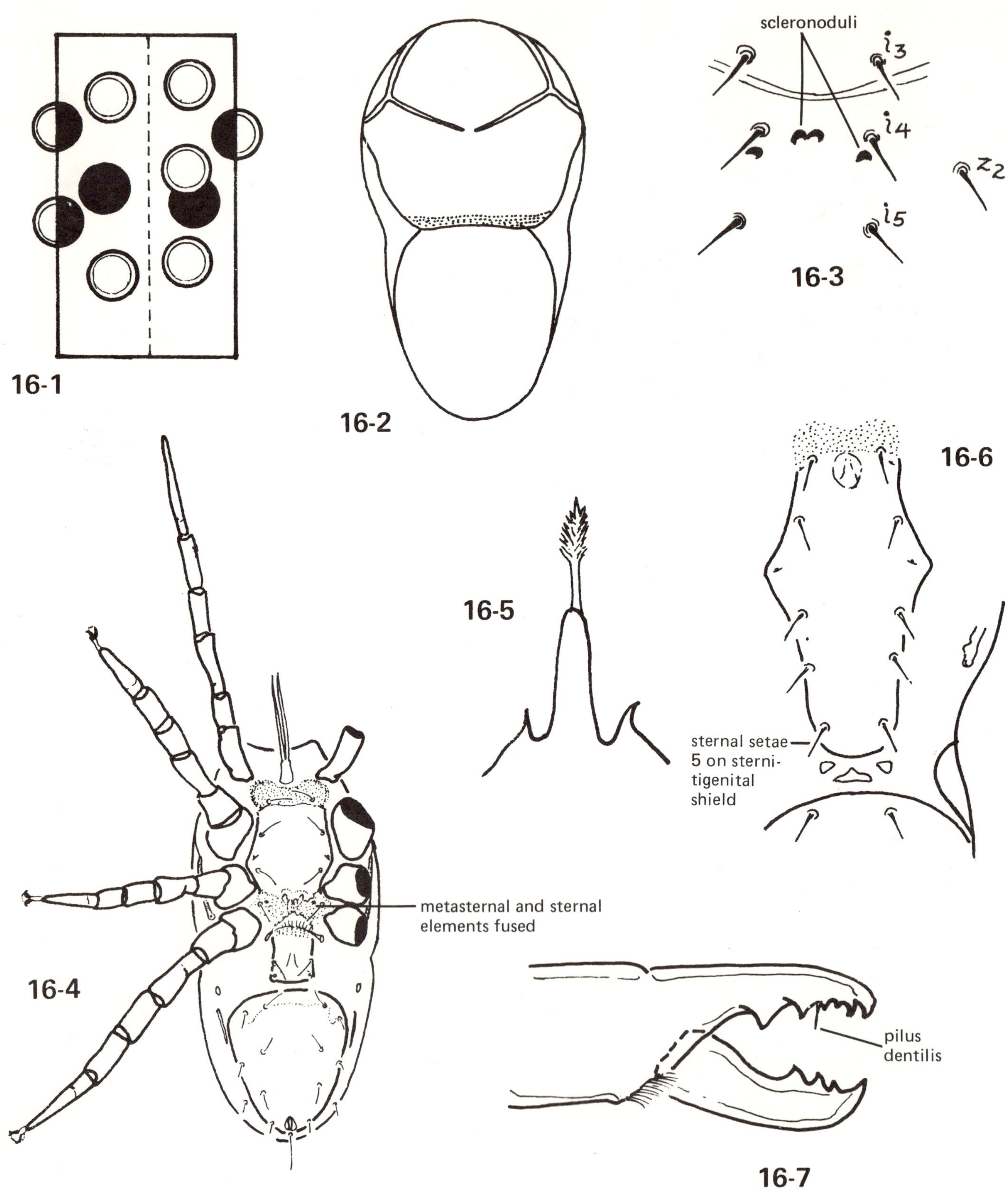

16-1 to 16-7; family RHODACARIDAE. **16-1**; chaetotaxy of genu IV: **16-2**; *Rhodacarus roseus* Oudemans (Oregon, USA), dorsum of female: **16-3**; *R. roseus*, position of dorsal scleronoduli: **16-4**; *R. roseus*, venter of female: **16-5**; *R. roseus*, epistome: **16-6**; *R. roseus*, sternitigenital shield of male: **16-7**; rhodacarid female chelicera

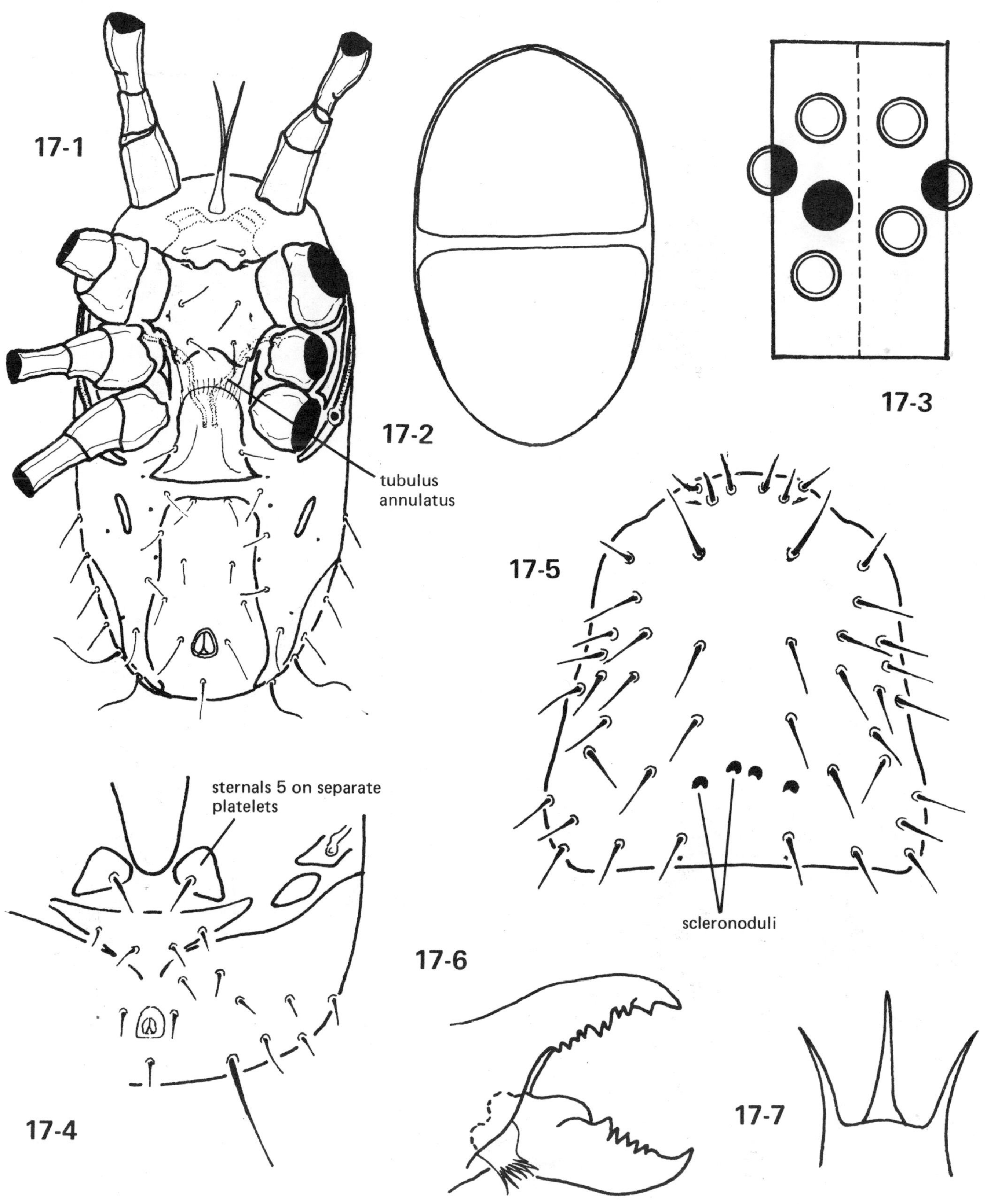

17-1 to 17-7; family DIGAMASELLIDAE. **17-1**; *Dendrolaelaps* sp. (Oregon, USA), venter of female: **17-2**; dorsum of female: **17-3**; chaetotaxy of genu IV: **17-4**; *Dendrolaelaps ulmi* Hirschmann, posteroventral aspect of male: **17-5**; *Dendrolaelaps* sp. (Oregon, USA), propodosoma of female: **17-6**; chelicera of female: **17-7**; epistome of female

PLATE 18

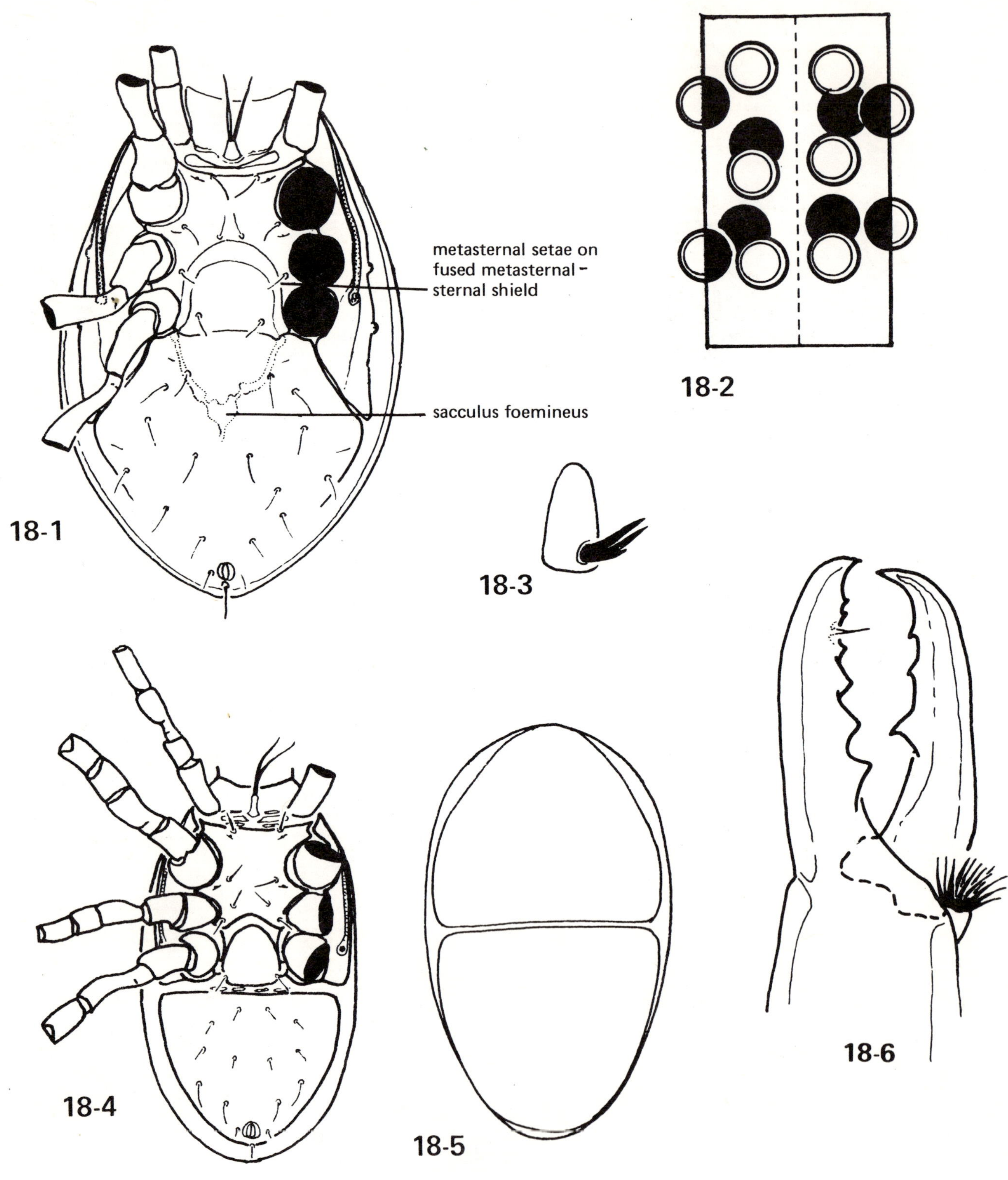

18-1 to 18-6; family OLOGAMASIDAE. **18-1**; *Gamasiphis* sp. (India), venter of female: **18-2**; typical chaetotaxy of tibia I: **18-3**; typical ologamasid palptarsus; **18-4 and 5**; *Gamasellus* sp. (Oregon, USA), ventral and dorsal aspects: **18-6**; *Euryparasitus emarginatus* (C.L. Koch), female chelicera

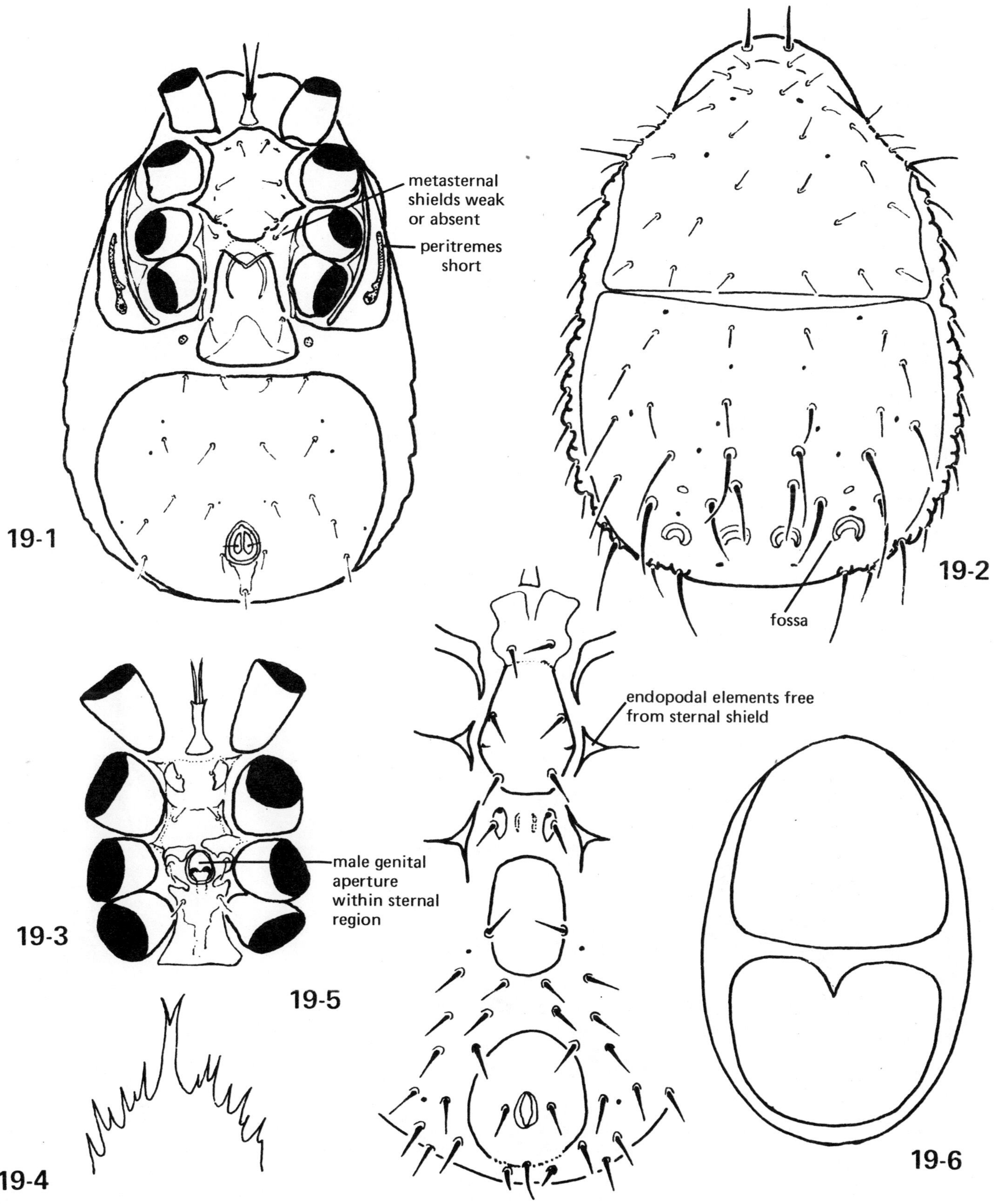

19-1 to 19-3; family ZERCONIDAE. **19-1**; *Zercon* sp. (Oregon, USA), venter of female: **19-2**; *Zercon* sp. (Norway), dorsum of female: **19-3**; *Zercon* sp. (Oregon, USA), podogastral region of male

19-4 to 19-6; family HALOLAELAPIDAE. **19-4**; *Halolaelaps* sp. (Oregon, USA), epistome of female: **19-5**; *Halolaelaps* sp. (Oregon, USA), venter of female: **19-6**; typical halolaelapid dorsal shield conformation

PLATE 20

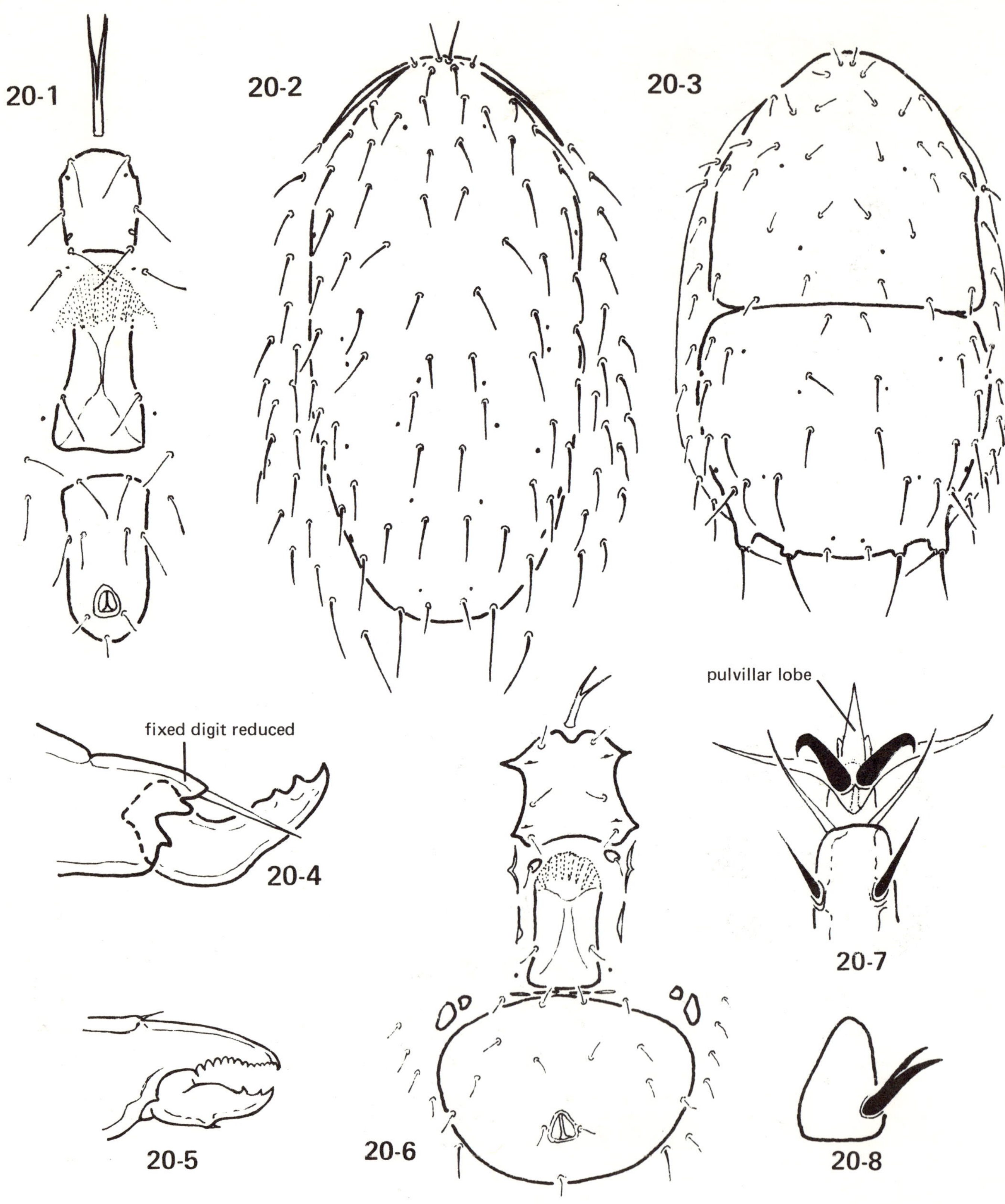

20-1 to 20-8; family ASCIDAE. **20-1**; *Blattisocius tarsalis* (Berlese) (Oregon, USA), venter of female: **20-2**; *B. tarsalis*, dorsum of female: **20-3**; *Asca* sp. (Oregon, USA), dorsum of female: **20-4**; *B. tarsalis*, chelicera of female: **20-5**; *Proctolaelaps pygmaeus* (Müller) (Oregon, USA), chelicera of female: **20-6**; *Lasioseius garambae* Krantz (Zaire), venter of female: **20-7**; *Cheiroseius* sp., tarsus II of female, dorsal aspect (after Lindquist and Evans 1965): **20-8**; typical ascid palptarsus

PLATE 21

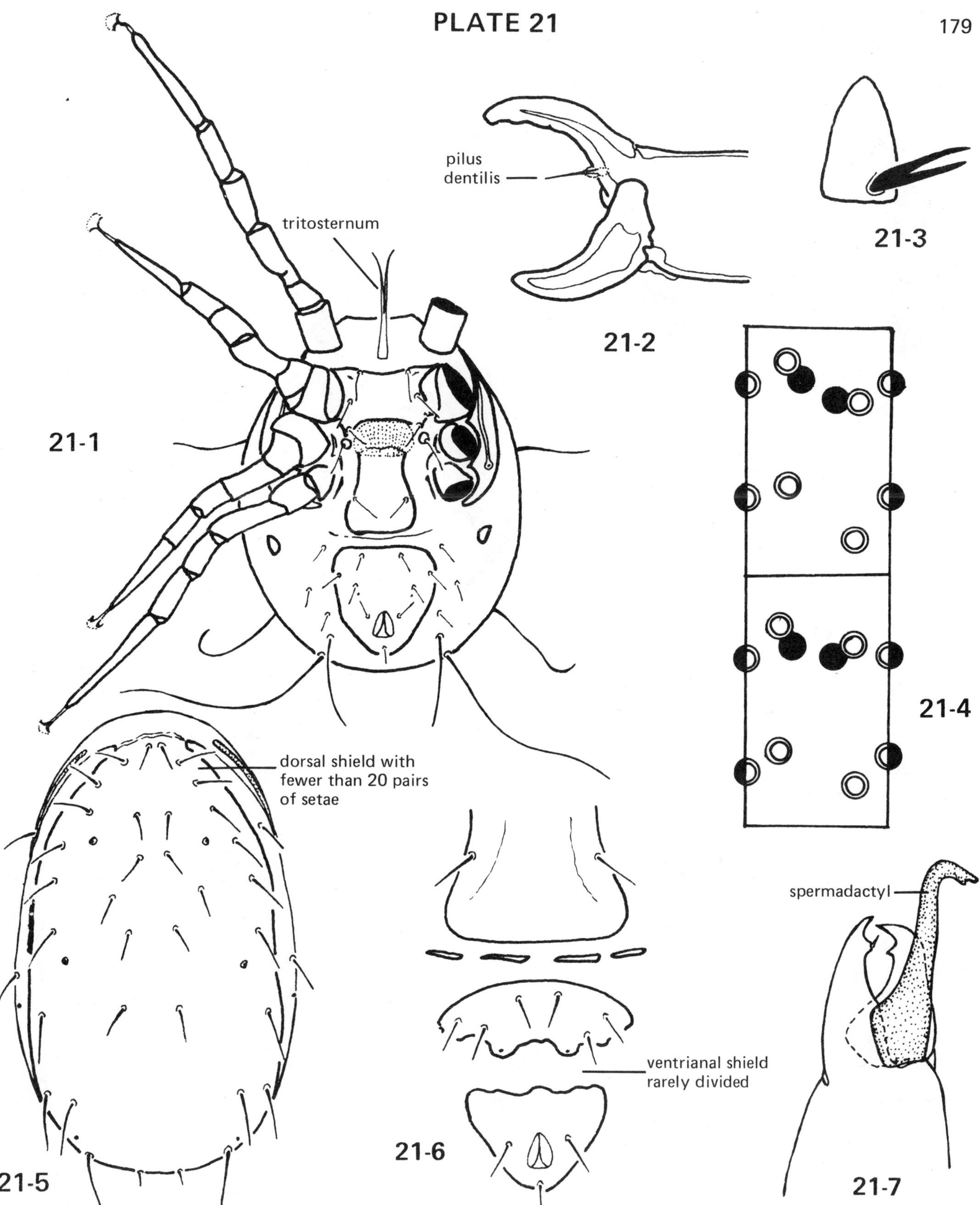

21-1 to 21-7; family PHYTOSEIIDAE. **21-1;** *Amblyseius* sp. (Oregon, USA), venter of female: **21-2;** *Amblyseius* sp., chelicera of female: **21-3;** phytoseiid palptarsus: **21-4;** chaetotaxy of genu (bottom) and tibia I of PHYTOSEIIDAE: **21-5;** *Typhlodromus pyri* Scheuten (Oregon, USA), dorsum of female: **21-6;** *Iphiseius degenerans* (Berlese) (Italy), epigynial and ventrianal region of female (after Chant 1959): **21-7;** *Typhlodromus pyri* Scheuten, chelicera of male

PLATE 22

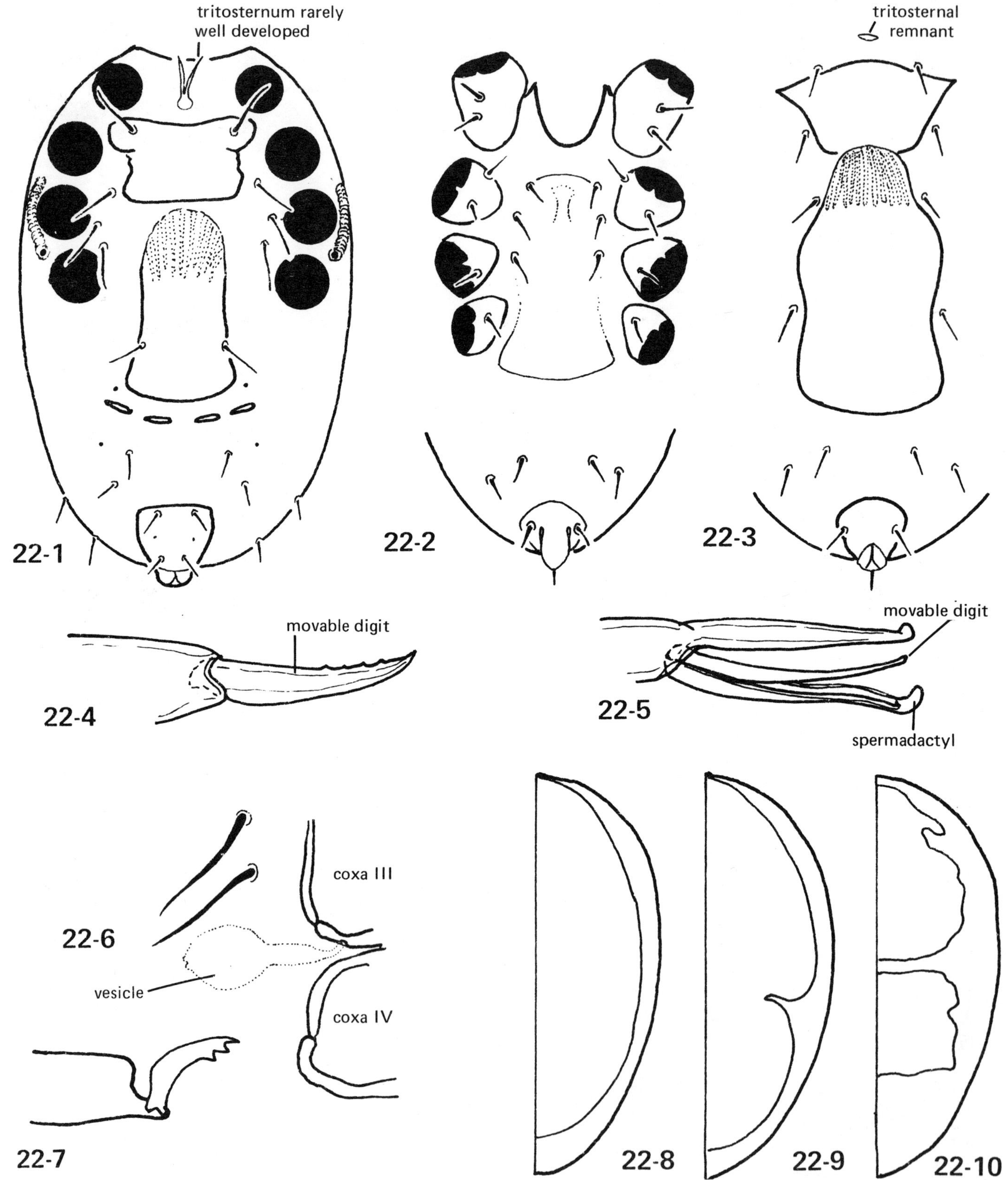

22-1 to 22-10; family OTOPHEIDOMENIDAE. **22-1**; *Hemipteroseius indicus* (Krantz and Khot), (India), venter of female: **22-2**; *Otopheidomenis zalelestes* Treat, venter of male (after Treat 1975): **22-3**; *Otopheidomenis* sp., venter of female: **22-4**; *O. zalelestes,* chelicera of female: **22-5**; *O. zalelestes,* chelicera of male: **22-6**; *Hemipteroseius indicus,* sacculus: **22-7**; *H. indicus,* chelicera: **22-8**; dorsal shield of *Nabiseius:* **22-9**; dorsal shield of *Otopheidomenis:* **22-10**; dorsal shields of *Hemipteroseius*

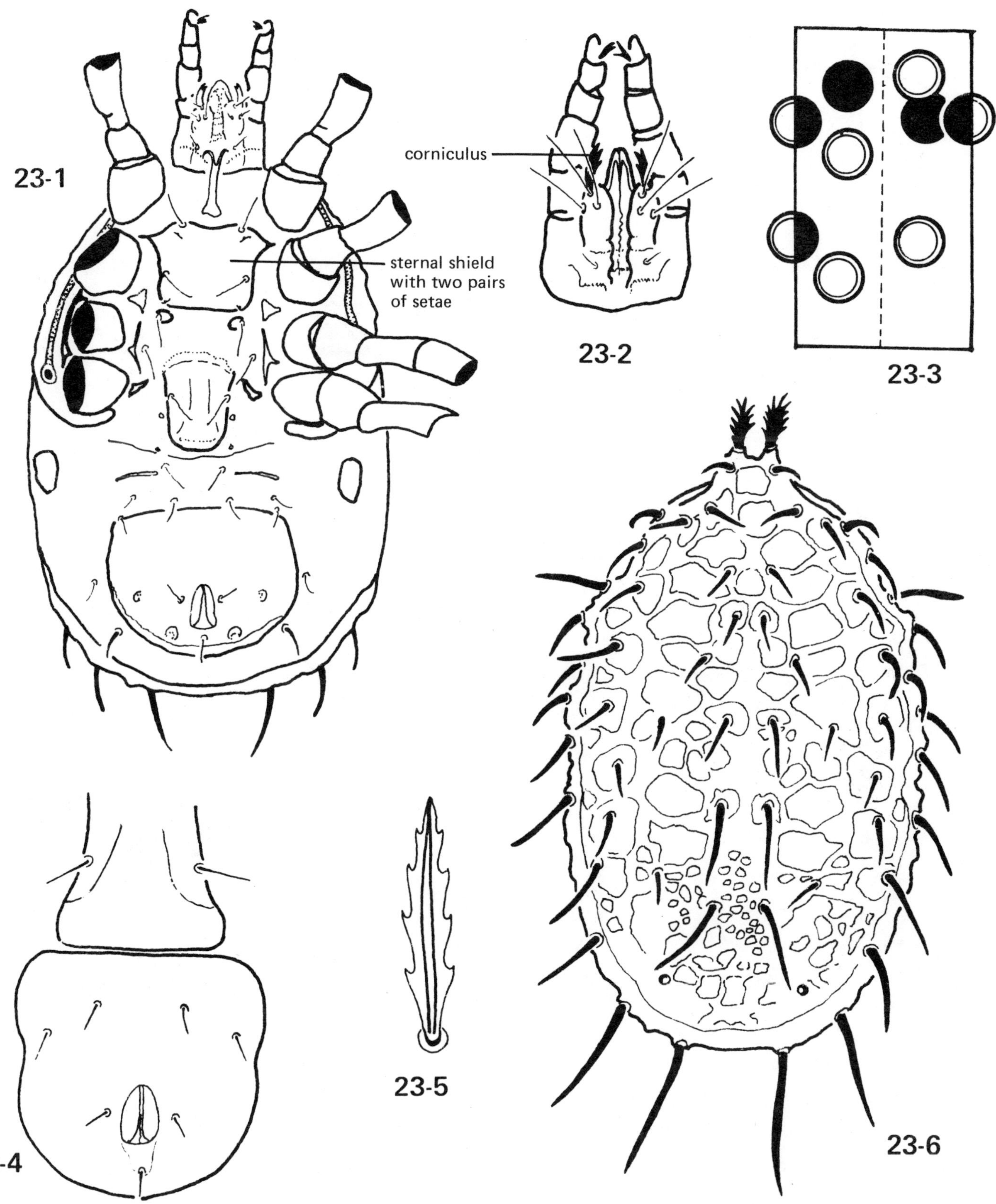

23-1 to 23-6; family AMEROSEIIDAE. **23-1**; *Ameroseius* sp. (Oregon, USA), venter of female: **23-2**; *Ameroseius* sp., venter of gnathosoma: **23-3**; chaetotaxy of tibia IV: **23-4**; *Kleemannia* sp. (Oregon, USA), epigynial and ventrianal shields of female: **23-5**; typical dorsal seta of *Ameroseius* sp.: **23-6**; *Ameroseius* sp., dorsum of female

PLATE 24

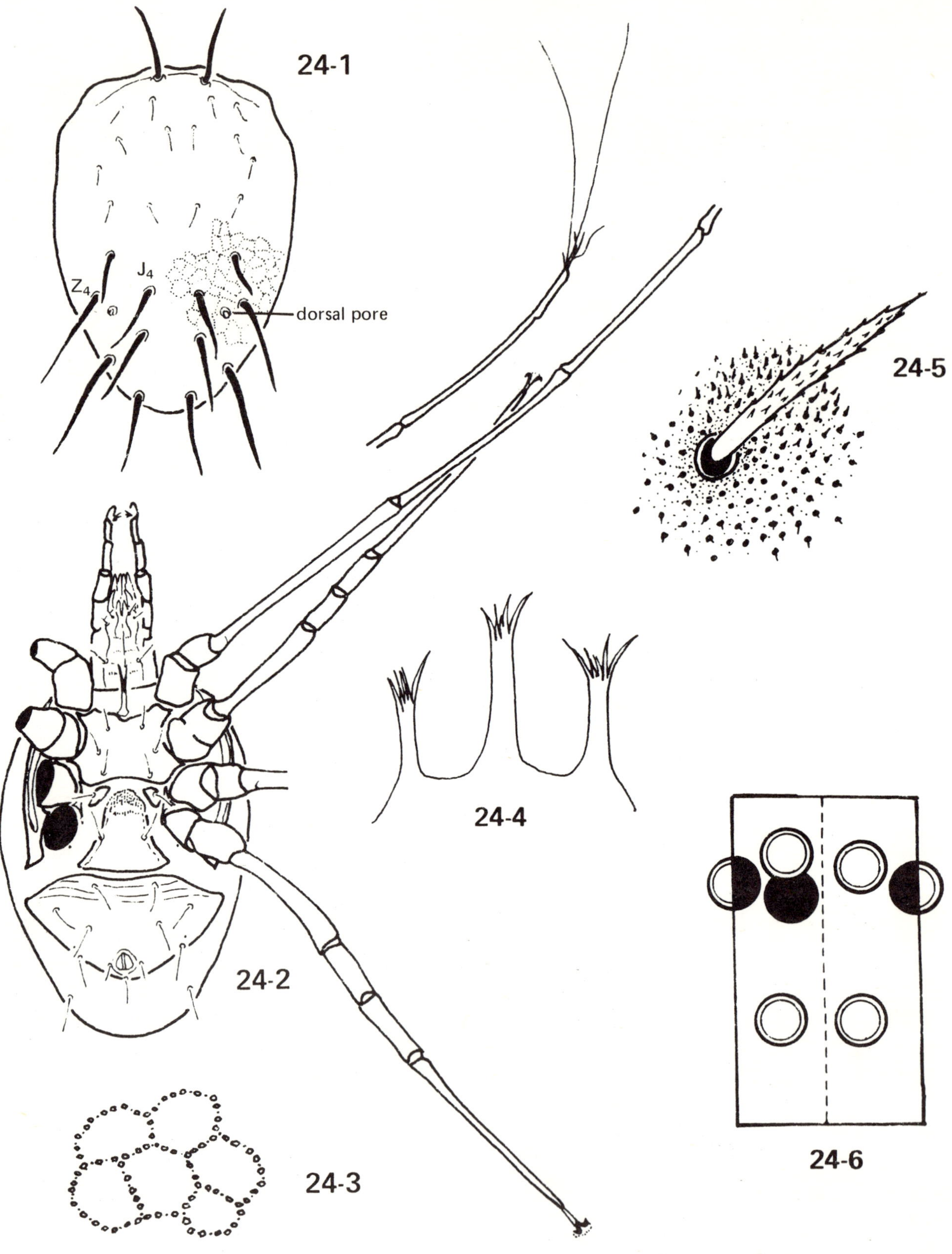

24-1 to 24-6; family PODOCINIDAE. 24-1; *Podocinum* sp. (Kansas, USA), dorsum of female: 24-2; *Podocinum* sp., venter of female: 24-3; *Podocinum* sp., ornamentation of dorsal shield: 24-4; *Podocinella* sp. (Mexico), epistome. 24-5; *Podocinella* sp., dorsal seta and surrounding ornamentation: 24-6; chaetotaxy of genu III

PLATE 25

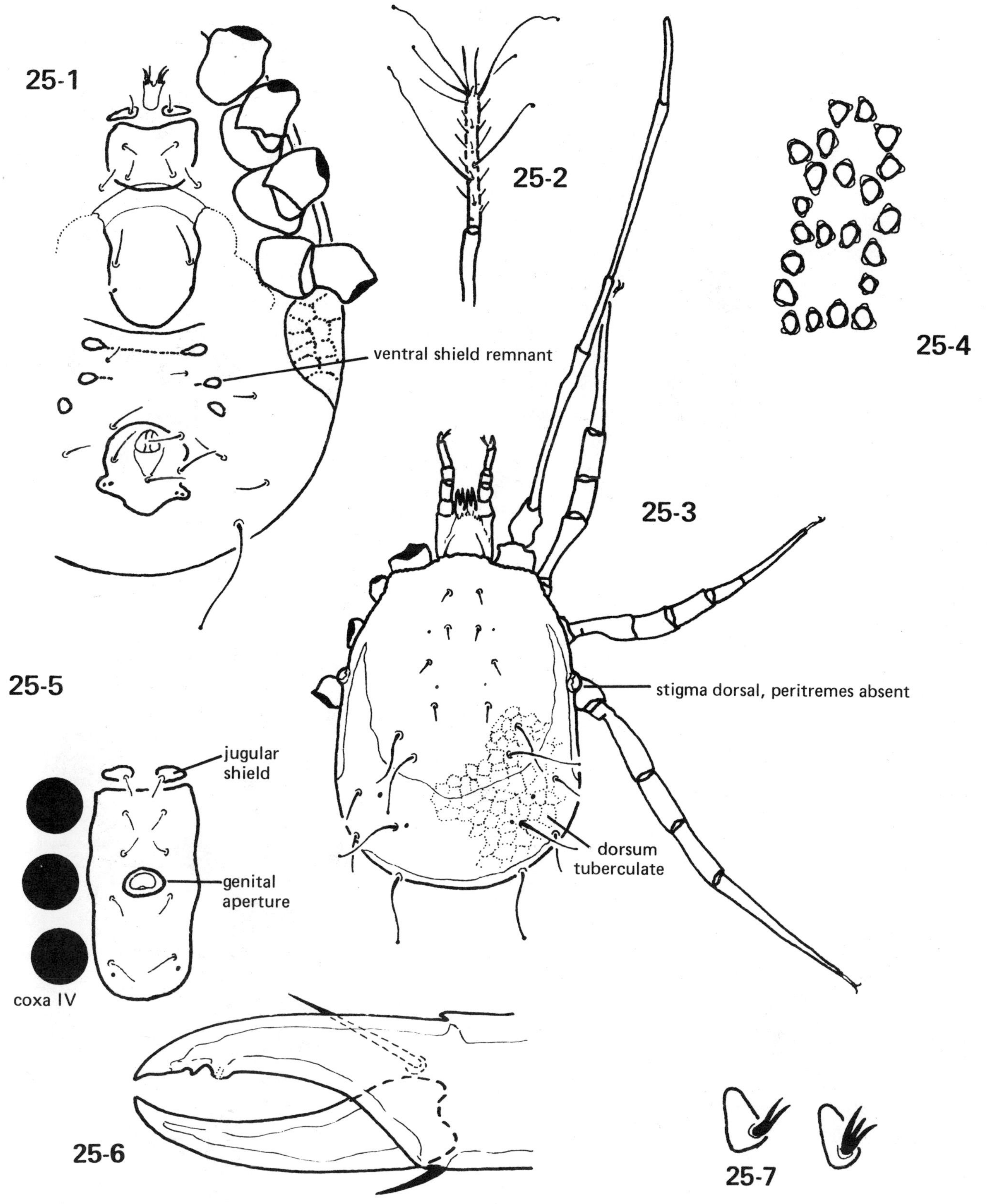

25-1 to 25-7; family EPICRIIDAE. **25-1**; *Epicrius* sp. (Oregon, USA), venter of female: **25-2**; *Epicrius* sp., tarsus I of female: **25-3**; *Epicrius* sp., dorsum of female: **25-4**; *Epicrius* sp., dorsal ornamentation: **25-5**; *Epicrius* sp., sternitigenital region of male: **25-6**; *Epicrius mollis* (Kramer), chelicera of female: **25-7**; epicriid palptarsal claw types

PLATE 26

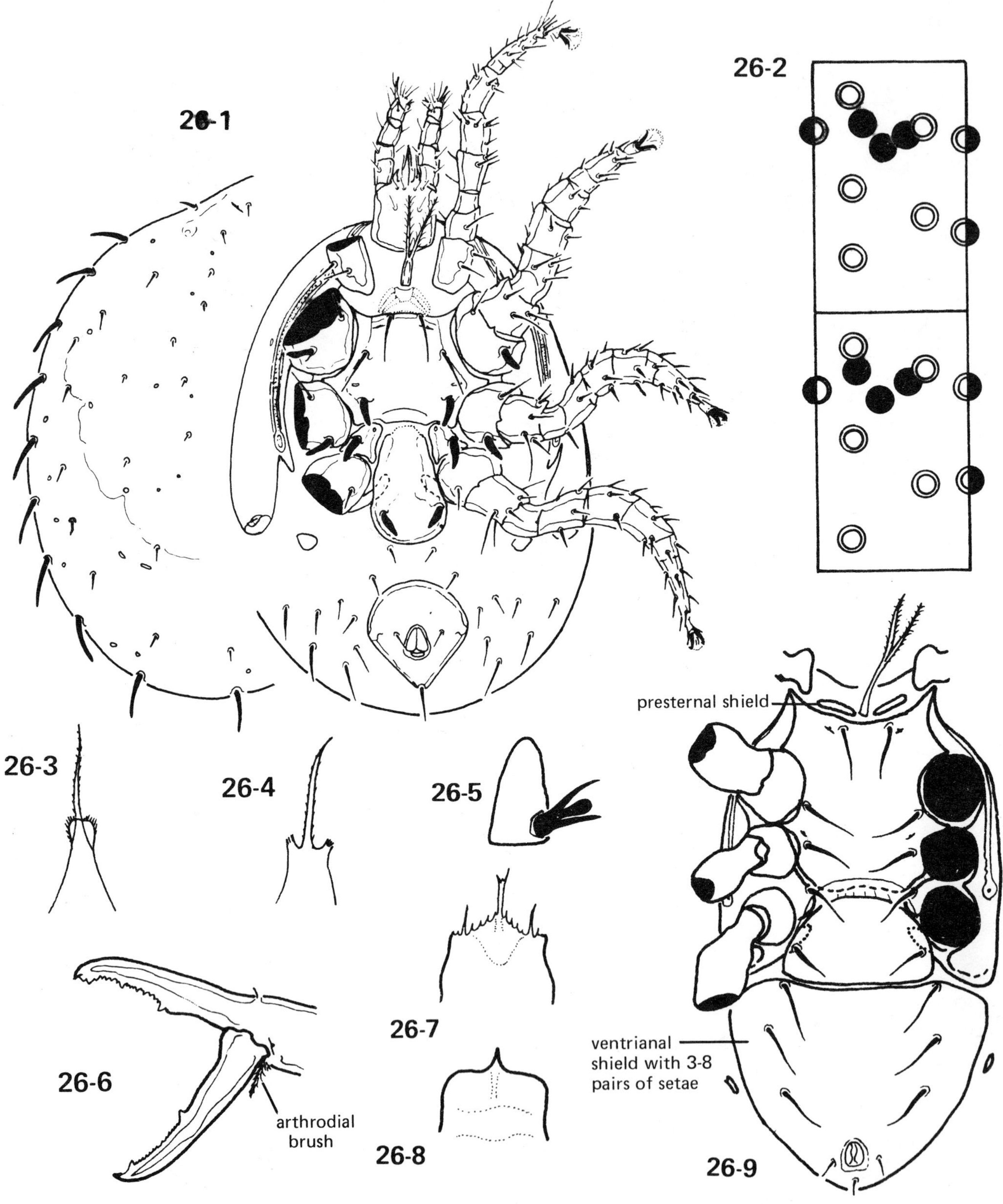

26-1; to 26-4; family EVIPHIDIDAE. **26-1;** *Eviphis* sp. (Africa), dorsum and venter of female: **26-2;** chaetotaxy of genu and tibia I of an eviphidid mite (after Evans 1963): **26-3;** *Eviphis stefaninianus* Berlese (Africa), epistome: **26-4;** *Scarabaspis rykei* Shoemake and Krantz (Zaire), epistome (after Shoemake and Krantz 1966)

26-5 to 26-9; family PARHOLASPIDIDAE. **26-5;** palpal tarsus of typical parholaspidid: **26-6;** *Parholaspulus lobatus* Krantz (Oregon, USA), chelicera of female: **26-7;** *Neoparholaspulus coalescens* Krantz (Louisiana, USA), epistome: **26-8;** *Calholaspis berlesei* Krantz (Maryland, USA), epistome: **26-9;** *Parholaspulus parvilobatus* Krantz (Oregon, USA), venter of female

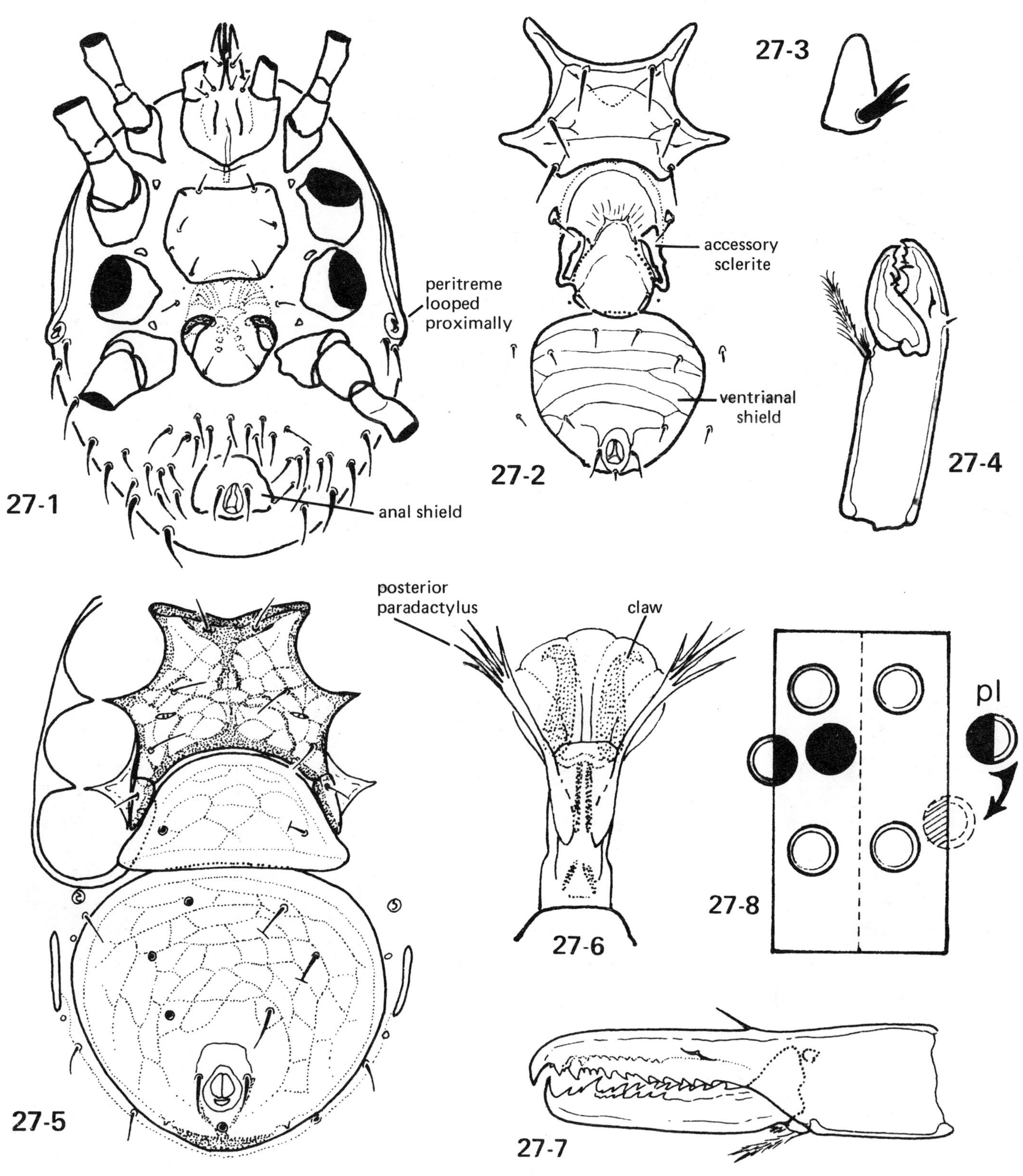

27-1 to 27-8; family MACROCHELIDAE. **27-1**; *Lordocheles rykei* Krantz (Zaire), venter of female: **27-2**; *Holocelaeno melisi* Krantz (Brazil), venter of female: **27-3**; typical macrochelid palptarsus: **27-4**; *Holostaspella bifoliata* (Trägårdh) (Oregon, USA), chelicera of female: **27-5**; *Holostaspella punctata* Krantz (Germany), venter of female: **27-6**; *Macrocheles* sp. (Trinidad), tarsus III: **27-7**; *Geholaspis (Longicheles) mandibularis* (Berlese) (Germany), chelicera of female: **27-8**; chaetotaxy of genu IV. An additional seta (pl) may occur in some species.

PLATE 28

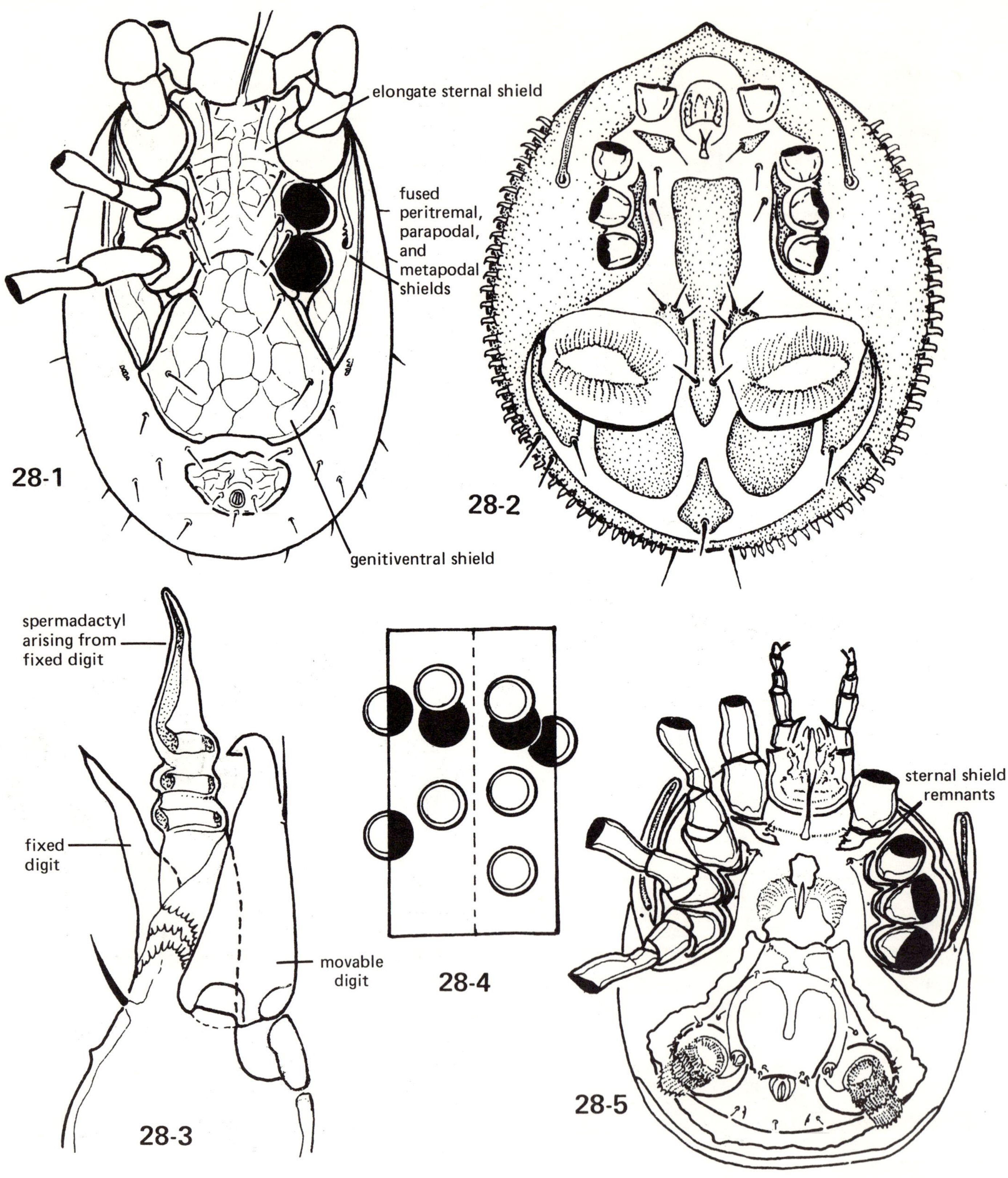

28-1; family PACHYLAELAPIDAE, *Pachylaelaps* sp. (Oregon, USA), venter of female

28-2; family DISCOZERCONIDAE, *Discozercon* sp., venter of female (after Baker et al. 1958)

28-3 to 28-5; family HETEROZERCONIDAE. **28-3**; *Heterozercon* sp. (Tanzania), chelicera of male: **28-4**; chaetotaxy of genua III and IV: **28-5**; *Heterozercon* sp. (Florida, USA), venter of female

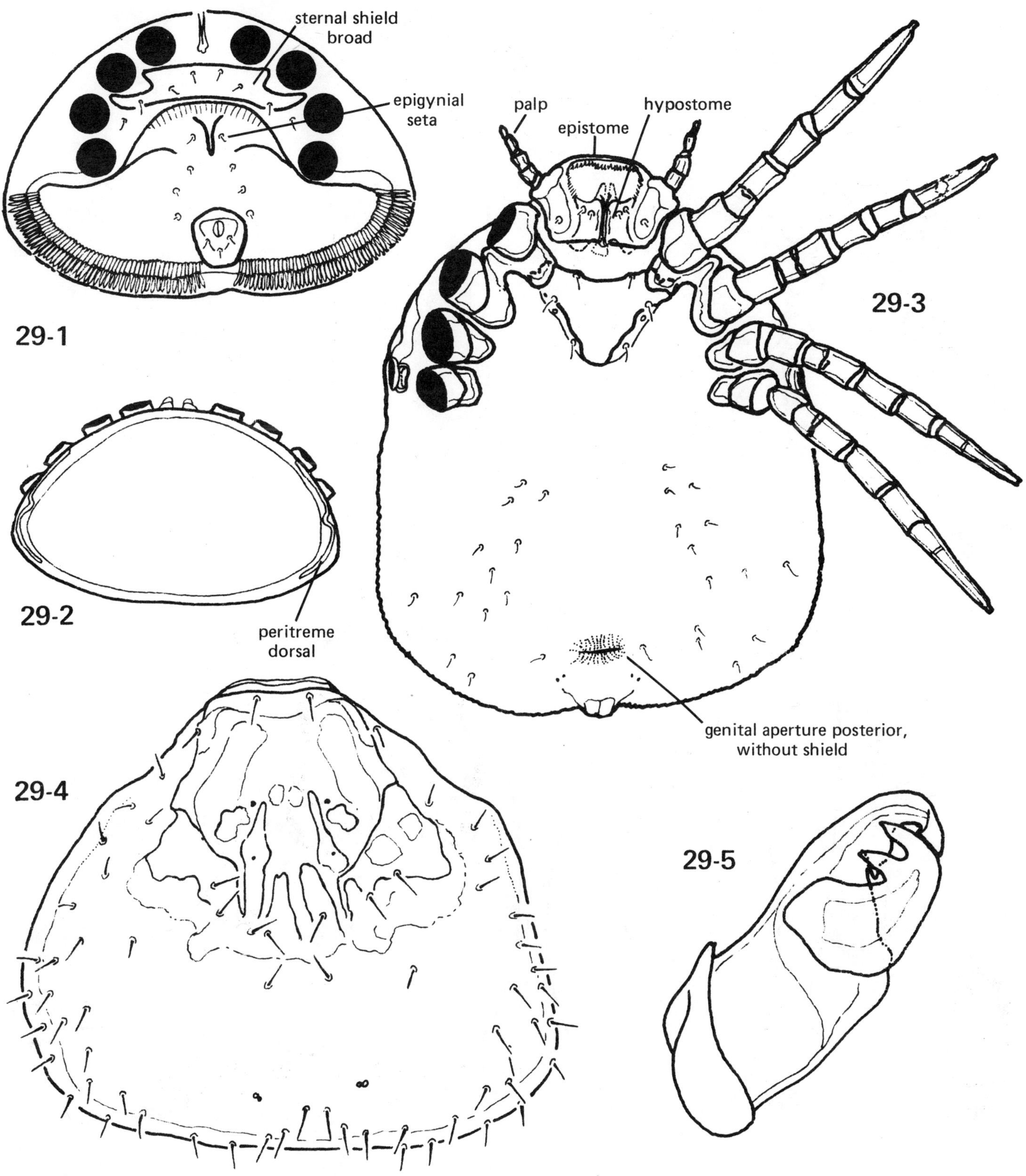

29-1 and 29-2; family OMENTOLAELAPIDAE, *Omentolaelaps mehelyae* Fain (Zaire). **29-1**; venter of female: **29-2**; dorsum of female (after Fain 1961)

29-3 to 29-5; family SPELAEORHYNCHIDAE, *Spelaeorhynchus praecursor* Neumann (South America). **29-3**; venter of female: **29-4**; dorsum of female: **29-5**; chelicera of female

PLATE 30

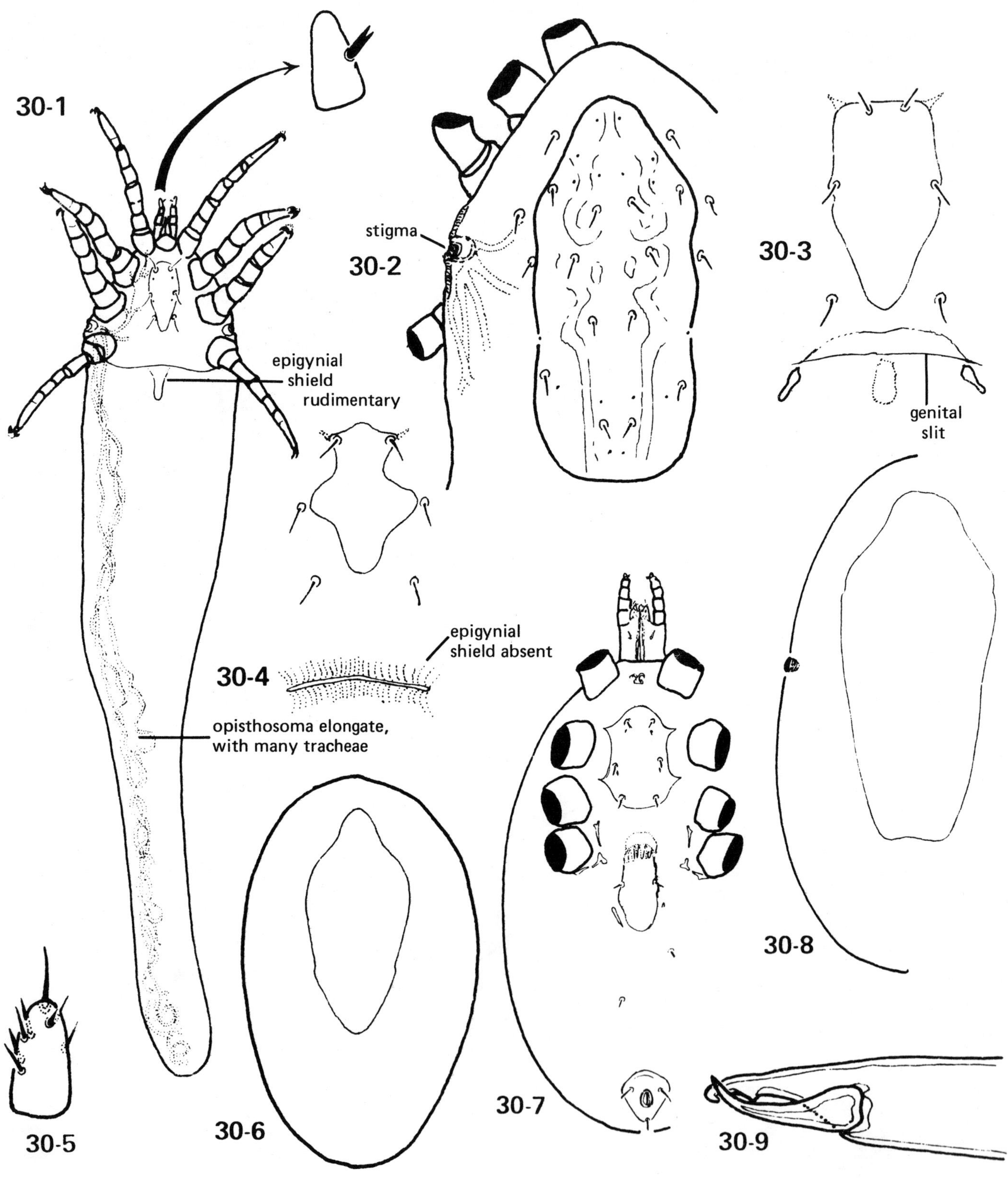

30-1 to 30-6; family HALARACHNIDAE. **30-1**; *Orthohalarachne attenuata* (Banks) (north Pacific), venter of female: **30-2**; *O. attenuata,* anterodorsal aspect of female: **30-3**; *Halarachne americana* Banks, sternal and genital region of female (after Newell 1947): **30-4**; *Pneumonyssoides caninum* (Chandler and Ruhe), sternal and genital region of female: **30-5**; *P. caninum,* palptarsus: **30-6**; *P. caninum,* dorsum

30-7 to 30-9; family ENTONYSSIDAE, *Ophiopneumicola* sp. (Washington, USA). **30-7**; venter of female: **30-8**; dorsum of female: **30-9**; chelicera of female

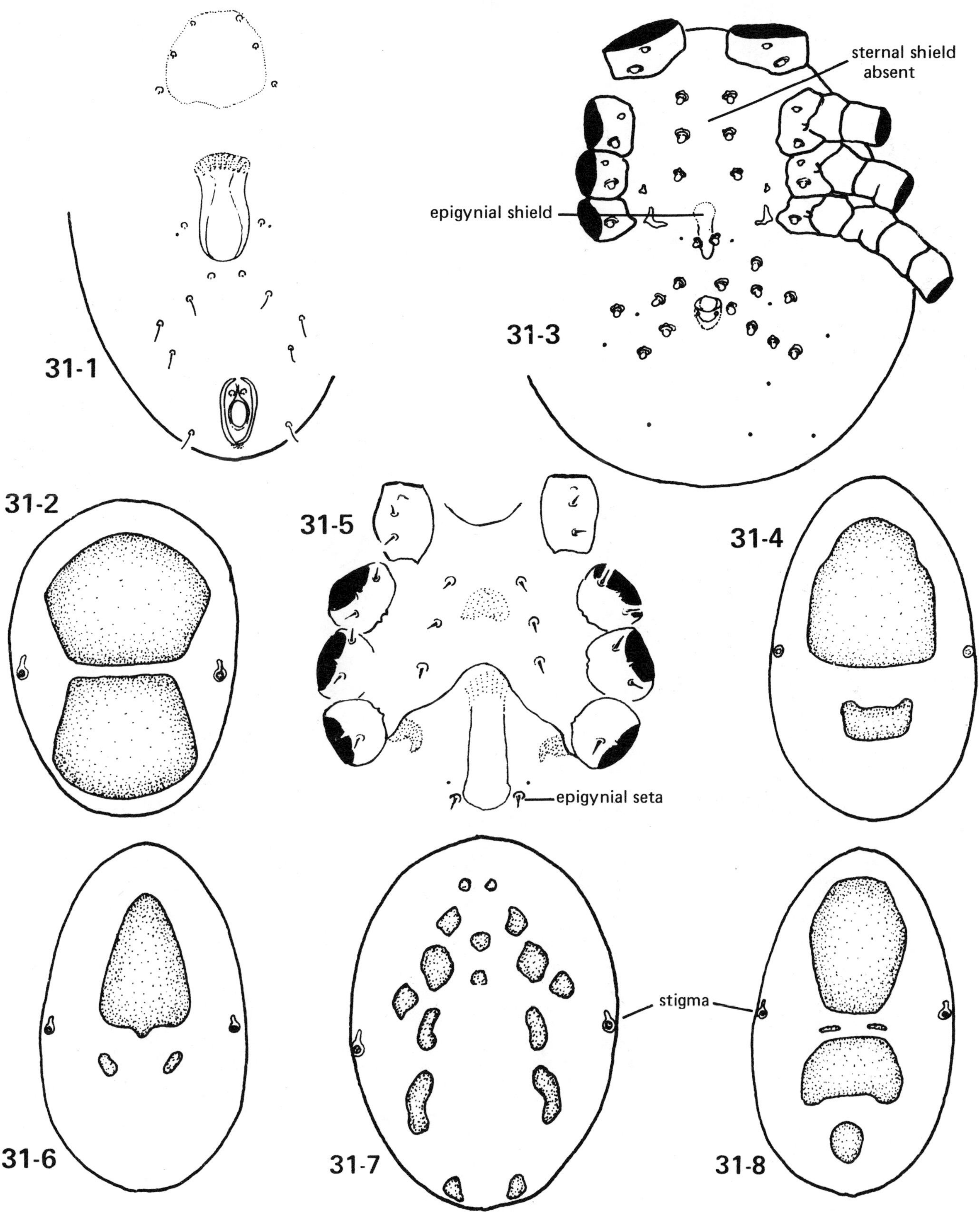

31-1 to 31-8; family RHINONYSSIDAE. **31-1**; *Neonyssus columbae* Crossley (Texas, USA), venter of female: **31-2**; *N. columbae,* dorsum of female: **31-3**; *Cas angrensis* (Castro) (Texas, USA), venter of female: **31-4**; *C. angrensis,* dorsum of female: **31-5**; *Ptilonyssus ohioensis* Fain and Johnston (Ohio, USA), sternal and genital region of female (after Fain and Johnston 1966): **31-6**; *Rhinoecius* sp., dorsum: **31-7**; *Larinyssus* sp., dorsum: **31-8**; *Ptilonyssoides* sp., dorsum

PLATE 32

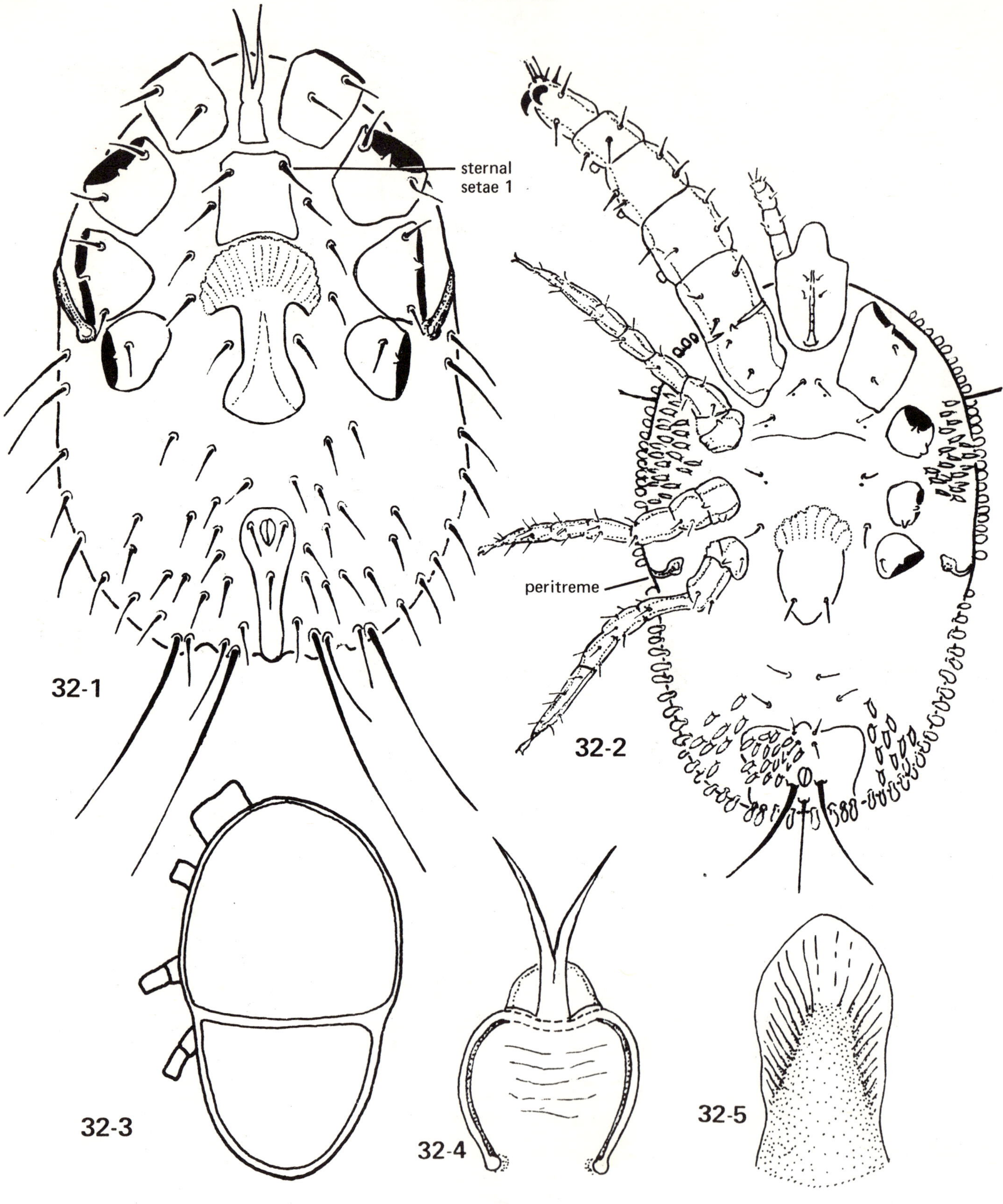

32-1; family MANITHERIONYSSIDAE, *Manitherionyssus heterotarsus* Vitzthum, venter of female (after Strandtmann and Wharton 1958)

32-2 to 32-5; family DASYPONYSSIDAE. 32-2; *Dasyponyssus neivai* Fonseca, venter of female (after Strandtmann and Wharton 1958): 32-3; *D. neivai,* dorsum (diagrammatic): 32-4; *Xenarthronyssus furmani furmani* Radovsky and Yunker, tritosternum and presternal region (after Radovsky and Yunker 1971): 32-5; *X. furmani furmani,* epistome (after Radovsky and Yunker 1971)

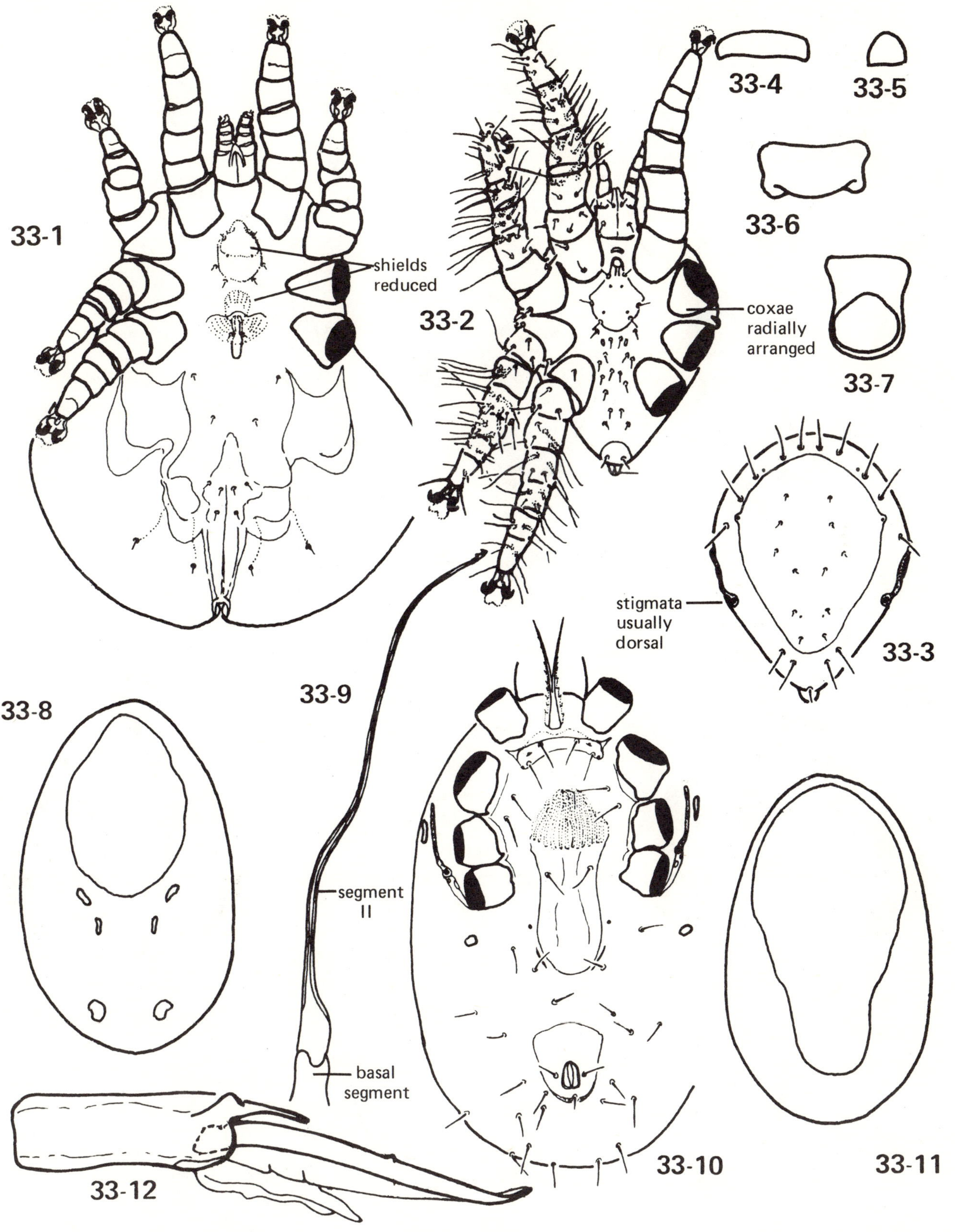

33-1 to 33-7; family SPINTURNICIDAE. **33-1**; *Periglischrus* sp. (Mexico), venter of female: **36-2**; *Spinturnix* sp. (Indiana, USA), venter of male: **33-3**; *Spinturnix* sp., dorsum of male: **33-4 to 33-7**; tritosternal bases of various spinturnicid species (after Rudnick 1960)

33-8 to 33-12; family DERMANYSSIDAE. **33-8**; *Dermanyssus triscutatus* Krantz (Alaska, USA), dorsum of female: **33-9**; *D. triscutatus,* chelicera of female: **33-10**; *D. triscutatus,* venter of female: **33-11**; *D. gallinae* (DeGeer) (Oregon, USA), dorsum of female: **33-12**; *D. gallinae,* chelicera of male (after Evans and Till 1965)

PLATE 34

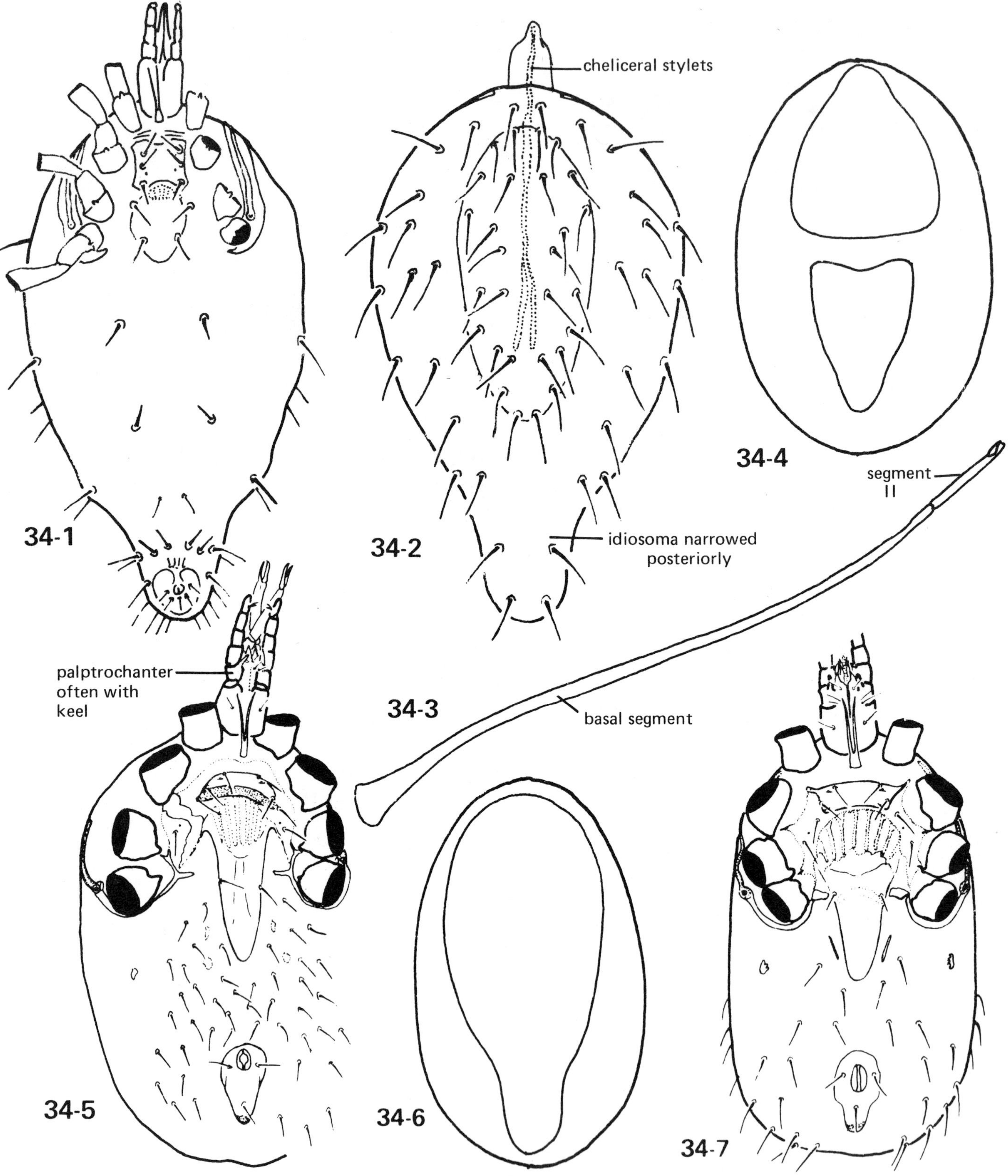

34-1 to 34-3; family HYSTRICHONYSSIDAE. **34-1**; *Hystrichonyssus turneri* Keegan, Yunker and Baker (Malaysia), venter of female (after Keegan, Yunker and Baker 1960): **34-2**; *H. turneri*, dorsum of female (after Keegan, Yunker and Baker 1960): **34-3**; *H. turneri*, chelicera of female

34-4 to 34-7; family MACRONYSSIDAE. **34-4**; *Steatonyssus* sp. (California, USA), dorsum of female: **34-5**; *Steatonyssus* sp., venter of female: **34-6**; *Ornithonyssus sylviarum* (Canestrini and Fanzago) (Oregon, USA), dorsum of female: **34-7**; *O. sylviarum*, venter of female

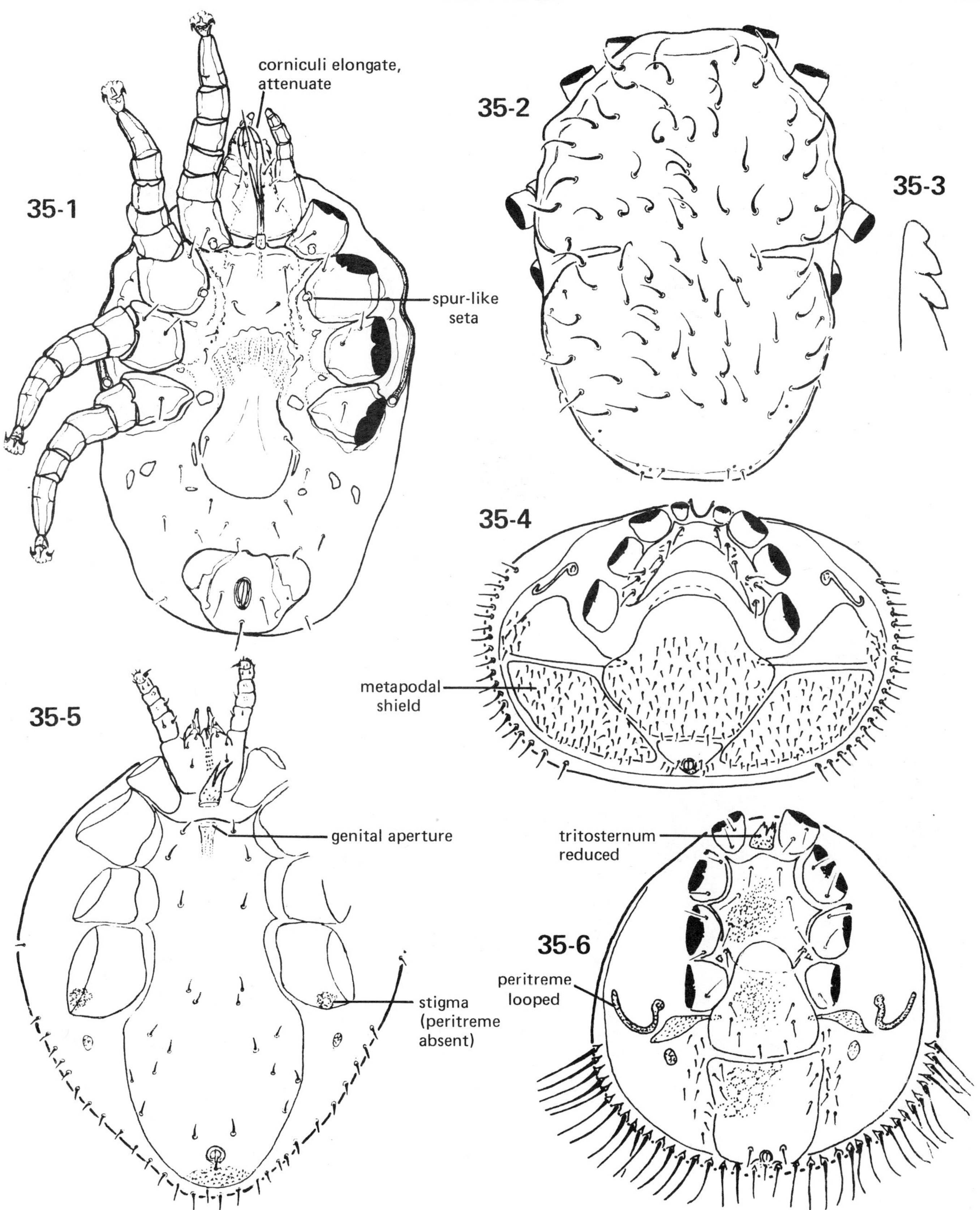

35-1 to 35-3; family IXODORHYNCHIDAE, *Ixodorhynchus liponyssoides* Ewing. **35-1**; venter of female: **35-2**; dorsum of female: **35-3**; chelicera of female

35-4 to 35-6; family VARROIDAE. **35-4**; *Varroa jacobsoni* Oudemans (Rep. China), venter of female (after Akratanakul 1976): **35-5**; *Euvarroa sinhai* Delfinado and Baker (Thailand), venter of male (from Akratanakul 1976): **35-6**; *E. sinhai*, venter of female (after Akratanakul 1976)

PLATE 36

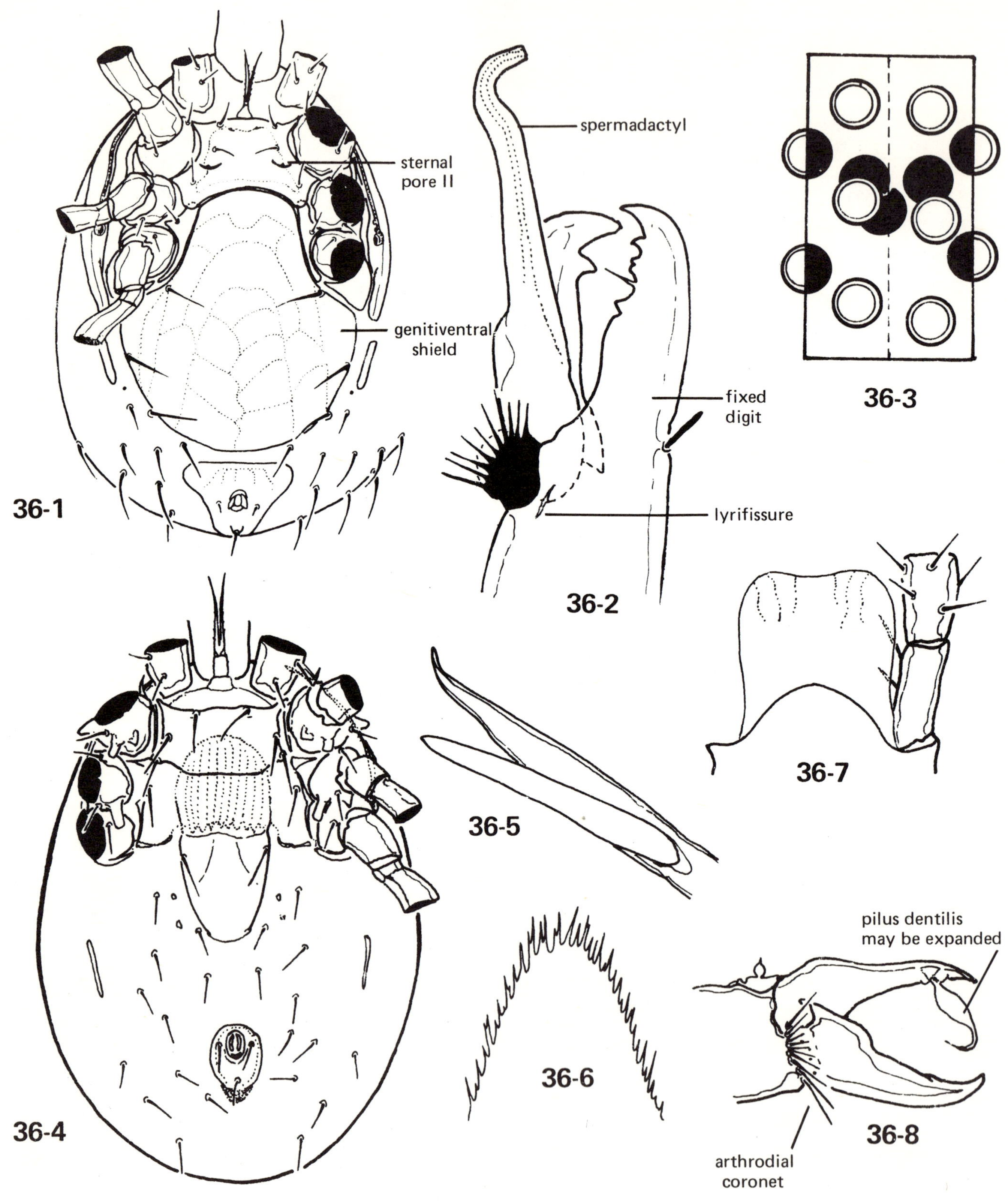

36-1 to 36-8; family LAELAPIDAE. **36-1**; *Hypoaspis* sp. (Oregon, USA), venter of female: **36-2**; *Hypoaspis krameri* (Canestrini), chelicera of male (after Evans and Till 1965): **36-3**; chaetotaxy of tibia I, typical for free-living LAELAPIDAE: **36-4**; *Hirstionyssus* sp. (Oregon, USA), venter of female: **36-5**; *Hirstionyssus* sp., chelicera of female: **36-6**; *Haemogamasus pontiger* (Berlese) (Oregon, USA), epistome: **36-7**; *Androlaelaps fahrenholzi* (Berlese) (Oregon, USA), epistome: **36-8**; *A. fahrenholzi*, chelicera of female

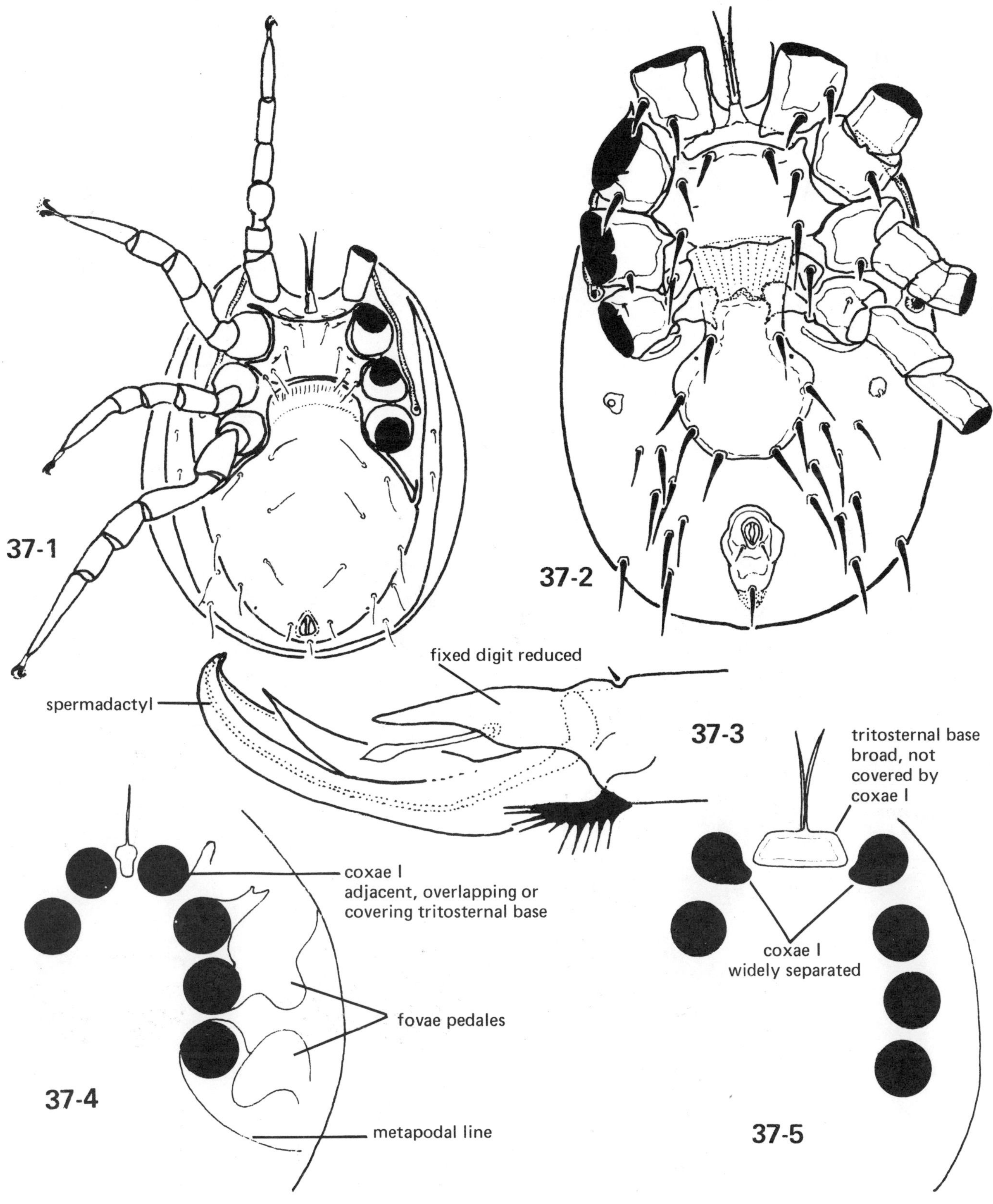

37-1 to **37-3**; family LAELAPIDAE. **37-1**; *Ololaelaps* sp. (Oregon, USA), venter of female: **37-2**; *Laelaps* sp. (Brazil), venter of female: **37-3**; *Laelaps hilaris* Koch, chelicera of male (after Evans and Till 1965)

37-4 and 37-5; Uropodina, venters of major uropodine types (diagrammatic) showing typical coxal and tritosternal characters. **37-4**; uropodoid type: **37-5**; polyaspidoid type

PLATE 38

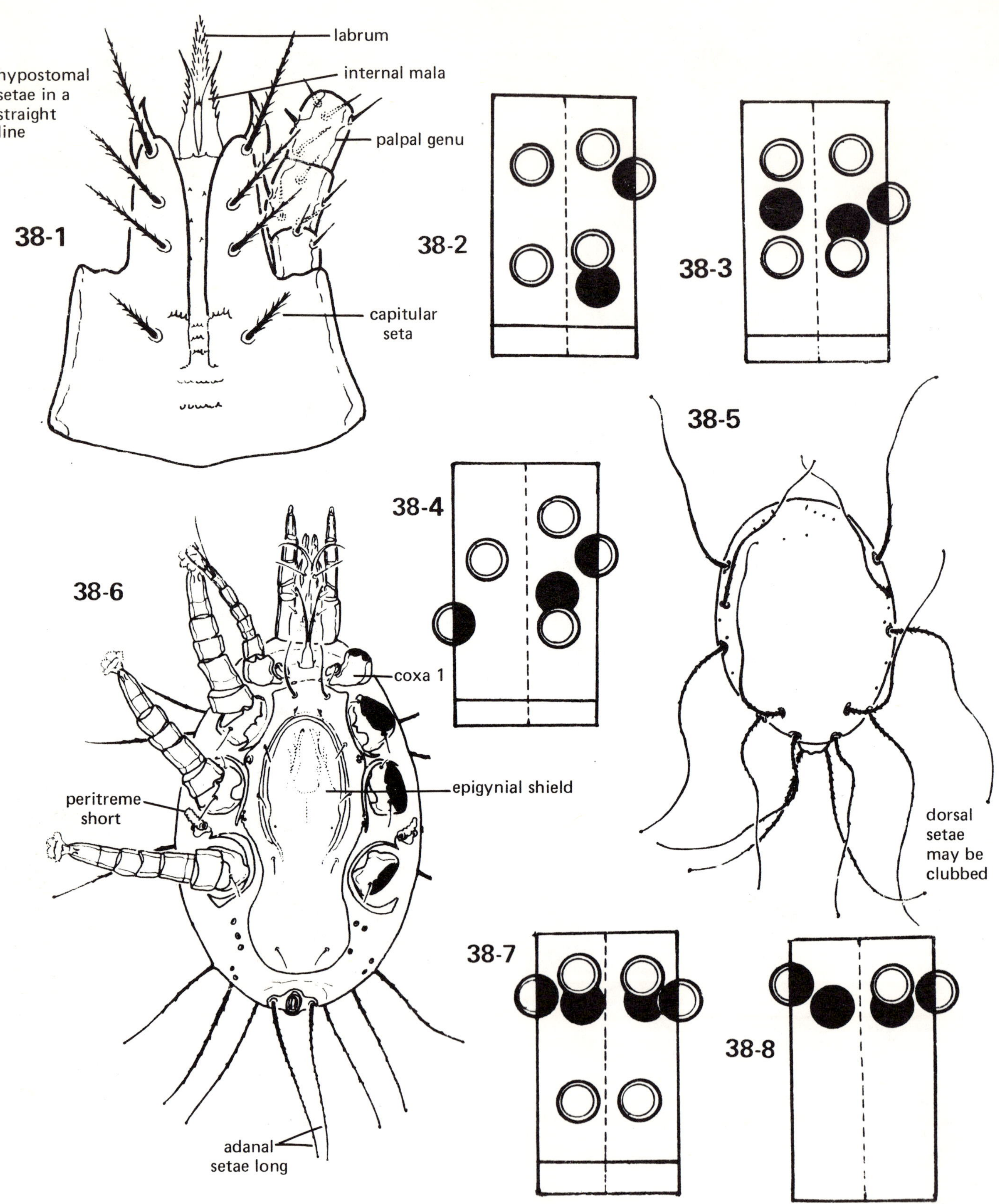

38-1; typical uropodine gnathosoma, ventral aspect

38-2 to 38-4; chaetotaxies of femur IV, Uropodina. **38-2**; Thinozerconoidea: **38-3**; Polyaspidoidea: **38-4**; Uropodoidea

38-5 to 38-8; family DIARTHROPHALLIDAE, *Diarthrophallus quercus* (Pearse and Wharton) (Mexico). **38-5**; dorsum of female: **38-6**; venter of female: **38-7**; chaetotaxy of femur IV: **38-8**; chaetotaxy of tibia I

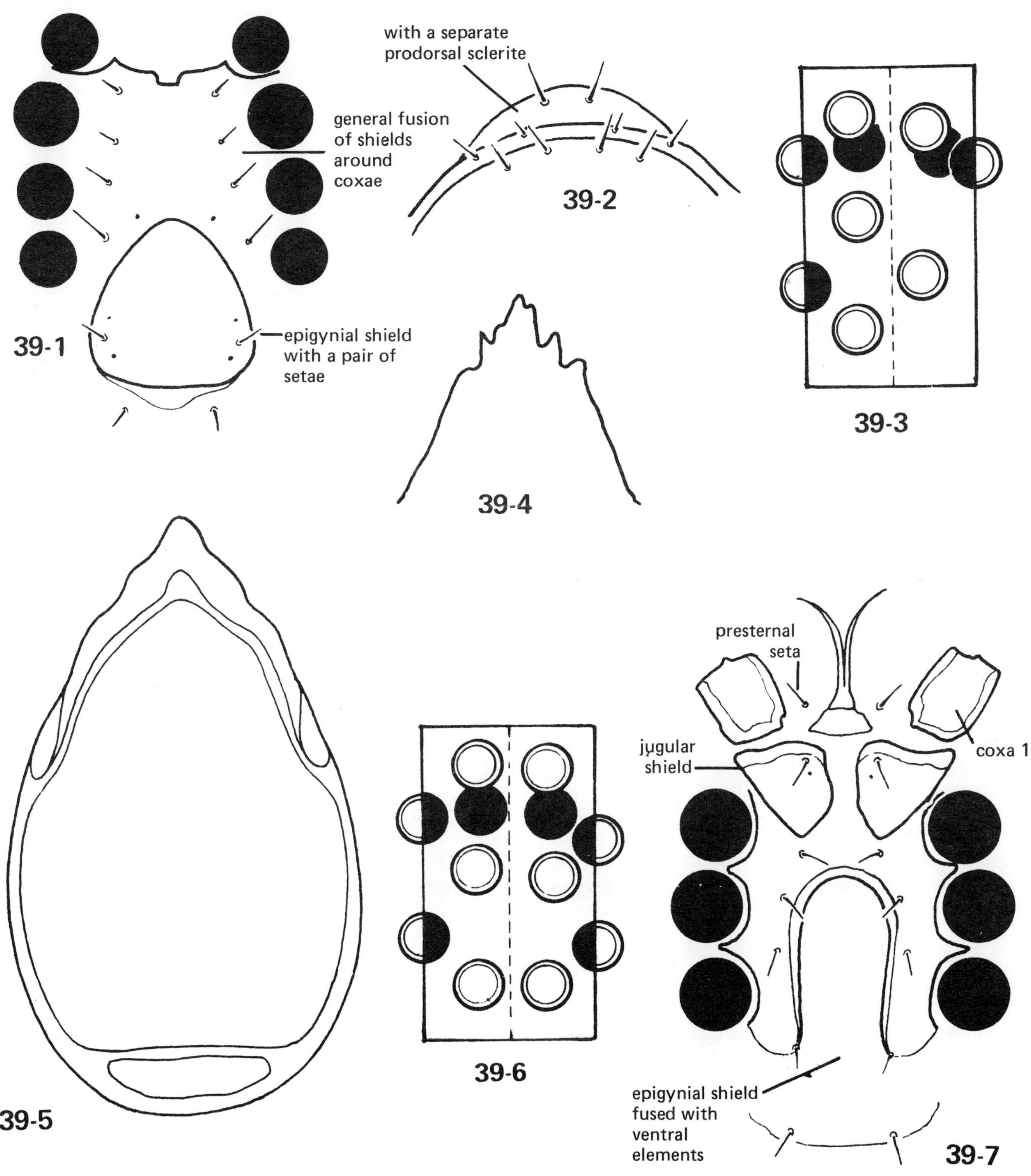

39-1 to 39-4; family PROTODINYCHIDAE. **39-1**; *Protodinychus* sp. (Canada), sternal and genital region of female: **39-2**; *Protodinychus punctatus* Evans (England), anterodorsal portion of idiosoma (after Evans 1957): **39-3**; chaetotaxy of genu I: **39-4**; *P. punctatus*, epistome

39-5 to 39-7; family THINOZERCONIDAE. **39-5**; *Thinozercon michaeli* Halbert (Ireland), dorsum of female (after Halbert 1915): **39-6**; chaetotaxy of genu I: **39-7**; *T. michaeli*, anteroventral aspect of female (after Trägårdh 1941)

PLATE 40

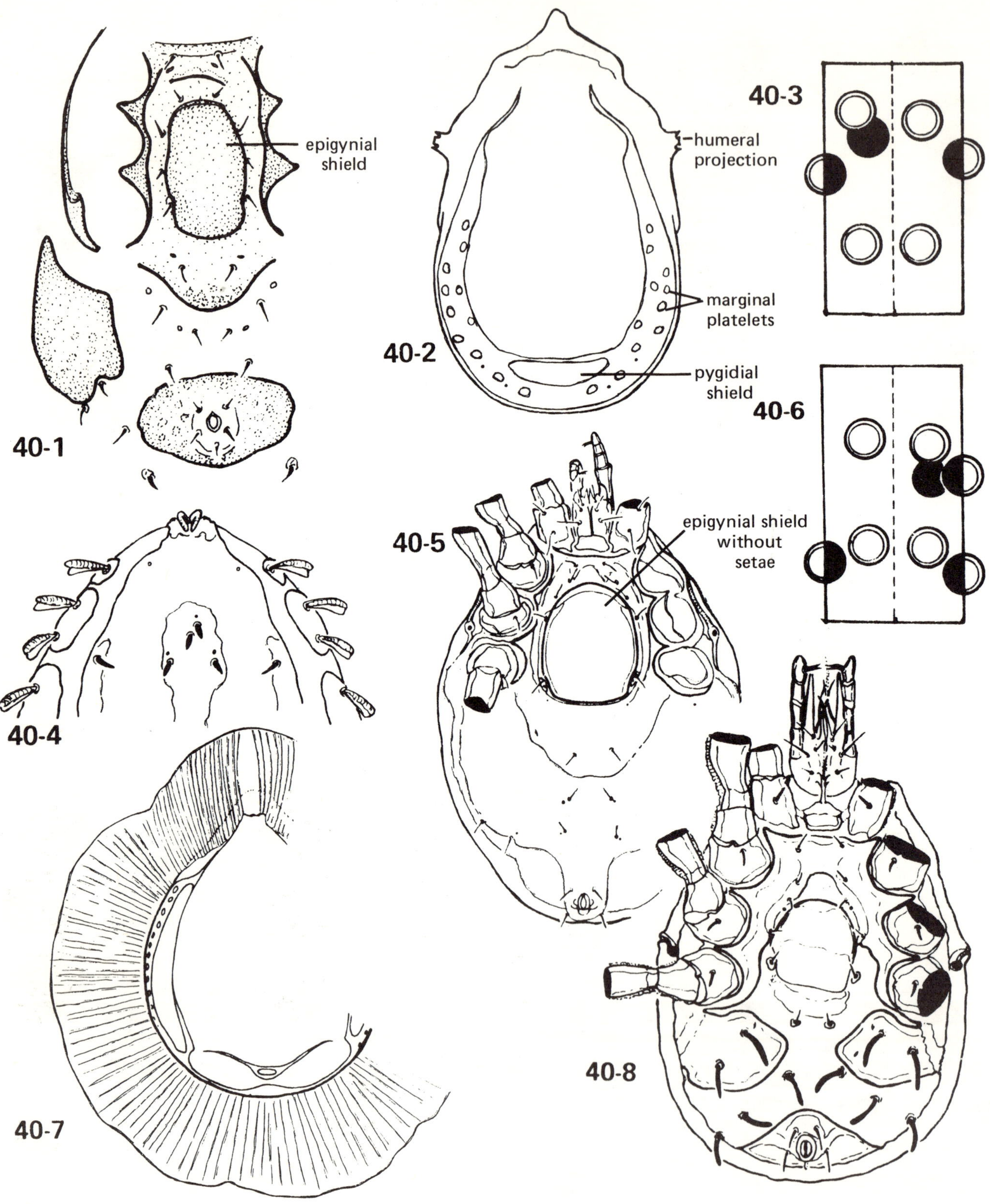

40-1 to 40-3; family DITHINOZERCONIDAE. **40-1**; *Apionoseius castrii* Hirschmann (Chile), venter of female (after Hirschmann 1973): **40-2**; *Caminella peraphora* Krantz and Ainscough (Oregon, USA), dorsum of female: **40-3**; chaetotaxy of genu IV of many dithinozerconids

40-4 to 40-8; family POLYASPIDIDAE. **40-4**; *Polyaspis* sp. (Oregon, USA), anterodorsal aspect of idiosoma of female: **40-5**; *Polyaspinus* sp. (Oregon, USA), venter of female: **40-6**; chaetotaxy of genu IV: **40-7**; *Trachytes* sp. (Oregon, USA), dorsum of female: **40-8**; *Polyaspis* sp. (Brazil), venter of female

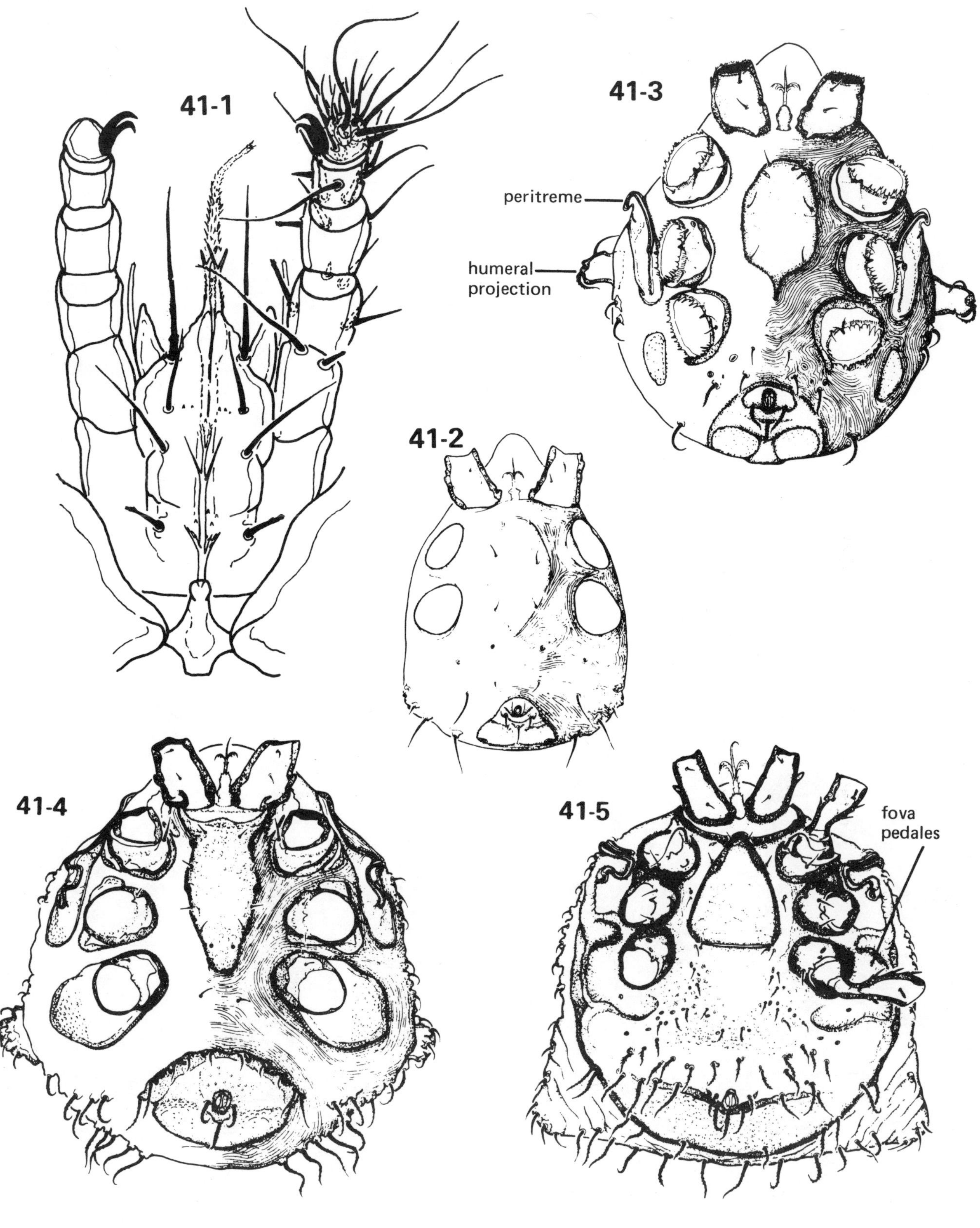

41-1 to 41-5; family UROPODIDAE. **41-1;** *Phaulodinychus mitis* (Leonardi) (Italy), venter of gnathosoma: **41-2;** *Eutrachytes maya* Krantz (Mexico), larva: **41-3;** *E. maya,* protonymph: **41-4;** *E. maya,* deutonymph: **41-5;** *E. maya,* female

PLATE 42

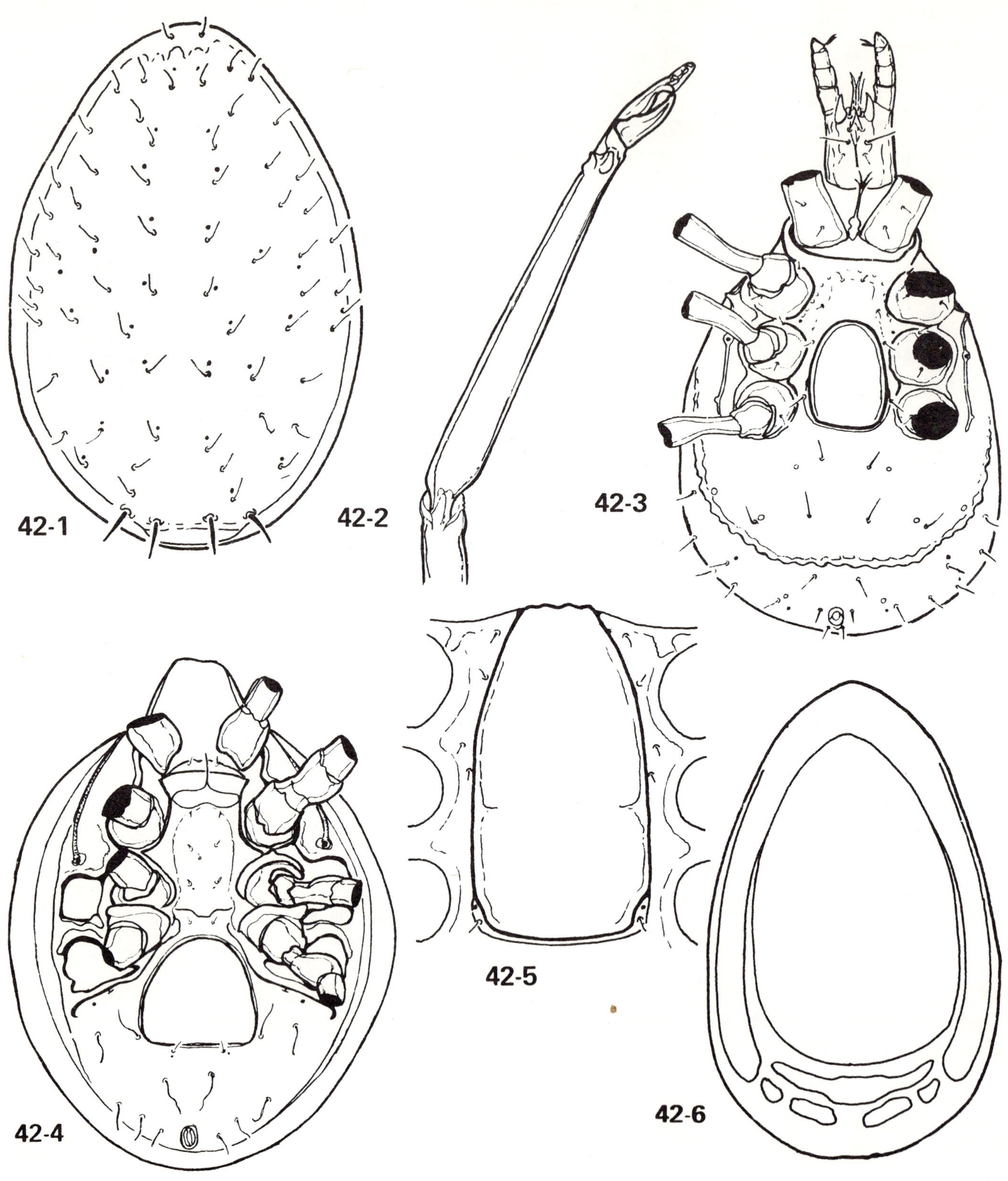

42-1 to 42-6; family UROPODIDAE. 42-1; *Dinychus* sp. (Oregon, USA), dorsum: 42-2; *Dinychus* sp., chelicera of female: 42-3; *Dinychus* sp., venter of female: 42-4; *Metagynella* sp. (South Africa), venter of female: 42-5; *Discourella* sp. (North America), sternitigenital region of female (after Johnston 1961): 42-6; *Discourella* sp., dorsum of female (after Johnston 1961)

PLATE 43

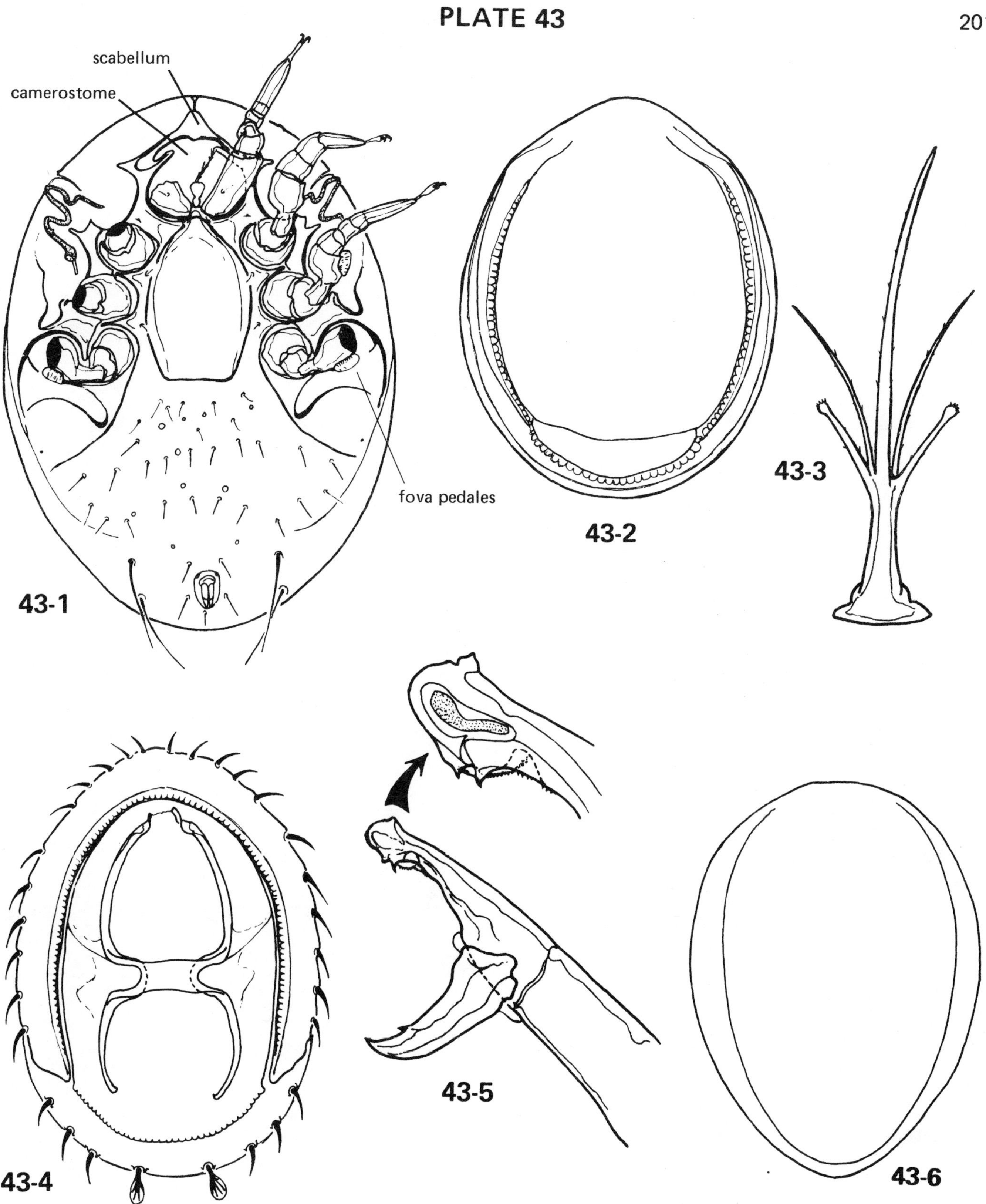

43-1 to 43-6; family UROPODIDAE. **43-1**; *Trichouropoda krantzi* Hirschmann (Zaire), venter of female: **43-2**; *Urodiaspis* sp. (Oregon, USA), dorsum: **43-3**; uropodid tritosternum: **43-4**; *Phaulodinychus krantzi* Hirschmann (Zaire), dorsum: **43-5**; *Phaulodinychus mitis* (Leonardi) (Italy), chelicera of female: **43-6**; *Fuscuropoda agitans* (Banks) (Maryland, USA), dorsum

PLATE 44

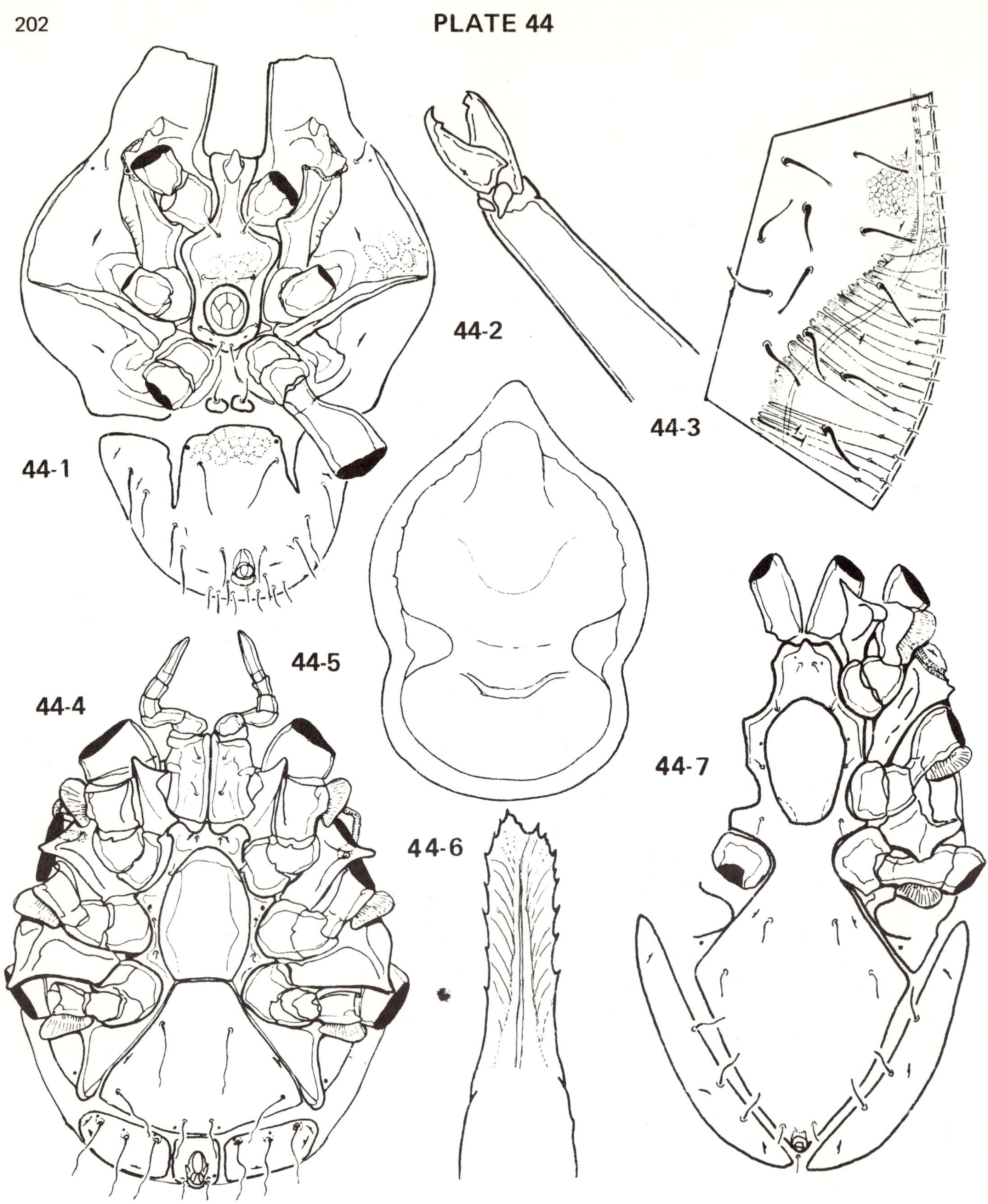

44-1 to 44-7; family TRACHYUROPODIDAE. **44-1;** *Coxequesoma panamensis* (Hirschmann) (Panama), venter of male: **44-2;** *C. panamensis,* chelicera of male: **44-3;** *Circocylliba krantzi* (Hirschmann) (Panama), dorsal ornamentation and setation: **44-4;** *C. krantzi,* venter of female: **44-5;** *Leonardiella* sp. (Zaire), dorsum: **44-6;** *Planodiscus burchelli* Elzinga and Rettenemeyer (Panama), epistome: **44-7;** *P. burchelli,* venter of female

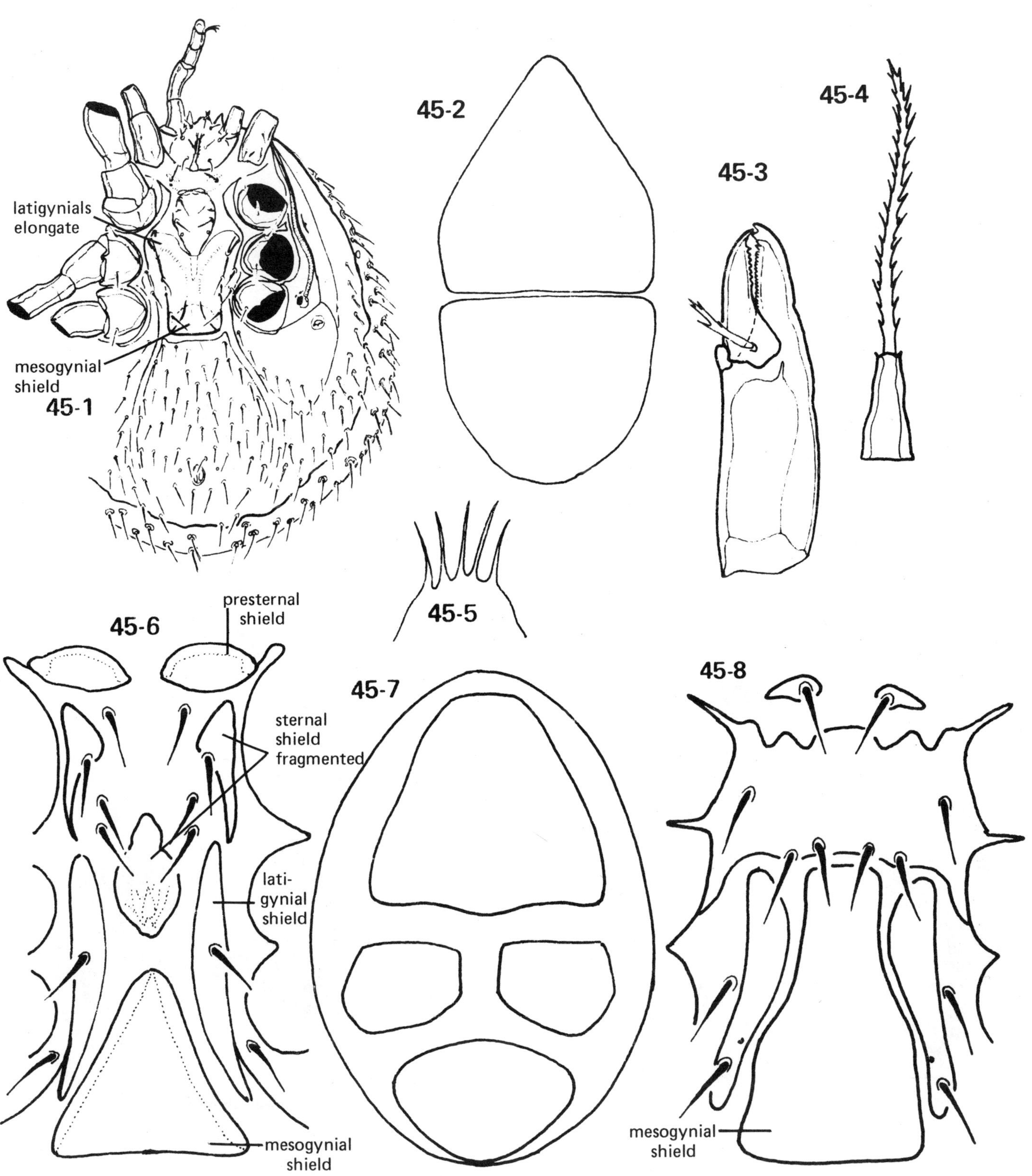

45-1 to 45-5; family CERCOMEGISTIDAE. **45-1;** *Cercomegistus evonicus* Kinn (Oregon, USA), venter of female: **45-2;** dorsum of female: **45-3;** chelicera of female: **45-4;** tritosternum: **45-5;** epistome

45-6; family ASTERNOSEIIDAE, *Asternoseius ciliatus* Berlese (Italy), sternitigenital region of female (based on drawing of type by J.B. Kethley)

45-7 and 45-8; family DAVACARIDAE, *Davacarus gressitti* Hunter (Antarctica). **45-7;** dorsum: **45-8;** sternitigenital region of female (based on drawing of type by J.B. Kethley)

PLATE 46

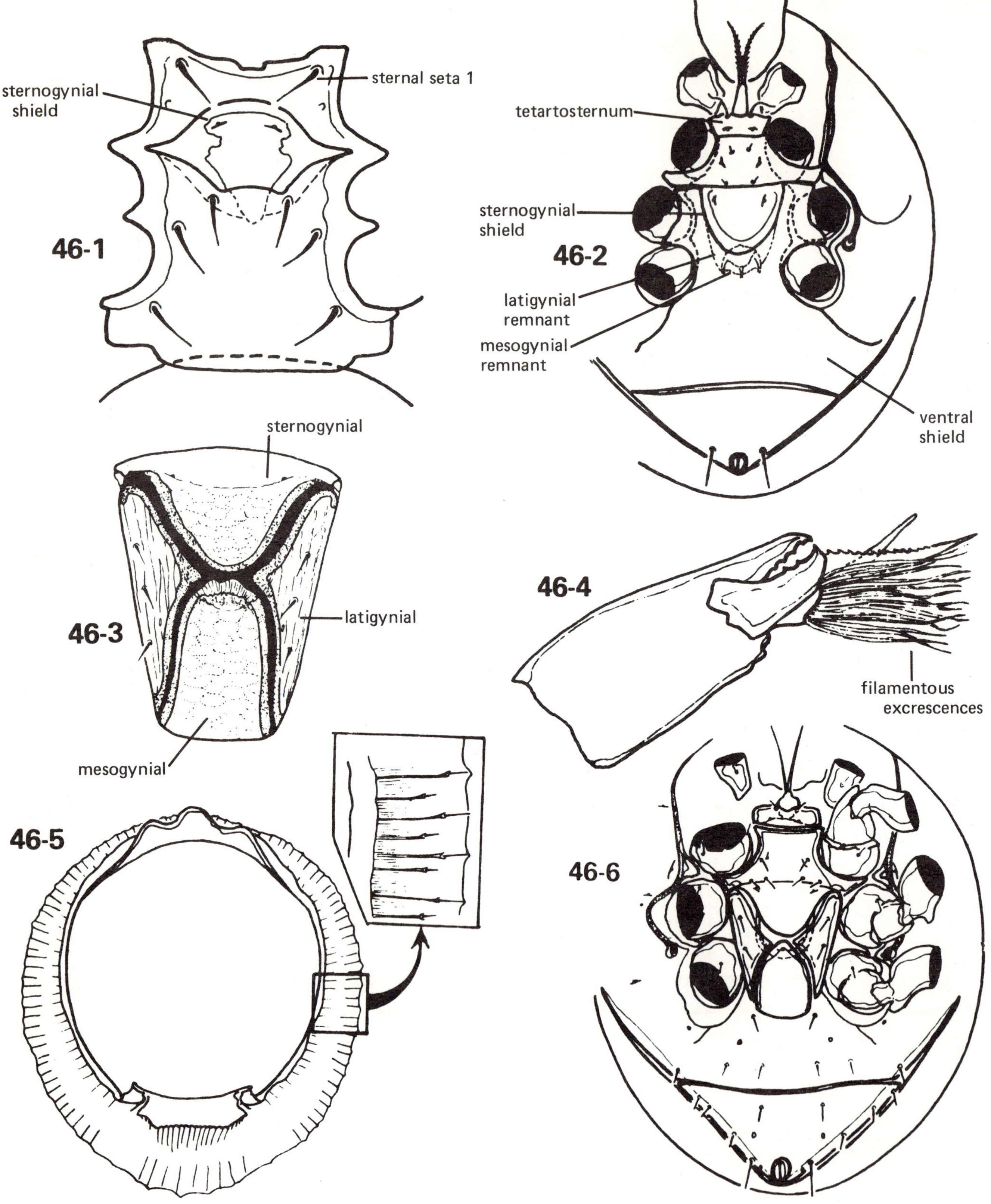

46-1; family SEIODIDAE, *Seiodes ursinus* Berlese, sternitigenital region of female (after Athias-Henriot 1959)

46-2; family FEDRIZZIIDAE, *Fedrizzia* sp. (New Guinea), venter of female

46-3 to 46-6; family KLINCKOWSTROEMIDAE, *Klinckowstroemia* sp. (Brazil). **46-3**; genital shields of female: **46-4**; chelicera of female: **46-5**; dorsum of female, with detail of marginal ornamentation: **46-6**; venter of female

PLATE 47

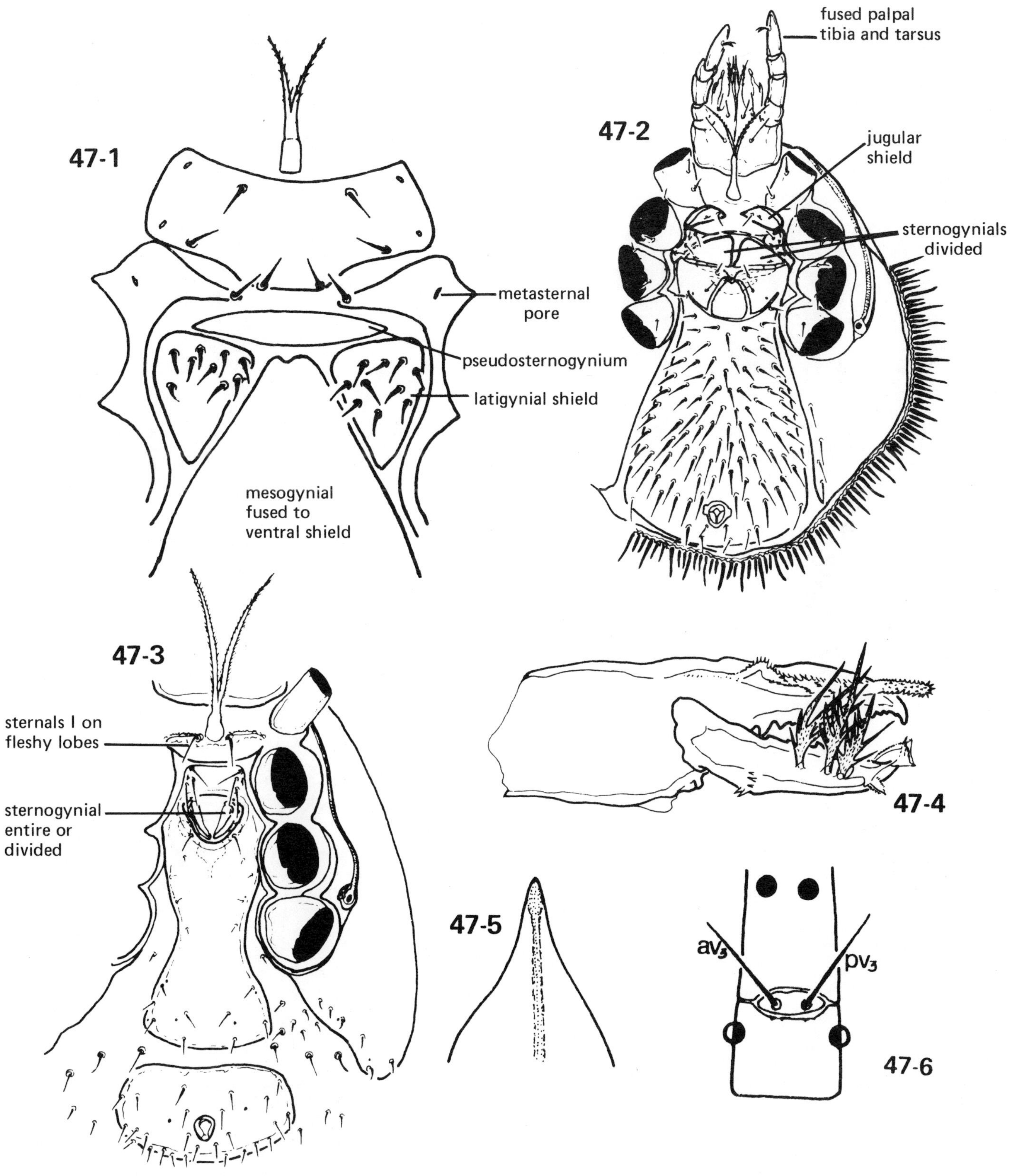

47-1; family PROMEGISTIDAE, *Promegistus armstrongi* Womersley (Australia), sternitigenital region of female (based on drawing of type by J.B. Kethley)

47-2; family PARAMEGISTIDAE, *Echinomegistus wheeleri* (Wasmann) (Kansas, USA), venter of female

47-3 to 47-6; family MEGISTHANIDAE, *Megisthanus floridanus* Banks (Florida, USA). **47-3**; venter of female: **47-4**; chelicera of female: **47-5**; epistome: **47-6**; proximoventral portion of tarsus IV of protonymph showing the intercalary sclerite with setae av_3 and pv_3 (after Evans 1965)

PLATE 48

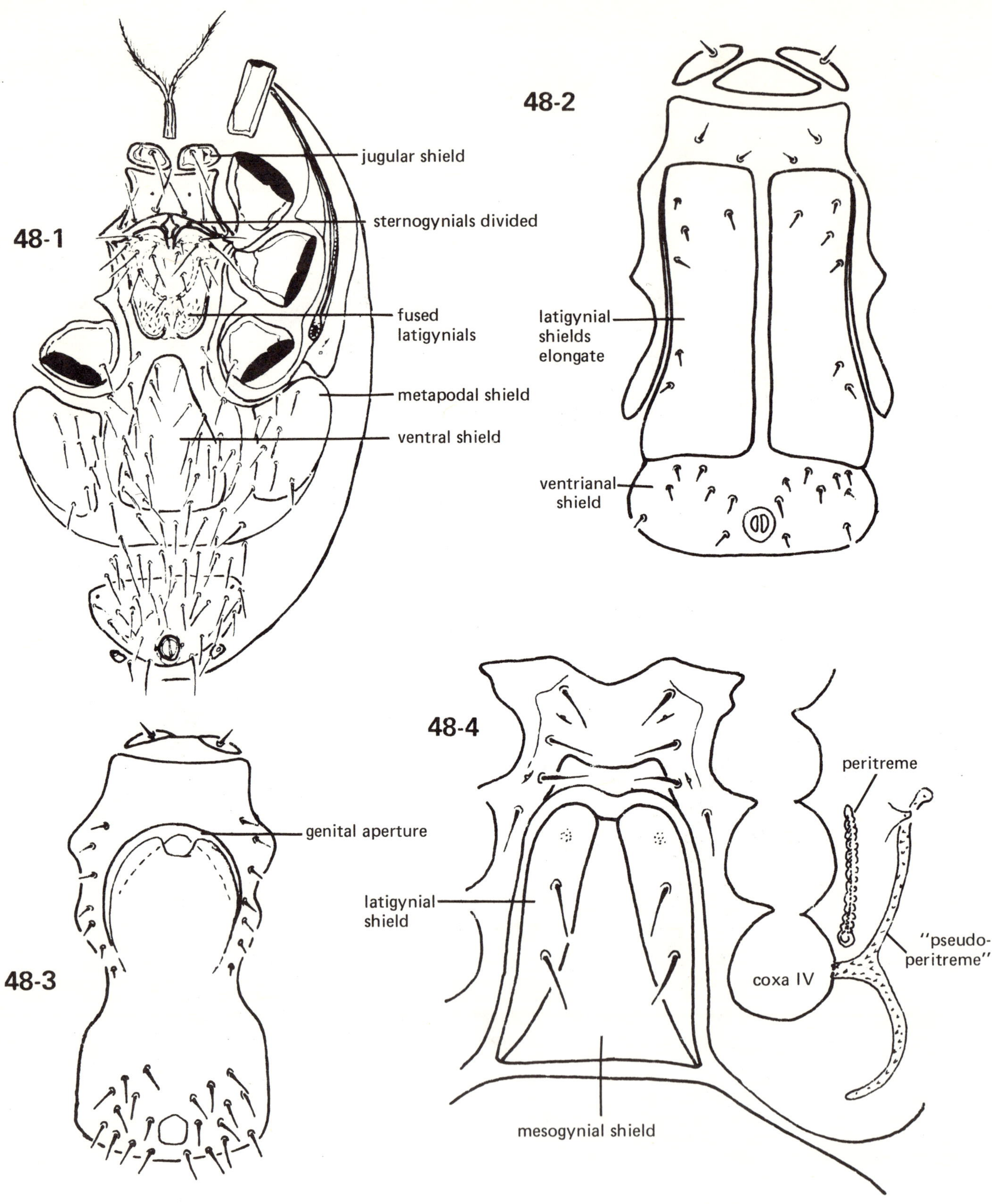

48-1; family HOPLOMEGISTIDAE, *Hoplomegistus* sp. (Brazil), venter of female

48-2 and 48-3; family MESSORACARIDAE, *Messoracarus mirandus* Silvestri (based on drawings of type by J.B. Kethley).
48-2; venter of female: **48-3**; venter of male

48-4; family AENICTEQUIDAE, *Aenicteques chapmani* Jacot (Philippines), sternitigenital region of female (based on drawing of type by J.B. Kethley)

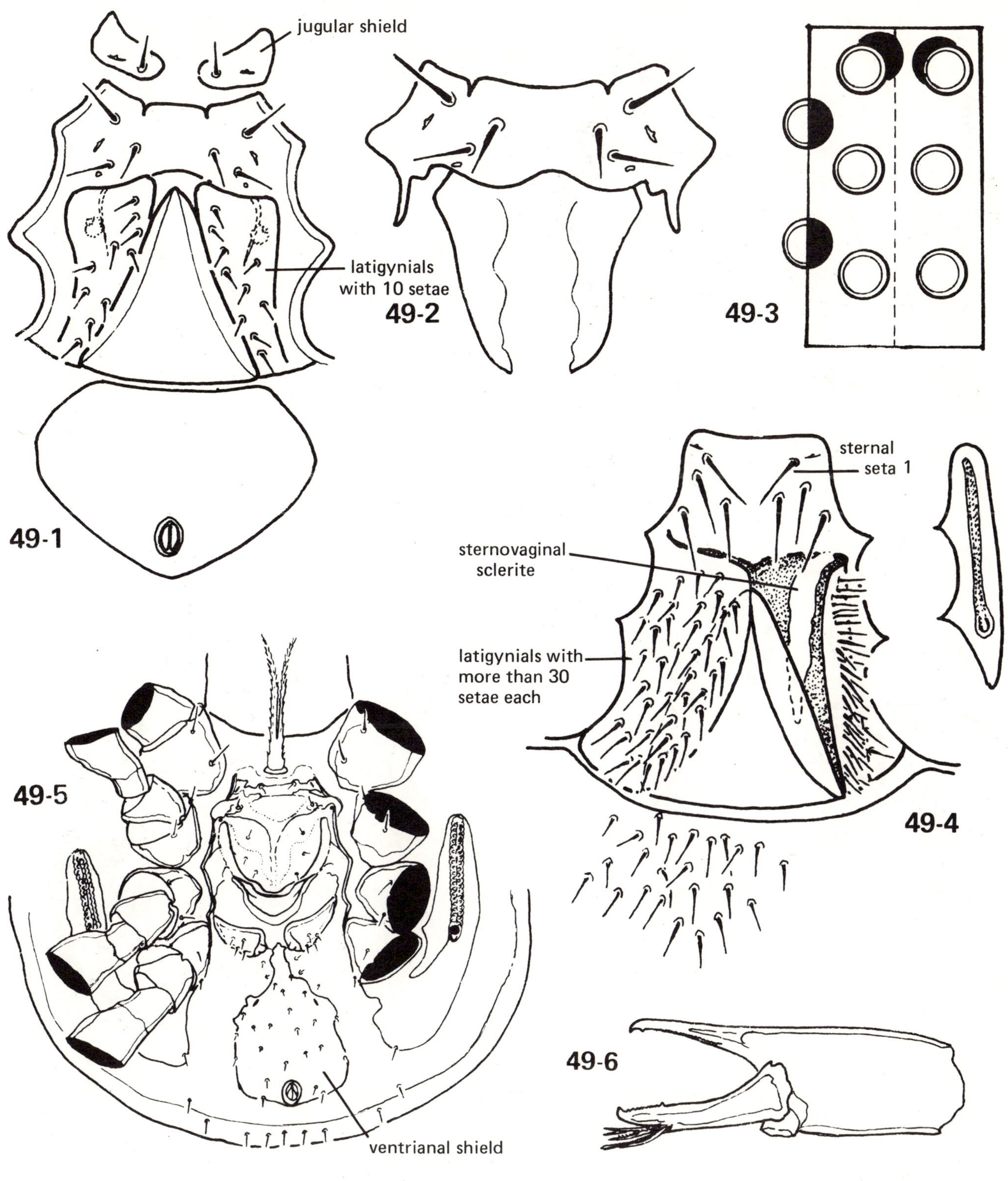

49-1 to 49-3; family PTOCHACARIDAE, *Ptochacarus silvestrii* Womersley (Australia). **49-1**; venter of female: **49-2**; sternal shield, with posterolateral sternovaginal shields exposed (based on drawing of type by J.B. Kethley): **49-3**; chaetotaxy of genu IV

49-4; family PHYSALOZERCONIDAE, *Physalozercon raffrayi* Wasmann, sternitigenital region of female, with one latigynial shield folded back to expose a sternovaginal shield (based on drawing of type by J.B. Kethley)

49-5 and 49-6; family ANTENNOPHORIDAE, *Antennophorus* sp. (British Columbia). **49-5**; venter of female: **49-6**; chelicera of female

PLATE 50

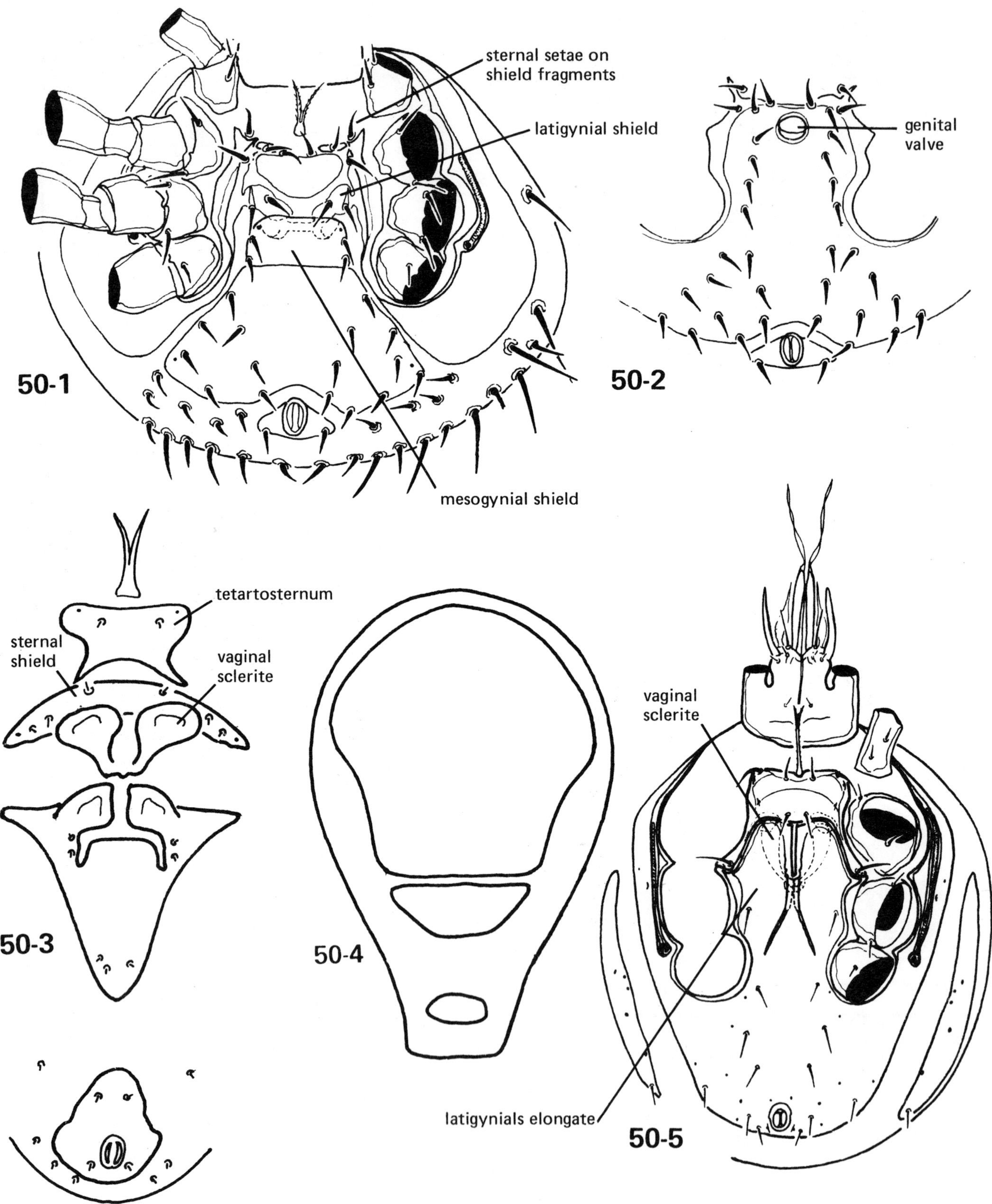

50-1 and 50-2; family PARANTENNULIDAE, *Micromegistus bakeri* Trägårdh (Ohio, USA). **50-1;** venter of female: **50-2;** venter of male

50-3 and 50-4; family NEOTENOGYNIIDAE, *Neotenogynium malkini* Kethley (Equador). **50-3;** venter of female (after Kethley 1973); **50-4;** dorsal shields of female (after Kethley 1973)

50-5; family SCHIZOGYNIIDAE, *Schizogynium* sp. (Panama), venter of female

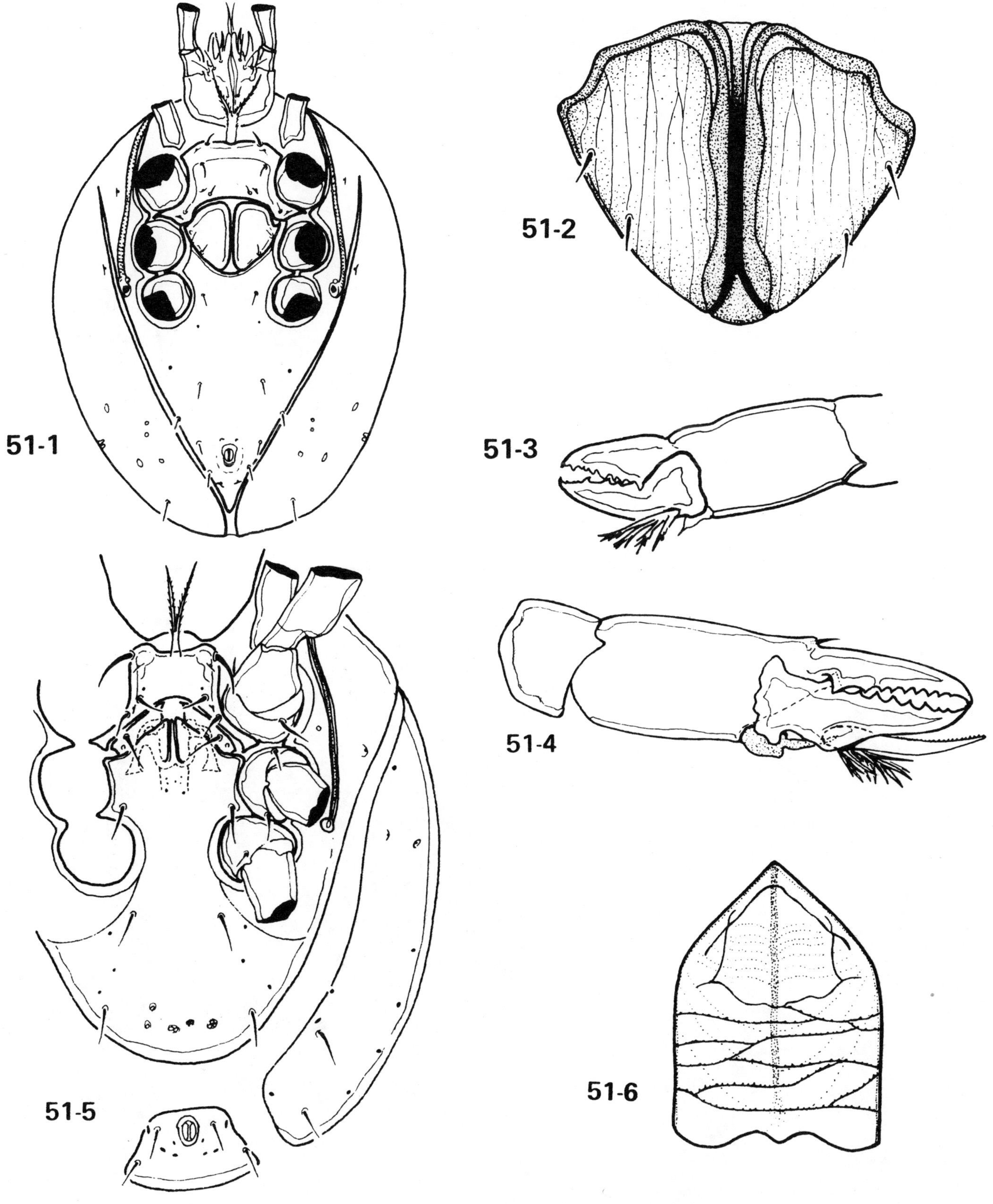

51-1 to 51-4; family DIPLOGYNIIDAE. **51-1**; ? genus (Ohio, USA), venter of female: **51-2**; genital shields of female: **51-3**; chelicera of female: **51-4**; *Lobogyniella trägårdhi* Krantz (Oregon, USA), chelicera of female

51-5 and 51-6; family EUZERCONIDAE. **51-5**; *Euzercon* sp. (Arkansas, USA), venter of female: **51-6**; *Euzercon* sp. (Mexico), epistome

PLATE 52

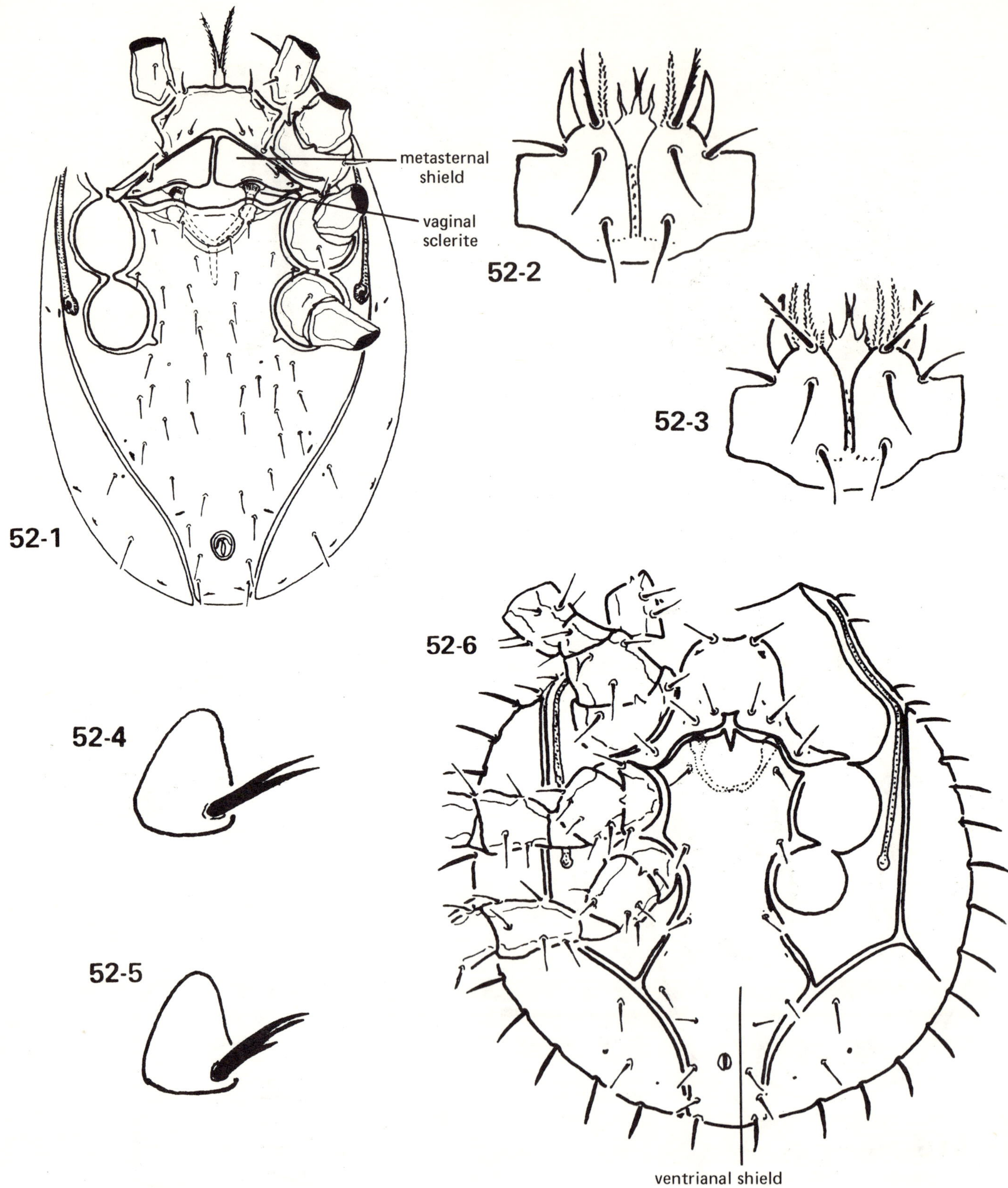

52-1; family CELAENOPSIDAE (Colorado, USA), venter of female.

52-2 to 52-6; family MEGALOCELAENOPSIDAE. **52-2**; *Pelorocelaenopsis camini* Funk (Zaire), venter of female gnathosoma: **52-3**; *P. camini*, venter of male gnathosoma (after Funk 1974): **52-4**; *P. camini*, palptarsal claw: **52-5**; *Megalocelaenopsis oudemansi* Funk (Panama), palptarsal claw: **52-6**; *M. oudemansi*, venter of female (after Funk 1974)

SUBORDER IXODIDA

The Ixodida, or ticks, are a small group of acarines (approximately 800 species described to date) which feed obligately on the blood of mammals, reptiles and birds (Doss et al. 1974). Their unique morphology, coupled with their great size (2,000 to over 30,000 μ in length), serves to distinguish the ticks from most other Acari (Hoogstraal 1973b).

While ticks illustrate the interstadial heteromorphy typical of other Acari, certain singular characters are present and distinct throughout ontogeny. A hypostome armed with retrorse teeth (Plate 54-7, p. 222) serves to anchor the tick to its host. A complex sensory setal field, *Haller's organ* (Fig. 30), is located on the dorsal aspect of tarsus I in all postembryonic stages, providing sites for contact or olfactory chaemoreception (Balashov 1972). Other distinguishing features are listed later in this section.

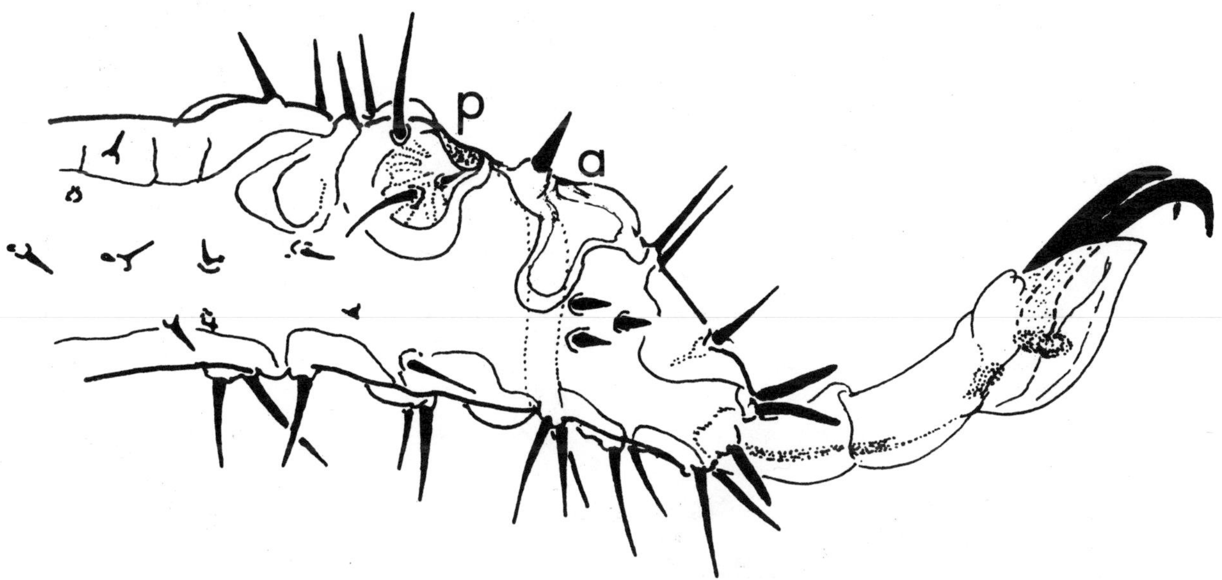

Fig. 30. Distal portion of tarsus I of *Dermacentor variabilis* (Say) (Massachusetts, USA), showing position of Haller's organ (a = anterior pit, p = posterior pit).

Larval ticks possess a small podonotal shield or *scutum* which is retained or enlarged in postlarval stadia, or a mesonotal shield which is lost at the larval molt. Tick larvae lack lateroventral stigmata and, like other acarines, are hexapod (Fig. 31). Nymphs and adults are octopod and possess stigmal plates behind or anterolateral to coxae IV. The morphology of the ixodid stigma is referred to in an earlier chapter (see page 22). Elongate peritremes are absent.

Unlike the Gamasida, differences between nymphal and adult Ixodida may not always be obvious, nor is sexual dimorphism clearly marked throughout the suborder. Adults possess a genital opening in the region of coxae I-II which, along with certain ventral grooves, are absent in the nymph. Nymphal hard ticks (IXODIDAE) closely resemble adult females in that both nymphs and females have an abbreviated podonotal scutum which allows for idiosomal expansion during feeding. Male ixodids have an entire dorsal shield (Plate 55-1, p. 223) and consequently are unable to consume large blood meals. Nymphal and adult soft ticks (ARGASIDAE) lack a true dorsal scutum and are difficult to distinguish other than through comparative size or the presence or absence of a genital opening. Sexual dimorphism in soft ticks is essentially confined to internal structures (Balashov 1972).

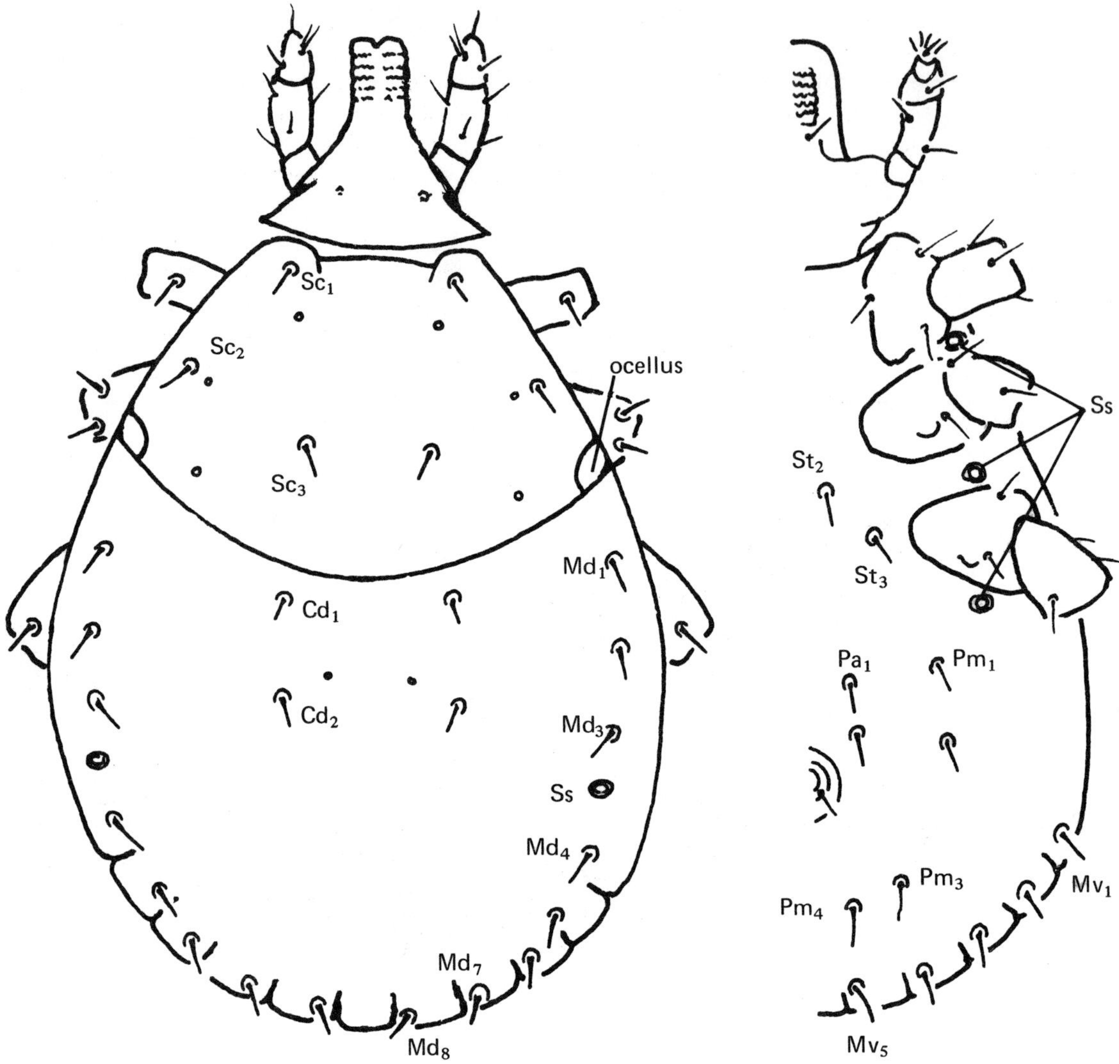

Fig. 31. Larva of *Dermacentor variabilis* (Say), with setal notations of Clifford and Anastos (1960) (Ss = sensilla sagittiformia).

In addition to having a highly modified hypostome, distinctive stigmal plates, and a Haller's organ on tarsus I, members of the Ixodida share the following characters:

1. Palp with only three or four segments. Claws are absent on the fused palpal tibiotarsus, which often is reduced (Plate 54-7, p. 222).

2. Chelicera 2-segmented, the distal movable segment comprised of an inner sclerotized element on which a strongly dentate external member is articulated. A dentate dorsal process is attached internally on the inner element. A membranous mantle surrounds the digit, leaving only the cutting external teeth exposed (Plate 55-6, p. 223).

3. Epistome, corniculi and tritosternum absent. One or two pairs of ocelli may be present on the lateral margins of the dorsal shield (Plate 55-4, p. 223), or in the pleural integument.

The Ixodida occur throughout the world but are most frequently encountered in tropical and subtropical realms (Hoogstraal 1973b). Nonetheless, ticks may be significant pests of man and animals in temperate forest, steppe and prairie habitats, as well as in the harsher tundra of the far north.

Attachment to the host by Ixodida is accomplished by piercing of the skin with the chelicerae, and anchoring to the site by inserting the hypostome into the wound (Arthur 1957). A cement-like substance is produced by the salivary glands of most ixodid genera during feeding (Chinery 1962 and 1965). This material is deposited around the mouthparts and extends into the host tissue to secure the tick to its feeding site (Chinery 1973). Subcutaneous infestations of *Ixodes* and *Hyalomma* species have been observed on some hosts.

Attraction and attachment responses of ticks are triggered, at least in part, by stimuli received through thin-walled contact and olfactory sensory setae located on the apex of the palpal tibiotarsus, on leg tarsus and pretarsus I, and in Haller's organ (Balshov 1972). Mechanoreception (sudden vibration), photoreception (sudden shadows), and host warmth also appear to play a part in attraction and attachment. Many tick species react positively to carbon dioxide emanating from a prospective host (Garcia 1962 and 1969, Nevill 1964).

The ocelli of at least some tick genera play an important role not only in perception of light and shadow, but also in perception of the host itself. *Hyalomma asiaticum* P. and E. Schulze utilizes the ocelli to orient to large vertebrate hosts and may pursue them over long distances.

Lees (1948) noted that host finding and attachment are ensured by 1) orientation responses leading to a definite distribution in vegetation layers and favorable waiting posture, 2) direct response to an approaching host by taking up a "questing" posture and 3) reactions associated with crawling to the host and choosing attachment sites. Chance of contact with a host often is enhanced by the tick's habit of climbing to the tops of grass blades or twigs in animal runways (Lees 1946). Desiccation of questing ticks on these exposed sites may occur within a few hours, although some genera are highly resistant to rapid water loss (*Dermacentor* and *Hyalomma,* for example). Ticks which are unsuccessful in immediately finding a host may regain satisfactory water balance by moving down to areas of higher relative humidity for short periods of time.

Tick drop-off appears to be subject to a predictable periodicity in certain ixodid and argasid species. Drop-off times observed in all three instars of the rabbit tick, *Haemaphysalis leporispalustris* (Packard), and in larvae and nymphs of the argasid kangaroo tick, *Ornithodoros gurneyi* Warburton, are regulated by a circadian rhythm entrained by the daily light cycle (George 1971, Doube 1975). Darkness plays the principal role in setting the phase of the rhythm. Similar rhythms have been suggested for species of *Argas, Hyalomma* and *Rhipicephalus* (Hadani and Rechav 1969). Diapause expressed through delay in oviposition has been noted in *Argas arboreus* Kaiser, Hoogstraal and Kohls (Khalil 1974).

Unfed ticks often engage in cannibalism or homoparasitism, feeding on the haemolymph and/or gut contents of engorged individuals of members of their own species. Examples of cannibalism are common amongst nymphs and males of the argasid genera *Ornithodoros* and *Argas,* and are regarded by Beklemishev (1948) as vestigial manifestations of ancestral entomophagy. An example of what appears to be obligate homoparasitism in the IXODIDAE was described by Moorhouse (1966). He observed male *Ixodes holocyclus* Neumann feeding on the haemolymph of engorged and partially engorged females. The female hosts apparently were unaffected by the male's feeding activities. Furthermore, males were never found feeding on vertebrate animals and may normally procure their small blood meal exclusively from females.

The only recorded arthropod parasites of ticks are two wasps of the family Encyrtidae (Smith et al. 1946). *Ixodiphagus texanus* Howard is widely distributed in the United States and normally parasitizes *Haemaphysalis leporispalustris.* It also attacks larvae and nymphs of the American dog tick, *Dermacentor variabilis* (Say) in the eastern United States. A second encyrtid wasp, *Hunterellus hookeri* How., occurs in warmer parts of the world and parasitizes the brown dog tick, *Rhipicephalus sanguineus* Latr.

A number of bird and rodent species feed on ticks (Wilkinson 1975), a phenomenon which may explain the observed link between the bird-rabbit tick-lagomorph cycle and the rodent-dog tick-mammal cycle of spotted fever rickettsiae in North America.

Ixodid and argasid ticks may cause injury to their hosts through exsanguination, secondary infection or irritation at attachment sites, or through the transmission of disease organisms (Philip 1963, Hoogstraal 1966, 1967; Balashov 1972, Southcott 1976, Shaw et al. 1976). Ticks serve as reservoirs and vectors for many infective viruses, rickettsias, bacteria, sporozoans and spirochaetes, and may effect transmission of these organisms through their bite or through contact of infective coxal fluids or excrement with the broken skin of the host (Hoogstraal 1973b). Ixodid ticks have been observed to produce motor paralysis in man and animals following an extended period of engorgement. The condition is thought to be produced by a neurotoxin injected by the feeding tick (Philip 1969). Tick paralysis may result in death if the feeding ticks are allowed to remain attached to the host.

Three families of Ixodida are presently recognized, and these are relegated to a single superfamily, the Ixodoidea.

Superfamily Ixodoidea
(Plates 53-55, pp. 221-223)

DIAGNOSIS: *With the characters of the suborder; weakly sclerotized but with a thick leathery cuticle; with or without a dorsal scutum, occasionally with a leathery false scutum. Gnathosoma (capitulum) anterior or anteroventral in position, with the hypostome modified as a holdfast organ; palpi simple, telescoped or normally extended. With a Haller's organ on the dorsum of tarsus I; all tarsi with claws.*

The hard ticks, or IXODIDAE (Plates 54-55, pp. 222-223), number approximately 650 species accommodated in 13 genera. Ixodids generally attach to three different host species during their various active life stages (larva, single nymph and adult). Each stage feeds only once on each host, generally for several days. Some ticks, however, do not leave the original host animal until after adult emergence, feeding, and mating have taken place (one-host ticks). Others drop off the initial host following nymphal development and acquire a second host only after molting to the adult stage on the ground (two-host ticks). The Texas cattle fever tick, *Boophilus annulatus* Say, is a one-host tick, while *Hyalomma marginatum* Koch is an example of a two-host tick.

Male ixodids may feed sparingly and for greatly extended periods (several months), increasing their body weight up to 1½-2 times. However, males of many nest or burrow-inhabiting species of *Ixodes* do not feed and have not been found on hosts (Balashov 1972). Ixodid females are slow feeders, often requiring several days to complete a blood meal. They may oviposit up to 18,000 eggs but average 2-8,000. Mortality of hatching larvae and subsequent developmental stages is high, owing primarily to their inability to find suitable hosts (Hoogstraal 1973b).

While most ixodid genera show little specificity for particular animal groups, some appear to illustrate more than an ecological dependency. For example, all but three of the 26 described species of *Aponomma* feed on large snakes and varanid lizards. *Dermacentor* species are primarily mammal parasites, as are members of the genus *Rhipicephalus.* On the other hand, the tropical and subtropical genus *Amblyomma,* which is common on mammals, frequently is recovered in the nymphal stage on migrating birds. *A. tuberculatum* Marx parasitizes land tortoises, while other species occur on amphibians. Finally, two *Amblyomma* species are parasites of sea snakes and lizards, attaching to their sea-going hosts while the latter are on land (Hoogstraal 1973b).

The soft ticks, or ARGASIDAE (Plates 53-54, pp. 221-222), are a relatively small group (140 species in five genera) which usually feed on hosts that return periodically to a particular shelter. Argasids are especially evident as nest, cave or burrowing inhabitants in semitropical and tropical realms where they parasitize birds and small mammals (Cooley and Kohls 1944). Unlike the ixodids, argasids feed intermittently and rapidly (2 minutes-2 hours), and often on a single host. Most species live in arid or semiarid regions and are able to tolerate long periods of drying. Many argasids have access to a host for only short periods of time (as little as a few days in the case of certain nesting birds and roosting bats), and have adapted to long-term survival on a single blood meal. *Ornithodoros papillipes* Birula, a burrow-inhabiting rodent parasite, may survive for up to 11 years without a blood meal (Pavlovsky and Skrynnik 1960). Other argasids are known to live without food for periods up to 10 years (Balashov 1972).

Unlike other argasid genera, *Otobius* and *Antricola* infest wandering hosts. This characteristic probably is shared by *Nothoaspis reddelli* Keirans and Clifford, a species found in bat guano in Mexico (Keirans and Clifford 1975). The habitat of *N. reddelli* is overlapped by *Antricola* species which parasitize cave-dwelling bats in Central and North America (Hoogstraal 1973b).

Because of their sedentary habits, argasids generally are considered less important than the more unrestricted ixodids in disease transmission. Nevertheless, species of *Ornithodoros* are known to transmit a number of arboviruses, while the spinose ear tick, *Otobius megnini* (Dugès) transmits both bacterial and rickettsial diseases. *Argas, Otobius* and *Ornithodoros* species may cause extreme irritation and trauma in their hosts (Hoogstraal 1963a), and are sometimes responsible for host deaths. As with hard ticks, argasids may attack man and sometimes cause severe symptoms. For example, *Ornithodoros (Alectorobius) muesebecki* Hoogstraal, a bird parasite of the Arabian Gulf, attacks man and causes fever, headache, itch and erythema. A similar syndrome occurs in guano diggers attacked by *O. (Alectorobius) amblus* Chamberlain, a related Peruvian bird parasite (Hoogstraal et al. 1970).

The family NUTTALLIELLIDAE (Plate 53, p. 221) is represented by a single species, *Nuttalliella namaqua* Bedford, collected in South and South-West Africa and in Tanzania from a variety of bird and mammal habitats (Keirans et al. 1976). Based on available data, *N. namaqua* appears to be a parasite of hyrax *(Procavia)* or other small rock-inhabiting mammals. Only the female has been described.

N. namaqua shares a number of ixodid and argasid characters (see key which follows this section), and its true position in the suborder may not become clear until males and immatures are examined.

Useful References

Arthur, D.R. (1957). The capitulum and feeding mechanism of *Dermacentor parumapertus.* Parasitol. **47**: 169-184. [IXODIDAE]

Arthur, D.R. (1960). Ticks. A Monograph of the Ixodoidea. Part 5. On the genera *Dermacentor, Anocentor, Cosminomma, Boophilus* and *Margaropus.* Cambridge Univ. Press: 251 pp. [IXODIDAE]

Arthur, D.R. (1962). Ticks and Diseases. Pergamon Press, Oxford: 445 pp. + xvi.

Balashov, Y.S. (1972). Bloodsucking ticks (Ixodoidea)-vectors of disease of man and animals. Misc. Publ. Ent. Soc. Amer. **8**(5):161-376.

Beklemishev, V.N. (1948). On interrelationships between the systematic position of the agent and the vector of transmissible diseases of terrestrial vertebrates and man. Med. Parasitol. Parazit. Bolezn. **17**:385-400.

Chinery, W.A. (1962). Cytochemical observation on the salivary glands of an ixodid tick. Trans. Roy. Soc. Trop. Med. Hyg. **56**:268.

Chinery, W.A. (1965). Studies on the various glands of the tick *Haemaphysalis spinigera* Neumann 1897. III. The salivary glands. Acta Trop. **22**:321-349. [IXODIDAE]

Chinery, W.A. (1973). The nature and origin of the "cement" substance at the site of attachment and feeding of adult *Haemaphysalis spinigera* (Ixodidae). J. Med. Ent. **10**(4):355-362.

Clifford, C.M. and G. Anastos (1960). The use of chaetotaxy in the identification of larval ticks (Acarina: Ixodidae). J. Parasitol. **46**(5):567-578.

Clifford, C.M., G.M. Kohls and D.E. Sonenshine (1964). The systematics of the subfamily Ornithodorinae (Acarina:Argasidae). I. The genera and subgenera. Ann. Ent. Soc. Amer. **57**(4):429-437.

Cooley, R.A. (1946). The genera *Boophilus, Rhipicephalus* and *Haemaphysalis* (Ixodidae) of the New World. Nat. Inst. Health Bull. 187:54 pp.

Cooley, R.A. and G.M. Kohls (1944). The Argasidae of North America, Central America, and Cuba. Amer. Midl. Natur. Monogr. 1:152 pp.

Cooley, R.A. and G.M. Kohls (1945). The genus *Ixodes* in North America. Nat. Inst. Health Bull. 184:246 pp.

Doss, M.A., M.M. Farr, K.F. Roach and G. Anastos (1974). Ticks and tickborne diseases II. Hosts. Index-Catalogue Med. Vet. Zool. Spec. Publ. 3, Parts 1-3: 1268 pp. + xvi.

Doube, R.M. (1975). Regulation of the circadian rhythm of detachment of engorged larvae and nymphs of the argasid kangaroo tick, *Ornithodoros gurneyi.* J. Med. Ent. **12**(1):15-22. [ARGASIDAE]

Filippova, N.A. (1961). Larvae and nymphs of the subfamily Ornithodorinae (Ixodoidea, Argasidae) in the fauna of the Soviet Union. Parazit. Sborn. Zool. Inst. Akad. Nauk SSSR **20**:148-184.

Garcia, R. (1962). Carbon dioxide as an attractant for certain ticks (Acarina:Argasidae and Ixodidae). Ann. Ent. Soc. Amer. **55**:605.

Garcia, R. (1969). Reaction of the winter tick, *Dermacentor albipictus* (Packard) to CO_2. J. Med. Ent. **6**(3):286. [IXODIDAE]

George, J.E. (1963). Responses of *Haemaphysalis leporispalustris* to light. Advances in Acarology, Cornell Univ. Press, Ithaca **1**:425-430.

George, J.E. (1971). Drop-off rhythms of engorged rabbit ticks *Haemaphysalis leporispalustris* (Packard, 1896) (Acari:Ixodidae). J. Med. Ent. **8**(5):461-479.

Hadani, A. and Y. Rechav (1969). Tick-host relationships. I. The existence of a circadian rhythm of "drop-off" engorged ticks from their hosts. Acta Trop. **26**(2):173-179.

Hoogstraal, H. (1956). African Ixodidae. Ticks of the Sudan. Bur. Med. Surg., U.S. Navy **1**:1101 pp.

Hoogstraal, H. (1966). Ticks in relation to human diseases caused by viruses. Ann. Rev. Ent. **11**:261-308.

Hoogstraal, H. (1967). Ticks in relation to human diseases caused by *Rickettsia* species. Ann. Rev. Ent. **12**:377-420.

Hoogstraal, H. (1973a). Ticks. **In** Parasites of Laboratory Animals, R.J. Flynn, ed. Iowa State Univ. Press, Ames: 398-424.

Hoogstraal, H. (1973b). Acarina (ticks). **In** Viruses and Invertebrates, A.J. Gibbs, ed. North-Holland Publ. Co., Amsterdam: 89-103.

Hoogstraal, H., R.M. Oliver and S.M. Giurgis (1970). Larva, nymph, and life cycle of *Ornithodoros (Alectorobius) muesebecki* (Ixodoidea:Argasidae), a virus-infected parasite of birds and petroleum industry employees in the Arabian Gulf. Ann. Ent. Soc. Amer. **63**(6):1762-1767.

Keirans, J.E. and C.M. Clifford (1975). *Nothoaspis reddelli,* new genus and new species (Ixodoidea:Argasidae), from a bat cave in Mexico. Ann. Ent. Soc. Amer. **68**(1):81-85.

Keirans, J.E., C.M. Clifford, H. Hoogstraal and E.R. Easton (1976). Discovery of *Nuttalliella namaqua* Bedford (Acarina:Ixodoidea:Nuttalliellidae) in Tanzania and redescription of the female based on scanning electron microscopy. Ann. Ent. Soc. Amer. **69**(5):926-932.

Khalil, G.M. (1974). The subgenus *Persicargas* (Ixodoidea:Argasidae:*Argas*) 19. Preliminary studies on diapause in *A. (P.) arboreus* Kaiser, Hoogstraal and Kohls. J. Med. Ent. **11**(3):363-366.

Lees, A.D. (1946). The water balance in *Ixodes ricinus* (L.) and certain other ticks. Parasitol.**37**:1-20.

Lees, A.D. (1948). The sensory physiology of the sheep tick, *Ixodes ricinus* L. J. Exp. Biol. **25**:145-207.

Moorhouse, D.E. (1966). Observations on copulation in *Ixodes holocyclus* Neumann and the feeding of the male. J. Med. Ent. **3**(2):168-171. [IXODIDAE]

Nevill, E.M. (1964). The role of carbon dioxide as stimulant and attractant to the sand tampan, *Ornithodoros savigni* (Audouin). Onderst. J. Vet. Res. **31**:59-68. [ARGASIDAE]

Nuttall, G.H.F., C. Warburton, W.F. Cooper and L.E. Robinson (1908). Ticks. A Monograph of the Ixodoidea. Part 1. The Argasidae. Cambridge Univ. Press, London: 104 pp.

Pavlovsky, E.N. and A.N. Skrynnik (1960). Comparative data on the biology of some species of ticks of the genus *Ornithodoros.* Dokl. Akad. Nauk SSSR **133**:734-736. [ARGASIDAE]

Philip, C.B. (1963). Ticks as purveyors of animal ailments: A review of pertinent data and of recent contributions. Advances in Acarology, Cornell Univ. Press. Ithaca **1**:285-325.

Philip, C.B. (1969). Tick paralysis. Rocky Mt. Laboratory, Hamilton, Montana, Circ. 4: 2 pp. (mimeo).

Pospelova - Shtrom, M.V. (1969) On the system of classification of ticks of the family Argasidae Can., 1890. Acarologia **11**(1):1-22.

Ricketts, H.T. (1907). A summary of investigations of the nature and means of transmission of Rocky Mountain spotted fever. Trans. Chicago Path. Soc. **7**:73-82.

Roberts, F.H.S. (1970). Australian ticks. CSIRO, Melbourne: 267 pp.

Shaw, R.D., J.A. Thorburn and H.G. Wallace (1976). Cattle Tick Control, 2nd Ed. Wellcome Foundation Ltd., London: 65 pp.

Smith, C.N., M.M. Cole and H.K. Gouck (1946). Biology and control of the American dog tick. USDA Tech. Bull. 905:74 pp. [IXODIDAE]

Sonenshine, D.E., C.M. Clifford and G.M. Kohls (1962). The identification of larvae of the genus *Argas* (Acarina:Argasidae). Acarologia **4**(2):193-214.

Southcott, R.V. (1976). Arachnidism and allied syndromes in the Australian region. Rec. S. Austral. Children's Hospital **1**(1):97-186.

Usakov, U.Y. (1961). Intraspecific parasitism (homovampirism) in ixodid ticks. Zool. Zhur.**40**:608-609.

Wilkinson, P.R. and M.B. Garvie (1975). Notes on the role of ticks feeding on lagomorphs and ingestion of ticks by vertebrates in the epidemiology of Rocky Mountain spotted fever. J. Med. Ent. **12**(4):480. [IXODIDAE]

SUBORDER IXODIDA

(Plates 53-55, pp. 221-223)

KEY TO THE FAMILIES AND GENERA[1]

1. Gnathosoma (capitulum) anteriorly situated in all known stages, visible from above; with a podonotal or holonotal scutum or pseudoscutum. Stigmal plates or stigma-like organs posterior to coxae IV. 2

— Gnathosoma anteriorly situated only in larva, inferior and not visible from above in nymphs and adults; median scutum present in larva, leathery pseudoscutum rarely present in postlarval forms. Stigmata lateral and anterior to coxae IV . . . . (Plates 53-54, pp. 221-222) Family ARGASIDAE ... 3

2. Integument strongly convoluted, without setae; with a leathery papillate podonotal pseudoscutum. Palpi apparently consisting of only three segments; hypostome with few denticles and indeterminant dental formula. Known only from female, Africa (monotypic) . . . . (Plate 53, p. 221) Family NUTTALLIELLIDAE, Genus *Nuttalliella*

— Integument normally striate, setae present; with a sclerotized podonotal or holonotal scutum. Palpi with four segments; palpal tibiotarsus reduced, often inserted ventrally on penultimate article. Hypostome strongly dentate or, where dentition is weak, at least with a distinctive dental formula. Larva with or without dorsolateral pores, or *sensilla sagittiformia* (Fig. 31, Ss, p. 212) . . . . (Plates 54-55, pp. 222-223) Family IXODIDAE ... 7

3. With a smooth leathery podonotal pseudoscutum; medial extension of proximal palpal articles form a ventral sheath for minute hypostome. Known only from male. From bat guano in cave, Mexico (monotypic). . . . . .Genus *Nothoaspis*

— Scutum absent in nymphs and adults, median scutum present in larva; palpal articles not as above 4

4. Hypostome vestigial in adults; integument of nymph (the stage ordinarily seen) strongly spinose; adult integument rugose, without spines; adult idiosoma constricted behind coxae IV. Associated with large mammals or lagomorphs in western North America (2 species) . . . . Genus *Otobius*

— Hypostome well developed, at least in female; integument leathery, often folded, occasionally tuberculate but without spines. Idiosoma rounded or ovoid. 5

5. Hypostome of female broad basally and scoop-like, vestigial in male; integument distinctly tuberculate. Pulvilli of larva modified into extremely large pads. From bats and bat guano in caves, North and Central America (4 species). . . . . Genus *Antricola*

— Hypostome of female not as above, male hypostome normally developed; integument various. Pulvillar pads of larva various but not as above 6

6. Nymphs and adults with a distinct sutural line separating dorsal and ventral surfaces; dorsum often with radiating rows of small discoid protuberances. Parasites of birds (subgenera *Argas* and *Persicargas*), bats (subgenera *Chiropterargas* and *Carios*), and various reptiles and mammals (subgenus *Secretargas*) in both the Old and New World (140 species) . . . . Genus *Argas*

---

[1]The genera *Rhipicentor* and *Cosmiomma* (IXODIDAE) are not included.

— Without a sutural line between dorsal and ventral surfaces of adults or nymphs; dorsum not as above, typically granular or mammillated. Parasites of birds and mammals in tropical and subtropical realms (90 species) . . . . Genus *Ornithodoros*

7. All stages with a distinctive groove curved anterior to anal opening; scutum inornate, eyes absent. Venter of male covered by a series of large, almost contiguous shields. Larva with two pairs of posthypostomal setae. Parasites of a variety of vertebrate hosts throughout much of the world (250 species) . . . . Genus *Ixodes*

— Without a preanal groove, with or without a postanal groove; eyes often present. Venter of male either without shields or only partially covered. Larva with only one pair of posthypostomal setae. 8

8. Eyes present, occasionally obsolescent 9

— Eyes absent, at least in nymphs and adults 16

9. Palpi short and broad, palpal segments II-III more or less subequal in length 10

— Palpi long, palpal segment II considerably longer than penultimate segment III 15

10. Leg segments greatly swollen, especially those of legs IV; male with a median preanal shield armed with two long spines extending posterior to anal opening, without *festoons* (Plate 55-4, p. 223). Palpi of larva with three segments, with five marginal setae anterior to each sensillum sagittiforme. Parasites of giraffes, zebras and horses in Africa (3 species). . . . . Genus *Margaropus*

— Leg segments normally formed, not greatly swollen; male without median preanal shield, festoons present or absent. Larval palpi and marginal setae various. 11

11. Gnathosomal base or *basis capituli* (Plate 54-7, p. 222) of male and female hexagonal in dorsal outline, usually inornate. Larva with four or five marginal setae anterior to each sensillum sagittiforme 12

— Basis capituli quadrangular in dorsal outline, usually ornate. Larva with two or three marginal setae anterior to each sensillum sagittiforme (if more than three, then eyes are raised and protruding) 13

12. Palpal segments II and III telescoped; coxae I terminating posteriorly in a pair of short, blunt spurs. Larval palpi and dorsal marginal setae as in the genus *Margaropus* (see couplet 10a). Parasites of large mammals in subtropical and tropical realms (5 species) . . . . Genus *Boophilus*

— Palpal segments II and III not telescoped; coxae I terminating posteriorly in a pair of long parallel spurs. Larval palpi with three articles, with four marginal setae anterior to each sensillum sagittiforme. Parasites of mammals and, rarely, birds or reptiles in temperate and tropical realms (63 species) . . . . Genus *Rhipicephalus*

13. Ornate species, eyes and postanal groove distinct 14

— Inornate forms, eyes and postanal groove present but indistinct or obsolescent; males with seven festoons. Larva with four palpal articles, with three marginal setae anterior to each sensillum sagittiforme. . . . . Genus *Anocentor*

14. Males with accessory adanal shields. Parasitic on man and animals in India and southeast Asia . . . . Genus *Nosomma*

— Male without accessory adanal shields. Parasites of mammals, primarily in holarctic realms (31 species) . . . . Genus *Dermacentor*

15. Eyes located on scutal margin. Larva with 4-segmented palpi; with two marginal setae anterior to each sensillum sagittiforme. Mostly large, brightly ornamented forms parasitic on mammals, birds, reptiles and amphibians in tropical and subtropical realms, occasionally holarctic (100 species). Genus *Amblyomma*

— Eyes located away from scutal margin and set in distinct "sockets", raised and protruding. Larval palpi with four segments, with four (occasionally five) setae anterior to each sensillum sagittiforme. Large, leathery ticks parasitic on a wide range of mammals, birds and reptiles in the Palearctic, Oriental and Ethiopian realms (21 species) . . . . Genus *Hyalomma*

16. Palpi of adults short and broad, approximately twice as long as wide; palpal segment II angulate, often strongly produced externally. Palpi of larva with three segments, with two marginal setae anterior to each sensillum sagittiforme. Parasites of a variety of wild and domestic mammals and, less commonly, of birds. Cosmopolitan (150 species) . . . . Genus *Haemaphysalis*

— Palpi of female elongate, more than three times as long as broad; palpal segment II without external angulation. Palpi of male various. 17

17. Palpal segment I hypertrophied, elongate-triangular and extended internally, adjacent to and almost as long as segment II; palpi of male short, segments II-III subequal. Dorsum strongly punctate (♂) or punctate-setate (♀). Parasitic on burrowing rodents and a woolly hare in Nepal (monotypic) . . . . Genus *Anomalohimalaya*

— Palpal segment I normally developed, considerably shorter than segment II; palpi of male elongate, similar to those of female, male idiosoma frequently very broad and subcircular. Often ornamented species, parasitic primarily on snakes and lizards in the Old World and on marsupials in Australia (26 species) . . . . Genus *Aponomma*

PLATE 53

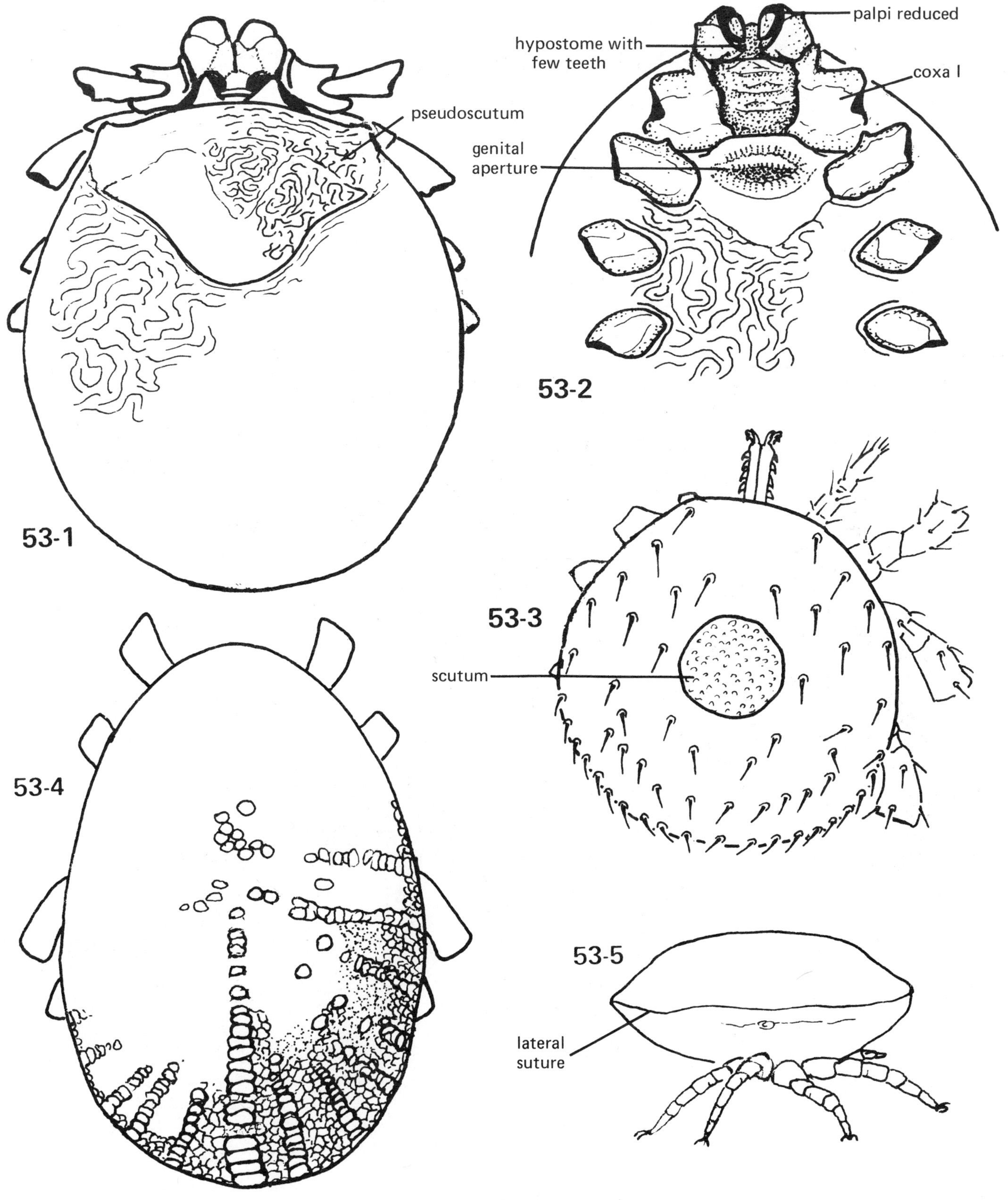

53-1 and 53-2; family NUTTALLIELLIDAE, *Nuttalliella namaqua* Bedford (Tanzania). **53-1;** dorsum of female (after an SEM photo, Keirans et al. 1976): **53-2;** anteroventral aspect (after an SEM photo, Keirans et al. 1976)

53-3 to 53-5; family ARGASIDAE, *Argas (Persicargas) persicus* (Oken). **53-3;** dorsum of larva: **53-4;** dorsum of female: **53-5;** lateral aspect of partially engorged female

PLATE 54

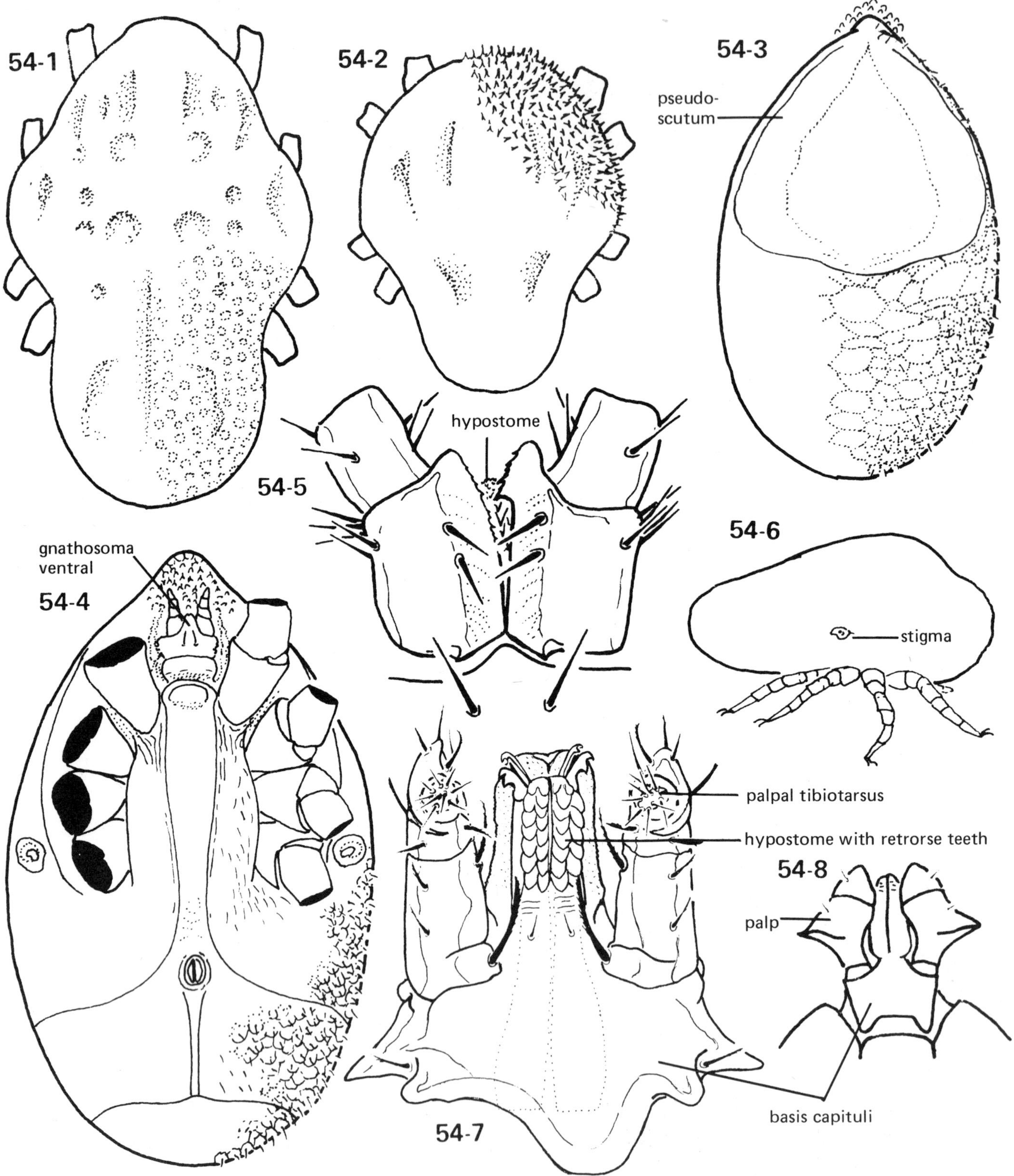

54-1 to 54-6; family ARGASIDAE. **54-1**; *Otobius lagophilus* Cooley and Kohls, dorsum of male (after Johnston 1968): **54-2**; *O. megnini* (Dugès), nymph: **54-3**; *Nothoaspis reddelli* Keirans and Clifford, dorsum of male (after Keirans and Clifford 1975): **54-4**; *N. reddelli,* venter of male (after Keirans and Clifford 1975): **54-5**; *N. reddelli,* ventral hypostomal sheath formed by proximal palpal articles (after an SEM photo, Keirans and Clifford 1975): **54-6**; *Ornithodoros* sp., lateral aspect of partially engorged female

54-7 and 54-8; family IXODIDAE. **54-7**; capitulum of an ixodid tick (Zaire), ventral aspect. The median hypostome is armed with retrorse teeth: **54-8**; *Haemaphysalis leporispalustris* (Packard), capitulum of male, dorsal aspect

PLATE 55

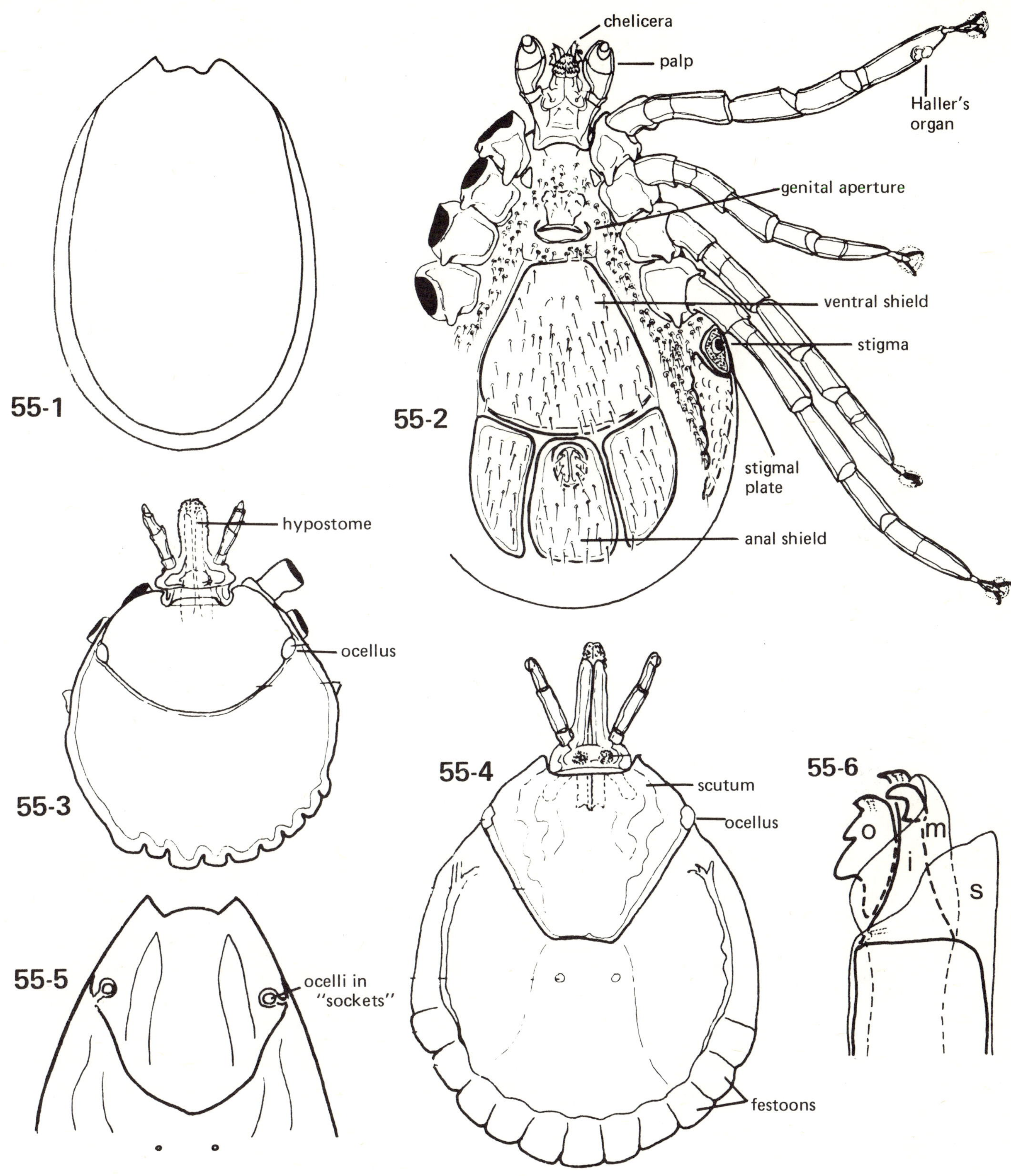

55-1 to 55-6; family IXODIDAE. **55-1**; *Ixodes* sp. (Oregon, USA), dorsum of male: **55-2**; *Ixodes* sp., venter of male: **55-3**; *Amblyomma cajennense* (Fabr.) (Mexico), dorsum of larva: **55-4**; *A. cajennense,* dorsum of female: **55-5**; *Hyalomma asiaticum* P. & E. Schulze, scutum of female (after Balashov 1972): **55-6**; *H. asiaticum,* chelicera of female (o = outer digit, i = inner digit, m = mantle, s = sheath) (from Balashov 1972)

SUBORDER ACTINEDIDA

The Actinedida is a large and complex group of terrestrial, aquatic and marine predators, phytophages, saprophages, and parasites. Some parasitic species may not exceed 100 μ in length, while certain predators may attain a size of 10,000 μ or more. The diversity of morphological characters in the suborder suggests that the Actinedida is in reality a composite of several subordinal entities. The supercohort Endeostigmatides, for example, might well be considered a separate suborder, as might the remarkable subcohort Parasitengonae. The great number of exceptions and alternate features listed in existing descriptions of the suborder amply illustrate the heterogeneity of the group. The rarity of unifying characteristics within the Actinedida often necessitates identification of its members through the process of elimination, i.e., through recognition of the lack of a particular attribute or combination of attributes found in other suborders.

Typically, the Actinedida are weakly or incompletely sclerotized forms which rarely show formulized incremental ontogenetic development of idiosomal shields. Where an internal respiratory system exists, it may debouch via paired stigmata at or near the bases of the chelicerae (Figs. 9c, 10a and b, pp. 22-23), at the humeral angles of the propodosoma (Fig. 9b) or, rarely, into the progenital region. Some actinedids have no distinct respiratory system of any kind (the Eriophyoidea, for example). Similar diversity exists in the structure of the chelicerae (stylettiform, chelate or reduced) and the palpi (simple, fang-like or with a palptibial claw). Although the empodial element of legs II-III may be claw-like or sucker-like in certain exceptional families, empodia II-III generally are pad-like, membranous, or rayed. Genital and anal openings often are approximate or contiguous on the venter of the opisthosoma. Two or three pairs of sucker-like acetabula may flank the genital slit (Plate 57-4, p. 312). Sexual dimorphism rarely is pronounced in the Actinedida. Minor differences in genital valve size or in internal genital structure often provide the only clues in differentiating between males and females.

Ontogenetic development in many actinedid groups is paurometabolous and, aside from the unique hexapod condition of the larva, all stages are similar in appearance as well as in habitat. Structural development in successive instars tends to be gradual and subtle (Plate 56, p. 225), although certain character state changes may be obvious. For example, in many families which have genital acetabula, the number of acetabular pairs changes from stage to stage, increasing from one pair in the protonymph to two in the deutonymph, and finally to three in the tritonymph and adult (Fig. 32, p. 226). Incremental acetabular increase of this type is referred to as *Oudeman's rule.* In some families, a maximum of two pairs of acetabula occurs in all instars subsequent to the protonymph. The presence of acetabula in postlarval instars usually is indicative of the presence of urstigmata in the larva. Incremental ontogenetic changes may also occur in numbers of genital setae and leg solenidia (Fig. 33, p. 226), as well as in overall size.

In contrast to most Actinedida, transformation from the larval to postlarval instars in the subcohort Parasitengonae is accompanied by radical changes both in appearance and in behavior. The heteromorphic larva (Plate 99-1, p. 354) is parasitic, while the nymph and adult are predaceous. Additional characters which apply to the Actinedida are listed below:

1. Coxae of legs fused to the venter so as to form distinctive independent coxal fields (Plate 62-1, p. 317), or insensibly amalgamated with the venter or with each other (Plate 87-2, p. 342).

2. Usually with a podocephalic canal (Fig. 16, p. 34) which may be internal or external. Ocelli may be present on the propodosoma.

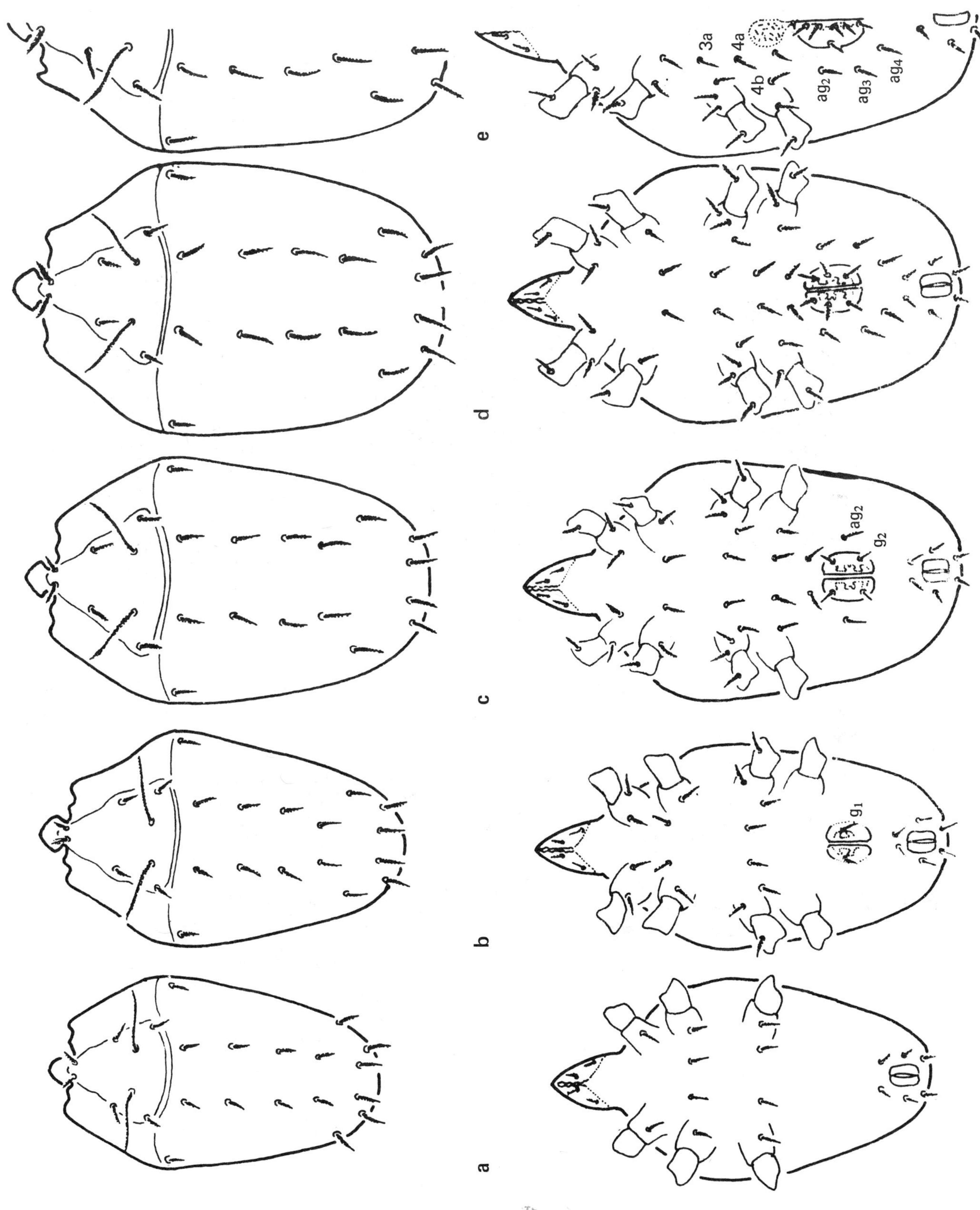

56; ontogenetic development of *Stereotydeus mollis* Womersley and Strandtmann (PENTHALODIDAE), showing dorsal (top) and ventral aspects of each instar: a, larva; b, protonymph; c, deutonymph; d, tritonymph; e, adult male (after Pittard 1971)

3. Specialized sensilla (= trichobothria) often present on the propodosoma (Plate 58-10, p. 313) and, less commonly, on the legs. Solenidia (Fig. 14b, p. 28) may be found on femora, genua, tibia and tarsi of the legs, and on the terminal segments of the palpi.

4. Larvae often with a pair of urstigmata, or organs of Claparède, in the region of coxae I-II (Plate 99-3, p. 354). Urstigmata may be greatly reduced in some families.

5. Epistome generally absent. Rutella may be present in some groups (Plate 60-4, p. 315.

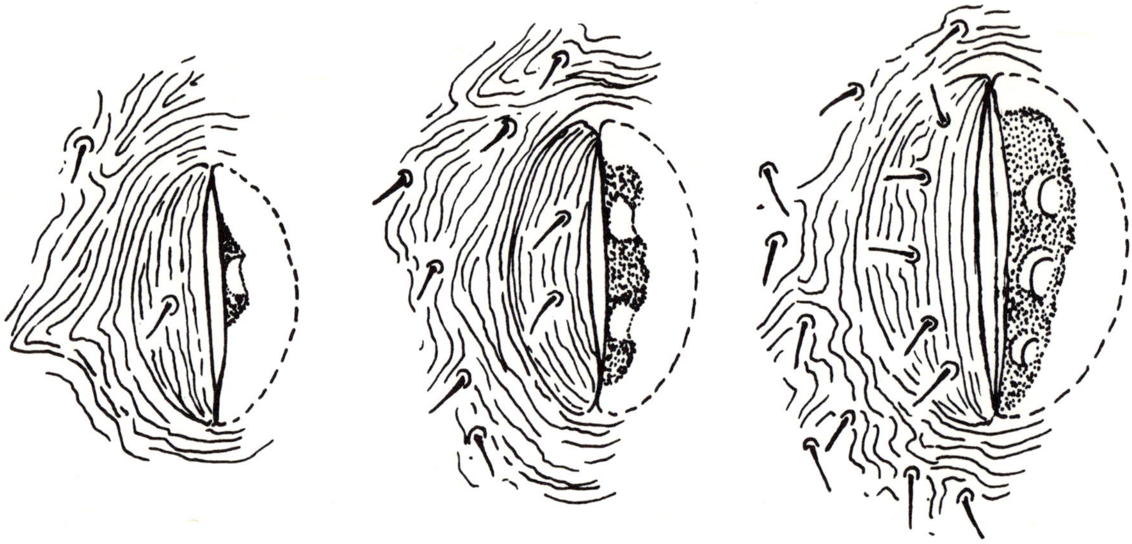

Fig. 32. Genital field of protonymph (left), deutonymph (center) and tritonymph (right) of *Bdella semiscutata* Sig Thor (BDELLIDAE), illustrating incremental ontogenetic changes in acetabular number and genital setae (after Grandjean 1938).

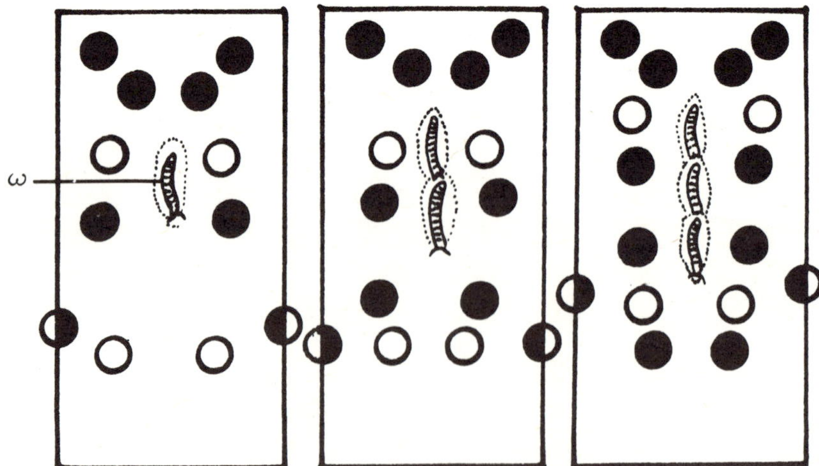

Fig. 33. Tarsus I (dorsal) of protonymph (left), deutonymph (center) and tritonymph (right) of *Stereotydeus mollis* Womersley and Strandtmann (PENTHALODIDAE), illustrating incremental ontogenetic changes in numbers of solenidia and setae. Ventral setae are indicated by solid circles (derived from Pittard 1971).

The Actinedida are cosmopolitan in distribution and are virtually unlimited in habitat. Their remarkable morphological variety is reflected in the great number of families (c. 120) and superfamilies (28) in which they are accommodated (see key to the suborder, page 295). The higher classification of the suborder used in this treatment includes two supercohorts, five cohorts, four subcohorts and, in the subcohort Parasitengonae, two phalanxes (Fig. 34, p. 228).

Useful References

Cook, D.R. (1974). Water mite genera and subgenera. Mem. Amer. Ent. Inst. 21:860 pp. + vii.

Cunliffe, F. (1955). A proposed classification of the trombidiforme mites. Proc. Ent. Soc. Wash. **57**(5):209-218.

Hammen, L. van der (1972). A revised classification of the mites (Arachnidea, Acarida) with diagnoses, a key, and notes on phylogeny. Zool. Meded. **47**(22):273-292.

Jeppson, L.R., H.H. Keifer and E.W. Baker (1975). Mites Injurious to Economic Plants. Univ. Calif. Press, Berkeley: 614 pp. + 74 plates + xix.

Lindquist, E.E. (1976). Transfer of the Tarsocheylidae to the Heterostigmata, and reassignment of the Tarsonemina and Heterostigmata to lower hierarchic status in the Prostigmata (Acari). Can. Ent. **108**:23-48.

Schweizer, J. and C. Bader (1963). Die Landmilben der Schweiz (Mittelland, Jura und Alpen). Trombidiformes Reuter. Mem. Soc. Helvet. Sci. Nat. **84**(2):209-378 + vi.

Vercammen-Grandjean, P. - H. (1973). Sur les statuts de la famille Trombidiidae Leach, 1815 (Acarina Prostigmata). Acarologia **15**(1):102-114.

Wharton, G.W. and H.S. Fuller (1952). A manual of the chiggers: The biology, classification, distribution, and importance to man of the larvae of the family Trombiculidae (Acarina). Mem. Ent. Soc. Wash. **4**:185 pp.

Zumpt, F. (1961). The arthropod parasites of vertebrates in Africa south of the Sahara (Ethiopian region). Vol. 1. (Chelicerata). S. Afr. Inst. Med. Res. **9**(1):1-457.

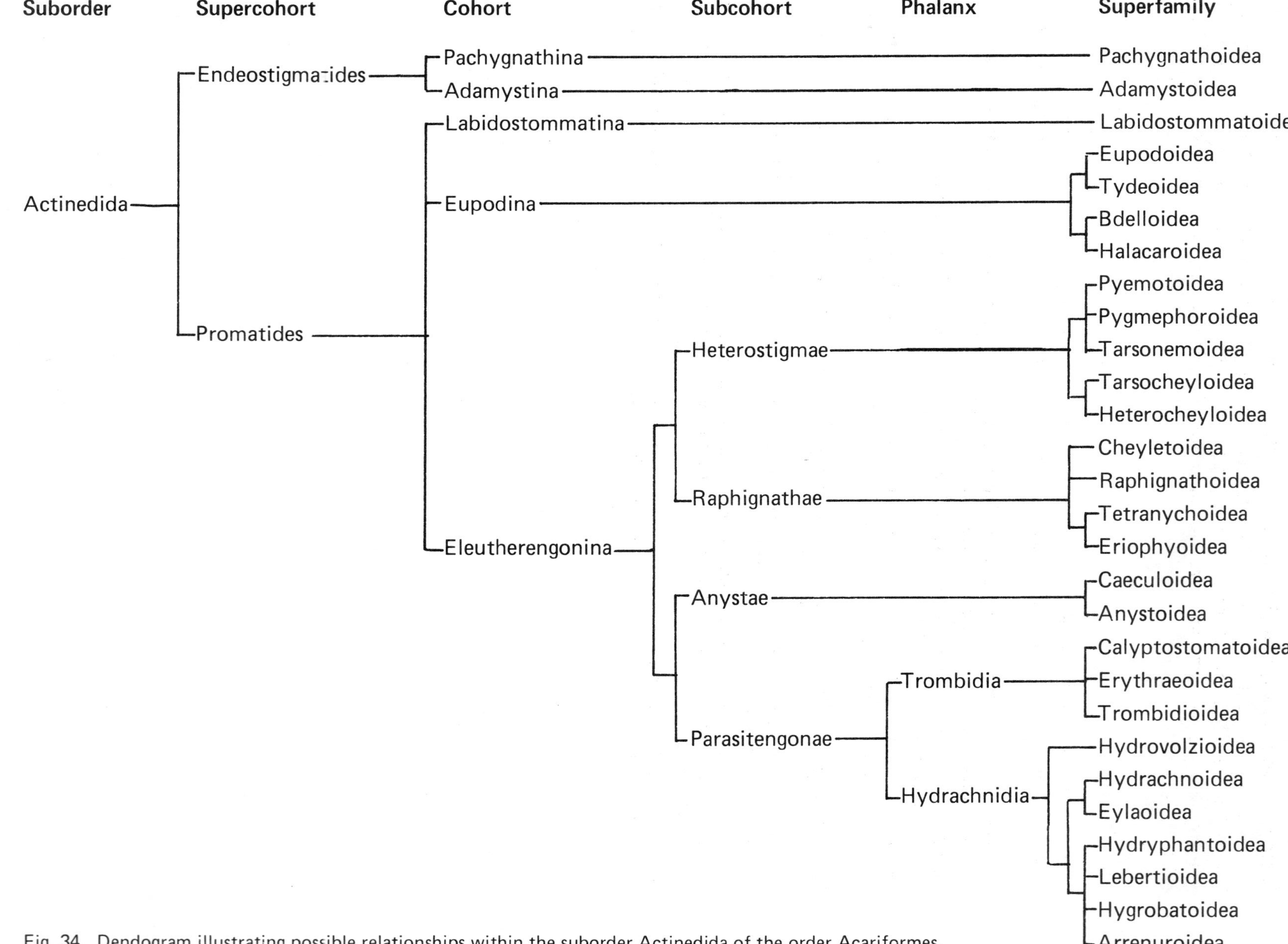

Fig. 34. Dendogram illustrating possible relationships within the suborder Actinedida of the order Acariformes.

Cohort Pachygnathina

The Pachygnathina is one of two cohorts assigned to the supercohort Endeostigmatides (Fig. 34, p. 228). The group is composed of soft-bodied, generally small (< 500 μ) species which usually lack external respiratory openings, and which possess sclerotized gnathosomal rutella (occasionally absent or with surrogate maxillary setae). Solenidia of tarsus I, when present, are erect.

While it is convenient to group the various members of the Pachygnathina in a single taxon on the basis of particular characteristics, there are strong indications that certain of the families may in fact have affinities with the suborder Oribatida rather than with the Actinedida. The family PACHYGNATHIDAE, for example, might easily be grouped with the primitive oribatid supercohort Macropylides (page 453), while the ubiquitous family NANORCHESTIDAE shares more structural similarities with the eleutherengonine Actinedida. For purposes of this review, and pending further study of its member groups, the Pachygnathina is retained as a single cohortal entity.

Superfamily Pachygnathoidea
(Plates 57-61, pp. 312-316)

DIAGNOSIS: With the characters of the cohort; idiosoma with little or no sclerotization, occasionally showing indications of segmentation; chelicerae variable but generally with opposed chelae, with one or two setae on the extended fixed digit; palpi simple, with five (exceptionally four) articles. With one or two pairs of sensilla on the propodosoma (if absent, then with an unpaired anteromedian propodosomal seta, pf); occasionally with stigmata at the cheliceral bases. Eyes and naso present or absent; tarsi with claws and/or empodia. With two-three pairs of genital acetabula.

The Pachygnathoidea is a cosmopolitan assemblage of nine families comprised of free-living species in litter, moss, humus, sand, and interstitial waters.

The PACHYGNATHIDAE (Plate 57, p. 312) are comparatively large, ovoid species which inhabit forest and garden litter. Specimens have also been taken in litter from semidesert vegetation and from beach wrack in northwestern United States. Pachygnathids are common in North America and Europe but have also been collected in South America (Thor and Willmann 1941) and Japan (Shiba 1969). Their feeding habits are unknown. The fossil species *Protacarus crani* Hirst, considered by some authors to be a member of the superfamily Eupodoidea, appears to be a pachygnathid (Hirst 1923). *Alycus roseus* Koch (= *Pachygnathus dugesi* Grandjean), a European species, recently was the subject of an intensive morphological study by van der Hammen (1969).

The ALICORHAGIIDAE (Plate 57, p. 312) is a small family which includes two genera and barely a half dozen tiny (ca. 300 μ) species, all of which are inhabitants of moss, soil or leaf mold in Europe, Japan, and Africa. Some alicorhagiids are unusual in that they possess a pair of tracheae which debouch into the region between coxae I-II. *Stigmalychus veretrum* Theron, Meyer and Ryke, provisionally placed in the ALICORHAGIIDAE, also is atypical in that it has a normally produced pair of stigmata and peritremes behind the chelicerae (Theron et al. 1970). The feeding habits of alicorhagiids are not known, but it is likely that they ingest solid particles of vegetable debris or mold. Species of the monogeneric family SPHAEROLICHIDAE (Plate 58, p. 313) occur in vegetable debris in Europe, North America and Japan. Few sphaerolichids have been described (Grandjean 1939, Thor and Willmann 1941), and their feeding habits are unknown.

Members of the family LORDALYCHIDAE (Plate 58, p. 313) are common inhabitants of moss and lichens in North America, Europe, South Africa and Japan. Representativves of the family OEHSERCHESTIDAE[1] (Plate 59, p. 314) are relatively broad but small (2-300 μ) species found in soil or humus in Java (Thor and Willmann 1941) and Africa (Theron 1974). *Grandjeanicus uncus* Theron is unusual in having a greatly elongate, basally modified pilose seta on the venter of tarsus I (Plate 59-4). Its function is unknown.

Species of NANORCHESTIDAE (Plate 59, p. 314) are found throughout the world in moss, humus and soil. A species of *Speleorchestes* has been taken from ant nests in Oregon where it may be feeding on fungi. *Nanorchestes antarcticus* Strandtmann is common to the Antarctic realm (Strandtmann 1964), and is one of the most southern terrestrial arthropods known. It has also been collected in subalpine habitats in Japan (Shiba 1969). *N. amphibius* Topsent and Trouessart is a littoral species which often occupies minute cracks in stratified rock of intertidal localities in Europe. Specimens of what appear to be *N. amphibius* have been found in rock cracks in the Oregon intertidal zone. Other nanorchestids occur in soils of dry savannah regions in South Africa (Theron and Ryke 1969, Theron 1975a). Baker and Wharton (1952) mention a species of *Speleorchestes* from Mexico which possesses paired stigmata opening at the cheliceral bases.

The greatly elongated annulate species of the family NEMATALYCIDAE (Plate 60, p. 315) are found in pasture and deep grassland soils of North America, in sandy soils of northern Brazil, and in coastal sands of southern Europe and North Africa. Nematalycids are aberrant forms which are well adapted to survival in deep interstitial habitats in a variety of soil types. Collections of *Psammolycus delamarei* Schubart were made at depths of up to 25 cm in sandy soils of Amazonia (Schubart 1973). Other species have been taken in far deeper strata. *Gordialycus tuzetae* Coineau, Fize and Delamare Deboutteville was collected at three meters in sands near Montpellier, France (Coineau et al. 1967). *Nematalycus nematoides* Strenzke occupies a similar habitat (Strenzke 1954). A nematalycid (possibly *Cunliffea strenzkei* (Cunliffe)) occurs at considerable depths in grassland soils of western United States.

The TERPNACARIDAE (Plate 60, p. 315) includes three described genera, the members of which occur in dry vegetable debris and soil. One undescribed species of *Terpnacarus* from Oregon occurs in wood dust generated by the activities of subterranean termites. Legs IV of most terpnacarids are adapted for jumping, but those of the highly ornate genus *Alycosmesis* are normally developed. *A. retiformis* Theron, *A. corallium* Th. and *A. granuliformis* Th. occur in *Eucalyptus* and *Acacia* debris in South Africa (Theron 1975b).

The PEDICULOCHELIDAE (Plate 61, p. 316) was once thought to be a primitive group of Acaridida, but is considered by most modern workers to be a pachygnathoid family. *Pediculochelus raulti* Lavoipierre was collected from African honey bees and from an anthophorid bee in South Africa (Lavoipierre 1946). The chelate chelicerae of *P. raulti* suggest a phoretic or symbiotic relationship of the mite with its bee hosts rather than a parasitic one (Price 1973). Other pediculochelids have been collected from chickens in the Philippines and from a murid rodent in Florida. *P. lavoipierri* Price and *P. parvulus* Pr. were collected from grassland soil in California (Price op. cit.).

---

[1]The HYBALICIDAE of Theron 1974. The generic name *Hybalicus* is referrable to the pachygnathoid family LORDALYCHIDAE. The name OEHSERCHESTIDAE is based on Jacot's (1939) genus *Oehserchestes,* a congener of *Hybalicus* sensu Theron (Kethley, personal communication).

Useful References

Baker, E.W. and G.W. Wharton (1952). Nanorchestidae Grandjean, 1937. **In** Introduction to Acarology. Macmillan Co., New York, N.Y.: 197-199.

Coineau, Y., A Fize and M.C. Delamare Deboutteville (1957). Découverte en France des Acariens Nematalycidae Strenzke à l'occasion des travaux d'aménagement du Languedoc-Roussillon. C.R. Acad. Sci. Paris **265** (Sér. D):685-688.

Cunliffe, F. (1956). A new species of *Nematalycus* Strenzke with notes on the family (Acarina, Nematalycidae). Proc. Ent. Soc. Wash. **58**(6):353-355.

Grandjean, F. (1936). Le genre *Pachygnathus* Dugès (*Alycus* Koch) (Acariens) 1re partie. Bull. Mus. nat. Hist. natur. Paris 8(2):398-405.

Grandjean, F. (1939). Quelques genres d'acariens appartenant au groupe des Endeostigmata. Ann. Sci. Zool., Ser. 11(2):1-122. [ALICORHAGIIDAE, LORDALYCHIDAE, SPHAEROLICHIDAE, TERPNACARIDAE]

Grandjean, F. (1942). Quelques genres d'acariens appartenant au groupe des Endeostigmata (2 Ser.), Première partie. Ann. Sci. nat. Zool., Ser. 11(4):85-135. [PACHYGNATHIDAE]

Grandjean, F. (1943). Quelques genres d'acariens apartenant au groupe des Endeostigmata (2 Ser.), Deuxieme partie. Ann. Sci. nat. Zool., Ser. 11(5):1-59. [PACHYGNATHIDAE]

Hammen, L. van der (1969). Notes on the morphology of *Alycus roseus* C.L. Koch. Zool. Med. **43**(15):177-202.

Hirst, S. (1923). On some arachnid remains from the Old Red Sandstone (Rhynie Chert Bed, Aberdeenshire). Ann. Mag. Nat. Hist. 9, **12**:455-474. [PACHYGNATHIDAE]

Lavoipierre, M. (1946). A new acarine parasite of bees. Nature **158**(4004):130. [PEDICULOCHELIDAE]

Price, D.W. (1973). Genus *Pediculochelus* (Acarina:Pediculochelidae), with notes on *P. raulti* and descriptions of two new species. Ann. Ent. Soc. Amer. **66**(2):302-307.

Schubart, H.O.R. (1973). The occurrence of Nematalycidae (Acari, Prostigmata) in Central Amazonia with a description of a new genus and species. Acta Amazonica **3**(3):53-57.

Shiba, M. (1968). Prostigmatic mites from Japan (III). On some endeostigmatic mites. Matsuyama Shinonome Junior Coll. Repts. **3**(2):219-228. [LORDALYCHIDAE, SPHAEROLICHIDAE, TERPNACARIDAE]

Shiba, M. (1969). Taxonomic investigations on free-living mites in the subalpine forest on Shiga Heights IBP area. II. Prostigmata. Bull. Nat. Sci. Mus. Tokyo **12**(1):65-115. [NANORCHESTIDAE, PACHYGNATHIDAE, ALICORHAGIIDAE]

Strandtmann, R.W. (1964). Insects of Campbell Island. Prostigmata:Eupodidae, Penthalodidae, Rhagidiidae, Nanorchestidae, Tydeidae, Ereynetidae. Pac. Insects Mono.: 148-156.

Strenzke, K. (1954). *Nematalycus nematoides* n. gen. n. sp. (Acarina, Trombidiformes) aus dem Grundwasser der algerischen Küste. Vie et Milieu **4**(4):638-647.

Theron, P.D. (1974). Hybalicidae, a new family of endeostigmatic mites (Acari:Trombidiformes). Acarologia **16**(3):397-412.

Theron, P.D. (1975a). Two new species of the family Nanorchestidae (Acari:Endeostigmata) from pasture soil in South Africa. Wetenskap. Byd. Potchefstroom Univ. C.H.O. Reeks B, Natuurw. 63:1-9.

Theron, P.D. (1975b). Three new species of the genus *Alycosmesis* (Acari:Terpnacaridae) from South Africa. J. Ent. Soc. S. Afr. **38**(2):289-296.

Theron, P.D., M.K.P. Meyer and P.A.J. Ryke (1970). The family Alicorhagiidae Grandjean (Acari:Trombidiformes) with descriptions of a new genus and species from South African soils. Acarologia **12**(4):668-676.

Theron, P.D. and P.A.J. Ryke (1969). The family Nanorchestidae Grandjean (Acari, Prostigmata) with descriptions of new species from South African soil. J. Ent. Soc. S. Afr. **32**(1):31-59.

Theron, P.D. and P.A.J. Ryke (1975). Five new species of the family Lordalychidae (Acari:Endeostigmata) from South Africa. Acarologia **17**(4):631-651.

Thor, S. and C. Willmann (1941). Acarina. Prostigmata 6-11 (Eupodidae, Penthalodidae, Penthaleidae, Rhagidiidae, Pachygnathidae, Cunaxidae). Das Tierreich 71a:1-186 + xxxvi.

Womersley, H. (1944). Australian Acarina, families Alycidae and Nanorchestidae. Trans. Roy. Soc. S. Austral. **68**(1):133-143.

Cohort Adamystina

Although their systematic position is unclear, the peculiar species comprising the monofamilial cohort Adamystina illustrate strong affinities with the Pachygnathina (Coineau 1974). Yet the presence of obscure but distinguishable postcheliceral peritremes and stigmata might argue against placement in the Endeostigmatides, which generally lack these structures. While the cheliceral morphology of adamystines is reminiscent of certain pachygnathines (Plate 59-3, p. 314), it also suggests that seen in the promatide cohort Labidostommatina (Plate 62-2, p. 317). Insertion of the anterior trichobothria on an anteromedian naso, on the other hand, is characteristic of certain of the promatid Eupodina (Plate 63-3, p. 318). Pending clarification of these and other shared characters, the Adamystina are included here in the Endeostigmatides, the major taxon with which they appear to have the greatest affinity. Their singular morphology, however, prompts the erection of a separate cohort to accommodate them.

The Adamystina may be distinguished from other endeostigmatide mites by the presence of a distinct or subtle idiosomal constriction at the level of coxae IV and by the absence of rutella. Adamystines have two pairs of propodosomal trichobothria, with the more anterior pair inserted on a well developed naso. Stigmata and peritremes are present at the cheliceral bases. Sensory setae of tarsus I are present and erect.

Superfamily Adamystoidea
(Plate 61, p. 316)

DIAGNOSIS: *With the characters of the cohort; idiosoma either covered by two large subequal shields, or with a single propodosomal shield; lateral eyes present, occasionally with a row of anterolateral lens-like structures which extend posteroventrally. Chelicerae with opposed chelae, with two or more dorsal setae on the extended dorsal jaw (cheliceral femur-genu-tibia-tarsus); palpi simple, with four articles; tarsi of legs with claws and empodia. With two pairs of genital acetabula.*

Members of the family ADAMYSTIDAE are free-living species found in soil and litter in North America and Europe (Cunliffe 1957, Hunter and Crossley 1968, Coineau 1974, 1977). *Saxidromus knoepffleri* Coineau and *Adamystis doumengei* C. were found on a limestone substrate in South Africa. The feeding habits of adamystids are unknown.

Useful References

Coineau, Y. (1974). Un type nouveau d'Acariens Prostigmates libres: les Saxidromoidea nouvelle super-famille. C.R. Acad. Sci. Paris **278** (Sér. D): 1059-1062.

Coineau, Y. (1977). Les Adamystidae, une étonnante famille d'Acariens Prostigmates primitifs. Proc. 4th Int. Congr. Acarology, Saalfelden (in press).

Cunliffe, F. (1957). Notes on the Anystidae with a description of a new genus and species, *Adamystis donnae,* and a new subfamily Adamystinae (Acarina). Proc. Ent. Soc. Wash. **59**:173-175.

Hunter, P.E. and D.A. Crossley Jr. (1968). *Adamystis sarae,* a new species of soil mite from cornfield litter in South Carolina (Acari:Anystidae). J. Georgia Ent. Soc. **3**(4):181-183.

Cohort Labidostommatina

A single superfamily, the Labidostommatoidea, is accommodated in the Labidostommatina. These are heavily armored, ornamented predatory mites of moderate size (500-2000 μ) which differ radically from other actinedids in general facies. The ventral armature is essentially complete, with the flattened coxae and genital and anal valves forming discrete shields or areas of sclerotization contiguous with the surrounding sclerites. The dorsum is virtually covered by a holonotal shield which may have a strongly alveolate or simple reticulate pattern. Legs I often are longer and more robust than legs II-IV. Sexual dimorphism is distinct in that the male genital and anal valves occupy separate fields (Plate 62-4, p. 317) while those of the female are amalgamated into a single posteroventral field.

Superfamily Labidostommatoidea
(Plate 62, p. 317)

DIAGNOSIS: *With the characters of the cohort; cheliceral bases free, not fused to rostrum or to each other, with well developed opposable digits, fixed digit with two setae; palpi simple. With two pairs of sensilla on the propodosoma; stigmata open at base of chelicerae. Lateral eyes generally present, often with one or more pairs of conspicuous lens-like pustules just posterior to eyes; anteromedian eye may be distinct or obscure. Tarsus I with claws, tarsi II-IV with claws and a claw-like empodium; with a pair of erect solenidia on tarsus I often accompanied by a simple, bifid or barbed famulus. With 2 pairs of genital acetabula.*

The Labidostommatoidea is represented by a single family, the LABIDOSTOMMATIDAE.[1] Members of the family occur in humus, lichens, soil or moss, subsisting on other small arthropods. Specimens have also been taken from caves in Europe. Labidostommatids are cosmopolitan in distribution but appear to have achieved their greatest success in the tropics. The genera *Sellnickiella* and *Dicastriella* are known only from tropical habitats (Feider and Vasiliu 1970, Feider et al. 1974), while *Labidostomma* and *Eunicolina* are most often encountered in temperate regions (Grandjean 1942). *Labidostomma pacifica* (Ewing), a common species in northwestern United States, is a familiar faunal component of forest litter. It has been observed to stalk its prey with the well developed legs I held as pincers, utilizing legs II-IV for locomotion.

Useful References

Atyeo, W.T. and D.A. Crossley (1961). The Labidostommidae of New Zealand (Acarina). Rec. Dominion Mus. **4**(4):29-48.

Coineau, Y. (1964a). Une espèce nouvelle francaise de Labidostomidae *Eunicolina travei.* Vie et Milieu **15**(1):153-175.

Coineau, Y. (1964b). Un nouveau *Labidostoma* à pustules multiples *Labidostoma jacquemarti* Coineau (Labidostomidae Acar. Prostigmata). Rev. Biol. Écol. Sol **1**(3):543-552.

Feider, Z. and N. Vasiliu (1968). *Labidostomma nepalense* (Nicoletiellidae) une nouvelle espèce asiatique. Acarologia **10**(4):630-644.

Feider, Z. and N. Vasiliu (1969). Révision critique de la famille Nicoletiellidae. Proc. 2nd Int. Congr. Acarology, Sutton Bonnington: 202-207.

Feider, Z. and N. Vasiliu (1970). Six espèces de Nicoletiellides d'Amerique du Sud. Acarologia **12**(2):282-309.

[1] Family NICOLETIELLIDAE of Feider and Vasiliu (1969).

Feider, Z., N. Vasiliu and M. Colugar (1974). Nouvelle contribution à l'étude des nicoletiellides d'Amerique du Sud. The Hungarian Soil Zoological Expedition in Chile and de [sic] collection of Prof. R. Schuster from Brasil. Acarologia **16**(3):413-427.

Grandjean, F. (1942). Observations sur les Labidostommidae. Bull. Mus. nat. Hist. natur. Paris Ser. 2, **14**(2): 118-125; (3):185-192; (5):319-326; (6):414-418.

Greenberg, B. (1952). New Labidostommidae with keys to the New World species (Acarina). J. New York Ent. Soc. **60**:195-209.

Schuster, R. and I.J. Schuster (1969). Gesteilte Spermatophoren bei Labidostomiden (Acari, Trombidiformes). Naturw. **56**(3):145.

Shiba, M. (1969). Taxonomic investigations on free-living mites in the subalpine forest of Shiga Heights IBP Area. II. Prostigmata. Bull. Nat. Sci. Mus. **12**(1):65-115.

Thor, S. (1931). Bdellidae, Nicoletiellidae, Cryptognathidae. Das Tierreich **56**:66-77.

Cohort Eupodina

The Eupodina is a large, diverse, and probably heterogenous group of predators, parasites, fungivores and phytophages. One superfamily, the Halacaroidea, contains primarily marine forms. Eupodines have postcheliceral peritremes and stigmata, although these structures have been secondarily lost in the Halacaroidea. The movable cheliceral digit of some eupodines may form part of a large chelate structure reminiscent of that seen in gamasid predators (Plate 64-2, p. 319). In others, the movable digits form needle-like extensions of the fused or contiguous basal fixed digits (Plate 63-4). The palpi may be simple or raptorial, and have one, three, four or five articles. Solenidia of tarsus I, when present, may be erect or appressed to the segment.

Four superfamilies are included in the Eupodina—the Eupodoidea, Tydeoidea, Bdelloidea and Halacaroidea. They are characterized in the sections to follow.

Superfamily Eupodoidea
(Plates 63-65, pp. 318-320)

DIAGNOSIS: Soft-bodied or weakly sclerotized, with a naso bearing a pair of setae, lateral eyes often present; with a pair of propodosomal sensilla. Movable cheliceral digit chelate and opposed to fixed digit, or needle-like and inserted subapically on cheliceral base (= fixed cheliceral digit), both digits occasionally distorted. Peritremes present or absent; palpi simple, generally four-segmented. With or without claws on tarsi I, with one or more specialized solenidia lying flush with tarsal surface in small dorsal depressions. With two or three pairs of genital acetabula.

Four families are included in the Eupodoidea—the EUPODIDAE, RHAGIDIIDAE, PENTHALODIDAE and PENTHALEIDAE. Eupodoids may be predaceous, fungivorous, or phytophagous, and one species may be parasitic on a snake.

The EUPODIDAE (Plate 63, p. 318) are mostly confined to damp soil, humus or moss in temperate, boreal, and antiboreal latitudes. Most of the known species of the type genus *Eupodes* have well developed femora IV (Thor and Willmann 1941) and are accomplished jumpers. *E. minutus* (Strandtmann), a species taken from moss and lichens and from bird nest material in the Antarctic realm (Strandtmann 1967, 1970), is an exception in that the femora of all legs are normally developed. Members of the genus *Linopodes* often are found under stones or debris in open situations. When threatened, *Linopodes* is capable of rapid backward movement, an ability which no doubt serves as a defensive response. Species of *Linopodes* may be encountered in mushroom houses where they are considered economic pests. Hussey et al. (1969) review evidence which indicates that *L. antennaepes* Banks, a common mushroom house inhabitant, is not fungivorous but rather is predaceous on other mites in mushroom beds. *Benoinyssus najae* Fain, the sole representative of the eupodid subfamily Benoinyssinae, was found in the nasal fossae of the cobra *Naja melanoleuca* in central Africa (Fain 1958).

Members of the predaceous family RHAGIDIIDAE (Plate 64, p. 319) are soft-bodied, rapidly moving mites which are found in essentially the same habitats as eupodids. Rhagidiids have strongly developed chelate chelicerae which are well suited for predation. They are a common holarctic group, although several species are known from other localities including South Africa (Meyer and Ryke 1960) and the Antarctic (Strandtmann 1970). The often colorful, delicately ornamented species of the family PENTHALODIDAE (Plate 64, p. 319) also may be predaceous (Baker 1946, Baker and Wharton 1952), although little is known of their biology. Penthalodids are found in moss and litter primarily in temperate and subtropical habitats. *Stereotydeus mollis* Womersley and Strandtmann, however, occurs commonly in Antarctica (Pittard 1971).

The winter grain mite, also known as the blue oat or pea mite, *Penthaleus major* (Dugès), is one of the economically important phytophagous species of the family PENTHALEIDAE (Plate 65, p. 320). *P. major* is widely distributed throughout temperate and tropical realms (Jeppson et al. 1975) where it attacks a variety of plants including clover (Swan 1934), peas (Campbell 1941) and grain (Chada 1956). Cotton, peanuts and ornamental flowers may also serve as hosts. *Halotydeus destructor* (Tucker), the red-legged earth mite or black sand mite, is a common pest of vegetable and legume crops, tobacco and grasses in Australia, South Africa, Rhodesia and Malawi (Meyer and Ryke 1960, Wallace and Mahon 1971, Jeppson et al., op. cit.). Species of both *Halotydeus* and *Penthaleus* have been taken from non-plant habitats (Thor and Willmann 1941), suggesting that not all penthaleid species are necessarily phytophagous.

Superfamily Tydeoidea
(Plates 65-67, pp. 320-322)

DIAGNOSIS: *Soft bodied or sclerotized, eyes present or absent; with a pair of propodosomal sensilla and, in certain EREYNETIDAE, a second pair on the posterodorsal aspect of the idiosoma. Movable cheliceral digit needle-like, cheliceral bases fused or closely contiguous along their mesal edges. Peritremes distinct or obscure; palpi simple, with one, three, or four segments. Tarsus I with or without claws, occasionally with only a median empodium; tarsi II-IV with claws and empodia, with one or several erect solenidia on tarsus I. With two pairs of genital acetabula, or acetabula absent.*

The Tydeoidea includes four families of predaceous, phytophagous and parasitic species from throughout the world. The elongate, oddly segmented members of the family PARATYDEIDAE (Plate 65, p. 320) occur in soil and litter in Europe, in North and Central America and in South Africa (Baker 1949, 1950, Theron et al. 1969), where they apparently prey on other minute arthropods. Collections at hand include specimens from bird nests, moss, and desert plant litter in the western United States. An undescribed paratydeid recently was collected in Oregon from trapped litter in a lichen colony on a Douglas fir tree branch over 30 meters above the forest floor. Price (1973) recovered a species of *Neotydeus* from pine forest soil in California.

While some members of the EREYNETIDAE (Plate 67, p. 322) are predators in moss, lichens, and leaf litter, others may be intimately associated with terrestrial gastropods (Baker 1970a and b) or found in the respiratory system of birds, mammals or amphibians (Fain 1962, 1963). *Ricardoella limacum* (Schrank), the slug mite, is found on the surface and within the mantle cavity of snails and slugs. Heavy infestations of *R. limacum* on slugs may occur under confined laboratory conditions, often resulting in death of the host. *R. limacum* has been demonstrated to be haematophagous on gastropods, forming a food canal or *stylostome* in the host tissue for injection of saliva and sucking up of food material (Baker 1973). Other members of the Ereynetinae are free-living predators.

Ereynetids of the subfamily Lawrencarinae are parasitic in the nasal passages of amphibians. *Lawrencarus eweri* (Fain) and *Xenopacarus africanus* Fain, Baker and Tinsley are African species which appear to be haemotaphagous in the nasal cavities of their frog and toad hosts (Lawrence 1952, Baker 1973). Other species of these genera, as well as members of the genus *Batracarus,* may share the haematophagous habit. Lawrencarines lack an active third nymphal stage. Members of the ereynetid subfamily Speleognathinae inhabit the nasal mucosa of a wide variety of birds and mammals throughout the world (Clark 1960, Fain 1963, Pence 1973, Fain and Hyland 1975). *Boydaia sturnellae* Clark, a North American speleognathine, appears to be capable of feeding on whole blood in the turbinates of its meadow lark host (Clark op. cit.), as does *B. nigra* Fain (Baker op. cit.). Speleognathines have no active nymphal stadia in their ontogenetic development and move directly from the hexapod larval form to the adult. Fain (1972) has observed the vestigial immobile nymphal forms developing within the larval skin.

The TYDEIDAE (Plate 66, p. 321) is a large cosmopolitan family of weakly to heavily sclerotized striate or reticulate mites which have been recorded as predators, plant feeders and scavengers (Oatman 1963, Baker 1965, Brickhill 1968, Gerson 1968). Several species are phoretic in the tympana of noctuid moths (Treat 1975), and one species is a symbiont of trigonid bees in Brazil (Flechtmann and Camargo 1974). *Tydeus molestus* (Moniez) has been reported to bite man and domestic animals in Europe, causing dermal irritation (Thor 1933). Tydeids are encountered in moss, litter, straw, soil or humus, fungi, bird nests, on plants, and in stored products (Marshall 1970). Although a number of tydeid species have been categorized as to general life style, little is actually known about the interactions of tydeids with their habitats. Many may be unspecialized feeders, a supposition which would account for the difficulty in analyzing tydeid behavior. *Tydeus californicus* (Banks) certainly fits the unspecialized category, having been considered a predator by Baker and Wharton (1952) and a phytophage by Fleschner and Arakawa (1953). *T. caudatus* (Dugès) has recently been observed in Oregon in the outer scales of filbert buds where it may be grazing on various fungi.

The tydeid *Lorryia formosa* Cooreman damages young fruit tissue of citrus, although it is considered to be fungivorous (Jeppson et al. 1975). *Pronematus ubiquitus* (McGregor) is an omnivorous species, apparently feeding on honeydew, fungi and dead arthropods in citrus trees (Jeppson et al., op. cit.). Eggs of *P. ubiquitus* are borne on filamentous stalks, suggesting adaptation for prevention of egg predation by early emerging *P. ubiquitus* larvae. The remarkable geographical and habital range of *L. bedfordiensis* Evans (Europe, North and South America, and Japan; on plants, in plant litter, fungi, stored grain, and moss) (Shiba 1969, Marshall 1970) suggests the existence of an *L. bedfordiensis* species complex.

The IOLINIDAE (Plate 66, p. 321) includes less than a half dozen tiny ($< 500\ \mu$) species of mites associated with orthopterans. *Iolina nana* Pritchard was described from the wing bases of a cockroach, *Blaberus craniifer,* in the United States (Pritchard 1956). Members of a second genus, *Proctotydaeus,* are found on locusts in Java and in the Galapagos Islands (Fain and Evans 1966). The relationship of iolinids with their hosts is not clear.

Superfamily Bdelloidea
(Plate 68, p. 323)

DIAGNOSIS: *Soft-bodied to strongly sclerotized forms, lateral eyes present or absent; with one or two pairs of propodosomal sensilla. Movable cheliceral digit sickle-shaped, terminally situated on conical fixed digit base; cheliceral bases separated and hinged to rostrum, capable of lateral scissors-like motion. Palpi with three-five segments, palpal tibiotarsus terminating in tactile setae, a distinctive solenidion, or a strong claw; podocephalic canal usually distinct, internal or external above coxae I. With claws on all tarsi, empodia pad- or claw-like; solenidia of tarsus I erect; trichobothria present on various leg segments. With two or three pairs of genital acetabula (rarely none).*

Two free-living families are included in the Bdelloidea—the BDELLIDAE and the CUNAXIDAE. Both are cosmopolitan in distribution.

The BDELLIDAE is a large family of active red, brown or green mites which prey on small arthropods or arthropod eggs (Atyeo 1960). The elongate rostrum characteristic of most bdellid genera accounts for the commonly applied appellation, "snout mites". Some bdellids are of great size (3-4,000 μ), but most species measure between 700-1200 μ. They are found in a variety of situations from dry exposed desert to cool moist forest habitats. *Neomologus littoralis* (L.), a large bright red species, often is found in enormous numbers on intertidal rock outcrops. *Bdella lignicola* Canestrini occurs in stored products where it preys on other mites (Hughes 1961). Some bdellid species may be efficacious in regulating spider mite populations (Snetsinger 1956) or economically important collembolans (Wallace and Mahon 1971, Wallace 1974, Wallace and Walters 1974). *Bdella depressa* Ewing is an effective predator of the clover mite, *Bryobia praetiosa* Koch, in the western United States (Jeppson et al. 1975). *Odontoscirus virgulatus* (C. & F.) preys on the red-legged earth mite, *Halotydeus destructor* (Tucker) in South Africa (Womersley 1933), but its effect on prey populations is not clear. Substantial numbers of the bdellid *Biscirus uncinatus* (Kramer) were collected in Canada from balsam fir trees heavily infested by the balsam woolly aphid. It is likely that this large predator utilizes the aphids for food.

Bdellids often secure their arthropod prey to the underlying substrate by means of silken threads (*Spinibdella cronini* (Baker and Balock) is an example). The silk spinning propensity is also brought into use by certain bdellids during molting. A silken cocoon is produced to provide shelter during these short periods of vulnerability (Wallace and Mahon 1972).

Members of the family CUNAXIDAE are small (400-600 μ) red or reddish brown species which feed on a variety of small arthropods in soil, litter, moss, or straw. Their palpi usually are raptorial and are armed with internal spines or spurs which better adapt them for grasping prey. Some are aerial predators which prey on phytophagous insects and mites (Muma 1961, Smiley 1975). The effect of cunaxids on prey populations is unknown, although Ewing (1917) refers to *Cunaxoides parvus* (Ewing) as an enemy of the oyster-shell scale. *Cunaxoides oliveri* (Schruft) is reported as a predator of the eriophyid mite *Calepitrimerus vitis* (Nal.) on grapes (Shruft 1971). Cunaxids often are found on the bark of fruit trees infested with spider mites in Oregon but their effect on spider mite populations has not been measured. Occasional records also exist of cunaxids in stored products (Hughes 1961).

Superfamily Halacaroidea
(Plate 69, p. 324)

DIAGNOSIS: *Soft-bodied or leathery, with or without lateral eyes, sometimes with an anteromedian pigmented region; propodosomal sensilla absent; with one to four dorsal shields. Movable digit of chelicerae sickle-shaped, elongate-dentate or reduced to a membranous structure (in the latter case, the terminus of the fixed digit may be hook- or sickle-shaped). Palpi simple or raptorial, three- or four-segmented. Podocephalic canal internal, occasionally visible in clear specimens; stigmata and peritremes absent. With strong claws on all legs, empodia present or absent; often with an erect solenidion near the terminus of tarsus I. Genital acetabula usually obscure, adults may have three pairs.*

A single family, the HALACARIDAE, is included in the Halacaroidea. The family consists of approximately 350 species of predaceous, parasitic and algivorous mites from a wide range of marine and freshwater habitats. Most of the known species have been collected from intertidal and shallow subtidal sand and algal substrates, and from plant-animal communities which thrive in these zones (André 1946, Newell 1957, Angelier 1953, Travé 1972, Krantz 1973, Bartsch 1975). However, many species have been taken at depths greater than 100 meters (Trouessart 1897, Lohmann 1907, Newell 1947, 1967a), and some have even been recovered from bottom samples collected at depths exceeding 4000 meters (Newell 1967b, Sokolov and Yankowskaya 1969). Members of several genera occur in brackish waters, and others have moved into fresh water habitats (Petrova 1971). Some have gained access to continental waters via subterranean migration (Delamare Debouteville 1960), with some species being found in lakes and streams which are far removed from the sea (*Lobohalacarus, Porohalacarus* and *Limnohalacarus* are examples). Species of *Soldanellonyx* have been taken from driven wells and subterranean pools (Walter 1917, Viets 1939, Imamura 1968, 1970). *S. chappuisi* Walter and *S. monardi japonicus* Imamura were collected in lava caves in Japan (Imamura 1971). Larvae and nymphs of *Enterohalacarus minutipalpus* Viets were removed from the digestive tract of a deep sea urchin in the southwest Pacific (Viets 1938).

The HALACARIDAE have succeeded in occupying virtually all of the major marine habitats and many freshwater niches as well. Despite their great diversity in habitat and life style, however, the halacarids present a remarkable degree of morphological uniformity.

While halacarids inhabit oceans and seas throughout the world, certain genera appear to be dominant at different latitudes, and even at different depths in the same latitude. *Metarhombognathus, Scaptognathus,* and *Arhodeoporus* are more or less confined to temperate seas while *Copidognathus, Rhombognathus,* and *Agauopsis* are cosmopolitan. *Halacarus* and *Agauopsis* species tend to be mutually exclusive throughout a range of substrates in the Adriatic, Baltic and North Seas (Krantz 1973), with *Agauopsis* being primarily intertidal and *Halacarus* favoring subtidal situations.

Useful References

André, M. (1946). Halacariens marins. Faune de France **46**:1-152.

Angelier, E. (1953). Halacariens des sables littoraux Méditerranéens. Vie et Milieu **4**(2):281-289.

Atyeo, W.T. (1960). A revision of the mite family Bdellidae in North and Central America. Univ. Kansas Sci. Bull. **40**(8):345-499.

Atyeo, W.T. (1963). The Bdellidae (Acarina) of the Australian Realm. Bull. Univ. Nebraska State Mus. **4**(8):113-210.

Baker, E.W. (1946). New species of North and Central American mites of the family Penthaleidae (Acarina). J. Wash. Acad. Sci. **36**(12):421-425. [PENTHALODIDAE]

Baker, E.W. (1949). Paratydeidae, a new family of mites (Acarina). Proc. Ent. Soc. Wash. **51**(3):119-122.

Baker, E.W. (1950). Further notes on the family Paratydeidae (Acarina) with a description of another new genus and species. J. Wash. Acad. Sci. **40**(9):289-291.

Baker, E.W. (1965). A review of the genera of the family Tydeidae (Acarina). Advances in Acarology **2**:95-133.

Baker, E.W. and A. Hoffmann (1948). Acaros de la familia Cunaxidae. Anales Esc. Nac. Ciencias Biol. **5**(3-4): 229-273.

Baker, E.W. and G.W. Wharton (1952). The suborder Trombidiformes Reuter, 1909. **In** An Introduction to Acarology. MacMillan Co., New York: 146-258. [TYDEIDAE and PENTHALODIDAE]

Baker, R.A. (1970a). The food of *Ricardoella limacum* (Schrank)—Acari-Trombidiformes and its relationship with pulmonate molluscs. J. Nat. Hist. **4**:521-530. [EREYNETIDAE]

Baker, R.A. (1970b). Studies on the life history of *Ricardoella limacum* (Schrank) (Acari:Trombidiformes). J. Nat. Hist. **4**:511-519. [EREYNETIDAE]

Baker, R.A. (1973). Notes on the internal anatomy, the food requirements and development in the family Ereynetidae (Trombidiformes). Acarologia **15**(1):43-52.

Bartsch, I. (1972). Ein Beitrag zur Systematik, Biologie und Ökologie der Halacaridae (Acari) aus dem Litoral der Nord- und Ostsee. I. Systematik und Biologie. Abh. Verh. naturw. Ver. Hamburg (NF) **16**:155-230.

Bartsch, I. (1975a). Eir Beitrag zur Rhombognathiden-Fauna (Halacaridae, Acari) der Bretagne-Küste. Acarologia **17**(1):53-80.

Bartsch, I. (1975b). Beitrag zur Halacariden-fauna (Halacaridae, Acari) der Bretagneküste. Beschreibung von fünf Arten aus dem Sandlückensystem. Acarologia **17**(4):652-667.

Brickhill, C.D. (1958). Biological studies of two species of tydeid mites from California. Hilgardia **27**(20):601-620.

Campbell, R.E. (1941). Further notes on the blue oat or pea mite *Penthaleus major* (Dugès) in California. Calif. Dept. Agr. Bull. 30:312-314. [PENTHALEIDAE]

Chada, H.L. (1956). Biology of the winter grain mite and its control in small grains. J. Econ. Ent. **49**(4):515-520. [PENTHALEIDAE]

Clark, G.M. (1960). Three new nasal mites (Acarina:Speleognathidae) from the grey squirrel, the common grackle, and the meadowlark in the United States. Proc. Helminth. Soc. Wash. **27**(1):103-110. [EREYNETIDAE]

Delamare Debouteville, C. (1960). Arthropodes autres que les Crustacés: Tardigrades, Pycnogonides, Acariens, Collemboles. **In** Biologie des Eaux Souterraines Littorales et Continentales. Hermann, Paris: 264-275. [HALACARIDAE]

Fain, A. (1958). Un nouvel acarien Trombidiforme parasitant les fosses nassales d'un Serpent au Ruanda-Urundi. Rev. Zool. Bot. Afr. **57**(1-2):177-183. [EUPODIDAE]

Fain, A. (1962). Les acariens parasites nasicoles de Batraciens. Revision des Lawrencarinae Fain, 1957 (Ereynetidae, Trombidiformes). Bull. Inst. roy. Sci. nat. Belg. **38**(25):1-69.

Fain, A. 1963. Chaetotaxie et classification des Speleognathinae. Bull. Inst. roy. Sci. nat. Belg. **39**:1-80. [EREYNETIDAE]

Fain, A. (1970). Nomenclature des poils idiosomaux et description de trois especes nouvelles dans la famille Ereynetidae (Trombidiformes). Acarologia **12**(2):313-325.

Fain, A. (1972). Développement postembryonnaire chez les acariens de la sous-famille Speleognathinae (Ereynetidae:Trombidiformes). Acarologia **13**(4):607-614.

Fain, A. and G.O. Evans (1966). The genus *Proctotydaeus* Berl. (Acari:Iolinidae) with descriptions of two new species. Ann. Mag. Nat. Hist. **9**(13):149-157.

Fain, A. and K.E. Hyland (1975). Speleognathinae collected from birds in North America (Acarina:Ereynetidae). J. New York Ent. Soc. **83**(3):203-208.

Flechtmann, C.H.W. and C.A. Camargo (1974). Acari associated with stingless bees...in Brazil. Abstract, 4th Int. Congr. Acarology, Saalfelden.

Fleschner, C.A. and K.Y. Arakawa (1953). The mite *Tydeus californicus* on citrus and avocado leaves. J. Econ. Ent. **45**:1092.

Gerson, U. (1968). Five tydeid mites from Israel (Acarina:Prostigmata). Israel J. Zool. **17**:191-198.

Grandjean, F. (1938). Observations sur les Bdelles (Acariens). Ann. Soc. Ent. France **107**:1-24.

Hughes, A.M. (1961). The Mites of Stored Food. Minis. Agr., Fish. and Food Tech. Bull. 9:287 pp. + vi.

Hussey, N.W., W.H. Read and J.J. Hesling (1969). Other arthropod pests of mushrooms. **In** The Pests of Protected Cultivation. Amer. Elsevier Publ. Co., New York: 338-351. [EUPODIDAE]

Imamura, T. (1968). Results of the speleological survey in South Korea 1966. IX. Halacaridae (Acari) found in a limestone cave of South Korea. Bull. Nat. Sci. Mus. Tokyo **11**(3):281-284.

Imamura, T. (1970). Subterranean water mites (Limnohalacarinae and Hydrachnellae) of the Tsushima Islands. Bull. Nat. Sci. Mus. Tokyo **13**(2):249-262. [HALACARIDAE]

Imamura, T. (1971). The fauna of lava caves around Mt. Fuji-san. V. Limnohalacarinae (Acari). Bull. Nat. Sci. Mus. Tokyo **14**(3):333-336. [HALACARIDAE]

Jeppson, L.R., H.H. Keifer and E.W. Baker (1975). Biological enemies of mites (Chapter 5); Tydeidae, Tuckerellidae, Pyemotidae, Penthaleidae, Astigmata, and Cryptostigmata (Chapter 11). **In** Mites Injurious to Economic Plants. Univ. California Press, Berkeley: 75-90, 307-326.

Krantz, G.W. (1973). Four new predatory species of Halacaridae (Acari:Prostigmata) from Oregon, with remarks on their distribution in the intertidal mussel habitat (Pelecypoda:Mytilidae). Ann. Ent. Soc. Amer. **66**(5):975-985.

Lawrence, R.F. (1952). A new parasitic mite from the nasal cavities of the South African toad *Bufo regularis* Reuss. Proc. Zool. Soc. London **121**:747-752. [EREYNETIDAE]

Lohmann, H. (1907). Die Meeresmilben der Deutschen Südpolar-Expedition 1901-1903. Deutsch. Südpolar-Exped. **9** (Zool. I) (5):361-413. [HALACARIDAE]

Marshall, V.G. (1970). Tydeid mites (Acarina:Prostigmata) from Canada. I. New and redescribed species of *Lorryia.* Ann. Soc. Ent. **15**(1):17-52.

Meyer, M.K.P. and P.A.J. Ryke (1959). Cunaxoidea (Acarina:Prostigmata) occurring on plants in South Africa. Ann. Mag. Nat. Hist. **2**(13):369-384.

Muma, M.H. (1975). Mites associated with citrus in Florida. Univ. Fla. Agr. Exp. Sta. Bull. 640A:92 pp. + iv.

Newell, I.M. (1947). A systematic and ecological study of the Halacaridae of eastern North America. Bull. Bingham Ocean. Coll. **10**(3):1-232.

Newell, I.M. (1967a). Prostigmata: Halacaridae (marine mites). Antarctic Res. Ser. **10**:81-95.

Newell, I.M. (1967b). Abyssal Halacaridae (Acari) from the southeast Pacific. Pac. Insects **9**(4):693-708.

Oatman, E.R. (1963). Mite species on apple foliage in Wisconsin. Advances in Acarology **1**:21-24. [TYDEIDAE]

Pence, D.B. (1973). The nasal mites of birds from Louisiana. VII. The Ereynetidae (Speleognathinae). J. Parasitol. **59**(2):364-368.

Petrova, A. (1971). On halacarid migration into freshwaters. Proc. 3rd Int. Congr. Acarology, Prague: 169-171.

Pittard, D.A. (1971). A comparative study of the life stages of the mite, *Stereotydeus mollis* W. & S. (Acarina). Pac. Insects Monogr. **25**:1-14. [PENTHALODIDAE]

Price, D.W. (1973). Abundance and vertical distribution of microarthropods in the surface layers of a California pine forest soil. Hilgardia **42**(4):121-147. [PARATYDEIDAE]

Prichard, A.E. (1956). A new superfamily of trombidiform mites with the description of a new family, genus, and species (Acarina:Iolinoidea:Iolinidae:*Iolina nana*). Ann. Ent. Soc. Amer. **49**(3):204-206.

Schruft, C. (1971). *Haleupalus oliveri* new species, a "thorn-palped" mite on grape vines (*Vitis* sp.) Acari: Cunaxidae. Deutsch. Ent. Z. **18**:377-382.

Shiba, M. (1969). Taxonomic invesigations on free-living mites in the subalpine forest on Shiga Heights IBP Area. II Prostigmata. Bull. Nat. Sci. Mus. Tokyo **12**(1):65-115.

Smiley, R.L. (1975). A generic revision of the mites of the family Cunaxidae (Acarina). Ann. Ent. Soc. Amer. **68**(2):227-244.

Snetsinger, R. (1956). Biology of *Bdella depressa,* a predaceous mite. J. Econ. Ent. **49**(6):745-746.

Sokolov, I.A. and A.I. Yankowskaya (1969). New genus and species of Halacaridae (Acari) from the abyssal region of the Kuril-Kamchatka trench. Proc. 2nd Int. Cong. Acarology, Sutton Bonnington: 113-114.

Strandtmann, R.W. (1967). Terrestrial Prostigmata (Trombidiform mites). Antarctic Res. Ser. **10**:51-80. [EUPODIDAE, TYDEIDAE]

Strandtmann, R.W. (1970). Acarina: eupodiform Prostigmata of South Georgia. Pac. Insects Monogr. **23**:89-106.

Strandtmann, R.W. (1971). The eupodoid mites of Alaska (Acarina:Prostigmata). Pac. Insects **13**(1):75-118.

Theron, P.D., M.K.P. Meyer and P.A.J. Ryke (1969). Two new genera of the family Paratydeidae (Acari: Prostigmata) from South African soils. Acarologia **11**(4):697-710.

Thor, S. (1931). Bdellidae, Nicoletiellidae, Cryptognathidae. Das Tierreich **56**:1-65 + xiii.

Thor, S. (1933). Acarina. Tydeidae, Ereynetidae. Das Tierreich **60**:1-82 + xi.

Thor, S. and C. Willmann (1941). Acarina. Prostigmata 6-11 (Eupodidae, Penthalodidae, Penthaleidae, Rhagidiidae, Pachygnathidae, Cunaxidae). Das Tierreich **71a**:1-186 + xxxvi.

Travé, J. (1972). Premières données sur les Halacariens (Acariens) interstitiels de Grèce. Biol. Gallo-Hellenica **4**(1):61-70.

Treat, A.E. (1975). Adult Prostigmata. **In** Mites of Moths and Butterflies. Cornell Univ. Press, Ithaca, N.Y.: 239-270. [TYDEIDAE]

Viets, K. (1938). Eine neue, in tiefsee-echiniden schmarotzende Halacaridengattung und Art (Acari). Zeit. Parasit. **10**(2):210-216. [HALACARIDAE]

Viets, K. (1939). Halacariden (Acari) aus süditalienischen Hohlengewässern. Arch. Hydrobiol. **35**(4):625-630.

Wallace, M.M.H. (1974). An attempt to extend the biological control of *Sminthurus viridis* (Collembola) to new areas in Australia by introducing a predatory mite, *Neomolgus capillatus* (Bdellidae). Austral. J. Zool. **22**:519-529.

Wallace, M.M.H. and J.A. Mahon (1971). The ecology of *Sminthurus viridis* (Collembola). III. The influence of climate and land use on its distribution and that of an important predator, *Bdellodes lapidaria* (Acari: Bdellidae). Austral. J. Zool. **19**:177-188.

Wallace, M.M.H. and J.A. Mahon (1972). The taxonomy and biology of Australian Bdellidae (Acari). I. Subfamilies Bdellinae, Spinibdellinae and Cytinae. Acarologia **14**(4):544-580.

Wallace, M.M.H. and J.A. Mahon (1976). The taxonomy and biology of Australian Bdellidae (Acari). II. Subfamily Odontoscirinae. Acarologia **18**(1):65-123.

Wallace, M.M.H. and M.C. Walters (1974). The introduction of *Bdellodes lapidaria* (Acari:Bdellidae) from Australia into South Africa for the biological control of *Sminthurus viridis* (Collembola). Austral. J. Zool. **22**:505-517.

Walter, C. (1917). Schweizerische Süsswasserformen der Halacariden. Rev. suisse Zool. **25**:411-423.

Womersley, H. (1933). On some Acarina from Australia and South Africa. Trans. Roy. Soc. S. Austral. **57**:108-112. [BDELLIDAE]

Cohort Eleutherengonina

The Eleutherengonina is a vast assemblage of mite groups which comprise 21 of the 28 actinedid superfamilies recognized in Figure 34. Few of these superfamilial aggregations are entirely limited to predation, and most of them illustrate highly specialized life styles in at least some of their constituent families. Many members of the subcohort Heterostigmae have established intimate phoretic or parasitic relationships with insects and other animals. Some have evolved as fungivores or phytophages while others are predators, often in singular habitats. Phytophagy is a major evolutionary trend in the subcohort Raphignathae although several families in this group are parasitic or predaceous. Larvae of the subcohort Parasitengonae are parasitic on invertebrates or vertebrates, but later active instars retain the predatory habit.

Eleutherengonines generally have a distal palptibial seta which is modified into a claw-like spine, and which often extends beyond the reduced, laterally displaced tarsus so that it is terminal on the palp (Plate 79-2, p. 334). The palpal *thumb-claw process,* however, is reduced or lost in some members of the subcohort Raphignathae and in certain Heterostigmae. The movable cheliceral digits are hook-like or stylettiform and are inserted apically or subapically on the fused or closely contiguous cheliceral bases. Chelate chelicerae of the type seen in other actinedid cohorts are unknown in the Eleutherengonina. When present, solenidia on tarsi I are erect rather than closely appressed to the segment. Podocephalic canals may often be distinguished anterodorsally.

The Eleutherengonina is considered in this treatment to include four subcohorts—the Heterostigmae, Raphignathae, Anystae, and Parasitengonae. The latter group is further divided into two phalanxes. The hierarchic scheme used here is based, in part, on that suggested by Lindquist (1976).

Subcohort Heterostigmae

The subcohort comprises five superfamilies of small to moderate-sized acarines (200-600 μ) with a longitudinal series of dorsal hysterosomal shields (basically four in number) which are often weakly defined. Adult females—and rarely males—have a pair of anterolateral propodosomal stigmata *(heterostigmata)* and associated tracheae but peritremes are absent. A pair of capitate sensilla arise on the propodosoma of most females and occasionally on that of males. Eyes are absent. The palpi often are greatly reduced, comprising only a few indistinct segments, and are closely appressed to a capsulate gnathosoma. In some groups, however, the palpi may be conspicuous 3-segmented appendages with a well-developed terminal claw. Tarsi II-III each have a stalked membranous empodium which usually is flanked by paired claws. These structures may be lost on tarsus IV. Tarsus I has no empodium but a single claw or pair of claws may be present. Genital acetabula are absent. Sexual dimorphism is pronounced throughout the subcohort.

The system of superfamilial classification followed here for the Heterostigmae is based largely on the concepts elaborated by Mahunka (1970b) and Lindquist (1976).

Superfamily Pyemotoidea
(Plates 70-72, pp. 325-327)

DIAGNOSIS: Gnathosoma capsulate or somewhat elongate, visible from above, usually with closely appressed reduced palpi; chelicerae stylettiform, partially retractile. Legs II-IV of females similar in shape and size; trochanter IV triangular, occasionally reduced to a narrow band, genu IV typically with two setae.

Four families are included in the Pyemotoidea. The PYEMOTIDAE (Plate 71, p. 326) are primarily insect associates and may commonly be found as predators of insects (Krczal 1959, Cross 1965, Cross and Moser 1971). Species of *Pyemotes,* the major genus of the family, attack and often kill preimaginal Homoptera, Coleoptera, Diptera, and Hymenoptera. *P. tritici* (Lagrèze-Fossat & Montagne), a member of the *P. ventricosus* species group (Moser 1975), is a common predator of larval and pupal Lepidoptera in grain storages, attacking other granary insects as well. *P. tritici* injects a toxin into its prey which causes paralysis and eventual death (Krczal 1959). The same species may produce severe skin lesions, asthma, nausea, and other symptoms in man (Treat 1975, Southcott 1976). Other members of this group also are known to produce a toxin (e.g., *P. zwoelferi* Krczal and *P. herfsi* (Oud.)). Moser (op. cit.) has separated species of *Pyemotes* on the basis of their toxicity to humans and to southern pine beetle larvae.

Pyemotes species of the *scolyti* group are inhabitants of bark beetle galleries, feeding on the eggs, larvae, and nymphs of their beetle hosts. Adult beetles are utilized for phoretic transfer of heteromorphic females. Cross and Moser (1975) observed that heteromorphy in the *P. scolyti* species group extends to both males and females, and that phoretic females, or *phoretomorphs,* are shorter and broader than "normal" females. In addition, legs I of phoretomorphs are greatly thickened and are armed with a more robust empodial claw.

While members of the *P. scolyti* group tend to be restricted in their host relationships (Moser et al. 1971, Cross and Moser 1975), species of the *P. ventricosus* group have wide host ranges. They are frequently encountered in laboratory insect cultures and stored foods.

The DOLICHOCYBIDAE (Plate 70, p. 325) are considered the most primitive pyemotoids, primarily because of their relatively freely-projecting developed palpi and the presence of bifid claws on tarsi I. Other primitive characteristics noted in the family are referred to by Lindquist (1976). *Dolichocybe keiferi* Krantz, *D. piceae* Rack and *D. hippocastani* R. are found under bark of trees in Europe and North America (Rack 1967). Other members of the family, however, are generally encountered on insects. *Pavania endroedyi* Mahunka and *P. equisetosa* were recovered from insects in Ghana (Mahunka 1975d). An unusual species of mite taken from beneath the elytra of the beetle *Agonoderus pallipes* in Kansas, USA, may represent still another primitive family of the Pyemotoidea. The gravid female of *Crotalomorpha* (CROTALOMORPHIDAE ms. of J.H. Camin) is illustrated in Plate 72.

Certain members of the genus *Acarophenax* (ACAROPHENACIDAE, Plate 71, p. 326) may be beneficial in that they parasitize graminivorous beetles of the genera *Tribolium* and *Cryptolestes* (Newstead and Duvall 1918, Cross and Krantz 1964). Species of *Paracarophenax* also are associated with beetles (Rack 1959, Mahunka 1975a), as are mites of the monogeneric family CARABOACARIDAE (Plate 72-2, p. 327).

Superfamily Pygmephoroidea
(Plates 72-73, pp. 327-328)

DIAGNOSIS: *Gnathosoma circular, oval, triangular, or elongate, generally not visible—or not easily visible—from above, with closely appressed reduced palpi; chelicerae stylettiform, partially retractile. Male gnathosoma reduced, tubiform, nonfunctional. Legs II-IV of females not similar in structure; trochanter IV always quadrangular, distinctly longer than wide, genu IV typically with only one seta.*

Pygmephoroids often are found in association with insects and small mammals. Many species, however, are free-living in a variety of organic habitats, and some have evolved close associations with plants. Three subfamilies are recognized in the superfamily.

The PYGMEPHORIDAE (Plate 73, p. 328) is an assemblage of mostly very small mites (200-300 μ) which may be found in soil, litter, organic substrates or on plants, or in nests of insects, birds and mammals (Cooper 1940, Krczal 1959, Cross 1965, Mahunka 1969a and b, 1970a, 1975c, 1976a). Species of *Pygmephorus* and *Pediculaster* are contaminants in mushroom houses (Gurney and Hussey 1967). Many of the species described to date have been collected from insects and small mammals on which the mites are phoretic. *Geotrupophorus gozmanyi* Mahunka and *Pediculaster geotrupi* M. are phoretic on geotrupine scarabs (Mahunka 1970b). *Parapygmephorus pappi* M. and *P. crossi* M. occur on the body and hairs of halictid bees (Mahunka 1974a), and species of the genus *Pediculaster* occur on Diptera (Rack 1974a, 1976) and on Lepidoptera (Cross 1965). An underwing moth, *Amphipyra pyramidoides,* serves as host for *Propygmephorus treati* Cross in North America (Treat 1975).

Pseudopygmephorus and *Bakerdania* species occur on the hair of small mammals (Krczal 1959, Sasa 1961, Mahunka 1973, 1974b, 1975b). Species of the same genera are recorded from mammal nests (Mahunka 1975c), along with representatives of *Pediculaster* and *Rackia.*

Siteroptes cerealium Kirchner is a known vector of *Fusarium poae,* the fungal pathogen which causes carnation bud rot (Cooper 1940). The same species also may be implicated in silver top of grasses in North America. *S. cerealium* has been successfully reared on *Fusarium* cultures in the laboratory. Various *Siteroptes* species, including *S. cerealium,* are found in mammal nests (Mahunka 1975c) and in litter (Mahunka 1969c), and some are found in association with bark beetles. The female of *S. reniformis* Krantz (Plate 73-3, p. 328) and certain other species of *Siteroptes* possess paired eversible saclike structures behind coxae IV in which fungal spores may be found. *Trochometridium tribulatum* Cross has similar sporothecae debouching on coxal plates IV (Lindquist, personal communication).

The discovery of phoretomorphy in the PYGMEPHORIDAE has led to problems in recognition of certain genera (Moser and Cross 1975). For example, a species of *Siteroptes* associated with bark beetles produces two distinct female types, one of which is referrable to the genus *Pediculaster.* Based on observations by Gurney and Hussey (1967), three distinctive morphs may develop in populations of *Pediculaster mesembrinae* (Can.), each of which identifies with what is currently recognized as a different genus *(Pediculaster, Siteroptes* and *Bakerdania).* These and related findings by Rack (1974b) and by Cross and Moser (1975) raise questions as to the validity of existing classifications in the Pygmephoroidea and Pyemotoidea (Mahunka 1970b).

The MICRODISPIDAE (Plate 73, p. 328) are mostly soil and litter forms which often possess greatly enlarged and/or ornate dorsal setae (Cross 1965, Mahunka 1975e). The majority of known species occur in the tropics.

The SCUTACARIDAE (Plate 72, p. 327) is a large and ubiquitous family represented in soil, forest litter, humus, manure, compost, and in nests of birds and small mammals (Karafiat 1959). Many species are associated with insects, being found either on the insects themselves or in nesting cells of their primary associates—bees and wasps (Batra 1965, Mahunka 1969d, Delfinado and Baker 1976, Delfinado et al. 1976). Scutacarids also are found on ants and in ant nests (Mahunka 1970c), and on beetles (Mahunka 1975a, Kurosa 1976).

Norton and Ide (1974) have studied the behavior of *Scutacarus baculitarsus agaricus* N. & I., a phoretic associate of a phorid fly, *Megaselia dakotensis.* They found that only the female mite is phoretic, with the non-feeding male serving only in a reproductive capacity. Female *S. baculitarsus agaricus* attach to their dipteran host by gripping the soft integument with claws I, which close against a proximal pretarsal sclerite. Males were seen to carry quiescent female calyptostases prior to final female ecdysis. Based on the scanty data available, a similar phoretic regime probably exists throughout the family, and throughout the Pygmephoroidea as well.

Superfamily Tarsonemoidea
(Plates 74-75, pp. 329-330)

DIAGNOSIS: Gnathosoma capsulate or somewhat elongate, generally visible from above, with closely appressed reduced palpi; chelicerae stylettiform, partially retractile. Legs IV of females either absent or smaller and narrower than legs II-III, with only one pair (legs I) or legless in some groups; males with three (legs I-III) or four pairs of legs; when present, genu IV in both sexes with one seta.

The Tarsonemoidea are considered to comprise two families which have evolved complex relationships with insects, with other mites, and with plants. The TARSONEMIDAE (Plate 74, p. 329) is a large family of insectophilous, nidicolous, algivorous, fungivorous, and phytophagous mites grouped in over a dozen genera. Species of *Steneotarsonemus, Hemitarsonemus,* and *Polyphagotarsonemus* are important phytophages, with the cyclamen mite, *S. pallidus* (Banks) and the broad mite, *P. latus* (Banks) being of prime economic significance (Moznette 1917, Gadd 1946, Schaarschmidt 1959, Taksdal 1973, Jeppson et al. 1975). Both species may produce teratosities in plant tissues on which they feed, presumably through injection of salivary toxins. Similar tissue toxaemia is reported in other tarsonemid mite-plant relationships (Jeppson et al., op. cit.). Several species of *Floridotarsonemus, Metatarsonemus, Lupotarsonemus, Tarsonemus,* and the strongly ornamented genus *Daidalotarsonemus* are recorded by Attiah (1970) from citrus trees in Florida. The female of *Fungitarsonemus pulvirosus* Attiah, an inhabitant of citrus orchards, carries debris on its dorsum which obscures the mite and probably offers a degree of protection from predators (Muma 1975). Other phytophagous and fungivorous tarsonemids are listed by Hussey et al. (1969) and by Jeppson et al. (op. cit.).

Tarsonemid associates of insects may be phoretic and/or predaceous. *Iponemus* species prey on the eggs of bark beetles and live as commensals in the beetle galleries (Lindquist and Bedard 1961, Lindquist 1969a). Adults are phoretic on the adult beetles and are easily carried from gallery to gallery. Species of the genera *Tarsonemus, Heterotarsonemus, Ununguitarsonemus,* and *Pseudotarsonemoides* also are associated with bark beetles (Lindquist 1969b, Smiley and Moser 1974). *Tarsonemus destructor* Smiley & Landwehr preys on the eggs of several species of TETRANYCHIDAE and TENUIPALPIDAE on Monterey pine in California (Smiley & Landwehr 1976). *Tarsonemus indoapis* Lindquist and *T. apis* (Rennie) are honeybee associates (Lindquist 1968). Male *T. indoapis* apparently are restricted to the region of the tentorial pits of the bee's head. Species of *Coreitarsonemus, Amcortarsonemus,* and *Asiocortarsonemus* are parasites in the odor glands of coreid bugs primarily in the tropics (Fain 1970, 1971).

Two of the three recognized species of the tarsonemid genus *Acarapis—A. dorsalis* Morgenthaler and *A. externus* M.—occur on the bodies of honeybees, apparently causing no injury to their hosts. Both species probably are cosmopolitan (Delfinado 1963). A third species, *A. woodi* (Rennie), is a parasite in the tracheal system of *Apis mellifera* where it causes a series of symptoms known as the Isle of Wight disease or, simply, acarine disease. Acarine disease has proven a serious problem in parts of Europe and Asia, and has been reported from South America as well (Nascimento et al. 1971). Observations on the feeding apparatus of

A. woodi by Hirschfelder and Sachs (1952) indicate that *A. woodi* can pierce the thoracic integument of their honeybee hosts under laboratory conditions. It seems likely that *A. dorsalis* and *A. externus* feed in a similar manner.

A type of social parasitism exists between members of the genus *Steneotarsonemus* and certain gall mites of the family ERIOPHYIDAE (page 262). *S. fulgens* Beer invades the galls of *Phyllocoptes didelphis* Keifer in *Populus,* causing the eriophyid mites to abandon the galls (Beer 1963). A similar situation exists between *S. nitidus* Beer and the gall mite *Phytoptus laevis* (Nalepa) on *Betula.*

Certain tarsonemids have been implicated in cases of human dermatitis. *Daidalotarsonemus hewitti* Mahunka was shaken from the clothing of a person afflicted with a skin rash (Mahunka 1974c). Other tarsonemids have been isolated from the surface and subsurface skin tissues of persons with dermal abnormalities, including species of *Tarsonemus* and *Tarsonemoides* (Hewitt 1973). Samšinák et al. (1976) feel that these associations may be the result of accidental or incidental contamination.

As mentioned for certain of the *SCUTACARIDAE* (page 245), male tarsonemids often carry a mature female calyptostase (or adult female) on their dorsum prior to copulation, using modified legs IV (Plate 74-6 and 7, p. 329) and the caudal genital papilla to hold the female in place (Jeppson et al. 1975).

All of the known members of the PODAPOLIPIDAE (Plate 75, p. 330) are internal or external arthropod parasites which often illustrate bizarre reductions in numbers of legs and in gnathosomal structures. Podapolipids occur under the elytra and in the vaginal membranes of beetles, in the tracheae and air sacs of orthopterans and hymenopterans, and externally on grasshoppers and cockroaches. *Podapolipoides* species feed externally on locusts and grasshoppers (Husband 1972b) while *Locustacarus trachealis* Ewing and *L. masoni* Husband occur in the respiratory systems of their locust hosts, apparently piercing the tracheal walls to feed (Wehrle and Welch 1925, Husband and Sinha 1970, Husband 1974a). *L. buchneri* (Stammer) also is a tracheal parasite but occurs in apid bees rather than in orthopterans. Species of *Tetrapolipus, Eutarsopolipus, Podapolipus, Coccipolipus, Dorsipes* and *Chrysomelobia* feed externally on a variety of beetles (Husband and Sinha op. cit., Feldman-Muhsam and Havivi 1972, Husband 1972a, Baker and Eickwort 1975, Eickwort 1975). Many podapolipids are associated with carabid beetles in Europe (Regenfuss 1968). Members of seven carabid genera serve as hosts for over 20 species of *Dorsipes* and *Eutarsopolipus. Ovacarus clivinae* Stannard and Vaishampayan and *O. peellei* Husband also are beetle parasites, but appear to be restricted to the membranous vaginal membranes and oviduct sacs of their carabid beetle hosts (Stannard and Vaishampayan 1971, Husband 1974b).

The unusual life history of *Locustacarus buchneri* Stammer probably reflects that of other podapolipids, at least in basic features. Female mites overwinter in the tracheae of queen bees *(Bombus bimaculatus)* and in the spring begin to feed on bee haemolymph through the tracheal walls. Within a week, the greatly enlarged female deposits up to 50 eggs which soon hatch into diminutive non-feeding hexapod males and motile, hexapod "females". Mating occurs immediately and the mated larviform females migrate directly to worker bees. They attach to the tracheal walls, molt to the so-called "adult" female stage, and begin to feed and enlarge. Adult females are sac-like and have only one pair of legs (Plate 75-2, p. 330). They are considerably larger than the tiny males and larviform females. Several generations may thus be passed in the bee's nest, with larviform females infesting successive worker broods. New queens leaving the nest in late summer carry with them enough larviform females to initiate a new cycle following winter hibernation (Stammer 1951). Males of *Ovacarus* species are atypical for the family in that they are much larger than larviform females, and apparently are capable of feeding (Husband 1974b).

Members of the genus *Chrysomelobia,* subelytral parasites of leaf beetles in Europe and North America, possess four pairs of legs in adult male and female stadia, resembling members of the family TARSONEMIDAE (Plate 74) in certain respects. However, leg segmentation and setation closely approximate those of other podapolipid genera (Eickwort 1975, Baker and Eickwort 1975).

Husband (personal communication) suggests that clues to evolutionary trends in the PODAPOLIPIDAE may lie in the morphological variation and position of the male aedeagal apparatus and, consequently, the mating stance of representative types. The aedeagus ranges from a terminal posteriorly directed organ as in the relatively unspecialized *TARSONEMIDAE* (e.g., *Eutarsopolipus, Archipolipus,* and *Ovacarus*), to a bizarre structure which arises anterodorsally and extends to or beyond the gnathosoma (e.g., *Rhynchopolipus* and *Locustacarus,* Plate 75-6 and 7). Retroconjugate or end-to-end mating as practiced by the former group is thought to indicate a lower phylogenetic level of development than that attained by the latter group which mate in an inverted proconjugate position.

Superfamily Tarsocheyloidea
(Plate 70, p. 325)

DIAGNOSIS: *Gnathosoma elongate, subconical, easily seen from above, with one or two pairs of dorsal setae; palpi well developed, usually projecting well beyond gnathosomal apex, with three segments and a distinctive terminal tibiotarsal claw, palptarsal "thumb" absent; chelicerae blade-like or stylettiform, partially retractile. Prodorsal shield of females (and probably all males) with a pair of anterolateral capitate sensilla. Legs I-IV similarly developed, each with paired tarsal claws; tarsi II-IV with empodia; adults and nymphs with a pair of platelets between coxae IV, each with two setae; coxal sclerites I-II separated at midline by striated integument.*

A single cosmopolitan family, the TARSOCHEYLIDAE, is included in the superfamily. The type genus *Tarsocheylus* is represented by a single species, *T. paradoxus* Berlese. Other species are accommodated in the genus *Hoplocheylus* (Atyeo and Baker 1964). Tarsocheylids are found in soil, sand, humus, forest or rodent nest litter, in rotting wood or trunk cavities of deciduous trees, or under the elytra of passalid beetles (Lindquist 1976). *H. pickardi* Smiley and Moser was taken from under pine bark infested with bark beetles *(Dendroctonus)* in Louisiana (Smiley and Moser 1968).

Tarsocheyloids appear to have strong affinities with members of the pyemotoid genera *Dolichocybe, Dolichomotes,* and *Pavania* (DOLICHOCYBIDAE, Plate 70). The form and setation of the gnathosoma and of the freely projecting palpi in *Pavania* (Plate 70-5) are especially reminiscent of those seen in the TARSOCHEYLIDAE (Lindquist 1976).

Superfamily Heterocheyloidea
(Plate 70, p. 325)

DIAGNOSIS: *Gnathosoma elongate, subconical, easily seen from above, without dorsal setae; palpi well developed, projecting well beyond gnathosomal apex, with two or three segments which may not be freely articulated; tibiotarsal claw reduced, rounded apically, palptarsal "thumb" absent; chelicerae blade-like, hooked. Prodorsal shield of female with a pair of posteromarginal "ampulliform organs", represented in males and immature instars by a pair of normal scapular setae. Legs I somewhat less robust than legs II-IV; claws absent on all legs, with a stalked discoid empodium and a heavy spinose ventridistal subinguinal seta on tarsi II-IV; with an elongate rectangular single shield medially between fused coxae III-IV, bearing a single pair of setae; coxal sclerites I-II united medially.*

The heterocheyloids are moderately large (400-600 μ), elongate species which usually are distinctly broadened in the region between coxae II-III. All are relegated to the monogeneric family HETEROCHEYLIDAE.

Over 20 species of *Heterocheylus* have been described, and all of them are subelytral associates of passalid beetles. Their known distribution and habitats overlap those of their beetle hosts; i.e. decaying wood in warm and temperate forests throughout the world except in Europe and western North America (Lindquist and Kethley 1975). The relationship between *Heterocheylus* species and passalids may involve parasitism, although only females are routinely found on the beetles. A fairly high degree of host specificity has been noted by Schuster and Lavoipierre (1970), at least at the generic level of the hosts.

Heterocheyloids are considered to have close phylogenetic ties with the Tarsocheyloidea (Lindquist and Kethley 1975) and, in fact, both taxa have at times been accommodated in the same family (Atyeo and Baker 1964). Both occur on passalid beetles (Lindquist 1976) although the tarsocheyloids utilize non-passalid habitats as well.

Useful References

Athias, F. (1973). Scutacaridae de la savane de Lamto (Côte d'Ivoire) (Acariens:Tarsonemida). 2. Imparipedinae, avec description d'une nouvelle espece. Acarologia **15**(1):129-137.

Attiah, H.H. (1970). New tarsonemid mites associated with citrus in Florida (Acarina:Tarsonemidae). Florida Ent. **53**(4):179-201.

Atyeo, W.T. and E.W. Baker (1964). Tarsocheylidae, a new family of prostigmatic mites (Acarina). Bull. Univ. Nebraska State Mus. **4**(11):243-256.

Baker, T.C. and G.C. Eickwort (1975). Development and bionomics of *Chrysomelobia labidomerae* (Acari: Tarsonemina; Podapolipidae), a parasite of the milkweed leaf beetle (Coleoptera:Chrysomelidae). Can. Ent. **107**:627-638.

Beer, R.E. (1954). A revision of the Tarsonemidae of the western hemisphere. Univ. Kansas Sci. Bull. **36**, 2(16):1091-1387.

Beer, R.E. (1958). The genus *Tarsonemella* Hirst, with description of a new species (Acarina, Tarsonemidae). J. Kansas Ent. Soc. **31**(2):188-192.

Beer, R.E. (1963). Social parasitism in the Tarsonemidae, with description of a new species of tarsonemid mite involved. Ann. Ent. Soc. Amer. **56**(2):153-160.

Beer, R.E. and A. Nucifora (1965). Revisione dei generi della famiglia Tarsonemidae (Acarina). Boll. Zool. Agr. Bachic., Ser. 2, **7**:19-43.

Cooper, K.W. (1940). Relations of *Pediculopsis graminum* and *Fusarium poae* to bud rot of carnation. Phytopath. **30**(10):853-859. [PYGMEPHORIDAE]

Cross, E.A. (1965). The generic relationships of the family Pyemotidae (Acarina:Trombidiformes). Univ. Kansas Sci. Bull. **45**(2):29-275.

Cross, E.A. and G.W. Krantz (1964). Two new species of the genus *Acarophenax* Newstead and Duvall 1918 (Acarina:Pyemotidae). Acarologia **6**(2):287-295. [ACAROPHENACIDAE]

Cross, E.A. and J.C. Moser (1971). Taxonomy and biology of some Pyemotidae (Acarina:Tarsonemoidea) inhabiting bark beetle galleries in North American conifers. Acarologia **13**(1):47-64.

Cross, E.A. and J.C. Moser (1975). A new dimorphic species of *Pyemotes* and a key to previously described forms (Acarina:Tarsonemoidea). Ann. Ent. Soc. Amer. **68**(4):723-732. [PYEMOTIDAE]

Delfinado, M.D. (1963). Mites of the honeybee in southeast Asia. J. Apic. Res. **2**(2):113-114. [TARSONEMIDAE]

Delfinado, M.D. and E.W. Baker (1976). New species of Scutacaridae (Acarina) associated with insects. Acarologia **18**(2):264-301.

Delfinado, M.D., E.W. Baker and M.J. Abbatiello (1976). Terrestrial mites of New York - III. The family Scutacaridae. J. New York Ent. Soc. **84**(2):106-145.

Eickwort, G.C. (1975). A new species of *Chrysomelobia* (Acari:Tarsonemina; Podapolipidae) from North America and the taxonomic position of the genus. Can. Ent. **107**:613-626.

Fain, A. (1970). *Coreitarsonemus* un nouveau genre d'acariens parasitant la glande odoriférante des Hémiptères Coreidae (Tarsonemidae:Trombidiformes). Rev. Zool. Bot. Afr. **82**(3-4):315-334.

Fain, A. (1971). Note sur la repartition geographique des Coreitarsoneminae parasites des Hemipteres coreides avec description de taxa nouveaux (Acarina:Trombidiformes). Bull. Ann. Soc. Roy. Ent. Belg. **107**:81-88. [TARSONEMIDAE]

Feldman-Muhsam, B. and Y. Havivi (1972). Two new species of the genus *Podapolipus* (Podapolipodidae, Acarina), redescription of *P. aharonii* Hirst, 1921 and some notes on the genus. Acarologia **14**(4):657-674.

Gadd, C.H. (1946). Observations on the yellow tea mite *Hemitarsonemus latus* (Banks) Ewing. Bull. Ent. Res. **37**:157-162. [TARSONEMIDAE]

Giordani, G. (1967). Laboratory research on *Acarapis woodi* Rennie, the causative agent of acarine disease of the honeybee. Note 5. J. Apic. Res. **6**(3):147-157. [TARSONEMIDAE]

Gurney, B. and N.W. Hussey (1967). *Pygmephorus* species (Acarina:Pyemotidae) associated with cultivated mushrooms. Acarologia **9**(2):353-358. [PYGMEPHORIDAE]

Hammen, L. van der (1970). *Tarsonemoides limbatus* nov. spec., and the systematic position of the Tarsonemoidea (Acarida). Zool. Verh. 108:1-35.

Hewitt, R., G.I. Barrow, D.C. Miller, F. Turk and S. Turk (1973). Mites in the personal environment and their role in skin disorders. Brit. J. Dermatol. **89**:401-409. [TARSONEMIDAE]

Hirschfelder, H. and H. Sachs (1952). Recent research on the acarine mite. Bee World **33**(12):201-209. [TARSONEMIDAE]

Husband, R.W. (1971). A new genus and species of mite (Acarina:Podapolipidae) associated with *Canthon humectus*. Ann. Ent. Soc. Amer. **64**(6):1297-1301.

Husband, R.W. (1972a). A new genus and species of mite (Acarina:Podapolipidae) associated with the coccinellid *Cycloneda sanguinea*. Ann. Ent. Soc. Amer. **65**(5):1099-1104.

Husband, R.W. (1972b). *Podapolipoides porteri* n. sp.: description with review of genus (Acarina, Podapolipidae). Trans. Amer. Microscop. Soc. **91**(3):422-428.

Husband, R.W. (1974a). Lectotype designation for *Locustacarus trachealis* Ewing and a new species of *Locustacarus* (Acarina:Podapolipidae) from New Zealand. Proc. Ent. Soc. Wash. **76**(1):52-59.

Husband, R.W. (1974b). *Ovacarus peelei,* a new species of mite (Acarina:Podapolipidae) associated with the carabid *Pasimachus elongatus.* Great Lakes Ent. **7**(1):1-7.

Husband, R.W. and R.N. Sinha (1970). A revision of the genus *Locustacarus* with a key to genera of the family Podapolipidae (Acarina). Ann. Ent. Soc. Amer. **63**(4):1152-1162.

Hussey, N.W., W.H. Read and J.J. Hesling (1969). Order Acarina: Mites. **In** The Pests of Protected Cultivation. Amer. Elsevier Publ. Co., New York: 190-228.

Jeppson, L.R., H.H. Keifer and E.W. Baker (1975). Tarsonemidae Kramer. **In** Mites Injurious to Economic Plants. Univ. Calif. Press, Berkeley: 285-305.

Karafiat, H. (1959). Systematik und Ökologie der Scutacariden. Beitr. Syst. Ökol. mitteleurop. Acarina **1**(4):627-712.

Karl, E. (1965a). Untersuchungen zur Morphologie und Ökologie von Tarsonemiden gärtnerischer Kulturpflanzen. I. *Tarsonemus pallidus* Banks. Biol. Zbl. **84**:47-80.

Karl, E. (1965b). Untersuchungen zur Morphologie und Ökologie von Tarsonemiden gärtnerischer Kulturpflanzen. II. *Hemitarsonemus latus* (Banks), *Tarsonemus confusus* Ewing. *T. talpae* Schaarschmidt, *T. setifer* Ewing, *T. smithi* Ewing und *Tarsonemoides belemnitoides* Weis-Fogh. Biol. Zbl. **84**:331-357.

Krczal, H. (1959). Systematik und Ökologie der Pyemotiden. Beitr. Syst. Ökol. mitteleurop. Acarina **1**(3): 385-625.

Kurosa, K. (1976). The scutacarid mites of Japan. V. Three new *Archidispus* associated with stenolophine ground beetles. Kontyû **44**(2):172-183. [SCUTACARIDAE]

Lindquist, E.E. (1968). An unusual new species of *Tarsonemus* (Acarina:Tarsonemidae) associated with the Indian honey bee. Can. Ent. **100**(9):1002-1006.

Lindquist, E.E. (1969a). Review of holarctic tarsonemid mites (Acarina:Prostigmata) parasitizing eggs of ipine bark beetles. Mem. Ent. Soc. Canada **60**:111 pp.

Lindquist, E.E. (1969b). New species of *Tarsonemus* (Acarina:Tarsonemidae) associated with bark beetles. Can. Ent. **101**(12):1291-1314.

Lindquist, E.E. (1976). Transfer of the Tarsocheylidae to the Heterostigmata, and reassignment of the Tarsonemina and Heterostigmata to lower hierarchic status in the Prostigmata (Acari). Can. Ent. **108**:23-48.

Lindquist, E.E. and W.D. Bedard (1961). Biology and taxonomy of mites of the genus *Tarsonemoides* (Acarina:Tarsonemidae) parasitizing eggs of bark beetles of the genus *Ips*. Can. Ent. **83**:982-999.

Lindquist, E.E. and J.B. Kethley (1975). The systematic position of the Heterocheylidae Trägårdh (Acari: Acariformes:Prostigmata). Can. Ent. **107**:887-898.

Mahunka, S. (1964). Neue Scutacariden aus Angola (Acari:Tarsonemini). Cult. Companhia Diamantes Angola, Lisboa 68:117-137.

Mahunka, S. (1969a). The scientific results of the Hungarian Soil Zoological Expeditions to South America. 9. Acari: Pyemotidae and Scutacaridae from the Guayaramerin region in Bolivia. Acta Zool. Acad. Sci. Hung. **15**(1-2):63-90.

Mahunka, S. (1969b). The scientific results of the Hungarian Soil Zoological Expeditions to South America. 13. Acari: Pygmephoridae and Scutacaridae from the material of the second expedition (Brazil and Bolivia). Acta Zool. Acad. Sci. Hung. **15**(3-4):333-370.

Mahunka, S. (1969c). Pyemotidae and Scutacaridae IV. Ergebnisse der Zoologischen Forschungen von Dr. Z. Kaszab in der Mongolei (Acari). Reichenbachia **12**(10):83-112. [PYGMEPHORIDAE,SCUTACARIDAE]

Mahunka, S. (1969d). *Imparipes (I.) eickworti* sp. n., a new scutacarid mite (Acari, Tarsonemina) from *Dialictus umbripennis* Ellis (Hym.). Parasit. Hung. **2**:153-158. [SCUTACARIDAE]

Mahunka, S. (1970a). Ergebnisse der zoologischen Forschungen von Dr. Z. Kaszab in der Mongolei. 227. Acari: Pygmephoroidea. Ann. Hist.-natur. Mus. Nat. Hung. (Zool.) **62**:343-362.

Mahunka, S. (1970b). Considerations on the systematics of the Tarsonemina and the description of new European taxa (Acari:Trombidiformes). Acta Zool. Acad. Sci. Hung. **16**(1-2):137-174.

Mahunka, S. (1970c). Two new scutacarid mites (Acari, Tarsonemina) from *Pogonomyrmex occidentalis* (Hymenoptera) in the United States of America. Parasit. Hung. **3**:87-96.

Mahunka, S. (1972). Tetüatkák - Tarsonemina. Fauna Hung. **18**(16):215 pp.

Mahunka, S. (1973). *Pygmephorus* species (Acari, Tarsonemida) from North American small mammals. Parasit. Hung. **6**:247-259.

Mahunka, S. (1974a). Beiträge zur Kenntnis der an Hymenopteren lebenden Milben (Acari), I. Ann. Hist.-natur. Mus. Nat. Hung. **66**:389-394. [PYGMEPHORIDAE, SCUTACARIDAE]

Mahunka, S. (1974b). New data to the knowledge of *Pygmephorus*-species (Acari:Tarsonemida) living on small mammals in America. Parasit. Hung. **7**:197-200.

Mahunka, S. (1974c). *Daidalotarsonemus hewitti* sp. n. (Acari:Tarsonemida) from human skin in England. Parasit. Hung. **7**:191-196.

Mahunka, S. (1975a). Neue und auf Insekten lebenden Milben aus Australien und Neu-Guinea (Acari:Acarida, Tarsonemida). Ann. Hist.-natur. Mus. Nat. Hung. **67**:317-325. [ACAROPHENACIDAE, SCUTACARIDAE]

Mahunka, S. (1975b). Further data to the knowledge of Tarsonemida (Acari) living on small mammals in North America. Parasit. Hung. **8**:85-94. [PYGMEPHORIDAE]

Mahunka, S. (1975c). Beiträge zur Kenntnis der Tarsonemiden (Acari) von Kleinsäugernestern aus der Umgebung von Ljubljana (Jugoslawien). Parasit. Hung. **8**:75-83. [PYGMEPHORIDAE, MICRODISPIDAE, SCUTACARIDAE, TARSONEMIDAE]

Mahunka, S. (1975d). Auf Insekten lebende Milben (Acari:Acarida und Tarsonemida) aus Afrika. V. Acta Zool. Acad. Sci. Hung. **21**(1-2):39-72. [DOLICHOCYBIDAE, PYGMEPHORIDAE, SCUTACARIDAE]

Mahunka, S. (1975e). Äthiopische Tarsonemiden (Acari:Tarsonemida). I. Acta Zool. Acad. Sci. Hung. **21**(3-4):369-410. [MICRODISPIDAE, SCUTACARIDAE]

Mahunka, S. (1976). Äthiopische Tarsonemiden (Acari:Tarsonemida). II. Acta Zool. Acad. Sci. Hung. **22**(1-2):69-96. [PYGMEPHORIDAE, SCUTACARIDAE]

Mahunka, S. and G. Rack (1975). Bibliographica Tarsonemidologica I. (1971-74). Folia Ent. Hung. (ser. nova) **28**(1):117-126.

Mahunka, S. and G. Rack (1976). Bibliographica Tarsonemidologica II. (1975). Folia Ent. Hung. (ser. nova) **29**(1):43-48.

Moser, J.C. (1975). Biosystematics of the straw itch mite with special reference to nomenclature and dermatology. Trans. Roy. Ent. Soc. London **127**(2):185-191. [PYEMOTIDAE]

Moser, J.C. and E.A. Cross (1975). Phoretomorph: a new phoretic phase unique to the Pyemotidae (Acarina: Tarsonemoidea). Ann. Ent. Soc. Amer. **68**(5):820-822. [PYGMEPHORIDAE]

Moser, J.C., E.A. Cross and L.M. Roton (1971). Biology of *Pyemotes parviscolyti* (Acarina:Pyemotidae). Entomophaga **16**(4):367-379.

Moznette, G.F. (1917). The cyclamen mite. J. Agr. Res. **10**(8):373-390. [TARSONEMIDAE]

Nascimento, C.B., R.P. de Mello, M.W. Dos Santos, R.V. do Nascimento and D.J. de Souza (1971). Occorrencia de acariose em *Apis mellifera* no Brazil. Pesq. Agropec. Brasil. Sec. Vet. **6**:57-60. [TARSONEMIDAE]

Naudo, M.H. (1967). Contribution a l'etude des Acariens parasites d'Orthopteres malgaches. I. Le genre *Podapolipus* (Podapolipidae): diagnoses preliminaires d'especes nouvelles. Acarologia **9**(1):30-54.

Newstead, R. and H.M. Duvall (1918). Bionomic, morphological and economic report of acarids of stored grain and flour. Roy. Soc. Rept. Grain Pests (War) Cmttee. No. 2:48 pp. [PYEMOTIDAE]

Nixon, J.W. (1944). Cheese 'itch' and 'itchy' cargoes in reference to workmen's compensation. Proc. Roy. Soc. Med. London **87**:405-410. [PYEMOTIDAE]

Norton, R.A. and G.S. Ide (1974). *Scutacarus baculitarsus agaricus* n. subsp. (Acarina:Scutacaridae) from commercial mushroom houses, with notes on phoretic behavior. J. Kansas Ent. Soc. **47**(4):527-534.

Rack, G. (1959). *Acarophenax dermestidarum* sp. n. (Acarina, Pyemotidae), ein Eiparasit von *Dermestes*-Arten. Z. Parasitenkunde **19**:411-431. [ACAROPHENACIDAE]

Rack, G. (1967). Untersuchungen über die Biologie von *Dolichocybe* Krantz, 1957 und Beschreibung von zwei neuen Arten (Acarina, Pyemotidae). Mitt. Hamburg Zool. Mus. Inst. **64**:29-42. [DOLICHOCYBIDAE]

Rack, G. (1972). Pyemotiden an Gramineen in schwedischen landwirtschaftslichen Betreiben. Ein Beitrag zur Entwicklung von *Siteroptes graminum* (Reuter, 1900) (Acarina, Pyemotidae). Zool. Anz. Leipzig **188**(3/4):157-174. [PYGMEPHORIDAE]

Rack, G. (1974a). Zwei neue Arten der Gattung *Pediculaster* von australischen Dipteren (Acarina, Tarsonemida, Pygmephoridae). Acarologia **16**(3):500-505.

Rack, G. (1974b). Neue und bekannte Milbenarten der Überfamilie Pygmephoroidea aus dem Saalkreis bei Halle (Acarina, Tarsonemida). Ent. Mitt. Zool. Mus. Hamburg **4**:499-521.

Rack, G. (1976). Milben (Acarina) von europäischen Limoniinen (Diptera, Nematocera). Mitt. Hamburg. Zool. Mus. Inst. **73**:63-85. [PYGMEPHORIDAE]

Regenfuss, H. (1968). Untersuchungen zur Morphologie, Systematik und Ökologie der Podapolipidae (Acarina, Tarsonemini). Zeitschr. wissenschaftl. Zool. **177**(3/4):183-282.

Samšiňák, K., P. Palička, K. Zítek, L. Mališ and E. Vobrázková (1976). Are the mites of the genus *Tarsonemus* really parasites of man? Fol. Parasitol., Praha **23**:91-93. [TARSONEMIDAE]

Sasa, M. (1961). New Mites of the genus *Pygmephorus* from small mammals in Japan (Acarina, Pygmetidae). Japan J. Exp. Med. **31**:191-208. [PYGMEPHORIDAE]

Schaarschmidt, L. (1959). Systematik und Ökologie der Tarsonemiden. Beitrag Syst. Ökol. mitteleurop. Acarina **1**(5):713-823.

Schuster, R.O. and M.M.J. Lavoipierre (1970). The mite family Heterocheylidae Trägårdh. Occasional Papers California Acad. Sci. 85:42 pp.

Smiley, R.L. and V.R. Landwehr (1976). A new species of *Tarsonemus* (Acarina:Tarsonemidae), predaceous on tetranychid mite eggs. Ann. Ent. Soc. Amer. **69**(6):1065-1072.

Smiley, R.L. and J.C. Moser (1968). New species of mites from pine. Proc. Ent. Soc. Wash. **70**:307-317. [TARSOCHEYLIDAE]

Smiley, R.L. and J.C. Moser (1974). New tarsonemids associated with bark beetles (Acarina:Tarsonemidae). Ann. Ent. Soc. Amer. **67**(4):639-665.

Southcott, R.V. (1976). Arachnidism and allied syndromes in the Australian region. Rec. Adelaide Children's Hospital **1**(1):97-186. [PYEMOTIDAE]

Stammer, H.G. (1951). Eine neue Trachenmilbe, *Bombacarus buchneri* n.g. n.sp. (Acar. Podapolipodidae). Zool. Anz. **146**(7-8):137-150.

Stannard, L.J. and S.M. Vaishampayan (1971). *Ovacarus clivinae,* new genus and species (Acarina:Podapolipidae), an endoparasite of the slender seedcorn beetle. Ann. Ent. Soc. Amer. **64**:887-890.

Suski, Z.W. (1973). A revision of *Siteroptes cerealium* (Kirchner) complex (Acarina, Heterostigmata, Pyemotidae). Ann. Zool. Warsawa **30**(17):509-535.

Taksdal, G. (1973). Interactions between pest and host plant in an attack by the broad mite, *Hemitarsonemus latus* (Banks) (Acarina, Tarsonemidae), on passion fruit, *Passiflora edulis* Sims. Norsk. Ent. Tidsskr. **20**:301-304.

Treat, A.E. (1975). Adult Prostigmata. **In** Mites of Moths and Butterflies. Cornell Univ. Press, Ithaca, New York: 239-270. [PYEMOTIDAE, PYGMEPHORIDAE]

Wehrle, L.P. and P.S. Welch (1925). The occurrence of mites in the tracheal system of certain Orthoptera. Ann. Ent. Soc. Amer. **28**(1):35-44. [PODAPOLIPIDAE]

Subcohort Raphignathae

Four superfamilies of predatory, parasitic and phytophagous mites are accommodated in the Raphignathae. With some exceptions (e.g., the STIGMAEIDAE, HOMOCALIGIDAE, and EUPALOPSELLIDAE), postcheliceral stigmata *(prostigmata)* and laterally produced peritremes are present throughout the subcohort.[1] A palpal thumb-claw process usually occurs, but may be reduced. Prodorsal ocelli often are present, but distinctive prodorsal trichobothria are absent. The gnathosoma is well developed although palpi and/or chelicerae may be reduced in highly specialized parasitic families. The cheliceral bases may be free or fused. Legs I-IV are more or less equally developed (the family MYOBIIDAE (Plate 82, p. 337) is exceptional in having legs I modified for grasping). The legs typically have a pair of well developed claws, and empodia are present in many forms. *Tenent hairs* (Plate 87-5, p. 342) often may be found on empodia, and occasionally on the claws themselves. Genital acetabula and larval urstigmata are absent (the family POMERANTZIIDAE (Plate 76, p. 331) is an exception), and genital and anal openings often are contiguous. Sexual dimorphism is mostly confined to differences in genitalia and in overall dimensions. Males may be smaller and/or narrower posteriorly than females, and often have a terminal or dorsal aedaegus.

While the Raphignathae provides a convenient and useful grouping for the purposes of this treatment, basic behavioral and habital differences among its constituent superfamilies suggests that further division above the superfamilial level might be warranted. The Raphignathoidea is essentially a predatory group, while both the Tetranychoidea and Eriophyoidea are exclusively phytophagous. With the exception of the CHEYLETIDAE, which are mostly free-living predators (Summers and Price 1970), the superfamily Cheyletoidea is composed of parasitic families. The dendrogram on page 228 indicates possible affinities and overall relationships in the subcohort.

Superfamily Raphignathoidea
(Plates 76-80, pp. 331-335)

DIAGNOSIS: *Soft-bodied or well sclerotized, with a holonotal shield or a series of smaller weakly tanned, reticulate or dimpled dorsal sclerites, rarely without shields. Gnathosoma well developed, occasionally retractile, chelicerae each with a stylettiform, partially retractile movable digit arising from fused or separate cheliceral bases which are often broadly expanded basally, not fused to underlying rostral elements (Plate 79-4, pp. 334); palpi 5-segmented, palpal thumb-claw process present but often modestly developed or obsolete, occasionally absent; postcheliceral peritremes and prodorsal ocelli present or absent. Empodial elements of tarsi commonly with tenent hairs, claws generally present on all legs.*[2]

The nine families currently included in the Raphignathoidea comprise small (300-500 μ) species which occur in a variety of terrestrial and semiaquatic habitats throughout the world. Raphignathoids are predaceous, with the possible exception of the families CRYPTOGNATHIDAE and RAPHIGNATHIDAE, and certain members of the STIGMAEIDAE.

---

[1] Grandjean (1944) noted that the podocephalic canals in various stigmaeid and homocaligid genera are distinctive and might be mistaken for peritremes. *Stigmaeus* and *Eustigmaeus (=Ledermuelleria)* were mentioned by Grandjean in this context.

[2] Members of the genus *Pilonychiopus* (STIGMAEIDAE) are unique in that the true claws are replaced by a balloon-shaped distal tarsal lobe (Meyer 1969).

Mites of the family RAPHIGNATHIDAE (Plate 76, p. 331) often are found on or beneath tree bark (Atyeo et al. 1961) and in dry organic litter. One species has been noted as a resident in stored grain. *Raphignathus gracilis* (Rack) and *R. collegiatus* Atyeo, Baker and Crossley occur in deeper soil layers in California pine forests (Price 1973).

The family POMERANTZIIDAE (Plate 76, p. 331) includes small (± 400 μ) greatly elongate mites found in soil or humus in North America, southern Europe, and North Africa. Collecting data for the four known species of *Pomerantzia* indicate that the genus probably is endemic to deep soil strata (Price and Benham 1976). *P. prolata* Price was collected at soil depths exceeding 80 cm., and *P. benhami* P. was recovered from samples over 100 cm. from the surface. *Stigmocheylus brevisetus* Berlese, a soil form from Italy, has been tentatively placed in the POMERANTZIIDAE by Lindquist (1976).[1] A second species of the genus, *S. pilosus* (Soliman and Zaher), recently was described from Egypt (Soliman and Zaher 1975). The CRYPTOGNATHIDAE (Plate 77, p. 332) is a well-defined family of highly ornamented mites found in many parts of the world in leaf litter, tree bark, moss, and lichens. Cryptognathids are easily recognized by their elongate, partially retractile gnathosoma which is obscured basally by a prodorsal "hood" (Summers and Chaudri 1965, Luxton 1973). Although cryptognathids have been thought to be predatory, Luxton (op. cit.) has noted a green material in the hindgut which suggests that the family may be phytophagous or algivorous.

Members of the EUPALOPSELLIDAE (Plate 78, p. 333) are predaceous mites with distinctive elongate cheliceral stylets. *Eupalopsellus* species are especially common in very dry regions (sage, heather, brushland habitats), although *E. olandicus* has been collected from coastal areas in northern Europe. *Exothorhis caudata* Summers and *Saniosulus nudus* S. are predators in citrus crops (Summers 1960a). Both *S. nudus* and *Eupalopsis maseriensis* (Canestrini and Fanzago) prey on the citrus chaff scale, *Parlatoria pergandii* in Israel (Gerson 1966, 1967, Gerson and Blumberg 1969). *E. pinicola* Oudemans is found on apple trees in eastern Canada (Baker and Wharton 1952) and is a scale insect predator in Europe. Other eupalopsellids have been taken in moss or leaf mold.

The family STIGMAEIDAE (Plate 79, p. 334) is a large cosmopolitan group of genera which often may be identified by their distinctive dorsal shield configurations (Summers 1966b). At least three stigmaeid genera contain species which are predators of phytophagous mites in orchard crops. *Zetzellia mali* (Ewing) feeds on several species of spider mites in North America, Europe and Israel (Jeppson et al. 1975) and is regarded as a beneficial species. Several representatives of the stigmaeid genera *Agistemus* and *Mediolata* prey on spider mites and other orchard-associated arthropods in various parts of the world (Gonzalez-R. 1965, Summers 1966b, Jeppson et al. op. cit.).

Eustigmaeus and related genera are globate, strongly ornamented stigmaeid species which may be encountered in soil and litter habitats (Summers and Price 1961, Price 1973). Species of *Eustigmaeus* have been observed to feed on mosses (Gerson 1972). *E. frigida* (Habeeb), for example, completed its entire life cycle on a diet of *Heterophyllium.*

In contrast to the strongly sclerotized species of *Eustigmaeus, Apostigmaeus* and *Erynigiopus* have little more than vague prodorsal shields, while *Summersiella ancydactyla* Gonzalez-R., a predator in leaf cavities in New Zealand, is practically unsclerotized

[1]Summers (1966) considered *S. brevisetus* and *Pomerantzia charlesi* Baker (the sole representative of the genus at the time) as unusual raphignathoids which illustrated little affinity with other members of the superfamily. The presence of genital acetabula and the absence of empodia on tarsi II-IV of both genera suggests that the POMERANTZIIDAE could indeed be accommodated elsewhere in the suborder.

(Gonzalez-R. 1967). A similar lack of shields may be noted in the family BARBUTIIDAE (Plate 80, p. 335), small fusiform species which, until recently, were considered aberrant stigmaeids. *Barbutia perretae* Robaux occurs in forest litter in the western United States (Robaux 1975). *B. anguineus* (Berlese), the only other described species of the genus, was recovered from ground litter in Italy. It has since been collected in California (Summers 1964, Robaux op. cit.).

The HOMOCALIGIDAE (Plate 80, p. 335) are semiaquatic or aquatic species, the adults of which have what appear to be accessory respiratory sacs or tubes. *Homocaligus muscorum* Habeeb was collected from a pasture swamp in eastern North America (Habeeb 1962), and *H. scapularis* (Koch) from a variety of habitats in Europe, including a pond (Wood 1969). A second genus, *Annerossella,* is known from swampy and aquatic habitats in South Africa (Meyer and Ryke 1959) and Malaysia (Wood, op. cit.).

The CALIGONELLIDAE (Plate 78, p. 333) are commonly encountered in leaf and grass litter (Summers and Schlinger 1955), but other habitats have also been successfully utilized by these predators. *Molothrognathus fulgidus* Summers and Schlinger occurs under the bark of prune and almond trees in California. Unidentified species of *Molothrognathus* have been collected from cherry tree twigs in the Pacific Northwest and from grain storages in Canada. *Neognathus terrestris* (Summers and Schlinger) inhabits the soil substrata in pine forest soils of California, while specimens of what may be a new species of *Molothrognathus* occur frequently in salt grass and sage litter in the dry deserts of central Oregon.

The stilt-legged species of the family CAMEROBIIDAE (Plate 77, p. 332) are found in leaf mold, or on or under tree bark, in temperate and tropical realms (McGregor 1950, Southcott 1957, De Leon 1967, Gerson 1972). One species of *Neophyllobius* is common in sage and salt grass litter in central and eastern Oregon, and another has been recorded from mite-infested farm grain storages. *N. fissus* De Leon was found on a cocoa tree which was heavily infested by a false spider mite, *Tenuipalpus orilloi* Rimando, in Trinidad (De Leon, op. cit.).

Superfamily Cheyletoidea
(Plates 81-85, pp. 336-340)

DIAGNOSIS: Soft-bodied but often with weak dorsal shields. Gnathosoma well defined, comprised of fused rostral and cheliceral elements; movable cheliceral digits stylettiform, needle-like, retractile, approximate terminally; palpi variously developed, often reduced or telescoped, sometimes large and armed with a distinctive tibial claw; postcheliceral stigmata and peritremes usually distinct, eyes present or absent. Claws always present on tarsi II-IV, present or absent on tarsus I; empodia present or absent, often with tenent hairs.. Genital opening longitudinal.

The Cheyletoidea is a diverse assemblage of nine families, of which only one, the CHEYLETIDAE, includes non-parasitic species. Arthropods, reptiles, birds and mammals—including man—serve as hosts for parasitic cheyletoids.

The CHEYLETIDAE (Plate 81, p. 336) includes over 40 genera and 150 species which are found in a broad range of habitats including soil, litter, tree bark, and debris, and grain storages. Cheyletids have also been collected from ant and termite nests, caves, bird nests, bat roosts, and in association with bark beetles.

While most cheyletids are considered to be free-living predators, some species are associated with other animals in various ways. *Chelacaropsis moorei* Baker and species of *Hemicheyletus* are found in the fur of mammals, where they subsist as predators of ectoparasitic arthropods sharing the same habitat (Lawrence 1954, Volgin 1960). Similarly, species of *Cheyletus, Eucheyletia* and *Cheletonella* are sometimes associated with mammals (Whitaker and Wilson 1974, Reisen et al. 1976). *Pavlovskicheyla platydemae* Thewke and Enns appears to be a true ectoparasite of the tenebrionid beetle *Platydema ruficorne* (Thewke and Enns 1975), with the mites attaching to the underside of the elytra of the beetle host. *Acaropsis docta* Berlese and a species of *Cheletonella* were observed to feed on the skin of house-martins in England (Woodroffe 1956), although neither genus is considered truly parasitic.

Species of *Cheyletus, Cheletomorpha, Cheyletia* and *Acaropsis* often occur as predators in grain storages infested by graminivorous mites (Hughes 1961). *Cheyletus eruditus* (Schrank) is considered by some observers to be effective in controlling grain mite populations (Pulpan and Verner 1965), but somewhat seasonal in its effectiveness (Solomon 1946). *C. eruditus* was noted to reduce laboratory populations of the granary mite, *Acarus siro* L., when prey and predator populations were equal or imbalanced in favor of the predator, at 20°C and 80% rh (Solomon 1969). *A. siro* populations quickly outnumbered *C. eruditus* at lower temperatures. House dust mites (Acaridida, PYROGLYPHIDAE) are effectively controlled by *C. eruditus* in laboratory populations (Lukoschus, personal communication). *Cheletomorpha lepidopterorum* (Shaw) has been observed to consume an average of four grain mites/day at 20° and 80% rh in the laboratory.

The family CHEYLETIELLIDAE (Plate 81, p. 336) includes eight genera of mites which are parasitic on birds and small mammals (Smiley 1970). Species of the genera *Neocheyletiella, Ornithocheyla, Ornithocheyletia* and *Bakericheyla* are bird parasites, while *Nihelia, Eucheyletiella* and *Criokeron* species parasitize small mammals. *Bakericheyla chanayi* (Berlese and Trouessart) constructs silken "nests" on the skin of its fringillid bird hosts (Furman and Sousa 1969).

Cheyletiella yasguri Smiley, a cheyletiellid which causes a superficial mange on dogs (Smiley 1965, van Bronswijk and de Kreek 1976), is also a hyperparasite of dog-infesting louse-flies in Iran (Vercammen-Grandjean and Rak 1968) and a causative agent of pruritic dermatitis in man (Rack 1971, Hewitt et al. 1971, Keh 1973). *C. parasitivorax* (Mégnin), often confused with *C. yasguri,* is a parasite of rabbits. Cases of eczema in humans as the result of contact with cats ostensibly infested by this species were reported by Olsen and Roth (1947). Later, Keh (1975) described cases of human eczema following contact with cats infested by *C. blakei* Smiley. Based on the emerging evidence of host specificity in the genus (Gething and Walton 1972, van Bronswijk and de Kreek, op. cit.), *C. blakei* rather than *C. parasitivorax* may have been involved in earlier episodes of human eczema described by Olsen and Roth.

Mites of the cosmopolitan family MYOBIIDAE (Plate 82, p. 337) are ectoparasites of marsupials, rodents, bats and insectivores, and cling to their hosts by means of highly modified legs I. Feeding is more or less confined to the hair follicle bases and often involves haematophagy. Female myobiids ingest blood during reproductive periods (Lukoschus, personal communication) but may feed on other extracellular tissue fluids as well (Yunker 1973). *Myobia musculi* (Schrank), *Radfordia ensifera* (Poppe) and *R. affinis* (P.) are common myobiid parasites of laboratory rodents, and cause dermatitis, alopecia and trauma in their hosts (Baker et al. 1956, Yunker op. cit.).

Dusbábek (1969) has illustrated a high degree of zoogeographical correlation between myobiid mites and their hosts. He feels that myobiids may have evolved originally as parasites of marsupials, later transferring to more advanced forms. Moreover, Fain (1975) has determined that parallelism exists between the evolution of myobiids and host mammals. Fain distinguishes three major myobiid categories: a primitive *marsupial group,* with the terminal segments of legs I free and well developed (Plate 82-3), a *rodent group* with the terminal three segments of legs I fused (Plate 82-1), and an *insectivore-bat group* with intermediate leg characters. Bat myobiids are commonly correlated in degree of primitiveness with their bat hosts (Dusbábek 1973).

The HARPYRHYNCHIDAE (Plate 84, p. 339) are parasitic in the cuticle of the skin and feathers of birds (Fritsch 1954, Zumpt 1961, Fain 1972, 1976). Members of the family SYRINGOPHILIDAE (Plate 83, p. 338) feed on the fluids of their bird hosts through the walls of the feather quills in which they reside. Kethley (1970, 1973) recognizes 18 syringophilid genera which parasitize 10 orders of birds. The mite genera appear to be more or less confined to particular avian orders. Syringophilid species are highly specific as to host and to feather type (primaries, secondaries, coverts), although more than one species may often be found in a single feather calmus. Thickness and size of a given quill largely determines the mite species complex inhabitating it (Kethley 1971). Adequate stylet length is a critical factor for survival in a chosen feeding site, as is utilization of the quill volume for maximum development of female dispersants at host molting time in the fall, and nestling development in the spring.

The family OPHIOPTIDAE (Plate 82, p. 337) includes circular mites which live in small pits in the scales of their snake hosts. Sambon (1928) reported that *Ophioptes parkeri* Sambon formed these pits in the scales of its colubrid host, a conclusion also reached by Ewing (1933) for *O. tropicalis* on another colubrid species. It has since been determined that the female makes the "pit" and deposits an egg therein. The apodous larvae feed, expand, and molt to apodous nymphs which then give rise to males and females (Fain 1964). Members of the PSORERGATIDAE (Plate 83, p. 338) are minute (130-200 μ) round mites which live in surface and subsurface skin layers of hair follicles of their mammal hosts. Psorergatids often cause dermatitis, mange or follicular infection which is aggravated by rubbing or biting by the host. The sheep itch mite, *Psorergates ovis* Womersley, is a major pest of sheep in the United States and in Australia (Baker et al. 1956). *P. bos* Johnston, an itch mite of cattle in the southwestern United States, precipitates symptoms in its host which are similar to those caused by *P. ovis* on sheep (Johnston 1964, Roberts and Meleney 1965). Other psorergatids have been collected from insectivores, bats, rodents, carnivores and primates in Africa, Asia, North and South America, Europe and Australia (Fain et al. 1966, Lukoschus et al. 1967, Lukoschus et al. 1974).

The minute, annulate, usually worm-like species of the family DEMODICIDAE (Plate 84, p. 339) are parasites which may cause epithelial destruction, hyperplasia and granuloma in their mammal hosts. Demodicids are known to invade hair follicles and Meibomian (holocrine) glands or ducts, and may penetrate the epidermis. Records also exist of invasion of auricular tissues, subdermal fascia, the digestive tract, and even the circulatory system (Nutting 1975). *Demodex caprae* may cause metaplasia of follicular epithelial cells of infested goats, followed by hyperkeratinization and eventual cratering of skin after extrusion of mites (Lebel and Nutting 1971). Mange is not uncommon in domestic animals attacked by *D. canis* Leydig (Yunker 1973). Demodectic mange in dogs and cats may be caused or aggravated by a *Staphylococcus* strain which occurs with the mite (Hirst 1919). *D. equi* Railliet causes pruritis and alopecia in horses, and subsurface skin pustules containing demodicid mites are routinely found in cattle, sheep and goats.

D. longissimus Desch, Nutting and Lukoschus and *D. molossi* D., N. and L. are greatly elongated (up to 900 μ) species which attach inside the large posterior Meibomian gland in Neotropical bats (Desch et al. 1972). They attach in such a way that the elongate opisthosoma hangs out over the eyeball of the host. Immobile larvae and nymphs have greatly enlarged hooked legs III with which they secure themselves to the host tissue (Plate 84-4). Lemurs are parasitized by species of *Stomatodex* and *Chinodex,* and man serves as the natural host for *Demodex folliculorum* (Simon) and *D. brevis* Akbulatova (Desch and Nutting 1972). Both *Demodex* species may be collected from the skin of the forehead, nose, and alar regions of the face, but each exploits a different niche. *D. folliculorum* is found in simple hair follicles above the level of the sebaceous glands while *D. brevis* colonizes sebaceous glands of the deeper vellus hairs. Both apparently are low-grade pathogens (Desch and Nutting op. cit.), although a dermatitis of the scalp has been attributed to *Demodex* mites (Miskjian 1951).

Species of the highly evolved family CLOACARIDAE (Plate 85, p. 340) are parasitic in the mucosa of the cloaca of turtles (Camin et al. 1967, Fain 1968). Cloacarids are thought to be venereally transmitted and are strongly host specific. *Caminacarus theodori* Fain and *Theodoracarus testudinis* F. both tend to invade the mucosa itself, while other species attach at the mucosal surface by means of their peculiar hooked palpi.

Superfamily Tetranychoidea
(Plates 85-88, pp. 340-343)

DIAGNOSIS: Soft-bodied and with striate or reticulate pattern, with or without a sejugal furrow and weakly defined shields. Gnathosoma with cheliceral bases fused, separate from ventral rostrum; movable cheliceral digits greatly elongated, whiplike, retractile, approximate terminally; palpi with five segments, often with a thumb-claw process, sometimes simple and with the palptarsal segment terminal; with postcheliceral stigmata and peritremes, peritremes emergent in some genera; eyes present or absent. Claws and empodia on all tarsi, with tenent hairs on claws and often on empodia as well. Genital opening transverse.

The five recognized families of Tetranychoidea include over 450 phytophagous species, many of which are of economic importance. The majority of species are contained in two families—the TETRANYCHIDAE or spider mites, and the TENUIPALPIDAE or false spider mites.

Species of the family TETRANYCHIDAE (Plates 85-87, pp. 340-342) are found throughout those parts of the world where higher plants flourish. They occur on virtually every major food crop and ornamental plant, often causing serious injury or death of the host. Spider mites are green, yellow, orange or red in color, and may commonly be found associated with fine silk webbing which they spin from large unicellular glands located in the palpi (Grandjean 1948, Alberti and Storch 1974). Tetranychids feed by inserting their whiplike cheliceral stylets through or between the palisade cells and into the underlying parenchyma (Jeppson et al. 1975). They may simultaneously inject a toxic salivary substance into the plant, although little experimental evidence is presently available on this point. Feeding symptoms may be confined to simple stippling of injured tissue, or the host plant may suffer leaf and fruit drop or twig die-back. Overall plant vigor and plant growth may be affected as well.

Grasses and other low-growing plants commonly are infested by tetranychids of the subfamily Bryobiinae. The clover mite, *Bryobia praetiosa* Koch, is found on grasses in many parts of the world (Pritchard and Baker 1955, Meyer 1974, Jeppson et al. 1975). Clover mites complete several generations during the growing season and may overwinter in any stage. Large numbers of mites often invade buildings in early spring to molt and oviposit.

B. praetiosa is regarded by some specialists as a complex of closely related races which differ somewhat in host plant preference, life history and behavior. The complex includes: *B. kissophila* van Eyndhoven, a multivoltine species on ivy which overwinters in all stages; *B. ribis* Thomas, a univoltine form on gooseberry which overwinters only in the egg stage; *B. rubrioculus* (Scheuten), a multivoltine inhabitant of fruit trees which overwinters in the egg stage. *B. rubrioculus,* the brown mite, is an important pest of pome and stone fruits in both the Old and New Worlds (Morgan 1960, Meyer op. cit., Jeppson et al., op. cit.).

Bryobiines of the genus *Monoceronychus* are more or less confined to grasses, while members of the genera *Petrobia, Aplonobia, Hystrichonyssus* and *Pseudobryobia* (among others) occur on a variety of low-growing plants. The brown wheat mite, *Petrobia latens* (Müller), and the legume mite, *P. apicalis* (Banks), are serious economic pests in many parts of the world. *Tetranycopsis horridus* (Canestrini and Fanzago), a pest of filberts in Europe, has recently appeared in a filbert planting in California (Jeppson et al. 1975).

The subfamily Tetranychinae includes species which infest all higher plant groups. A substantial number of tetranychines are economically important forms, but only a few can be mentioned here. *Tetranychus (T.) urticae* Koch, the two-spotted spider mite, is a widespread major pest of deciduous fruit trees. *T. (Armenychus) mcdanieli* McGregor also is an important fruit tree pest, especially in the northwestern United States. Cotton and low-growing vegetable crops often are infested by populations of the carmine spider mite, *T. (T.) cinnabarinus* (Boisduval), the desert spider mite, *T. (T.) desertorum* Banks, the strawberry spider mite, *T. (T.) turkestani* (Ugarov and Nikolski), and the tumid spider mite, *T. (T.) tumidus* Banks (Pritchard and Baker 1955, Boudreaux 1956, Jeppson et al. 1975). *Panonychus ulmi* (Koch), the European red mite, may be found in association with *Tetranychus* species on deciduous fruit trees, and is considered a major economic problem in European, North American and Japanese orchards (Groves and Massee 1951, Ehara 1975a, Jeppson et al., op. cit.). The citrus red mite, *P. citri* (McGregor), attacks citrus fruit trees and other plants in North America, South Africa and Japan. It has been highly injurious in California orchards, particularly in coastal and intermediate citrus-growing areas (Jeppson 1963, Jeppson et al., op. cit.).

Banks grass mite, *Oligonychus (Reckiella) pratensis* (Banks) has proven to be a serious pest of grasses in North and Central America, Hawaii and Africa (Meyer 1974, Jeppson et al. 1975). Wheat, corn, bluegrass, sorghum and sugar cane comprise a partial list of host plants for this species. Coniferous and deciduous trees are major hosts for species of the subgenera *O. (Oligonychus), O. (Wainsteinella)* and *O. (Homonychus).* The spruce spider mite, *O. (O.) ununguis* (Jacobi), is a major pest of conifers throughout the world. An epidemic outbreak of *O. (O.) ununguis* in Montana and Idaho in 1957 severely damaged over 800,000 acres of Douglas fir forests (U.S. Forest Service Report, 1958). The same species is reported on chestnut in Japan (Ehara 1975a). The southern red mite, *O. (O.) ilicis* (McGregor), is injurious on holly, rhododendron and other hosts in eastern United States (Weidhaas and Reeves 1963, Jeppson et al., op. cit.). It is also a widely distributed pest of coffee in Brazil.

Members of the genus *Eotetranychus* feed primarily on trees and shrubs, and some species are considered major economic pests. *E. carpini carpini* (Oudemans) may become a problem in unsprayed apple orchards although natural enemies tend to keep it in check. The yellow spider mite, *E. carpini borealis* (Ewing), is an injurious pest of apple and pear in the northwestern United States. *E. willamettei* (McGregor) causes extensive injury of grapes in California, and *E. pruni* (Oud.) damages maple in western Oregon.

A comprehensive list of injurious tetranychid mites may be found in Jeppson et al. (1975). An excellent survey of ecological factors affecting host injury by important tetranychid species is presented by Huffaker et al. (1970), McMurtry et al. (1970) and van de Vrie et al. (1972).

The false spider mites or TENUIPALPIDAE (Plate 88, p. 343) comprise 14 genera of small, somewhat flattened red or green species which appear to be best adapted to subtropical or tropical climates. Some species, however, occur well into the northern extremes of the temperate zone. Feeding generally occurs on the midrib or a vein on the underside of the host leaf (Baker and Pritchard 1960). Some species are found in leaf sheaths on grass, or in plant galls. Members of the genera *Brevipalpus, Tenuipalpus* and *Cenopalpus* are of particular importance as plant pests (Jeppson et al. 1975).

Citrus trees serve as hosts for *Brevipalpus californicus* (Banks), *B. lewisi* McGregor, *B. obovatus* Donnadieu and *B. phoenicis* (Geijskes) in commercial orchards. *B. chilensis* Baker is a serious pest of grapes in Chile. Other hosts for *Brevipalpus* species include ornamentals, berry crops, orchids and olives. *Tenuipalpus* species infest grasses, tea, orchids and pomegranates, among other hosts. *T. pacificus* Baker attacks orchids in European greenhouses (Dosse 1954). *Cenopalpus lanceolatisetae* (Attiah) and *C. pulcher* (Canestrini and Fanzago) are pests of deciduous fruit trees in Europe, Asia and North Africa (Jeppson et al. 1975).

The elongate bright red tenuipalpid species of the genus *Dolichotetranychus* are found primarily in grasses (Pritchard and Baker 1958). *D. vandergooti* is only known to occur on orchids (Collyer 1973a). *Larvacarus transitans* (Ewing), a gall-inhabiting tenuipalpid, has reduced palpi and, as with members of the genus *Phytoptipalpus,* has only three pairs of legs in postlarval stadia. Extreme palpal reduction also has been observed in *Obuloides rajamohani* Baker and Tuttle, a tenuipalpid inhabitant of eriophyid galls in India (Baker and Tuttle 1975a).

The families TUCKERELLIDAE, ALLOCHAETOPHORIDAE and LINOTETRANIDAE (Plate 88, p. 343) are monogeneric groups containing few species. Some species of the genus *Tuckerella* occur in soil, probably in association with underground plant parts. *T. knorri* Baker and Tuttle was collected from the roots of *Pandanus* and *Achras* in Thailand, and *T. hypoterra* McDaniel, Morihara and Lewis was found in pasture soil in South Dakota (Baker and Tuttle 1975b, McDaniel et al. 1975). An unidentified *Tuckerella* species was collected from plant roots in California, and *T. litoralis* Collyer was found in moss on boulders bordering a salt water harbor (Collyer 1969). Other species of the genus are aerial forms (Jorgensen 1967, Zaher and Rasmy 1969, McDaniel et al., op. cit., Ehara 1975b).

Allochaetophora californica McGregor feeds on Bermuda grass, and *Linotetranus cylindricus* Berlese, the type species of the genus, was collected in moss (McGregor 1950, Baker and Pritchard 1953). A second species, *L. protractulus* Athias-Henriot, was recovered from soil and litter in Algeria (Athias-Henriot 1961).

Superfamily Eriophyoidea[1]
(Plates 89-90, pp. 344-345)

DIAGNOSIS: Soft-bodied or armored, annulate and worm-like. Gnathosoma comprised of median rostrum enclosed by simple 4-segmented lateral palpi, movable cheliceral digits whip-like, lying in an anterodorsal groove of the palpal bases;

[1]Newkirk and Keifer (1971) reviewed the higher classification of the Eriophyoidea and concluded that its members are referrable to three distinct family categories on the basis of morphological and habital characters. Newkirk and Keifer's classification is followed here, with certain nomenclatorial emendations suggested by Lindquist (1974).

without stigmata, peritremes, tracheae, or eyes. With only two pairs of anteriorly inserted legs with few setae, often lacking some segments; tarsi each with a feathered or rayed empodium; true claws absent but with a dorsodistal claw-like seta. Genital opening transverse. Small to extremely small species (90-350 μ).

The Eriophyoidea is virtually worldwide in distribution, occurring on a multitude of primarily perennial hosts. Although eriophyoid injury may be economically significant, eriophyoids rarely kill their host plants. Like the TETRANYCHIDAE (with which they appear to have phylogenetic affinities), eriophyoids feed by inserting the cheliceral stylets into plant cells and sucking up their liquid contents (Jeppson et al. 1975). Unlike the tetranychids, however, eriophyoids illustrate a high degree of host specificity, at least to the generic level of their hosts. Host reaction to eriophyoid feeding often is bizarre, involving formation of galls and other tissue teratosities which the mites then utilize as feeding niches. Eriophyoids are referred to as gall mites, blister mites, bud mites or rust mites, depending on host reaction to their feeding. Those which cause no observable host injury are considered to be leaf vagrants. Symptomatology may be useful, at least to some extent, in defining family categories in the Eriophyoidea. For example, the family RHYNCAPHYTOPTIDAE[1] (Plate 89, p. 344) consists entirely of leaf vagrants which do little or no damage to their hosts. The ERIOPHYIDAE (Plate 90), on the other hand, commonly precipitate gall formation and other host abnormalities, as do many members of the family SIERRAPHYTOPTIDAE[2] (Plate 89).

Rhyncaphytoptids are unusual in having a greatly enlarged gnathosoma which is abruptly bent ventrad. The chelicerae also are exceptionally elongate, which suggests that these mites are capable of feeding on deeper tissues than are other eriophyoids. Yet the only noticeable symptoms produced by rhyncaphytoptids involve surface leaf cells. Many rhyncaphytoptid nymphs and adults produce a flocculent, waxy exudate which covers the mite's dorsum. Presumably, the exudate protects the exposed vagrant mites from excessive water loss. *Catarhinus tricholaenae* Keifer, a grass-inhabiting rhyncaphytoptid from Brazil, causes leaf rust on corn (Jeppson et al. 1975). Other rust-producing rhyncaphytoptids are found in the genus *Diptacus.*

Leaf injury produced by members of the family ERIOPHYIDAE may be of several types, and may involve surface or mesophyll tissues (Keifer 1952). *Aculus comatus* (Nalepa) causes rusting of filbert leaves in Europe and North America, but may also give rise to leaf edgerolling (Krantz 1973). *A. cornutus* (Banks), the peach silver mite, causes leaf spotting in the spring followed by leaf curling and silvering later in the season. *A. fockeui* (Nalepa and Trouessart) may be intraspecifically related to *A. cornutus,* but is found instead on plums and cherries where it produces leaf chlorosis. Severe leaf edgerolling is caused by *Eriophyes granati* (Canestrini) on pomegranate, and by *E. caryae* Keifer on pecans.

Eriophyes brachytarsus K. causes a localized leaf puckering, or *purse galls* Plate 90-8), on California black walnut. Other eriophyid species induce a secondary development of papillar outgrowths on leaves or petioles which afford shelter for the mites. Papillar development may be patchy or generalized, and may appear on either leaf surface. Papillar masses are known as *erinea* (Plate 90-10), and constitute "open galls" in which pocketing is minimal. Several species of *Eriophyes, Acalitus* and *Colomerus* induce erineal formation in their hosts. The injurious grape erineum mite, *Colomerus vitis* (Pgst.) appears to consist of three strains or races, only one of which precipitates development of erinea (Smith and Stafford 1948, Stafford and

---

[1]Based on rules of priority, the family name DIPTILOMIOPIDAE Keifer 1944 should supersede RHYNCAPHYTOPTIDAE Roivainen 1953 *sensu* Newkirk and Keifer 1971 (Lindquist 1974). The latter name is retained here only on the strength of its current general usage.

[2]The family NALEPELLIDAE Roivainen 1953 of Newkirk and Keifer 1971. The name SIERRAPHYTOPTIDAE Keifer 1944 has priority (Lindquist 1974).

Doutt 1974). The other two strains feed on bud tissues or cause leaf curl in susceptible grape varieties.

True galls are produced by host reactions to feeding by numerous eriophyid mite species. Galls may assume a variety of shapes (nail, bead, finger, bladder, etc.), but all are highly developed purse galls which have virtually closed, and in which the abnormal growth of papillar material provides a substrate in which mites may feed with little danger from predators or from desiccation (Plate 90-8). Galls may be produced on the upper sides of leaves, on flowers, stems, buds or petioles. All are formed prior to maturation of the affected plant part. Jeppson et al. (1975) list several gall-inducing ERIOPHYIDAE, and Mani (1964) provides illustrations of several gall types.

Pears, apples, quince and other pomaceous plants may be affected by *Eriophyes pyri* (Pgst.) or one of the *E. pyri* complex of blister mites. Unlike galls, blisters are formed within the parenchymal leaf tissues (Plate 90-11). Ingress occurs early in the spring following hypertrophy and necrosis of surface cells previously fed upon by early emerging mites. The collapse of necrotic cells leaves a tiny opening through which a female mite finally passes. *E. pyri* may also damage bud tissues in the fall (Jeppson et al. 1975). But injury by other eriophyids may be caused by surface feeding or by bud gall formation. The citrus bud mite, *Eriophyes sheldoni* Ewing, feeds in leaf buds and fruit buttons of lemons and causes severe fruit and leaf malformation (Boyce and Korsmeier 1941). Members of the genus *Cecidophyopsis* induce bud gall formation in a variety of hosts including yew, currant, and filbert (Jeppson et al. 1975, Krantz in press) (Fig. 35, p. 264). In addition to leaf and bud injury, eriophyid mites are responsible for development of "witches broom" (adventitious twig development), flower galls, shortening of internodes, and plant stunting.

At least some eriophyids apparently are capable of spinning silk. Keifer (1976) describes a thin webbing on the dorsal surfaces of leaves infested by *Aculops knorri* K., which evidently is spun by the mites.

The SIERRAPHYTOPTIDAE are primarily rust and bud mites which often infest monocotyledenous hosts. None is an erineum-forming species. Comparatively few sierraphytoptids are of economic significance, although several feed on economically important hosts. Species of *Trisetacus* attack coniferous hosts, feeding chiefly on bud tissues. The juniper bud mite, *T. quadrisetus* (Thomas), lives and feeds inside the berries of its host. Others cause bud swelling ("big bud") or bud proliferation in larch, spruce, and Douglas fir. *Phytoptus avellanae* Nalepa produces big bud symptoms in hazelnut and commercial filberts, often causing significant bud losses (Fig. 35). Leaf stunting and deformation of English ivy is caused by *P. hedericola* (Keifer).

With few exceptions, eriophyoids (and, more specifically, the ERIOPHYIDAE), are the only mites which have been demonstrated to transmit plant viruses (Slykhuis 1969, 1972, Oldfield 1970). Grasses serve as hosts for diseases transmitted by *Eriophyes tulipae* Keifer (wheat streak and wheat spot mosaic) and *Abacarus hystrix* (Nalepa) (ryegrass and *Agropyron* mosaic). Currant reversion, a serious viral disease of black currants, is transmitted by *Cecidophyopsis ribis* (Westwood and Nalepa). Other woody perennials are affected by viruses transmitted by *Eriophyes ficus* Cotte (fig mosaic), *Eriophyes insidiosus* Wilson and Keifer (peach mosaic), and *Aculus fockeui* (Nalepa and Trouessart) (latent plum virus). Additional mite-virus relationships are listed by Oldfield (op. cit.).

Eriophyoids in northern temperate and subarctic latitudes often produce special overwintering females called *deutogynes.* Deutogynes differ morphologically from normal females or *protogynes* (Plate 90) and generally are difficult to identify. They may appear early or late

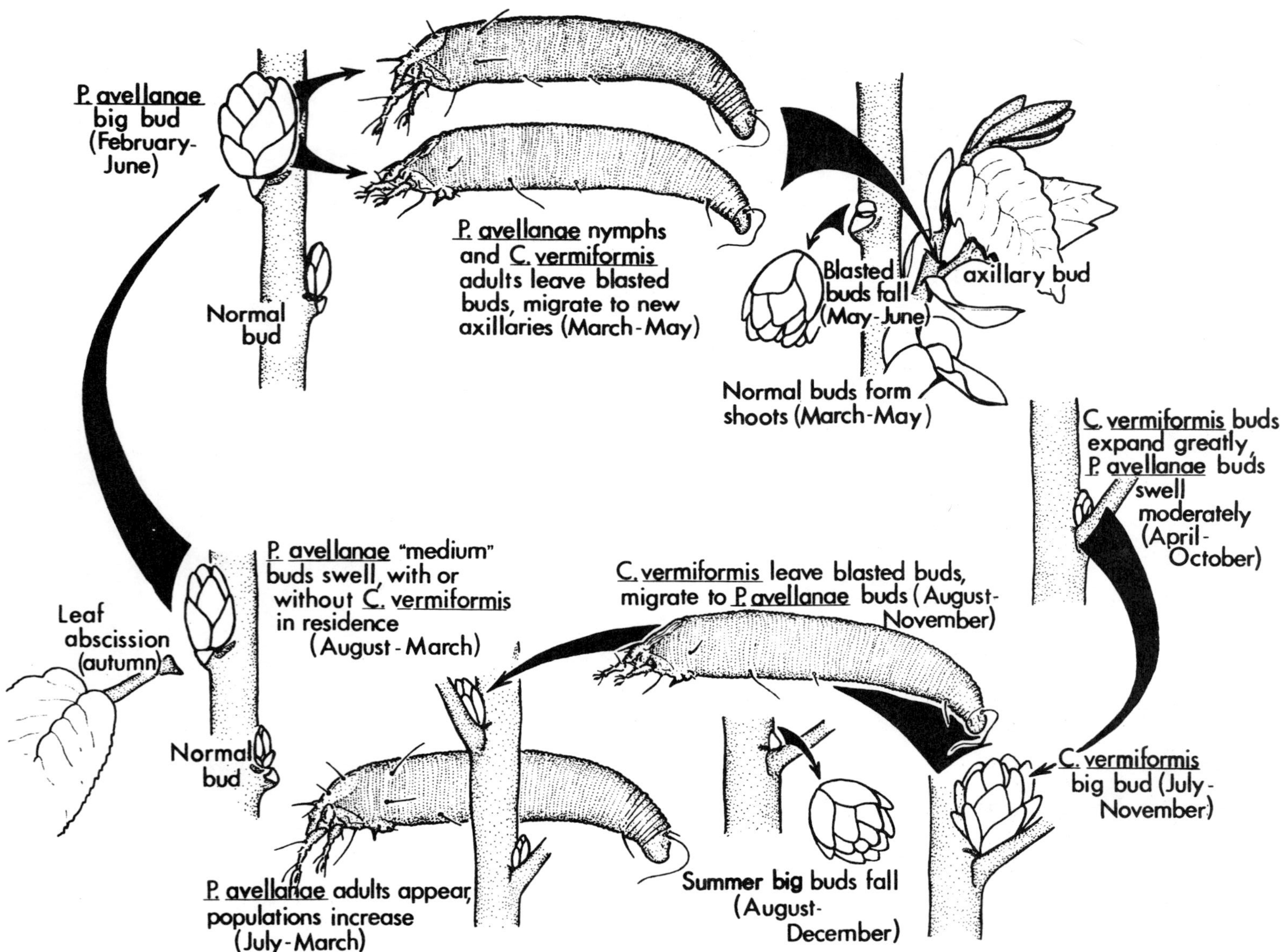

Fig. 35. Dynamics of the filbert bud mites *Phytoptus avellanae* Nalepa (SIERRAPHYTOPTIDAE) and *Cecidophyopsis vermiformis* (Nalepa) (ERIOPHYIDAE) in commercial filberts (Oregon, USA).

in the growing season, depending on host physiology, but always seek an aestivation or hibernation niche. According to Oldfield and Newell (1973), fertilization of deutogynes by spermatophores occurs prior to hibernation in the peach silver mite, *Aculus cornutus* (Banks). Prehibernation fertilization of deutogynes may occur in other species as well. Detailed information on deuterogyny is presented by Keifer (1942, 1952) and by Jeppson et al. (1975).

Useful References

Ah, H.-S., J.C. Peckham and W.T. Atyeo (1973). *Psorergates glaucomys* sp.n. (Acari:Psorergatidae), a cystogenous mite from the southern flying squirrel *(Glaucomys v. volans),* with histopathologic notes on a mite-induced dermal cyst. J. Parasitol. **59**(2):369-374.

Alberti, G. and V. Storch (1974). Über Bau und Funktion der Prosoma-Drüsen von Spinnmilben (Tetranychidae, Trombidiformes). Z. Morph. Tiere **79**:133-153.

Anon. (1958). Forest insect conditions in 1957: a status report. Div. Forest Insects Res.: 20 pp.

Athias-Henriot, C. (1961). Nouveaux acariens phytophages d'Algérie (Actinotrichida, Tetranychoidea: Tetranychidae, Linotetranidae). Ann. École Nat. d'Agr. d'Alger **3**(3):1-10.

Atyeo, W.T., E.W. Baker and D.A. Crossley, Jr. (1961). The genus *Raphignathus* Dugès (Acarina, Raphignathidae) in the United States with notes on the Old World species. Acarologia **3**(1):14-20.

Baker, E.W. (1949a). Pomerantziidae, a new family of mites. J. Wash. Acad. Sci. **30**(8):269-271.

Baker, E.W. (1949b). A review of the mites of the family Cheyletidae in the United States National Museum. Proc. U.S. Nat. Mus. **99**(3238):267-320.

Baker, E.W., T.M. Evans, D.J. Gould, W.B. Hull and H.L. Keegan (1956). A Manual of Parasitic Mites of Medical or Economic Importance. Nat. Pest Control Assoc. Tech. Publ.: 170 pp.

Baker, E.W. and A.E. Pritchard (1953). The family categories of tetranychoid mites, with a review of the new families Linotetranidae and Tuckerellidae. Ann. Ent. Soc. Amer. **46**(2):243-258.

Baker, E.W. and A.E. Pritchard (1960). The tetranychoid mites of Africa. Hilgardia **29**(11):455-574.

Baker, E.W. and D.M. Tuttle (1975a). A new genus of Tenuipalpidae (Acarina) from India. USDA Coop. Econ. Insect Rept. **25**(22):453-455.

Baker, E.W. and D.M. Tuttle (1975b). A new species of *Tuckerella* (Acarina:Tuckerellidae) from Thailand. USDA Coop. Econ. Insect Rept. **25**(17):337-340.

Baker, E,W., D.M. Tuttle and M.J. Abbatiello (1975). The false spider mites of northwestern and north central Mexico (Acarina:Tenuipalpidae). Smithsonian Contrib. Zool. 194:23 pp.

Baker, E.W. and G.W. Wharton (1952). The Suborder Trombidiformes Reuter, 1909. **In** An Introduction to Acarology. Macmillan Co. New York: 146-258.

Blauvelt, W.E. (1945). The internal morphology of the common red spider mite (*Tetranychus telarius* Linn.). Cornell Univ. Agr. Exp. Sta. Mem. 270:3-46.

Boudreaux, H.B. (1956). Revision of the two-spotted spider mite (Acarina, Tetranychidae) complex, *Tetranychus telarius* (Linnaeus). Ann. Ent. Soc. Amer. **49**(1):43-48.

Boyce, A.M. and R.B. Korsmeier (1941). The citrus bud mite, *Eriophyes sheldoni* Ewing. J. Econ. Ent. **34**(6):745-756.

Boyle, W.W. (1956). On the mode of dissemination of the two-spotted spider mite, *Tetranychus telarius* (L.) (Acarina:Tetranychidae). Proc. Hawaiian Ent. Soc. **16**(2):261-268.

Briones, M.L. and B. McDaniel (1976). The eriophyid plant mites of South Dakota. South Dakota Agr. Exp. Sta. Tech. Bull. 43:123 pp. + viii.

Bronswijk, J.E.M.H. van and E.J. de Kreek (1976). *Cheyletiella* (Acari:Cheyletiellidae) of dog, cat and domesticated rabbit, a review. J. Med. Ent. **13**(3):315-327.

Camin, J.H., W.W. Moss, J.H. Oliver, Jr. and G. Singer (1967). Cloacaridae, a new family of cheyletoid mites from the cloaca of aquatic turtles (Acari:Acariformes:Eleutherengona). J. Med. Ent. **4**(3):261-272.

Coates, T.J.D. (1974). The influence of some natural enemies and pesticides on various populations of *Tetranychus cinnabarinus* (Boisduval), *T. lombardinii* Baker and Pritchard and *T. ludeni* Zacher (Acari: Tetranychidae) with aspects of their biologies. Dept. Agr. Serv. Ent. Mem. 42:40 pp.

Collyer, E. (1969). Two species of *Tuckerella* (Acarina:Tuckerellidae) from New Zealand. N. Z. J. Sci. **12**(4):811-814.

Collyer, E. (1973a). A new species of the genus *Dolichotetranychus* (Acari:Tenuipalpidae) from New Zealand. N. Z. J. Sci. **16**:747-749.

Collyer, E. (1973b). New species of the genus *Tenuipalpus* (Acari:Tenuipalpidae) from New Zealand, with a key to the world fauna. N. Z. J. Sci. **16**:915-955.

De Leon, D. (1967). Some mites of the Caribbean area. Part 1. Acarina on plants in Trinidad, West Indies. Allen Press Inc., Lawrence, Kansas: 1-46. [CAMEROBIIDAE, TENUIPALPIDAE, CHEYLETIDAE]

Desch, C. and W.B. Nutting (1972). *Demodex folliculorum* (Simon) and *D. brevis* Akbulatova of man: redescription and reevaluation. J. Parasitol. **58**(1):169-177. [DEMODICIDAE]

Desch, C., W.B. Nutting and F. Lukoschus (1972). Parasitic mites of Surinam VII: *Demodex longissimus* n. sp. from *Carollia perspicillata* and *D. molossi* n. sp. from *Molossus molossus* (Demodicidae:Trombidiformes); Meibomian complex inhabitants of Neotropical bats (Chiroptera). Acarologia **14**(1):35-53.

Dosse, G. (1954). *Tenuipalpus orchidarum* Parfitt nun auch in deutschen Gewachshäusern. Zeitschr. ang. Ent. **36**:304-315. [TENUIPALPIDAE]

Dusbábek, F. (1969). To the phylogeny of genera of the family Myobiidae (Acarina). Acarologia **11**(3):537-584.

Dusbábek, F. (1973). A systematic review of the genus *Pteracarus* (Acariformes:Myobiidae). Acarologia **15**(2):240-288.

Ehara, S. (1975a). A guide to the tetranychid mites of agricultural importance in Japan. **In** Approaches to Biological Control. K. Yasumatsu and H. Mori, eds. JIBP Synthesis, Tokyo **7**:15-23.

Ehara, S. (1975b). Description of a new species of *Tuckerella* from Japan (Acarina:Tuckerellidae). Int. J. Agr. **1**(2):1-5.

Ehara, S. and T. Wongsiri (1975). The spider mites of Thailand (Acarina:Tetranychidae). Mushi **48**(13):149-185.

Ewing, H.E. (1933). A new pit-producing mite from the scales of a South American snake. J. Parasit. **20**(1): 53-56. [OPHIOPTIDAE]

Fain, A. (1960). Les acariens psoriques parasites des chauves-souris. XIII. La famille Demodicidae Nicolet. Acarologia **2**(1):80-87.

Fain, A. (1964). Les Ophioptidae acariens parasites des ecailles des serpents (Trombidiformes). Bull. Inst. roy. Sci. nat. Belg. **40**(15):1-57.

Fain, A. (1968). Notes sur les Acariens de la famille Cloacaridae Camin et al. parasites du cloaque et des tissus profonds des tortues (Cheyletoidea:Trombidiformes). Bull. Inst. roy. Sci. nat. Belg. **44**(5):1-33.

Fain, A. (1972). Notes sur les Acariens des familles Cheyletidae et Harpyrhynchidae producteurs de gale chez les oiseaux ou les mammifères. Acta Zool. Path. Antverp. **56**:37-60.

Fain, A. (1973). Notes sur la nomenclature des poils idiosomaux chez les Myobiidae, avec description de taxa nouveaux (Acarina:Trombidiformes). Acarologia **15**(2):279-309.

Fain, A. (1975). Observations sur les Myobiidae parasites des rongeurs. Évolution parallele hotes-parasites (Acariens:Trombidiformes). Acarologia **16**(3):441-475.

Fain, A. (1976). Notes sur les Harpyrhynchidae. Description de quatre espèces nouvelles (Acarina:Prostigmates). Acarologia **18**(1):124-132.

Fain, A., F. Lukoschus and P. Hallmann (1966). Le genre *Psoregates* chez les muridés. Description de trois espèces nouvelles (Psorergatidae:Trombidiformes). Acarologia **8**(2):251-274.

Fleschner, C.A., M.E. Badgley, D.W. Ricker and J.C. Hall (1956). Air drift of spider mites. J. Econ. Ent. **49**(5):624-627. [TETRANYCHIDAE]

Fritsch, W. (1954). Die Milbengattung *Harpyrhynchus* Megnin, 1878. Zool. Anz. **152**:177-198.

Furman, D.P. and O.E. Sousa (1969). Morphology and biology of a nest-producing mite, *Bakericheyla chanayi* (Acarina:Cheyletidae). Ann. Ent. Soc. Amer. **62**(4):858-863.

Geijskes, D.C. (1939). Beiträge zur Kenntnis der europaischen Spinnmilben (Acari, Tetranychidae), mit besonderer Berucksichtigung der niederlandischen Arten. Meded. Landbouwhoogeschool **42**(4):1-68.

Gerson, U. (1966). A redescription of *Eupalopsis maseriensis* (Canestrini and Fanzago) (Acarina:Eupalopsellidae). Israel J. Zool. **15**:148-154.

Gerson, U. (1967). The natural enemies of the chaff scale, *Parlatoria pergandii* Comstock, in Israel. Entomophaga **12**:97-109. [EUPALOPSELLIDAE]

Gerson, U. (1972a). A redescription of *Ledermuelleria frigida* Habeeb (Acarina:Prostigmata:Stigmaeidae). Acta Arachnol. **24**(1):15-28.

Gerson, U. (1972b). A new species of *Camerobia* Southcott, with a redefinition of the family Camerobiidae (Acari:Prostigmata). Acarologia **13**(3):502-508.

Gerson, U. and D. Blumberg (1969). Biological notes on the mite *Saniosulus nudus*. J. Econ. Ent. **62**(3):729-730. [EUPALOPSELLIDAE]

Gething, M.A. and G.S. Walton (1972). Possible host specificity of *Cheyletiella* mites. Vet. Rec. **90**(18):512. [CHEYLETIELLIDAE]

Gonzalez-Rodriquez, R.H. (1965). A taxonomic study of the genera *Mediolàta, Zetzellia* and *Agistemus* (Acarina:Stigmaeidae). Univ. Calif. Publ. Ent. **41**:1-64.

Gonzalez-Rodriquez, R.H. (1967). *Summersiella,* a new stigmaeid mite from New Zealand (Acarina:Prostigmata). Pan-Pac. Ent. **43**(3):236-239. [STIGMAEIDAE]

Grandjean, F. (1944). Observations sur les Acariens de la famille des Stigmaeidae. Arch. Sci. phys. natur. 5, **26**:103-131.

Grandjean, F. (1948). Quelques caractères des tetranyques. Bull. Mus. nat. Hist. natur. Paris **20**(2):517-524. [TETRANYCHIDAE]

Groves, J.R. and A.M. Massee (1951). A synopsis of the world literature on the fruit tree red spider mite *Metatetranychus ulmi* (C.L. Koch, 1835) and its predators. Commonw. Inst. Ent., London: 180 pp.

Habeeb, H. (1962). A new species of *Homocaligus* (Acarina, Raphignathidae). Leaflet Acadian Biol. **27**:1-3.

Hewitt, M., G.S. Walton and M. Waterhouse (1971). Pet animal infestations and human skin lesions. Brit. J. Dermatol. **85**:215-225. [CHEYLETIELLIDAE]

Hirst, S. (1919). The genus *Demodex* Owen. Brit. Mus. (Nat. Hist.) Studies on Acari 1:1-44. [DEMODICIDAE]

Hirst, S. (1922). Mites injurious to domestic animals. Brit. Mus. (Nat. Hist.) Econ. Ser. 13:107 pp.

Huffaker, C.B., M. van de Vrie and J.A. McMurtry (1970). Ecology of tetranychid mites and their natural enemies: a review. II. Tetranychid populations and their possible control by predators: an evaluation. Hilgardia **40**(11):391-458.

Hughes, A.M. (1961). The mites of stored food. Min. Agr. Fish. and Food Tech. Bull. 9:287 pp. + vi. [CHEYLETIDAE]

Jameson, E.W. (1948). Myobiid mites (Acarina:Myobiinae) from shrews (Mammalia:Soricidae) of eastern North America. J. Parasitol. **34**(4):336-342. [MYOBIIDAE]

Jameson, E.W. (1955). A summary of the genera of Myobiidae (Acarina). J. Parasitol. **41**(4):407-416.

Jameson, E.W. and F. Dusbábek (1971). Comments on the myobiid genus *Protomyobia*. J. Med. Ent. **8**(1):33-36. [MYOBIIDAE]

Jeppson, L.R. (1963). Interrelationships of weather and acaricides with citrus mite infestations. Advances in Acarology **1**:9-13. [TETRANYCHIDAE]

Johnston, D.E. (1964). *Psorergates bos,* a new mite parasite of domestic cattle (Acari-Psorergatidae). Ohio Agr. Exp. Sta. Res. Circ. 129:7 pp.

Jorgensen, C.D. (1967). A new species of *Tuckerella* (Acarina:Tuckerellidae) from Nevada. Ent. News **78**:141-146.

Keh, B. (1973). Dermatitis in man traced to dog infested with *Cheyletiella yasguri* Smiley (Acari:Cheyletiellidae) in California. California Vector Views **20**(10):77-79.

Keh, B. (1975). Intense pruritis in man and concurrent infestation of *Cheyletiella blakei* Smiley (Acari: Cheyletiellidae) on cats in a home in California. California Vector Views **22**(1):1-4.

Keifer, H.H. (1942). Eriophyid studies XII. Bull. Calif. Dept. Agr. **31**:117-129.

Keifer, H.H. (1946). A review of North American economic eriophyid mites. J. Econ. Ent. **39**(5):563-570.

Keifer, H.H. (1952). The eriophyid mites of California (Acarina:Eriophyidae). Bull. Calif. Insect Surv. **2**(1):123 pp.

Keifer, H.H. (1976). Eriophyid studies, C-12. Bull. California Dept. Agr.: 24 pp.

Kethley, J.B. (1970). A revision of the family Syringophilidae (Prostigmata:Acarina). Contrib. Amer. Ent. Inst. **5**(6):1-76.

Kethley, J.B. (1971). Population regulation in quill mites (Acarina:Syringophilidae). Ecol. **52**(6):1113-1118.

Kethley, J.B. (1973). A new genus and species of quill mites (Acarina:Syringophilidae) from *Colinus virginianus* (Galliformes:Phasianidae). Fieldiana Zool. **65**(1):1-8.

Krantz, G.W. (1973). Observations on the morphology and behavior of the filbert rust mite, *Aculus comatus* (Prostigmata:Eriophyoidea) in Oregon. Ann. Ent. Soc. Amer. **66**(4):709-717.

Krantz, G.W. (1977). The role of *Phytocoptella avellanae* and *Cecidophyopsis vermiformis* (Eriophyoidea) in big bud of filbert. Proc. 4th Int. Congr. Acarology, Saalfelden (in press).

Lawrence, R.F. (1954). The known African species of Cheyletidae and Pseudocheylidae (Acarina:Prostigmata). Ann. Natal Mus. **13**:65-77.

Lawrence, R.F. (1959). New mite parasites of African birds (Myobiidae, Cheyletidae). Parasitol. **49**(3-4):416-438.

Lebel, R.R. and W.B. Nutting (1971). Population dynamics of a parasitic mite *Demodex caprae* (Trombidiformes:Demodicidae). Proc. 3rd Int. Congr. Acarology, Prague: 517-521.

Lindquist, E.E. (1974). Nomenclatural status and authorship of some family-group names in the Eriophyoidea (Acarina:Prostigmata). Can. Ent. **106**:209-212.

Lindquist, E.E. (1976). Transfer of the Tarsocheylidae to the Heterostigmata, and reassignment of the Tarsonemina and Heterostigmata to lower hierarchic status in the Prostigmata (Acari). Can. Ent. **108**:23-48. [POMERANTZIIDAE]

Lukoschus, F.S., F. Dusbábek and E.W. Jameson, Jr. (1972). Parasitic mites of Surinam IV. *Archemyobia philander* spec. nov. (Myobiidae:Trombidiformes) from *Philander opossum.* Acarologia **14**(2):179-189.

Lukoschus, F., A. Fain and M.M.J. Beaujean (1967). Beschreibung neuer *Psorergates* Arten (Psorergatidae: Trombidiformes). Tijdschr. Ent. **110**:133-181.

Lukoschus, F.S., R.H.G. Jongman and W.B. Nutting (1972). Parasitic mites of Surinam. XII. *Demodex melanopteri* sp. n. (Demodicidae:Trombidiformes) from the Meibomian glands of the Neotropical bat *Eptesicus melanopterus.* Acarologia **14**(1):54-58.

Lukoschus, F.S., J.M.W. Louppen and T.C. Maa (1974). *Psorergates squamipes,* n. sp. (Acari:Psorergatidae), a skin mite from *Anourosorex squamipes* (Insectivora) in Taiwan. Pac. Insects **16**:51-56.

Luxton, M. (1973). Mites of the genus *Cryptognathus* from Australia, New Zealand and Niue Island. Acarologia **15**(1):53-75. [CRYPTOGNATHIDAE]

Mani, M.S. (1964). Ecology of plant galls in E.P. Felt's Plant Galls and Gall Makers. Comstock Publ. Co., Ithaca, New York: 434 pp. [ERIOPHYIDAE]

McDaniel, B., D.K. Morihara and J.K. Lewis (1975). A new species of *Tuckerella* from South Dakota and a key with illustrations of all known described species. Acarologia **17**(1):274-283. [TUCKERELLIDAE]

McGregor, E.A. (1950). Mites of the genus *Neophyllobius.* Bull. S. Calif. Acad. Sci. **49**(2):55-70. [CAMEROBIIDAE]

McMurtry, J.A., C.B. Huffaker and M. van de Vrie (1970). Ecology of tetranychid mites and their natural enemies: a review. I. Tetranychid enemies: their biological characters and the impact of spray practices. Hilgardia **40**(11):331-390.

Meyer, M.K.P. (1969). Some stigmaeid mites from South Africa (Acari:Trombidiformes). Acarologia **11**(2): 227-271. [STIGMAEIDAE]

Meyer, M.K.P. (1974). A revision of the Tetranychidae of Africa (Acari) with a key to the genera of the world. Dept. Agr. Tech. Serv. Mem. 36:291 pp.

Meyer, M.K.P. and P.A.J. Ryke (1959). Mites of the superfamily Raphignathoidea (Acarina:Prostigmata) associated with South African plants. Ann. Mag. Nat. Hist. **2**(13):209-234.

Miskjian, H.G. (1951). Demodicidosis (*Demodex* infestation of the scalp). Arch. Derm. Syph. **63**:282-283. [DEMODICIDAE]

Morgan, C.V.G. (1960). Anatomical characters distinguishing *Bryobia arborea* M. & A. and *B. praetiosa* Koch (Acarina:Tetranychidae) from various areas of the world. Can. Ent. **92**(8):595-604.

Mumcuoglu, Y. and E. Stix (1974). Milben in der Luft. Rev. suisse Zool. **81**(3):673-677.

Newkirk, R.A. and H.H. Keifer (1971). Eriophyid studies C-5. Revision of types of *Eriophyes* and *Phytoptus.* Bull. California Dept. Agr.: 24 pp.

Norris, J.D. (1958). Observations on the control of mite infestations in stored wheat by *Cheyletus* spp. (Acarina, Cheyletidae). Ann. Appl. Biol. **46**(3):411-422.

Nutting, W.B. (1975). Pathogenesis associated with hair follicle mites (Acari:Demodicidae). Acarologia **17**(3):493-507.

Oldfield, G.N. (1970). Mite transmission of plant viruses. Ann. Rev. Ent. **15**:343-380.

Oldfield, G.N. and I.M. Newell (1973). The role of the spermatophore in the reproductive biology of protogynes of *Aculus cornutus* (Acarina:Eriophyidae). Ann. Ent. Soc. Amer. **66**(1):160-163.

Oldfield, G.N., I.M. Newell and D.K. Reed (1972). Insemination of protogynes of *Aculus cornutus* from spermatophores and description of the sperm cell. Ann. Ent. Soc. Amer. **65**(5):1080-1084. [ERIOPHYIDAE]

Olsen, S.J. and H. Roth (1947). On the mite *Cheyletiella parasitivorax,* occurring on cats, as a facultative parasite of man. J. Parasitol. **33**:444-445. [CHEYLETIDAE]

Price, D.W. (1973). Abundance and vertical distribution of microarthropods in the surface layers of a California pine forest soil. Hilgardia **42**(4):121-148.

Price, D.W. (1974). Notes on the genus *Pomerantzia* Baker, with a description of a second species from California (Acarina:Pomerantziidae). Proc. Ent. Soc. Wash. **76**(4):419-427.

Price, D.W. and G.S. Benham, Jr. (1976). Vertical distribution of pomerantziid mites (Acarina:Pomerantziidae). Proc. Ent. Soc. Wash. **78**(3):309-313.

Pritchard, A.E. and E.W. Baker (1955). A revision of the spider mite family Tetranychidae. Pac. Coast Ent. Soc. Mem. Ser. **2**:472 pp.

Pritchard, A.E. and E.W. Baker (1958). The false spider mites (Acarina:Tenuipalpidae). Univ. Calif. Publ. Ent. **14**(3):175-274.

Pulpan, J. and P.H. Verner (1965). Control of tyroglyphoid mites in stored grain by the predatory mite *Cheyletus eruditus* (Schrank). Can. J. Zool. **43**:417-432. [CHEYLETIDAE]

Rack, G. (1971). *Cheyletiella yasguri* Smiley, 1965 (Acarina, Cheyletiellidae), ein facultativ menschenpathogener Parasit des Hundes. Z. Parasitenk. **36**:321-334.

Reisen, W.K., M.L. Kennedy and N.T. Reisen (1976). Winter ecology of ectoparasites collected from hibernating *Myotis velifer* (Allen) in southwestern Oklahoma (Chiroptera:Vespertilionidae). J. Parasitol. **62**(4):628-635. [CHEYLETIDAE]

Robaux, P. (1975). Observations sur quelques Actinedida (= Prostigmates) du sol d'Amérique du Nord. V. Barbutiidae, une nouvelle famille d'Acariens (Acari:Raphignathoidea) et description d'une nouvelle espèce appartenant au genre *Barbutia.* Acarologia **17**(3):480-488.

Roberts, I.H. and W.P. Meleney (1965). Psorergatic acariasis in cattle. J. A. V. M. A. **146**(1):17-23.

Sambon, L.W. (1928). *Ophioptes parkeri.* A new species and genus of Cheyletid inhabiting the scales of reptiles. Ann. Trop. Med. & Parasit. **22**(1):137-142. [OPHIOPTIDAE]

Schevtchenko, V.G. (1970). The origin and morpho-functional analysis of quadruped mites (Acarina, Eriophyoidea). **In** Studies on the Evolutional Morphology of Invertebrates. Leningrad Univ., Publ.: 152-183. [in Russian]

Schevtchenko, V.G. (1971). The phylogenetic relationships and basic trends in the evolution of the four-legged mites (Acariformes, Tetrapodili). Proc. 13th Int. Congr. Ent., Moscow **1**:295. [in Russian]

Slykhuis, J.T. (1969). Mites as vectors of plant viruses. **In** Viruses, Vectors and Vegetation, Contrib. 613:121-141.

Slykhuis, J.T. (1972). Transmission of plant viruses by eriophyid mites. **In** Principles and Techniques in Plant Virology, Kado and Agarwal, eds. Van Nostrand—Reinhold Co.: 204-225.

Smiley, R.L. (1965). Two new species of the genus *Cheyletiella.* Proc. Ent. Soc. Wash. **67**(2):75-79. [CHEYLETIELLIDAE]

Smiley, R.L. (1970). A review of the family Cheyletiellidae (Acarina). Ann. Ent. Soc. Amer. **63**(4):1056-1078.

Smiley, R.L. and J.C. Moser (1968). New species of mites from pine (Acarina:Tarsocheylidae, Eupalopsellidae, Caligonellidae, Cryptognathidae, Raphignathidae, and Neophyllobiidae). Proc. Ent. Soc. Wash. **70**(4):307-317.

Smith, L.M. and E.M. Stafford (1948). The bud mite and the erineum mite of grapes. Hilgardia **18**(7):317-334. [ERIOPHYIDAE]

Soliman, Z.R. and M.A. Zaher (1975). *Hemitarsocheylus* a new genus from the family Tarsocheylidae with a description of a new species. Acarologia **17**(1):103-105. [POMERANTZIIDAE]

Solomon, M.E. (1946). Tyroglyphid mites in stored products. Ecological studies. Ann. Appl. Biol. **33**(1):82-97. [CHEYLETIDAE]

Solomon, M.E. (1969). Elements of predator-prey interactions of storage mites. Acarologia **11**(3):484-501. [CHEYLETIDAE]

Somsen, H.W. and W.H. Sill, Jr. (1970). The wheat curl mite, *Aceria tulipae* Keifer, in relation to epidemiology and control of wheat streak mosaic. Kansas Agr. Exp. Sta. Res. Publ. 162:24 pp. [ERIOPHYIDAE]

Stafford, E.M. and R.L. Doutt (1974). Insect grape pests of northern California. California Agr. Exp. Sta. Circ. 566:75 pp. [ERIOPHYIDAE]

Sternlicht, M. and D.A. Griffiths (1974). The emission and form of spermatophores and the fine structure of adult *Eriophyes sheldoni* Ewing (Acarina, Eriophyoidea). Bull. Ent. Res. **63**:561-565. [ERIOPHYIDAE]

Southcott, R.V. (1956). Notes on the acarine genus *Ophioptes,* with a description of a new Australian species. Trans. Roy. Soc. S. Australia **79**:142-147. [OPHIOPTIDAE]

Southcott, R.V. (1957). Description of a new Australian raphignathoid mite, with remarks on the classification of the Trombidiformes (Acarina). Proc. Linn. Soc. N.S.W. **81**(3):306-312. [CAMEROBIIDAE]

Summers, F.M. (1960a). *Eupalopsis* and eupalopsellid mites (Acarina:Stigmaeidae, Eupalopsellidae). Florida Ent. **43**(3):119-138.

Summers, F.M. (1960b). Several stigmaeid mites formerly included in *Mediolata* redescribed in *Zetzellia* Ouds. and *Agistemus,* new genus (Acarina). Proc. Ent. Soc. Wash. **62**(4):233-247.

Summers, F.M. (1964). Three uncommon genera of the mite family Stigmaeidae (Acarina). Proc. Ent. Soc. Wash. **66**(3):184-192. [BARBUTIIDAE]

Summers, F.M. (1966a). Key to families of the Raphignathoidea (Acarina). Acarologia **8**(2):226-229.

Summers, F.M. (1966b). Genera of the mite family Stigmaeidae Oudemans (Acarina). Acarologia **8**(2):230-250.

Summers, F.M. and W.M. Chaudri (1965). New species of the genus *Cryptognathus* Kramer (Acarina:Cryptognathidae). Hilgardia **36**(7):313-326.

Summers, F.M. and D.W. Price (1961). New and redescribed species of *Ledermuelleria* from North America (Acarina:Stigmaeidae). Hilgardia **31**(10):369-382.

Summers, F.M. and D.W. Price (1970). Review of the mite family Cheyletidae. Univ. California Publ. Ent. **61**:153 pp. [CHEYLETIDAE, CHEYLETIELLIDAE]

Summers, F.M. and E.I. Schlinger (1955). Mites of the family Caligonellidae (Acarina). Hilgardia **23**(12):539-561.

Summers, F.M. and R.L. Witt (1972). Nesting behavior of *Cheyletus eruditus* (Acarina:Cheyletidae). Pan-Pac. Ent. **48**(4):261-269.

Thewke, S.E. and W.R. Enns (1975). A new species of *Pavlovskicheyla* (Acarina:Cheyletidae) from the elytra of *Platydema ruficorne* (Coleoptera:Tenebrionidae) from Missouri. Acarologia **17**(4):671-682.

Tuttle, D.M. and E.W. Baker (1968). Spider mites of the southwestern United States and a revision of the family Tetranychidae. Univ. Arizona Press, Tucson: 143 pp. + vii.

Vercammen-Grandjean, P.H. and H. Rak (1968). *Cheyletiella yasguri* Smiley, 1965, un parasite de Canides aux États-Unis et hyperparasite d'Hippoboscide en Iran (Acarina:Cheyletidae). Ann. Parasit. Hum. et Comp. **43**(3):405-412. [CHEYLETIELLIDAE]

Volgin, V.I. (1960). On the taxonomy of predatory mites of the family Cheyletidae. II. Genus *Cheyletiella* Can. Akad. Nauk Zool. Inst. Parsit. Sborn. **19**:237-248. [CHEYLETIELLIDAE]

Volgin, V.I. (1969). Acarina of the family Cheyletidae, world fauna. Akad. Nauk SSSR, Zool. Inst. 101: 432 pp. [CHEYLETIDAE, CHEYLETIELLIDAE]

Vrie, M. van de, J.A. McMurtry and C.B. Huffaker (1972). Ecology of tetranychid mites and their natural enemies: a review. III. Biology, ecology, and pest status, and host-plant relations of tetranychids. Hilgardia **41**(13):343-432.

Wainstein, B.A. (1960). Tetranychoid mites of Kazakhstan (with revision of the family) Kazakh. Akad. Sel'sk. Nauk. Nauch.-Issled. Inst. Zash. Rast. Trudy **5**:1-276.

Weidhaas, J.A. and R.M. Reeves (1963). The occurrence and importance of tetranychid and eriophyid mites on woody plants in New York. Advances in Acarology **1**:25-29.

Whitaker, J.O., Jr. and N. Wilson (1974). Host and distribution lists of mites (Acari), parasitic and phoretic, in the hair of wild mammals of North America, north of Mexico. Amer. Midl. Natur. **91**(1):1-67.

Wood, T.G. (1969). The Homocaligidae, a new family of mites (Acari: Raphignathoidea), including a description of a new species from Malaya and the British Solomon Islands. Acarologia **11**(4):711-729.

Wood, T.G. (1973). Revision of Stigmaeidae (Acari:Prostigmata) in the Berlese collection. Acarologia **15**(1): 76-95.

Woodroffe, G.E. (1956). Some insects and mites associated with bat-roosts, with a discussion of the feeding habits of the cheyletids (Acarina). Ent. Month. Mag. 92:138-141.

Yunker, C.E. (1973). Mites. **In** Parasites of Laboratory Animals. R.J. Flynn, ed. Iowa State Univ. Press, Ames: 425-492.

Zaher, M.A. and A.H. Rasmy (1969). A new species of the genus *Tuckerella* from UAR (Acarina:Tuckerellidae). Acarologia **11**(4):730-732.

Zumpt, F. (1961). The Arthropod Parasites of Vertebrates in Africa South of the Sahara (Ethiopian Region). Vol. 1. (Chelicerata). Publ. S. Afr. Inst. Med. Res. **9**(1):1-457.

Subcohort Anystae

The Anystae includes five predaceous and parasitic families grouped in two superfamilies. Members of the subcohort have short, often sickle-like movable cheliceral digits arising terminally from broad unfused bases. Fixed cheliceral extensions are absent or virtually so. A well defined palpal thumb-claw process is present throughout the subcohort but in some groups it may be reduced. Prodorsal ocelli and postcheliceral stigmata and peritremes usually are distinct, and a naso may be present.

Many of the predatory Anystae are xerophiles, ranging over arid and often exposed substrates in their search for prey. Parasitic species of the anystoid family PTERYGOSOMATIDAE often utilize xerophilous lizards or arthropods as hosts. Newell (1971) noted that development of pterygosomatids is comparable to that seen in the highly evolved subcohort Parasitengonae (page 276), where the protonymph and tritonymph are represented by quiescent calyptostases and the larva is parasitic. Based on these characteristics, along with the shared traits of an extensive host range and possession of certain trombid chaetotactic attributes, Newell postulated that the PTERYGOSOMATIDAE may be ancestral to the Parasitengonae. The dendrogram on page 228 suggests instead a common ancestral stock for the two taxa. Pterygosomatid-like ontogenetic development may also occur in the family ANYSTIDAE where calyptostasis apparently occurs during nymphal ontogeny (Baker and Wharton 1952). Anystid larvae, however, are predaceous rather than parasitic.

Superfamily Caeculoidea
(Plate 91, p. 346)

DIAGNOSIS: Heavily sclerotized forms, with eight dorsal shields arranged in a characteristic pattern. Cheliceral bases unfused, apparently capable of limited lateral motion; palptibia always with a single well defined terminal claw and ancillary subterminal spines; with two pairs of lateral eyes and an anteromedian prodorsal eye associated with a naso. Legs I with strong, internal spinose setae, tibia and tarsus I each with a solenidion and a vestigial actinopilous seta in an integumental sink. Claws present on all legs; with three pairs of genital acetabula.

The Caeculoidea includes only one family, the CAECULIDAE or rake-legged mites. Caeculids are large (1000-3000 μ) heavily armored, wrinkled, slow-moving species which typically inhabit dry rocky niches in desert and mountain habitats. Here they lie in wait for other more agile arthropods, using the strong inner spines of legs I to capture and/or hold their prey. Coineau (1974) has reviewed in depth the biology, morphology and behavior of caeculids.

Laboratory observations by Crossley and Merchant on a species of *Caeculus* suggested that caeculids may feed on fungus (Crossley and Merchant 1971). The mites never were observed to feed on collembolans or their eggs, but reproduced in cultures where only fungal hyphae were available as a food source. Feeding in fungal cultures was indicated by recovery of ^{134}Cs from mites confined on fungi treated with this radioactive material.

Superfamily Anystoidea

DIAGNOSIS: Soft-bodied, sometimes with weak prodorsal shield or shields, dorsal setal bases may also be sclerotized. Cheliceral bases freely articulated, capable of scissors-like movement over the rostrum; palptibia with one, two or three claws, palpal tarsus strongly developed or reduced; one or two pairs of eyes may be present, often with an anteromedian naso. Legs may be hypertrichous or normally setate, tibia and tarsus I with or without reduced chemosensory setae in integumentary pits, tarsi occasionally subdivided distally into several distinct segments. Claws present or absent on all tarsi; with or without empodia. With two or three pairs of genital acetabula, or acetabula absent.

Both predaceous and parasitic species are found among the four families comprising the Anystoidea. The ANYSTIDAE (Plate 92, p. 347) includes moderately large (500-1500 μ) bright orange or red, long-legged rapidly moving mites which often are recovered from plants on which they search for prey. *Anystis baccarum* (L.) and other species of the genus are aerial predators of phytophagous insects and mites in Europe, Australia, Africa and North America, but their efficacy as control agents on economic crops may be offset by their slow rate of increase (Jeppson et al. 1975). *A. agilis* (Banks), the common whirlagig mite, has been identified as a predator of the citrus thrips in California (Mostafa et al. 1975). The more elongate species of the genera *Anandia, Bechsteinia* and *Chausseria* occur as predators in soil and on low-lying plants (Meyer and Ryke, 1960). A species of *Tarsotomus* is a regular inhabitant of farm storage bins in Oregon, USA.

Lee (1961) and Southcott (1976) noted that *Anystis* sp. (prob. *baccarum*) pierces human skin and has no difficulty in feeding to repletion on subdermal fluids. Southcott described the effect of an anystid bite as consisting of a stinging sensation followed in some cases by swelling and localized hemorrhaging.

Like the ANYSTIDAE, the TENERIFFIIDAE (Plate 93, p. 348) are predators which are capable of rapid movement. Teneriffiids have been collected in habitats ranging from intertidal sand banks to the undersides of rocks in the Tyrol at altitudes exceeding 1000 meters (Irk 1939, Eller and Strandtmann 1963). A species of *Parateneriffia* has been collected in sage and saltbrush litter in the high desert country of central Oregon in company with other mites which probably serve as its prey. *P. uta* (Tibbetts), *P. luxoriensis* (Hirst) and other members of the genus seem to share a predilection for arid desert habitats (McDaniel et al. 1976). Species of *Teneriffia,* however, are encountered only in seashore situations. *T. quadripapillata* Thor, the type species for the family, was collected on a beach at Teneriffe. *T. marina* (Hirst) is an intertidal predator in Malaysia, and *T. mexicana* McDaniel, Morihara and Lewis occurs in a variety of intertidal substrates in Baja California (McDaniel et al., op. cit.).

The PSEUDOCHEYLIDAE (Plate 92, p. 347) are infrequently collected predators which usually occur in leaf litter. *Neocheylus natalensis* Trägårdh was collected from moss in South Africa (Trägårdh 1906, Meyer and Ryke 1960). Many members of the family PTERYGOSOMATIDAE (Plate 93, p. 348) are ectoparasites of lizards, attaching under scales or between the toes of their hosts (Hirst 1925). The broad short species of the genus *Pterygosoma* are common on agamid lizards in North and East Africa, and in India (Jack 1961). Gekkonid and iguanid lizards are parasitized by species of *Geckobia* and *Geckobiella* (Lawrence 1953, Davidson 1958). *Geckobiella texana* (Banks) has been recovered from iguanids of the genus *Sceloporus* in Texas and in California, while *Geckobia* species are common on geckos of the genus *Hemidactylus* in southeast Asia. Specimens of *Hirstiella bakeri* Cunliffe have been found on *Ctenosaura hemilopha* in the San Diego Zoo, California.

Species of the pterygosomatid genus *Pimeliaphilus* are confined to arthropods. Two species attach to the conjunctival tissues of scorpions where they feed on the blood of their hosts (Beer 1960, Cunliffe 1949). *P. podapolipophagus* is an ectoparasite of cockroaches (Cunliffe 1952), often proving troublesome in laboratory rearing operations. Heavy infestations of this mite may cause roach mortality within a few hours after initial exposure. Several species of triatomine kissing bugs (family Reduviidae) also are attacked by species of

Pimeliaphilus (Newell and Ryckman 1966). The bugs may be killed by the feeding of large mite populations.

Prasad (1975) described a pterygosomatid, *Bharatoliaphilus punjabensis,* from a dove *(Streptopelia decaocta)* in India. The description was based on a single specimen, and a natural association with the bird carrier was not definitely established.

Useful References

André, M. (1935). Notes sur le genre *Caeculus* Dufour (Acariens) avec descriptions d'espèces nouvelles africaines. Bull. Soc. Hist. Nat. Afr. Nord **26**:79-127. [CAECULIDAE]

Beer, R.E. (1960). A new species of *Pimeliaphilus* (Acarina:Pterygosomidae) parasitic on scorpions, with discussion of its postembryonic development. J. Parasit. **46**(4):433-440.

Coineau, Y. (1963). Contribution a l'étude des Caeculidae. Première série: Développement postlarvaire de *Allocaeculus catalanus* Franz 1954.—Première partie: la chetotaxie du corps. Acarologia **5**(2):189-212.

Coineau, Y. (1964). Contribution a l'étude des Caeculidae. Première série: Développement postlarvaire de *Allocaeculus catalanus* Franz 1954. Deuxième partie: la chétotaxie des pattes. Acarologia **6**(1):47-72.

Coineau, Y. (1966). Contribution a l'étude des Caeculidae. Deuxième série: Développement postlarvaire de *Microcaeculus hispanicus* Franz 1952. Acarologia **8**(2):23-44.

Coineau, Y. (1970). A propos de l'oeil anterieur du naso des Caeculidae. Acarologia **12**(1):109-118.

Coineau, Y. (1974). Éléments pour une monographie morphologique, écologique et biologique des Caeculidae (Acariens). Mém. Mus. nat. Hist. natur. (N.S.)A, 81:299 pp. + vi.

Crossley, D.A. and V. Merchant (1971). Feeding by caeculid mites on fungus demonstrated with radioactive tracers. Ann. Ent. Soc. Amer. **64**(4):760-762.

Cunliffe, F. (1949). *Pimeliaphilus isometri,* a new scorpion parasite from Manila, P.I. Proc. Ent. Soc. Wash. **51**:123-124. [PTERYGOSOMATIDAE]

Cunliffe, F. (1952). Biology of the cockroach parasite, *Pimeliaphilus podapolipophagus* Trägårdh, with a discussion of the genera *Pimeliaphilus* and *Hirstiella.* Proc. Ent. Soc. Wash. **54**:153-169. [PTERYGOSOMATIDAE]

Davidson, J.A. (1958). A new species of lizard mite and a generic key to the family Pterygosomidae (Acarina, Anystoidea). Proc. Ent. Soc. Wash. **60**(2):75-78.

Eller, R. and R.W. Strandtmann (1963). Notes on Teneriffiidae (Acarina:Prostigmata). Southwest. Nat. **8**(1):23-31.

Franz, H. (1952). Revision der Caeculidae Berlese 1894 (Acari). Bonner Zool. Beitr. **2**(1-2):91-124.

Grandjean, F. (1943). Le développement postlarvaire d'"*Anystis*" (Acarien). Mém, Mus. nat. Hist. natur. (N.S.) **18**(2):33-77. [ANYSTIDAE]

Grandjean, F. (1944). Observations sur les Acariens du genre *Caeculus.* Arch. Sci. Phys. Nat., 5, **26**:33-46.

Hirst, S. (1925). On the parasitic mites of the suborder Prostigmata (Trombidioidea) found on lizards. J. Linn. Soc. London Zool. **36**:173-200. [PTERYGOSOMATIDAE]

Irk, V. (1939). Drei neue Milbenarten aus dem Tiroler Hochgebirge. Zool. Anz. **128**(7-8):216-223. [TENERIFFIIDAE]

Jack, K.M. (1961). New species of Near Eastern agamid scale-mites (Acarina, Pterygosomidae) with notes on the developmental stages of *Geckobia hemidactyli* Law., 1936. Parasitol. **51**:241-256.

Lawrence, R.F. (1935). The prostigmatic mites of South African lizards. Parasitol. **27**(1):1-45. [PTERYGOSOMATIDAE]

Lawrence, R.F. (1936). The prostigmatic mites of South African lizards. Parasitol. **28**(1):1-39. [PTERYGOSOMATIDAE]

Lawrence, R.F. (1953). Two new scale-mite parasites of lizards. Proc. U.S. Nat. Mus. **103**(3312):9-18. [PTERYGOSOMATIDAE]

McDaniel, B., D. Morihara and J.K. Lewis (1976). The family Teneriffiidae Thor, with a new species from Mexico. Ann. Ent. Soc. Amer. **69**(3):527-537.

Meyer, M.K.P. and P.A.J. Ryke (1960). Acarina of the families Anystidae, Pseudocheylidae and Cheyletidae (Prostigmata) found associated with plants in South Africa. J. Ent. Soc. S. Afr. **23**(1):177-193.

Mostafa, A.R., P. de Bach and T.W. Fisher (1975). Anystid mite: citrus thrips predator. California Agr., March: 1 p.

Mulaik, S. (1945). New mites in the family Caeculidae. Bull. Univ. Utah **35**(17):1-23.

Newell, I.M. (1971). The protonymph of *Pimeliaphilus* (Pterygosomatidae) and its significance relative to the calyptostases in the Parasitengona. Proc. 3rd Int. Congr. Acarology, Prague: 789-795.

Newell, I. and R.E. Ryckman (1966). Species of *Pimeliaphilus* (Acari:Pterygosomidae) attacking insects, with particular reference to the species parasitizing Triatominae (Hemiptera:Reduviidae). Hilgardia **37**(12): 403-436.

Oudemans, A.C. (1936). Neues über Anystidae (Acari). Arch. für Naturg. N.F. **5**:364-446.

Prasad, V. (1975). A new genus and species of pterygosomatid mite (Acarina:Pterygosomatidae) from India. Int. J. Acarology **2**(1):14-17.

Schuster, R. and I.J. Schuster (1966). Über das Fortpflanzungsverhalten von Anystiden-Männchen (Acari, Trombidiformes). Naturw. **6**:162.

Strandtmann, R.W. (1965). Additional notes on Teneriffiidae (Acarina:Prostigmata) with two previously unpublished plates by A.C. Oudemans. J. Kansas Ent. Soc. **38**(3):258-261.

Thor, S. (1911). Eine neue Acarinenfamilie (Teneriffiidae) und zwei neue Gattungen, die eine von Teneriffa, die andere aus Paraguay. Zool. Anz. **38**:171-179.

Subcohort Parasitengonae

The ten superfamilies comprising the Parasitengonae are highly evolved terrestrial and aquatic forms which have unique postembryonic development. The larval instar is parasitic and, with the exception of the homeomorphic larval Calyptostomatoidea, bears little or no resemblance to the predatory nymphal and adult forms (Plates 97-98, pp. 352-353). As in certain anystoids, the protonymphal and tritonymphal stases are calyptostatic and are passed within the integument of the preceding instar (Johnston and Wacker 1967). All stages have either sickle-like movable chelae which arise from free or partially fused cheliceral bases, or long retractile cheliceral stylets. An opposed fixed digit is absent. A palpal thumb-claw often is present, although the palp is modified as a simple or chelate raptorial structure in the post-larval stadia of most aquatic families. One or two pairs of prodorsal eyes generally are present, and sensillary trichobothria may also be found. Stigmata, when present, are located between the cheliceral bases. They open internally into paired air sacs which may or may not be contiguous anteriorly with chambered peritremes (Plate 97-2, p. 352). With few exceptions, larval Parasitengonae have paired urstigmata on or between coxae I or II (Plate 99-3, p. 354).

While the Parasitengonae is a fairly homogenous grouping in regard to basic ontogeny, it is convenient to consider them as two entities based on life style and on certain key morphological characters. Accordingly, the subcohort has been divided into two phalanxes: 1) the Trombidia, and 2) the Hydrachnidia.

1. Phalanx Trombidia

The three superfamilies comprising the Trombidia are essentially terrestrial forms although some families tend to favor semiaquatic or aquatic habitats. All stages have a palpal thumb-claw process, and pronounced hypertrichy of postlarval instars usually occurs throughout the phalanx. Prodorsal trichobothria (1-2 pairs) are inserted medially, generally in a distinct prodorsal shield in larval forms and in a reduced, often elongate-ossiform prodorsal sclerite or *crista metopica* in nymphs and adults (Plate 94-6, p. 349).

Superfamily Calyptostomatoidea
(Plate 94, p. 349)

DIAGNOSIS: *Broad unsclerotized forms with moderate hypertrichy, often with distinctive integumental patterns, dorsal setae lacking on sites of muscle insertions; with a pair of hair-like median prodorsal sensilla in pronounced adjacent raised insertions, crista metopica absent; with a pair of large ocular pits antero-laterally. Gnathosoma retractile; chelicerae greatly elongated, flanked by reduced palpi with minute tibial claws. Adults with two pairs of genital acetabula, empodia absent. Larva homeomorphic, with urstigmata and a well defined anal opening; coxae I-II contiguous; parasites of dipteran insects.*

A single monogeneric family, the CALYPTOSTOMATIDAE, is included in the superfamily. Adult and nymphal calyptostomatids are slow-moving, large, generally red or orange-red mites which inhabit wet stream banks or moss in Europe (Turk 1945), Japan (Shiba 1969), Australia (Vercammen-Grandjean 1975a), Africa, and North America. They often secrete themselves under stones or in debris so that they are difficult to see, despite their size. Calyptostomatid larvae (Plate 94-3, p. 349) parasitize adult Tipulidae and Limoniidae (Diptera) (Vistorin-Theis 1975), and are more readily collected than the cryptic postlarval stages. The larva of *Calyptostoma velutinus* (Müller) is encountered on limoniids in Europe (Rack 1976), and is also reported to be a tipulid parasite in Ehime Prefecture, Japan (Shiba 1969).

Superfamily Erythraeoidea
(Plates 94-95, pp. 349-350)

DIAGNOSIS: *Ovoid or elongate forms with moderate to extreme hypertrichy, dorsal setae sometimes lacking on muscle insertion sites; with two pairs of prodorsal sensilla in all stages, usually on a crista metopica in postlarval instars; generally with a pair of lateral eyes. Gnathosoma normally developed or greatly reduced, capable in the latter instance of complete withdrawl into the idiosoma, chelicerae of nymphs and adults greatly elongated, retractile; genital acetabula and empodia absent. Larvae heteromorphic, without urstigmata or an anal opening; coxae I-II distinctly separated; parasites of arthropods.*

Three families are presently recognized in the Erythraeoidea and one of these, the PROTERYTHRAEIDAE, includes only fossil forms (Vercammen-Grandjean 1973a).

The ERYTHRAEIDAE (Plate 94, p. 349) is a large cosmopolitan family of primarily red or reddish-brown species which are predatory in postlarval instars on both ground and aerial arthropods in many habitats. Vercammen-Grandjean (1973a) recognizes 33 erythraeid genera grouped in five subfamilies. Larvae of many of these groups are known to be ectoparasites of insects or arachnids while others are found in a free-living state. However, not all erythraeid larvae fit easily into one or the other category; e.g. a species of the leptine genus *Leptus* which apparently was successfully reared on lizards in the southwest Pacific (Baker and Wharton 1952). Larvae of the balaustiine genus *Balaustium* have been observed to feed on pollen (Grandjean 1946), a habit which may be shared by larval forms of the genus *Microsmaris.* Larvae of *B. putmani* Smiley feed on *Panonychus ulmi* (Koch), among other hosts, and must be considered predatory on these miniscule forms.

Larvae of *Erythraeus* (Lawrence 1940), *Caeculisoma* (Southcott 1961b) and *Charletonia* (Southcott 1965) are common parasites of locusts. *Caeculisoma huxleyi* Southcott and three species of *Charletonia* are recorded from moths (Treat 1975), while *Rainbowia imperator* Hirst (Hirst 1928) and certain *Erythroides* species parasitize Homoptera in Australia (Southcott 1946). Aphids and other heteropterans serve as hosts for *Bochartia kuyperi* (Oudemans) and for species of *Balaustium* (Womersley 1934, Evans et al. 1961). *Bochartia mentonensis* André utilizes the ant *Phagiolepis pygmaea* as its host (André 1929a). *Myrmicothrombium brevicristatum* Womersley has also been found with ants, but larval host relationships are obscure (Southcott 1957).

Callidosoma larvae often parasitize Lepidoptera. Treat (1975) mentions seven species of *Callidosoma* taken from several families of moths. Another callidosomatine erythraeid, *Momorangia vallata* Southcott, was taken 20 miles off the west coast of New Zealand from a far-ranging specimen of the noctuid *Agrotis ypsilon.* Larvae of the widespread genus *Leptus* also parasitize lepidopterans (Treat, op. cit.), but they have been found on many other hosts including spiders (Lawrence 1940, Shiba 1969), opilionids (Evans et al. 1961), scorpions (André 1953), and flies.

Adult and nymphal erythraeids often occur on plants, in humus, or on open ground where they prey on small arthropods. Species of *Balaustium* are considered of special significance as predators of phytophagous arthropod pests. *B. aonidaphagus* (Ebeling) was found to prey on citrus red scale in California. Postlarval instars of *Balaustium* species have been collected from balsam fir infested with aphids, and may have been feeding on these insects. All active stages of *B. putmani* Smiley feed on European red mite, San José scale, *Aphis pomi,* and newly hatched leafhopper nymphs in California orchards (Putman 1970). However, the importance of *Balaustium* species as aerial predators is minimized by their low rate of increase (Jeppson et al. 1975).

Rack (1973) reports *Balaustium murorum* (Hermann) as a common invader of buildings in Europe and North America, and some of these invasions may involve attacks on man. Newell (1963) also observed that the normally predaceous adults of *Balaustium* will bite man, causing a sharp stinging sensation followed by itching and development of lesions. The same species also feed on the leaves and pollen of plants, all of which indicates that *Balaustium* spp. are capable of phytophagy and haematophagy in addition to insect parasitism and predation.

Larvae of the family SMARIDIDAE (Plate 95, p. 350) are poorly known, and relatively few observations have been made on host relationships. Many have been found in humus, moss, or on the bark of trees, suggesting that larval parasitism is not obligate throughout the family. Womersley and Southcott (1941) recorded a species of psocid as host for the larva of *Smaris prominens* (Banks), but the larvae of another smaridid, *Sphaerotarus leptophilus* Womersley, refused psocids and other insects offered to them in the laboratory (Southcott 1960). Adult and nymphal smaridids are predators with essentially the same habits as postlarval erythraeids.

Superfamily Trombidioidea
(Plates 95-99, pp. 350-354)

DIAGNOSIS: Ovoid or elongate hypertrichous forms; with one or two pairs of prodorsal sensilla in all stages, generally on a crista metopica in postlarval instars; with one or two pairs of lateral eyes which are sometimes weak or absent. Gnathosoma normally developed, with broad unfused cheliceral bases giving rise to blade-like movable digits which may be variously toothed; cheliceral and gnathosomal elements not retractile; with or without genital acetabula and empodia. Larvae heteromorphic, with urstigmata and an anal opening; coxae I-II usually contiguous; parasites of invertebrates and vertebrates.

Vercammen-Grandjean (1973a) listed eight families in the Trombidioidea. An additional family, the CHYZERIDAE, has subsequently been recognized (V.-G., personal communication). Like the Erythraeoidea, trombidioids are predaceous in adult and nymphal stadia, and generally parasitic as larvae. The TROMBELLIDAE (Plate 96, p. 351) are a distinctive group of primarily Old World forms which often have a strongly granular or tuberculate integument. Adults of the colorful *Trombella favosa* (Berlese) have been collected in several localities in central Africa. *T. otiorum* Berlese was recovered from moss and other substrates in Europe (Feider 1955). The tanaupodine trombellid genus *Rhinothrombium* is commonly encountered in forest litter in the northwestern United States. *R. nemoricola* Berlese is a well known European counterpart. Other trombellid species have been described from southeast Asia and Australia.

Trombellid larvae are described only for the genera *Durenia* and *Audyana,* with the remaining 19 genera known only as adults (Mullen 1974). Larvae of three species of *Durenia* have been recovered from mosquitoes, one of which is a recent new record for the United States (Mullen 1975). *D. singaporensis* Vercammen-Grandjean and Audy parasitizes *Aedes curtipes,* while *A. quasiunivittatus* serves as host for *D. bukavuensis* Grandjean in Congo-Kinshasa (Vercammen-Grandjean 1955, V-G. and Audy 1959).

Larvae of the family JOHNSTONIANIDAE (Plate 95, p. 350) are known to be parasitic only on certain Diptera (Tipulidae, Limoniidae, and Ceratopogonidae) and on aquatic Coleoptera (Newell 1957, Vercammen-Grandjean and Cochrane 1974, Rack 1976). An unconfirmed report of a johnstonianid larva on a mosquito, *Aedes triseriatus,* is noted by Mullen (1974). Larvae of *Centrotrombidium dichotomicoxala* V.-G. and Cochrane were found parasitizing the midge *Culicoides crepuscularis* in New York State. Three species of *Culicoides* are attacked by *Centrotrombidium culicoides* (V.-G.) in Scotland (Vercammen-Grandjean 1957). Adult johnstonianids have been collected from many habitats, but mostly in humid or semiaquatic situations (Feider 1955, Newell op. cit.). One species, *Johnstoniana helvetica* Cooreman, was found in a damp grotto in the Jura of Switzerland (Cooreman 1959), while an unidentified species of *Diplothrombium* was collected from under a waterfall in the coastal mountains of Oregon. Based on their morphology and habits, Newell has postulated that johnstonianid stock may have provided an evolutionary stepping-stone between terrestrial trombidioids and the water mites of the Phalanx Hydrachnidia (page 286). Unlike the semiaquatic johnstonianids, members of the hygrophilous family STYGOTHROMBIIDAE have adapted to a truly aquatic existence (André 1949b and c). Larval relationships in the family are unclear at present.

The NEOTROMBIDIIDAE (Plate 96, p. 351) is a small family comprising only four genera. Adult neotrombidiids are found in forest litter and under bark of trees, while larvae are parasitic on Coleoptera and Diptera (Singer 1971, Lindquist and Vercammen-Grandjean 1971). The monotypic genus *Monunguis* is described only from the larva, a hyperparasite of streblid flies parasitic on bats in Mexico.[1] Larvae of *Neotrombidium* parasitize a variety of coleopterans including members of the families Cerambycidae, Tenebrionidae, Elateridae and Cleridae. *N. beeri* Singer, *N. tenuipes* (Womersley) and *N. bengalense* Lindquist and V.-G. are subelytral forms.

The monogeneric family PODOTHROMBIIDAE (Plate 97, p. 352) comprises about a dozen described species which occur in soil or under stones, primarily in the Palaearctic realm (Feider 1955). An undescribed adult *Podothrombium* was collected in *Alnus* litter in western Oregon. Larval associations in the genus are unclear.

The TROMBIDIIDAE (Plates 97-98, pp. 352-353) is considered by Vercammen-Grandjean (1973a) to include six subfamilies and over 50 genera of mites often characterized by a bright red or orange color and a heavy, satiny pelage of setae which completely obscures the idiosoma. Trombidiid larvae are known to be parasitic on Orthoptera, Lepidoptera, Coleoptera, Diptera, Heteroptera, and on a variety of arachnids. Inasmuch as the larval instar has been identified in only 19 of the 50+ recognized trombidiid genera (Mullen 1974), much still remains to be learned about host relationships in the family.

Larvae of *Trombidium* species parasitize various species of Lepidoptera, sometimes in such numbers as to impair the flight of the host (Robaux 1971). *Isothrombium oparbellae* André, another trombidiine trombidiid, is parasitic on a solfugid, *Oparbella fagei* (André 1949a). Larvae of the trombidiine *Parathrombium teres* André parasitize capsid bugs (André 1929b), while *P. egregium* Bray. is parasitic on a spider wasp of the genus *Pompilus* in France. An unusual record of apparent parasitism of a mite by a trombidiine larva *(Hoplothrombium quinquescutatum* Ewing) is all the more remarkable in that the parasitized mite, an oribatid, was found in the stomach of a toad (Ewing 1925, Thor and Willmann 1947).

[1] Adults of *Monunguis* have recently been found (Lindquist, personal communication).

Larvae of the trombidiine genera *Dolichothrombium* and *Angelothrombium* parasitize locusts, and adults and nymphs are predaceous on subterranean termites (Newell and Tevis 1960). The adults of these two genera are referred to as giant red velvet mites because of their great size (1,000-13,000 μ), their color, and their extreme idiosomal hypertrichy. Newell and Tevis tell of an unusual incident in which a species of *Dolichothrombium* was observed from the air near Tucson, Arizona, following a heavy rainstorm. A 5-acre eruption of mites from the interstices of the sandy soil was seen as a bright red bloom on the desert floor. Another species of *Dolichothrombium* sent to Oregon State University was collected in India in great numbers under virtually identical circumstances of weather and terrain.

Larvae of *Allothrombium aphidis* (De Geer) and *A. fulginosum* (Hermann) are aphid parasites in northwestern United States and Europe. Another allothrombiine, *A. neapolitanum* (Oudemans), parasitizes phalangids. *Feiderium (Parafeiderium) culicoidium* Vercammen-Grandjean and Cochrane and *Attractothrombium dictyostracum* V.-G. and C., representatives of two of the four described feideriine trombidiid genera, parasitize midges in North America (Vercammen-Grandjean and Cochrane 1974). Larvae of the eutrombidiine species *Eutrombidium rostratus* (Scopoli) are parasites of Orthoptera throughout most of the world. The life cycle of *E. rostratus* has been described by Severin (1944).

Few microtrombidiine trombidiid larvae have been described or linked with particular arthropod hosts. A microtrombidiine may parasitize the mosquito *Mansonia titillans* in Panama (Michener 1946), and what appeared to be a *Microtrombidium* species was recovered from aphids in the northwestern United States.

Adults and nymphs of the TROMBICULIDAE (Plates 98-99, pp. 353-354) prey on small arthropods and are most often encountered in soil and ground litter. One species is a commensal of termites in East Africa (Vercammen-Grandjean 1965a). Relatively few trombiculid species are easily identified from postlarval forms (Thor and Willmann 1947, Feider 1955) but the larvae, or *chiggers,* have been studied both extensively and intensively by many acarologists. With few exceptions, trombiculid larvae are parasitic on vertebrates and may cause injury to their hosts through their bite or through transmission of disease organisms. Their importance as parasites of man helps account for the enormous bibliography dealing with chiggers (see lists in Fuller 1952, André 1965, Audy 1968, and Traub and Wisseman 1974).

Every major group of terrestrial vertebrates is attacked by trombiculid larvae, and the results of such attacks often include the development of a distinctive host symptomatology. The formation of a feeding tube or *stylostome* (Fig. 36) at the site of chigger attachment is characteristic of all chiggers (André 1927, Cross 1964), entailing injection of salivary secretions and consequent lysis of host tissues. Dermatitis, or *trombidiosis,* often occurs in man as the result of attacks by chiggers of the genera *Eutrombicula, Neotrombicula* and *Schöngastia. N. autumnalis* (Shaw), the harvest mite of Europe, is particularly annoying to man but also causes dermatitic symptoms in domestic animals, rodents and birds. Species of *Eutrombicula* are common causal agents of human trombidiosis in the Western Hemisphere, with *E. alfreddugesi* (Oudemans) being particularly troublesome (Williams 1946). Members of the genus *Schöngastia* are pests of man in the Orient

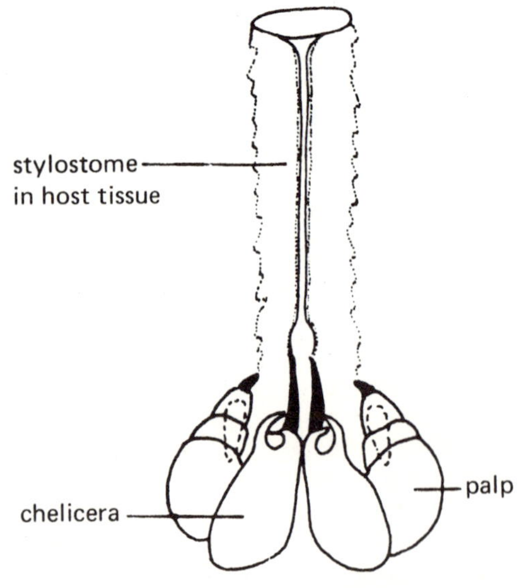

Fig. 36. Stylostome of a trombiculid larva.

and in Australia (Womersley and Heaslip 1943, Womersley 1952). Traub and Wisseman (1974) make the interesting observation that the chigger species involved in human trombidiosis normally attack birds or reptiles rather than mammals. The implication is that the intense itching reaction experienced by man, the accidental host, reflects lack of host adaptation. By the same token, resultant scratching of the bite is inimical to mite survival. The itching reaction in man is relatively mild following attack by *Leptotrombidium* species, which normally infest other mammals (rodents).

Members of *Leptotrombidium deliense* species group serve as the major vectors of scrub typhus, or tsutsugamushi fever, in the Asiatic-Pacific region (Audy 1968). Scrub typhus is an infectious reckettsial disease which proved to be a major problem for both Allied and Japanese military forces during World War II. Traub and Wisseman (1974) have provided a comprehensive review of scrub typhus ecology which should be consulted for additional information on this subject.

While most chiggers feed ectodermally on exposed sites, many species have adapted to the intranasal niche. Members of the genera *Doloisia* and *Microtrombicula* live as intranasal parasites of bats (Yunker and Brennan 1962, Vercammen-Grandjean 1965a) and a variety of rodents (Nadchatram 1970a). *Schoutedenichia* species, and members of the genera *Walchia* and *Schöngastiella* (Gahrliepiinae), are nasal parasites of small mammals (Vercammen-Grandjean 1975b). The nasal fossae, tracheae and lungs of marine iguanids and sea snakes are infested by species of *Vatacarus* (Southcott 1957, Vercammen-Grandjean 1965b). Other intranasal chiggers are listed by Nadchatram (op. cit.).

Based on body color, host and habitat preferences, Nadchatram (1970b) separated the chiggers of Malaysia into ecological groups, and drew some interesting conclusions from his results. Nidicolous, usually pallid, forms were found to be strongly habitat-specific (*Babiangia, Leptotrombidium* and *Herpetacarus,* for example) while the yellow, orange and red ground surface and open niche forms appeared more host-specific (*Microtrombicula, Neoschöngastia* and *Siseca,* for example). Nadchatram concluded that, from an epidemiological standpoint, special attention should be focused on the pallid, nidicolous species as potential pests of man.

While trombiculid larvae are typically parasites of terrestrial vertebrates, a few have opted for other hosts. *Eutrombicula poppi* Vercammen-Grandjean has been collected from a sea snake, an unusual host for an acarine ectoparasite (Vercammen-Grandjean 1971). Since other *Eutrombicula* species parasitize terrestrial snakes, and since the sea snake may well spend part of its life ashore, the choice of host by *E. poppi* is not necessarily incongruous. *Microtrombicula (Eltonella) eltoni* Audy and *Neotrombicula scorpionis* have gone even farther afield in their choice of hosts. Both are parasites of scorpions (Audy 1956, Nadchatram and Dohany 1974). Vercammen-Grandjean et al. (1970) recorded a species of *Schöngastia* as parasitizing a snail in Africa.

As mentioned earlier, relatively little is known about postlarval trombiculids. The systematics of the family as now constituted (Vercammen-Grandjean 1973a) is based primarily on larval forms. Adults of a number of trombiculid species have been described, but identification of postlarval instars remains difficult or impossible for most genera.

Members of the LEEUWENHOEKIIDAE (Plate 99, p. 354) are distinguished from trombiculids mainly on larval morphological characters. Leeuwenhoekiids have obvious structural and behavioral affinities with the TROMBICULIDAE and, in fact, are considered a trombiculid subfamily by some specialists (Wharton and Fuller 1952, Nadchatram 1970b). I have followed Vercammen-Grandjean (1973a) in recognizing Womersley's family ranking for this group.

Larval leeuwenhoekiids are parasites of a variety of birds, reptiles and mammals, and occasionally are found even on arthropods (*Odontacarus paradoxa* (André) was taken from a scorpion, *Buthus gibbosus,* on Crete (André 1943)). Larvae of *Apolonia tigipioensis* Torres and Braga attack chickens in Brazil, often killing young birds (Torres and Braga 1939). Attacks by *A. tigipioensis* on humans also have been recorded. Larvae of the genus *Whartonia* are bat ectoparasites (Nadchatram 1974).

While none is recorded from an intranasal habitat, certain leeuwenhoekiids are intradermal parasites. *Hannemania* species, for example, invade the skin of amphibians. *H. hylae* (Ewing) is an intradermal parasite of tree frogs in North America (Ewing 1926).

Nymphs of the leeuwenhoekiid *Womersia strandtmanni* Wharton apparently are capable of immobilizing their prospective prey by means of an attractant substance which is released from the anal region of the mite (Huber 1978). Collembolans were observed to suffer paralysis following brief contact with *W. strandtmanni,* after which the mite fed upon its disabled victims.

Useful References

André, M. (1927). Digestion "extra-intestinale" chez le Rouget (*Leptus autumnalis* Shaw). Bull. Mus. nat. Hist. natur. Paris **33**:509-516. [TROMBICULIDAE]

André, M. (1929a). Sur une nouvelle forme larvaire d'Acarien (Erythraeidae), parasite d'une fourmi (*Phagiolepis pygmaea* Latr.). Bull. Mus. nat. Hist. natur. Paris 2, **1**:255-259.

André, M. (1929b). Note complémentaire sur *Parathrombium teres* M. André. Bull. Soc. Zool. France **14**:644. [TROMBIDIIDAE]

André, M. (1943). Une espèce nouvelle de *Leeuwenhoekia* parasite de scorpions. Bull. Mus. nat. Hist. natur. Paris, 2, **15**:294-298. [LEEUWENHOEKIIDAE]

André, M. (1949a). Nouvelle forme larvaire de Thrombidion *Isothrombium oparbellae* (n.g., n.sp.) parasite d'un solifuge. Bull. Mus. nat. Hist. natur. Paris, 2, **21**:354-357. [TROMBIDIIDAE]

André, M. (1949b). Une nouvelle espèce de Thrombidion (Stygothrombidiidae) recueillie, en France, dans un cours d'eau phréatique. Bull. Mus. nat. Hist. natur. Paris, 2, **21**:67-71.

André, M. (1949c). Les *Stygothrombium* (Acariens) de la faune francaise. Bull. Mus. nat. Hist. natur. Paris, 2, **21**:680-689.

André, M. (1953). Un espèce nouvelle de *Leptus* (Acarien) parasite de scorpions. Bull. Mus. nat. Hist. natur. Paris, 2, **25**:150-154. [ERYTHRAEIDAE]

Audy, J.R. (1956). Trombiculid mites infesting birds, reptiles and arthropods in Malaya, with a taxonomic revision and descriptions of a new genus, two new subgenera, and six new species. Bull. Raffles Mus., Singapore **28**:27-80.

Audy, J.R. (1968). Red mites and Typhus. Univ. London, Athlone Press: 191 pp. + x. [TROMBICULIDAE]

Baker, E.W., T.M. Evans, D.J. Gould, W.B. Hull and H.L. Keegan (1956). A Manual of Parasitic Mites of Medical or Economic Importance. Natl. Pest Control Assoc. Tech. Publ.: 170 pp.

Baker, E.W. and G.W. Wharton (1952). The Suborder Trombidiformes Reuter, 1909. **In** An Introduction to Acarology. Macmillan Co., New York: 146-258.

Brennan, J.M. and E.K. Jones (1959). Keys to the chiggers of North America with synonymic notes and descriptions of two new genera (Acarina:Trombiculidae). Ann. Ent. Soc. Amer. **52**(1):7-16.

Cooreman, J. 1959. Notes sur quelques acariens de la faune cavernicole (2me serie). Inst. roy. Sci. natur. Belg. **35**(34):1-40. [JOHNSTONIANIDAE]

Cross, H.F. (1964). Observations on the formation of the feeding tube by *Trombicula splendens* larvae. Acarologia **6**(fasc. h.s.):255-261.

Crossley, D.A. (1960). Comparative external morphology and taxonomy of nymphs of the Trombiculidae (Acarina). Univ. Kansas Sci. Bull. **40**(6):135-321.

Evans, G.O., J.G. Sheals and D. Macfarlane (1961). Mites associated with insects and other invertebrates. **In** The Terrestrial Acari of the British Isles. Volume I. Introduction and Biology. British Mus., London: 166-185.

Ewing, H.E. (1925). A contribution to our knowledge of the taxonomy of chiggers (Trombidiidae), including the descriptions of a new genus, six new species and a new variety. Amer. J. Trop. Med. **5**:251-265. [TROMBICULIDAE]

Ewing, H.E. (1926). The life history and biology of the Tree-Toad Chigger, *Trombicula hylae* Ewing. Ann. Ent. Soc. Amer. **19**:261-267.

Feider, Z. (1955). Fauna Republicii Populare Romîne. Arachnida, Acarina, Trombidoidea. Acad. Rep. Pop. Romîne **5**(1):1-186.

Fuller, H.S. (1952). The mite larvae of the family Trombiculidae in the Oudemans collection: taxonomy and medical importance. Zool. Verhand. **18**:261 pp.

Grandjean, F. (1947). Étude sur les Smarisidae et quelques autres Érythroïdes (Acariens). Arch. Zool. exp. gén. **85**(1):1-126. [SMARIDIDAE]

Hirst, S. (1928). On some new Australian mites of the families Trombidiidae and Erythraeidae. Ann. Mag. Nat. Hist. 10, **1**(4):563-571.

Huber, I. (1978). Prey attraction and immobilization by allomone from nymphs of *Womersia strandtmanni* (Acarina:Trombiculidae). Acarologia **20**(1):112-115.

Jeppson, L.R., H.H. Keifer and E.W. Baker (1975). Biological enemies of mites. **In** Mites Injurious to Economic Plants. Univ. California Press, Berkeley: 75-90. [ERYTHRAEIDAE]

Johnston, D.E. and R.R. Wacker (1967). Observations on postembryonic development in *Eutrombicula splendens* (Acari-Acariformes). J. Med. Ent. **4**(3):306-310. [TROMBICULIDAE]

Lawrence, R.F. (1940). New larval forms of South African mites from arthropod hosts. Ann. Natal Mus. **9**(3):401-408. [ERYTHRAEIDAE]

Lindquist, E.E. and P.-H. Vercammen-Grandjean (1971). Revision of the chigger-like larvae of the genera *Neotrombidium* Leonardi and *Monunguis* Wharton, with a redefinition of the subfamily Neotrombidiinae Feider in the Trombidiidae (Acarina:Prostigmata). Can. Ent. 103:1557-1590.

Michener, C.D. (1946). The taxonomy and bionomics of some Panamanian trombidiid mites (Acarina). Ann. Ent. Soc. Amer. **39**(3):349-380. [TROMBIDIIDAE]

Mullen, G.R. (1974). Acarine parasites of mosquitoes. II. Illustrated larval key to the families and genera of mites reportedly parasitic on mosquitoes. Mosquito News **34**(2):183-195.

Mullen, G.R. (1975). Acarine parasites of mosquitoes. I. A critical review of all known records of mosquitoes parasitized by mites. J. Med. Ent. **12**(1):27-36.

Nadchatram, M. (1970a). A review of intranasal chiggers with descriptions of twelve species from east New Guinea (Acarina, Trombiculidae). J. Med. Ent. **7**(1):1-29.

Nadchatram, M. (1970b). Correlation of habitat, environment and color of chiggers, and their potential significance in the epidemiology of scrub typhus in Malaya (Prostigmata:Trombiculidae). J. Med. Ent. **7**(2):131-144.

Nadchatram, M. and A.L. Dohany (1974). A pictorial key to the subfamilies, genera and subgenera of southeast Asian chiggers (Acari, Prostigmata, Trombiculidae). Bull. Inst. Med. Res. Malaysia 16:1-67.

Newell, I.M. (1957). Studies on the Johnstonianidae (Acari, Parasitengona). Pac. Sci. **11**:396-466.

Newell, I.M. (1963). Feeding habits in the genus *Balaustium* (Acarina, Erythraeidae), with special reference to attacks on man. J. Parasitol. **49**(3):498-502.

Newell, I.M. and L. Tevis, Jr. (1960). *Angelothrombium pandorae* n.g., n.sp. (Acari, Trombidiidae), and notes on the biology of the giant red velvet mites. Ann. Ent. Soc. Amer. **53**(3):293-304.

Putman, W.L. (1970). Life history and behavior of *Balaustium putmani* (Acarina:Erythraeidae). Ann. Ent. Soc. Amer. **63**(1):76-81.

Rack, G. (1973). *Balaustium murorum* (Hermann, 1804) (Acarina, Erythraeidae) in Häusern. Eine wenig beachtete sommerliche Milbenplage. Anz. Schädl. Pflanzen-Umwelt. **46**:129-132.

Rack, G. (1976). Milben (Acarina) von europäischen Limoniinen (Diptera, Nematocera). Mitt. Hamburg. Zool. Mus. Inst. **73**:63-85. [CALYPTOSTOMATIDAE]

Robaux, P. (1967). Contribution a l'étude des acariens Thrombidiidae d'Europe. Mém. Mus. nat. Hist. natur. Paris 46:124 pp.

Robaux, P. (1970). Étude des larves de Thrombidiidae III. La larve de *Johnstoniana errans* (Johnston) 1852. Redescription de l'adulte et de la nymphe. Acarologia **12**(2):339-356.

Robaux, P. (1971). Recherches sur le développement et la biologie des Acariens Thrombidiidae. PhD Thesis, Faculty of Science, Univ. Paris. C.N.R.S. Reg. No. 5616.

Robaux, P. (1973). Importance de l'étude des caractères morphologiques, de la biologie et de l'écologie a toutes les stases, pour établir la phylogénèse des Acariens voisins des Thrombidions. Acarologia **15**(1):121-128.

Severin, H.C. (1944). The grasshopper mite *Eutrombidium trigonum* (Hermann) an important enemy of grasshoppers. S. Dakota Agr. Exp. Sta. Tech. Bull. 3:36 pp. [TROMBIDIIDAE]

Shiba, M. (1969). Taxonomic investigations on free-living mites in the subalpine forest of Shiga Heights IBP Area. II. Prostigmata. Bull. Nat. Sci. Mus. Tokyo **12**(1):65-115. [CALYPTOSTOMATIDAE]

Singer, G. (1971). *Neotrombidium* Leonardi (Acarina:Trombidioidea). Part 1: revision of the genus, with description of two new species. Acarologia **13**(1):119-142.

Smiley, R.L. (1966). Further descriptions of two erythraeids predaceous upon cotton bollworm eggs (Acarina:Erythraeidae). Proc. Ent. Soc. Wash. **68**(1):25-28.

Southcott, R.V. (1946). Studies on Australian Erythraeidae (Acarina). Proc. Linn. Soc. N. S. W. **71**(1-2):6-48.

Southcott, R.V. (1957). On *Vatacarus ipoides* n. gen., n. sp. (Acarina:Trombidioidea). Trans. Roy. Soc. S. Austral. **80**:165-176. [TROMBICULIDAE]

Southcott, R.V. (1960). Notes on the genus *Sphaerotarsus* (Acarina, Smaridiidae). Trans. Roy. Soc. S. Austral. **83**149-161.

Southcott, R.V. (1961a). Studies on the systematics of the Erythraeoidea (Acarina), with a critical revision of the genera and subfamilies. Austral. J. Zool. **9**(3):367-610.

Southcott, R.V. (1961b). Notes on the genus *Caeculisoma* (Acarina:Erythraeidae) with comments on the biology of the Erythraeoidea. Trans. Roy. Soc. S. Austral. **84**:163-178.

Southcott, R.V. (1963). The Smaridiidae (Acarina) of North and Central America and some other countries. Trans. Roy. Soc. S. Austral. **86**:159-245.

Southcott, R.V. (1965). Revision of the genus *Charletonia* Oudemans (Acarina:Erythraeidae). Austral. J. Zool. **14**:687-819.

Thor, S. and C. Willmann (1947). Trombidiidae. Das Tierreich **71b**:187-541.

Torres, S. and W. Braga (1939). *Apolonia tigipioensis* g. et. sp. n. (Trombiculinae) parasito de *Gallus gallus domesticus.* Bol. Soc. Brasil. Med. Vet. **9**:28-34. [LEEUWENHOEKIIDAE]

Traub, R. and C.L. Wisseman, Jr. (1974). The ecology of chigger-borne rickettsiosis (scrub typhus). J. Med. Ent. **11**(3):237-303. [TROMBICULIDAE]

Treat, A.E. (1975). Protelean parasites: Family Erythraeidae and Family Trombidiidae (Chapters 9 and 10). **In** Mites of Moths and Butterflies. Cornell Univ. Press, Ithaca, N.Y.: 191-238.

Turk, F.A. (1945). Studies of Acari, V. - Notes on and descriptions of new and little-known British Acari. Ann. Mag. Nat. Hist. **12**(11):785-820. [CALYPTOSTOMATIDAE]

Vercammen-Grandjean, P.-H. (1955). Un genre nouveau: *Durenia,* dans la sous-famille des Trombellinae (Trombidiidae-Acarina). Rev. Zool. Bot. Afr. **52**:252-260. [TROMBELLIDAE]

Vercammen-Grandjean, P.-H. (1957). Un nouveau Trombidiidae larvaire parasite de divers culicoides originaires d'Ecosse: *Evansiella culicoides* n. g., n. sp. (Acarina). Ann. Mag. Nat. Hist. **10**:263-286. [JOHNSTONIANIDAE]

Vercammen-Grandjean, P.-H. (1965a). *Tenotrombicula mintneri* n. g., n. sp., an interesting commensal of African termites (Trombiculidae:Acarina). Acarologia **7** (fasc. suppl.):259-265.

Vercammen-Grandjean, P.-H. (1965b). *Iguanacarus,* a new subgenus of chigger mite from nasal fossae of the marine iguana in the Galapagos Islands, with a revision of the genus *Vatacarus* Southcott (Acarina, Trombiculidae). Acarologia **7** (fasc. suppl.):266-274.

Vercammen-Grandjean, P.-H. (1967). Revision of *Hoplothrombium quinquescutatum* Ewing, 1925 (Trombidiidae:Acarina). Opusc. Zool. 96:1-7.

Vercammen-Grandjean, P.-H. (1968). Chigger mites of the Far East (Acarina:Trombiculidae & Leeuwenhoekiidae). U.S. Army Med. Res. Dev. Command, Washington, D.C. Spec. Study: 135 pp.

Vercammen-Grandjean, P.-H. (1971). *Eutrombicula (E.) poppi,* a new chigger from a sea snake (Acarina, Trombiculidae). Opusc. Zool. 114:1-3.

Vercammen-Grandjean, P.-H. (1973a). Sur les statuts de la famille des Trombidiidae Leach, 1815 (Acarina Prostigmata). Acarologia **15**(1):102-114.

Vercammen-Grandjean, P.-H. (1973b). Tentative nepophylogeny of trombiculids. Folia Parasitol. (Praha) **20**:49-66.

Vercammen-Grandjean, P.-H. (1975a). Les organes de Claparède et les papilles genitales de certains Acariens sont-ils des organes respiratoires? Acarologia **17**(4):624-630. [CALYPTOSTOMATIDAE]

Vercammen-Grandjean, P.-H. (1975b). Some larval Trombiculidae of the Ethiopian region (Acari). Rev. Zool. Afr. **89**(2):397-439.

Vercammen-Grandjean, P.-H. and M. André (1966). Introduction à la notion de "complex" appliquée à la systématique des Trombiculidae (Acarina). Acarologia **8**(1):62-70.

Vercammen-Grandjean, P.-H. and J.R. Audy (1959). Une seconde espèce appartenant au genre *Durenia* Vercammen 1955 et originaire de Malaisie: *Durenia singaporensis* n. sp. (Acarina:Trombidiidae). Biol. Jaarb. **44**:98-101. [TROMBELLIDAE]

Vercammen-Grandjean, P.-H., P.L.G. Benoit and J.J. van Mol (1970). Terrestrial snail a new host for trombiculid larvae. Acta Trop. **27**(2):177.

Vercammen-Grandjean, P.-H. and A. Cochrane (1974). On three new species of larval Trombidiformes parasitizing American midges (Acarina:Trombidiidae & Johnstonianidae). J. Kansas Ent. Soc. **47**(1):66-79.

Vercammen-Grandjean, P.-H. and R. Langston (1971). The chigger mites of the world. Vol. III. *Guntherana* complex. Section A. Genus *Guntherana.* G.W. Hooper Foundation, San Francisco: 153 pp. + vii.

Vercammen-Grandjean, P.-H. and R. Langston (1975). The chigger mites of the world. Vol. III. *Leptotrombidium* complex. Section C. Iconography. G.W. Hooper Foundation, San Francisco: 298 pp.

Vercammen-Grandjean, P.-H. and R. Langston (1976a). The chigger mites of the world. Vol. III. *Leptotrombidium* complex. Section A. *Leptotrombidium* sensu stricto. G.W. Hooper Foundation, San Francisco: pp. 1-612.

Vercammen-Grandjean, P.-H. and R. Langston (1976b). The chigger mites of the world. Vol. III. *Leptotrombidium* complex. Section B. *Trombiculindus, Hypotrombidium* and *Ericotrombidium,* plus heterogenera. G.W. Hooper Foundation, San Francisco: 613-1061.

Vistorin-Theis, G. (1975). Entwicklungszyklus der Calyptostomiden (Acari, Trombidiformes). Acarologia **17**(4):683-692.

Wharton, G.W. and H.S. Fuller (1952). A manual of the chiggers. Mem. Ent. Soc. Wash. **4**:185 pp.

Williams, R.W. (1946). A contribution to our knowledge of the bionomics of the common North American chigger, *Eutrombicula alfreddugesi* (Oudemans) with a description of a rapid collecting method. Amer. J. Trop. Med. **26**:243-250.

Womersley, H. (1934). A revision of the trombid and erythraeid mites of Australia with descriptions of new genera and species. Rec. S. Austral. Mus. **5**(2):179-254.

Womersley, H. (1952). The scrub-typhus and scrub-itch mites (Trombiculidae, Acarina) of the Asiatic-Pacific region. Res. S. Austral. Mus. **10**:1-673.

Womersley, H. and W.G. Heaslip (1943). The Trombiculinae (Acarina) or itch-mites of the Austro-Malayan and Oriental regions. Trans. Roy. Soc. S. Austral. **67**:68-142.

Womersley, H. and R.V. Southcott (1941). Notes on the Smarididae (Acarina) of Australia and New Zealand. Trans. Roy. Soc. S. Austral. **65**(1):61-78.

Yunker, C.E. and J.M. Brennan (1962). Endoparasitic chiggers: II. Rediscovery of *Doloisia synoti* Oudemans, 1910, with descriptions of a new subgenus and two new species (Acarina:Trombiculidae). Acarologia **4**(4):570-576.

2. Phalanx Hydrachnidia

The Hydrachnidia or water mites are a large, distinctive, widely distributed group classified in seven superfamilies, 45 families and 287 genera (Cook 1974, Smith and Oliver 1976). While the larval stage typically has a palpal thumb-claw process similar to those seen in the Trombidia, adults and nymphs generally lack a terminal tibial claw, and the palp may be a simple setate appendage. In some groups, the palptibia is produced distidorsally so as to oppose or cradle the terminal tarsal segment (Fig. 37b).[1] The palptarsus itself may be pointed or variously produced as a raptorial structure. The movable cheliceral digits are sickle-like and well situated for operating in concert with the grasping palpi (Plate 108-2, p. 363). Idiosomal hypertrichy of the type seen in the Trombidia is absent, but leg hypertrichy is common in actively swimming species. Leg setae adapted as swimming hairs are located ventrodistally on the more distal leg segments (Plate 109-2, p. 364). Shorter secondary swimming setae also may be found on the ventral surfaces of postcoxal segments. Prodorsal trichobothria and cristae metopica are absent, but a series of distinctive glands, or *glandularia,* may be found associated with various dorsal setae (Plate 103-1, p. 358). A profusion of small genital acetabula often flank the genital opening, or *gonopore,* in both males and females.

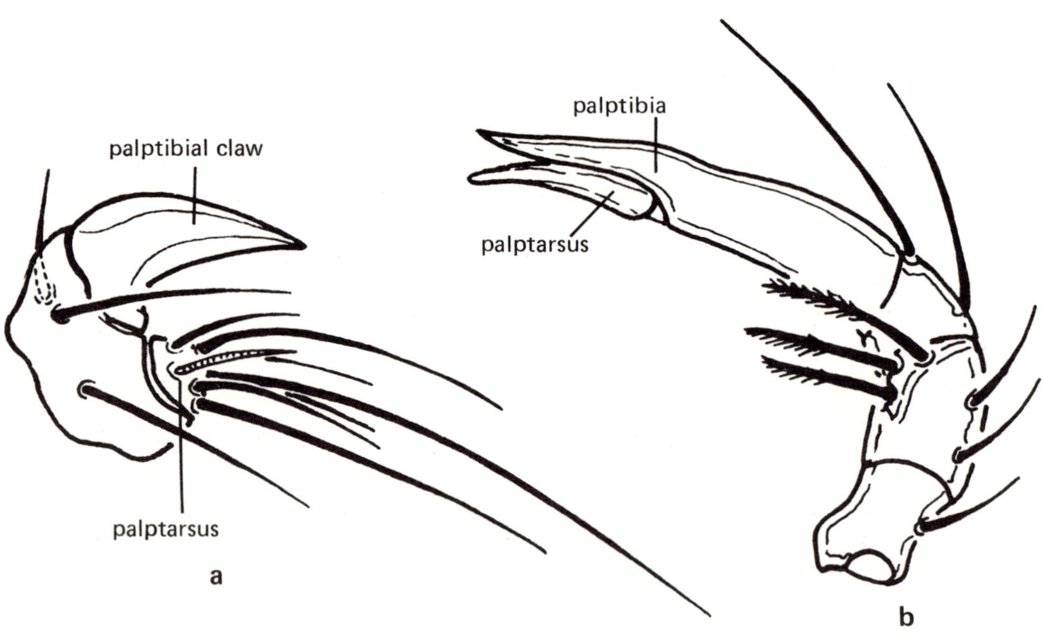

Fig. 37. Palpal types commonly observed in the Hydrachnidia: **a**; *Piona* sp. (PIONIDAE), larva, showing thumb-claw process (after Prasad and Cook 1972): **b**; *Hydrodoma monticola* (Piersig) (HYDRODROMIDAE), male, showing distidorsally produced palptibia (after Cook 1974).

[1]Cook (1974) uses the term "uncate" to describe this condition and defines it as follows. "In water mites, referring to a condition of the palp in which the ventral portion of the fourth segment [tibia] is greatly expanded and the fifth segment [tarsus] is able to fold against this expanded portion."

The Hydrachnidia are found throughout the world (Antarctica is an exception) in lakes and ponds, streams, seepage waters, wells, hot springs, cave pools and interstitial spaces in bottom sediments. Some have adapted to intertidal waters, but most species are confined to fresh water habitats. Adult and nymphal Hydrachnidia are predaceous on a variety of aquatic insects and their eggs, and on other small invertebrates (Mullen 1975b). Larvae are typically parasitic on insects, but may sometimes be associated with mollusks, gastropods, decapods, or freshwater sponges.

Based on morphological considerations, Cook (1974) feels that the Hydrachnidia have evolved from the terrestrial Trombidia and probably are a monophyletic group (with the possible exclusion of the superfamily Hydrovolzioidea).

Specialization for survival in restricted aquatic niches under conditions where environmental factors are more or less constant has resulted in distinctive and occasionally startling modifications in the morphology of many postlarval Hydrachnidia, modifications which tend to obscure natural family groupings. Since relatively few larval forms are known (Cook op. cit.), the familial classification of Hydrachnidia has been based largely on postlarval stases. The resulting system is both cumbersome and highly artificial. Hopefully, discovery of key larval forms and further study of larval host associations will make possible a more "streamlined" treatment in the future.

Superfamily Hydrovolzioidea
(Plate 100, p. 355)

DIAGNOSIS: *Dorsum with an anterior and a posterior shield, the anterior wider than long; with a series of marginal shields flanking the unpaired median sclerites; dorsal glandularia, if present, on platelets which are adjacent to but separate from setal platelets; prodorsal eyes present or absent, without a median eye. Palptibia not produced distidorsally. Genital acetabula absent in all stages; genital opening (gonopore) covered by genital valves. Larvae aerial (surface swimmers), without urstigmata.*

The Hydrovolzioidea is represented by a single family, the HYDROVOLZIIDAE. Hydrovolziids are known primarily from the Asiatic-Pacific region and from Europe, but some species have been identified from Africa and North America. Postlarval instars are predators in streams, springs and seepage waters. *Stygovolzia uenoi* Imamura was taken from a driven well in Japan (Imamura 1957). *Bharatavolzia musicola* Cook and most species of the genus *Acherontacarus* are interstitial forms (Cook 1974). *Hydrovolzia placophora* Monti and *H. cancellata* Walter are cold water species usually found at high altitudes or latitudes in Europe (Monti 1905, Walter 1906), and are considered glacial relicts.

Hydrovolziid larvae are confined to the surface water film. Known species are parasites of certain Hemiptera (Mesoveliidae) and Diptera (Empididae) (Mitchell 1954, Efford 1966). *Hydrovolzia gerhardi* Mitchell parasitizes terminal nymphs and adults of the water bug *Microvelia americana* in North America.

Superfamily Hydrachnoidea
(Plate 100, p. 355)

DIAGNOSIS: *Spherical species with varying degrees of dorsal sclerotization, integument strongly papillate; dorsal glandularia and associated setae each on a shared platelet; lateral eyes separately encapsulated, median eye present or absent; palptibia produced distidorsally. With three or more pairs of genital acetabula; females with a tubular ovipositor. Larvae aquatic (sub-surface swimmers), gnathosoma greatly enlarged, over 1/3 length of idiosoma, palpi short and without thumb-claw process; with swimming hairs on legs; legs with five segments, excluding coxae. Dorsal shield large, covering most of the idiosoma in unengorged individuals; urstigmata present.*

Hydrachnoids are typically bright red mites which are found in standing waters and sluggish streams throughout much of the world. A single family, the HYDRACHNIDAE, is included in the superfamily.

Hydrachnid females are unique amongst the Hydrachnidia in that they oviposit via a tubular ovipositor and secrete the eggs in intercellular air spaces of higher aquatic plants (Cook 1974). The larvae are subaquatic but tend to remain in the upper water layers. They are parasitic on certain Coleoptera (Dytiscidae) and on a variety of Hemiptera which live in subsurface situations (Smith and Oliver 1976). Hydrachnid larvae remain attached to their hosts for long periods of time (as long as 10 months) and may overwinter on the host insect. The larva increases greatly in size during this time, assuming a tear-drop shape prior to molting to the active deutonymphal stage (Fig. 21, p. 67).

Superfamily Eylaoidea
(Plate 101, p. 356)

DIAGNOSIS: Spherical or elongate species; dorsal sclerotization variable, often without shields or with a few small sclerites, integument papillate and with or without glandularia; lateral eyes, when present, on a common median ocular shield. Buccal opening with a distinct membranous fringe. Palpi simple, not specialized for grasping; genital acetabula present, often stalked, widely scattered in genital region or in discrete acetabular plates. Larvae aerial, with or without swimming hairs, legs with six segments, excluding coxae. Dorsal shield large, covering half the idiosoma, with eight pairs of setae; urstigmata present.

The Eylaoidea is represented by three families of red mites which live in standing waters, streams, swamps and temporary ponds. One species, *Eylais thermalis* Uchida (EYLAIDAE), is found in hot springs in Formosa. Others are interstitial in habitat. *Piersigia limophila* Protz (PIERSIGIIDAE), an inhabitant of rotting vegetation in marshes and swamps (Imamura and Mitchell 1967), may represent an early step toward a truly aquatic type from a terrestrial trombid progenitor.

Larvae of the EYLAIDAE, LIMNOCHARIDAE and PIERSIGIIDAE differ considerably from one another both in form and in host preferences. Eylaid larvae are known to parasitize Hemiptera of the families Belostomatidae and Corixidae (Lanciani 1969, Smith and Oliver 1976), and Coleoptera of the families Dytiscidae, Haliplidae and Hydrophilidae (Lanciani, op. cit.). *Limnochares* larvae, on the other hand, favor coenagrionid and libellulid dragonflies, and also attack hydrometrids, gerrids and veliids (Hemiptera). Piersigiid larvae have been collected on hydrophilid beetles of the genera *Enochrus* (Imamura and Mitchell 1967) and *Anacaena* (Smith and Oliver, op. cit.).

Superfamily Hydryphantoidea
(Plates 102-104, pp. 357-359)

DIAGNOSIS: *Sclerotization highly variable, ranging from a simple prodorsal shield to a series of greatly expanded sclerites which virtually cover the idiosoma, glandularia normally situated; lateral eyes, when present, on separate shields, median eye present or absent. Palpi simple or palptibia produced distidorsally; genital acetabula present, with as few as two pairs, occasionally on long stalks; genital valves usually present in both sexes. Larvae typically aerial, with or without swimming hairs, legs with six postcoxal segments. Dorsal shield small, covering less than 1/3 the length of the idiosoma, with one, two or four pairs of setae; urstigmata present.*

Cook (1974) lists six families in the Hydryphantoidea: the HYDRYPHANTIDAE, HYDRODROMIDAE, TERATOTHYADIDAE, RHYNCHOHYDRACARIDAE, CTENOTHYADIDAE and THERMACARIDAE. The superfamily is worldwide and occurs in a great variety of aquatic habitats. Hydryphantoid nymphs and adults usually are non-swimming or weak-swimming large red mites which prey on aquatic insects. The most widely distributed family is the HYDRYPHANTIDAE (Plate 102, p. 357), a taxon representing 10 subfamilies and over 40 genera. *Trichothyas* and related thyadine hydryphantids often inhabit seepage waters and may be encountered in wet moss or in litter along stream banks. Temporary ponds also are favored by many of these species. Nymphs and adults of two *Thyas* species were found to prey on mosquito eggs after refusing other candidate prey (Mullen 1975b). Eggs were holed, slit or torn open along one side, and the contents were removed. *Hydryphantes ruber* (DeGeer) was noted to feed on a fourth instar larva of *Aedes stimulans.*

Members of the genus *Euwandesia* (Wandesiinae) occur interstitially in river sand (Hopkins and Schminke 1970). Like marine interstitial mites (page 238), *Euwandesia* species are elongate forms well suited to an arenicolous existence. Species of *Partnuniella,* another wandesiine genus, are known from thermal waters in the New World (Viets 1938). The hot springs habitat is shared by species of the genus *Thermacarus* (see below).

Larvae of hydryphantoids parasitize a variety of insects including Hemiptera, Odonata, Diptera, Plecoptera and Trichoptera. All attach to the thoraces of their hosts (Smith and Oliver 1976). Mullen (1974) reported a number of thyadine hydryphantids as parasites of mosquitoes. Thyadines have also been recorded recently from eight genera of Chironomidae (Smith and Oliver, op. cit.). Larvae of *Hydrodroma* (HYDRODROMIDAE) share the chironomid niche, attaching to Tanypodinae and Chironominae.

Larvae of *Thermacarus nevadensis* Marshall (THERMACARIDAE, Plate 104, p. 359) run rapidly on the surface of thermal waters, but engorged larvae are found crawling on the bottom (Mitchell 1962). In an earlier paper (1960), Mitchell noted that *T. nevadensis* larvae attach to humans but do not engorge. However, there are indications that *T. nevadensis* larvae were responsible for multiple bites reported by swimmers in a Harney County (Oregon) hot spring in 1971. Field observations on larvae in 1974 revealed that they will attach to any object passed through the water surface, but quickly detach and swim away in most cases. Although attachment often was found to be more prolonged on human skin, no bites were experienced. The natural host for the larva of *T. nevadensis* is unknown.

Superfamily Lebertioidea
(Plates 104-106, pp. 359-361)

DIAGNOSIS: *Shields and integument various, from unsclerotized and smooth to strongly papillate, areolate or granular and/or with extensive or complete idiosomal armature; lateral eyes generally present, separated, not on a single median shield, with or without a median eye. Palpal tarsus produced terminally into a point, or a series of points, palpi never chelate or produced distidorsally. With three pairs of genital acetabula (occasionally six pairs), usually in two parallel rows medially in the gonopore and covered by large well developed genital valves. Larvae aquatic, with or without swimming hairs; legs with five segments, excluding coxae. Dorsal shield various but typically more than 1/3 length of idiosoma; urstigmata present.*

The families LEBERTIIDAE, SPERCHONTIDAE, TEUTONIIDAE, RUTRIPALPIDAE, ANISITSIELLIDAE, OXIDAE and TORRENTICOLIDAE are included in the Lebertioidea by Cook (1974). The superfamily is worldwide but few (if any) families are recorded as occurring in Australia. Lebertioids occur in all types of water from cold streams in the temperate zone to seepage waters in the tropics. One species, *Nilotonia thermophila* (Lundblad) (ANISITSIELLIDAE) inhabits hot springs in the Old World (Cook, op. cit.). The SPERCHONTIDAE and most genera of the OXIDAE (Plate 105, p. 360) are especially prevalent in the New and Old World tropics, whereas the TEUTONIIDAE, RUTRIPALPIDAE and LEBERTIIDAE are more or less restricted to the holarctic realms.

Relatively few lebertioid larvae have been identified, and these are known only from nematocerous Diptera (Smith and Oliver 1976). Larvae of *Sperchon* and *Sperchonopsis* have been observed on Chironomidae, as have representatives of the genera *Lebertia* (LEBERTIIDAE), *Oxus*, *Frontipoda* (OXIDAE), and *Torrenticola* (TORRENTICOLIDAE). Larvae of *Sperchon* also are known from black flies (Simuliidae). Most lebertioid larvae attach to the thorax of their chironomid hosts but some are encountered on the abdomen as well (Smith and Oliver, op. cit.). Larvae attach to the host pupa beneath the surface water film, transferring to the emerging adult at host ecdysis.

Superfamily Hygrobatoidea
(Plates 106-110, pp. 361-365)

DIAGNOSIS: *Shields and integument various, from unsclerotized and smooth or papillate to entirely sclerotized forms; lateral eyes, when present, on separate lateral shields or beneath the integument, not on a median shield. Palpal tarsus variously produced terminally, palptibia not produced distidorsally. Genital acetabula generally present, three or more pairs on acetabular plates or free in the genital field, never in the gonopore (acetabula absent in marine family PONTARACHNIDAE) genital valves absent. Larvae aquatic, often with swimming hairs; legs with five segments, excluding coxae. Dorsal shield large, covering most of the idiosoma, typically with four pairs of setae; urstigmata present.*

Nine families are contained in the Hygrobatoidea (Cook 1974). The PONTARACHNIDAE (Plate 106, p. 361) is unusual in that it has adapted completely to marine littoral habitats. Its relationships with other water mites are unclear, but it is traditionally placed with the hygrobatoids. Pontarachnids occur in most coastal waters, except for the Arctic and Antarctic littoral and the waters flanking South America. A species of *Litarachna* was taken in numbers from wharf piling in a heavy surf zone on the Caribbean coast of Yucatan, Mexico where it apparently was preying on other microinvertebrates in the same niche.

The eight remaining families of Hygrobatoidea are freshwater groups and include the majority of described water mite taxa. In addition to the marine pontarachnids, Cook (1974) lists the LIMNESIIDAE, OMARTACARIDAE, ASTACOCROTONIDAE, UNIONICOLIDAE, FELTRIIDAE, PIONIDAE, HYGROBATIDAE and ATURIDAE in the superfamily. An additional family, the AXONOPSIDAE,[1] is recognized by Smith (1976). Smith has observed that the AXONOPSIDAE plus the five last named families recognized by Cook constitute a relatively homogenous phyletic group which is definable on the basis of both larval and adult characters. Smith refers to this infragroup as the *pioniform water mites,* and defines them as hygrobatoids whose larvae exhibit a degree of coxal plate fusion, lack lateral blades on the urstigmata (Prasad and Cook 1972), and have nine setae on tibia I. Adult pioniform water mites lack genital valves and do not have distidorsally produced palptibiae.

The AXONOPSIDAE (Plate 109, p. 364) is a cosmopolitan assemblage especially adapted to surface or interstitial waters of streams. The FELTRIIDAE (Plate 110-1), on the other hand, is essentially an holarctic group which favors mosses and matted aquatic plants in cold streams (Cook 1974). Both the ATURIDAE and PIONIDAE (Plates 108-109) are highly diverse in habitat preferences, although neither group is represented in thermal waters. Permanent fresh water habitats are favored by members of the families UNIONICOLIDAE (Plate 108, p. 363) and HYGROBATIDAE (Plate 110). Among the non-pioniform freshwater hygrobatoids, only the primarily tropical family LIMNESIIDAE (Plate 107, p. 362) is polygeneric. *Astacocroton molle* Haswell, the type and only species of the ASTACOCROTONIDAE, occurs in the gill chambers of a freshwater decapod in Australia (Cook 1974).

Böttger (1970) observed that postlarval instars of *Piona* spp. (PIONIDAE) prey on cladocerans, but also attack chironomid larvae and mayfly nymphs. Laird (1947) and Mullen (1976b) have reported that mosquito eggs or larvae also serve as prey for hygrobatoids. A species of *Limnesia* was reported to prey on eggs and first instar larvae of *Anopheles farauti* and *Culex pullus* in the southwest Pacific (Laird, op. cit.). Nymphs and adults of *Piona* were found to feed on mosquito larvae both in the laboratory and in the field (Mullen, op. cit.). Larvae are subdued within 1-3 minutes following initial piercing of the prey integument.

Larvae of seven hygrobatoid families and 21 genera are known to be parasitic on nematocerous Diptera (Mullen 1974, Smith and Oliver 1976). Larvae of *Hygrobates* and *Atractides* (HYGROBATIDAE), *Unionicola* (UNIONICOLIDAE), and *Piona* (PIONIDAE) also parasitize Trichoptera, as does a species of *Albia* (AXONOPSIDAE). Most hygrobatoid larvae attach to the abdomen of their insect hosts. Larvae of *Limnesia* (LIMNESIIDAE), however, appear to be restricted to the thorax, and some species of UNIONICOLIDAE may be found on leg femora (Smith and Oliver, op. cit.).

Many members of the UNIONICOLIDAE are associated with freshwater mollusks, snails or sponges during one or more life stages (Mitchell 1955). Most species of the genus *Unionicola* deposit their eggs in a molluscan or sponge host, having invaded the host to feed or to pass the calyptostatic nymphal stases. *U. (Vietsatax) parasiticum* Uchida and Imamura, and most members of the subgenera *Parasitatax* and *Pentatax,* are associated with clams during all life stages. Several members of the subgenus *Polyatax* are found on snails (Cook 1974).

Some species of PIONIDAE are exceptional in that the larval instar may be partially or entirely suppressed, with the resultant omission of a parasitic stage. Larvae of *Piona* may

---

[1]Subfamilies Axonopsinae, Albiinae, and Frontipodopsinae of the family ATURIDAE as recognized by Cook (1974).

appear only briefly following a long embryonic development, immediately entering the quiescent protonymphal stage and subsequently molting to an active deutonymph. Other species of *Piona* and certain *Forelia* and *Pionacerus* species may exhibit complete larval repression, with the larva forming a protonymphal calyptostase within the egg membrane and finally ecdysing as an active deutonymph (Smith 1976).

Superfamily Arrenuroidea
(Plates 110-114, pp. 365-369)

DIAGNOSIS: Typically heavily sclerotized forms, occasionally soft-bodied or with only muscle attachment platelets; lateral eyes well separated, not on a median shield. Palpi various, tending toward "uncate" type; palptarsus often simple or weakly split terminally. Genital acetabula either in gonopore, on plates flanking the gonopore, or both; genital valves absent. Larvae aquatic, often with swimming hairs; legs with five segments, excluding coxae. Dorsal shield large, covering most of the idiosoma, with four or five pairs of setae; urstigmata present.

A total of 17 families are included in the Arrenuroidea by Cook (1974). Cook notes that the superfamily is greatly oversplit, probably because the described taxa illustrate such a diversity of unusual characters. Newly described genera, especially those from ground waters, are so divergent that they create as many classificatory problems as they resolve. Consequently, many arrenuroid families are based on a single genus (LAVERSIIDAE, ARENOHYDRACARIDAE, ACALYPTONOTIDAE, MIDEIDAE, KANTACARIDAE and NIPPONACARIDAE) or, at most, contain two or three small genera. A broader understanding of larval arrenuroids may eventually lead to a more realistic familial classification. At present, the larval stage is known in only five families.

The Arrenuroidea is worldwide in distribution, although most of the described taxa are largely or completely holarctic. Arrenuroids inhabit standing water, streams, springs, seepage and interstitial waters. They apparently do not occur in hot springs, although some have been found in other unusual habitats. For example, *Horeolanus orphanus* Mitchell (BOGATIIDAE) was collected from water emerging from a field drainage tile (Mitchell 1955). *Micruracaropsis phytotelmaticola* Viets (ARRENURIDAE) was taken from water trapped in an epiphytic bromeliad in Surinam (Viets 1939).

Arrenuroid larvae of seven genera in six families parasitize nematocerous Diptera. Larvae of the genus *Arrenurus* are found also on Odonata and Trichoptera (Oliver and Smith 1976). Most of the known dipteran associations involve various subfamilies of Chironomidae, although members of the genus *Arrenurus* are recorded parasites of chaoborids, ceratopogonids, and at least 37 species of mosquitoes (Mullen 1975a, Mullen 1976, Smith and Oliver, op. cit.). Thoracic parasites are found in the arrenuroid families MIDEOPSIDAE, ATHIENEMANNIIDAE and LAVERSIIDAE (Plates 110, 112), while larvae of the families KRENDOWSKIIDAE and MIDEIDAE (Plates 110, 113) attach to the host's abdomen. Larvae of *Arrenurus* (ARRENURIDAE, Plate 111, p. 366) occur both on the thorax and abdomen (Smith and Oliver, op. cit.).

Useful References

Barr, D. (1972). The ejaculatory complex in water mites (Acari:Parasitengona): morphology and potential value for systematics. Roy. Ontario Mus. Life Sci. Contr. 81:87 pp.

Böttger, K. (1962). Zur Morphologie und Ethologie der einheimischen Wassermilben *Arrenurus (Megaluracarus) globator* (Müll.), 1776; *Piona nodata nodata* (Müll.), 1776 und *Eylais infundibulifera meridionalis* (Thor), 1899 (Hydrachnellae, Acari). Zool. Jahrb. Abt. Syst. Ökol. Geogr. **89**:501-584.

Böttger, K. (1970). Die Ernährungsweise der Wassermilben (Hydrachnellae, Acari). Feeding in water mites. Int. Rev. Ges. Hydrobiol. **55**:895-912.

Cassagne-Méjean, F. (1966). Contribution a l'étude des Arrenuridae (Acari, Hydrachnellae) de France. Acarologia **8**, fasc. suppl.: 186 pp.

Cook, D.R. (1974). Water mite genera and subgenera. Mem. Amer. Ent. Inst. 21:860 pp. + vii.

Efford, I. (1963). The parasitic ecology of some water mites. J. Anim. Ecol. **32**:141-156.

Efford, I. (1966). Observations on the life history of three stream-dwelling water mites. Acarologia **8**(1):86-93.

Hopkins, C.L. and H.K. Schminke (1970). A species of *Euwandesia* (Acari:Hydrachnellae) from New Zealand. Acarologia **12**(2):357-359. [HYDRYPHANTIDAE]

Imamura, T. (1954). Studies on water-mites from Hokkaido. J. Hokkaido Gakugei Univ. Sect. B., Suppl. 1:148 pp.

Imamura, T. (1957). A new genus of subterranean water-mites from Kyoto. J. Fac. Sci. Hokkaido Univ., Ser. VI, Zool. **13**:49-53. [HYDROVOLZIIDAE]

Imamura, T. (1970). Subterranean water mites (Limnohalacarinae and Hydrachnellae) of the Tsushima Islands. Bull. Nation. Sci. Mus., Tokyo **13**(2):249-262. [MIDEOPSIDAE, UCHIDASTYGACARIDAE, HUNGAROHYDRACARIDAE]

Imamura, T. and R.D. Mitchell (1967). The ecology and life cycle of the water mite, *Piersigia limophila* Protz. Ann. Zool. Japon. **40**(1):37-44. [PIERSIGIIDAE]

Jalil, M. and R.D. Mitchell (1972). Parasitism of mosquitoes by water mites. J. Med. Ent. **9**(4):305-311.

Laird, M. (1947). Some natural enemies of mosquitoes in the vicinity of Palmalmal, New Britain. Trans. Roy. Soc. N. Z. **76**:453-476. [LIMNESIIDAE]

Lanciani, C. (1969). Three species of *Eylais* (Acari:Eylaidae) parasitic on aquatic Hemiptera. Trans. Amer. Microsc. Soc. **88**:356-365.

Lanciani, C. (1971). Host exploitation and synchronous development in a water mite parasite of the marsh treader *Hydrometra myrae* (Hemiptera:Hydrometridae). Ann. Ent. Soc. Amer. **64**:1254-1259. [HYDRYPHANTIDAE]

Lundblad, O. (1927). Die Hydracarinen Schwedens I. Zool. Bidr. Uppsala **11**:185-540.

Lundblad, O. (1941). Die Hydracarinenfauna Südbrasiliens und Paraguays. Erster Teil. Svensk. Vetenskap. Handl. 3, **19**(7):1-183.

Lundblad, O. (1942). Die Hydracarinenfauna Südbrasiliens und Paraguays. Zweiter Teil. Svensk. Vetenskap. Handl. 3, **20**(2):1-175.

Lundblad, O. (1943a). Die Hydracarinenfauna Südbrasiliens und Paraguays. Dritter Teil. Svensk. Vetenskap. Handl. 3, **20**(5):1-148.

Lundblad, O. (1943b). Die Hydracarinenfauna Südbrasiliens und Paraguays. Vierter Teil. Svensk. Vetenskap. Handl. 3, **20**(8):1-171.

Lundblad, O. (1949). Hydrachnellae. Exploration du Parc National Albert. Mission H. Damas (1935-1936). Bruxelles. 18:1-87.

Lundblad, O. (1956). Zur Kenntnis süd-und mitteleuropäischer Hydrachnellen. Ark. Zool. **10**(1):1-306.

Lundblad, O. (1961). Die Hydracarinen Schwedens. II. Ark. Zool. **14**(1):1-635.

Lundblad, O. (1969). Indische Wassermilben, hauptsächlich von Hinterindien. Ark. Zool. **22**(10):289-443.

Mitchell, R.D. (1954). A description of a water-mite, *Hydrovolzia gerhardi* new species, with observations on the life history and ecology. Nat. Hist. Misc. (Chi. Acad. Sci.) 134:1-9. [HYDROVOLZIIDAE]

Mitchell, R.D. (1955a). Anatomy, life history and evolution of the mites parasitizing mussels. Univ. Michigan Misc. Publ. Zool. 89:27 pp. + plates. [UNIONICOLIDAE]

Mitchell, R.D. (1955b). Two water-mites from Illinois. Trans. Amer. Microsc. Soc. **74**(4):333-342. [PIERSIGIIDAE]

Mitchell, R.D. (1957a). Locomotor adaptations of the family Hydryphantidae. Abh. naturw. Ver. Bremen **35**(1):75-100.

Mitchell, R.D. (1957b). Major evolutionary lines in water mites. Syst. Zool. **6**(3):137-148.

Mitchell, R.D. (1960). The evolution of thermiphilous water mites. Evol. **14**:361-377.

Mitchell, R.D. (1962). The growth of *Thermacarus nevadensis* Marshall (Acari, Hydrachnellae). Zool. Anz. **168**:269-275. [THERMACARIDAE]

Mitchell, R.D. (1964). An approach to the classification of water mites. Proc. 1st. Int. Congr. Acarology, Fort Collins. Acarologia **6** (fasc. h. s.): 75-79.

Monti, R. (1905). Genere e specie nouvi di Idrachnidae. Rend. Istit. Lomb. Sci. e Lett. 2, **38**:168-176.

Mullen, G.R. (1974). Acarine parasites of mosquitoes. II. Illustrated larval key to the families and genera of mites reportedly parasitic on mosquitoes. Mosquito News **34**(2):183-195.

Mullen, G.R. (1975a). Acarine parasites of mosquitoes. I. A critical review of all known records of mosquitoes parasitized by mites. J. Med. Ent. **12**(1):27-36.

Mullen, G.R. (1975b). Predation by water mites (Acarina:Hydrachnellae) on the immature stages of mosquitoes. Mosquito News **35**(2):168-171.

Mullen, G.R. (1976). Water mites of the subgenus *Truncaturus* (Arrenuridae, *Arrenurus*) in North America. Search (Agr., Ent. (Ithaca) 17) **6**(6):3-35.

Prasad, V. and D.R. Cook (1972). The taxonomy of water mite larvae. Mem. Amer. Ent. Inst. 18:326 pp. + ii.

Smith, I.M. (1976). A study of the systematics of the water mite family Pionidae (Prostigmata:Parasitengona). Mem. Ent. Soc. Canada 98:249 pp.

Smith, I.M. and D.R. Oliver (1976). The parasitic associations of larval water mites with imaginal aquatic insects, especially Chironomidae. Can. Ent. **108**:1427-1442.

Soar, C.D. and W. Williamson (1925). The British Hydracarina. Ray Society, London **1**:216 pp. + x + plates.

Soar, C.D. and W. Williamson (1927). The British Hydracarina. Ray Society, London **2**:215 pp. + viii + plates.

Soar, C.D. and W. Williamson (1929). The British Hydracarina. Ray Society, London **3**:184 pp. + viii + plates.

Sparing, I. (1959). Die Larven der Hydrachnellae, ihre parasitische Entwicklung und ihre Systematik. Parasit. Schriftenreihe **10**:1-168.

Stout, V. (1953). *Eylais waikawae* n. sp. (Hydracarina) and some features of its life history and anatomy. Trans. Roy. Soc. N. Z. **81**:389-416. [EYLAIDAE]

Viets, K. (1935). Die Wassermilben von Sumatra, Java und Bali nach den Ergebnissen der Deutschen Limnologischen Sunda-Expedition. Arch. Hydrobiol. Suppl. Bd. **13**:484-738; **14**:1-113.

Viets, K. (1936). Wassermilben oder Hydracarina (Hydrachnellae und Halacaridae). Tierwelt Deutschl. **31-32**:1-574.

Viets, K. (1938). Über die verscheidenen Biotope der Wassenmilben, besonders über solche mit abnormalen Lebensbedingungen und über einige neue Wassermilben aus Thermalgewässern. Verh. Int. Ver. Limnol. Paris 8:209-224.

Viets, K. (1939). Eine neue, die erste Süsswassermilbe (Hydrachnellae, Acari) aus tropischen Pflanzengewässern. Zool. Anz. **128**:69-77. [ARRENURIDAE]

Viets, K. (1956). Die Milben der Süsswassers und des Meeres. Zweiter und Dritter Teilen. Fischer Verlag, Jena: 870 pp.

Walter, C. (1906). Neue Hydrachnidenarten aus der Schweiz. Zool. Anz. **30**:570-575.

SUBORDER ACTINEDIDA
(Plates 57-114, pp. 312-369)

KEY TO THE FAMILIES

1. Females (rarely males) with a pair of anterolateral prodorsal stigmata *(heterostigmata,* Plate 74-2, p. 329) and associated trachea; peritremes absent. Opisthosoma generally showing indications of segmentation. Palpi often 2-segmented, greatly reduced and closely appressed to a small, capsulate gnathosoma (if not reduced, then with only three distinct palpal segments which may or may not be articulated). Sexual dimorphism pronounced . . . . .Cohort ELEUTHERENGONINA, Subcohort HETEROSTIGMAE ... 25

— Females and males with stigmata between cheliceral bases or on posterodorsal margin of gnathosoma, or stigmata absent; peritremes often present in stigmate forms, extending along anterior propodosomal margins. Opisthosoma generally without indications of segmentation. Gnathosoma variously shaped, generally conspicuous and with well developed articulated palpi. Sexual dimorphism often obscure, largely confined to primary genital structures 2

2. With a palpal thumb-claw process (Plates 76-5 and 97-3; pp. 331, 352) in all stages. Primarily terrestrial forms 48

— Palpi simple, chelate, or produced distidorsally in postlarval instars, rarely fang-like, occasionally absent; larvae of aquatic species may have a palpal thumb-claw process. Terrestrial, aquatic and marine species; free-living or parasitic 3

3. Aquatic species (one family, the PONTARACHNIDAE, is intertidal), often modified for swimming. Larvae occasionally suppressed; parasitic on insects, sometimes associated with mollusks, decapods, sponges or gastropods . . . . Cohort ELEUTHERENGONINA, Subcohort PARASITENGONAE, Phalanx HYDRACHNIDIA[1] ... 77

— Primarily terrestrial groups, rarely aquatic (one family, the HALACARIDAE, comprises mostly marine forms) 4

4. Strongly armored forms, generally bright orange or yellow in life. Usually with a pair of lens-like pustules laterally on idiosoma, just posterior to ocelli; with two pairs of prodorsal sensilla. Coxae heavily sclerotized, expanded, contiguous. . . . . Cohort LABIDOSTOMMATINA, Superfamily LABIDOSTOMMATOIDEA (Plate 62, p. 317) Family LABIDOSTOMMATIDAE

— Soft-bodied or armored, but without a pair of lens-like pustules or coxal development as above. With one or two pairs of prodorsal sensilla, or sensilla absent 5

5. Prodorsal sensilla present, hair-like or variously produced but differing in form from other prodorsal setae, generally inserted in distinctive bothridia which are quite unlike insertions of surrounding setae (Plates 58-1 and 62-5, pp. 313, 317) (the family NEMATALYCIDAE lacks sensilla, but has instead an unpaired anteromedian prodorsal seta (*pf,* Plate 60-2, p. 315)). Femora of legs usually subdivided (Plate 64-1, p. 319) 6

---

[1] The portion of this key which deals with the water mites (couplets 77-131) is derived from Cook (1974).

— Prodorsal sensilla absent, prodorsal setae paired, similar in size and form, or absent. Femora of legs generally undivided 35

6. Tarsus I with one or more recumbent solenidia lying in a specialized membranous depression. Fixed digit of chelicera usually reduced, occasionally distorted or well developed . . . . .Cohort EUPODINA, Superfamily EUPODOIDEA ... 7

— Solenidia of tarsus I erect or bent, never recumbent, arising from small circular membranous depressions (Plage 58-6, p. 313); sometimes absent. Chelicerae various. 10

7. With small and sometimes distorted cheliceral digits; recumbent tarsal solenidia elbowed basally, usually not produced proximal to insertions (Plate 63-6 and 7, p. 318) 8

— With large opposed cheliceral shears; each recumbent solenidion produced proximally behind its insertion (*rhagidial organs,* Plate 64-1, p. 319). Free-living predators. . . . .(Plate 64, p. 319) Family RHAGIDIIDAE

8. Idiosoma ornamented in a tuberculate pattern, often with a distinctive "V" or "Y" suture dorsally. With a sclerotized roof-like prodorsal projection over the gnathosoma. Free-living . . . .(Plate 64, p. 319) Family PENTHALODIDAE

— Soft-bodied, with or without a roof-like prodorsal projection over gnathosoma. Free-living or phytophagous 9

9. Anal aperture ventral; femur IV often greatly enlarged. Predators and fungivores, rarely associated with vertebrates . . . .(Plate 63, p. 318) Family EUPODIDAE

— Anal aperture dorsal or terminal; femur IV normally developed. Phytophagous species. . . . . (Plate 65, p. 320) Family PENTHALEIDAE

10. Chelicerae free, broad basally and narrowed apically, often chelate or hooked terminally, inserted on elongate rostral "snout" in such a way that considerable lateral motion is possible. With trichobothria on various leg segments. Free-living. . . . . .Cohort EUPODINA, Superfamily BDELLOIDEA ... 11

— Chelicerae free or fused but without potential for great lateral motion, inserted on or fused with short rostrum. Leg trichobothria absent 12

11. Palpi long, antenniform, often elbowed, typically with strong distal setae. With three pairs of genital acetabula. . . . .(Plate 68, p. 323) Family BDELLIDAE

— Palpi extending beyond gnathosoma or barely equal to chelae in length, terminating in a tarsal claw (*Parabonzia* is an exception). With two or three pairs of genital acetabula (rarely none) . . . . (Plate 68, p. 323) Family CUNAXIDAE

12. Cheliceral digits opposed, cheliceral bases free . . . . Supercohort ENDEOSTIGMATIDES ... 13

— Fixed cheliceral digits reduced, cheliceral bases fused or contiguous along their mesal edges, movable digits short, needle-like. Free-living or parasitic forms . . . . .Cohort EUPODINA, Superfamily TYDEOIDEA ... 22

13. Idiosoma covered by an entire or divided dorsal shield, often with a constriction at level of coxae IV. Rutella absent; extended fixed cheliceral member with two or more dorsal setae; palp with four articles . . . . .Cohort ADAMYSTINA, Superfamily ADAMYSTOIDEA (Plate 61, p. 316) Family ADAMYSTIDAE

— Idiosoma often ornamented but not covered by an entire or divided dorsal shield. Rutella present, sometimes reduced to fleshy lobes; extended fixed chela with one or two setae; palp with five (exceptionally four) articles. . . . . .Cohort PACHYGNATHINA, Superfamily PACHYGNATHOIDEA ... 14

14. Idiosoma greatly elongated, annulate, hysterosoma approximately twice the length of propodosoma; sensilla absent but with an unpaired anteromedian prodorsal seta *(pf)*. Free-living in soil, sand or in interstitial subterranean waters. . . . . (Plate 60, p. 315) Family NEMATALYCIDAE

— Idiosoma various, sometimes elongate but not approaching the dimensions cited above; with one or two pairs of sensilla, seta *pf* absent 15

15. Tarsi I-IV each with two claws and empodium 16

— Some or all of the tarsi lacking claws, empodia, or both 18

16. Adults with two pairs of genital acetabula. Fixed chela often somewhat reduced. With one or two pairs of prodorsal sensilla, eyes absent . . . . (Plate 58, p. 313) Family LORDALYCHIDAE

— Adults with three pairs of genital acetabula. Fixed chelae normally produced 17

17. With one pair of prodorsal sensilla. . . . .(Plate 60, p. 315) Family TERPNACARIDAE

— With two pairs of prodorsal sensilla (one pair may be capitate). . . . . (Plate 57, p. 312) Family PACHYGNATHIDAE

18. Claws lacking on all tarsi, claw-like or pulvillar empodia always present 19

— Claws present on tarsi II-IV, present or absent on tarsus I, with or without empodia . . 21

19. Empodia weakly developed, arising from a membranous pulvillar pad; with one pair of clavate prodorsal sensilla. . . . . (Plate 61, p. 316) Family PEDICULOCHELIDAE

— Empodia distinct, claw-like; with one or two pairs of setate sensilla 20

20. With one pair of prodorsal sensilla. . . . . . . (Plate 57, p. 312) Family ALICORHAGIIDAE

— With two pairs of prodorsal sensilla. . . . . (Plate 59, p. 314) Family NANORCHESTIDAE

21. With one pair of prodorsal sensilla; tarsus I without true claws, with or without empodium. . . . .(Plate 59, p. 314) Family OEHSERCHESTIDAE

— With two pairs of prodorsal sensilla; tarsus I with true claws, empodial claw lacking. . . . . (Plate 58, p. 313) Family SPHAEROLICHIDAE

22. Hysterosoma divided transversely by a postpedal suture. Empodia reduced, claw-like; tarsus I with more than one solenidion. Free-living. . . . .(Plate 65, p. 320) Family PARATYDEIDAE

— Hysterosoma without postpedal suture. Empodia pad- or hair-like; tarsus I with one solenidion 23

23. With an *ereynetal organ* opening in the distal portion of tibia I, consisting of an inverted sac-like structure in a narrow duct which opens at or near the insertion of a simple or highly modified seta. With or without genital acetabula, dorsal opisthosomal sensilla, net-like ornamentation on the legs, or distal tarsal concavities on tarsi I-II. Free-living, associated with gastropods, or nasal parasites of vertebrates . . . . (Plate 67, p. 322) Family EREYNETIDAE

— Without ereynetal organ. Genital acetabula, opisthosomal sensilla, leg ornamentation and tarsal concavities absent 24

24. Female genital opening transverse, male with a large terminal aedeagus. Without claws or empodium on tarsus I. Associated with Orthoptera (locusts) and Blattaria. . . . .(Plate 66, p. 321) Family IOLINIDAE

— Female genital opening longitudinal; male genital area posteroventral, intromittent organ indistinct. Tarsus I with or without claws and empodium. Free-living, occasionally phytophagous or associated with insects . . (Plate 66, p. 321) Family TYDEIDAE

25. Chelicerae fused into a subtrapezoidal stylophore which is not integrated with the rostral elements; palpi pronounced, 3-segmented, extending well beyond gnathosomal extremities 26

— Cheliceral stylophore fused with rostral elements into a capsulate gnathosoma; palpi reduced, usually with no more than two non-articulating segments, extending to or slightly beyond gnathosomal extremities. 27

26. Gnathosoma without dorsal setae; female with a pair of posteromedial prodorsal *ampulliform organs.* Tarsi II-IV without claws, empodia discoid, tarsus I without claws or empodium. Subelytral associates of passalid beetles . . . . Superfamily HETEROCHEYLOIDEA (Plate 70, p. 325) Family HETEROCHEYLIDAE

— Gnathosoma with two pairs of dorsal setae; female with a pair of lateral prodorsal capitate bothridial sensilla. Tarsi II-IV with claws and a stalked membranous empodium, tarsus I with sessile claws and no empodium. Free-living or associated with passalid beetles . . . . Superfamily TARSOCHEYLOIDEA (Plate 70, p. 325) Family TARSOCHEYLIDAE

27. Legs IV of female with separate femora and genua, often with claws and membranous empodia; legs IV of male 5-segmented 28

— Legs IV of female, when present, with fused femora and genua and without claws or empodia; legs IV of male, when present, usually 4-segmented, with a single sessile claw. . . . . Superfamily TARSONEMOIDEA ... 29

28. Trochanter IV subtriangular (similar to trochanter III) or, if compressed, wider than long. Gnathosoma of male never reduced to tubiform appendage. Insect or plant associates. . . . . .Superfamily PYEMOTOIDEA ... 30

— Trochanter IV quadrangular, generally longer than wide. Gnathosoma of male tubiform, unsuitable for food uptake . . . . Superfamily PYGMEPHOROIDEA ... 33

29. Female always with 4 pairs of legs, leg IV attenuate, 3-segmented. Male with legs IV always ventrally inserted, 3- or 4-segmented. Plant or fungus feeders, or insect associates . . . .(Plate 74, p. 329) Family TARSONEMIDAE[1]

— Female typically with 1-3 pairs of legs (if legs IV present, then they are 5-segmented). Male with 3-4 pairs of legs (if legs IV present, then they are 5-segmented, or reduced and dorsally inserted). Associated with insects. . . . .(Plate 75, p. 330) Family PODAPOLIPIDAE

30. Tarsus I with 2 claws. . . . . (Plate 70, p. 325) Family DOLICHOCYBIDAE

---

[1]Includes the family ACARAPIDAE of Mahunka (1970).

— Tarsus I with 1 claw 31

31. Coxal fields I-II with 4-6 pairs of setae; gnathosoma of male and female normally developed . . . . (Plate 71, p. 326) Family PYEMOTIDAE

— Coxal fields I-II with three pairs of setae, or less; gnathosoma either very large, or reduced and fused into propodosoma 32

32. Bothridial sensilla present; gnathosoma very large, wider than anterior margin of propodosoma; coxal fields I-II with three pairs of setae . . . . (Plate 72, p. 327) Family CARABOACARIDAE

— Bothridial sensilla absent; gnathosoma reduced, partly or completely fused into propodosoma; coxal fields I-II with fewer than three pairs of setae . . . . (Plate 71, p. 326) Family ACAROPHENACIDAE

33. Podonotal shield large, covering propodosoma completely; free margin of shield striated, forming a roof over gnathosoma; distance between insertions of legs II-III equal to that between legs III-IV. . . . . (Plate 72, p. 327) Family SCUTACARIDAE

— Podonotal shield generally smaller and not covering entire propodosoma, or absent (if covered in dorsal view, then shield without striated free margin); distance between insertions of legs II-III at least twice that of distance between legs III-IV 34

34. Propodosoma with 2-3 pairs of setae; femur I with 3-5 setae (if only three, then dorsal seta modified, not setiform) . . . .(Plate 73, p. 328) Family PYGMEPHORIDAE[1]

— Propodosoma with only one pair of setae; femur I with three setiform setae, dorsal seta unmodified. . . . . (Plate 73, p. 328) Family MICRODISPIDAE

35. Primarily marine, estuarine or arenicolous species, sometimes aquatic; typically with four dorsal and four ventral shields, although partial loss may occur; with three or four palpal segments. Predaceous, algivorous or parasitic . . . . .Cohort EUPODINA, Superfamily HALACAROIDEA (Plate 69, p. 324) Family HALACARIDAE

— Terrestrial species, with variable numbers of palpal segments and idiosomal shields; discrete ventral shields often absent 36

36. Gnathosoma elongate, enclosed within a dorsal sheath arising from the prodorsum; cheliceral bases free. Free-living species . . . . .Subcohort RAPHIGNATHAE, Superfamily RAPHIGNATHOIDEA (Plate 77, p. 332) Family CRYPTOGNATHIDAE

— Gnathosoma not as above, cheliceral bases fused into a stylophore. 37

37. Cheliceral stylophore not fused with rostral elements; movable cheliceral digits stylettiform. Often with fewer than four pairs of legs . . . . .Subcohort RAPHIGNATHAE (pars) ... 38

— Stylophore and rostral elements fused into a single structure; movable cheliceral digits may be stylettiform, or absent . . . . .Subcohort RAPHIGNATHAE, Superfamily CHEYLETOIDEA (pars) ... 43

38. Idiosoma greatly elongated, annulate; with only two pairs of legs in all stadia; without true claws. Phytophagous. . . . . Superfamily ERIOPHYOIDEA ... 41

— Idiosoma not as above; with more than two pairs of legs in all stadia 39

[1] Includes the family SITEROPTIDAE of Mahunka (1970).

39. Stylophore short, rounded, with abbreviated stylets. Legs stilt-like, generally longer than rounded idiosoma . . . . (Plate 77, p. 332) Family CAMEROBIIDAE

— Stylophore various, stylets elongate-spinose or whip-like. Legs not as above, shorter than elongate idiosoma 40

40. Stylophore narrow and elongate, with spinose cheliceral stylets; palptarsus elongate, clearly exceeding length of tibia. Free-living predators. . . . . .Superfamily RAPHIGNATHOIDEA (Plate 78, p. 333) Family EUPALOPSELLIDAE

— Cheliceral stylets whip-like, recurved basally; palptarsus not elongate, generally shorter than tibia. Phytophagous species . . . . Superfamily TETRANYCHOIDEA (Plate 88, p. 343) Family TENUIPALPIDAE

41. Gnathosoma large (50-70 μ in length), tapering, abruptly bent ventrad near base. Chelicerae and associated stylets long, extending the entire length of the elongate gnathosoma. Dorsal shield setae, when present, always oriented so as to point anteriorly. Leaf vagrants or rust mites. . .(Plate 89, p. 344) Family RHYNCAPHYTOPTIDAE

— Mouthparts various but not as above; cheliceral stylets short, corresponding in length to the small gnathosoma (Plate 90-4, p. 345). Dorsal setae various, present or absent . 42

42. With one or two anterior setae on the dorsal shield; female spermathecal tubes long (Plate 90-1, p. 345) or short (Plate 89-6); if short, then extending anteriorly from genital opening and then curving posteriorly. With a distiventral tibial spur and a pair of subdorsal hysterosomal setae usually present. Bud mites, rust mites and grass sheath mites, some gall mites. . . . . (Plate 89, p. 344) Family SIERRAPHYTOPTIDAE

— With two anterior dorsal shield setae, or setae absent; female spermathecal tubes always short and extending laterally or posterolaterally from genital opening. Distiventral tibial spur and subdorsal hysterosomal setae absent. Gall and erineum mites, bud mites, rust mites and leaf vagrants . . . .(Plate 90, p. 345) Family ERIOPHYIDAE

43. Legs I-IV strongly telescoped; leg setae, when present, only on tibiae and tarsi. Idiosomal setation lacking or limited to one or two pairs of minute setae. 44

— Legs II-IV normally developed, often short, legs I may be telescoped; setae present on pretibial segments of all legs. Idiosoma with more than two pairs of well developed setae 45

44. Palpi 3-segmented, chelicerae tiny, stylettiform. Elongate, annulate, worm-like species found in hair follicles, surface glands and ducts of vertebrates. . . . . (Plate 84, p. 339) Family DEMODICIDAE

— Palpi each comprised of a single elongate, retrorse fang-like segment, chelicerae absent. Opisthosoma reduced, virtually undefinable. Ovoid species parasitic in the cloacal mucosa of turtles. . . . . (Plate 85, p. 340) Family CLOACARIDAE

45. Legs I-IV similarly developed, ambulatory 46

— Legs I telescoped, adapted for clasping hairs of mammal hosts . . . . (Plate 82, p. 337) Family MYOBIIDAE

46. Femora I-IV each with a strong ventral spur; empodia pad-like, sometimes bilobed. Tiny (100-150 μ) circular mites parasitic in skin of mammals . . . . (Plate 83, p. 338) Family PSORERGATIDAE

— Without strong femoral spurs, empodia not as above. 47

47. Claws present, empodia each with a double row of tenent hairs. Elongate species found in the quills of birds . . . . (Plate 83, p. 338) Family SYRINGOPHILIDAE

— Claws absent, tarsi each with a terminal cup-like pretarsus from which a branched empodium arises. Circular forms, parasitic on snakes . . . .

48. Adults and nymphs often strongly hypertrichous; idiosomal setae not arranged in orderly rows, often forming a thick pelage; with one or two pairs of prodorsal sensilla usually inserted in a median crista metopica (Plate 94-6, p. 349). Larvae usually heteromorphic, parasitic; adults and nymphs free-living predators . . . . .Cohor ELEUTHERENGONINA, Subcohort PARASITENGONAE, Phalanx TROMBIDIA ... 66

— Body setae of adults and nymphs relatively few, generally arranged in transverse rows, rarely with hypertrichous patches (some PTERYGOSOMATIDAE); prodorsal sensilla usually absent (if present, then not inserted in a crista). Larvae homeomorphic, similar to nymphs and adults in habit 49

49. Cheliceral bases contiguous, fused, or partially fused, without potential for lateral motion, movable cheliceral digits stylettiform, aciculate or whip-like; with only one palptibial claw. Rarely with genital acetabula. 50

— Chelicerae freely articulated at bases so that a degree of lateral movement is possible, movable chela strongly hooked, aciculate, weakly chelate or sickle-shaped; palpal tibia with 1-3 claws. Often with genital acetabula . . . . .Subcohort ANYSTAE ... 62

50. Cheliceral bases forming a stylophore which is fused with the rostrum. Predators or parasites . . Subcohort RAPHIGNATHAE, Superfamily CHEYLETOIDEA (pars) . . . 51

— Cheliceral bases closely contiguous or fused with each other, rarely integrated with other gnathosomal elements. Phytophages or predators 53

51. Tarsi II-IV, and usually tarsi I, with claws and/or empodia 52

— Tarsi III-IV without claws or empodia, terminating instead in a series of long whip-like setae. Parasitic in skin and feathers of birds. . . . . (Plate 84, p. 339) Family HARPYRHYNCHIDAE

52. Palpal tarsus ornamented with sickle- and/or comb-like setae, palptibial claws oriented in horizontal plane for grasping prey. Aedeagus of male posteroventral or terminal. Primarily free-living predators (certain species are associated with insects or vertebrates) . . . . (Plate 81, p. 336) Family CHEYLETIDAE

— Palpal tarsus without specialized setae, often greatly reduced; palptibial claws usually curved ventrally, developed for fixation in fur or feathers, sometimes greatly enlarged; palpal femora may have well developed spur-like apophyses. Aedeagus of male dorsal. Parasites of birds and mammals. . . . . (Plate 81, p. 336) Family CHEYLETIELLIDAE

53. Chelicerae long, recurved and whip-like, arising from a stylophore; female genital aperture transverse, rarely appearing triangular. Phytophagous forms . . . . Superfamily TETRANYCHOIDEA (pars) ... 54

— Chelicerae short, aciculate; female genital aperture longitudinal. Free-living . . . . Superfamily RAPHIGNATHOIDEA (pars) ... 57

54. Lateral prodorsal eyes absent. Elongate species with indications of a postpedal furrow . . . . (Plate 88, p. 343) Family LINOTETRANIDAE

— Lateral prodorsal eyes present 55

55. Each claw with a subterminal series of nail-like tenent hairs (Plate 88-5, p. 343), empodia similarly ornamented. With a series of flagelliform or dendritic setae posteriorly 56

— Each claw reduced to a pair of tenent hairs (Plate 87-5, p. 342), or claw-like and ornamented with a fringe of tenent hairs (Plate 87-4); empodia claw- or pad-like, rarely obscure, often divided distally into a series of fine projections (Plate 87-6) or with a basal divided spur, empodia with or without tenent hairs. Without highly modified setae caudally . . . . (Plates 85-87, pp. 340-342) Family TETRANYCHIDAE

56. Dorsum of idiosoma with 44 fan-shaped setae; several dorsal setae of legs I-II similarly expanded. With a series of long flagelliform setae on the posterior margin of the idiosoma . . . . (Plate 88, p. 343) Family TUCKERELLIDAE

— Dorsal setae simple, with a series of short dendritic setae on the posterior margin of the idiosoma . . . . (Plate 88, p. 343) Family ALLOCHAETOPHORIDAE

57. Elongate species with five dorsomedian hysterosomal shields arranged end-to-end, eyes absent. Empodia absent, tarsus I with 6-8 solenidia. Adults with two-three pairs of genital acetabula. In soil. . . . . (Plate 76, p. 331) Family POMERANTZIIDAE

— Ovoid or round in shape, occasionally hemispherical (if elongate, then without dorsomedian shields), eyes present. Empodia present, tarsus I with one or two solenidia. Genital acetabula absent 58

58. With a narrow, arched transverse groove on the prodorsum extending laterally behind the large raised ocelli; groove opens internally into large paired flask- or funnel-shaped sacs (♀) or tubes (♂) which are easily seen in whole mount. Heavily sclerotized hemispherical mites in aquatic or semi-aquatic habitats . . . . (Plate 80, p. 335) Family HOMOCALIGIDAE

— Without prodorsal groove or internal structures as described above. Moderately or lightly sclerotized, more or less compressed species in terrestrial habitats 59

59. Small (± 400 μ), elongate, virtually unsclerotized species; fixed cheliceral digits fused basally, free distally. Tarsus I with two solenidia . . . . (Plate 80, p. 335) Family BARBUTIIDAE

— Mostly ovoid in shape; with a variety of shield arrangements dorsally, or shields absent (if absent, then peritremes extend into cheliceral bases); fixed cheliceral digits either free or totally coalesced into a stylophore.[1] Tarsus I with one or two solenidia 60

60. Coxae II-III contiguous . . . . (Plate 76, p. 331) Family RAPHIGNATHIDAE

— Coxae II-III narrowly or widely separated 61

61. Cheliceral bases fused to form a stylophore into which a pair of sinuous peritremes extend. Dorsal shields absent or weakly developed. . . . . (Plate 78, p. 333) Family CALIGONELLIDAE

— Cheliceral bases usually contiguous but not insensibly fused; peritremes absent. Dorsum completely or partially covered by shields . . . . (Plate 79, p. 334) Family STIGMAEIDAE

[1]Partial fusion of cheliceral bases may occur in the genus *Mediolata* (family STIGMAEIDAE). Where fusion occurs, tarsus I bears a single solenidion.

62. Heavily armored, with eight dorsal shields arranged in characteristic fashion. Movable cheliceral digit sickle-shaped, palptibia with a single large terminal claw. Legs I with several long, spinose internal setae inserted in horny protuberances. Free-living. . . . . Superfamily CAECULOIDEA (Plate 91, p. 346) Family CAECULIDAE

— Soft-bodied, occasionally with weakly sclerotized prodorsal shield or shields and dorsal setal bases. Movable cheliceral digits hooked or chelate distally, palptibia with 1-3 terminal claws. Legs hypertrichous or normally setate, without spinose internal setae. Free-living or parasitic . . . . Superfamily ANYSTOIDEA ... 63

63. Palptarsus long and prominent, usually longer than palptibia and extending well beyond palpal claw(s); movable chelae strongly hooked distally. Free-living . . . . (Plate 92, p. 347) Family ANYSTIDAE

— Palptarsus not prominent, often greatly reduced; chelicerae not as above 64

64. Claws of tarsus I broadly bipectinate, other tarsal claws may be similarly ornamented. Often with three pairs of weakly developed genital acetabula. Free-living . . . . (Plate 93, p. 348) Family TENERIFFIIDAE

— Claws of tarsus I not as above, sometimes absent. Genital acetabula absent 65

65. Claws each ornamented with one to several pairs of tenent hairs, empodia absent. Parasites of lizards and arthropods . . . (Plate 93, p. 348) Family PTERYGOSOMATIDAE

— Tarsi with or without claws, often with only a stalked membranous empodium. Free-living . . . . (Plate 92, p. 347) Family PSEUDOCHEYLIDAE

66. Movable chelae of nymph and adults short, curved, sometimes toothed distally; chelicerae and associated gnathosomal entities fixed, not retractible into idiosoma; with or without empodia. Larva heteromorphic, with urstigmata and an anal opening; with one pair of bothridial sensilla on the prodorsal shield (scutum), coxae I-II usually contiguous; parasites of vertebrates and invertebrates . . . . Superfamily TROMBIDIOIDEA ... 69

— Movable chelae of nymph and adults long, straight, either independently retractile or integrated with wholly retractile gnathosoma, empodia absent. Larva homeomorphic or heteromorphic; larval morphology variable, but usually with two pairs of bothridial sensilla on the prodorsal shield (scutum) (the CALYPTOSTOMATIDAE have only one pair) 67

67. Adults and nymph with a single pair of median prodorsal sensilla in raised insertions, crista absent. Gnathosoma retractile; with two pairs of genital acetabula. Larva homeomorphic, with urstigmata and an anal opening, coxae I-II contiguous; parasites of Diptera . . . . Superfamily CALYPTOSTOMATOIDEA (Plate 94, p. 349) Family CALYPTOSTOMATIDAE

— Adults and nymph with two pairs of sensilla in a median crista or platelet. Chelicerae or entire gnathosoma may be retractile; genital acetabula generally absent. Larva heteromorphic, urstigmata and anal opening absent, coxae I-II distinctly separated; parasites of arthropods, occasionally free-living . . . . Superfamily ERYTHRAEOIDEA ... 68

68. Gnathosoma of adults and nymph greatly narrowed, retractible into idiosoma; propodosoma generally narrowed and elongate anteriorly, with an ossiform crista or prodorsal scutellum. Body setae generally flattened, leaf-like, broad and serrate. . . . . (Plate 95, p. 350) Family SMARIDIDAE

— Gnathosoma normally developed, chelicerae independently retractible; propodosoma not narrowed anteriorly, generally with an ossiform crista. Body setae simple or pectinate, often somewhat broadened. Larva parasitic on insects and arachnids, or predatory on mites or insects .... (Plates 94-95, pp. 349-350) Family ERYTHRAEIDAE

69. With one pair of prodorsal sensilla (rarely with one or three bothridial hairs), idiosomal setae not tylochorous (i.e., not borne on individual platelets)............... 71

— With two pairs of prodorsal sensilla (if only one pair, then idiosomal setae are tylochorous, or dorsum has hummocks of long, matted setae) 70

70. Dorsal setae of all stages simple, tylochorous. Posterior bothridial sensilla of larva (and of some postlarval forms) capitate; larva with one or two claws on each tarsus, claw-like empodium present. Postlarval forms mostly in damp habitats, larva parasitic on certain Diptera.......... (Plates 95-96, pp. 350-351) Family JOHNSTONIANIDAE

— Dorsal setae simple or barbed, usually not tylochorous, often long and matted in postlarval stadia. Both sensillary pairs simple; larva with two claws and a claw-like empodium on each tarsus .... (Plate 96, p. 351) Family CHYZERIDAE

71. Palptibia ornamented externally with a comb of spines, with an additional spine cluster internally.................. (Plate 97, p. 352) Family PODOTHROMBIIDAE

— Without palptibial ornamentation as above, spines may be present externally but are lacking on the internal face.. 72

72. Crista metopica generally reduced or absent (if present, then sensilla usually arise in adjacent integument), sensilla attenuate, simple. Often highly ornamented species with strongly granular or tuberculate integumental patterns; setae simple, barbed or pedicellate. Larva with simple or bifid palptibial claw, tarsi of legs each armed with one claw; parasites of Diptera .... (Plate 96, p. 351) Family TROMBELLIDAE

— Crista metopica present; sensilla, dorsal setae, and larval morphology various........ 73

73. Idiosomal setae trifurcate or mushroom-shaped; with two or three pairs of genital acetabula. Larva with two anteromedian setae on prodorsal scutum, tarsi each with one claw; parasites of Coleoptera and of Diptera (Streblidae).................. ....(Plate 96, p. 351) Family NEOTROMBIDIIDAE

— Idiosomal setae simple, barbed or papillate, not as above. Larva with one or two anteromedian scutal setae, tarsi with one claw, or two claws and empodium 74

74. Anterior extension of crista metopica *(tectum)* with two, one, or no setae. Larva typically with a single prodorsal shield, mostly parasitic on vertebrates 76

— Tectum (when present) with many setae. Larva usually with more than one dorsal shield, parasitic on arthropods.. 75

75. Aquatic species; sensilla inserted in anterior portion of crista. Tarsi each with two claws and a claw-like empodium. Larval relationships obscure.................. .... Family STYGOTHROMBIIDAE

— Terrestrial species, sometimes found in damp habitats; sensilla variously inserted, often median in position. Empodia present or absent. Larvae parasitic on a variety of arthropods .... (Plates 97-98, pp. 352-353) Family TROMBIDIIDAE

76. Adults and nymph often broadly expanded in the humeral region, with two, one or no setae on the tectum. Larva with one anteromedian scutal seta, or none; legs I with seven segments; parasitic on vertebrates, occasionally on invertebrates . . . . (Plates 98-99, pp. 353-354) Family TROMBICULIDAE

— Adults and nymph normally oval in shape, with two setae on the tectum. Larva with two anteromedian scutal setae, legs I with six segments; mostly parasitic on birds, bats, and amphibians, occasionally on invertebrates . . . . (Plate 99, p. 354) Family LEEUWENHOEKIIDAE

77. Marine intertidal species, genital acetabula absent. . . . . Superfamily HYGROBATOIDEA (pars) (Plate 106, p. 361) Family PONTARACHNIDAE

— Aquatic species, with or without genital acetabula 78

78. Genital acetabula absent; dorsal *glandularia* (Plate 103-1, p. 358), when present, on platelets adjacent to but separate from setal platelets. Larva without urstigmata . . . . Superfamily HYDROVOLZIOIDEA (Plate 100, p. 355) Family HYDROVOLZIIDAE

— Genital acetabula present; dorsal glandularia, when present, each share a platelet with an associated seta. Larva with urstigmata 79

79. Palptibia produced distidorsally, lateral eyes separately encapsulated; female with a tubular ovipositor. Larva with gnathosoma greatly enlarged, over 1/3 length of idiosoma; palp without thumb-claw process; legs with some setae elongated as swimming hairs, with five postcoxal segments (femur undivided) . . . . .Superfamily HYDRACHNOIDEA (Plate 100, p. 355) Family HYDRACHNIDAE

— Palpi various, eyes on common ocular shield or separated; female without tubular ovipositor. Larval gnathosoma not as above; typically with a palpal thumb-claw process 80

80. Buccal opening surrounded by a distinct membranous fringe; lateral eyes, when present, on a common median ocular shield; palpi simple. Larva with or without swimming hairs, legs with six postcoxal segments (femur divided) . . . . Superfamily EYLAOIDEA ... 81

— Membranous fringe around buccal opening indistinct or absent; lateral eyes, when present, not incorporated in a median ocular shield; palpi simple or palptibia produced distidorsally 83

81. Genital acetabula widely scattered, not lying on or surrounded by acetabular plates . . 82

— Genital acetabula either on acetabular plates or contained in two pairs of ring-like sclerites . . . . (Plate 101, p. 356) Family PIERSIGIIDAE

82. Ocular shield typically as wide or wider than long, coxae III-IV joined only basally on their respective sides . . . . (Plate 101, p. 356) Family EYLAIDAE

— Ocular shield much longer than wide, coxae III-IV broadly joined . . . . Plate 101, p. 356) Family LIMNOCHARIDAE

83. With three (occasionally six) pairs of genital acetabula medially in the gonopore, usually in parallel rows and covered by large, well developed genital valves (occasionally reduced). Larva with or without swimming hairs, legs with five postcoxal segments; urstigmata present. . . . . Superfamily LEBERTIOIDEA ... 84

— Genital acetabula not arranged in median parallel rows, genital valves present or absent. Larval morphology various, always with urstigmata 90

84. Terminal palpal segment broad, spatulate; terminal setae also spatulate . . . . (Plate 106, p. 361) Family RUTRIPALPIDAE

— Palpi not spatulate terminally 85

85. Coxae IV broadened, each completely or partially surrounding a medially placed gland . . . . (Plate 106, p. 361) Family TEUTONIIDAE

— Coxae IV either without incorporated glands or, if glands are present, they are laterally placed 86

86. Dorsal and ventral shields present; dorsal sclerotization typically consisting of a large median shield and two or four adjacent anterior shields or several peripheral platelets; fusion of anterior shields may occur. Ventrally with a Y-shaped suture extending anteriorly from genital field to tips of coxae I. . . . . (Plate 105, p. 360) Family TORRENTICOLIDAE

— Without above combination of characters 87

87. Legs inserted far forward on body; palpal femur and genu without ventral setae or apophyses . . . . (Plate 105, p. 360) Family OXIDAE

— Legs normally inserted (if placed far forward, then palpal femur and genu with setae or apophyses) 88

88. Dorsal and ventral shields absent, ventrally with a Y-shaped suture extending anteriorly from genital field to tips of coxae I. Palpal femur with a long distiventral seta, palpal genu with several long dorsal setae . . . . (Plates 104-105, pp. 359-360) Family LEBERTIIDAE

— Without the above combination of characters 89

89. Tarsus IV with well developed claws; lateral eyes in capsules, never with an entire ventral shield . . . . (Plate 105, p. 360) Family SPERCHONTIDAE

— Tarsus IV without claws (if claws are present, then with an entire ventral shield (*Bharatonia* is an exception)) . . . . (Plate 106, p. 361) Family ANISITSIELLIDAE

90. Bright red or orange mites, generally with movable genital flaps in both sexes; palpi simple or palptibia produced distidorsally. Larva with or without swimming hairs, legs with six postcoxal segments; dorsal shield small, covering less than 1/3 the length of the idiosoma; shield with one, two, or four pairs of setae . . . . Superfamily HYDRYPHANTOIDEA ... 91

— Variously colored mites, genital flaps absent; palpi various; simple, produced distidorsally or chelate. Legs of larva with five postcoxal segments; dorsal shield large, covering more than 1/2 the length of the idiosoma; shield with four or five pairs of setae . . . . Superfamilies HYGROBATOIDEA and ARRENUROIDEA[1,2] 96

---

[1]Adult HYGROBATOIDEA and ARRENUROIDEA illustrate great morphological diversity, and many character states are found to occur in both superfamilies. While a family key to adults can be constructed which copes with these similarities, it is presently impractical to attempt a superfamilial separation based solely on adult characters. However, larvae of these superfamilies may easily be distinguished on the basis of coxal plate morphology. Coxal plates of arrenuroid larvae are completely separated from one another by distinct suture lines, while plates II-III (and often I-II) of hygrobatoid larvae are fused, at least medially.

[2]In couplets 96-131, families of the Hygrobatoidea are indicated by an asterisk.

91. Palpal femur with a distiventral row of more than 12 distinctive setae on the internal aspect; palpal tibia with a row of over 12 setae on the ventral axis . . . . (Plate 103, p. 358) Family CTENOTHYADIDAE

— Palpal femur without distinctive distal setae or, if present, numbering less than four; without row of palptibial setae as described above 92

92. Palpal femur either without distal setae or with only one unmodified hair 93

— Palpal femur with two or three modified distal setae. 94

93. Genital flaps present, with numberous acetabula; palp not produced distidorsally, rarely with one distiventral seta on the palpal femur. Dorsum with closely fitting porous platelets or more widely spaced reticulate platelets . . . .(Plate 104, p. 359) Family RHYNCOHYDRACARIDAE

— Genital flaps with only three pairs of acetabula, flaps occasionally reduced, or rarely lost; palptibia typically produced distidorsally (if not, then the idiosoma is unsclerotized and greatly elongated, or the acetabula are stalked), palpal femur without distiventral seta. Dorsum variously ornamented . . . . (Plate 103, p. 358) Family HYDRYPHANTIDAE

94. Palpal femur with two long, thickened and smooth distal bi- or trifurcate setae; with a collar of setae around the camerostome which extend medioventrally between coxae I. Leathery species found only in thermal waters . . . . (Plate 104, p. 359) Family THERMACARIDAE

— With two or three distal setae on palpal femur, setae pectinate or serrate, not distally furcated; without setal collar as above. Not found in thermal waters 95

95. Dorsum and venter with reticulate platelets which occupy most of the body surface and surround the glandularia; lateral eyes in capsules . . . . (Plate 103, p. 358) Family TERATOTHYADIDAE

— Dorsum and venter unsclerotized, without platelets; lateral eyes not encapsulated . . . . Family HYDRODROMIDAE

96. Coxae I-IV fused on their respective sides but widely separated medially; coxae I and IV each with a heavy peg-like seta ventrally. Parasitic in gill cavities of freshwater crabs . . . . (Plate 107, p. 362) Family ASTACOCROTONIDAE*

— Without above combination of characters 97

97. Idiosoma greatly elongated, unsclerotized, coxae inserted far anteriorly; genital acetabula numerous, located on two elongate acetabular plates in the female and on a single plate in the male. Palpal tibia with a heavy peg-like seta ventrally . . . . (Plate 107, p. 362) Family OMARTACARIDAE*

— Idiosomal and coxal combination not as above; with few or many acetabula. Palpal tibia with or without ventral seta 98

98. Palpal femur with a hair-like or peg-like ventral seta which is either sessile or on a tubercle. Tarsi IV usually without claws; rarely with idiosomal shields . . . . (Plate 107, p. 362) Family LIMNESIIDAE*

— Palpal femur without a hair-like or peg-like seta ventrally, but occasionally with more than one ventral seta 99

99. Tarsus I much shorter than tibia I; claws of tarsus I enlarged and able to fold against a spur-like dorsal projection of tarsus I to form a grasping organ . . . . (Plate 112, p. 367) Family MOMONIIDAE

— Tarsus and tibia I various, but not as above 100

100. Cheliceral claw armed with a row of small serrate teeth; with many small genital acetabula scattered over the venter, in addition to those on or surrounding the gonopore. . . . . (Plate 113, p. 368) Family HARPAGOPALPIDAE

— Cheliceral claw without serrations; acetabula generally confined either to the gonopore or to acetabular plates flanking the gonopore, rarely with a few acetabula on the ventral shield 101

101. Palpal tibia and tarsus fused into a sickle-shaped segment, suture visible between the shorter tibial and longer tarsal portions. Palpal femur with two or three ventral setae; with a pair of large posteromarginal glandularia ventrally . . . . (Plate 113, p. 368) Family NIPPONACARIDAE

— Without the above combination of characters 102

102. Palpal femur expanded, much larger than the combined remaining segments; palpal genu and tibia reduced, sometimes fused with each other; palpal tibia and tarsus partially fused, palpal femur with three or four medioventral setae which may be thickened. Dorsal and ventral shields present. (Plate 112, p. 367) Family BOGATIIDAE

— Palp without modifications described above 103

103. With a well developed ventral shield, dorsally with no more than a median plate; with 3-16 pairs of genital acetabula. . . . . . . (Plate 114, p. 369) Family ACALYPTONOTIDAE

— Dorsal and ventral shields present or absent but, if a ventral shield is present, dorsum is covered completely by a shield; genital acetabula various 104

104. Palptibia produced distidorsally; with dorsal and ventral shields. 105

— Palptibia not produced distidorsally; with or without shields 114

105. Epimeroglandularia 2 shifted anteromedially and located close together between coxae III-IV. . . . . (Plate 114, p. 369) Family CHAPPUISIDIDAE (pars)

— Epimeroglandularia 2 not shifted as above 106

106. Insertions of legs I-III beneath a rounded flap which extends anteriorly beyond coxae I . . . . (Plate 114, p. 369) Family UCHIDASTYGACARIDAE (pars)

— Without rounded coxal flap. 107

107. Epimera between coxae III-IV oriented so that, if extended posteriorly, they would meet in the genital field. . . . . (Plate 114, p. 369) Family NEOACARIDAE

— Coxal epimera not oriented as above. 108

108. Genital acetabula confined to gonopore, with well developed condyles associated with articulations of coxae IV. Epimeroglandularia 2 on medial margins of coxae III-IV . . . . (Plate 113, p. 368) Family KRENDOWSKIIDAE

— Without the above combination of characters 109

109. With seven or fewer pairs of genital acetabula in two parallel rows in the gonopore . . . . (Plate 112, p. 367) Family MIDEOPSIDAE (pars)

— Genital acetabula may or may not be confined to gonopore; if confined to gonopore, then either with more than seven pairs or arranged in more than two rows. 110

110. Palpal femur with two strong spinose setae ventrally or medioproximally; genital acetabula in male gonopore. . . (Plate 111, p. 366) Family HUNGAROHYDRACARIDAE

— Setae on palpal femur not as above; with or without acetabula in male gonopore. 111

111. With three pairs of genital acetabula . . . . . .(Plate 110, p. 365) Family KANTACARIDAE

— With considerably more than three pairs of genital acetabula 112

112. Posterior epimera of coxae IV obliterated; males with two pairs of genital acetabula in the gonopore itself, females without distinct acetabular plates . . . . (Plate 112, p. 367) Family ARENOHYDRACARIDAE

— Posterior epimera of coxae IV distinct; males either with many, or no, acetabula in the gonopore, females usually with distinct acetabular plates 113

113. Male gonopore without acetabula; acetabula in both sexes usually extending far laterally on distinct acetabular plates (females of *Africasia* are exceptions) . . . .(Plate 111, p. 366) Family ARRENURIDAE (pars)

— Male gonopore with numerous acetabula; acetabula of female on elongate acetabular plates which flank the gonopore and do not extend far laterally. . . . . (Plate 112, p. 367) Family ATHIENEMANNIIDAE

114. With three to five pairs of genital acetabula, all confined to the gonopore 115

— From three to several pairs of acetabula, typically on acetabular plates but occasionally in ventral integument (if some acetabula occur in the gonopore, then others are on the acetabular plates) 117

115. Palpal femur with two heavy distal setae; palpal tarsus long and narrow, subequal to palpal tibia . . . . (Plate 114, p. 369) Family CHAPPUISIDIDAE (pars)

— Palp not as described above. 116

116. Posterior pair of genital acetabula rounded or triangular, anterior two pairs oval . . . .(Plate 113, p. 368) Family UCHIDASTYGACARIDAE (pars)

— All genital acetabula more or less the same shape . . . . (Plate 112, p. 367) Family MIDEOPSIDAE (pars)

117. Dorsal and ventral shields present; genital field extending between widely separated coxae IV 118

— With or without sclerotization (if dorsal and ventral shields present, then coxae IV are not completely separated and the genital field is accordingly less extensive). 119

118. Insertions of trochanters IV more or less on a level with the epimeron separating coxae III-IV. . . . . (Plate 110, p. 365) Family MIDEIDAE

— Insertions of trochanters IV posterior to level of epimeron separating coxae III-IV; palpal tibia with a ventral peg-like seta; some ventral idiosomal setae branched . . . . (Plate 110, p. 365) Family LAVERSIIDAE

119. With dorsal and ventral shields; numerous genital acetabula on wing-like acetabular plates, without glandularia on coxae IV. Palpal tibia with a sessile spinose seta ventrally . . . .(Plate 111, p. 366) Family ARRENURIDAE, genus *Micruracaropsis*

— Without the above combination of characters 120

120. With three pairs of genital acetabula on well delineated acetabular plates flanking a gonopore which is completely incorporated into the ventral shield; epimera of coxae IV evident, coxae IV without glandularia . . . .(Plate 113, p. 368) Family UCHIDASTYGACARIDAE (pars)

— Without the above combination of characters 121

121. Body laterally compressed, distinctly higher than wide; dorsum with narrow median integumental strip which may contain small sclerites . . . . (Plate 109, p. 364) Family AXONOPSIDAE* (pars)

— Body not compressed as above, without a median unsclerotized strip 122

122. Coxae IV each with a coxoglandularium 123

— Coxoglandularia IV present, but in integument posterior to coxae IV 125

123. Typically soft-bodied (if dorsal and ventral shields are present, then a downcurved seta usually is present distally on tibia I).(Plate 110, p. 365) Family HYGROBATIDAE*[1]

— Dorsal and ventral shields present; tibia I without a downcurved distal seta; with 3-50 pairs of genital acetabula 124

124. With coxae I projecting well beyond anterior border of idiosoma; with eight or more pairs of genital acetabula . . . .(Plate 108, p. 363) Family ATURIDAE* (pars)

— Usually with fewer than eight pairs of acetabula (if more than eight pairs are present, then the anterior coxae extend only slightly or not at all beyond anterior border of idiosoma) . . . . (Plate 109, p. 364) Family AXONOPSIDAE* (pars)

125. Dorsal and ventral shields present 126

— Unsclerotized, or with scattered sclerites 129

126. With two pairs of glandularia located more or less in a transverse row between coxae IV and the genital field; insertions for legs IV partially covered by posterolaterally directed condyles; claws with distal clawlets . . . . (Plate 110, p. 365) Family FELTRIIDAE* (pars)

— Without the above combination of characters 127

127. With more than 10 pairs of genital acetabula; genu IV concave ventrally, the concavity lined with many peg-like setae (if concavity absent, then palpal tibia with a distal peg-like seta, and insertions of legs IV with associated projections) . . . .(Plate 109, p. 364) Family PIONIDAE* (pars)

— Without the above combination of characters 128

128. Insertions of legs IV with large projections which extend laterally or slightly posterolaterally (if projections are strongly posterolateral, then palpal tibia and tarsus with terminal spine-like projections) . . . .(Plate 108, p. 363) Family UNIONICOLIDAE* (pars)

---

[1]Only males of the hygrobatid genera *Atractidella, Corticacarus, Hygrobatopsis* and *Africacarus* possess dorsal and ventral shields. A series of separate platelets may be found posterodorsally in some species of *Aspidiobates*.

— Projections of leg IV insertions small or absent (if projections are present, they are directed posteriorly or strongly posterolaterally and the palpi are not terminally spinose as above) . . . . (Plate 108, p. 363) Family ATURIDAE* (pars)

129. With two pairs of glandularia in a more or less transverse line between coxae IV and the genital field; claws with clawlets. Palpal tibia usually without peg-like setae . . . . (Plate 110, p. 365) Family FELTRIIDAE* (pars)

— Without the above combination of characters 130

130. Without apodemes on the posterior aspect of fused coxal fields I-II . . . . (Plate 110, p. 365) Family HYGROBATIDAE* (pars)

— With pointed projecting apodemes on posterior aspect of each fused coxal field I-II. . . 131

131. Claws simple or with clawlets; posterior margins of coxae IV truncate or rounded. Chelicerae may be fused medially, or separate. . . . . (Plate 108, p. 363) Family UNIONICOLIDAE* (pars)

— Claws always with clawlets; posterior margins of coxae IV more or less angled, not truncate or rounded. Chelicerae always separate . . . . (Plate 109, p. 364) Family PIONIDAE* (pars)

PLATE 57

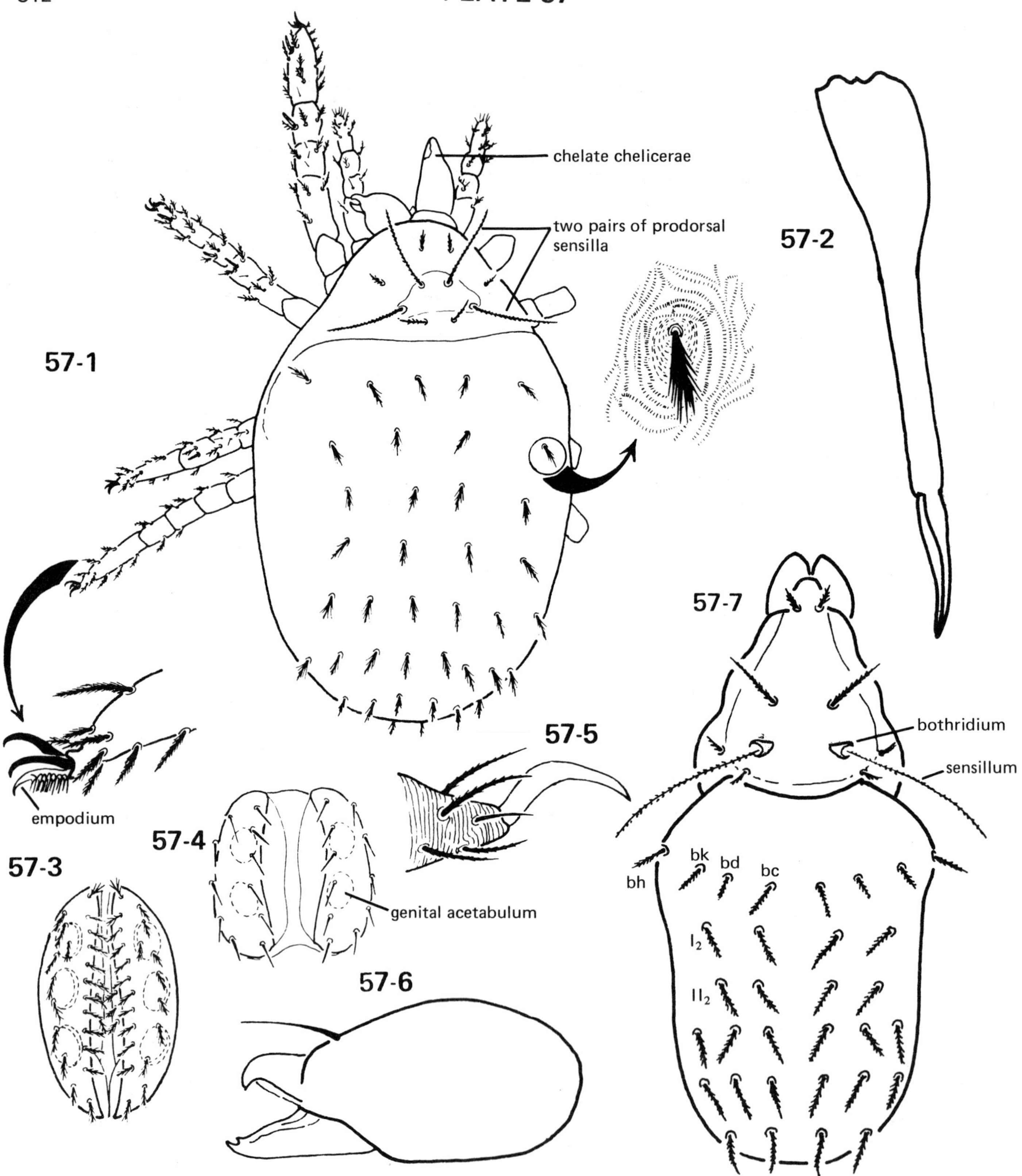

57-1 to 57-3; family PACHYGNATHIDAE. **57-1**; *Pachygnathus* sp. (Oregon, USA), dorsum of female with detail of tarsus IV and dorsal ornamentation: **57-2**; *Bimichaelia diadema* Grandjean, chelicera of female: **57-3**; *Pachygnathus* sp., genital region of female

57-4 to 57-7; family ALICORHAGIIDAE. **57-4**; *Alicorhagia* sp., genital region of female (after Grandjean 1939): **57-5**; *Alicorhagia* sp., apex of tarsus I (after Grandjean 1939): **57-6**; *Alicorhagia fragilis* Berlese (Japan), chelicera of female (after Shiba 1969): **57-7**; *A. fragilis,* dorsum of female (after Shiba 1969)

PLATE 58

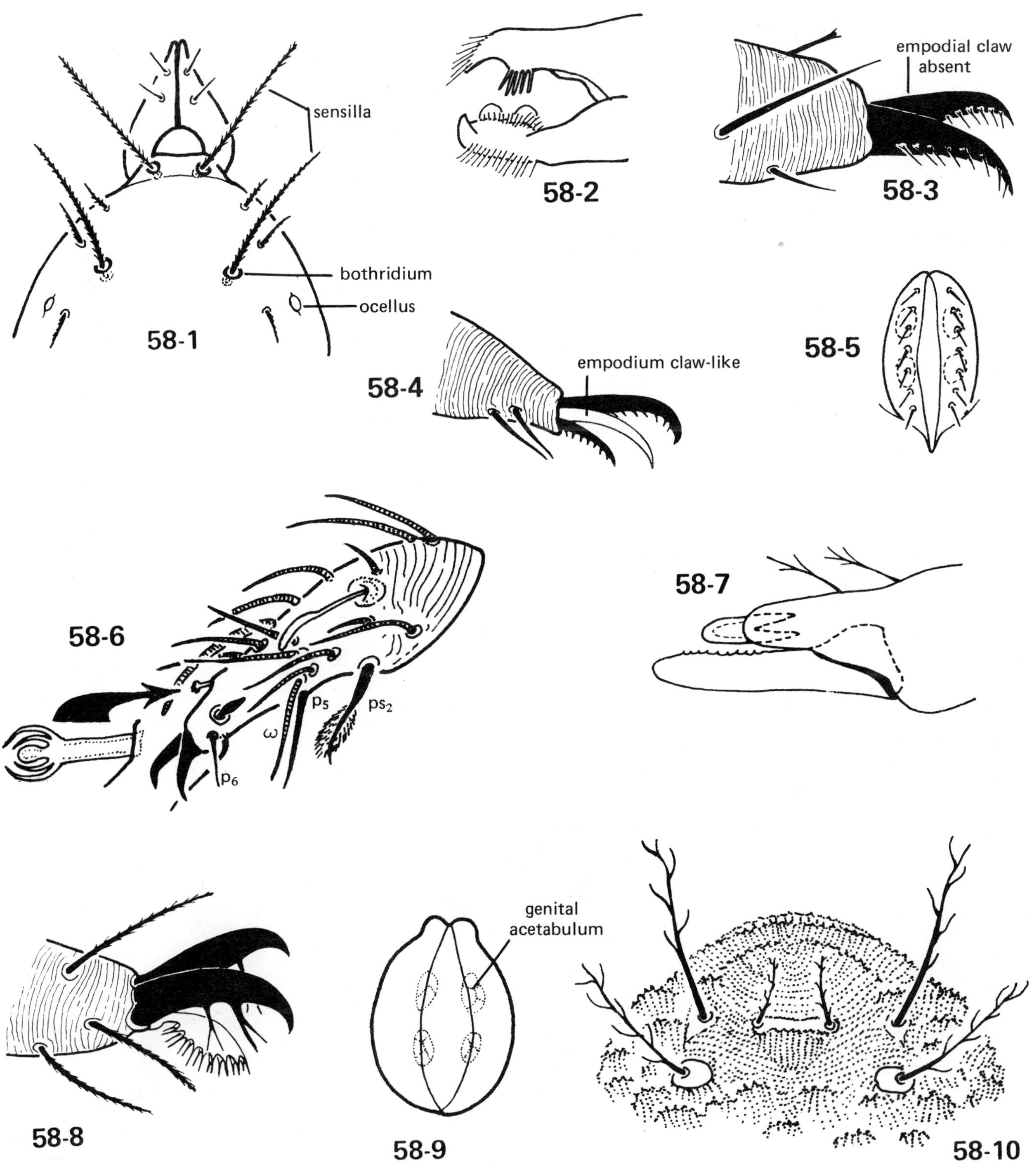

58-1 to 58-6; family SPHAEROLICHIDAE, *Sphaerolichus barbarus* Grandjean (France) (after Grandjean 1939). **58-1**; anterodorsal aspect of female: **58-2**; chelicera: **58-3**; pretarsus I: **58-4**; pretarsus III: **58-5**; genital region of female: **58-6**; tarsus I (antiaxial), with detail of famulus

58-7 to 58-10; family LORDALYCHIDAE, *Lordalychus peraltus* Grandjean. **58-7**; cheliceral extremities (after Grandjean 1939): **58-8**; pretarsus I (after Grandjean 1939): **58-9**; genital region of female (after Grandjean 1939): **58-10**; propodosoma of female (after Shiba 1968)

PLATE 59

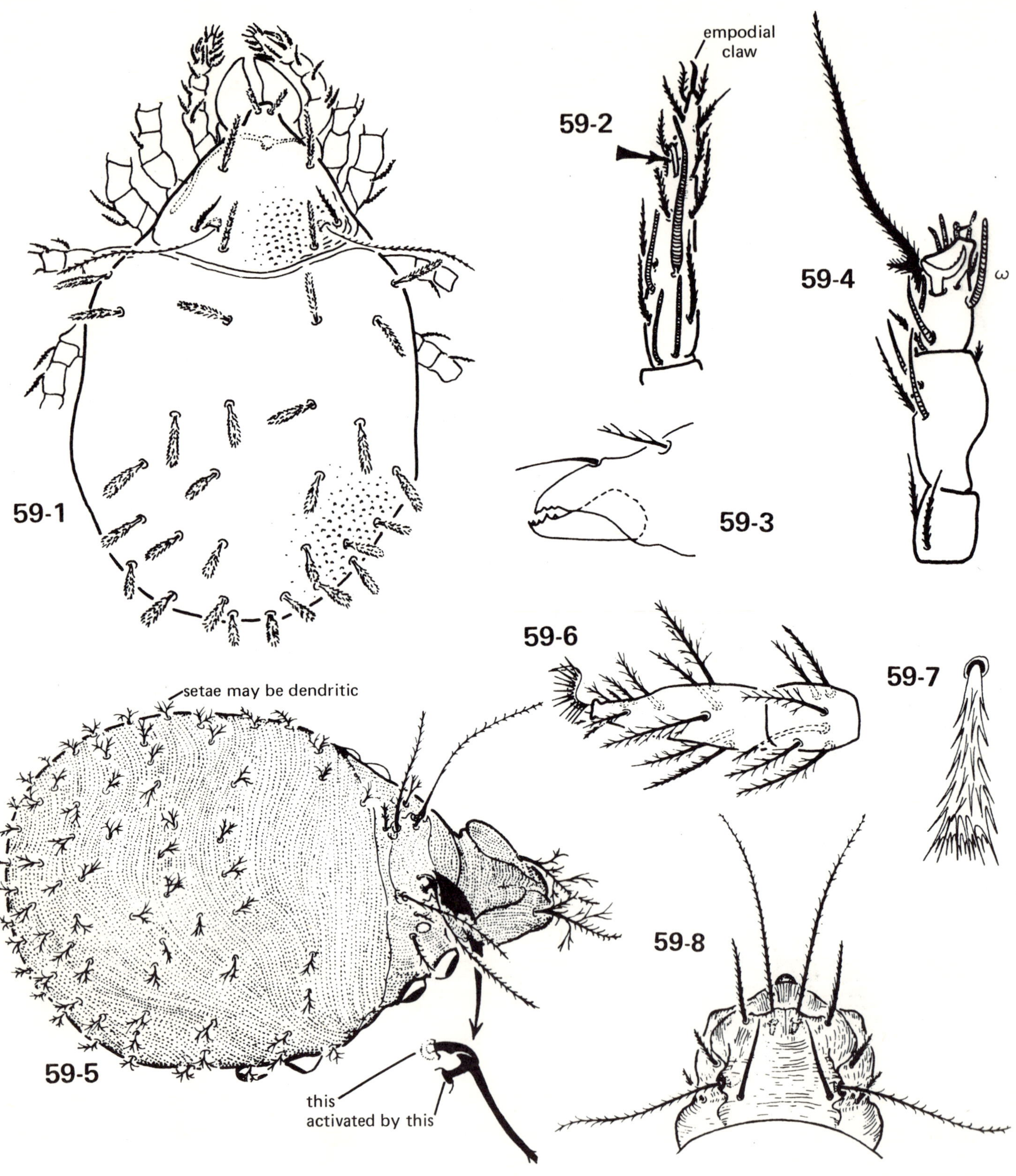

59-1 to 59-4; family OEHSERCHESTIDAE. **59-1**; *Oehserchestes dorysetatus* (Theron) (South Africa), dorsum of female (from Theron 1974): **59-2**; *O. dorysetatus,* tarsus I of female. Arrow designates the famulus (from Theron 1974): **59-3**; *O. dorysetatus,* chelicera of female (from Theron 1974): **59-4**; *Grandjeanicus uncus* Theron (South Africa), genu, tibia and tarsus I of protonymph (from Theron 1974)

59-5 to 59-8; family NANORCHESTIDAE; **59-5**; *Nanorchestes* sp. (Oregon, USA), dorsolateral aspect of female with detail of sensillary apparatus: **59-6**; *Nanorchestes* sp., tibia and tarsus I: **59-7**; *Speleorchestes* sp. (Oregon, USA), idiosomal seta: **59-8**; *Nanorchestes* sp., propodosoma

PLATE 60

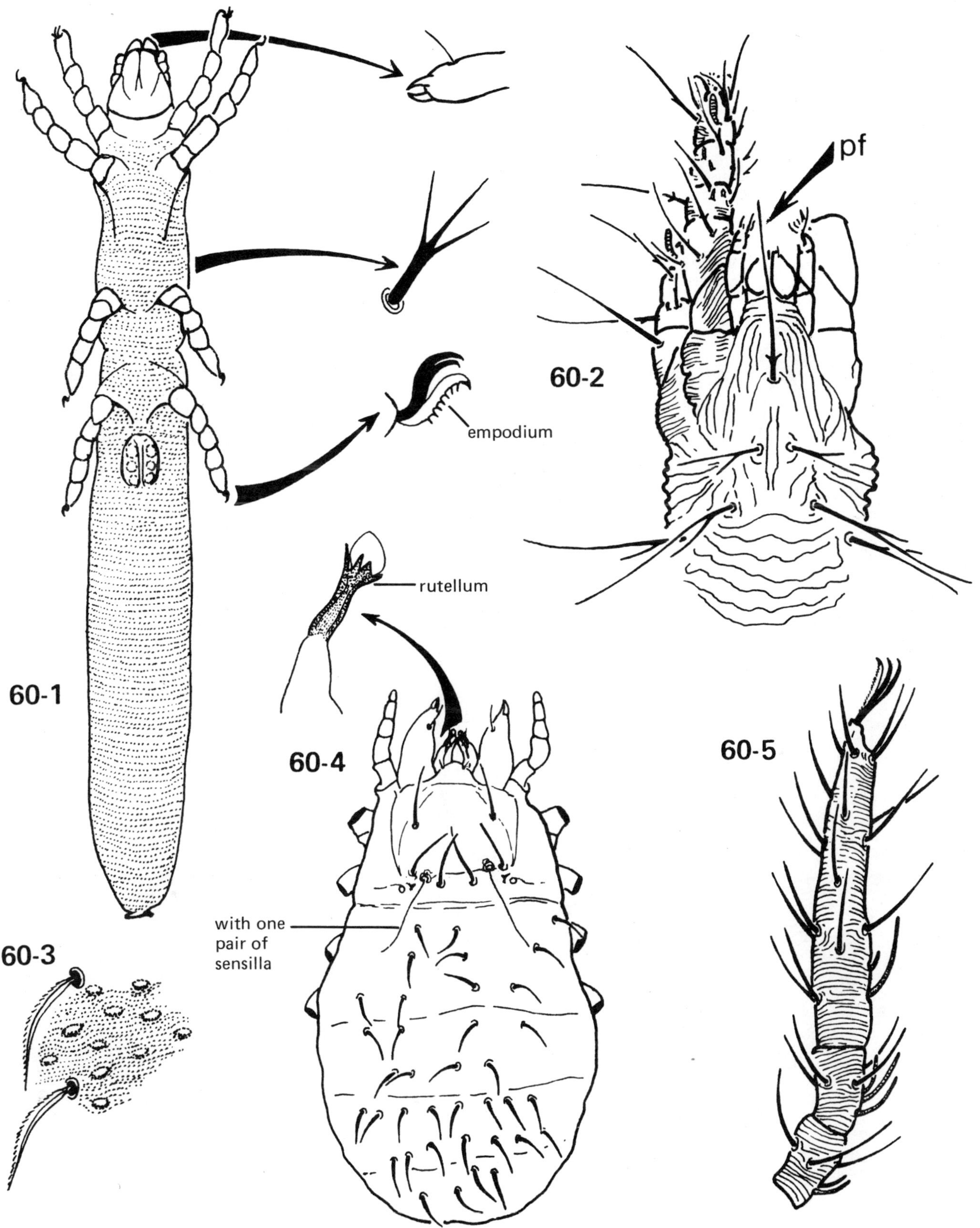

60-1 and 60-2; family NEMATALYCIDAE. **60-1**; *Nematalycus* (diagrammatic), venter of female with details of chelicera, dorsal seta and pretarsus IV: **60-2**; *Psammolycus delamarei* Schubart (Brazil), dorsum of prosoma (from Schubart 1973)

60-3 to 60-5; family TERPNACARIDAE, *Terpnacarus* sp. (Oregon, USA). **60-3**; detail of dorsal ornamentation: **60-4**; dorsum of female with detail of rutellum: **60-5**; genu, tibia and tarsus I of female

PLATE 61

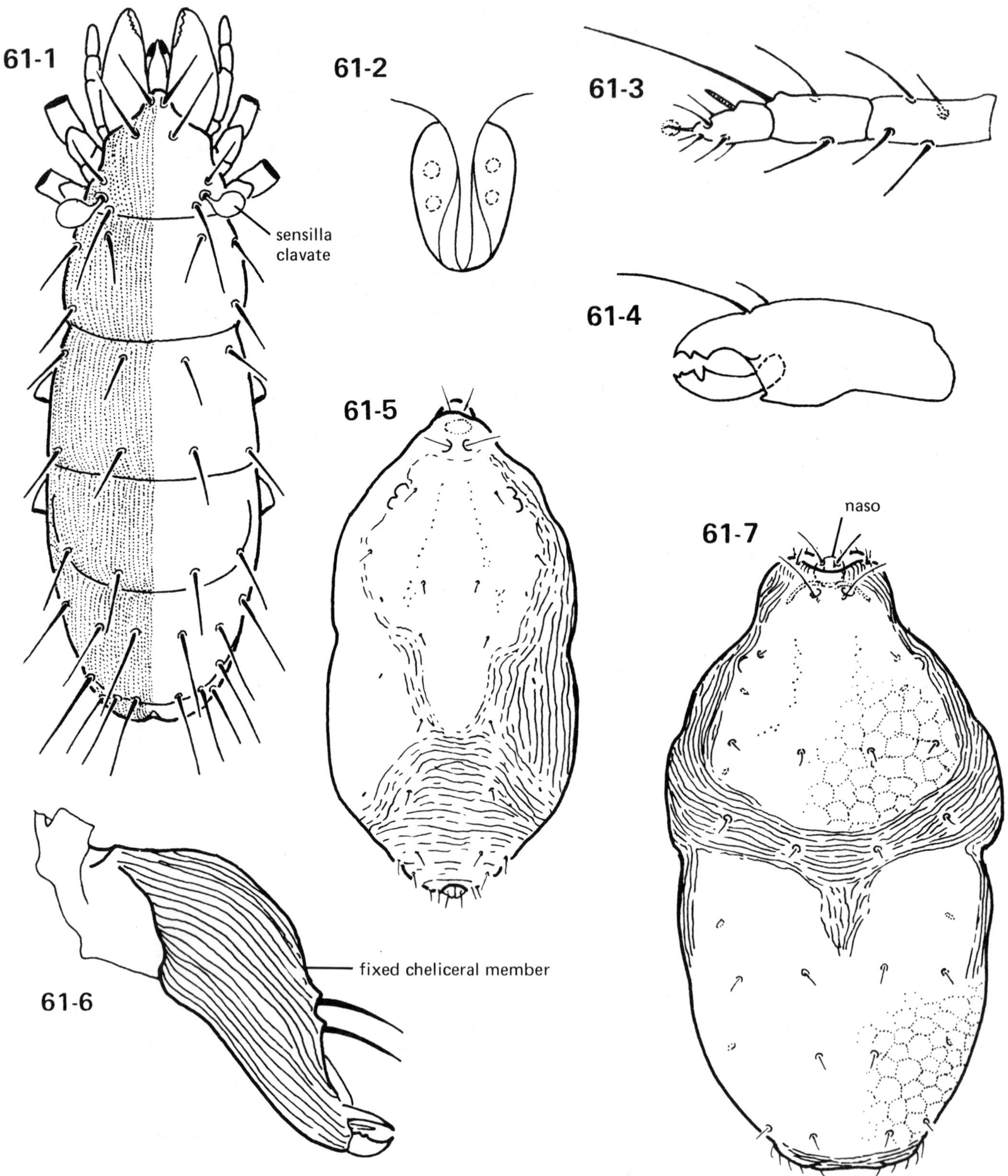

61-1 to 61-4; family PEDICULOCHELIDAE. **61-1**; *Pediculocheles* (diagrammatic), dorsum: **61-2**; *Pediculocheles*, genital region: **61-3**; *Pediculocheles lavoipierri* Price (California, USA), genu, tibia and tarsus I (after Price 1973): **61-4**; *P. lavoipierri*, chelicera of female (after Price 1973)

61-5 to 61-7; family ADAMYSTIDAE. **61-5**; *Adamystis fonsi* Coineau (France), dorsum (after Coineau 1974): **61-6**; *A. fonsi*, chelicera (antiaxial) (after Coineau 1974): **61-7**; *Saxidromus delamarei* Coineau (France), dorsum (after Coineau 1974)

PLATE 62

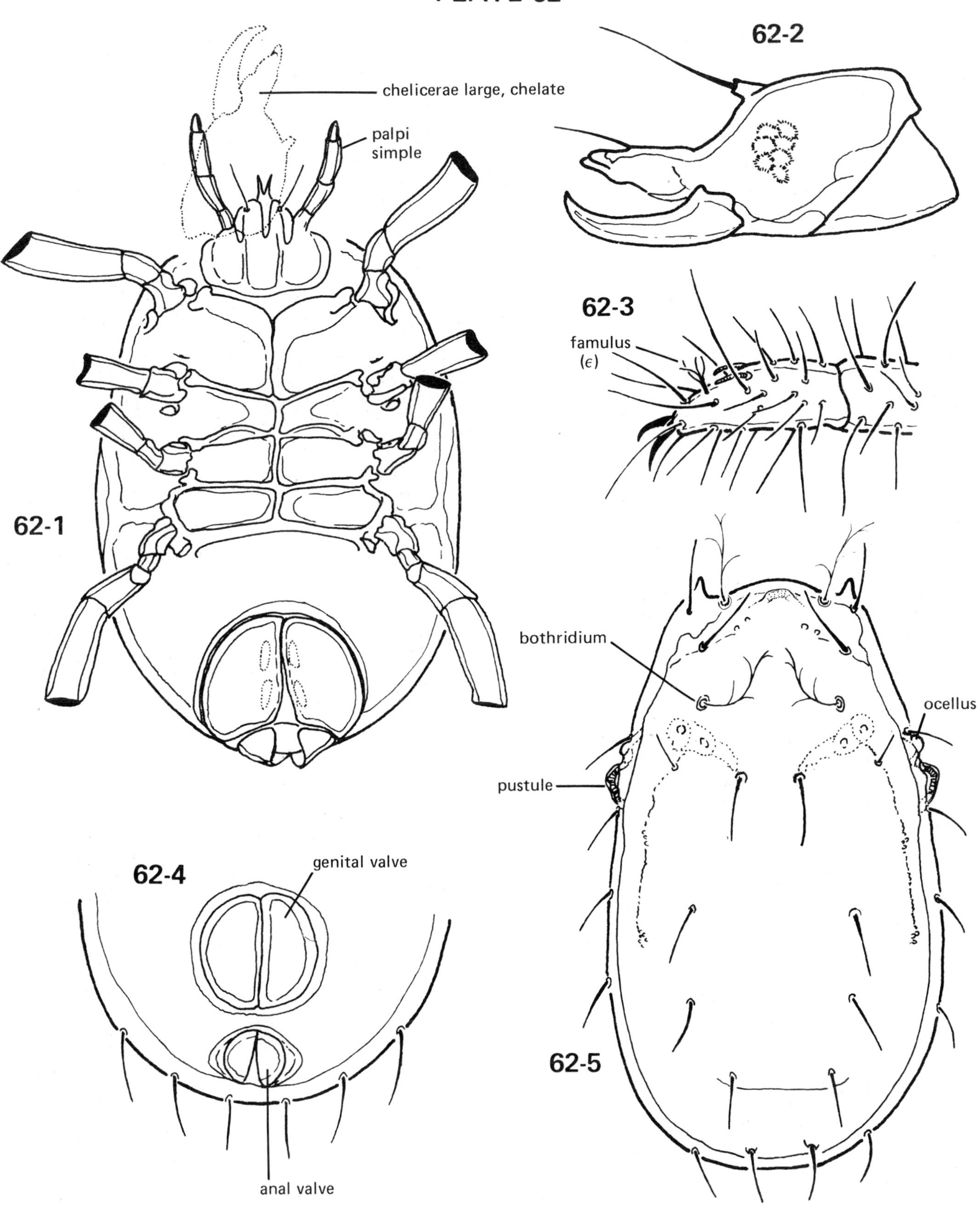

62-1 to 62-5; family LABIDOSTOMMATIDAE, *Labidostomma* sp. (Oregon, USA). **62-1**; venter of female: **62-2**; chelicera of female: **62-3**; tarsus I of female: **62-4**; genital-anal region of male: **62-5**; dorsum of female

PLATE 63

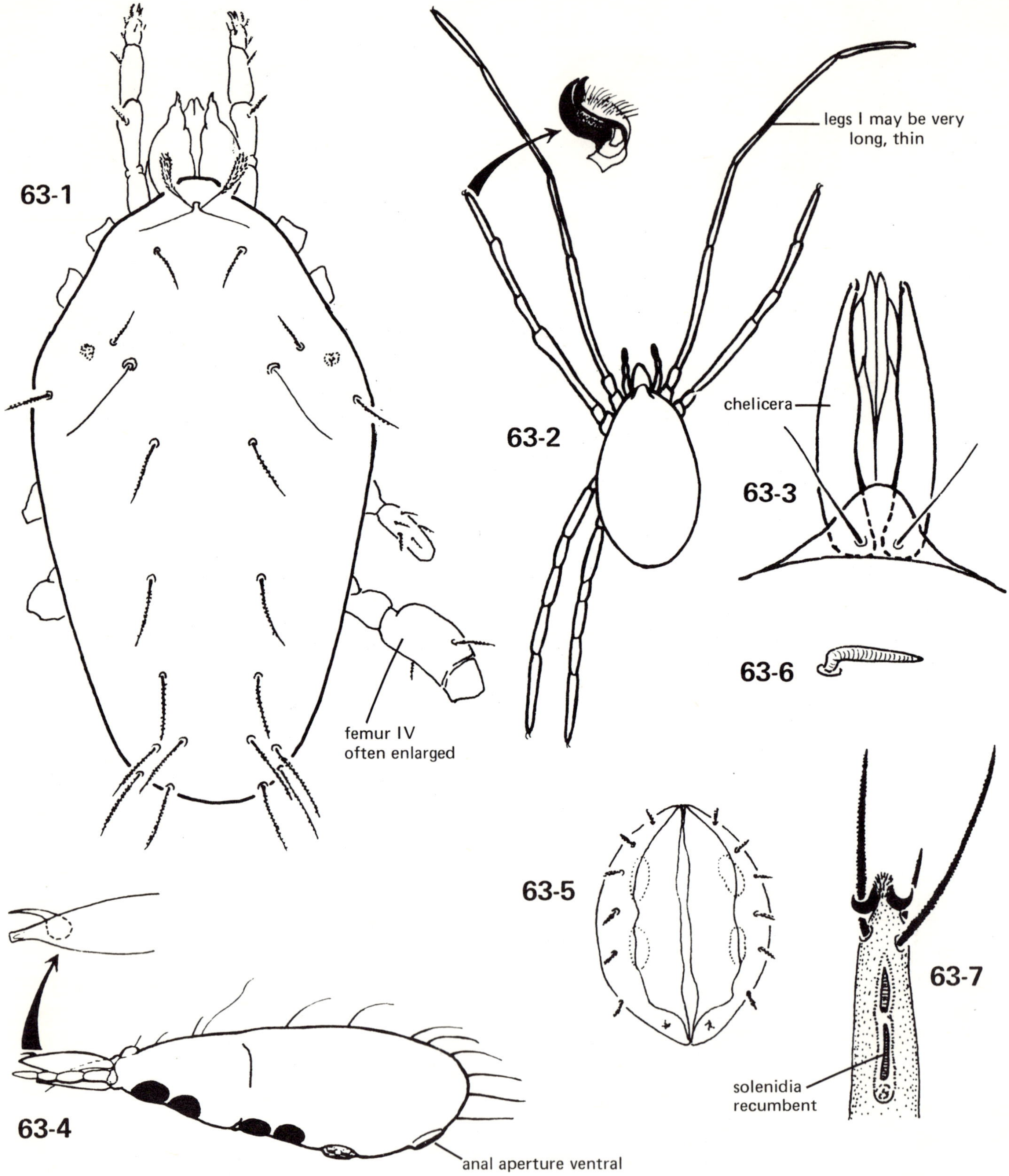

63-1 to 63-7; family EUPODIDAE. **63-1;** *Cocceupodes communis* Shiba (Japan), dorsum of female (after Shiba 1969): **63-2;** *Linopodes* sp. (Oregon, USA), dorsum, with detail of pretarsus II: **63-3;** *Linopodes* sp., dorsum of rostrum: **63-4;** *Eupodes* sp. (Oregon, USA), lateral aspect with detail of chelicera: **63-5;** *Cocceupodes communis* Shiba, genital region of female (from Shiba 1964): **63-6;** tarsal solenidion: **63-7;** *Eupodes longisetatus* Strandtmann (Japan), dorsal aspect of tarsus I (after Shiba 1969)

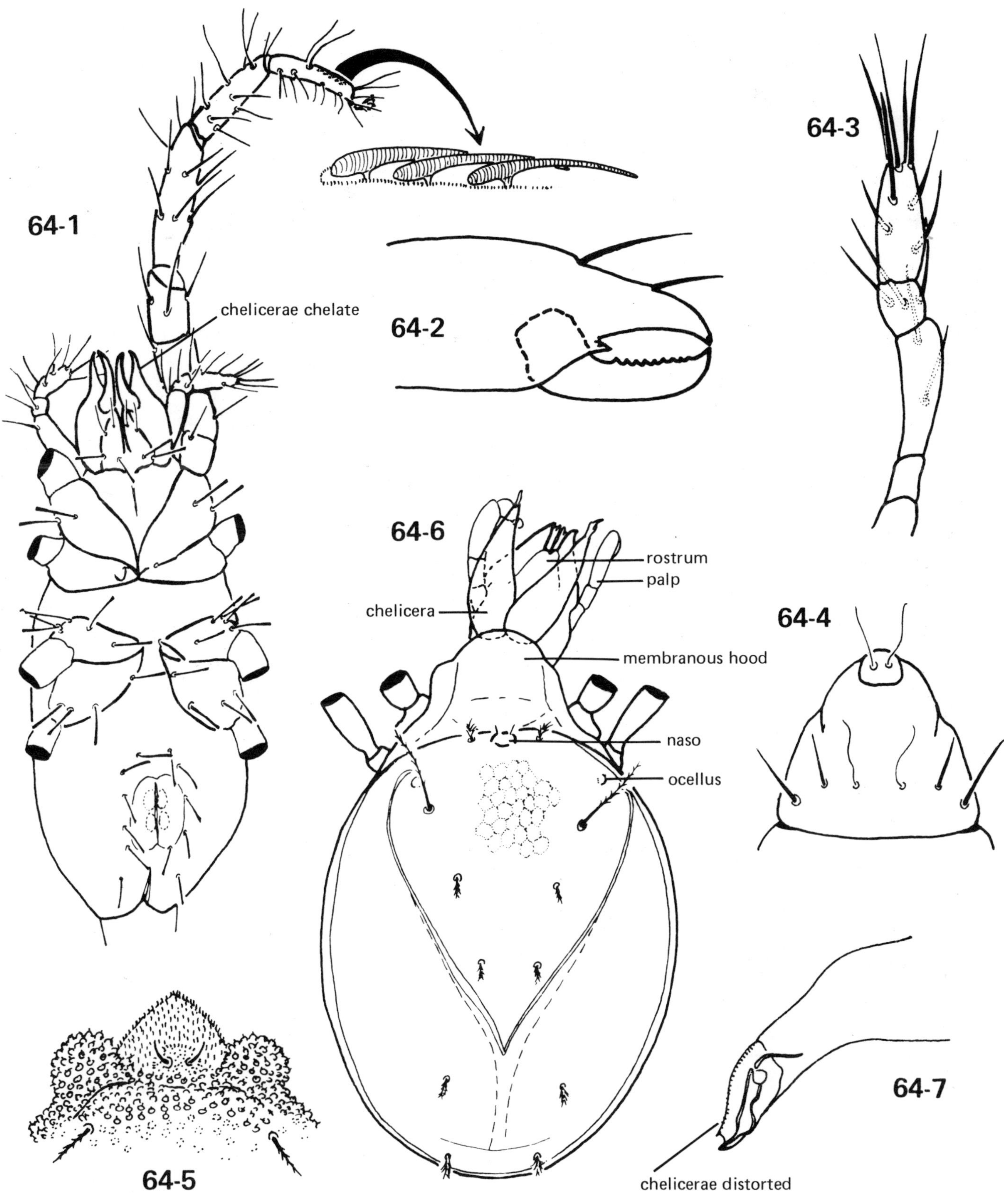

64-1 to 64-4; family RHAGIDIIDAE. **64-1;** *Rhagidia* sp. (Oregon, USA), venter of female with detail of tarsal solenidia (rhagidial organs): **64-2;** *Coccorhagidia clavifrons* (Canestrini) (Japan), chelicera (after Shiba 1969): **64-3;** *Rhagidia* sp., palp: **64-4;** *Rhagidia* sp., propodosoma

64-5 to 64-7; family PENTHALODIDAE. **64-5;** *Stereotydeus longipes* Strandtmann (South Georgia), anterior aspect of propodosoma (after Strandtmann 1970): **64-6;** *Penthalodes* sp. (Oregon, USA), dorsum of female: **64-7;** *Penthalodes* sp., chelicera of female

PLATE 65

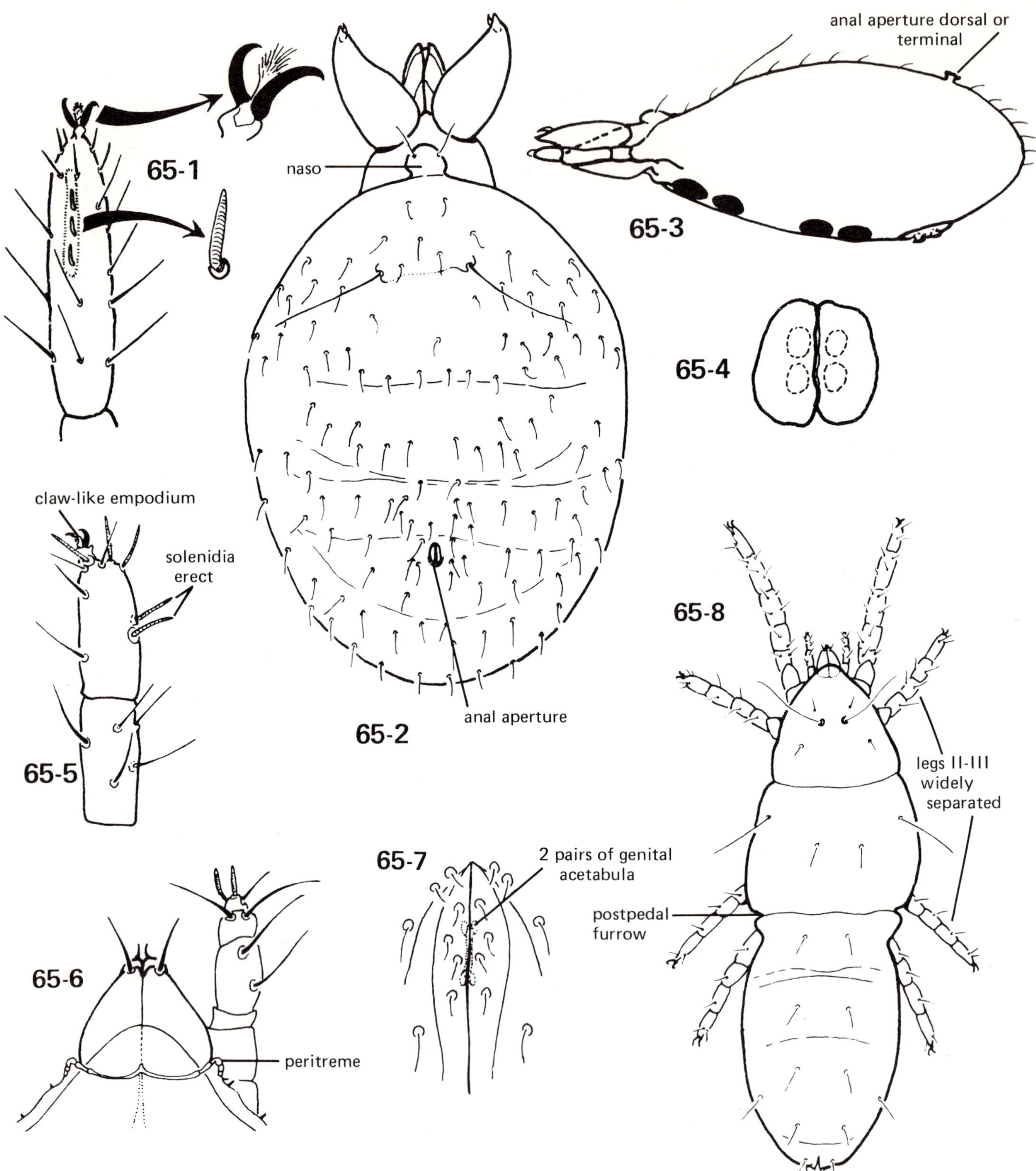

65-1 to 65-4; family PENTHALEIDAE, *Penthaleus major* (Dugès) (Oregon, USA). **65-1;** tarsus I with detail of pretarsus and solenidion: **65-2;** dorsum of female: **65-3;** lateral aspect of female: **65-4;** genital valves of female

65-5 to 65-8; family PARATYDEIDAE, *Paratydeus* sp. (Arizona, USA). **65-5;** tibia and tarsus I of female: **65-6;** anterodorsal aspect of female: **65-7;** genital region of female: **65-8;** dorsum of female

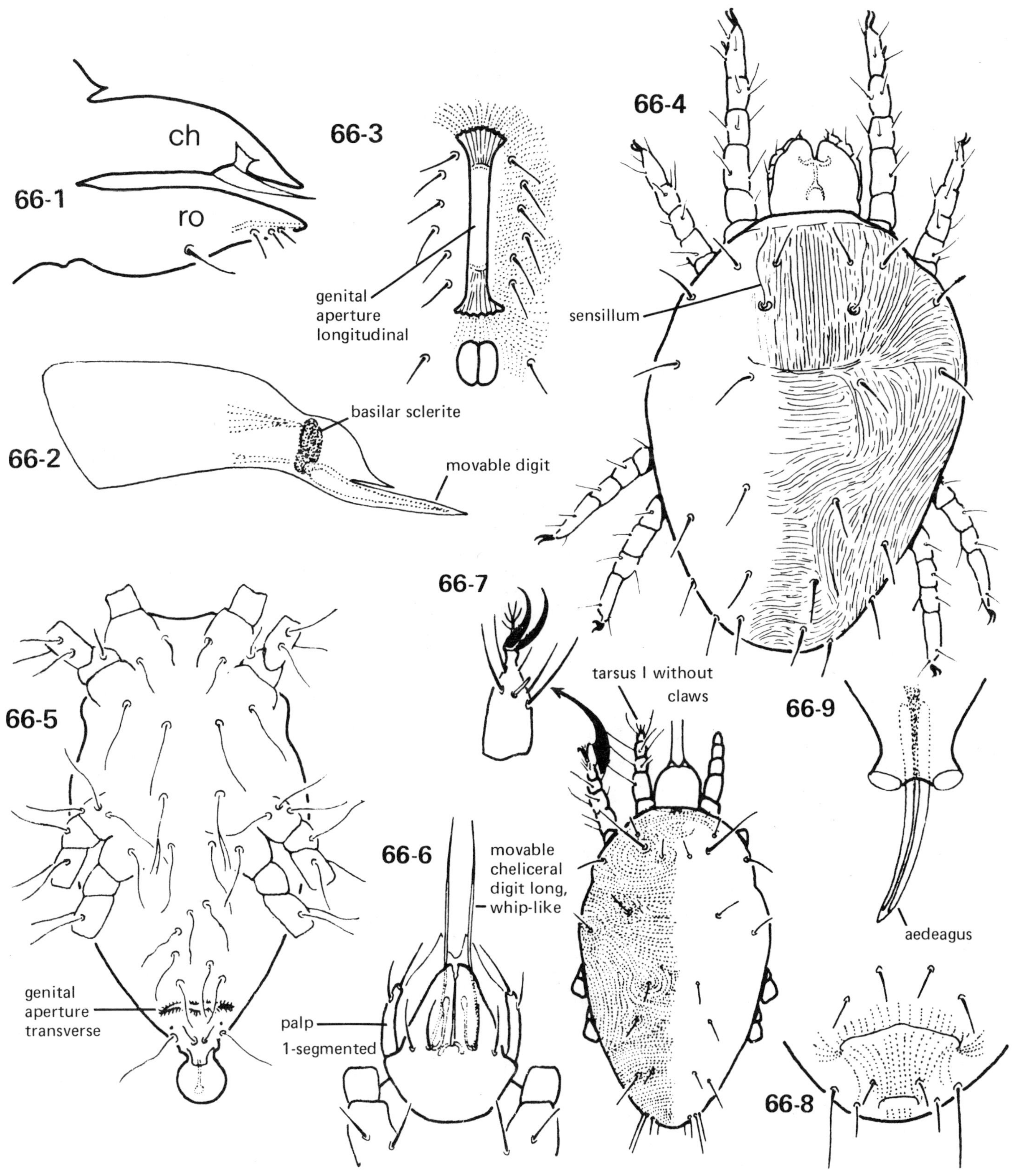

66-1 to 66-4; family TYDEIDAE. **66-1**; *Lorryia bedfordiensis* Evans (Canada), lateral aspect of gnathosoma (ch - chelicera, ro = rostrum) (after Marshall 1970): **66-2**; *L. montrealensis* Marshall (Canada), chelicera (antiaxial) (from Marshall 1970): **66-3**; *L. bedfordiensis*, genital region of female: **66-4**; *Paralorryia* sp. (Oregon, USA), dorsum of female

66-5 to 66-8; family IOLINIDAE. **66-5**; *Proctotydaeus schistocercae* Fain and Evans (Galapagos), venter of female (after Fain and Evans 1966): **66-6**; *Iolina nana* Pritchard (Massachusetts, USA), venter of gnathosoma (after Pritchard 1956): **66-7**; *I. nana*, dorsum of female with detail of tarsus II (after Pritchard 1956): **66-8**; *I. nana*, genital-anal region of female (after Pritchard 1956): **66-9**; *Proctotydaeus schistocercae*, male genitalia (from Fain and Evans 1966)

PLATE 67

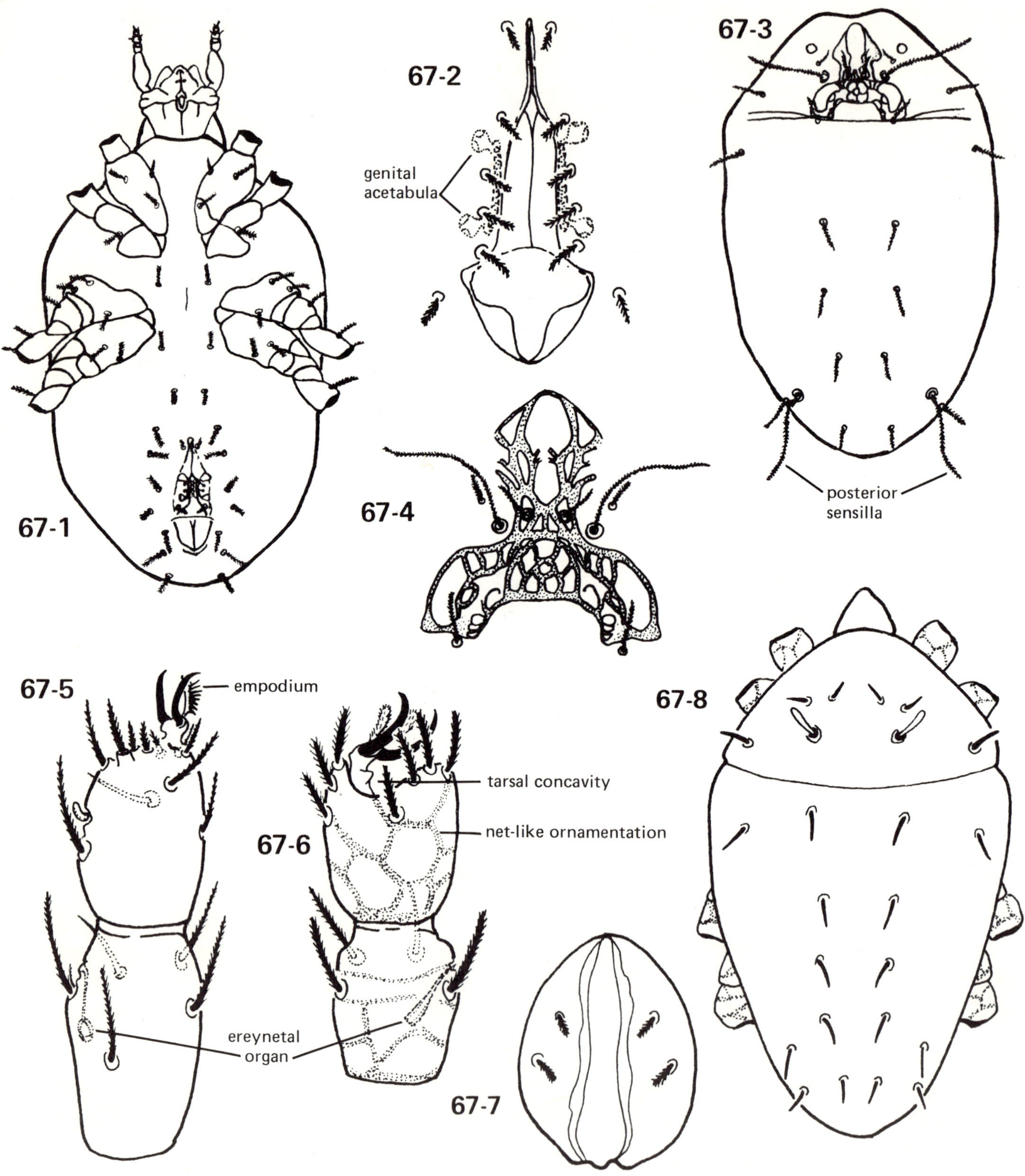

67-1 to 67-7; family EREYNETIDAE. **67-1**; *Ereynetes* sp. (Oregon, USA), venter of female: **67-2**; *Ereynetes* sp., genital-anal region: **67-3**; *Ereynetes* sp., dorsum of female: **67-4**; *Ereynetes* sp., detail of propodosomal ornamentation: **67-5**; *Ereynetes* sp., tibia and tarsus I: **67-6**; *Boydaia sturni* (Boyd), tibia and tarsus I: **67-7**; *Speleognathus* sp., genital region: **67-8**; *Speleognathus* (diagrammatic), dorsum

PLATE 68

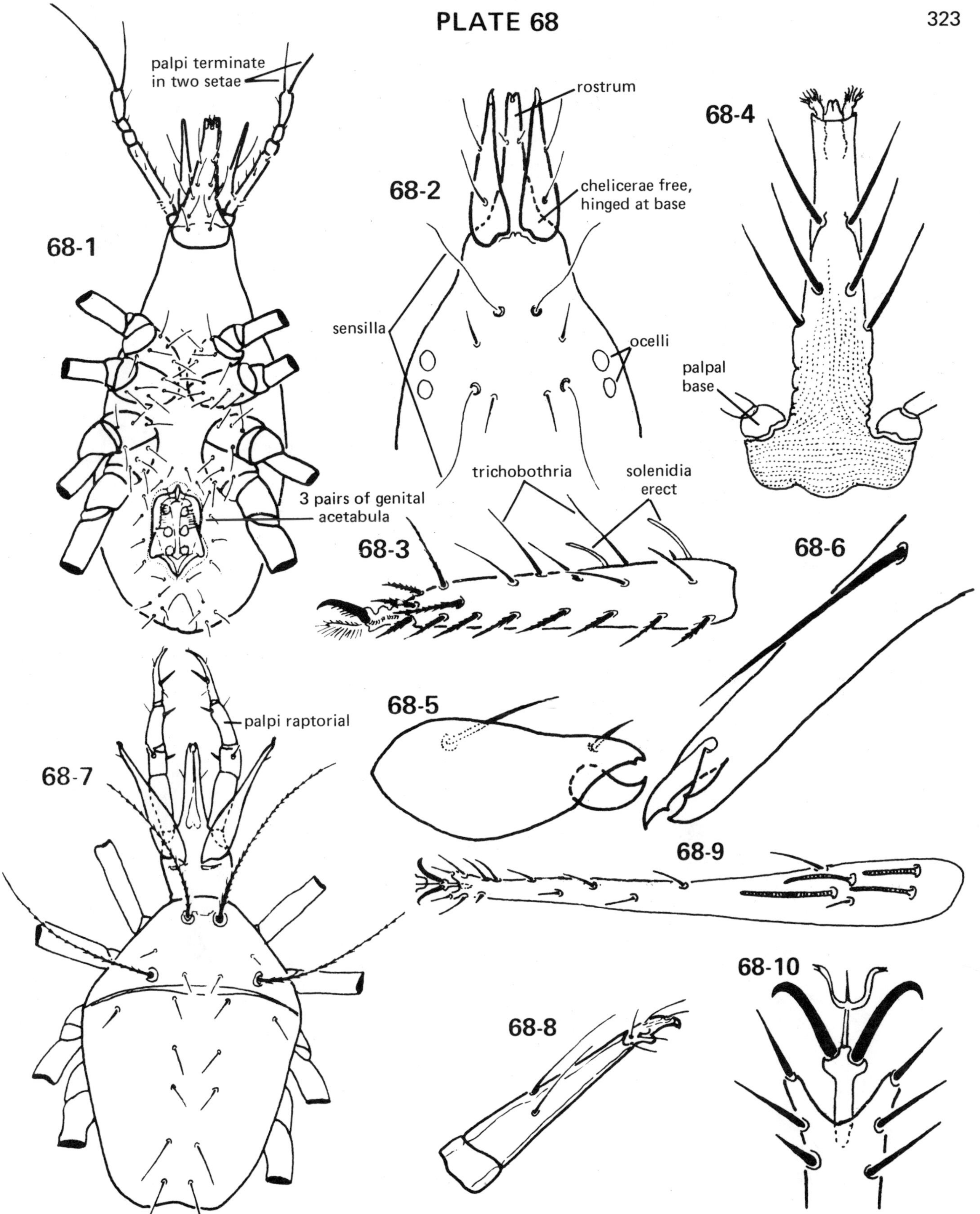

68-1 to 68-4; family BDELLIDAE. **68-1;** *Bdella* sp. (Oregon, USA), venter of male: **68-2;** *Bdella* sp., prosoma (dorsal): **68-3;** *Bdella* sp., tarsus I: **68-4;** *Bdella longicornis* (L.) (Japan), venter of gnathosoma (from Shiba 1969): **68-5;** *Cyta latirostris* (Hermann), chelicera: **68-6;** *Bdellodes (B.) longirostris,* chelicera (after Shiba 1969)

68-7 to 68-10; family CUNAXIDAE. **68-7;** *Cunaxa* sp. (Oregon, USA), dorsum of female: **68-8;** *Cunaxoides* sp. (Central Africa), palp: **68-9;** *Cunaxa setirostris* (Hermann), tarsus I (from Smiley 1975); **68-10;** *C. setirostris,* detail of pretarsus I

PLATE 69

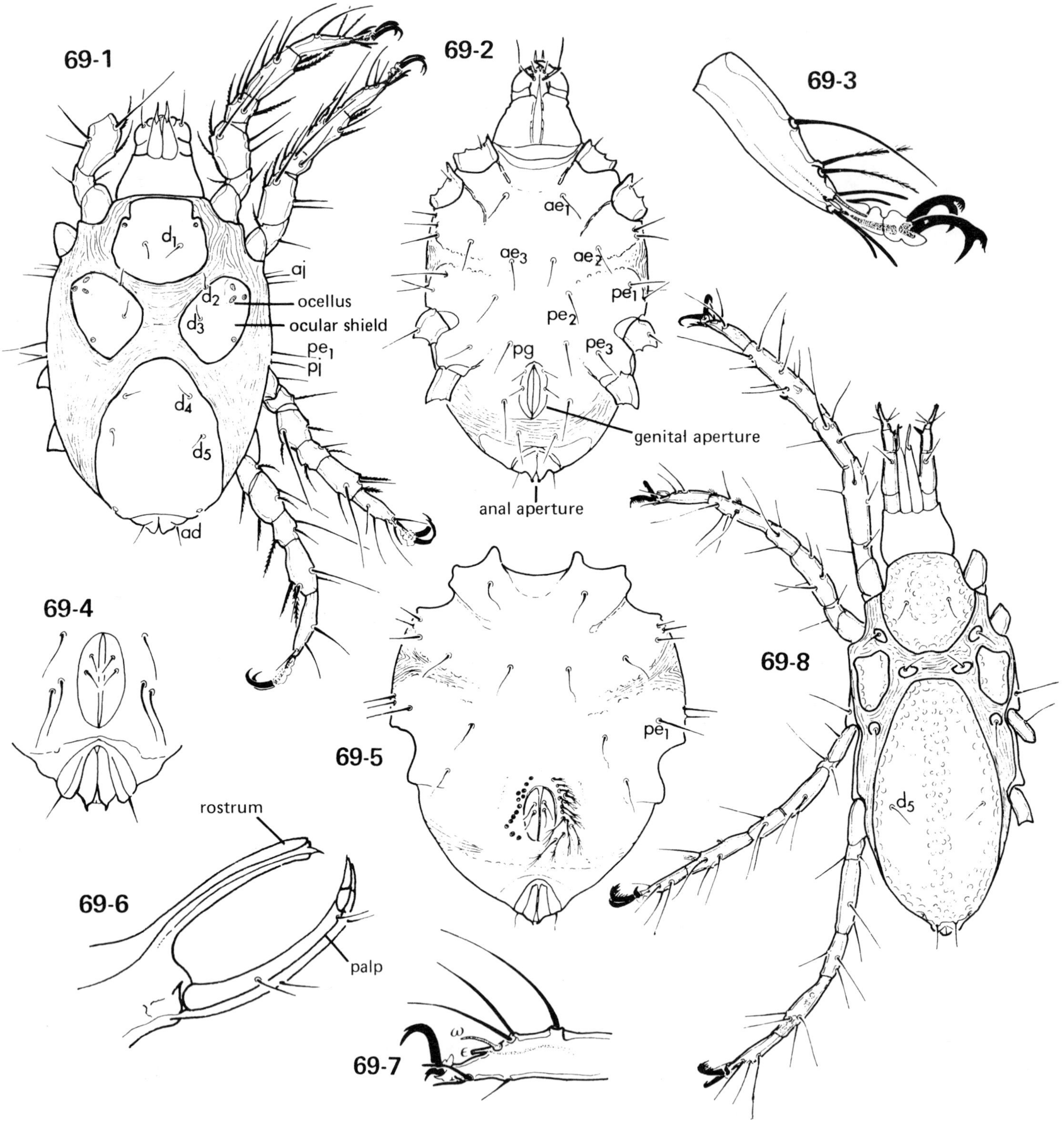

69-1 to 69-8; family HALACARIDAE. **69-1**; *Rhombognathus leurodactylus* Krantz (Oregon, USA), dorsum of female: **69-2**; *R. leurodactylus*, venter of female: **69-3**; *R. reticulatus* Krantz (Oregon, USA), tarsus I of male: **69-4**; *R. reticulatus*, genital-anal region of female: **69-5**; *R. reticulatus*, venter of male: **69-6**; *Lohmannella* sp., lateral aspect of gnathosoma: **69-7**; *Thalassarachna psammophila* Krantz (Oregon, USA), terminal portion of tarsus I of male, showing the famulus (ϵ) and solenidion (ω): **69-8**; *T. psammophila*, dorsum of female

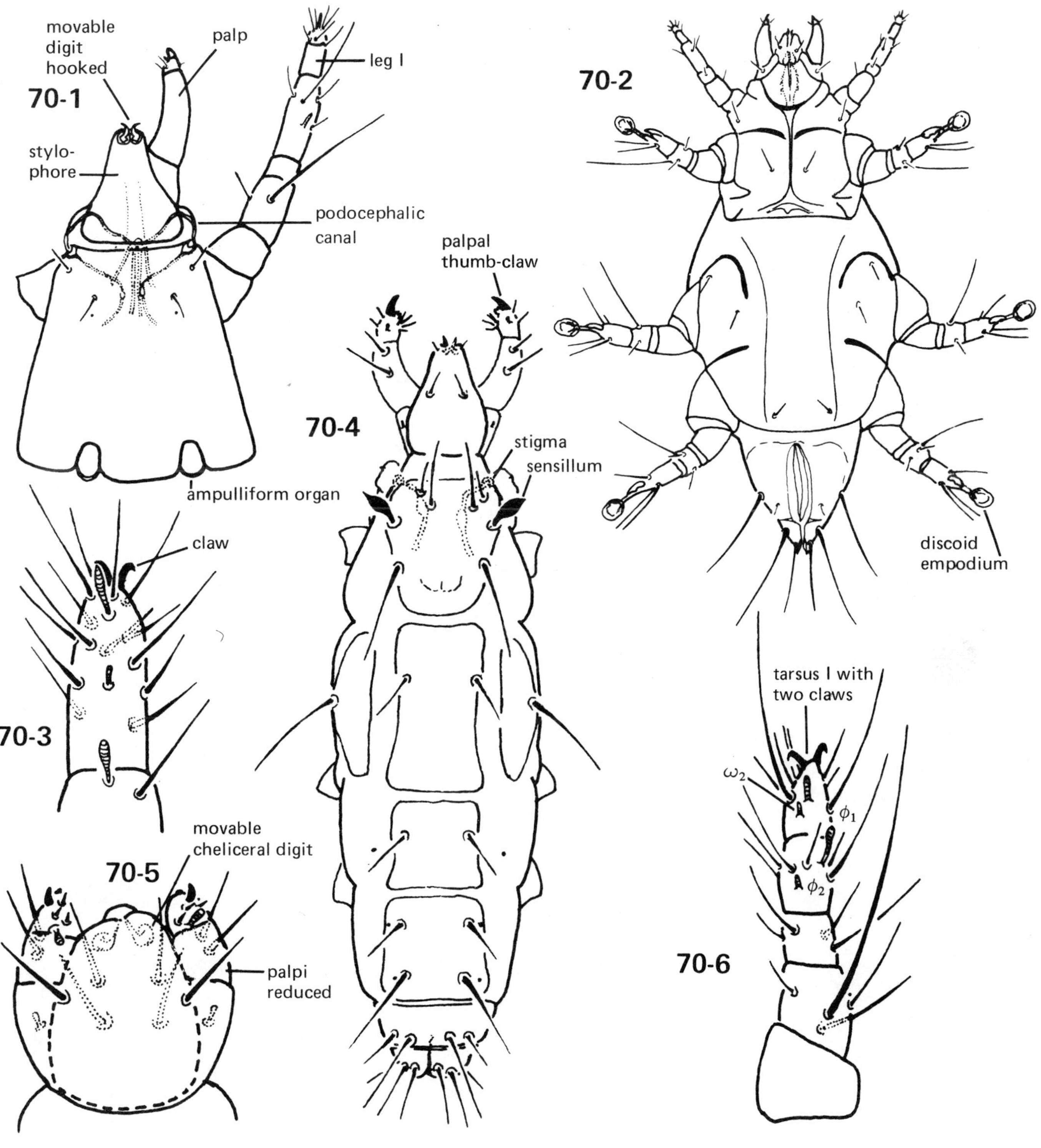

70-1 and 70-2; family HETEROCHEYLIDAE, *Heterocheylus* sp., (Georgia, USA). **70-1**; dorsum of proterosoma: **70-2**; venter of female

70-3 and 70-4; family TARSOCHEYLIDAE, *Hoplocheylus* sp. (after Lindquist 1976). **70-3**; dorsum of tarsus I of female: **70-4**; dorsum of female

70-5 and 70-6; family DOLICHOCYBIDAE. **70-5**; *Pavania* sp., dorsum of gnathosoma (after Lindquist 1976): **70-6**; *Dolichocybe* sp., dorsum of leg I (after Mahunka 1970)

PLATE 71

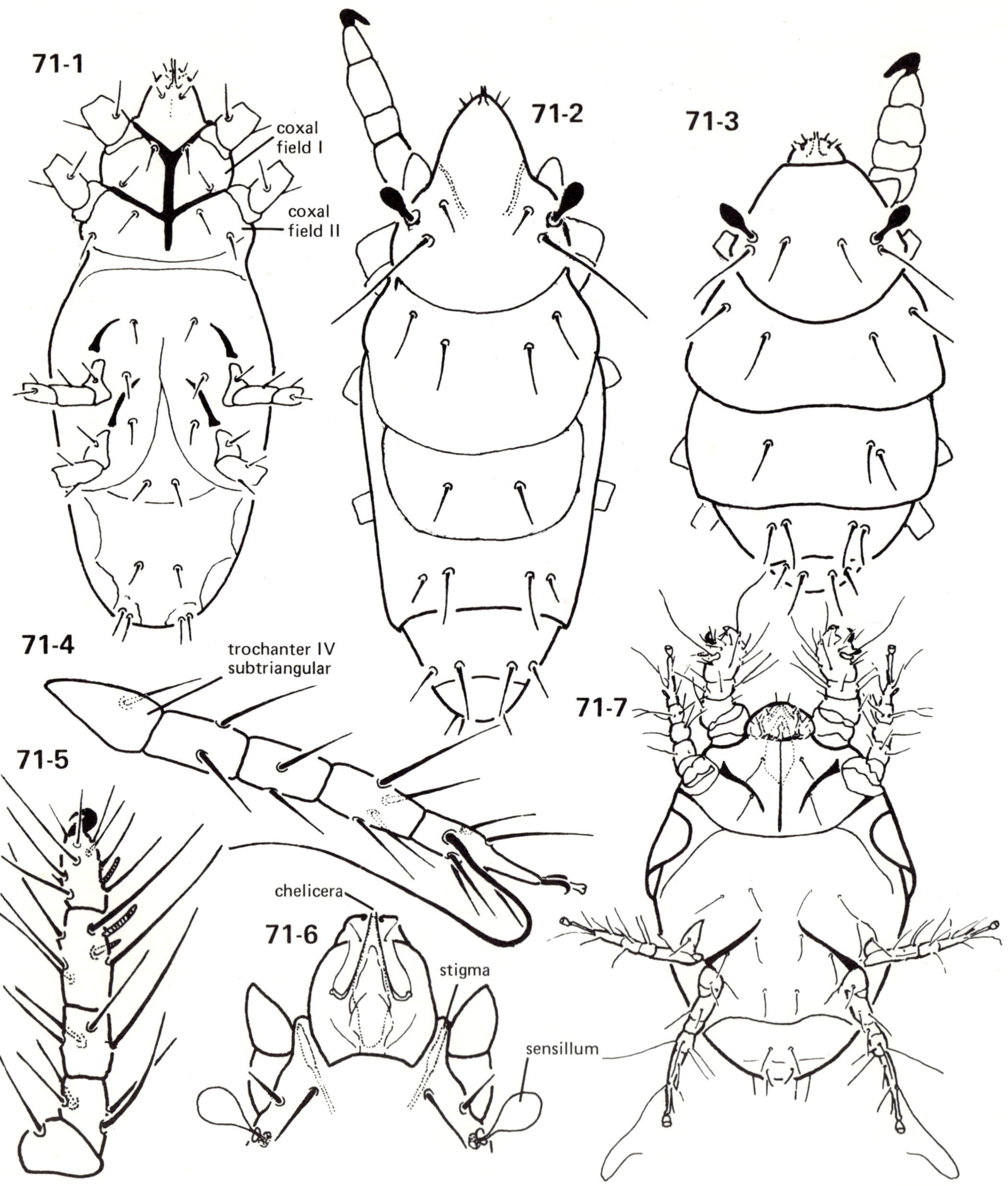

71-1 to 71-6; family PYEMOTIDAE. **71-7**; *Pyemotes dimorphus* Cross and Moser (New Hampshire, USA), venter of female (after Cross and Moser 1975): **71-2**; *P. dimorphus,* dorsum of female (after Cross and Moser 1975): **71-3**; *P. dimorphus,* dorsum of female phoretomorph (after Cross and Moser 1975): **71-4**; *Pyemotes* sp., leg IV of female (after Mahunka 1970): **71-5**; *Pyemotes* sp., leg I of female (after Mahunka 1970): **71-6**; *Pyemotes* sp., dorsum of gnathosoma and portion of propodosoma
71-7; family ACAROPHENACIDAE, *Acarophenax nidicolus* Cross and Krantz (Colorado, USA), venter of female (after Cross and Krantz 1964)

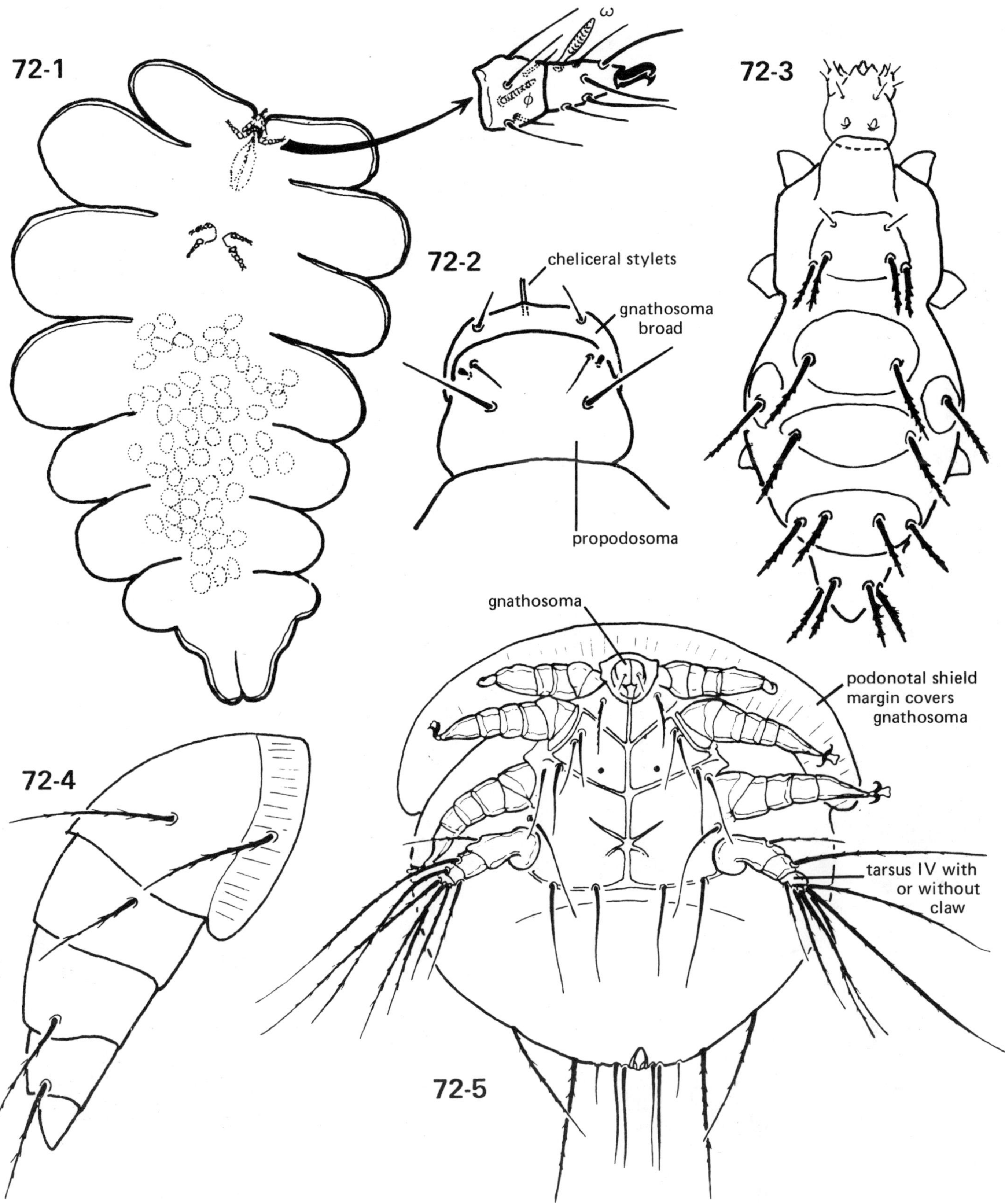

72-1; family CROTALOMORPHIDAE (ms. of J.H. Camin), *Crotalomorpha* (Kansas, USA), gravid female with detail of tibia and tarsus II

72-2; family CARABOACARIDAE, *Caraboacarus stammeri* Krczal, dorsum of proterosoma (after Mahunka 1970)

72-3 to 72-5; family SCUTACARIDAE. 72-3; *Imparipes* sp. (New York, USA), dorsum of larva (after Delfinado and Baker 1976): 72-4; *Scutacarus* sp., lateral aspect (diagrammatic) (after Karafiat 1959): 72-5; *Scutacarus* sp. (Oregon, USA), venter

PLATE 73

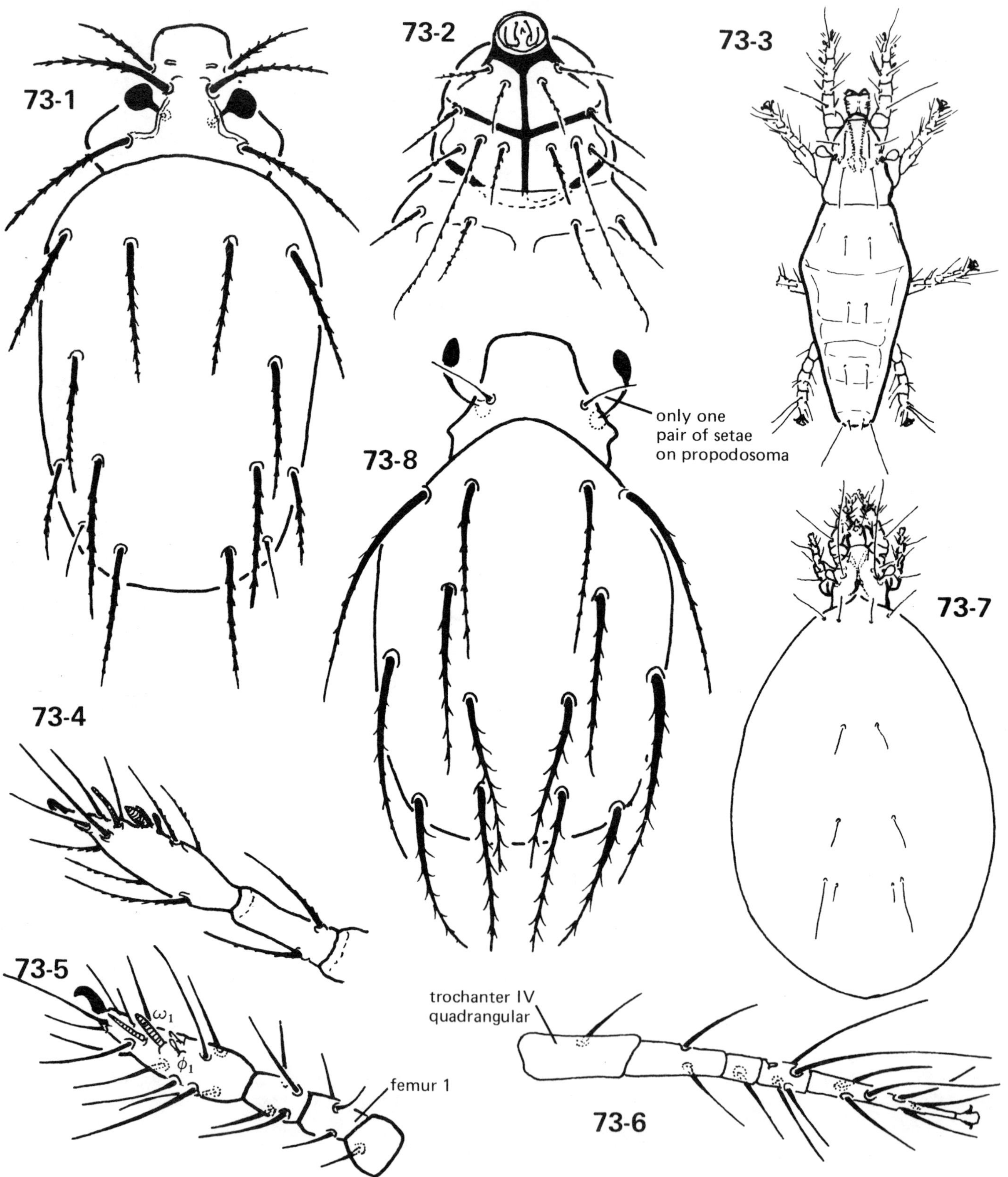

73-1 to 73-7; family PYGMEPHORIDAE. **73-1**; *Pediculaster mongolicus* Mahunka (Mongolia), dorsum (after Mahunka 1970): **73-2**; *P. mongolicus*, venter of proterosoma (after Mahunka 1970): **73-3**; *Siteroptes reniformis* Krantz (California, USA), dorsum of female (after Krantz 1957): **73-4**; *Pygmephorus* sp. (Oregon, USA), genu and tibiotarsus I: **73-5**; *Bakerdania* sp., leg I (after Mahunka 1970): **73-6**; leg IV of a typical pygmephorid (after Mahunka 1970): **73-7**; *Siteroptes cerealium* Kirchner (Oregon, USA), physogastric female (from Krantz 1959)

73-8; family MICRODISPIDAE, *Brennandania pilosa* Mahunka (Ghana), dorsum (after Mahunka 1975)

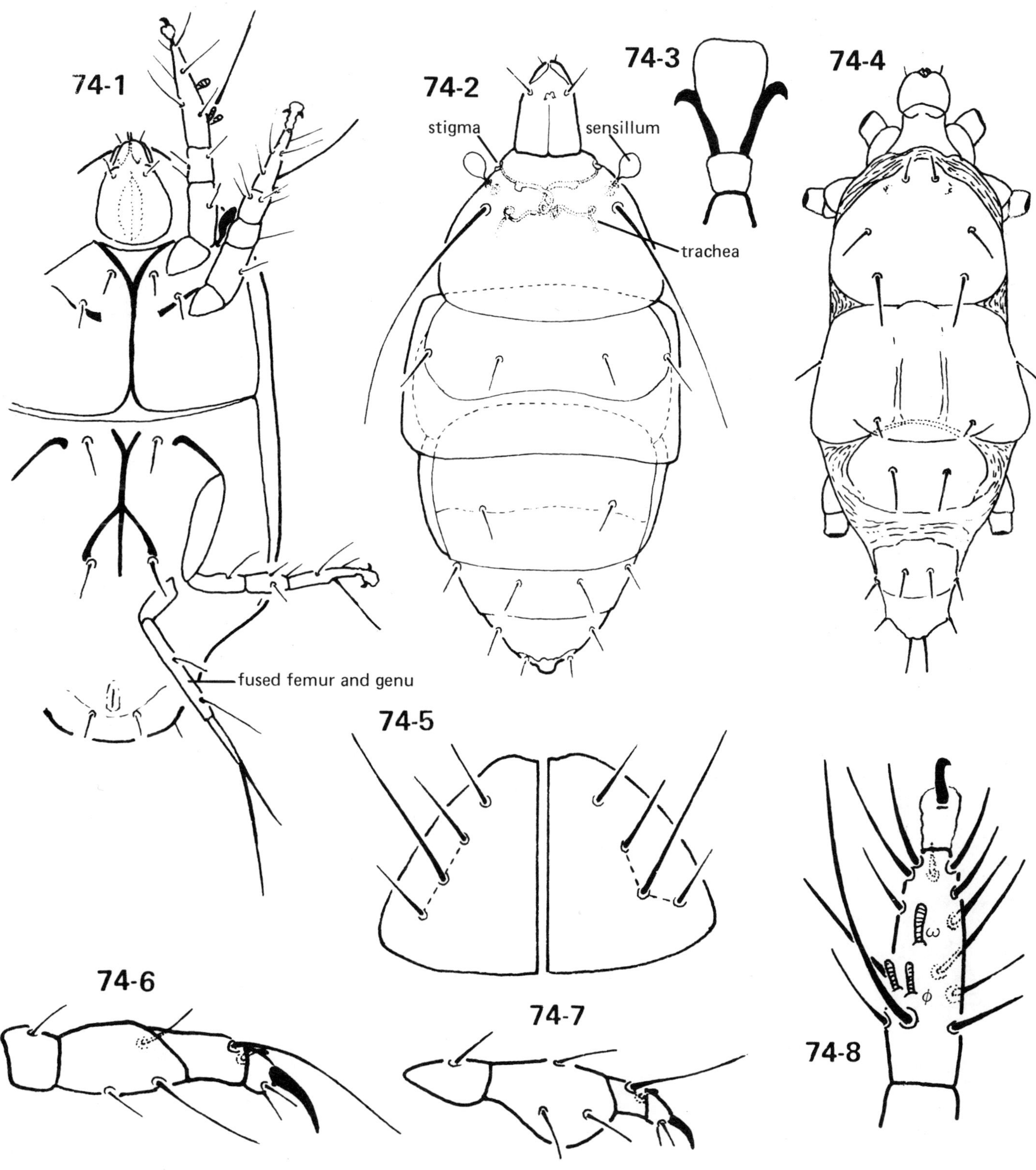

74-1 to 74-8; family TARSONEMIDAE. **74-1**; *? Tarsonemoides* sp. (Oregon, USA), venter of female: **74-2**; *Tarsonemus* sp. (Oregon, USA), dorsum of female: **74-3**; characteristic pretarsus II-III of a tarsonemid mite: **74-4**; *Steneotarsonemus* sp. (Oregon, USA), dorsum of larva: **74-5**; male propodosomal setal patterns of *Steneotarsonemus* (left) and *Tarsonemus* (right): **74-6 and 74-7**; leg IV of male, characteristic types: **74-8**; characteristic tibiotarsus I of a tarsonemid mite (after Mahunka 1970)

PLATE 75

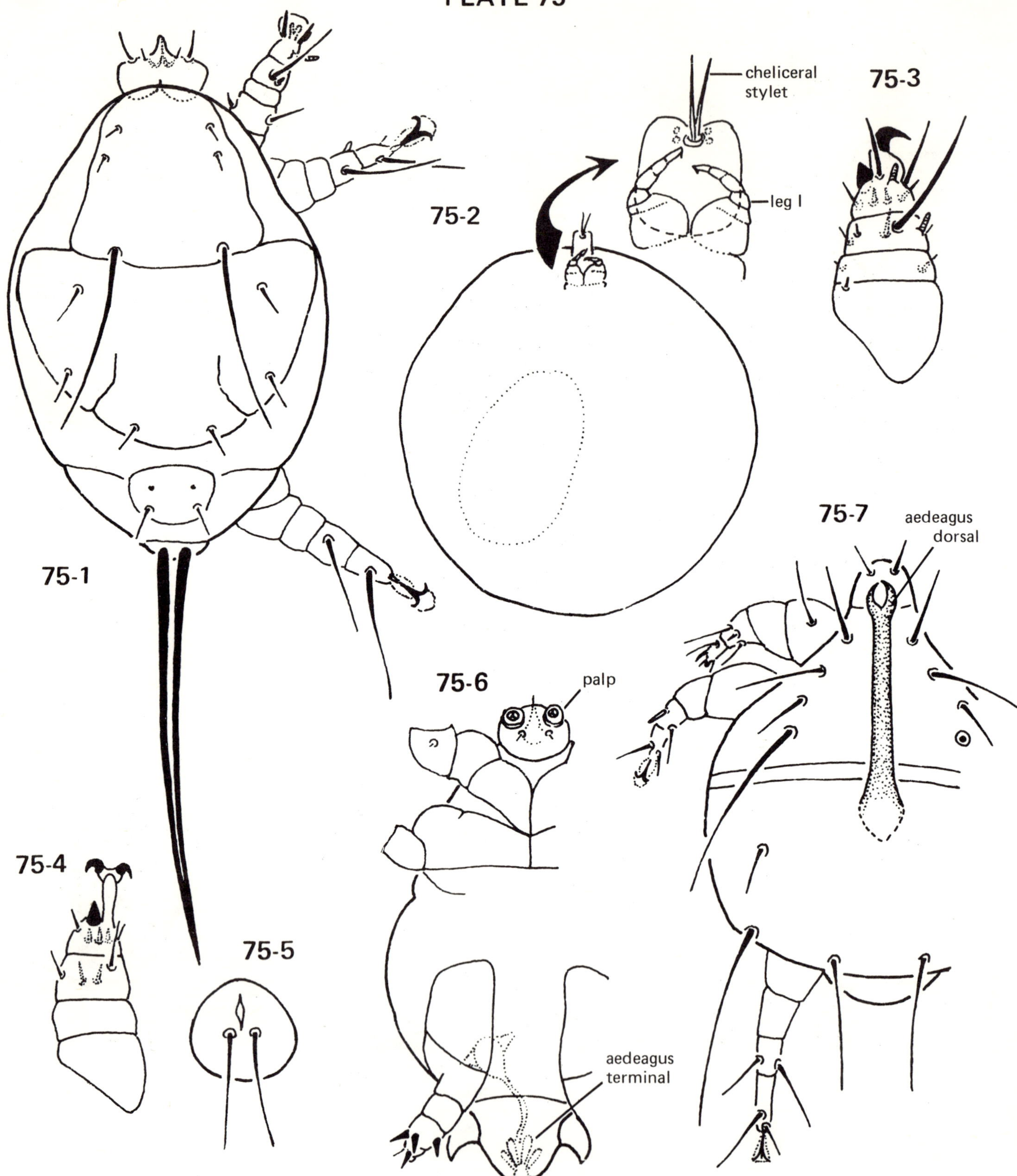

75-1 to 75-7; family PODAPOLIPIDAE. **75-1**; *Locustacarus buchneri* Stammer (Germany), dorsum of larviform female (after Husband and Sinha 1970): **75-2**; *L. buchneri,* adult female with detail of gnathosoma and single pair of legs (after Husband and Sinha 1970): **75-3**; *Eutarsopolipus* sp., leg I (after Mahunka 1970): **75-4**; *Eutarsopolipus* sp., leg III (after Mahunka 1970): **75-5**; *Ovacarus peelei* Husband (North Dakota, USA), opisthonotal shield (after Husband 1974): **75-6**; *O. peelei,* venter of male (after Husband 1974): **75-7**; *Locustacarus trachealis* Ewing (Kansas, USA), dorsum of male (after Husband and Sinha 1970)

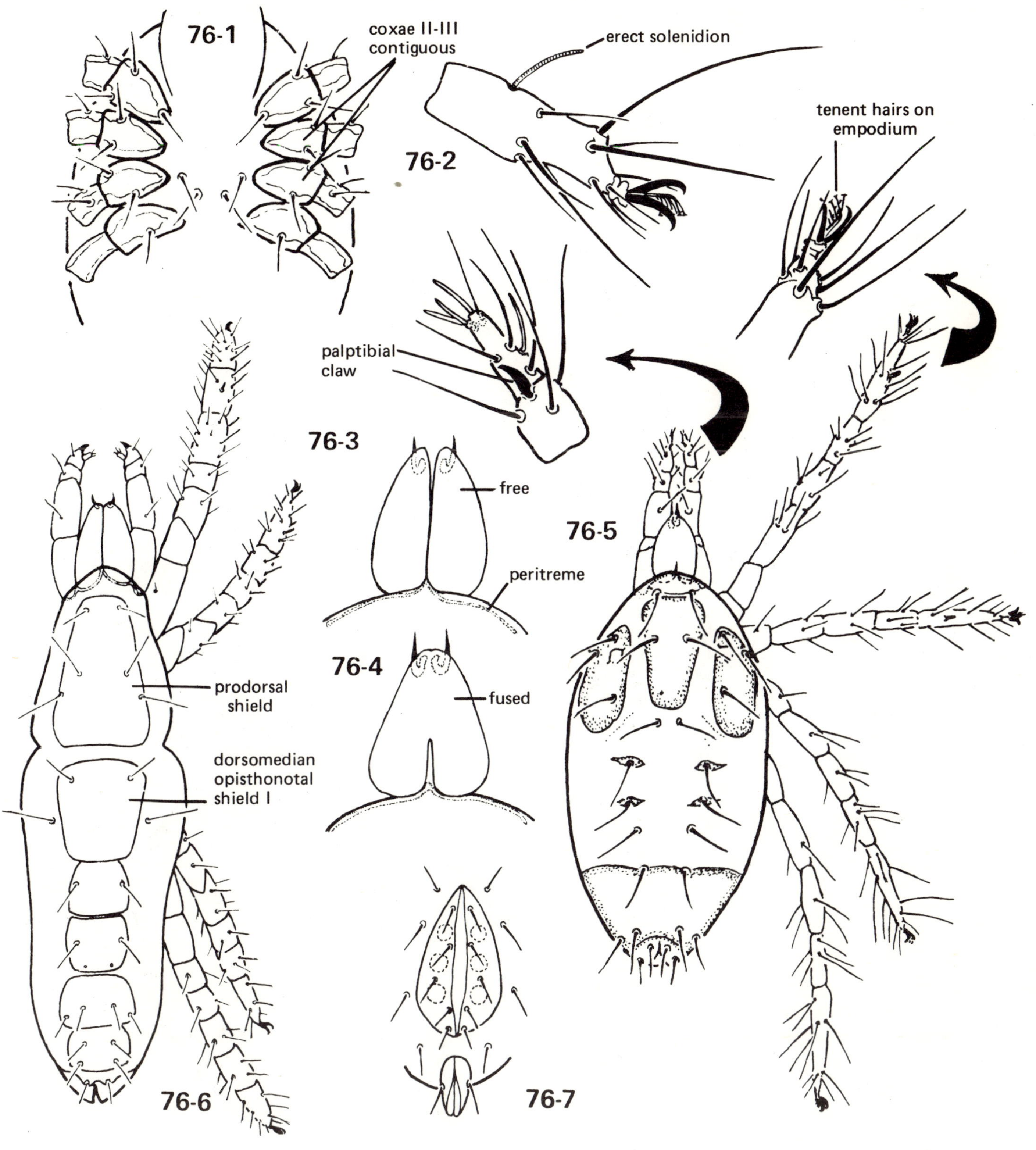

76-1 to 76-5; family RAPHIGNATHIDAE. **76-1**; raphignathid coxal arrangement: **76-2**; tarsus I of typical raphignathid: **76-3 and 76-4**; free and fused cheliceral bases in Raphignathoidea: **76-5**; *Raphignathus* sp. (Oregon, USA), dorsum of female with details of the palptibia and tarsus (left) and the terminus of tarsus I (right)

76-6 and 76-7; family POMERANTZIIDAE, *Pomerantzia charlesi* Baker (Georgia, USA) (after Baker 1949). **76-6**; dorsum of female: **76-7**; genital-anal region of female

PLATE 77

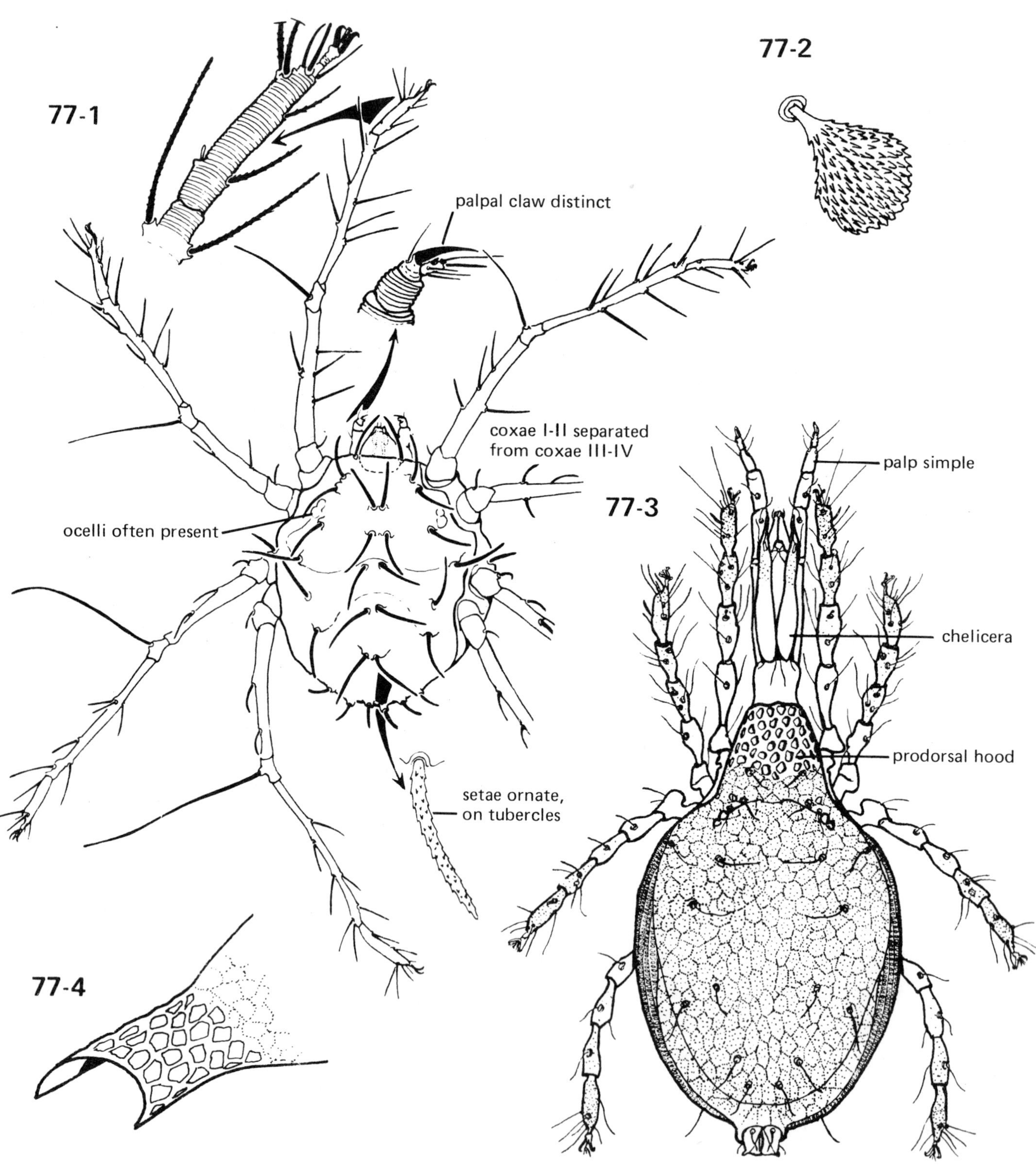

77-1 and 77-2; family CAMEROBIIDAE, *Neophyllobius* sp. (Oregon, USA). **77-1;** dorsum with details of tarsus I (top), the palptibia and tarsus, and dorsal seta: **77-2;** *Camerobia australis* Southcott (Australia), body seta (after Southcott 1957)

77-3 and 77-4; family CRYPTOGNATHIDAE, *Cryptognathus lagena* Kramer (Oregon, USA). **77-3;** dorsum (from Krantz 1958): **77-4;** prodorsal sheath (from Krantz 1958)

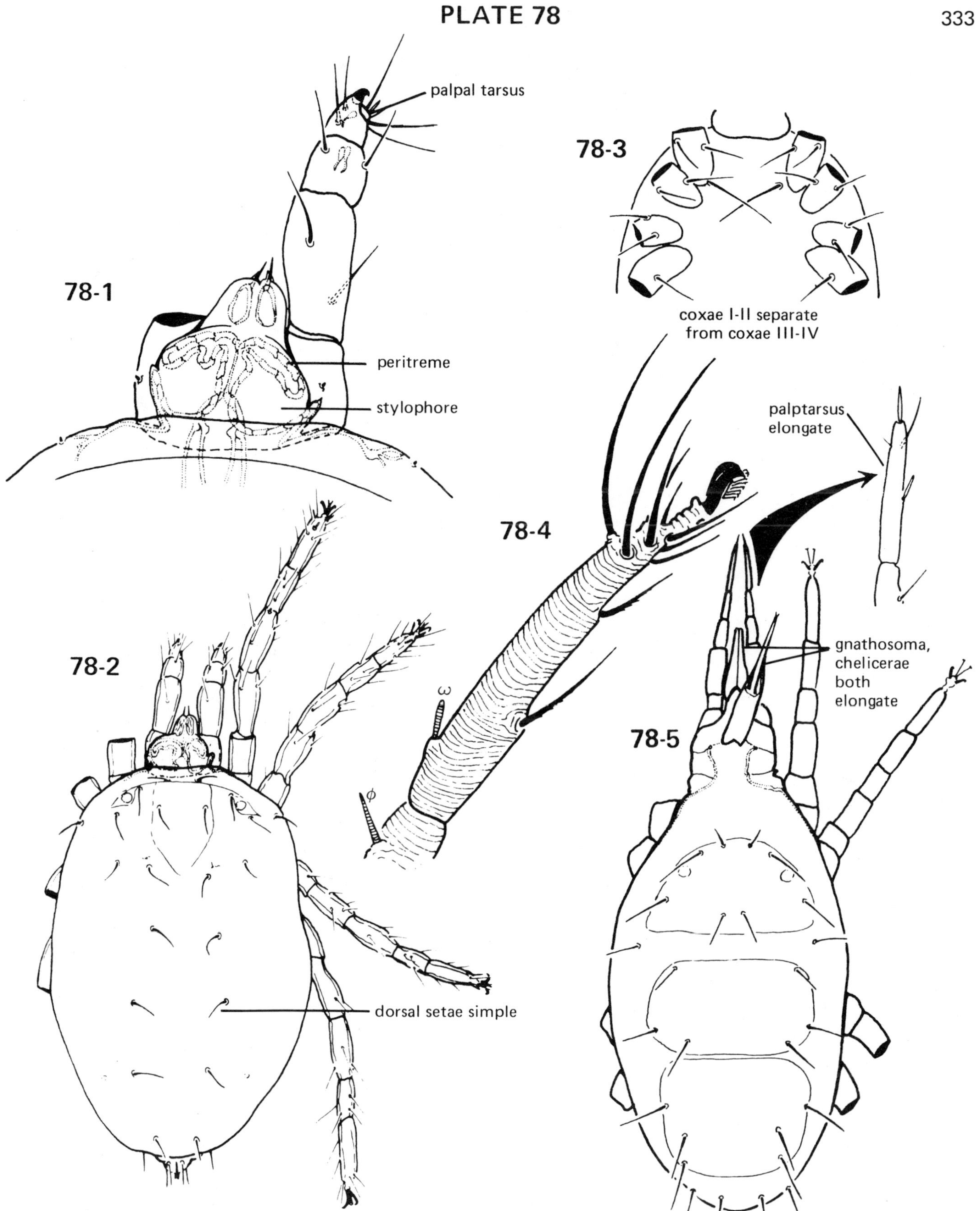

78-1 to 78-4; family CALIGONELLIDAE. **78-1**; *Coptocheles* sp. (Oregon, USA), gnathosoma and anterior margin of propodosoma: **78-2**; *Coptocheles* sp., dorsum of female: **78-3**; *Coptocheles* sp., coxal arrangement: **78-4**; tarsus I of caligonellid type
78-5; family EUPALOPSELLIDAE, *Eupalopsellus* sp. (Oregon, USA), dorsum with detail of palp

PLATE 79

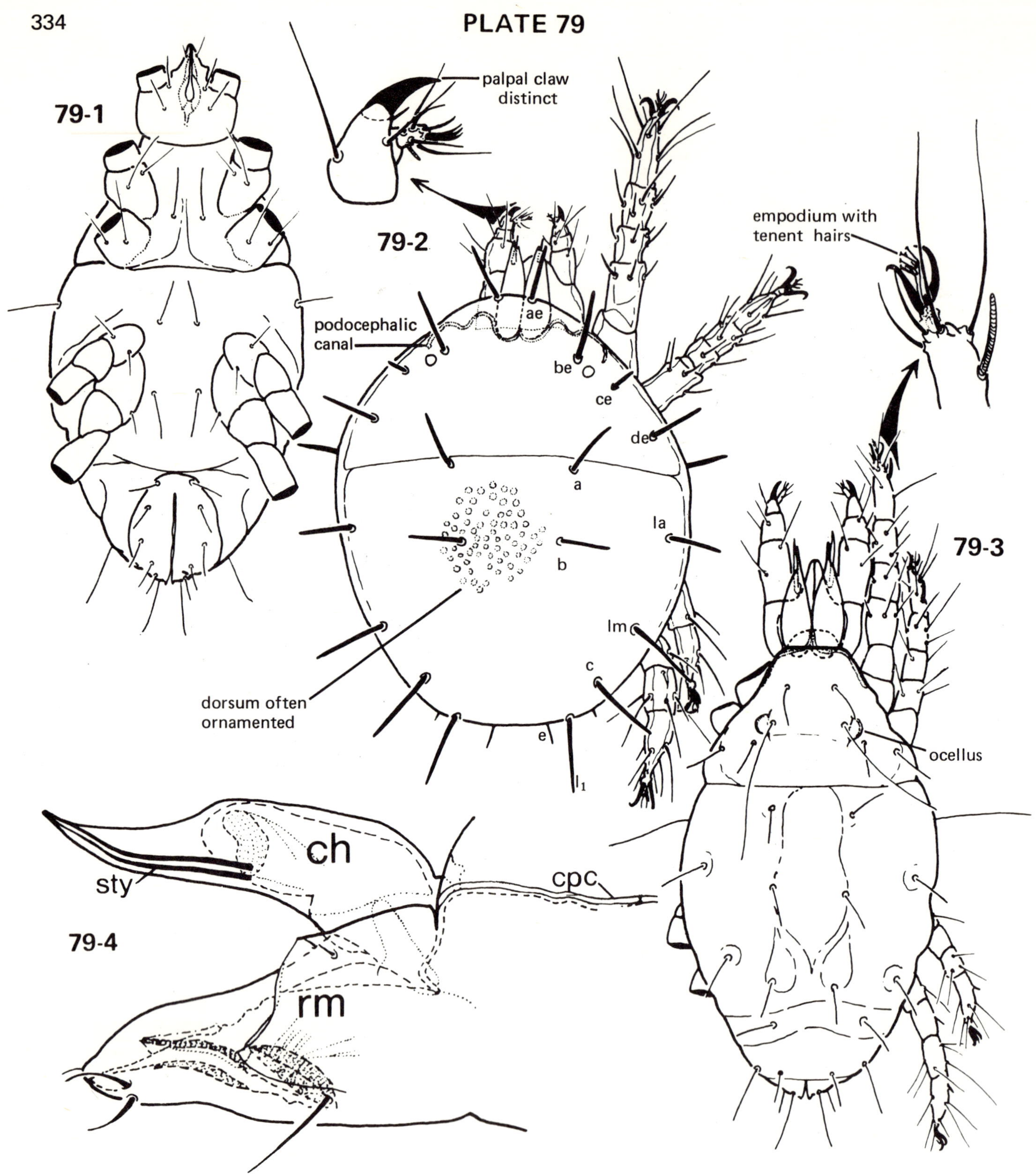

79-1 to 79-4; family STIGMAEIDAE. **79-1;** *Stigmaeus* sp. (Oregon, USA), venter of female: **79-2;** *Eustigmaeus* sp. (Oregon, USA), dorsum with detail of palpal thumb-claw complex: **79-3;** *Stigmaeus* sp., dorsum with detail of terminal portion of leg I: **79-4;** *Mediolata mariaefrancae* André (Belguim), lateral aspect of gnathosoma showing separation of rostrum (rm) and the partially fused cheliceral bases (ch) (cpc; podocephalic canal, sty; movable cheliceral digits) (after André 1977)

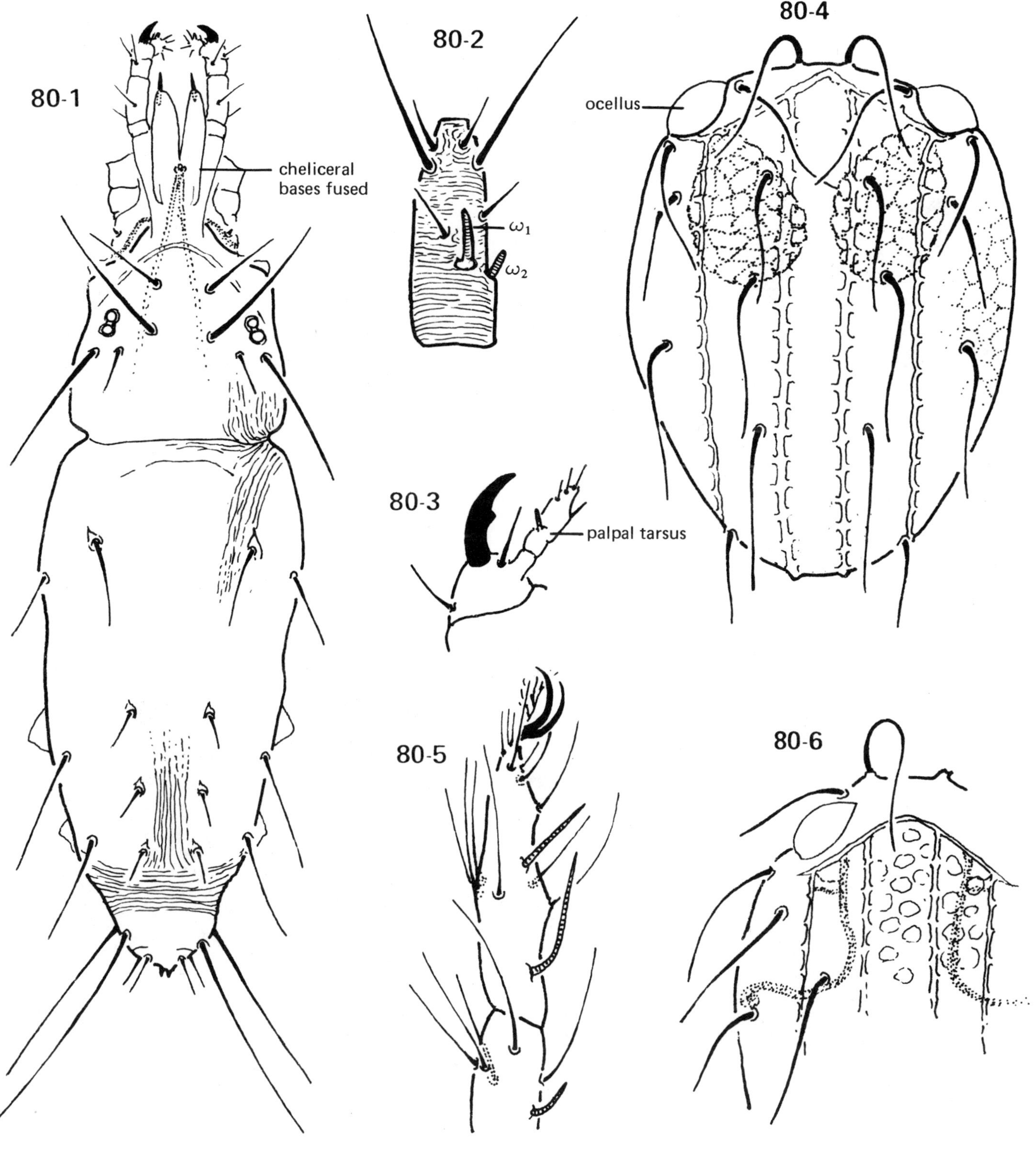

80-1 to 80-3; family BARBUTIIDAE. **80-1**; *Barbutia perretae* Robaux (California, USA), dorsum of male (after Robaux 1975): **80-2**; *B. perretae,* tarsus I of male (pretarsus deleted) (after Robaux 1975): **80-3**; *B. anguinea* (Berlese), palpal thumb-claw process (after Robaux 1975)

80-4 to 80-6; family HOMOCALIGIDAE, *Annerossella pacifica* Wood (Malaysia) (after Wood 1969). **80-4**; dorsum of female: **80-5**; tarsus and distal portion of tibia I of male: **80-6**; portion of male dorsum showing lateral internal tubes

PLATE 81

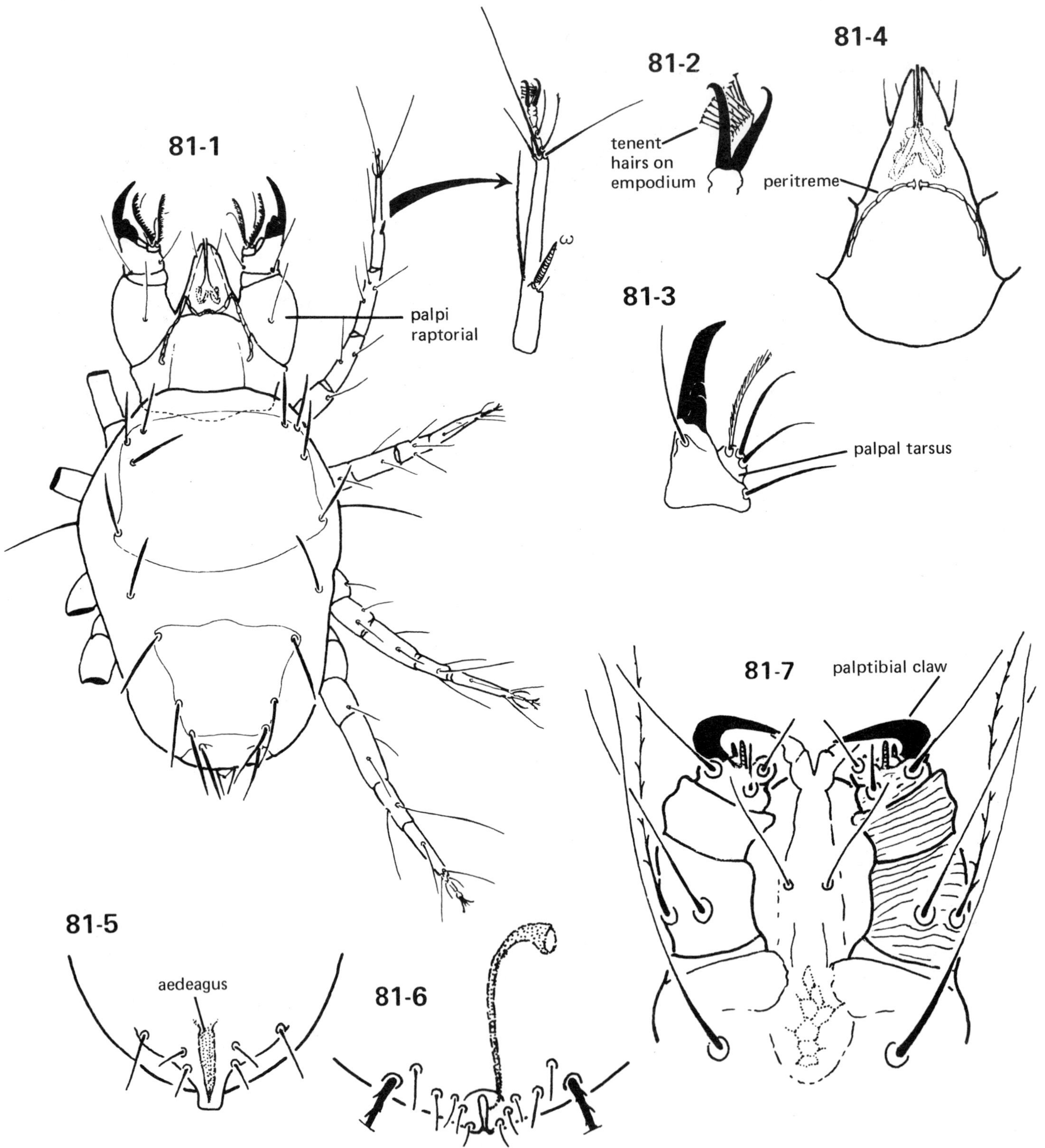

81-1 to 81-5; family CHEYLETIDAE. **81-1**; *Cheyletus malaccensis* Oudemans (Oregon, USA), dorsum with detail of tarsus I: **81-2**; *C. malaccensis,* pretarsus I: **81-3**; *C. malaccensis,* palpal thumb-claw process: **81-4**; *Acaropsis docta* Berlese (Oregon, USA), dorsomedian aspect of gnathosoma: **81-5**; *A. docta,* posteroventral aedeagal area of male

81-6 and 81-7; family CHEYLETIELLIDAE. **81-6**; *Cheyletiella parasitivorax* (Mégnin) (Oregon, USA), posterodorsal aedeagal area of male: **81-7**; *Ornithocheyletia* sp. (Oregon, USA), venter of gnathosoma

PLATE 82

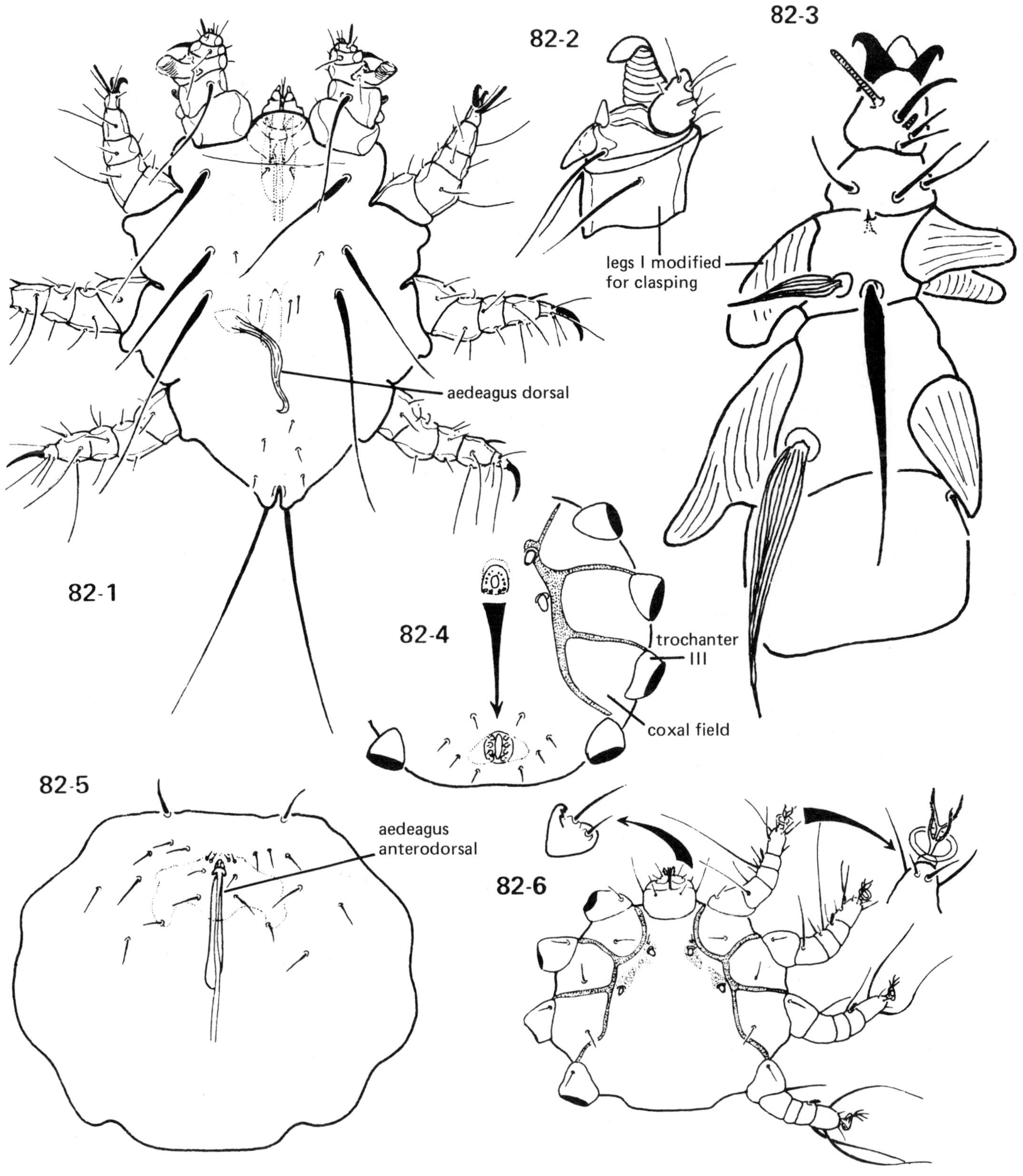

82-1 to 82-3; family MYOBIIDAE. **82-1**; *Protomyobia* sp. (Oregon, USA), dorsum of male: **82-2**; *Radfordia* sp. (Oregon, USA), leg I of female: **82-3**; *Archemyobia philander* Lukoschus, Dusbábek and Jameson (Netherlands), leg I of female (dorsal) (after Lukoschus, Dusbábek and Jameson 1972)

82-4 to 82-6; family OPHIOPTIDAE. **82-4**; position of genital aperture in females of *Afrophioptes* (anterior aperture) and of *Ophioptes* (arrow) (adapted from Fain 1964): **82-5**; *Ophioptes coluber* Radford (India), dorsum of male (after Fain 1964): **82-6**; *Ophioptes southcotti* Fain with details of palp and empodium I (after Fain 1964)

PLATE 83

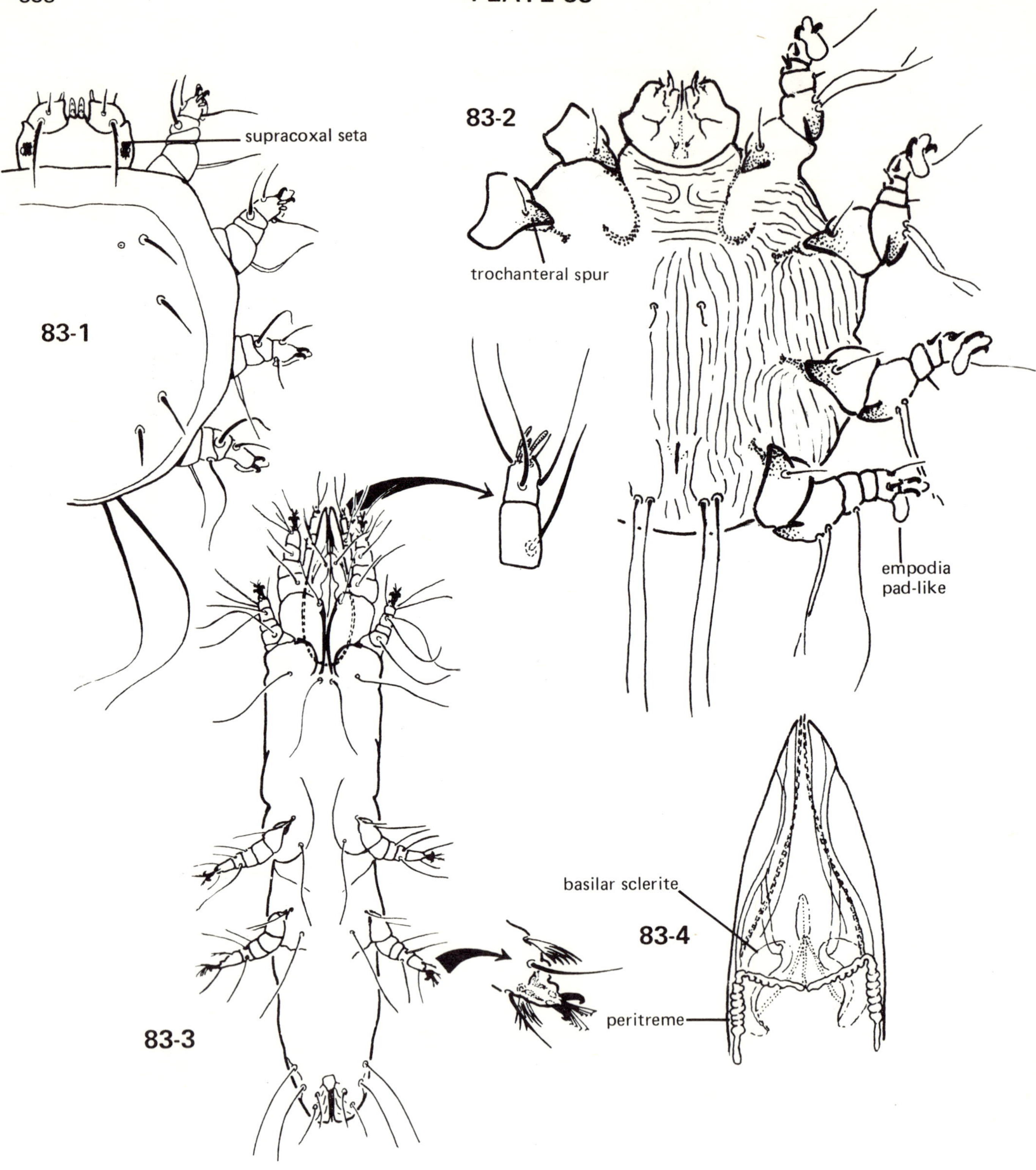

83-1 and 83-2; family PSORERGATIDAE. **83-1**; *Psorergates bos* Johnston (New Mexico, USA), dorsum of female (after Johnston 1964): **83-2**; *Psorergates talpae* Lukoschus (Netherlands), venter of female

83-3 and 83-4; family SYRINGOPHILIDAE, *Syringophilus* sp. (Delaware, USA). **83-3**; venter of female with details of palp and of distal aspect of tarsus IV: **83-4**; stylophore of female, showing position of chelicerae (heavy dotted lines) and reniform basilar sclerites.

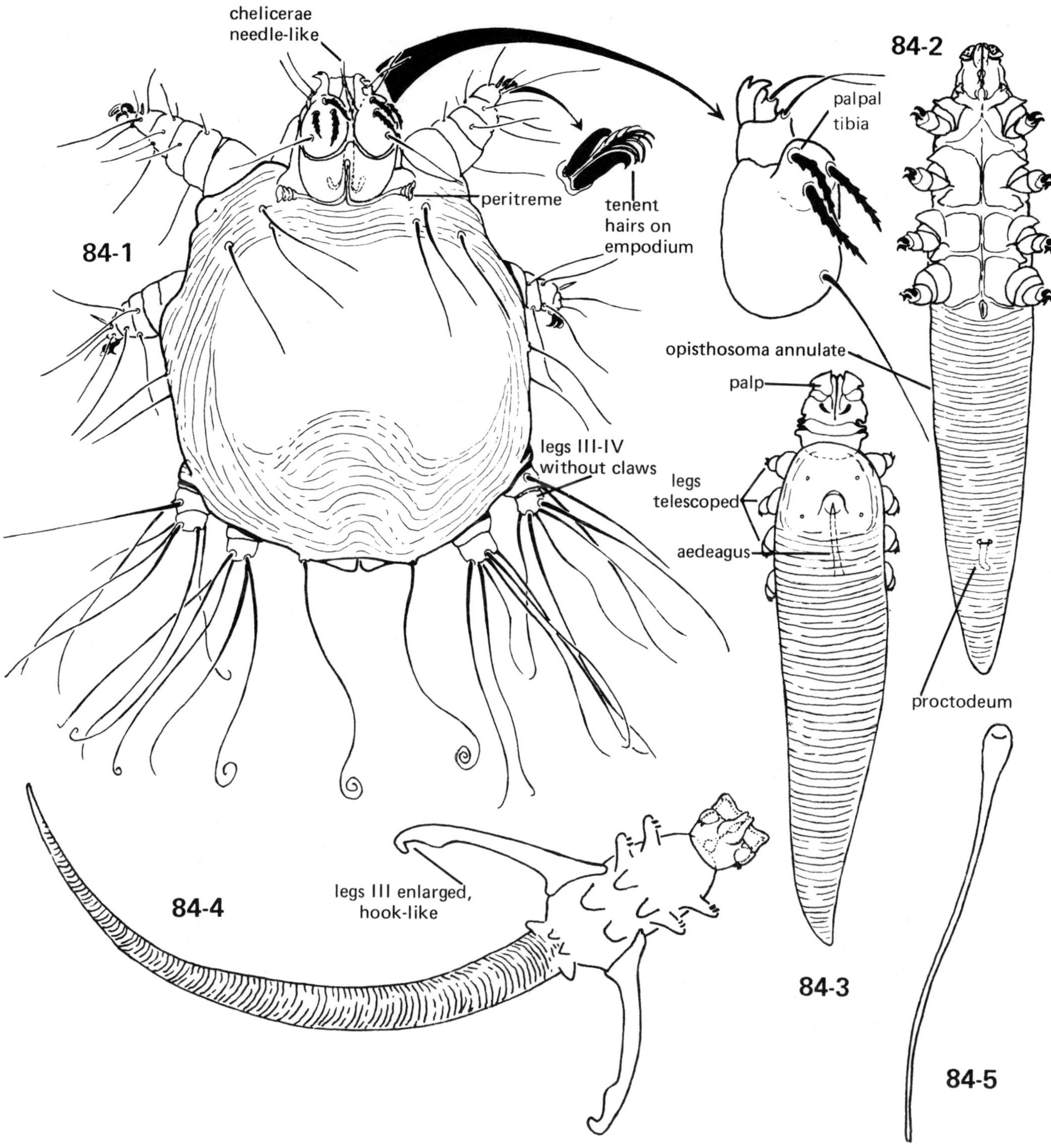

84-1; family HARPYRHYNCHIDAE, *Harpyrhynchus* sp. (California, USA), dorsum of female with details of pretarsus I and of the palp

84-2 to 84-5; family DEMODICIDAE. **84-2**; *Demodex* sp. (Oregon, USA), venter of female: **84-3**; *Demodex* sp., dorsum of male: **84-4**; *D. longissimus* Desch and Nutting (Surinam), venter of nymph (opisthosoma normally twice length illustrated) (after Desch and Nutting 1972): **84-5**; *D. longissimus*, ovum (after Desch and Nutting 1972)

PLATE 85

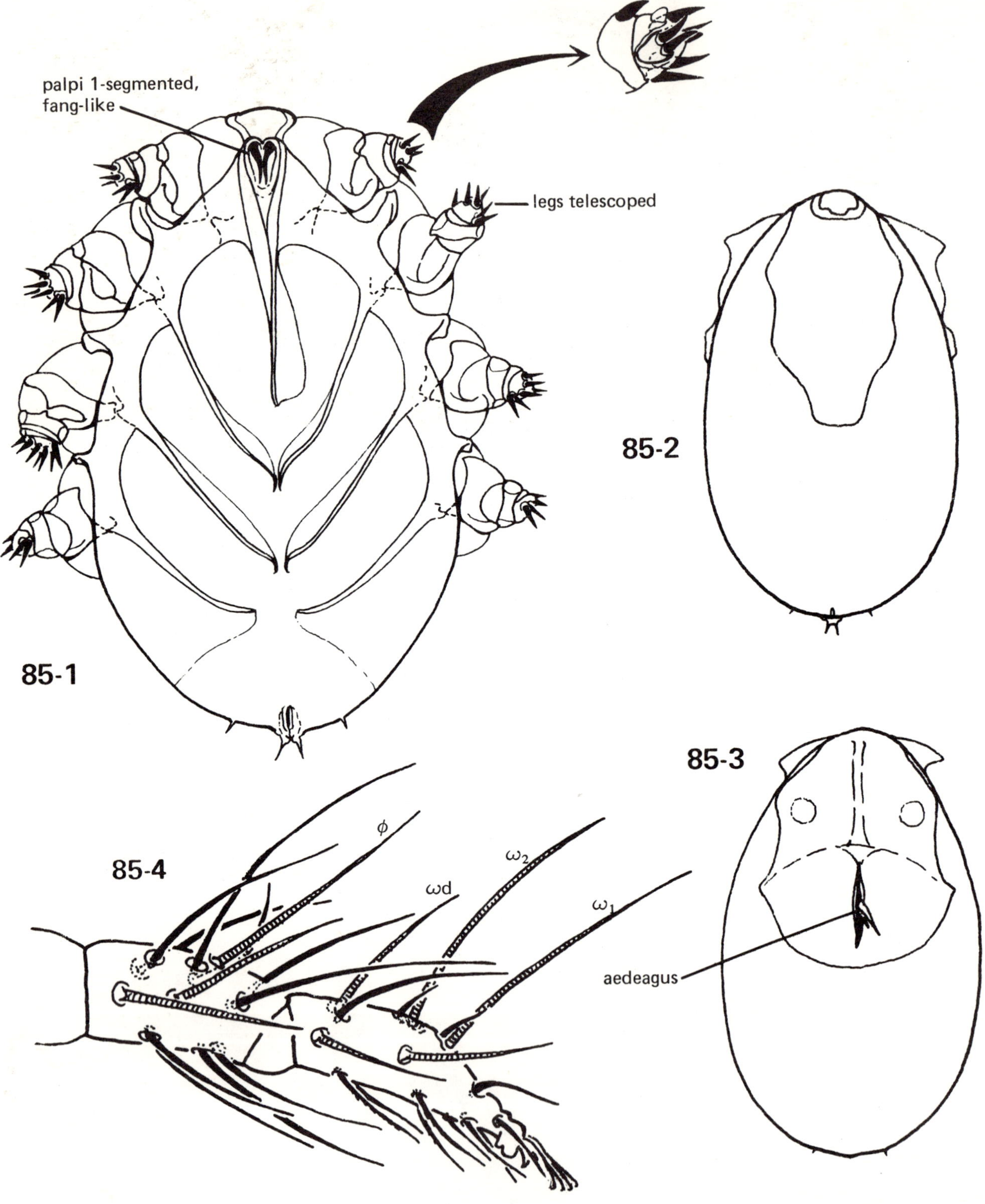

85-1 to **85-3**; family CLOACARIDAE, *Cloacarus faini* Camin, Moss, Oliver and Singer (Missouri, USA). **85-1**; venter of female with detail of leg I: **85-2**; dorsum of female: **85-3**; dorsum of male (after Camin et al. 1967)

85-4; family TETRANYCHIDAE, *Tetranychus urticae* Koch, tibia and tarsus of leg I of male (antiaxial), showing position of setae and solenidia (striated) (after Grandjean 1948)

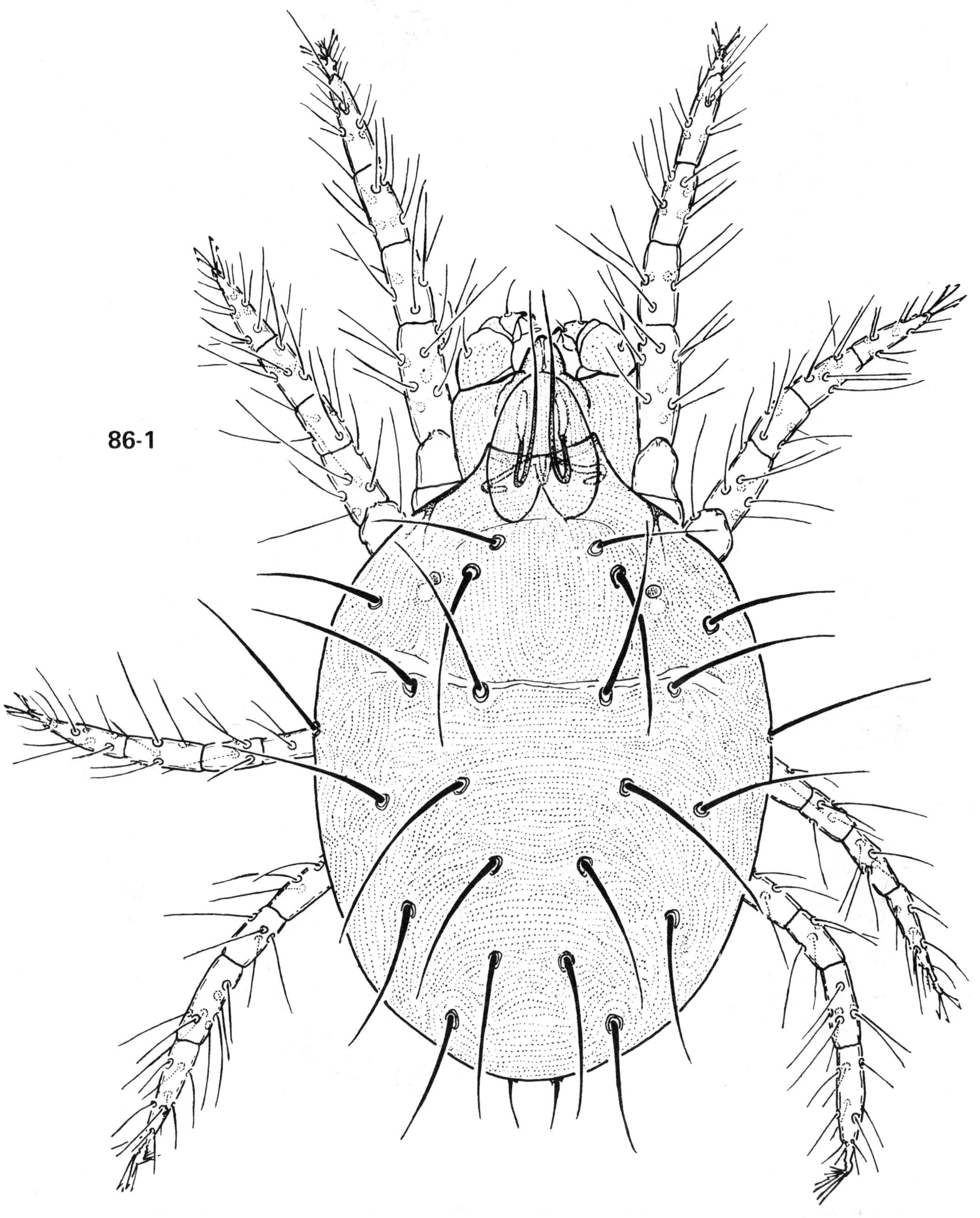

86-1; family TETRANYCHIDAE, *Tetranychus (Armenychus) mcdanieli* McGregor (Oregon, USA), dorsum of female

PLATE 87

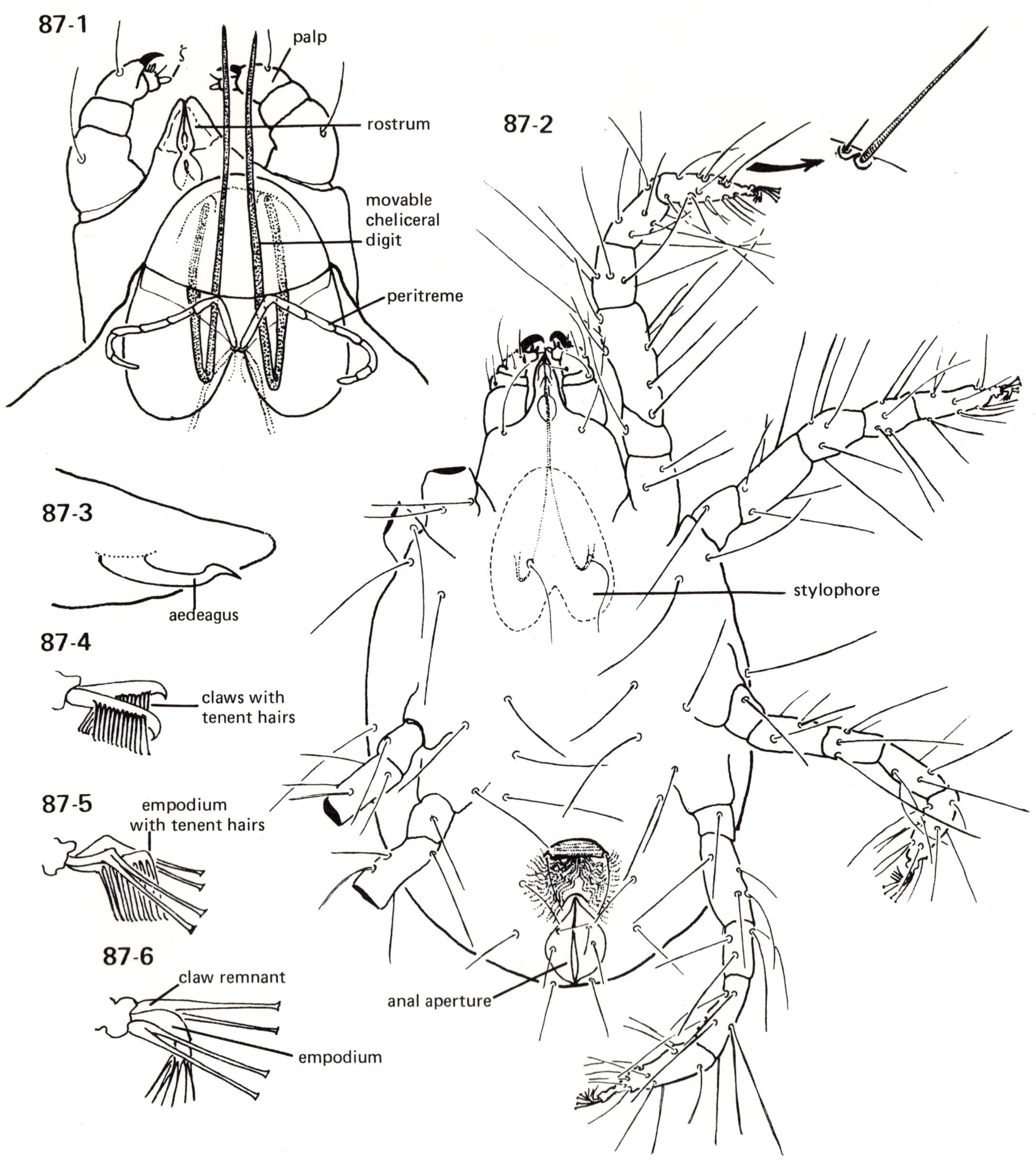

87-1 to 87-6; family TETRANYCHIDAE. **87-1**; *Tetranychus* sp (Oregon, USA), dorsal aspect of gnathosoma: **87-2**; *Tetranychus* sp., venter of female with detail of distal duplex setae of tarsus I: **87-3**; *Tetranychus* sp., aedeagus of male, lateral aspect: **82-4**; pretarsus of *Tetranycopsis:* **87-5**; *pretarsus* of *Petrobia:* **87-6**; pretarsus of *Tetranychus*

PLATE 88

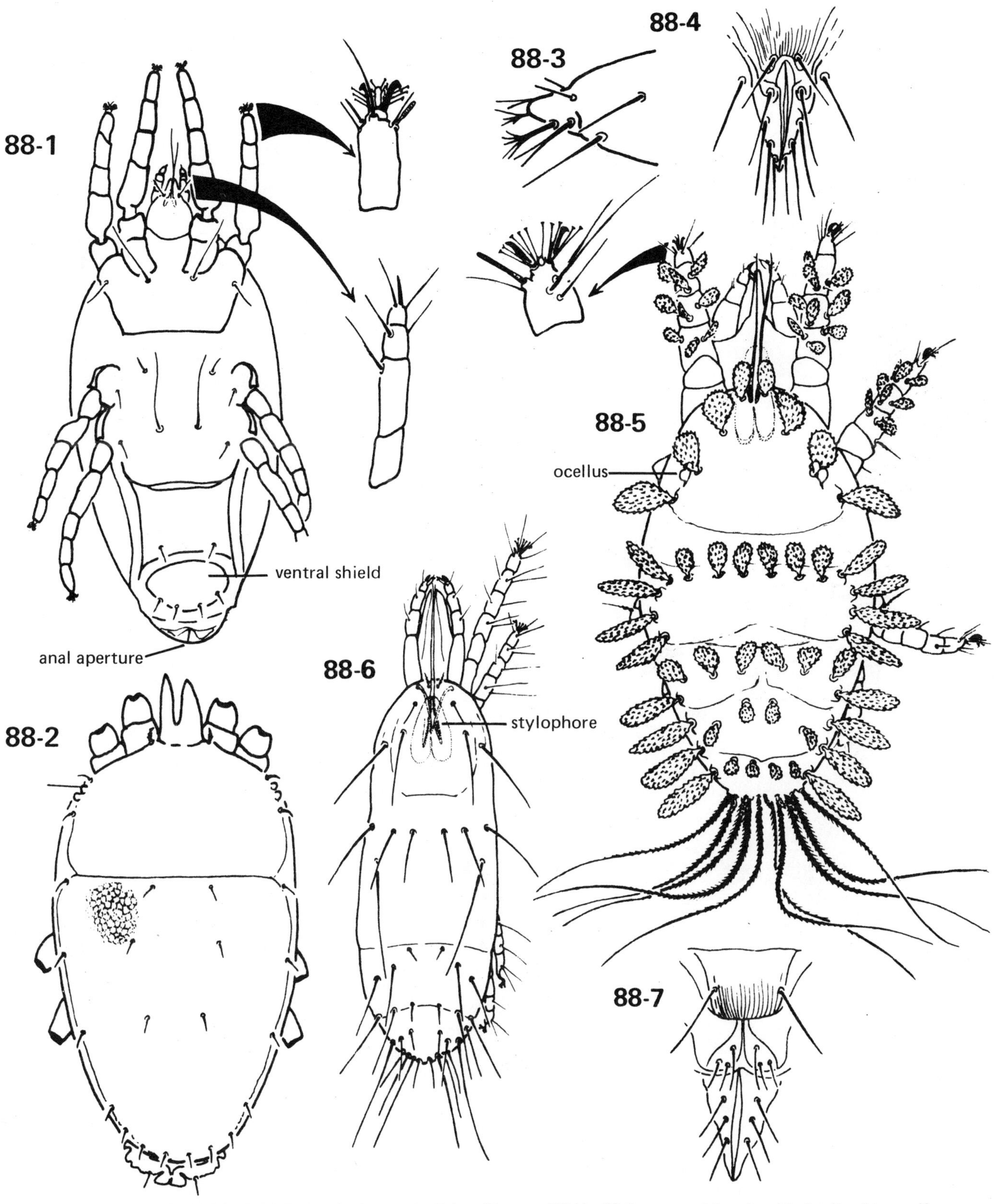

88-1 and 88-2; family TENUIPALPIDAE, *Brevipalpus essigi* Baker (Oregon, USA). **88-1**; venter of female with details of tarsus II (top) and palp: **88-2**; dorsum

88-3; family ALLOCHAETOPHORIDAE, *Allochaetophora californica* McGregor (California, USA), caudal aspect of nymph

88-4 and 88-5; family TUCKERELLIDAE, *Tuckerella* sp. (Hawaii, USA). **88-4**; genital-anal area of female: **88-5**; dorsum of female with detail of tarsus I

88-6 and 88-7; family LINOTETRANIDAE, *Linotetranus* sp. **88-6**; dorsum of female: **88-7**; genital-anal area of female

PLATE 89

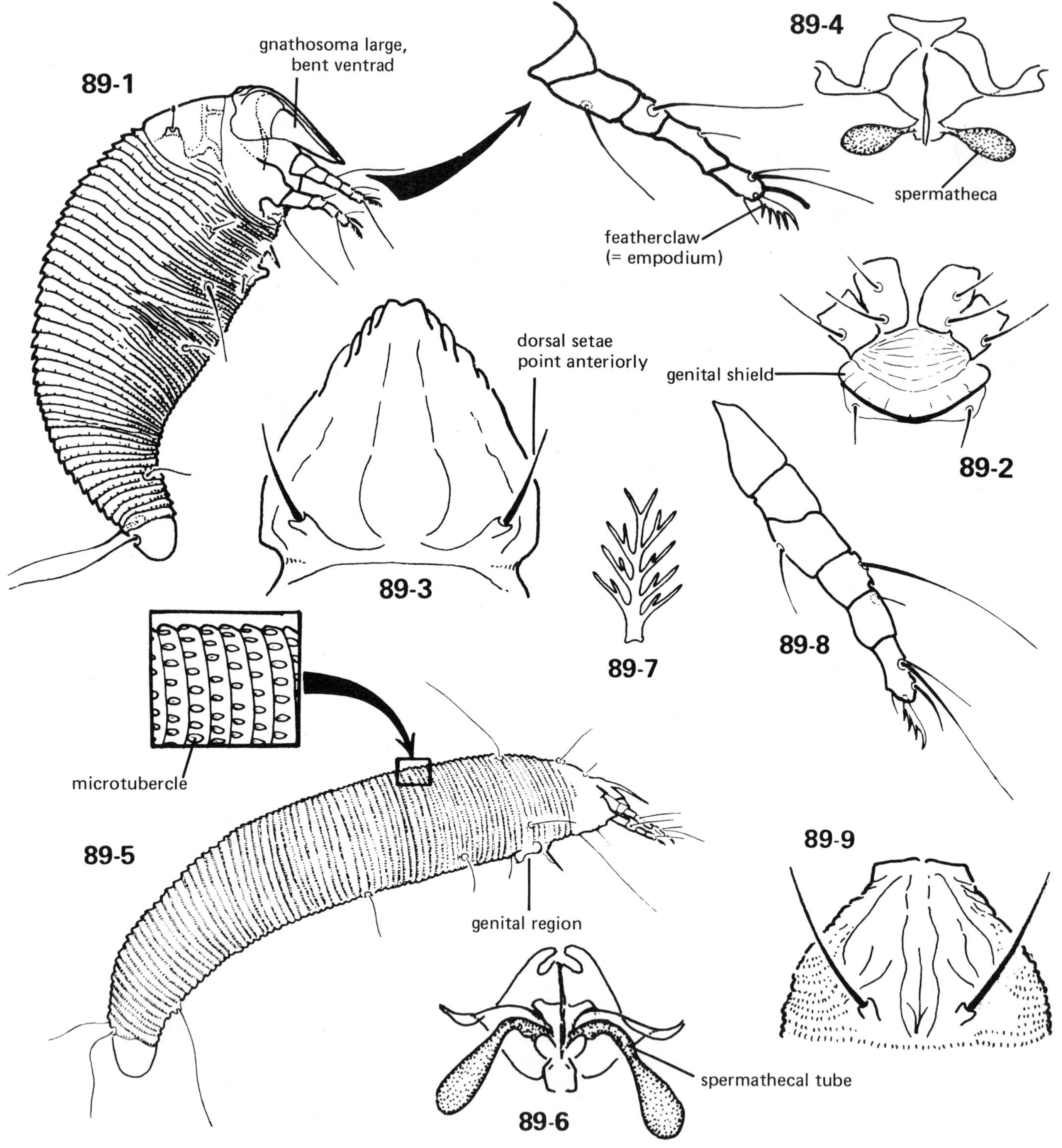

89-1 to 89-4; family RHYNCHAPHYTOPTIDAE. **89-1**; *Rhynchaphytoptus strigatus* Keifer (Maryland, USA), lateral aspect with detail of leg I: **89-2**; *R. strigatus*, genital region of female: **89-3**; *R. ficifoliae* Keifer, prodorsal shield (after Jeppson et al. 1975): **89-4**; *R. ficifoliae*, spermathecae and associated structures (after Jeppson et al. 1975).

89-5 to 89-8; family SIERRAPHYTOPTIDAE. **89-5**; *Phytoptus avellanae* Nalepa (Oregon, USA), lateral aspect of female with detail of idiosomal ornamentation: **89-6**; *P. avellanae*, spermathecae and associated structures (after Jeppson et al. 1975): **89-7**; *P. avellanae*, empodium: **89-8**; *P. avellanae*, leg I: **89-9**; *P. pseudoinsidiosus* (Wilson), prodorsal shield (after Jeppson et al. 1975)

PLATE 90

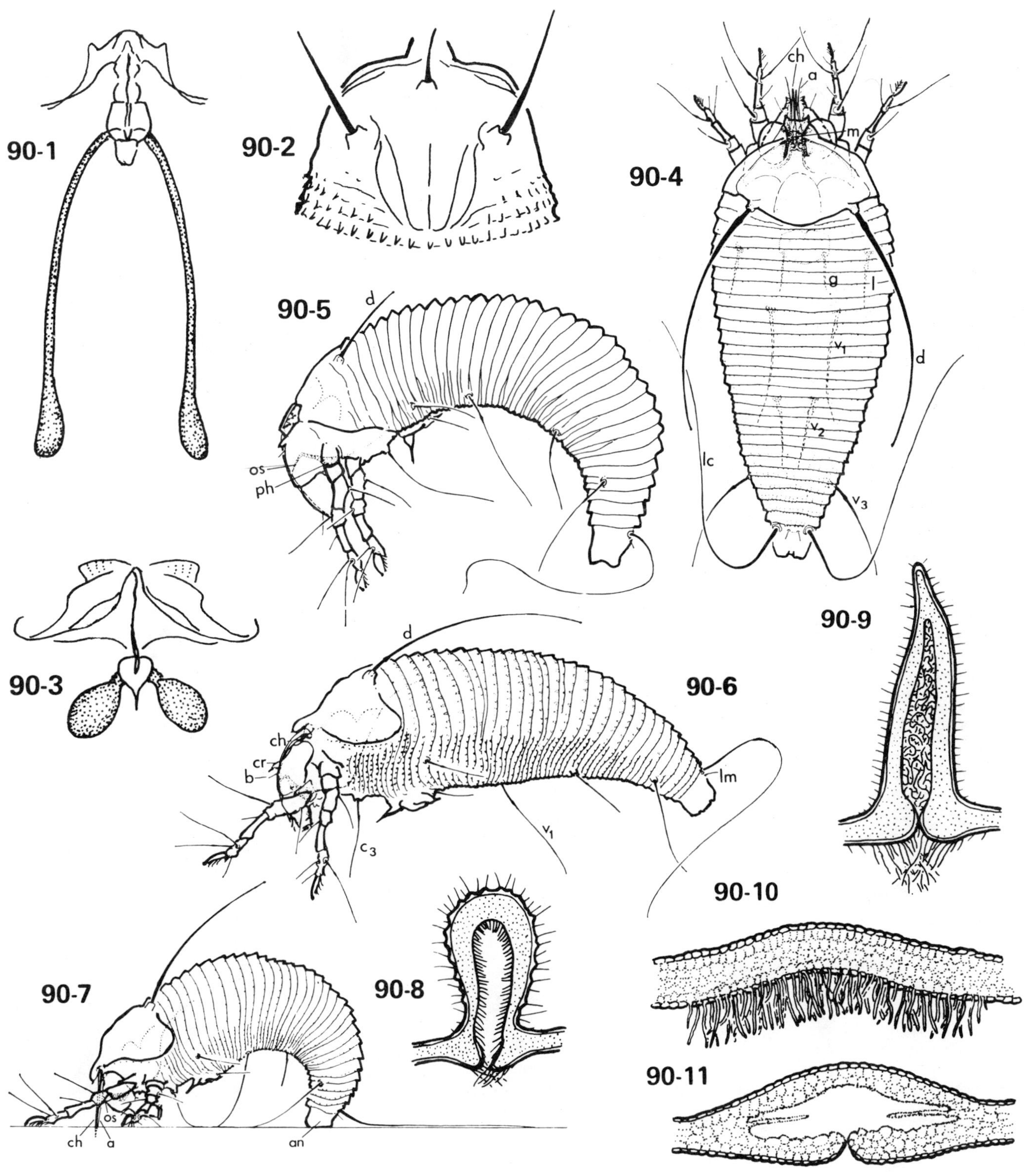

90-1 and 90-2; family SIERRAPHYTOPTIDAE (after Jeppson et al. 1975). **90-1**; *Trisetacus juniperinus* (Nalepa), spermathecal tubes and associated structures: **90-2**; *T. pini* (Nalepa), prodorsal shield

90-3 to 90-11; family ERIOPHYIDAE. **90-3**; *Aculops pelekassi* (Keifer), spermathecal tubes and associated structures: **90-4 to 90-7**; *Aculus comatus* (Nalepa) (Oregon, USA). **90-4**; dorsum of male: **90-5**; lateral aspect of deutogyne: **90-6**; lateral aspect of protogyne: **90-7**; feeding stance. Position of palpi is inferred (a = auxiliary stylet, an = anal sucker, ch = chelicera, cr = cheliceral retainer, m = cheliceral motivator, os = oral stylet, ph = pharynx; b,c,d,g,l,lc,lm and v are setae): **90-8 to 90-11**; plant host reactions to eriophyid feeding. **90-8 and 90-9**; closed gall types: **90-10**; erineum: **90-11**; mesophyllous blister

PLATE 91

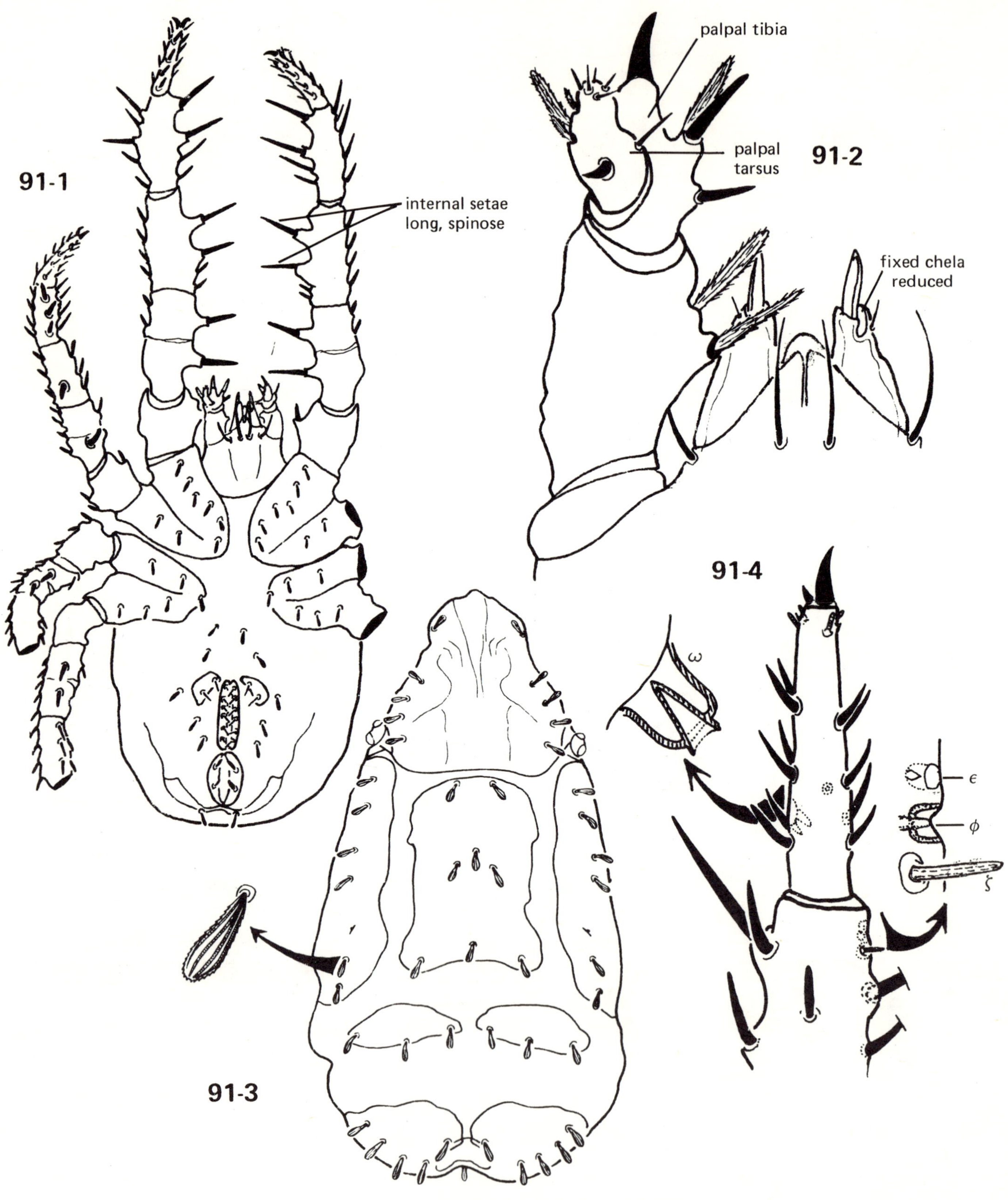

91-1 to 91-4; family CAECULIDAE. **91-1**; *Caeculus* sp. (Oregon, USA), venter: **91-2**; *Caeculus* sp., venter of gnathosoma: **91-3**; *Caeculus* sp., dorsum with detail of dorsal seta: **91-4**; *C. ? liguricus* Vitzthum (France), tibia and tarsus I, with details of the recessed tarsal solenidion (left) and the tibial complex of recessed famulus (ϵ) and solenidion (ϕ), flanked by a eupathidium (ζ) (after Grandjean 1944)

PLATE 92

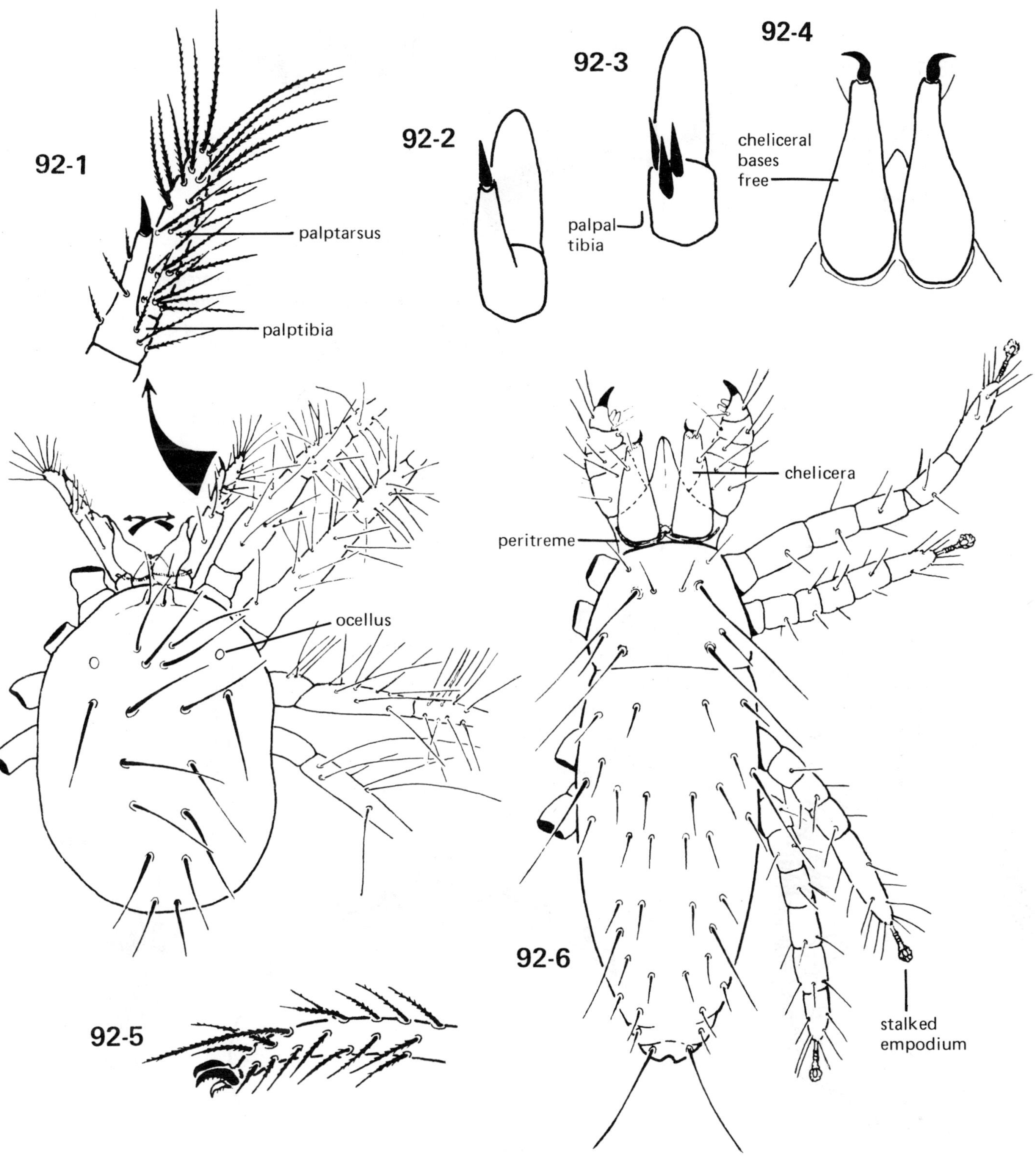

92-1 to 92-5; family ANYSTIDAE. **92-1**; *Bechsteinia* sp. (Oregon, USA), dorsum with detail of palpal thumb-claw process: **92-2**; thumb-claw process typical of the anystid subfamily Erythracarinae: **92-3**; thumb-claw process typical of the anystid subfamily Anystinae: **92-4**; cheliceral arrangement in ANYSTIDAE: **92-5**; *Anystis* sp. (Oregon, USA), distal portion of tarsus I

92-6; family PSEUDOCHEYLIDAE, *Pseudocheylus* sp., dorsum

PLATE 93

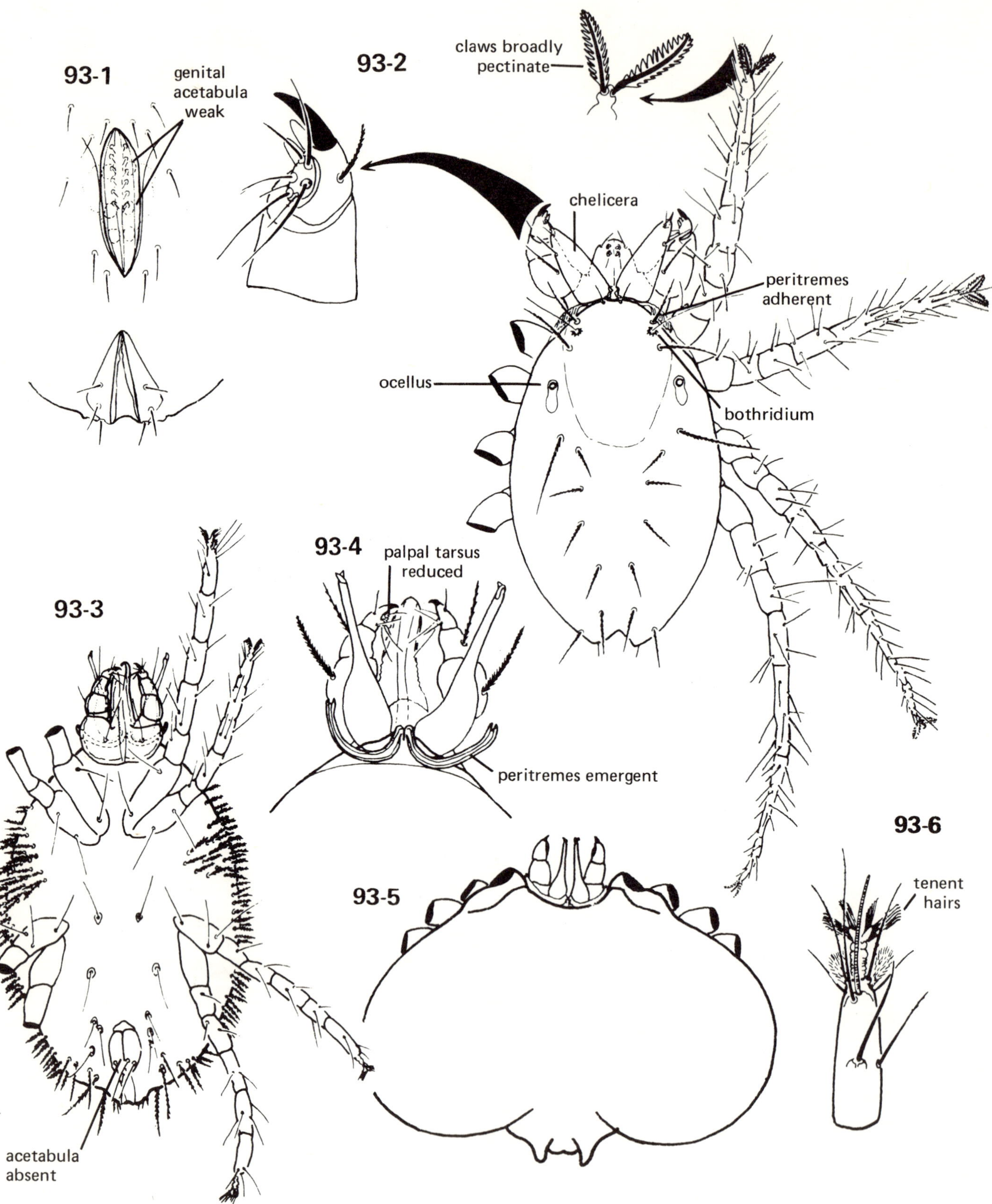

93-1 and 93-2; family TENERIFFIIDAE, *Teneriffia* sp. (Oregon, USA). **93-1**; genital-anal region of female: **93-2**; dorsum of female with detail of palpal thumb-claw process, and of pretarsus I

93-3 to 93-6; family PTERYGOSOMATIDAE. **93-3**; *Geckobiella texana* Banks (California, USA), venter of female: **93-4**; *G. texana*, dorsum of gnathosoma: **93-5**; *Geckobia*, dorsum (diagrammatic): **93-6**; *Geckobiella texana*, tarsus I (dorsum)

PLATE 94

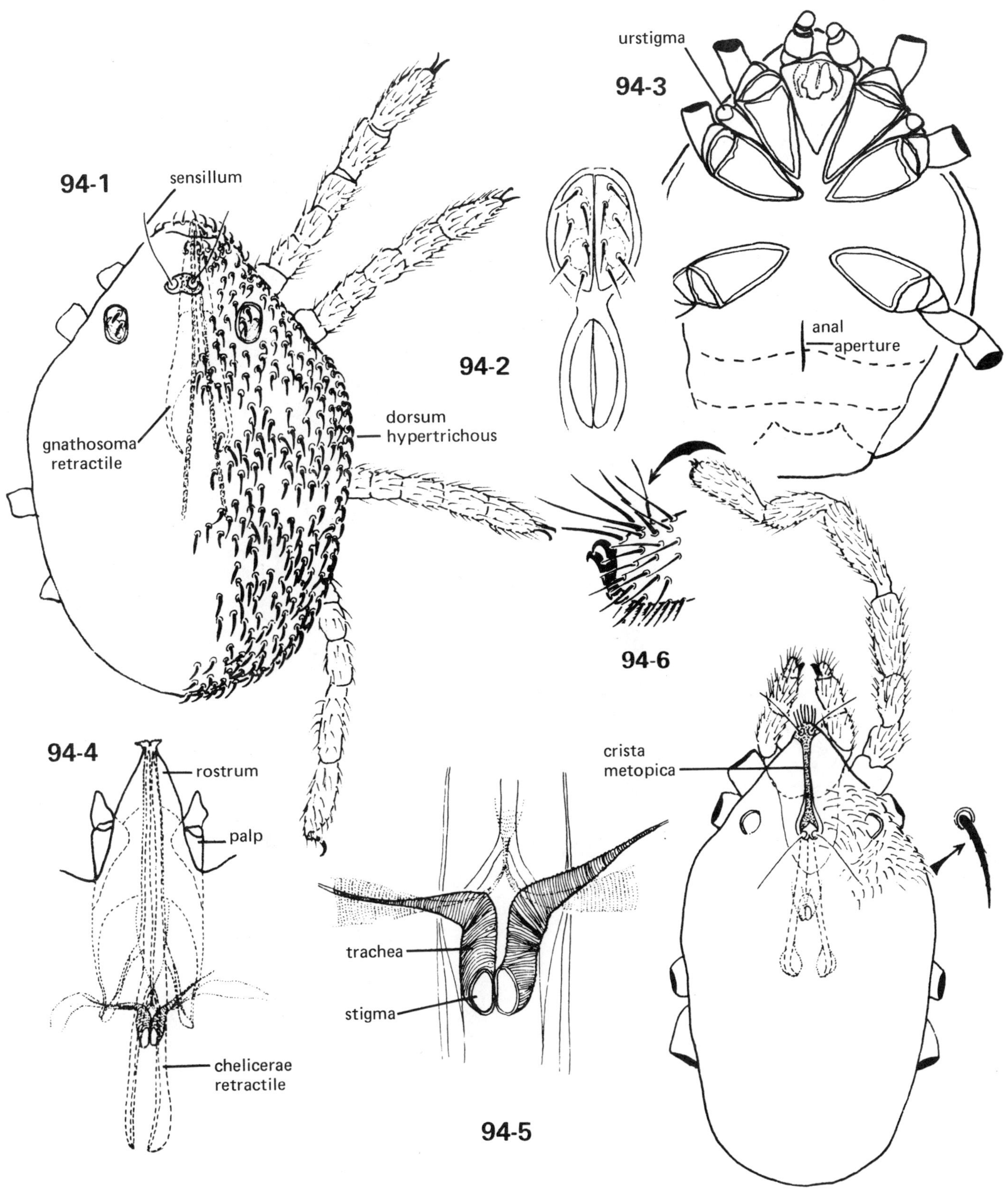

94-1 to 94-3; family CALYPTOSTOMATIDAE, *Calyptostoma* sp. (California, USA). **94-1;** dorsum: **94-2;** genitianal region: **94-3;** venter of larva

94-4 to 94-6; family ERYTHRAEIDAE. **94-4;** rostrum and associated respiratory structures: **94-5;** stigmata and associated tracheal trunks: **94-6;** *Balaustium* sp. (Oregon, USA), dorsum with details of terminal portion of tarsus (top) and of typical body seta

PLATE 95

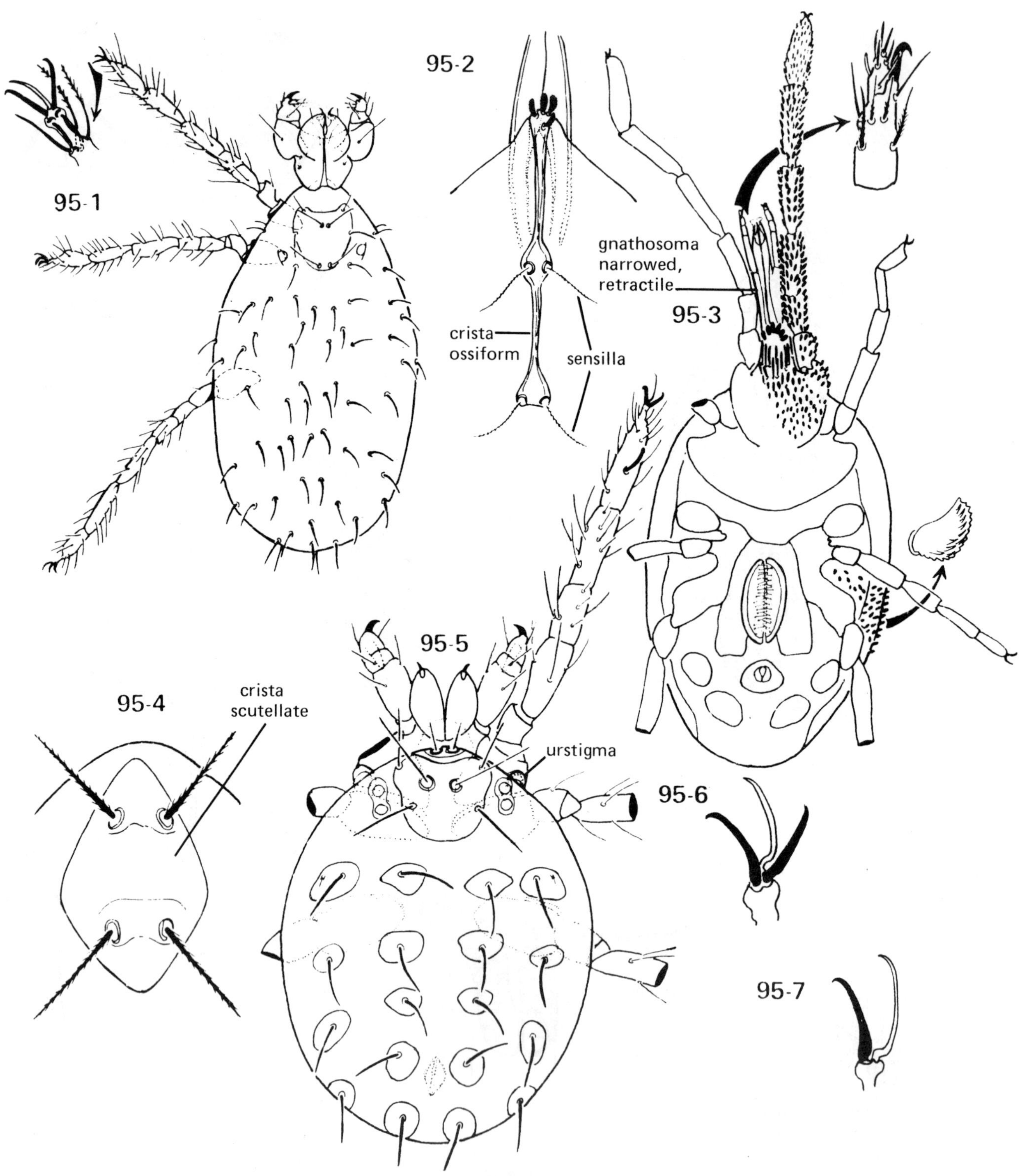

95-1; family ERYTHRAEIDAE, ?genus (Oregon, USA), dorsum of larva with detail of pretarsus I

95-2 to **95-4**; family SMARIDIDAE. **95-2**; *Fessonia* sp. (Oregon, USA), crista metopica of female: **95-3**; *Smaris* sp. (Oregon, USA), venter with details of palpal thumb-claw process (top) and of typical body seta: **95-4**; scutellate prodorsal shield of smaridid larva (diagrammatic)

95-5 to 95-7; family JOHNSTONIANIDAE. **95-5**; *? Diplothrombium* sp., dorsum of larva: **95-6 and 7**; pretarsi typical of johnstonianid larvae (true claws are solid, empodia are clear)

PLATE 96

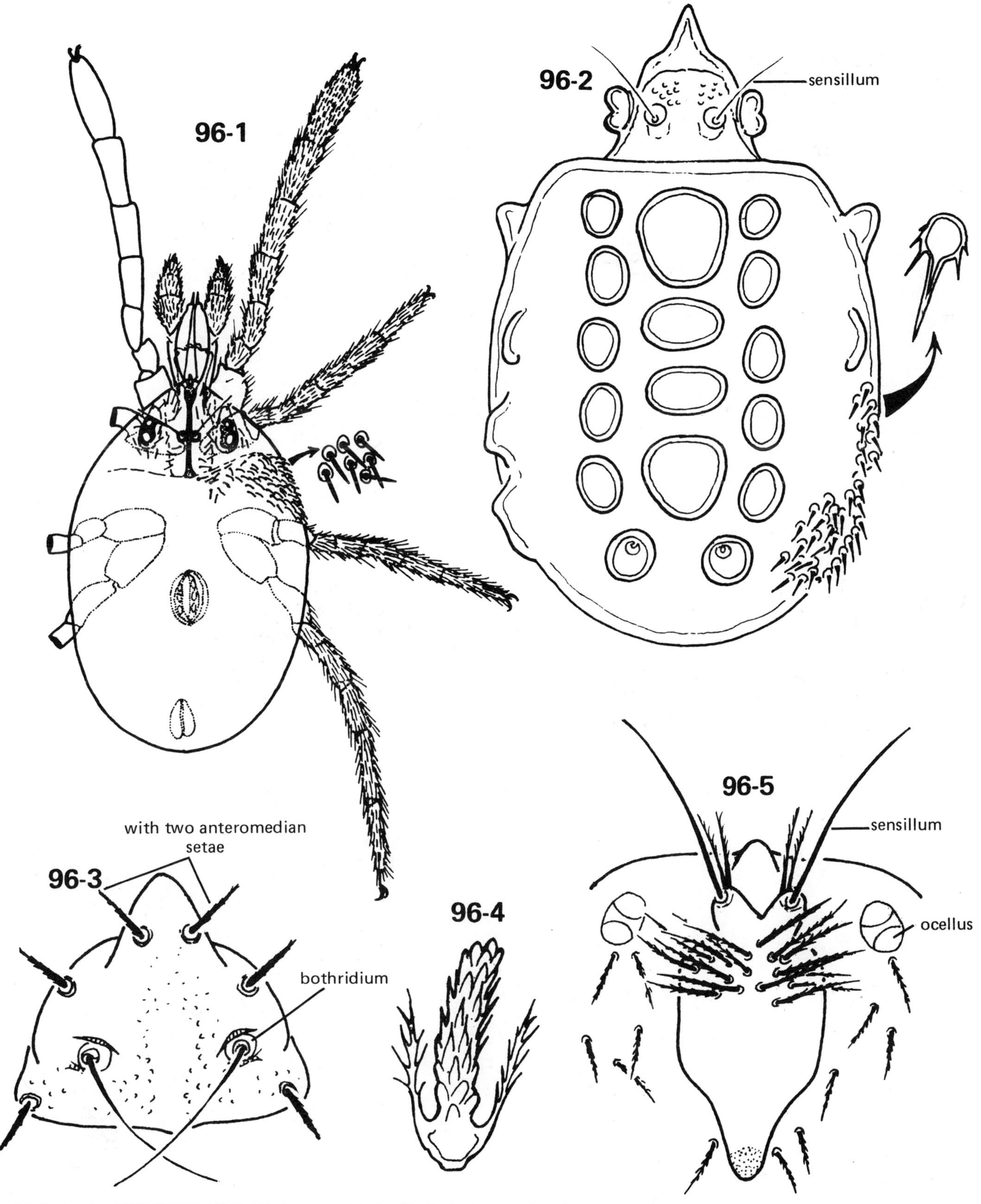

96-1; family JOHNSTONIANIDAE, dorsum with detail of tylochorous dorsal setae

96-2; family TROMBELLIDAE, *Trombella favosa* André (Zaire), dorsum of female with detail of typical dorsal seta

96-3 and 96-4; family NEOTROMBIDIIDAE. **96-3**; *Neotrombidium barringunense* Hirst (Australia), scutum of larva (after Lindquist and Vercammen-Grandjean 1971): **96-4**; *Neotrombidium* sp. (Zaire), typical dorsal seta

96-5; family CHYZERIDAE, *Chyzeria poivrei* Robaux (Chile), propodosoma of male (after Robaux 1969)

PLATE 97

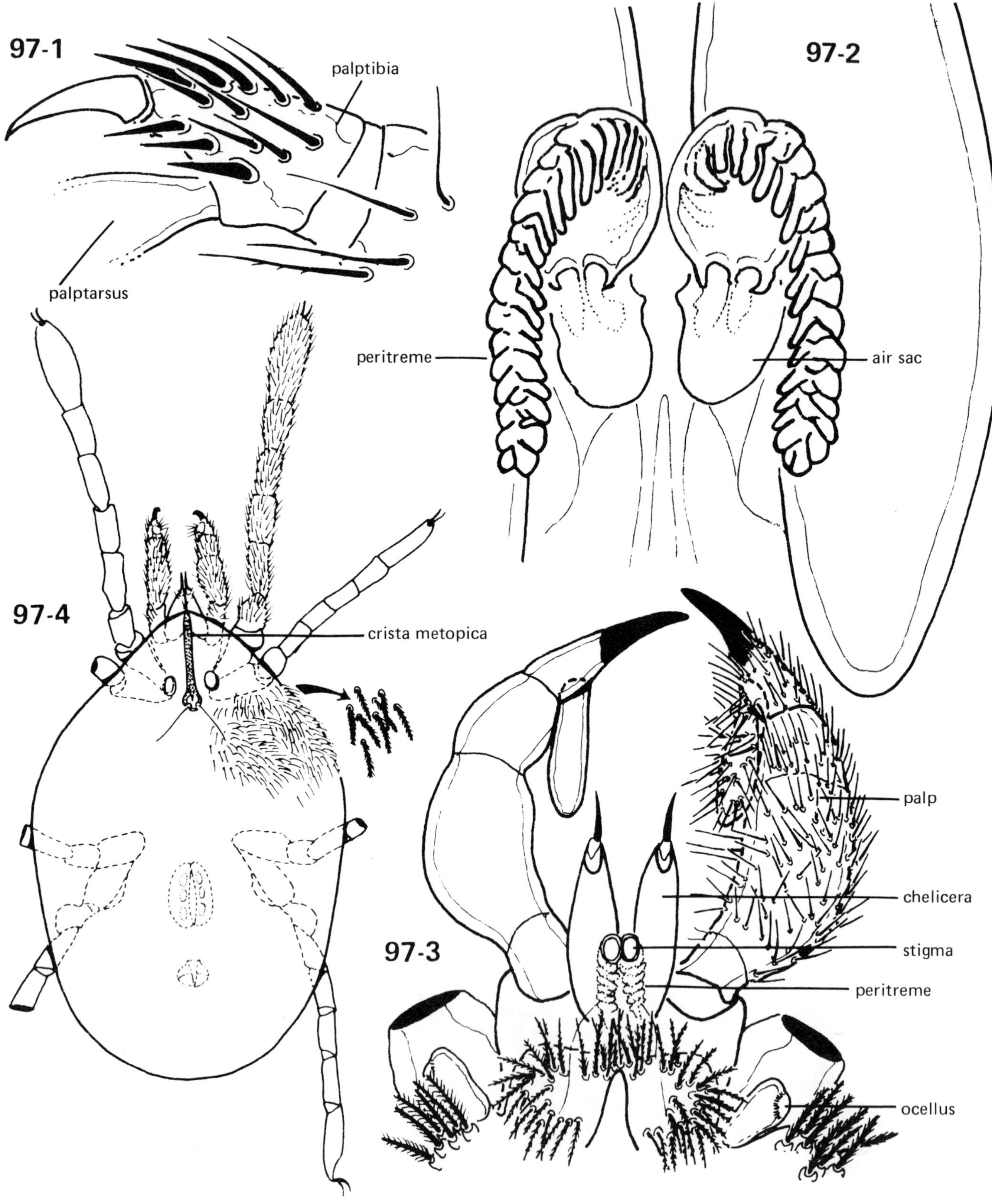

97-1; family PODOTHROMBIIDAE, *Podothrombium* sp., palpal tibia (axial) of female

97-2 to 97-4; family TROMBIDIIDAE. **97-2 and 97-3**; *Trombidium* sp. (Wyoming, USA). **97-2**; internal air sacs and peritremes of adult: **97-3**; proterosoma of adult: **97-4**; *Microtrombidium* sp. (Oregon, USA), dorsum of female with detail of typical dorsal setae

PLATE 98

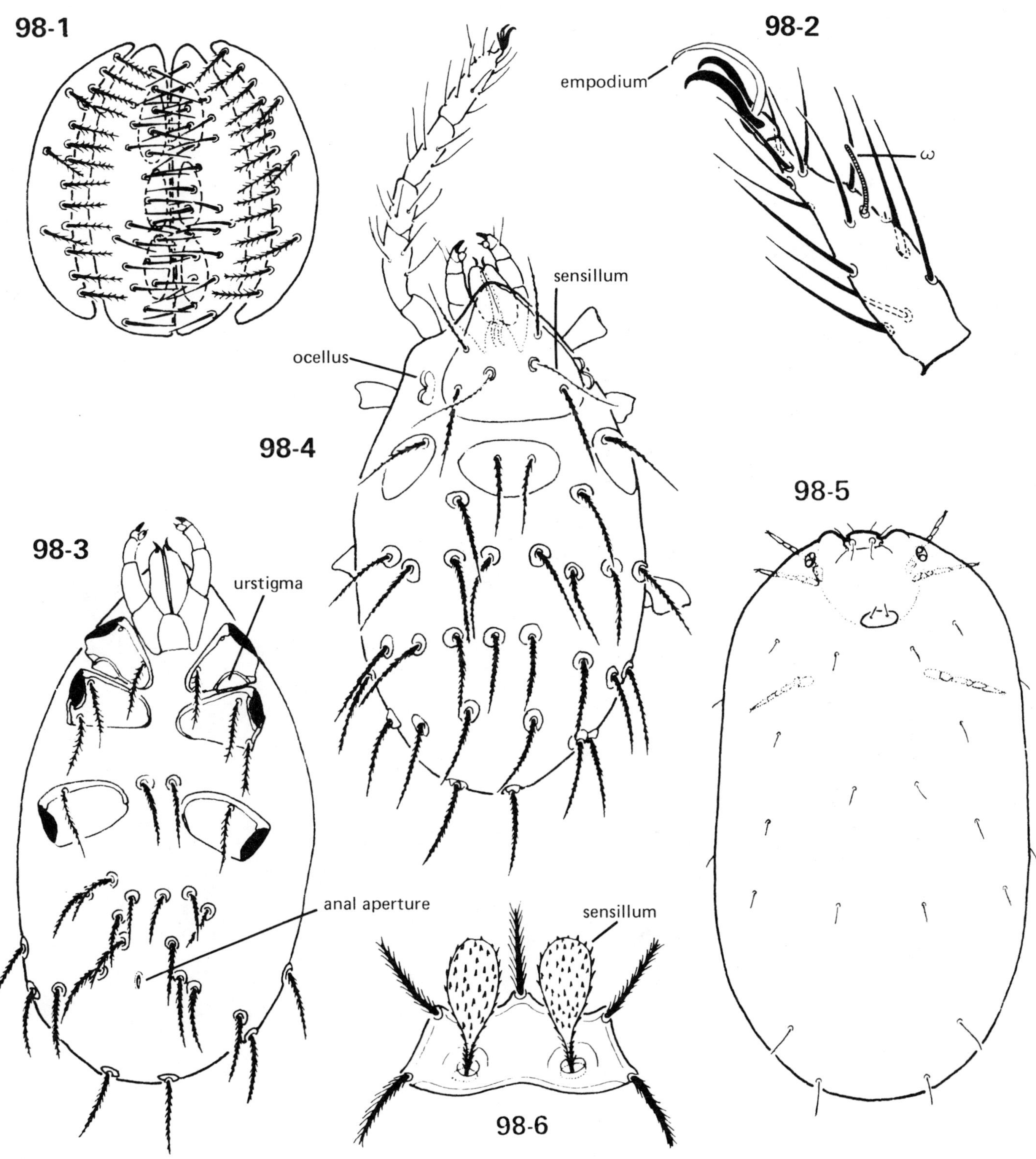

98-1 to 98-5; family TROMBIDIIDAE. **98-1**; *Microtrombidium* sp. (Oregon, USA), genital region of female: **98-2 to 98-4**; *Allothrombium* sp. (Washington, USA). **98-2**; tarsus I: **98-3**; venter of larva: **98-4**; dorsum of larva: **98-5**; dorsum of engorged trombidiid larva

98-6; family TROMBICULIDAE, *Euschöngastia* sp. (Oregon, USA), prodorsal shield (scutum) of larva

PLATE 99

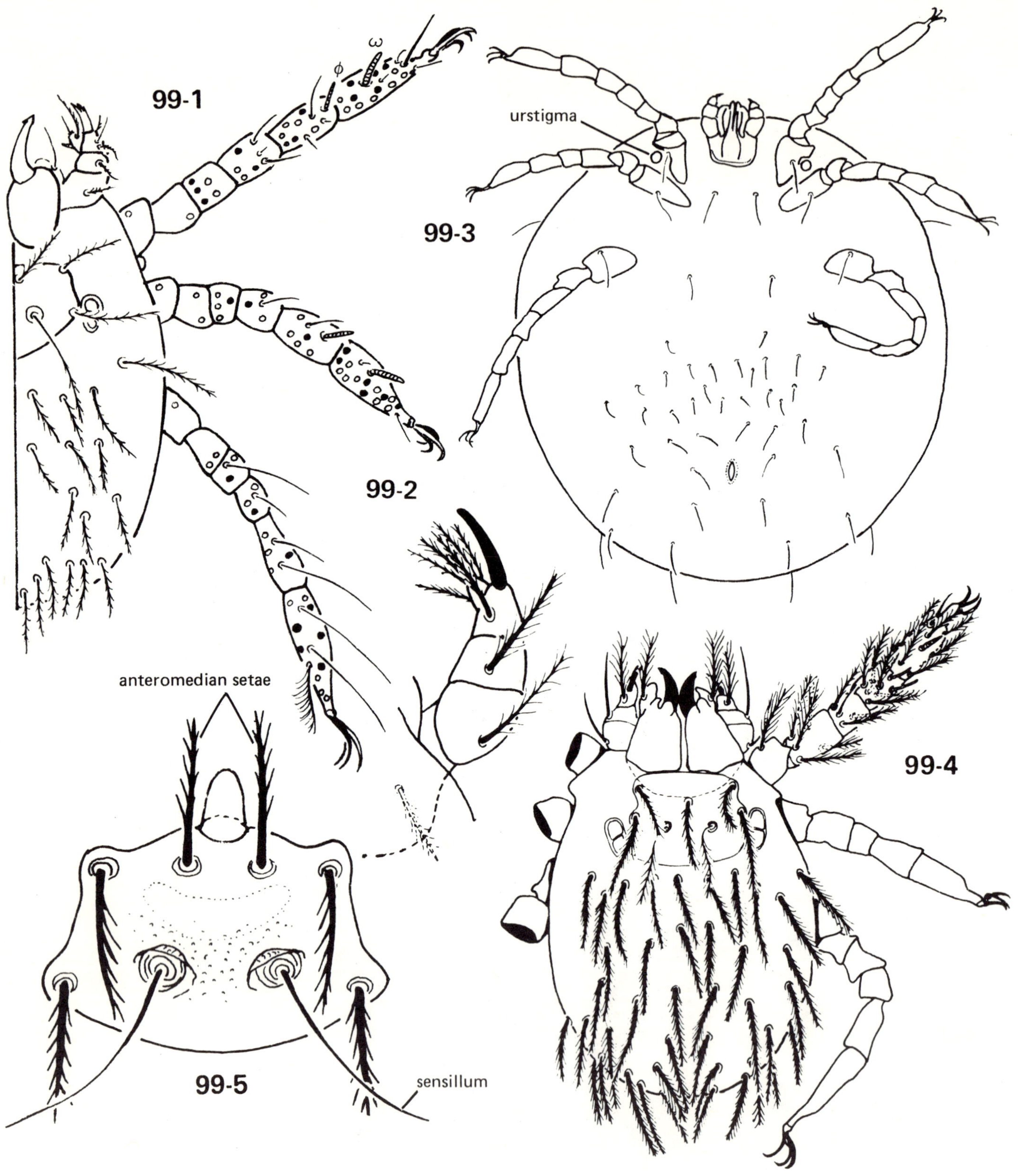

99-1 to 99-4; family TROMBICULIDAE. **99-1**; *Neotrombicula lipovskyi* Brennan and Wharton, dorsum of larva (tactile setae of legs are not included): **99-2**; *Sasacarus* sp. (Nevada, USA), larval palp: **99-3**; *Euschöngastia* sp. (Oregon, USA), venter of engorged larva: **99-4**; *Trombicula* sp. (South Korea), dorsum of larva

99-5; family LEEUWENHOEKIIDAE, *Leeuwenhoekia (Comatacarus) americana* (Ewing) (Oregon, USA), prodorsal shield of larva

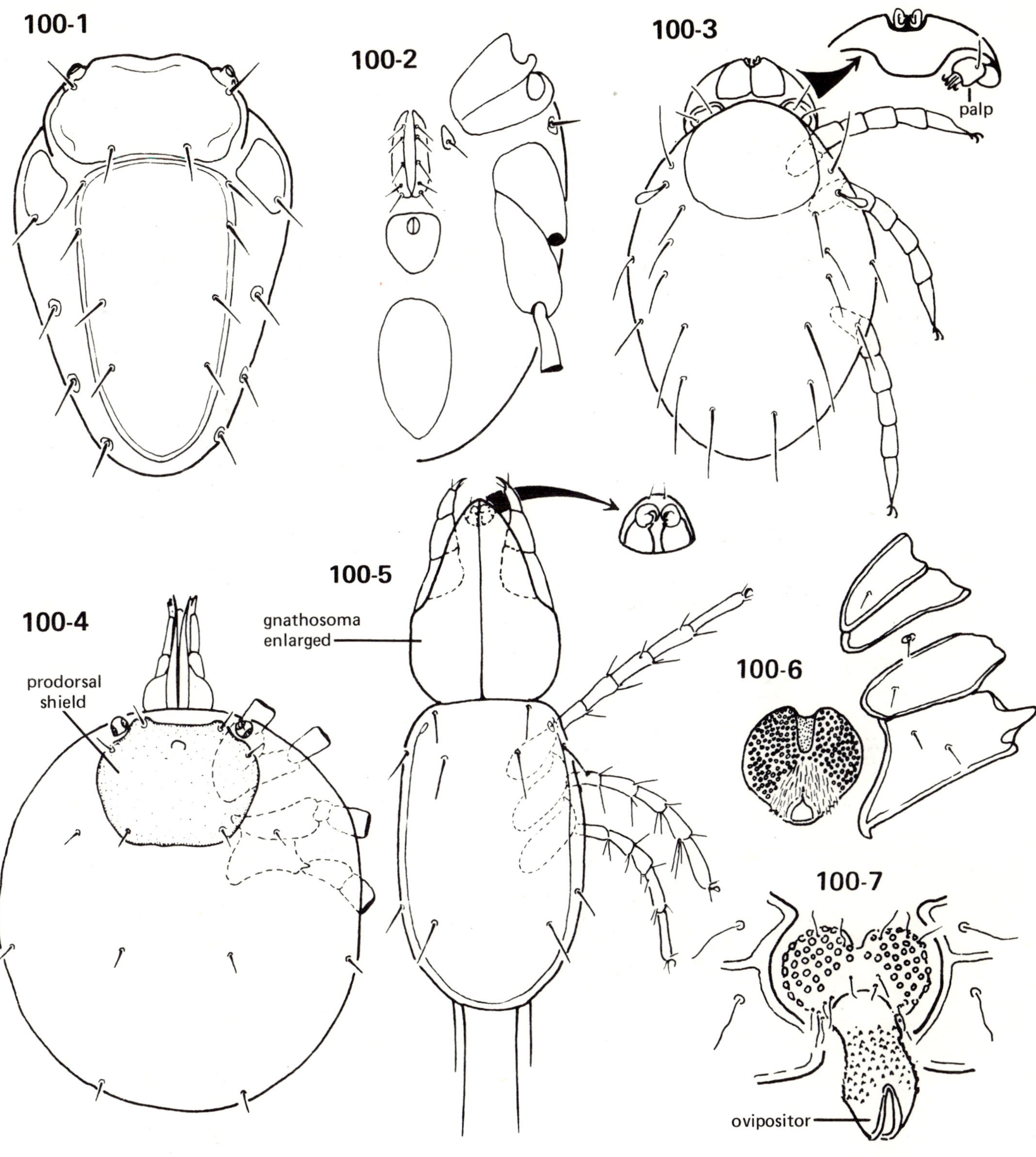

100-1 to 100-3; family HYDROVOLZIIDAE, *Hydrovolzia* sp. **100-1**; dorsum of adult: **100-2**; venter of adult: **100-3**; dorsum of larva with detail of gnathosoma (after Mitchell 1957)

100-4 to 100-7; family HYDRACHNIDAE. **100-4**; *Hydrachna* sp., dorsum of adult: **100-5**; *H. magniscutata* Marshall, dorsum of larva with detail of chelicerae (after Mitchell 1957): **100-6**; *Hydrachna* sp., genital field: **100-7**; *H. mysorensis* Cook, genital field of female, illustrating ovipositor (after Cook 1974)

PLATE 101

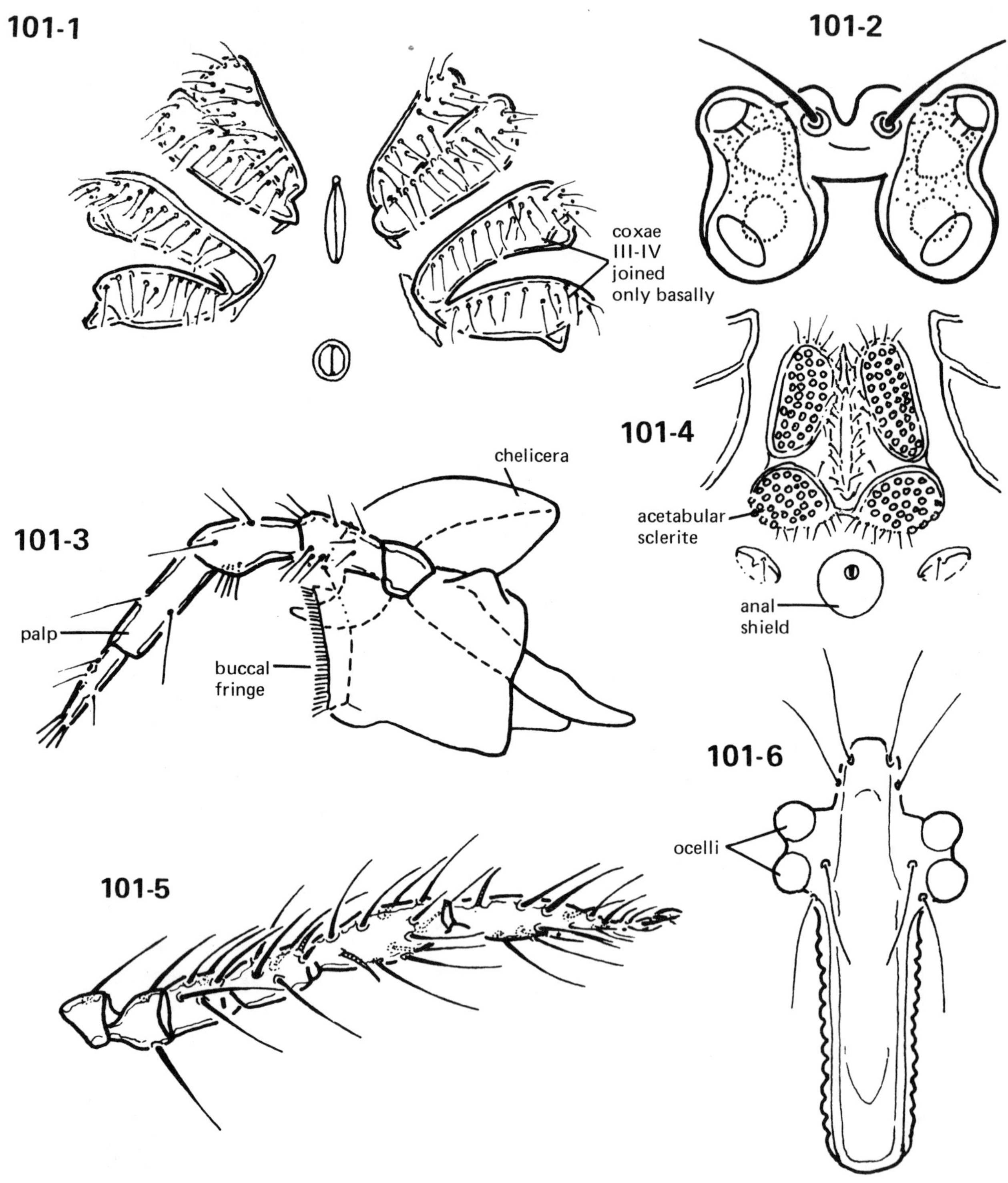

101-1 and 101-2; family EYLAIDAE (after Cook 1974). **101-1;** *Eylais* sp., genital and coxal fields of female: **101-2;** *E. (Proteylais) degenerata* Koenike, ocular shield of female

101-3 and 101-4; family PIERSIGIIDAE (after Cook 1974). **101-3;** *Piersigia limnophila* Protz (Michigan, USA), lateral aspect of capitulum, showing buccal fringe: **101-4;** *P. crusta* Mitchell, genital-anal region of female

101-5 and 101-6; family LIMNOCHARIDAE. **101-5;** *Limnochares americana* Lundblad, leg I of larva (after Prasad and Cook 1972): **101-6;** *L. (Cyclothrix) crinita* Koenike, ocular shield of female (after Cook 1974)

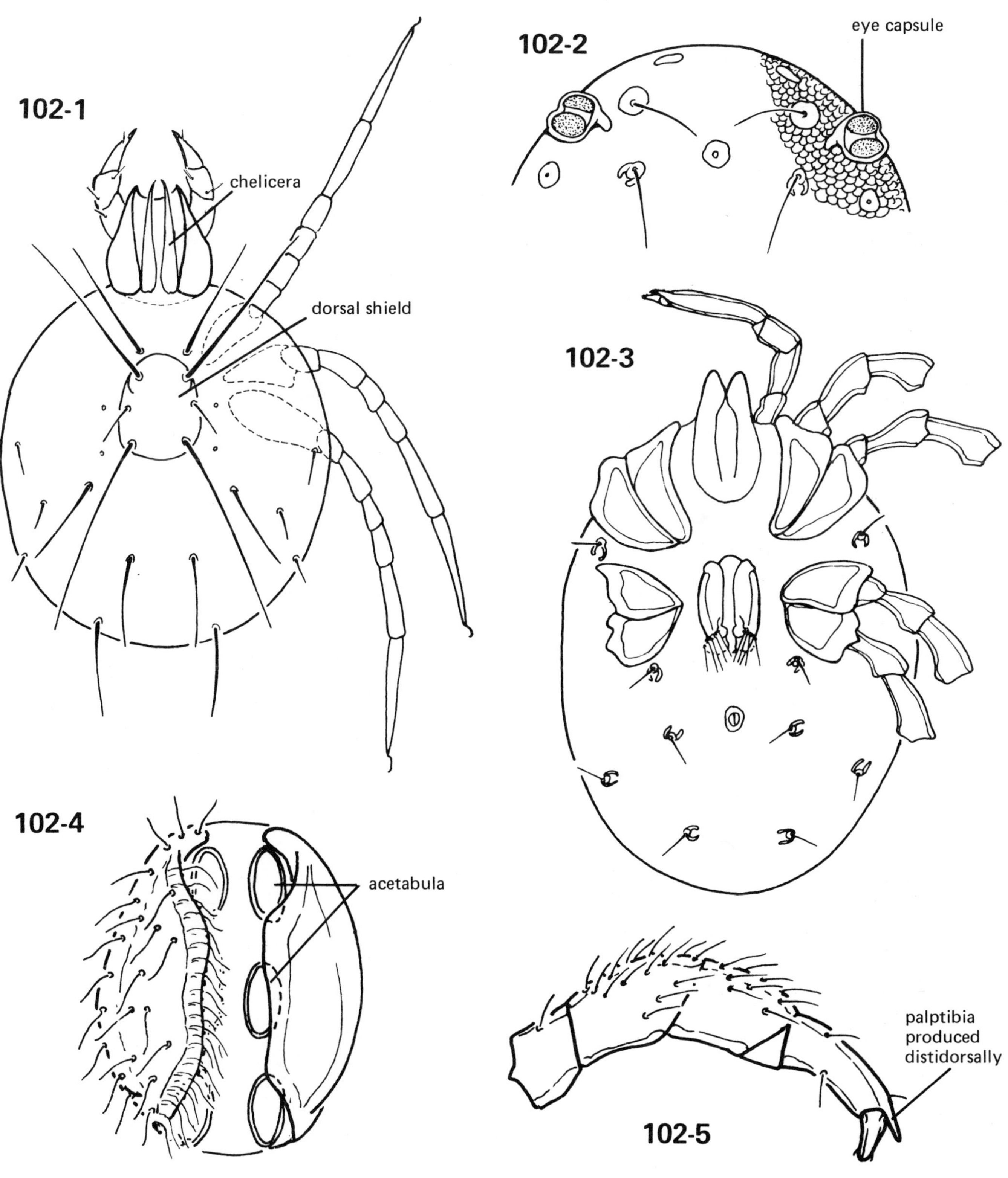

102-1 to 102-5; family HYDRYPHANTIDAE. **102-1 to 102-3**; *Thyas barbigera* Viets (Michigan, USA). **102-1**; dorsum of larva (after Mitchell 1957): **102-2**; anterodorsal portion of idiosoma: **102-3**; venter of adult: **102-4 and 102-5**; *Euthyas mitchelli* Cook. **102-4**; genital region of male (after Cook 1974): **102-5**; palp of male (after Cook 1974)

PLATE 103

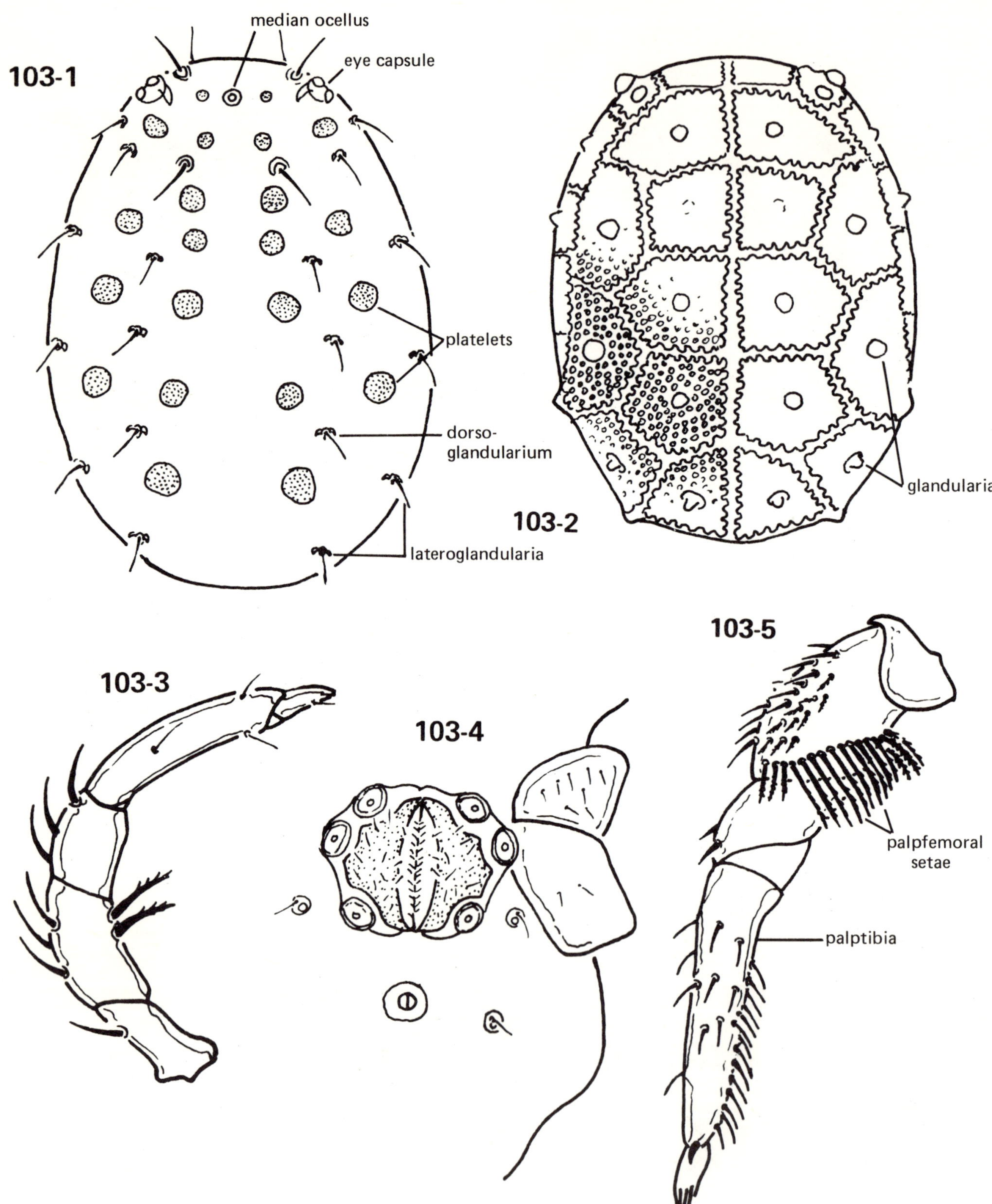

103-1; family HYDRYPHANTIDAE, *Thyas pachystoma,* dorsal aspect (lg = lateroglandularia, dg = dorsoglandularia) (after Cook 1974)

103-2 and 103-3; family TERATOTHYADIDAE (after Cook 1974). **103-2**; *Tetratothyasides lundbladi* Cook (Africa), dorsum of female: **103-3**; *T. (Hansvietsia) undulatus* Cook (Africa), palp of female

103-4 and 103-5; family CTENOTHYADIDAE (after Cook 1974). **103-4**; *Ctenothyas verrucosa* Lundblad (Java), genital-anal region of female: **103-5**; *T. verrucosa,* palp of female

PLATE 104

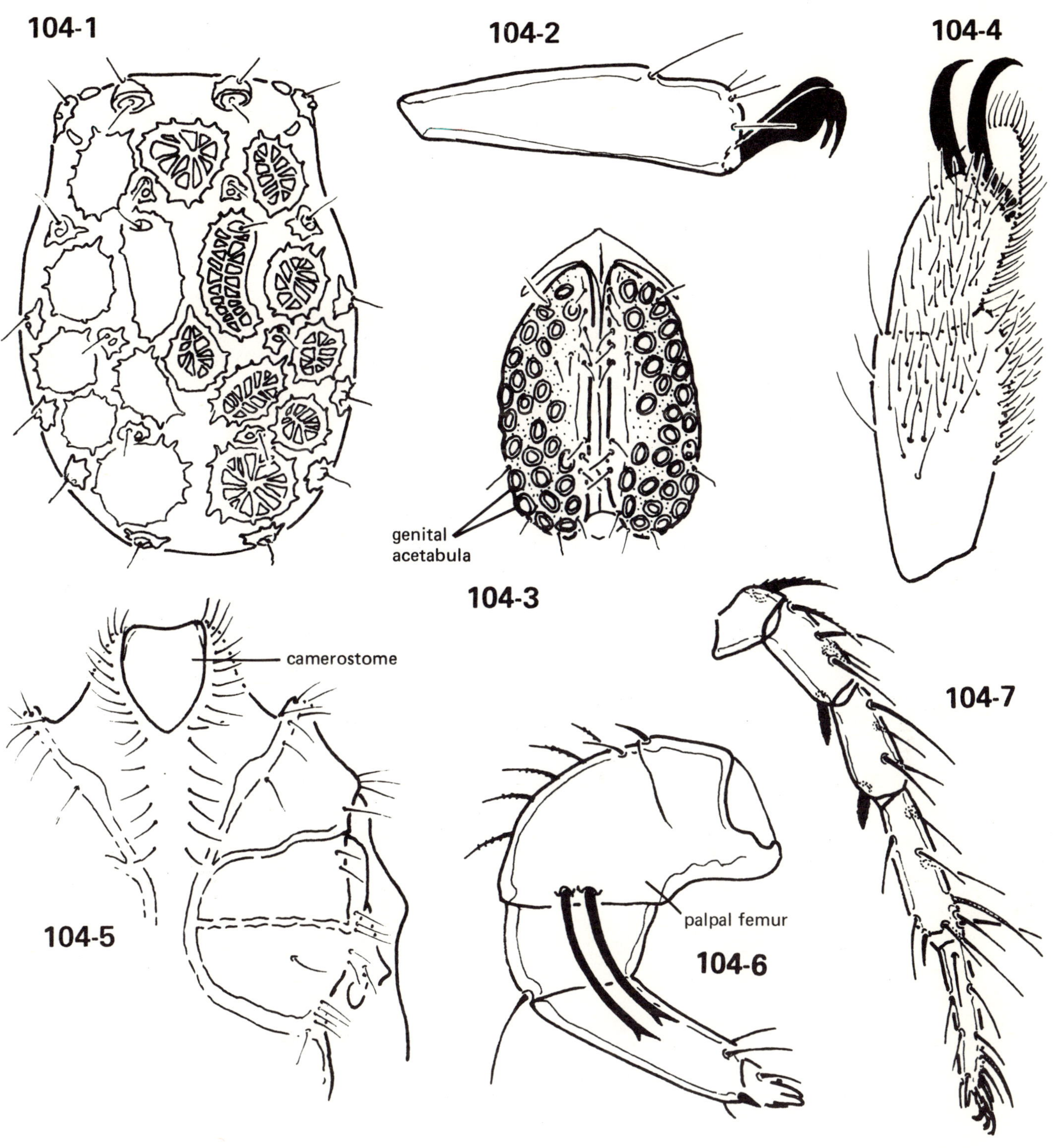

104-1 to 104-3; family RHYNCHOHYDRACARIDAE, *Clathrosperchon ornatus* Cook (after Cook 1974). **104-1**; dorsum of male: **104-2**; tarsus IV of female: **104-3**; genital field of female

104-4 to 104-6; family THERMACARIDAE, *Thermacarus nevadensis* Marshall (after Cook 1974). **104-4**; tibia and tarsus IV of male: **104-5**; anteroventral aspect of male: **104-6**; palp of male

104-7; family LEBERTIIDAE, *Lebertia quinquemaculosa* Marshall, leg I of larva (after Prasad and Cook 1972)

PLATE 105

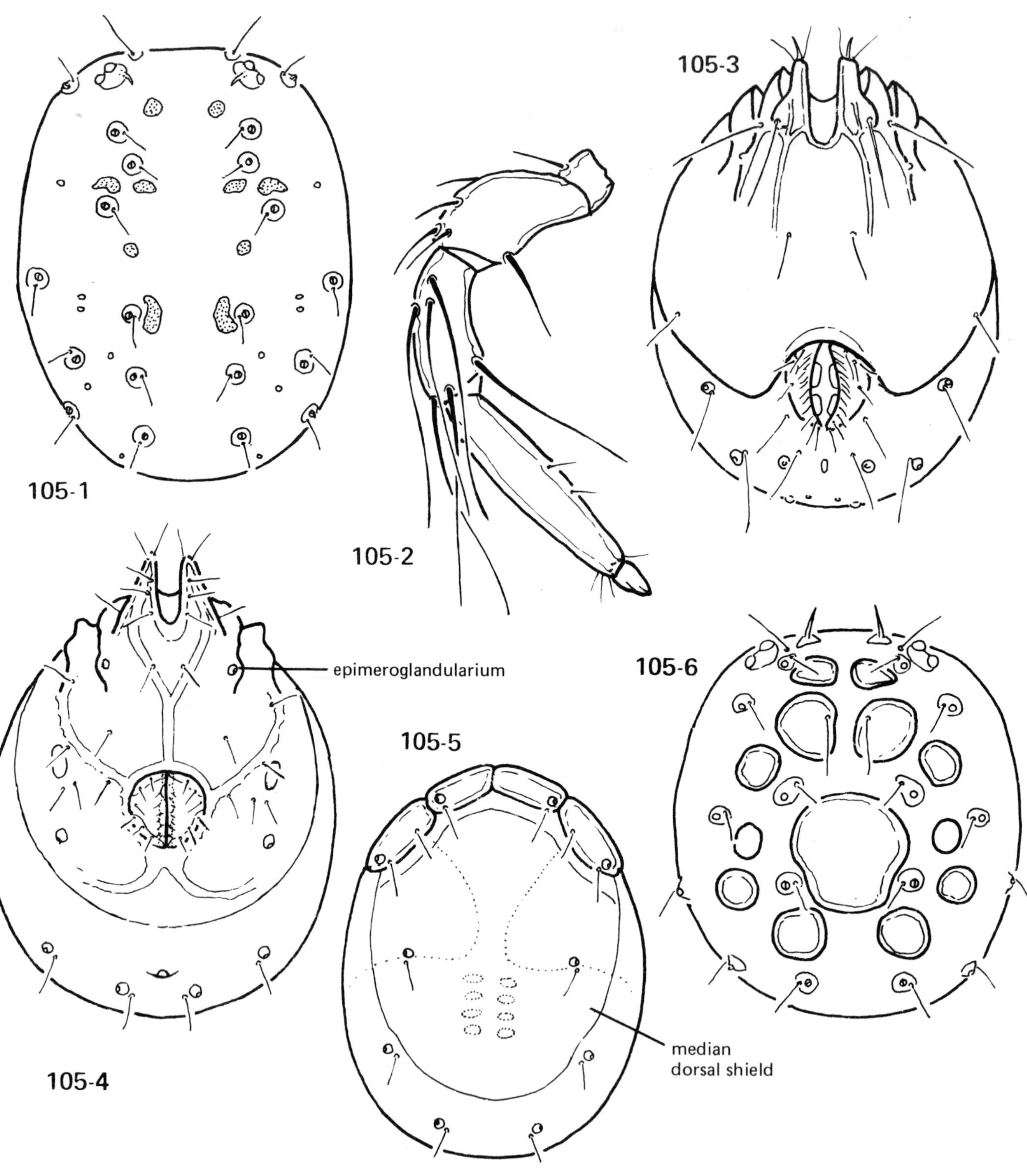

105-1 and 105-2; family LEBERTIIDAE (after Cook 1974). **105-1**; *Lebertia (Pseudolebertia) ventriscutata* Cook, dorsum of male: **105-2**; *L. (Hexalebertia) michiganensis* Cook, palp of female

105-3; family OXIDAE, *Africoxus curvisetus* (Viets), venter of male (after Cook 1974)

105-4 and 105-5; family TORRENTICOLIDAE (after Cook 1974). **105-4**; *Torrenticola (Monatractides) pinapalpus* Cook, venter of female: **105-5**; *T. tetraporella* Cook, dorsum of female

105-6; family SPERCHONTIDAE, *Sperchon (Hispidosperchon) nilgiris* Cook, dorsum of male

PLATE 106

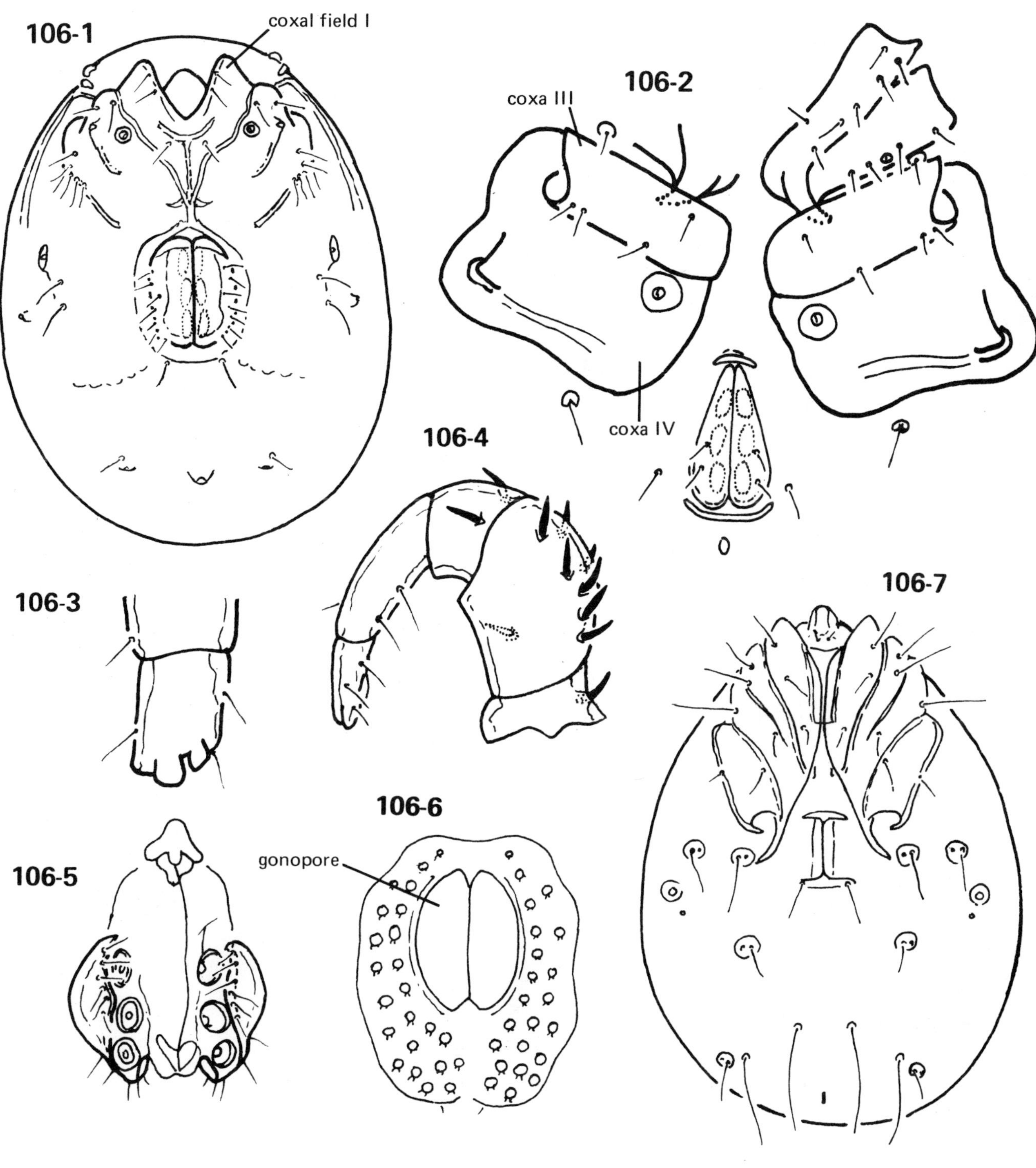

106-1; family ANISITSIELLIDAE, *Anisitsiellides monticolus* Lundblad, venter of female (after Cook 1974)

106-2; family TEUTONIIDAE, *Teutonia cometes* (Koch), coxal field and genital region of male

106-3 to 106-5; family RUTRIPALPIDAE, *Rutripalpus limicola* Sokolow (after Cook 1974). **106-3**; terminal segment of palp (dorsal): **106-4**; palp of female: **106-5**; genital region of female

106-6 and 106-7; family PONTARACHNIDAE (after Cook 1974). **106-6**; *Pontarachna pacifica pilosa* Sokolow, genital field of male (setal insertions only): **106-7**; *P. punctulum* Philippi, venter of female

PLATE 107

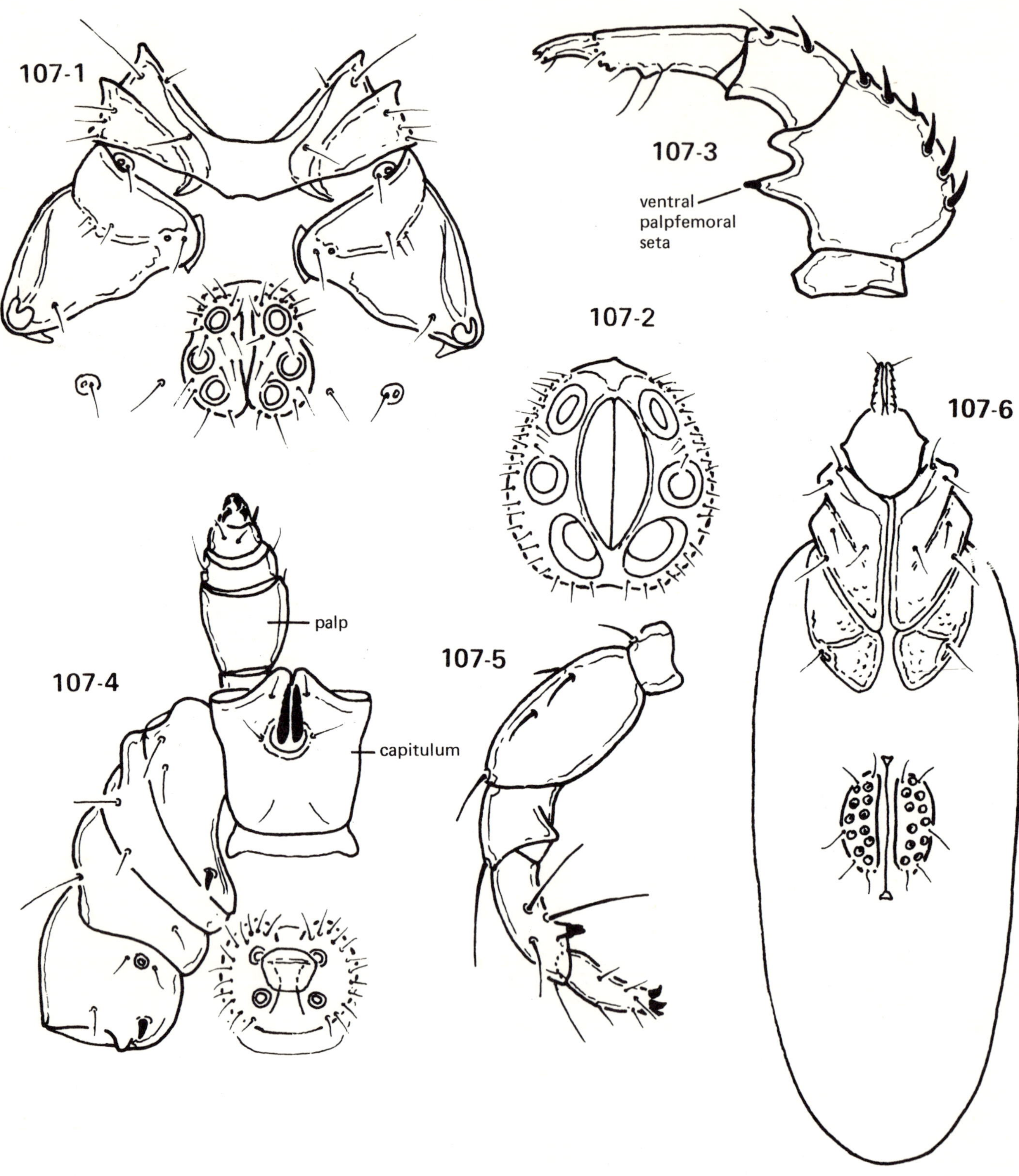

107-1 to 107-3; family LIMNESIIDAE (after Cook 1974). **107-1**; *Limnesia lembangensis* Piersig, coxal and genital fields of female: **107-2**; *L. lembangensis,* genital field of male: **107-3**; *Neotorrenticola violacea* Lundblad, palp of male

107-4; family ASTACOCROTONIDAE, *Astacocroton molle* Haswell, venter of male (after Cook 1974)

107-5 and 107-6; family OMARTACARIDAE, *Omartacarus* elongatus Cook (after Cook 1974). **107-5**; palp of female: **107-6**; venter of female

PLATE 108

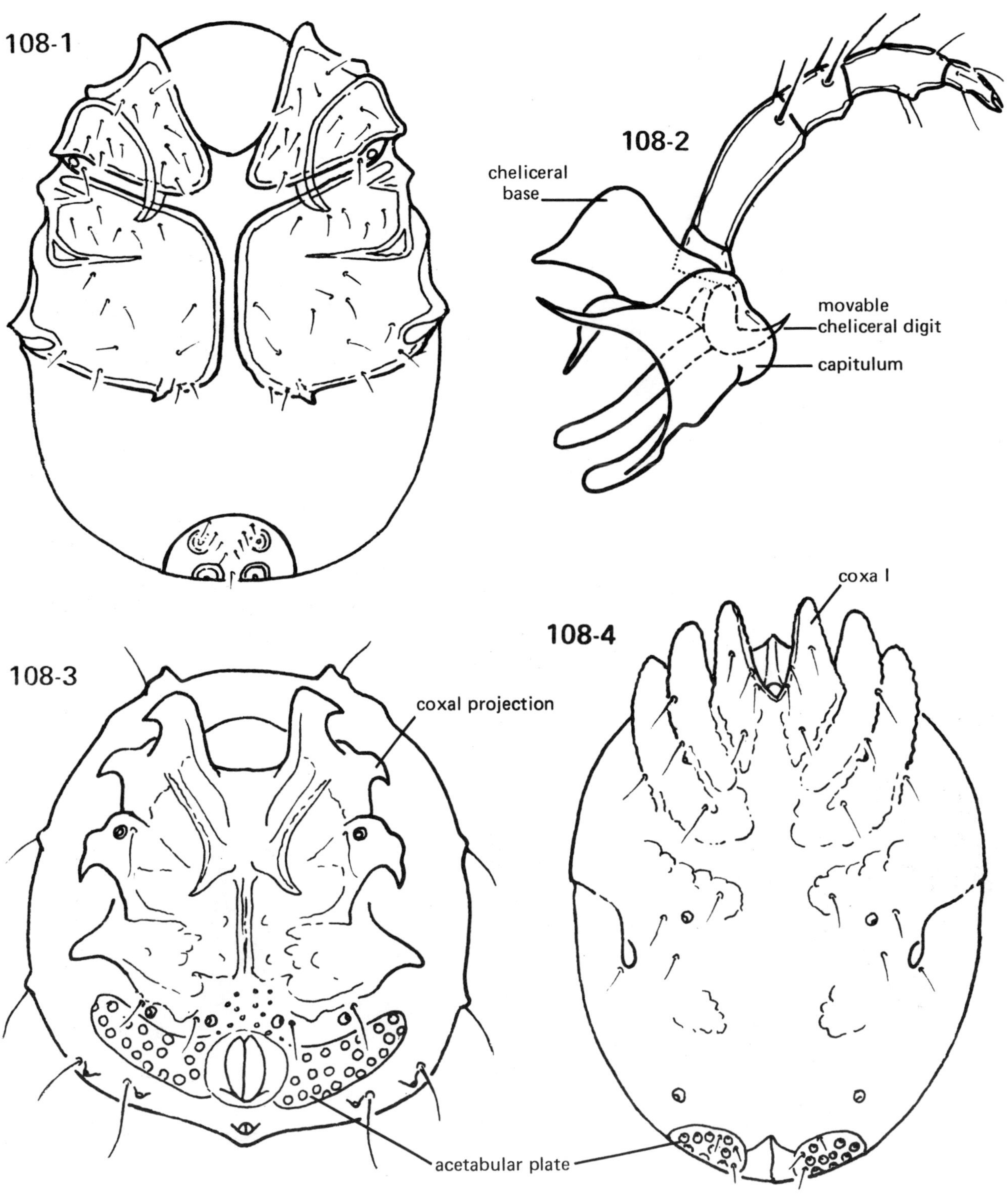

108-1 to 108-3; family UNIONICOLIDAE (after Cook 1974). **108-1;** *Vietsatax parasitica* Uchida, venter of male: **108-2;** *Koenikia concava* Wolcott, excised gnathosoma (lateral): **108-3;** *K. (Recifella) laminipes* Viets, venter of male
108-4; family ATURIDAE, *Phreatobrachypoda robusta* Cook, venter of female (after Cook 1974)

PLATE 109

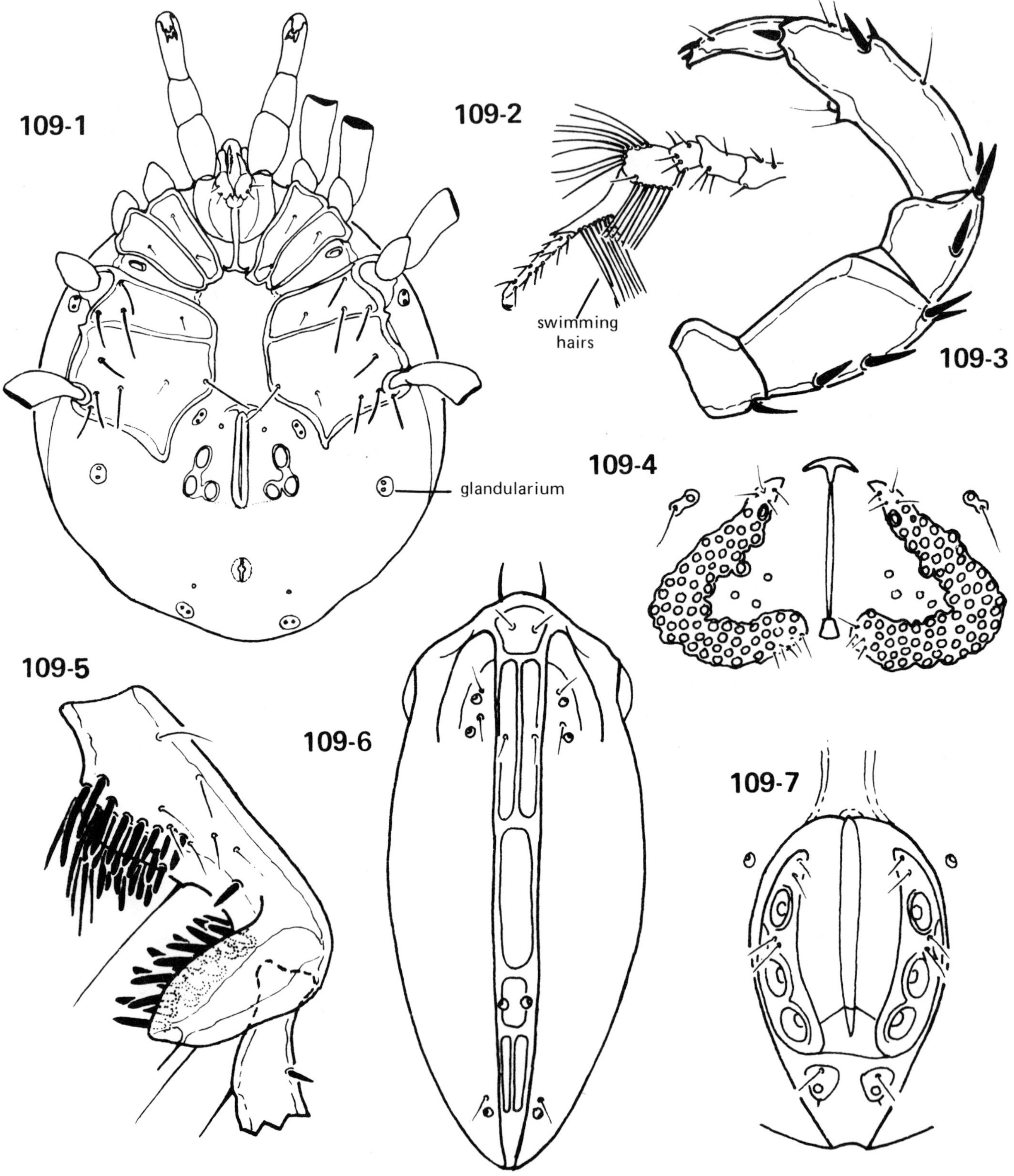

109-1 to 109-5; family PIONIDAE. **109-1;** *Pionopsis* sp. (Oregon, USA), venter of female: **109-2;** *Aceropsis* sp., leg IV of adult: **109-3;** *Pseudofeltria laversi* Cook, palp of female (after Cook 1974): **109-4;** *Piona angulata multiacetabulata* Cook, genital field of female (after Cook 1974): **109-5;** *Nautarachna (Pionella) queticoensis* Smith, genu IV (after Cook 1974)

109-6 and 109-7; family AXONOPSIDAE, *Karlvietsia angustipalpis* K.O. Viets (after Cook 1974). **109-6;** dorsum of female: **109-7;** genital field of female

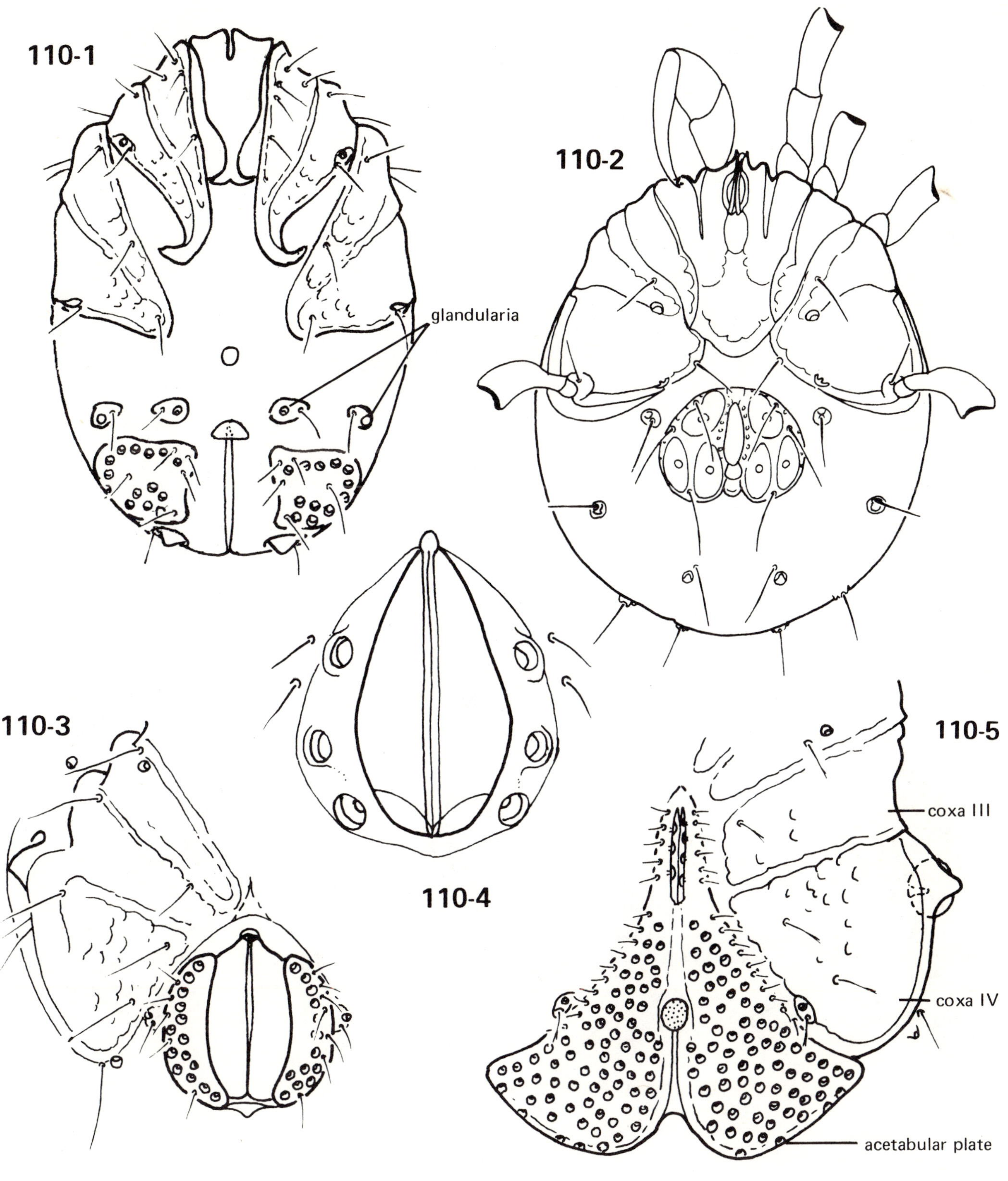

110-1; family FELTRIIDAE, *Feltria (Feltriella) multiscutata* Cook, venter of female (after Cook 1974)
110-2; family HYGROBATIDAE, *Hygrobates* sp. (Michigan, USA), venter of male
110-3; family MIDEIDAE, *Midea expansa* Marshall, coxal and genital fields of female (after Cook 1974)
110-4; family KANTACARIDAE, *Kantacarus matsumotoi* Imamura, genital field of female (after Cook 1974)
110-5; family LAVERSIIDAE, *Laversia berulophila* Cook, coxal and genital fields of male (after Cook 1974)

PLATE 111

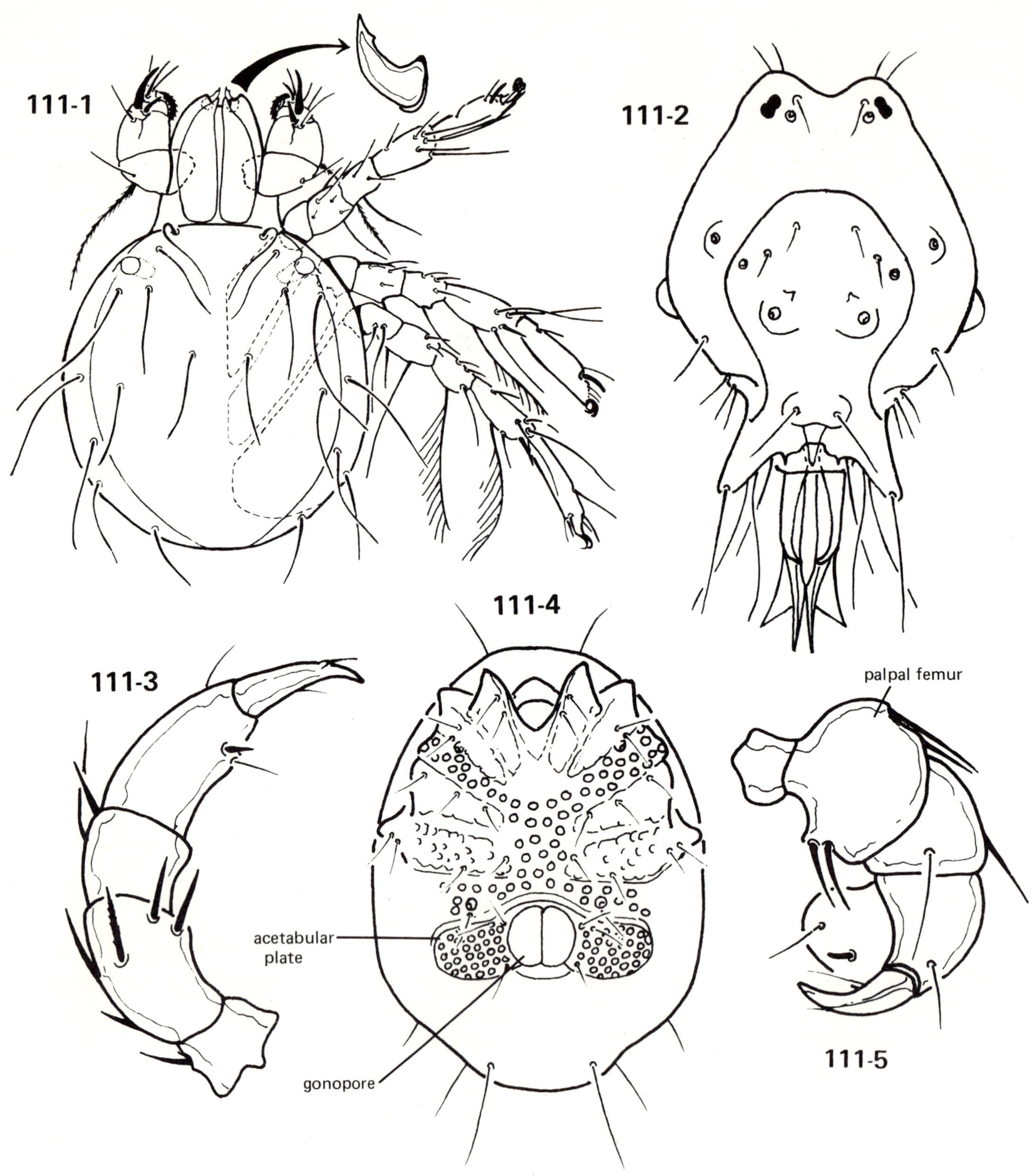

111-1 to 111-4; family ARRENURIDAE. **111-1**; *Arrenurus* sp. (Illinois, USA), dorsum of larva with detail of movable cheliceral digit: **111-2**; *A. dadayi* Cook, dorsum of male (after Cook 1974): **111-3**; *Micruracaropsis phytotelmaticola* (Viets), palp of male (after Cook 1974): **111-4**; *Arrenurus (Megaluracarus) liberiensis* Cook, venter of female

111-5; family HUNGAROHYDRACARIDAE, *Hungarohydracarus szalayi* Cook, palp of male (after Cook 1974)

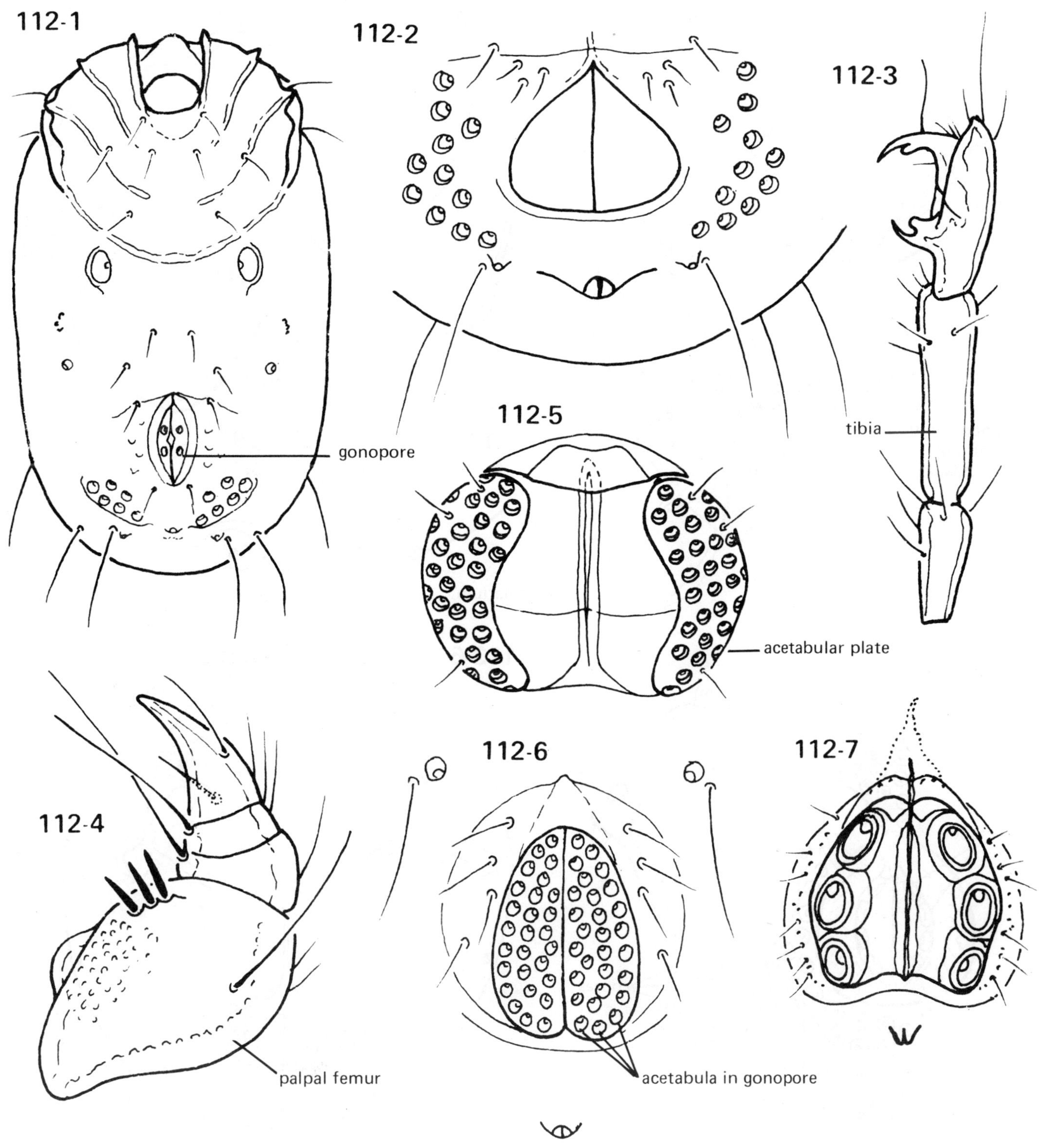

112-1 and 112-2; family ARENOHYDRACARIDAE, *Arenohydracarus minimus* Cook (after Cook 1974). **112-1;** venter of male: **112-2;** posteroventral aspect of female

112-3; family MOMONIIDAE, *Momoniella africana* Cook, distal segments of leg I, male (after Cook 1974).

112-4; family BOGATIIDAE, *Bogatia maxillaris* Motas and Tanasachi, palp of male (after Cook 1974)

112-5 and 112-6; family ATHIENEMANNIIDAE, *Chelomideopsis besselingi* (Cook) (after Cook 1974). **112-5;** genital field of female: **112-6;** genital field of male

112-7; family MIDEOPSIDAE, *Mideopsis longidens* Lundblad, genital field of female

PLATE 113

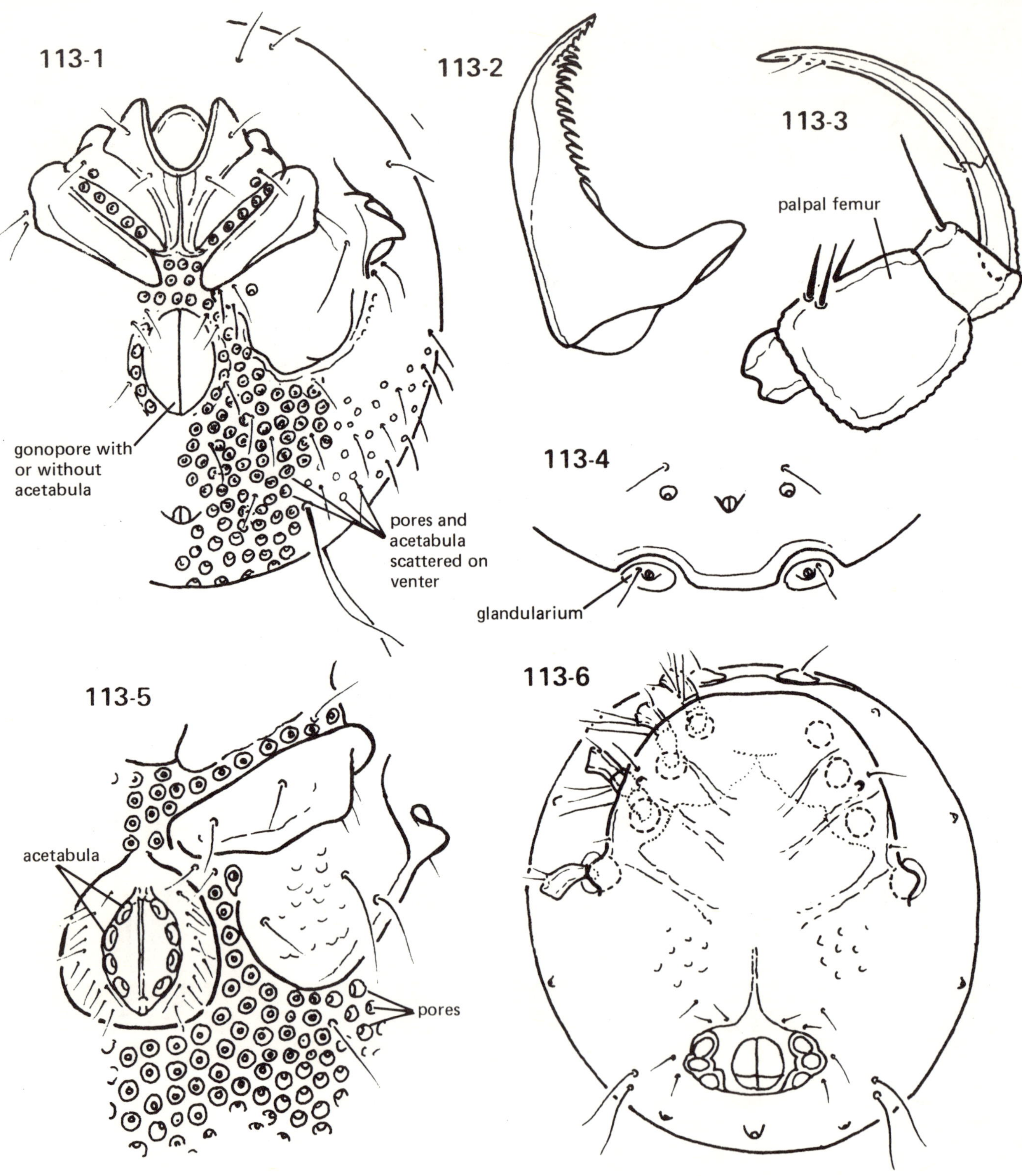

113-1 and 113-2; family HARPAGOPALPIDAE, *Harpagopalpus ?indicus* Cook (after Cook 1974). **113-1**; venter of female: **113-2**; movable cheliceral digit of female

113-3 and 113-4; family NIPPONACARIDAE, *Nipponacarus matsumotoi* Imamura (after Cook 1974). **113-3**; palp of male: **113-4**; posteroventral aspect of male

113-5; family KRENDOWSKIIDAE, *Geayia ovata* (Wolcott), coxal and genital fields of female (after Cook 1974)

113-6; family UCHIDASTYGACARIDAE, *Uchidastygacarus ovalis* Cook, venter of female (after Cook 1974)

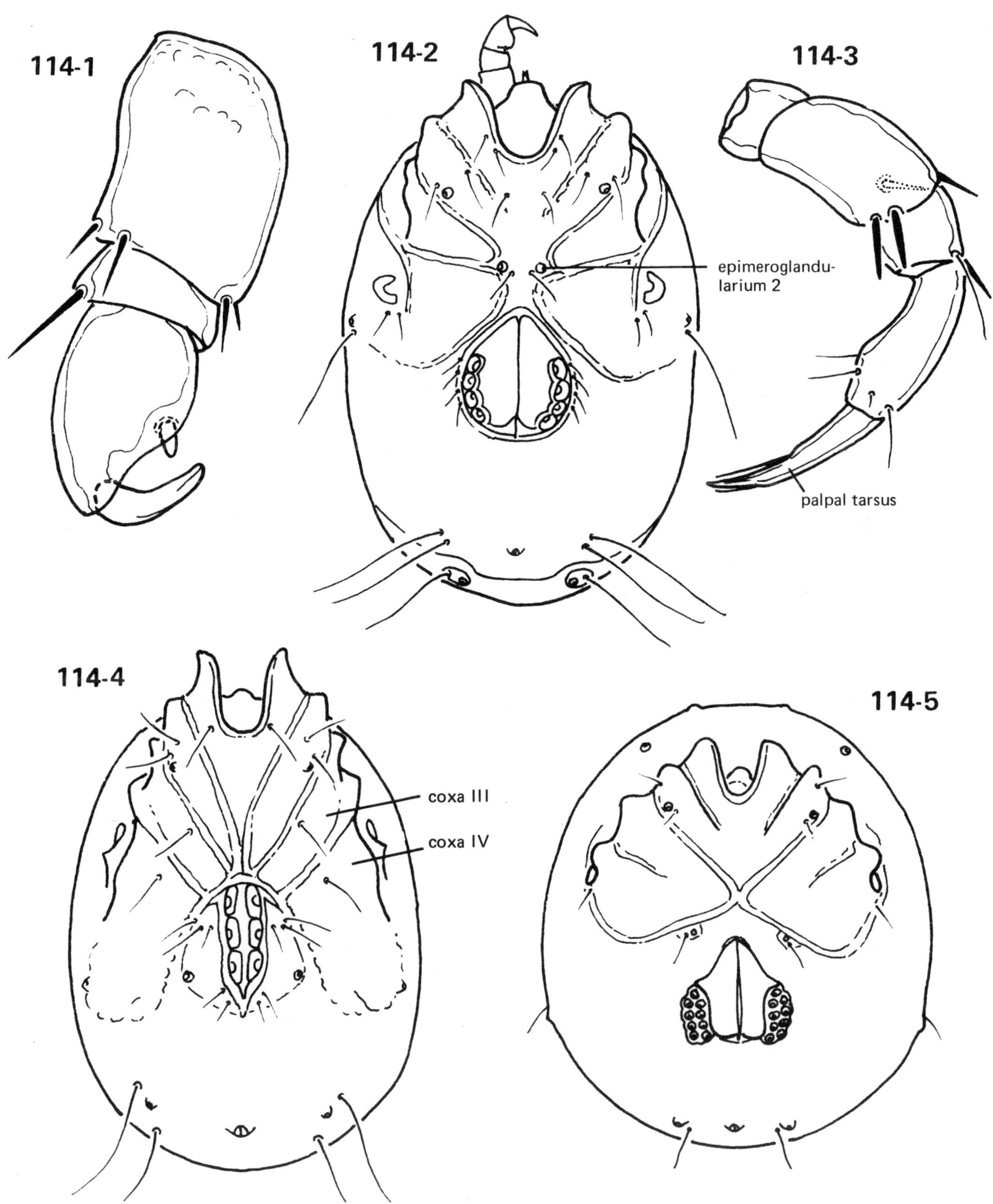

114-1; family UCHIDASTYGACARIDAE, *Uchidastygacarus (Imamurastygacarus) ovalis* Cook, palp of female (after Cook 1974)

114-2 and 114-3; family CHAPPUISIDIDAE (after Cook 1974). **114-2**; *Chappuisides eremitus* Cook, venter of female: **114-3**; *Tsushimacarus uenoi* Imamura, palp of female

114-4; family NEOACARIDAE, *Neoacarus occidentalis* Cook, venter of male (after Cook 1974)

114-5; family ACALYPTONOTIDAE, *Acalyptonotus violaceus* Walter, venter of female (after Cook 1974)

SUBORDER ACARIDIDA

The Acaridida is a large assemblage of primarily terrestrial mites which, with few exceptions, are non-predatory in habit. Many species are saprophagous, fungivorous or graminivorous (Türk and Türk 1959, Hughes 1976), and several groups are exclusively parasitic. Most Acaridida are slow-moving, weakly or incompletely sclerotized mites which range in size from 200 to 1800 μ. There are no indications of stigmata or peritremes, and tracheae occur only rarely.[1] Respiration apparently is integumental throughout the suborder. Mouthparts may occasionally be reduced or highly modified, but as a rule they are normally developed and distinct (Plate 3-2, p. 15). The chelicerae often are chelate-dentate and are never fused. A paraxial seta may be found on the fixed digit. The palpi are distinct, but are quite small and typically two-segmented. The chelicerae or palpi of certain parasitic groups are modified in various ways for attaching to their hosts. Mouthparts are greatly reduced and non-functional in the hypopus, a heteromorphic nymphal form which occurs in several acaridid families (Plate 121-4, p. 419). Leg pretarsi never bear true claws, but often have a terminal empodial claw surrounded basally by a stalked or sessile membranous pulvillus (Fig. 38a). The claw may be greatly reduced and virtually covered by an expanded discoid pulvillar pad which is joined to the tarsus on the dorsomesal pad surface. The claw-pad complex is referred to as the *ambulacral disc* (Fig. 38b). Some or all of the tarsi may be without ambulacra in certain parasitic forms.

Sexual dimorphism in the Acaridida usually is pronounced and may be expressed both in primary and secondary characters. Females have a terminal or posterodorsal bursa copulatrix (Plate 116-2, p. 414) or an elongate externally produced sperm duct (Plate 140-5, p. 438) which provides a locus for sperm acquisition from the male. Males usually have a sclerotized, often elongate aedeagus, and commonly possess special anal suckers and modified discoid tarsal setae (Plate 116-4 and 6, p. 414) with which they attach to the female during copulation. Legs III and/or IV may be enlarged or otherwise modified for the same purpose. The male aedeagal complex is greatly reduced in those groups where the female sperm duct is strongly produced (e.g., the CRYPTUROPTIDAE, Plate 140-5, p. 438), suggesting that the sperm duct functions as an intromittent organ.

The *oviporus*[2] of the female may be a naked transverse or longitudinal aperture, but is more often covered by flanking paragynial flaps and a posteromedian epigynial element (Plate 116-1). Two pairs of closely associated acetabula often occur in the paragynial region of post-protonymphal stadia, but they may be indistinct or absent in various parasitic, hygrophilous and halophilous groups. A pregenital sclerite or *epigynium* is present in both males and females of many acaridid families.

Ontogenetic development in the Acaridida (Plate 115, p. 372) generally is paurometabolous, with passage through successive instars involving few obvious changes aside from size increase, addition of acetabula, and certain setal additions to legs and idiosoma. However, a heteromorphic hypopus stage may follow the protonymphal instar in several acaridid families (see key on page 412), differing from preceding and succeeding instars both in morphology and behavior.

---

[1] A pair of genital tracheae has been observed in the female of *Gohieria fusca* (Oudemans) (GLYCYPHAGIDAE, Plate 118, p. 416), but they are absent in other Acaridida (van der Hammen 1972).

[2] The term "genital aperture" is misleading in the Acaridida since sperm transfer occurs via the terminal bursa copulatrix or sperm duct rather than via the genital aperture. Inasmuch as the aperture serves an ovipositional function, it is more appropriate to use the term *oviporus* in describing it.

PLATE 115

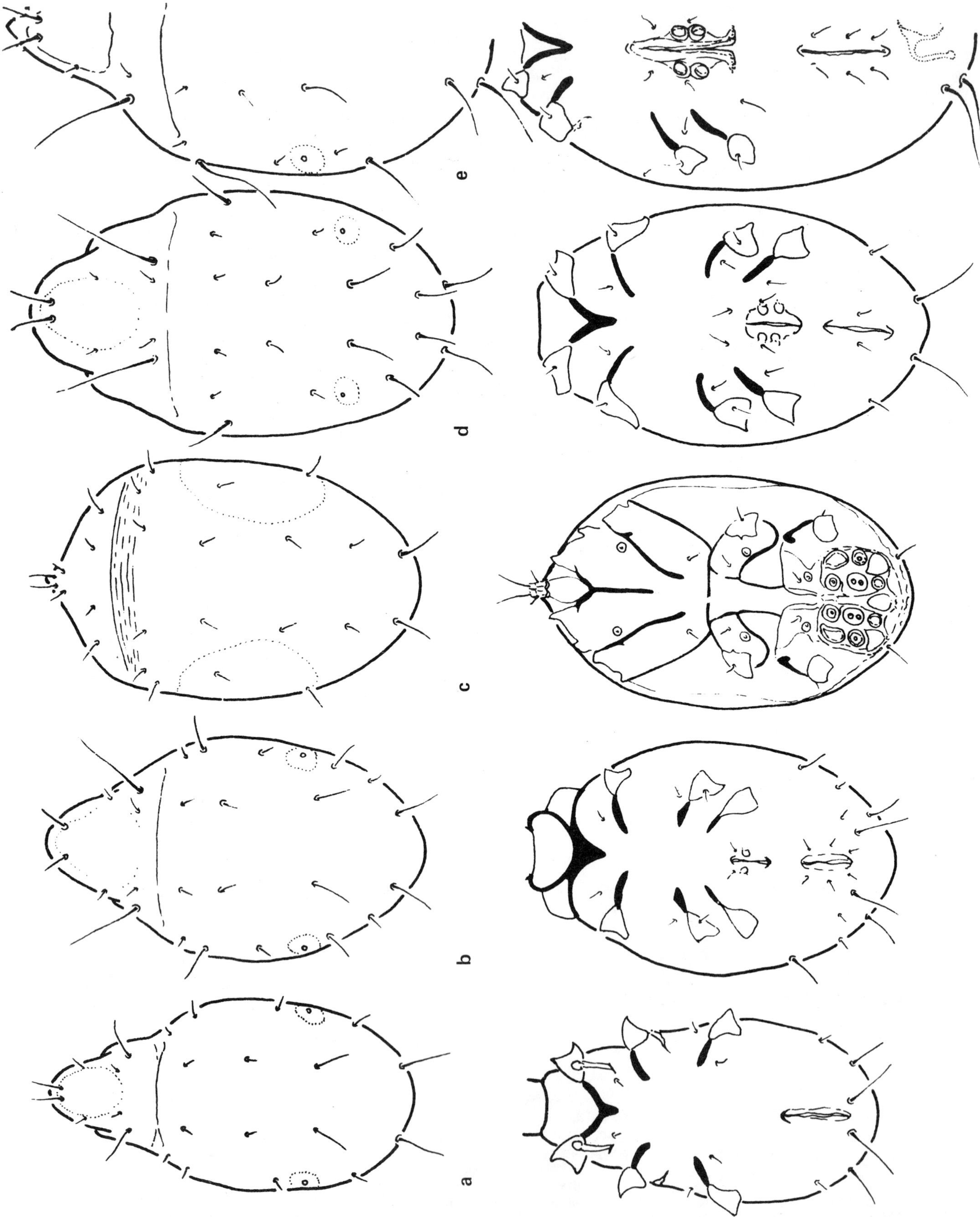

115; ontogenetic development of *Sancassania phyllophagianus* (Oseto and Mayo) (ACARIDAE) showing dorsal (top) and ventral aspects of each instar: a, larva; b, protonymph; c, deutonymph (hypopus); d, tritonymph; e, adult female (after Oseto and Mayo 1975)

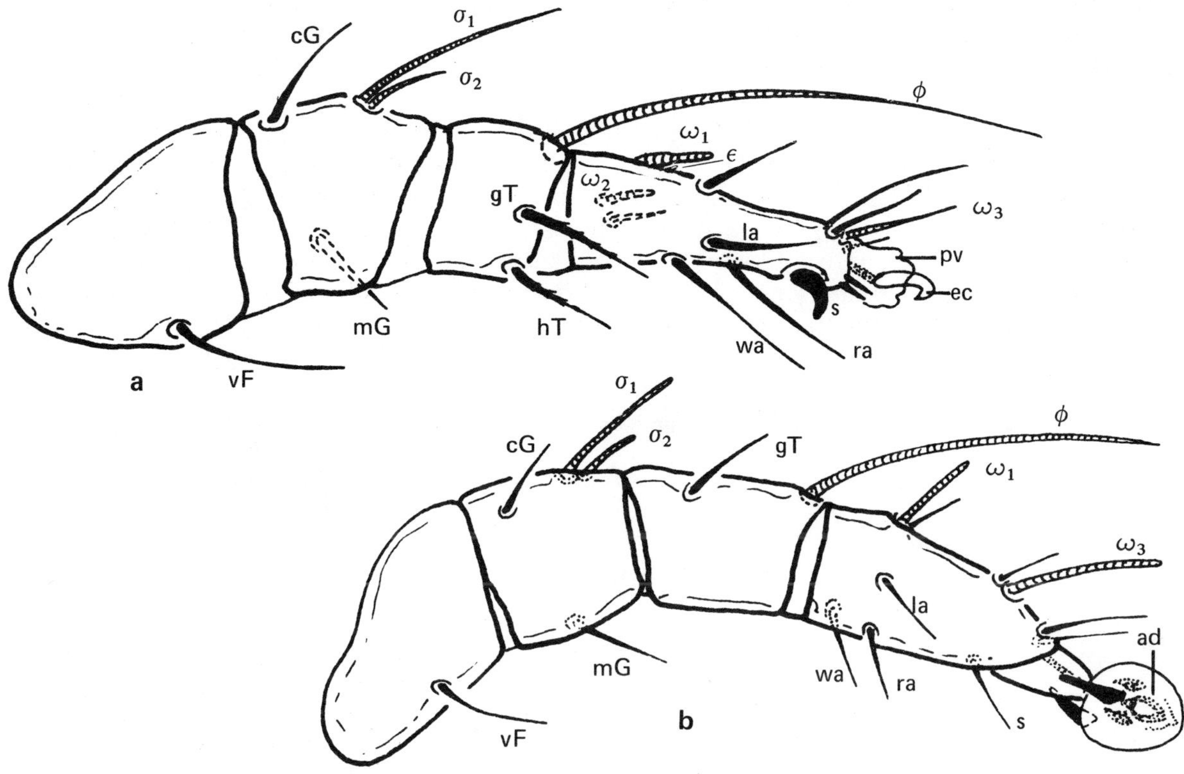

Fig. 38. Legs I (paraxial) of representative Acaridida: **a**; *Acarus siro L.* (ACARIDAE) (after Griffiths 1964) (ec = empodial claw, pv = pulvillus): **b**; *Pterolichus* (PTEROLICHIDAE) (after Atyeo and Gaud 1966) (ad = ambulacral disc). Solenidial and setal signatures are based on the system of Grandjean 1939.

Other important identifying features in the Acaridida include the following:

1. Coxae of legs fused to the venter, with the position of the coxal fields often demarcated by subintegumental epimera (Plate 115, p. 372).

2. With a short, external podocephalic canal often visible (Plate 116-5, p. 414). Eyes usually absent (present in some SAPROGLYPHIDAE, Plate 122, p. 420).

3. With paired laterodorsal opisthonotal glands (Plate 116-2, p. 414) in most families (absent in SARCOPTIDAE and others).

4. Without sensilla on the propodosoma. Solenidia often occur on the legs (Fig. 38a, p. 373) and on the terminal segments of the palpi.

5. Larvae generally with urstigmata between coxae I-II (Plate 115a, p. 372).

6. Epistome and rutella are absent.

The Acaridida are cosmopolitan and occupy a broad spectrum of nonpredatory niches. Many are intimately associated with vertebrate and invertebrate animals either as paraphages, phoretics, or parasites. The latter category includes feather and quill mites of birds, dermal and subdermal parasites of birds and mammals, gill parasites of Crustacea, and respiratory and visceral parasites of a variety of animal hosts. Approximately 60 families are presently included in the Acaridida and these are grouped in two supercohorts and eleven superfamilies.

The higher classification of the suborder Acaridida is relatively simple when compared to that of other major acarine suborders. Two supercohortal categories are recognized by most specialists, and infragroups above the superfamilial level have not been formalized. It may be that the degree of homogeneity in existing higher categories is such that further division is impractical. It seems more likely, however, that the static hierarchic system presented repeatedly in general treatments of the suborder (Fig. 39) reflects a paucity of analytic information at levels above the superfamily.

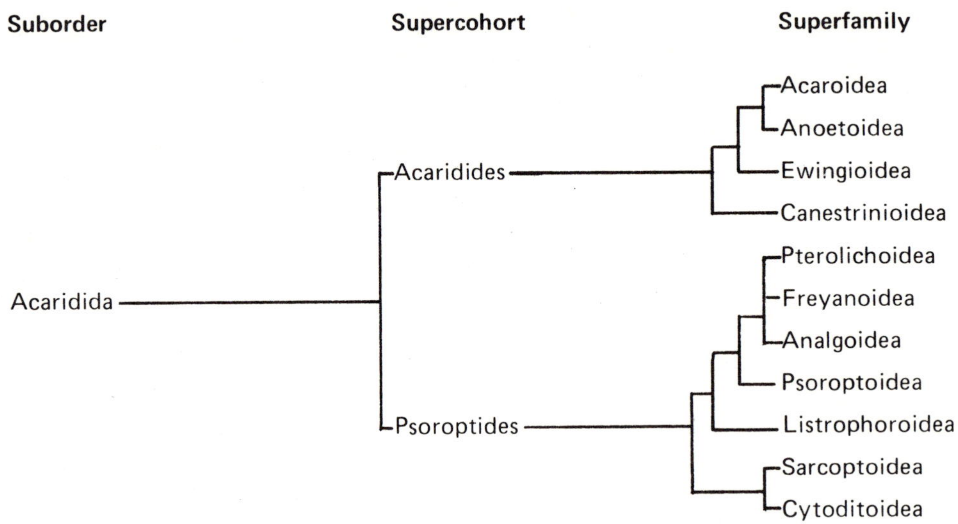

Fig. 39. Dendrogram illustrating possible relationships within the suborder Acaridida of the order Acariformes.

Useful References

André, M. (1949). Ordre des Acariens. Traite de Zoologie, Paris **4**:794-892.

Hammen, L. van der (1972). A revised classification of the mites (Arachnidia, Acarida) with diagnoses, a key, and notes on phylogeny. Zool. Meded. **47**(22):273-292.

Hughes, A.M. (1961). The Mites of Stored Food. Ministry Agr. Fish. Food Tech. Bull. 9:287 pp. + vi.

Türk, E. and F. Türk (1959). Systematik und Ökologie der Tyroglyphiden Mitteleuropas. Beitr. Syst. Ökol. mitteleurop. Acarina **1**(1):3-231.

Vitzthum, H.G. (1942). Acarina. Bronn's Klassen und Ordnungen des Tierreichs, Leipzig: 5, Sect. 4, Book 5 (6):801-912.

Yunker, C.E. (1955). A proposed classification of the Acaridiae (Acarina:Sarcoptiformes). Proc. Helminth. Soc. Wash. **22**(2):98-105.

Zakhvatkin, A.A. (1952). Division of the Acarina into orders and their position in the system of the Chelicerata. Mag. Parasitol. Moscow **14**:5-46.

Supercohort Acaridides

The Acaridides includes four superfamilies of mostly soft-bodied, primarily free-living species which typically possess membranous pretarsal pulvilli, empodial claws on at least some of the tarsi, and two pairs of distinct genital acetabula in post-protonymphal stases. Tibia I usually has two tactile setae and a solenidion (Figure 38a), although setae are absent on tibia I in the entomophilous superfamily Canestrinioidea.

Heteromorphic nymphs, or hypopodes, are known to occur in three of the four superfamilies included in the Acaridides (the Acaroidea, Anoetoidea and Canestrinioidea). Fain (1971) recognizes three hypopodial types based primarily on their organs of attachment. Hypopodes of the first type have a posteroventral sucker plate with a pattern of discoid suckers (Plate 115c, p. 372), and are generally entomophilous. Members of this group occur primarily in the acaroid and anoetoid families ACARIDAE, SAPROGLYPHIDAE, ANOETIDAE, HEMISARCOPTIDAE, and CHAETODACTYLIDAE. Hypopodes of the second type are pilicolous, attaching to the hairs of mammals. In place of a sucker plate, they have two pairs of claspers ornamented with numerous transverse ridges (Plate 120-1, p. 418). Hypopodes with claspers are found only in the acaroid family GLYCYPHAGIDAE, subfamily Labidophorinae. The third hypopodial group includes species which lack both claspers and sucker plates (Plate 120-2, p. 418), and which are found at or beneath the bases of hair follicles or in the connective tissues of birds and mammals. Hypopodes of the family GLYCYPHAGIDAE, subfamily Hypodectinae, are included in this category.

Additional information on hypopodes is presented under applicable superfamilial headings.

Superfamily Acaroidea
(Plates 116-125, pp. 414-423)

DIAGNOSIS: *Soft-bodied, often with a prodorsal shield, sejugal furrow present; chelicerae chelate-dentate, palpi simple. Tarsi of legs usually with a fleshy pulvillus and empodial claw, rarely with minute ambulacral discs. Female oviporus longitudinal, resembling an inverted "Y", with two pairs of genital acetabula flanking the opening; male with a pair of anal suckers, dorsal sucker-like setae often present on tarsus IV. Hypopodial instar often enountered.*

The Acaroidea includes twelve families of free-living and phoretic species from many terrestrial niches such as animal nests, food storages and plant tissues. Some have adapted to a semiaquatic, aquatic, or marine existence. Certain endofollicular hypopodes of the family GLYCYPHAGIDAE are considered parasitic since they are able to invade animal tissues and apparently derive sustenance from them. A number of acaroid species are economically important contaminants of foodstuffs or causal agents of dermal irritation in man.

The ACARIDAE (Plates 116-117, pp. 414-415) is a large assemblage of saprophagous, graminivorous, fungivorous and phytophagous species which often are encountered as contaminants of stored and processed foods. *Acarus,* the type genus, includes 10 species (Griffiths 1970), three of which comprise a species complex commonly found in stored grains and cereals and in adjacent habitats (Griffiths 1964). Of these, *A. siro* L. appears to be the dominant species in processed cereals. Baker et al. (1976) record it also as an inhabitant of bird's nests in New York. Graminivorous species of *Acarus* (*A. siro, A. farris* (Oudemans) and *A. immobilis* Griffiths) feed directly on the commodities which they infest and may cause considerable injury (Solomon 1946, Griffiths, op. cit.). Other *Acarus* species are nidicoles of and birds. Members of the genus *Tyrophagus* occupy a wide range of niches including ...ı nests, mushroom houses, vegetable crops, commercial cheeses, flower bulbs, grain ...ages, grasses and cereals (van den Bruel 1940, Robertson 1959, Johnston and Bruce 1965, ...ussey et al. 1969, Nemestóthy and Mahunka 1972, Jeppson et al. 1975, Baker et al. 1976). *T. putrescentiae* (Schrank), a common and cosmopolitan contaminant of foodstuffs, also is well known as a pest in laboratory animal cultures. *T. putrescentiae* and *T. longior* (Gervais) are recorded as causal agents of human intestinal and urogenital acariasis (Baker and Wharton 1952, Southcott 1976). Species of both *Tyrophagus* and *Acarus* may cause contact dermatitis in individuals who handle contaminated foodstuffs.

Members of the acarid genera *Suidasia, Thyreophagus, Lardoglyphus* and *Aleuroglyphus* often invade stored products. *Sancassania* species may also be found in food storages but more often are associated with scarab beetles, both as phorionts and as saprophages (Samšiňák 1970a). Phoretic hypopodes of *Sancassania* molt to succeeding stages following the death of their beetle carrier and feed saprophagically on it, finally reproducing to enormous numbers. *Sancassania* species are among the largest of the Acaridida, with adult females often exceeding 1200 μ in length.

The acarid bulb mites *Rhizoglyphus echinopus* (Fumouze and Robin) and *R. callae* Oudemans feed on ornamental and vegetable bulbs and tubers, and on the roots of a variety of cultivated crops (Hussey et al. 1969, Manson 1972, Jeppson et al. 1975). Damage to bulbs in storage by *R. echinopus* may follow invasion of bulb tissues by pathogenic fungi. Price (1976) demonstrated that *R. echinopus* feeds on *Verticillium* fungi in the laboratory, which suggests that tissue feeding by bulb mites is secondary to fungivory.

Species of the acarid genus *Naiadacarus* have adapted to an aquatic existence in water-filled treeholes in North America (Fashing 1974, 1975). *N. arboricola* Fashing is saprophagous, feeding on decomposing leaves and arthropods which fall into the treehole. Dispersal from treeholes is accomplished through hypopodial phoresy on syrphid flies (*Mallota* spp.) which are treehole inhabitants during their preimaginal development.

Acarid hypopodes generally are active forms which attach to insects (Zakhvatkin 1941, Türk and Türk 1959, Baker 1962a, Fain and Johnston 1974, Fashing 1975) and occasionally to other animals. The hypopus of *Acarus immobilis* Griffiths, however, is an inert non-phoretic stage with reduced legs and a rudimentary sucker plate (Griffiths 1970).

Like the ACARIDAE, members of the families GLYCYPHAGIDAE (Plates 117-120, pp. 415-418), CHORTOGLYPHIDAE (Plate 123, p. 421) and CARPOGLYPHIDAE (Plate 123, p. 421) infest stored grain and processed feeds (Hughes 1961), but are by no means confined to these habitats. *Carpoglyphus lactis* L. (CARPOGLYPHIDAE) is found in dried fruit, milk products, wine, caramel and flour (Hughes, op. cit.). It has also been reported from bee hives in central Europe and is considered an important contaminant of honey (Chmielewski 1971). Hypopodes of *C. lactis* have been taken from the head and tongue of a nymphalid butterfly, leading Vitzthum (1940) to surmise that *C. lactis* probably frequents flowers also. *Chortoglyphus arcuatus* (Troupeau) (CHORTOGLYPHIDAE) inhabits hay dust or fodder in stables. *Glycyphagus* and *Gohieria* species (GLYCYPHAGIDAE) share this habitat, but also are present in processed feed and food storages. *G. domesticus* (De Geer), a common invader of foodstuffs which have gone out of condition, is a causative agent of "grocer's itch", a contact dermatitis prevalent amongst food handlers (Southcott 1976). *G. domesticus* also is an intermediate host of a rodent tapeworm (Joyeaux and Baer 1945). Hypopodes of *Glycyphagus* are inert forms which lack a developed ventral sucker or clasper apparatus (Fig. 40, p. 377). They remain within the thin protonymphal exuvium and move easily from place to place on light air currents. Another glycyphagid, *Aeroglyphus robustus*

(Banks), feeds on grains and grain products of farm storages and elevators (Sinha 1966). However, adults of other species of *Aeroglyphus,* along with aeroglyphine glycyphagids of the genera *Melisia* and *Contramelisia,* are associated with passalid beetles and with xylocopid, bombid or eumenid bees (Cooreman 1959, Samšiňák 1963, Vomero 1972). *Hericia hericia* (Robin) and other members of the genus feed on sap exudates of various deciduous trees, often occurring in great numbers on the bark surface (Michael 1901).

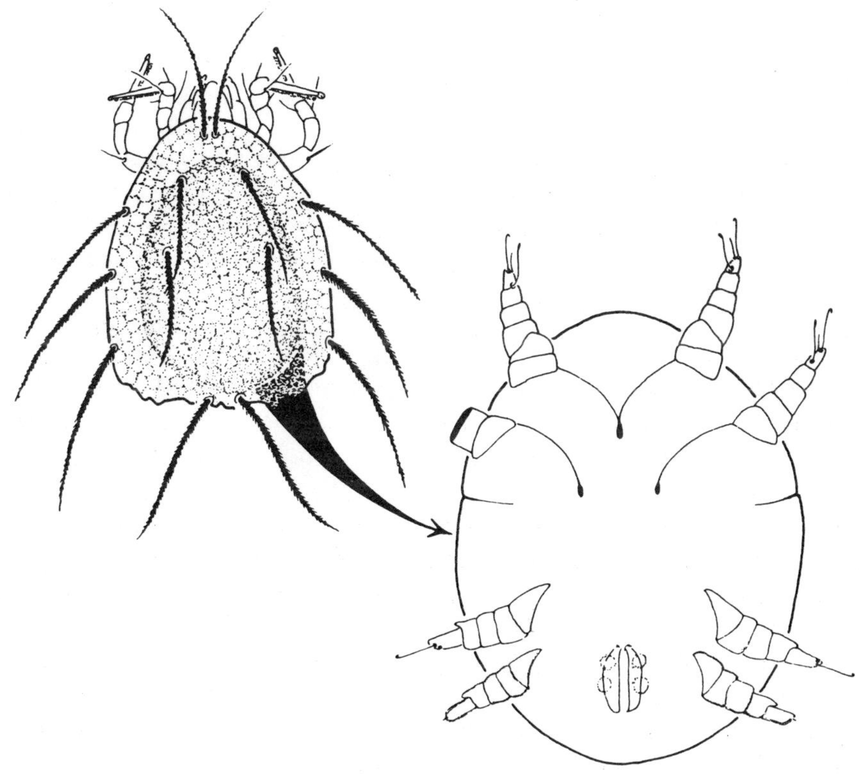

Fig. 40. Inert hypopus of *Glycyphagus destructor* (Schrank) within the protonymphal exuvium (left) and free (right).

Many glycyphagid species are found in the nests of wild mammals (Rupeš and Whitaker 1968, Fain 1969, Whitaker and Wilson 1974). Hypopodes of nine subfamilies and 24 genera of nest-inhabiting GLYCYPHAGIDAE attach at the bases of hair follicles or invade the endofollicular recesses of their nesting hosts. Other instars of these species are saprophages or fungivores which rely on the hypopodial instar for dispersal to new nest loci. The hypopus is the only known stage for most of the genera and species of mammal-associated glycyphagids. Consequently, comprehensive biological information on this group is limited to those few species which have been successfully maintained in culture or reared through an entire generation (Lukoschus et al. 1972).

Hypopodes of the glycyphagid subfamily Labidophorinae are distinctive in that they attach to the hairs of their hosts by means of ventral claspers (Plate 120-1, p. 418). Hypopodes of *Labidophorus* are common on mice, which may explain the prevalence of these mites in farm stored grain (Hughes 1961). Insectivores and rodents are hosts for the genera *Orycteroxenus, Dermacarus, Labidophorus, Zibethacarus* and *Rhynchocyonopus* (Fain 1969b and c, Rupeš et al. 1971). The five described species of *Marsupialichus* occur on edentates and marsupials (Pence 1973, Whitaker and Wilson 1974).

Rodents serve as hosts for the endofollicular hypopodes of the glycyphagid genera *Rodentopus* and *Cryptomyopus* (subfamily Ctenoglyphinae). *Ctenoglyphus plumiger* (Koch) is a grain-inhabiting ctenoglyphine (Hughes 1961), although the hypopodial instar strongly resembles those of endofollicular *Rodentopus* species (Chmielewski 1975). Nonhypopodial instars of *C. shoutedeni* Fain were recovered from rodent nests in Africa (Fain 1967a).

Two distinct hypopodial types have been observed in cultures of *Apodemopus apodemi* Fain (subfamily Lophuromyopinae) (Lukoschus et al. 1972). One was removed from host tissues, and the other was collected from rearing dishes following the protonymphal molt. The "tissue hypopus" is considerably longer and broader than the "free hypopus", and the host tissues which surround it exhibit abnormalities which suggest that extraintestinal tissue digestion has taken place. Similarly, histological sections of host tissues invaded by *Echimyopus dasypus* Fain, Lukoschus, Louppen and Mendez (subfamily Echimyopinae) illustrate hyperkeratosis and hypertrophy around the embedded mites, as well as injury to subepidermal connective tissues (Fain et al. 1973). Since hypopodes do not have functional mouthparts, the means of absorption of lysed tissues by endofollicular species is a matter for conjecture.

Subepidermal hypopodes in pigeons and various other birds have, until recent years, been considered the hypopodial form of the feather mite, *Falculifer rostratus* (Buchholz) (Psoroptidia, FALCULIFERIDAE), a common parasite of pigeons. Fain and Bafort (1967) discovered that the mite in question is actually a free-living nest inhabitant, *Hypodectes propus* (Nitzsch). *H. propus,* along with other species of *Hypodectes,* is included in the glycyphagid subfamily Hypodectinae.[1] Unlike related glycyphagids, female *Hypodectes* produce eggs which develop directly into hypopodes. Following tissue invasion and growth, the enlarged hypopodes leave the host tissue and develop directly to the adult stage. The morphology of the adult mouthparts is such that imaginal feeding is unlikely, meaning that food intake apparently can occur only during the hypopodial stage. The magnitude of food absorption by the hypopus is illustrated by its remarkable growth following host tissue invasion. The free hypopus is only 150 μ long but it attains a length of 1500 μ by the time it leaves the host. As with the endofollicular hypopodes, the means of food absorption by subepidermal hypodectine hypopodes is unclear (Fain 1967b and 1969a, Fain and Bafort 1967).

Members of the SAPROGLYPHIDAE (Plate 122, p. 420) are opaque, essentially unsclerotized saprophagous or fungivorous species which are best known as hypopodes (Fain 1972b). Non-hypopodial saproglyphids have been collected from a number of habitats, many of which are insect-associated (species of *Winterschmidtia, Olafsenia, Vidia* and *Calvolia* are examples). Other non-hypopodial saproglyphids are associated with plant materials *(Calvolia lordi* (Nesbitt), *C. heterocoma* (Michael), and *Oulenzia arboricola* Oudemans are examples). *Saproglyphus neglectus* Berlese was collected from rotting shelf fungi in Italy (Baker and Wharton, 1952).

Saproglyphid hypopodes of the genus *Procalvolia* inhabit stored grain and cheeses, while those of *Acalvolia* may be found in housedust (Fain 1972a). Hypopodes of most of the remaining 18 saproglyphid genera are phoretic on insects. Species of *Monobiacarus, Vidia, Crabrovidia, Zethovidia,* and *Zethacarus* are found on bees and solitary wasps. Hypopodes of *Vespacarus* are carried in a special acarinarium of solitary wasps of the genus *Parancistrocerus* (Mostafa 1970a and b), and those of *Kennethiella trisetosa* (Cooreman) are transmitted venereally from males to females of the wasp *Ancistrocerus antilope* (Cooper 1955).

---

[1] Family HYPODERIDAE of Fain (1969).

Bees provide the habitat and means of transport for members of the family CHAETODACTYLIDAE (Plate 121, p. 419). Adult *Chaetodactylus, Sennertia* and related genera inhabit the nests of wood-boring xylocopid and solitary megachilid and osmid bees both in tropical and temperate realms (Zakhvatkin 1941, Krombein 1962a, Baker 1962b, Fain 1966b). Fain (op. cit.) has noted that *Chaetodactylus* species have two types of hypopodes, one of which is immobile and has poorly developed legs and rudimentary ventral suckers. The immobile hypopus is a hibernating form while the well developed active hypopodial morph is migratory, attaching to the emerging adult bee as it leaves the brood cell. According to Popovici-Baznosanu (1913) *C. osmiae* (Dufour) may be truly parasitic on *Osmia* bees. Krombein (1962a) observed *C. krombeini* Baker to feed on osmid eggs and larvae.

A bee also serves as host for *Gaudiella minuta* Atyeo, Baker and Delfinado, the type and only species of the family GAUDIELLIDAE (Plate 124, p. 422). *G. minuta* was collected from a stingless bee, *Melipona quadrifasciata,* in Brazil (Atyeo, Baker and Delfinado 1974). *Platyglyphus malayanus* Kurosa (family PLATYGLYPHIDAE, Plate 124) was collected in a bee nest in a tree hole (Kurosa 1976). Its relationship to the bee inhabitants has not been determined.

Heterocoptes tarsii Fain was described from a single male specimen collected on a tarsier (*Tarsius* sp.) from Sarawak, and named as the type species of the family HETEROCOPTIDAE (Plate 124, p. 422) (Fain 1967d). The affinities of *H. tarsii* are questionable, but it is tentatively included in the Acaroidea pending description of additional material.

The ROSENSTEINIIDAE (Plate 125, p. 423) are small (300-400 μ), weakly sclerotized, sometimes scaly mites which are found either on bats (Ah and Hunter 1968, Fain 1963, 1972b) or free in bat caves (Strandtmann 1962, Fain 1968c). *Nycteriglyphus,* the largest genus in the family, occurs in the Far East, North America, Australia, Europe and Africa (Fain, op. cit.). *Mydopholeus capillus* McDaniel and Baker was collected from a free-tailed bat in Mexico and *Cheiromelichus malayi* Fain, a distinctive monotypic species with spur-like ventral setae, occurs on a Malaysian bat (Fain 1970a).

Members of the acaroid family HYADESIIDAE (Plate 125, p. 423) are encountered in tide pools, algae and mussel beds in coastal areas of Europe (Michael 1901, André 1931), the Subantarctic (Fain 1974a), Africa, the southwest Pacific and the Americas. *Hyadesia algivorans* (Michael) is found in algae bathed by fresh water but within reach of sea spray. An undescribed species of *Hyadesia* was taken in red coralline algae on the Caribbean coast of Yucatan, Mexico in a situation where the mites were perpetually submerged. The hyadesiid genera *Algophagus* and *Algophagopsis* are found only in fresh or brackish water (Fain 1974b, Fain and Johnston 1975).

Superfamily Anoetoidea
(Plates 125-126, pp. 423-424)

DIAGNOSIS: *Soft-bodied, with prodorsal shield, sejugal furrow present; movable cheliceral digit reduced, palpi modified as "strainers". With an empodial claw and sessile pulvillus on each tarsus. Oviporus transverse, flanked by a pair of sclerotized ring-like structures, with a second pair in the region of coxal fields IV, genital rings present also in male. Hypopodial instar often encountered.*

The Anoetoidea includes only one family, the ANOETIDAE or slime mites. Anoetids are virtually worldwide in distribution, favoring highly organic wet substrates. Non-hypopodial immatures and adults probably feed on bacteria and other microorganisms which are filtered from semiaquatic substrates by means of palpal "strainers" (Plate 126-2). However, *Histiostoma murchiei* Hughes and Jackson and *H. berghi* Jensen are reported to feed on earthworm eggs and leeches respectively. *Loxanoetus bassoni* Fain was encountered in great numbers in exudates and on the hair fringes of the ears of an elephant in South Africa (Fain 1970b).

The adaptation of anoetids to aquatic habitats reaches a climax in species such as *Histiostoma nigrelli* H. and J., *H. cyrtandrae* (Vitzthum) and *Mauduytia mallotae* Fashing, all of which live completely submerged in their aquatic niches (Hughes and Jackson 1958, Fashing 1973). All stages of *Anoetus gibsoni* Hunter and Hunter were collected from below the surface of liquid in a pitcher plant (Hunter and Hunter 1964). Other *Anoetus* species also are found in the pitcher plant habitat (Hirst 1928, Nesbitt 1954). Damp, rotting vegetables or insect cultures which have gone out of condition are common sources of burgeoning anoetid populations. Some anoetid species of the genera *Myianoetus* and *Histiostoma* are halophiles, favoring intertidal algal habitats or moist situations with a marine exposure (Fain 1976).

Anoetid hypopodes are common on insects and other arthropods (Scheucher 1959, Mahunka 1972a and b) but also are active inhabitants of damp organic substrates which shelter other anoetid instars (Hughes and Jackson 1958, Mahunka 1963, 1973a and b).

Superfamily Ewingioidea
(Plate 126, p. 424)

DIAGNOSIS: *Soft-bodied, without dorsal shields or sejugal furrow; chelicerae narrowed and chelate, palpi simple. With strong empodial claws or knob-like projections on tarsi I-II; tarsi III-IV uncate, with large empodial claws which oppose a ventral tarsal projection to form a grasping organ, pulvilli absent on all tarsi. Oviporus longitudinal, with two pairs of flanking acetabula (reduced in females); males with a pair of anal suckers, sucker-like dorsal setae may be present on tarsus IV. Hypopodes not known to occur.*

Only one family, the EWINGIIDAE, is included in the superfamily. Three genera have been described and all of them are found in the gills or on the protruded vasa deferentia of legs V of pagurid land crabs (Yunker 1970). The highly modified legs III-IV of ewingiids provide the means by which the mites adhere to the tissues of their hosts. *Ewingia coenobitae* Pearse and an undescribed species of *Askinasia* are New World forms found on *Coenobita diogenes* and *C. clypeatus* respectively. *Hoogstraalacarus tiwiensis* Yunker and *Askinasia aethiopicus* Y. live on *Coenobita rugosus* on the coast of Kenya, East Africa, while *A. sinusarabicus* Y. was collected from *C. jousseaumi* on the Red Sea coast in Egypt.

It is of interest to note that *Askinasia* species attach to the hairs of the vasa deferentia of their crab hosts, rather than to the gills (Yunker 1970).

Superfamily Canestrinioidea
(Plates 126-127, pp. 424-425)

DIAGNOSIS: *Soft-bodied but with a prodorsal shield which may be indistinct, idiosoma often with strong reticulate pattern, sejugal furrow present or absent; chelicerae narrowed, fixed digit somewhat reduced, occasionally absent, palpi simple. With a sucker-like stalked pulvillus on each tarsus, empodial claws minute; tibia I with solenidion ø, simple setae absent. Oviporus longitudinal, resembling a narrow inverted "Y", with two pairs of flanking acetabula; male with a pair of anal suckers. Hypopodial stage infrequently encountered.*

Two families are included here in the Canestrinioidea. Both are associated with insects.

Members of the CANESTRINIIDAE are weakly sclerotized, moderately large (400-600 μ) mites which are associated with several families of Coleoptera. Canestriniids generally favor protected sites beneath the elytra of their beetle hosts (Cooreman 1955) but some have been collected from the gular or coxal recesses. Many canestriniid species are tropical or subtropical although members of at least six genera occur in the Palearctic realm as associates of carabid beetles (Samšiňák 1971). Canestriniids are recorded from Europe, Africa, Asia, Central and South America on lucanid, carabid, tenebrionid, chrysomelid, passalid and scarabaeid beetles. Their feeding habits are unknown, but the intimate association of canestriniids with their hosts suggests a parasitic relationship.

Hemisarcoptes malus (Shimer), one of three described species of the family HEMISARCOPTIDAE, is a small (300 μ) mite which feeds on scale insects (Coccidae) and their eggs in North America, Europe and Africa (Tothill 1918, André 1942, Gerson 1967). *H. malus* appears to exert a significant degree of control on various coccids including *Parlatoria blanchardi* in North Africa and *Lepidosaphes ulmi* in Nova Scotia (Gerson, op. cit.). *H. coccophagus* Meyer preys on California red scale, *Aonidiella aurantii,* in South Africa (Meyer 1962) and on *Parlatoria pergandii* and *P. cinerea* in citrus orchards of Israel (Gerson 1964).

Hemisarcoptid hypopodes are phoretic on coccinellid beetles of the genus *Chilocorus,* which in turn are scale insect predators. The hypopus of *H. cooremani* (Thomas) was collected in Texas from beneath the elytra of *C. cacti,* and is presently the only known instar of this species. *Chilocorus bipustulatus* and *C. stigma* serve as carriers for *H. coccophagus* and *H. malus* respectively. Examination of additional species of *Chilocorus* may result in the discovery of other hemisarcoptid species possessing biological control potential (Gerson 1967).

Useful References

Ah. H.S. and P.E. Hunter (1968). *Nycteriglyphus vespertilio* n. sp. a new acarid mite associated with bats from Korea (Acarina:Rosensteiniidae). Acarologia **10**(2):269-275.

André, M. (1931). Sur le genre *Hyadesia* Mégnin 1889 (Sarcoptides hydrophiles). Bull. Mus. Hist. nat. Paris 2, **3**(6):496-506. [HYADESIIDAE]

André, M. (1942). Sur l'*Hemisarcoptes malus* Shimer (=*coccisugus* Lignières) (Acariens). Bull. Mus. Hist. nat. Paris 2, **14**(3):173-180. [HEMISARCOPTIDAE]

Atyeo, W.T., E.W. Baker and M.D. Delfinado (1974). *Gaudiella minuta,* A new genus and species of mite (Acarina:Acaridia) belonging to the new family Gaudiellidae. Wash. Acad. Sci. **64**(4):295-298.

Baker, E.W. (1962a). Some Acaridae from bees and wasps (Acarina). Proc. Ent. Soc. Wash. **64**(1):1-10.

Baker, E.W. (1962b). Natural history of Plummers Island, Maryland. XV. Descriptions of the stages of *Chaetodactylus krombeini,* new species, a mite associated with *Osmia lignaria* Say. Proc. Biol. Soc. Wash. **75**:227-236.

Baker, E.W. (1964). *Vidia cooremani,* a new species of Saproglyphidae from a crabronine wasp (Acarina). Ent. News **75**(2):43-46.

Baker, E.W. and D.A. Crossley Jr. (1964). A new species of mite, *Fusoherecia lawrenci,* from an artificial tree-hole (Acarina:Glycyphagidae). Ann. Natal Mus. **16**:1-6.

Baker, E.W., M.D. Delfinado and M.J. Abbatiello (1976). Terrestrial mites of New York II. Mites in bird's nests. J. New York. Ent. Soc. **84**(1):48-66. [ACARIDAE, GLYCYPHAGIDAE, PYROGLYPHIDAE]

Baker, E.W. and G.W. Wharton (1952). The Suborder Sarcoptiformes Reuter, 1909. **In** An Introduction to Acarology. MacMillan Co., New York: 320-386.

Bruel, W.E. van den (1940). Un ravageur de l'epinard d'hiver: *Tyrophagus dimidiatus* Herm. (*longior* Gerv.). Bull. Inst. Agron. Gembloux **9**(1-4):81-99. [ACARIDAE]

Chmielewski, W. (1971). Morfologia, biologia i ekologia *Carpoglyphus lactis* (L., 1758) (Glycyphagidae, Acarina). Prace Nauk Inst. Ochrony Roslin **13**(2):63-166. [CARPOGLYPHIDAE]

Chmielewski, W. (1975). Recently stated developmental stage of mites belonging to the genus *Ctenoglyphus*-hypopus of *C. plumiger* (Koch, 1835) (Acarina, Glycyphagidae). Zeszyty Problem. Post. Nauk Rolniczyck **171**:261-268.

Cooper, K.W. (1955). Venereal transmission of mites by wasps, and some evolutionary problems arising from the remarkable association of *Ensliniella trisetosa* with the wasp *Ancistrocerus antilope.* Trans. Amer. Ent. Soc. **80**:119-174. [SAPROGLYPHIDAE]

Cooreman, J. (1950). Étude de quelques Canestriniides (Acari) vivant sur des Chrysomelidae et sur des Carabidae (Insecta, Coleoptera). Bull. Inst. roy. Sci. nat. Belg. **26**(33):1-36.

Cooreman, J. (1954). Acariens Canestriniidae de la collection A.C. Oudemans, a Leiden. Zool. Meded. **33**(13):83-90.

Cooreman, J. (1955). Acari. Exploration du Parc National Albert. I. Mission G.F. de Witte, Fasc. 85:3-43. [CANESTRINIIDAE]

Cooreman, J. (1959). Note sur le genre *Aeroglyphus* Zachvatkine, 1941 (Acaridiae, Glycyphagidae). Bull. Inst. roy. Sci. nat. Belg. **35**(33):1-19.

Cunnington, A.M. (1976). The effect of physical conditions on the development and increase of some important storage mites. Ann. Appl. Biol. **82**:175-201. [ACARIDAE]

Fain, A. (1963). Les Tyroglyphides commensaux des chauves-souris insectivores. Description de cinq espèces nouvelles. Rev. Zool. Bot. Afr. **67**:33-58. [ROSENSTEINIIDAE]

Fain, A. (1965). Un nouveau type d'hypope, parasite cuticole de rongeurs Africains (Acari:Sarcoptiformes) Zeit. für Parasit. **26**:82-90. [GLYCYPHAGIDAE]

Fain, A. (1966a). Acariens cavernicoles du Congo I.-*Troglocoptes luciae* g. n., sp. n. provenant d'une grotte à Thysville (Acaridae:Sarcoptiformes). Rev. Zool. Bot. Afr. **73**(3-4):397-400.

Fain, A. (1966b). Notes sur la biologie des acariens du genre *Chaetodactylus* et en particulier de *C. osmiae,* parasite des abeilles solitaires *Osmia rufa* et *O. cornuta* en Belgique. Bull. Ann. Soc. roy. Ent. Belg. **102**(16):249-261.

Fain, A. (1967a). Acariens nidicoles et détriticoles d'Afrique au Sud du Sahara. II. *Ctenoglyphus schoutedeni* sp. n., vivant dans les nids de murides au Congo (Glycyphagidae:Sarcoptiformes). Rev. Zool. Bot. Afr. **75**(1-2):162-170.

Fain, A. (1967b). Les hypopes parasites des tissus cellulaires des oiseaux (Hypodectidae:Sarcoptiformes). Bull. Inst. roy. Sci. nat. Belg **43**(4):1-139. [GLYCYPHAGIDAE]

Fain, A. (1967c). Nouveaux hypopes vivant en association phoretique sur des rongeurs et des marsupiaux (Acarina:Glycyphagidae). Acarologia **9**(2):415-434.

Fain, A. (1967d). Un acarien remarquable recolte sur un tarsier (Heterocoptidae f.n.:Sarcoptiformes). Zool. Anz. **178**(1/2):90-94.

Fain, A. (1967e). Solenidiotaxy of leg I in the hypopi of the Acaridiae (Acari:Sarcoptiformes). Rev. Zool. Bot. Afr. **76**(3-4):244-248.

Fain, A. (1968a). Acariens nidicoles et détriticoles en Afrique au Sud du Sahara. III. Espèces et genres nouveaux dans les sous-familles Labidophorinae et Grammolichinae (Glycyphagidae:Sarcoptiformes). Acarologia **10**(1):86-110. [GLYCYPHAGIDAE]

Fain, A. (1968b). Un hypope de la famille Hypoderidae Murray 1877 vivant sous la peau d'un rongeur (Hypoderidae:Sarcoptiformes). Acarologia **10**(1):111-115. [GLYCYPHAGIDAE]

Fain, A. (1968c). Deux nouveaux acariens cavernicoles du Gabon (Sarcoptiformes). Biol. Gabon. **4**(2):195-205. [ROSENSTEINIIDAE]

Fain, A. (1969a). Adaptation to parasitism in mites. Acarologia **11**(3):429-449. [GLYCYPHAGIDAE]

Fain, A. (1969b). Les deutonymphes hypopiales vivant en association phoretique sur les mammiferes (Acarina:Sarcoptiformes). Bull. Inst. roy. Sci. nat. Belg. **45**(33):1-262.

Fain, A. (1969c). Morphologie et cycle évolutif des Glycyphagidae commensaux de la taupe *Talpa europaea* (Sarcoptiformes). Acarologia **11**(4):750-795.

Fain, A. (1970a). Trois nouveaux Nycteriglyphinae commensaux de chauves-souris (Acarina:Sarcoptiformes). Bull. Inst. roy. Sci. nat. Belg. **46**(28):1-13. [ROSENSTEINIIDAE]

Fain, A. (1970b). Un nouvel anoetide vivant dans la graisse de l'oreille d'un elephant (Acarina:Sarcoptiformes). Acta Zool. Path. Antverp. 50:173-177. [ANOETIDAE]

Fain, A. (1971). Évolution de certains groupes d'hypopes en fonction du parasitisme (Acarina:Sarcoptiformes). Acarologia **13**(1):171-175.

Fain, A. (1972a). Notes sur les hypopes des Sproglyphidae (Acarina:Sarcoptiformes). II. Redefinition des genres. Acarologia **14**(2):225-249.

Fain, A. (1972b). Notes sur deux nouveaux acariens commensaux de chiropteres (Sarcoptiformes:Rosensteiniidae). Acarologia **14**(2):219-224.

Fain, A. (1974a). Acariens récoltées par le Dr. J. Travé aux iles subantarctiques. I. Familles Saproglyphidae et Hyadesiidae (Astigmates). Acarologia **16**(4):684-708.

Fain, A. (1974b). Deux nouvelles espèces du genre *Hyadesia* recoltées par le Dr. J. Travé aux Iles Saint-Paul et Nouvelle-Amsterdam (Astigmates:Hyadesiidae). Acarologia **17**(1):153-159.

Fain, A. (1976). Acariens récoltées par le Dr. J. Travé aux iles subantarctiques. II. Familles Acaridae, Anoetidae, Ereynetidae et Tarsonemidae (Astigmates et Prostigmates). Acarologia **18**(2):302-328.

Fain, A. and J. Bafort (1967). Cycle évolutif et morphologie de *Hypodectes (Hypodectoides) propus* (Nitzsch) acarien nidicole à deutonymphe parasite tissulaire des pigeons. Bull. Acad. roy. Belg. **53**(5): 501-533. [GLYCYPHAGIDAE]

Fain, A. and D.E. Johnston (1974). Three new species of hypopi phoretic on springtails (Collembola) in England (Acari:Acaridiae). J. nat. Hist. **8**:411-420. [ACARIDAE, SAPROGLYPHIDAE]

Fain, A. and D.E. Johnston (1975). A new algophagin mite, *Algophagopsis pneumatica* gen. n., sp. n. living in a river (Astigmata:Hyadesiidae). Bull. Ann. Soc. roy. belg. Ent. 111:66-70.

Fain, A., F.S. Lukoschus, J.M.W. Louppen and E. Méndez (1973). *Echimyopus dasypus,* n. sp., a hypopus from *Dasypus novemcinctus* in Panama (Glycyphagidae, Echimyopinae:Sarcoptiformes). J. Med. Ent. **10**(6):552-555.

Fashing, N.J. (1973). The post-embryonic stages of a new species of *Mauduytia* (Acarina:Anoetidae). J. Kansas Ent. Soc. **46**:454-468.

Fashing, N. (1974). A new subfamily of Acaridae, the Naiadacarinae, from water-filled treeholes (Acarina: Acaridae). Acarologia **16**(1):166-181.

Fashing, N.J. (1975). Life history and general biology of *Naiadacarus arboricola* Fashing, a mite inhabiting water-filled treeholes (Acarina:Acaridae). J. nat. Hist. **9**:413-424.

Gerson, U. (1964). *Parlatoria cinerea,* a pest of citrus in Israel. FAO Plant Prot. Bull. **12**:82-85. [HEMISARCOPTIDAE]

Gerson, U. (1967). Observations on *Hemisarcoptes coccophagus* Meyer (Astigmata:Hemisarcoptidae), with a new synonym. Acarologia **9**(3):632-638.

Griffiths, D.A. (1964). A revision of the genus *Acarus* L., 1758 (Acaridae, Acarina). Bull. Brit. Mus. (Nat. Hist.) Zool. **11**(6):415-464 + plate.

Griffiths, D.A. (1970). A further systematic study of the genus *Acarus* L. 1758 (Acaridae, Acarina), with a key to species. Bull. Brit. Mus. (Nat. Hist.) Zool. **19**(2):85-118.

Hirst, S. (1928). A new Tyroglyphid mite (*Zwickia nepenthesiana*-sp. n.) from the pitchers of *Nepenthes ampullaria.* J. Malay, Brit. Asiat. Soc. **6**:19-22. [ANOETIDAE]

Hughes, A.M. (1976). The Mites of Stored Food and Houses. Second Edition. Min. Agr. Fish. Food, London: Tech. Bull. **9**:400 pp.

Hughes, R.D. and C.G. Jackson (1958). A review of the Anoetidae (Acari). Virginia J. Sci. **9**, N.S.(1):5-198.

Hunter, P.E. and C.A. Hunter (1964). A new *Anoetus* mite from pitcher plants (Acarina:Anoetidae). Proc. Ent. Soc. Wash. **66**(1):39-46.

Hussey, N.W., W.H. Read and J.J. Hesling. Other arthropod pests of mushrooms. **In** The Pests of Protected Cultivation. Amer. Elsevier Publ. Co., New York:338-342. [ACARIDAE]

Jeppson, L.R., H.H. Keifer and E.W. Baker (1975). Tydeidae, Tuckerellidae, Pyemotidae, Penthaleidae, Astigmata, and Cryptostigmata. **In** Mites Injurious to Economic Plants. Univ. California Press, Berkeley: 307-326. [ACARIDAE]

Johnston, D.E. and W.A. Bruce (1965). *Tyrophagus neiswanderi,* a new acarid mite of agricultural importance. Ohio Agr. Res. Dev. Center 977:3-17. [ACARIDAE]

Joyeaux, C. and G. Baer (1945). Morphologie, evolution et position systematique de *Catenotaenia pusilla* (Goeze, 1782), Cestode parasite de Rongeurs. Rev. Suisse Zool. 52(2):13-51. [GLYCYPHAGIDAE]

Knülle, W. (1959). Morphologische und Entwicklungsgeschichtliche untersuchungen zum phylogenetischen System der Acari: Acariformes Zachv. II. Acaridiae: Acaridae. Mitt. Zool. Mus. Berlin **35**(2):347-417.

Krombein, K.V. (1961). Some symbiotic relations between saproglyphid mites and solitary vespid wasps (Acarina, Saproglyphidae and Hymenoptera, Vespidae). J. Wash. Acad. Sci. (Oct.):89-92.

Krombein, K.V. (1962a). Natural history of Plummers Island, Maryland XVI. Biological notes on *Chaetodactylus krombeini* Baker, a parasitic mite of the megachilid bee, *Osmia (Osmia) lignaria* Say (Acarina: Chaetodactylidae). Proc. Biol. Soc. Wash. **75**:237-250.

Krombein, K.V. (1962b). Biological notes on acarid mites associated with solitary wood-nesting wasps and bees (Acarina:Acaridae). Proc. Ent. Soc. Wash. **64**(1):11-19.

Kurosa, K. (1976). Notes on Tarsonemini and Acaridiae (Acari) from Pasoh Forest Reserve and its vicinities, Malaya. Nature Life S.E. Asia **7**:61-81. [PLATYGLYPHIDAE]

Lukoschus, F.S., A. Fain and F.M. Driessen (1972). Life cycle of *Apodemopus apodemi* (Fain, 1965) (Glycyphagidae:Sarcoptiformes). Tijdsch. Ent. **115**(8):325-339.

Mahunka, S. (1963). Neue Anoetiden (Acari) aus Angola. Publ. Cult. Companhia Diamantes Angola 63:27-43.

Mahunka, S. (1972a). Neue, auf Tenebrioniden (Coleoptera) gesammelte Anoetidenarten (Acarina) von den Salamon-Inseln. Parasit. Hung. **5**:349-360.

Mahunka, S. (1972b). Untersuchungen über taxonomische und systematische Probleme bei der Gattung *Myianoetus* Oudemans, 1913 und der Unterfamilie Myianoetinae (Acari, Anoetoidea). Ann. Hist.-natur. Mus. Nat. Hung. **64**:359-372.

Mahunka, S. (1973a). *Xenanoetus grandiceps* sp. n., sowie weitere Angaben über die Anoetiden-Fauna der Mongolei (Acari). Fol. Ent. Hung. **26**(1):57-63.

Mahunka, S. (1973b). Neue und interessante Milben aus dem Genfer Museum V. *Ceylanoetus excavatus* gen. nov., sp. nov. und andere neue Anoetida Arten aus Ceylan (Acari). Acarologia **15**(3):506-513.

Mahunka, S. (1974). Beiträge zur Kenntnis der an Hymenopteren lebenden Milben (Acari). II. Fol. Ent. Hung. **27**(1):99-108. [ANOETIDAE]

Mahunka, S. (1975). Auf Insekten lebenden Milben (Acari:Acarida und Tarsonemida) aus Afrika. V. Acta Zool. Acad. Sci. Hung. **21**(1-2):39-72. [ACARIDAE, ANOETIDAE]

Manson, D.C.M. (1972). A contribution to the study of the genus *Rhizoglyphus* Claparède 1869 (Acarina: Acaridae). Acarologia **13**(4):621-650.

McDaniel, B. and E. W. Baker (1962). A new genus of Rosensteiniidae (Acarina) from Mexico. Fieldiana-Zool. **44**(16):127-131.

Meyer, M.K.P. (1962). Two new mite predators of red scale *(Aonidiella aurantii)* in South Africa. S. Afr. J. Agr. Sci. **5**:411-417. [HEMISARCOPTIDAE]

Michael, A.D. (1901). British Tyroglyphidae. Ray Soc., London **1**:291 pp. + xiii + plates. [GLYCYPHAGIDAE]

Michael, A.D. (1903). British Tyroglyphidae. Ray Soc., London **2**:183 pp. + vii + plates. [CHORTOGLYPHIDAE, FUSACARIDAE, CHAETODACTYLIDAE, GLYCYPHAGIDAE, CARPOGLYPHIDAE, ACARIDAE]

Mostafa, A.-R.I. (1970a). Saproglyphid hypopi (Acarina:Saproglyphidae) associated with wasps of the genus *Zethus* Fabricius. Acarologia **12**(1):168-192.

Mostafa, A.-R.I. (1970b). Saproglyphid hypopi (Acarina:Saproglyphidae) associated with wasps of the genus *Zethus* Fabricius. Acarologia **12**(2):383-401.

Nemestóthy, K. and S. Mahunka (1972). Kártevö és parazita Acaridae-fojok Magyarország faunájából. Parasit. Hung. **5**:361-374.

Nesbitt, H.H.J. (1946). Three new mites from Nova Scotian apple trees. Can. Ent. **78**:15-22. [SAPROGLYPHIDAE]

Nesbitt, H.H.J. (1954). A new mite, *Zwickia gibsoni* n. sp. from the pitchers of *Sarracenia purpurea.* Can. Ent. **86**:192-197.

Pence, D.B. (1972). The hypopi (Acarina:Sarcoptiformes:Hypoderidae) from the subcutaneous tissues of birds in Louisiana. J. Med. Ent. **9**(5):435-438. [GLYCYPHAGIDAE]

Pence, D.B. (1973). Notes on two species of hypopial nymphs of the genus *Marsupialichus* (Acarina:Glycyphagidae) from mammals in Louisiana. J. Med. Ent. **10**(4):329-332.

Popovici - Baznosanu, A. (1913). Étude biologique sur l'Acarien *Trichotarsus osmiae* Duf. Arch. Zool. Exp. Paris **52**:32-41. [CHAETODACTYLIDAE]

Price, D.W. (1976). Passage of *Verticillium albo-atrum* propagules through the alimentary canal of the bulb mite. Phytopath. **66**(1):46-50. [ACARIDAE]

Robertson, P. (1959). Revision of the genus *Tyrophagus* with a discussion on its taxonomic position in the Acarina. Austral. J. Zool. **7**(2):146-181. [ACARIDAE]

Robin, C. and M.P. Mégnin (1877). Mémoire sur les Sarcoptides plumicoles. J. Anat. Physiol. **13**:209-656 + plates. [GLYCYPHAGIDAE]

Rupeš, V. and J.O. Whitaker, Jr. (1968). Mites of the subfamily Labidophorinae (Acaridae, Acarina) in North America. Acarologia **10**(3):493-499. [GLYCYPHAGIDAE]

Rupeš, V., C.E. Yunker and N. Wilson (1971). *Zibethacarus,* n. gen., and three new species of *Dermacarus* (Acari:Labidophoridae). J. Med. Ent. **8**(1):17-22. [GLYCYPHAGIDAE]

Samšiňák, K. (1963). *Melisia* Lombardini, 1944- eine auf Insekten lebende Gattung der Unterfamilie Glycyphaginae (Acari). Acta Soc. Ent. Čechoslov. **60**(3):252-262. [GLYCYPHAGIDAE]

Samšiňák, K. (1964). Die auf *Procerus* lebenden Formen der Gattungen *Procericola* Cooreman, 1950 und *Photia* Oudemans, 1904. Vestn. Česchoslov. Spol. Zool. **28**:34-43. [CANESTRINIIDAE]

Samšiňák, K. (1965). Termitophile Milben aus der VR China, 2. Acaridoidea. Reichenbachia **5**(34):291-294. [ACARIDAE]

Samšiňák, K. (1970a). Zwei neue Arten der Gattung *Sancassania* Oudemans, 1916 (Acari:Acaridae). Zool. Anz. **184**(5/6):403-412.

Samšiňák, K. (1970b). Die auf *Blaps* (Col., Tenebrionidae) lebenden Milben der Gattung *Canestrinia* Berlese, 1881 (Acari). Ent. Mitt. Hamburg **4**(68):71-78.

Samšiňák, K. (1971). Die auf *Carabus*-Arten (Coleoptera, Adephaga) der palaearktischen Region lebenden Milben der Unterordnung Acariformes (Acari); ihre Taxonomie und Bedeutung für die Lösung zoogeographischer, entwicklungsgeschichtlicher und parasitophyletischer Fragen. Ent. Abh. **38**(6):145-234. [CANESTRINIIDAE]

Scheucher, R. (1959). Systematik und Ökologie der deutschen Anoetinen. Beitr. Syst. Ökol. mitteleurop. Acarina **1**(2):233-384.

Sinha, R.N. (1966). *Aeroglyphus robustus,* a pest of stored grain. J. Econ. Ent. **59**(3):686-688. [GLYCYPHAGIDAE]

Solomon, M.E. (1946). Tyroglyphid mites in stored products. Nature and amount of damage to wheat. Ann. Appl. Biol. **33**(3):280-289. [ACARIDAE]

Southcott, R.V. (1976). Arachnidism and allied syndromes in the Australian region. Rec. Adelaide Children's Hospital **1**(1):97-186.

Strandtmann, R.W. (1962). *Nycteriglyphus bifolum* n. sp., a new cavernicolous mite associated with bats (Chiroptera) (Acarina:Glycyphagidae). Acarologia **4**(4):623-631. [ROSENSTEINIIDAE]

Thomas, H.A. (1961). *Vidia (Coleovidia) cooremani,* a new subgenus and new species and notes on the life history (Acarina:Saproglyphidae). Ann. Ent. Soc. Amer. **54**(3):461-463. [HEMISARCOPTIDAE]

Tothill, J.D. (1918). The predaceous mites, *Hemisarcoptes malus* Shimer, and its relation to the natural control of the oyster-shell scale, *Lepidosaphes ulmi* L. Agr. Gazette Canada **5**(3):234-239.

Turk, F.A. (1948). Insecticolous Acari from Trinidad, B.W.I. Proc. Zool. Soc. London **118**:82-128. [CANESTRINIIDAE]

Türk, E. and F. Türk (1959). Systematik und Ökologie der Tyroglyphiden Mitteleuropas. Beitr. Syst. Ökol. mitteleurop. Acarina **1**(1):3-231.

Vitzthum, H.G. (1940). Die Deutonympha von *Carpoglyphus lactis* (L. 1763) (Acari:Tyroglyphidae). Zool. Anz. **129**:197-201. [CARPOGLYPHIDAE]

Volgin, V.I. (1971). The hypopus and its main types. Proc. 3rd Int. Congr. Acarology, Prague: 381-383.

Vomero, V. (1972). Una nuova specie del genere *Melisia* Lombardini 1944 associata a coleotteri passalidi (Acarina-Sarcoptiformes). Acarologia **14**(1):94-108. [GLYCYPHAGIDAE]

Whitaker, J.O., Jr. and N. Wilson (1974). Host and distribution lists of mites (Acari), parasitic and phoretic, in the hair of wild mammals of North America, north of Mexico. Amer. Midland Natur. **91**(1):1-67.

Willmann, C. (1939). *Winterschmidtia crassisetosa* spec. nov. (Winterschmidtiidae, Acari). Boll. Lab. Zool. Gen. Agrar., Portici 31:65-68. [SAPROGLYPHIDAE]

Woodring, J.P. (1966). North American Tyroglyphidae (Acari): I. New species of *Calvolia* and *Nanacarus,* with keys to the species. Proc. Louisiana Acad. Sci. **29**:76-84. [SAPROGLYPHIDAE]

Yunker, C.E. (1970). New genera and species of Ewingidae (Acari:Sarcoptiformes) from pagurids (Crustacea), with notes on *Ewingia cenobitae* Pearse 1929. Rev. Zool. Bot. Afr. **81**(3-4):237-254.

Zakhvatkin, A.A. (1941). Fauna of U.S.S.R. Arachnoidea **6**(1), Tyroglyphoidea (Acari). Zool. Inst. Acad. Sci. U.S.S.R., N.S. 28 [translated by Amer. Inst. Biol. Sci.: 573 pp. + v].

Supercohort Psoroptides

The Psoroptides includes a large number of primarily parasitic families which have membranous ambulacral discs or claw-like setae or spines on the tarsal apices. Sessile claws may rarely be present, but pad-like fleshy pulvilli are absent. Psoroptides either lack genital acetabula or have obscure acetabular remnants in post-protonymphal stases. Tibia I has only one tactile seta and a solenidion (Fig. 42, p. 390). A heteromorphic hypopus stage is absent.

Representatives of the seven psoroptide superfamilies are dermal, subdermal, follicular and internal parasites of birds and mammals. Many are found on feathers and some have moved within the feather calmus, feeding through the quill wall. Certain members of the family EPIDERMOPTIDAE are ectoparasites of birds, but oviposit on hippoboscid flies or on menoponid lice which share the same host. Some psoroptide species are nidicolous or free-living, but these are considered exceptional.

Superfamily Psoroptoidea
(Plates 128-130, pp. 426-428)

DIAGNOSIS: *Soft-bodied but commonly with discrete podonotal, opisthonotal and lateral shields, sejugal furrow absent; chelicerae chelate or terminally stylettiform, weakly developed; palpi simple. Pretarsi, when present, each consisting of a stalked terminal or subterminal sucker-like ambulacral disc; tarsi III-IV with or without discs, with either or both often terminating in long whip-like setae, leg IV sometimes greatly reduced. Oviporus an inverted "V", "Y" or "U" shape, often with two pairs of greatly reduced acetabula; males usually with anal suckers, often illustrating heteromorphic development of legs III, typically with sucker-like dorsal setae on tarsus IV.*

Five of the seven recognized psoroptoid families are mammal parasites and some are economically important pests of domestic animals. The PYROGLYPHIDAE is considered a free-living assemblage although records exist of pyroglyphid infestations of skin or feathers of various mammals or birds. The GUANOLICHIDAE also may be a free-living group, and is included in the Psoroptoidea on a provisional basis.

The family PSOROPTIDAE (Plate 128, p. 426) includes parasites of many mammal groups in most areas of the world. Psoroptids live at the follicular bases or in subcutaneous depressions adjacent to the follicle where they feed on tissue fluids. Many psoroptids have special spurs on the legs and gnathosoma (more rarely on the idiosoma) which aid them in adhering to their dermal substrate (Fain 1969a).

Psoroptes species often are serious pests of both domestic and wild animals, causing severe mange in horses, cattle, buffalo, sheep, goats, elk, rabbit and mountain sheep (Baker et al. 1956, Fain 1970a, Yunker 1973). The broad host range of *P. ovis* (Hering), the common scab mite, gave rise some years ago to a varietal concept in which different strains of *P. ovis* were considered to infest different hosts (Baker and Wharton 1952, Baker et al., op. cit.). This concept was refuted by Sweatman (1958) who found no host specificity amongst a range of ecological entities referrable to *P. ovis.* However, he did identify four additional species of *Psoroptes* on the basis of morphological and biological criteria. Populations of *Psoroptes* from particular hosts are considered as varietal, subspecific or specific entities in certain cases (Evans et al. 1961, Roberts and Meleney 1970, Yunker, op. cit.).

Psoroptic mange may appear on the body or head of the host as oozing itchy lesions which finally harden into scabby patches. These patches are not optimal feeding sites and consequently the mites tend to migrate outward into healthy tissue, constantly extending the lesion. Some strains of *P. ovis* appear to be highly pathogenic, precipitating massive lesions, hair loss and pruritis. Severe psoroptic mange in cattle and sheep may be fatal (Roberts and Meleney 1970). *P. cuniculi* (Delafond) in rabbits causes an ear canker which is characterized by scab formation in the pinna and a malodorous discharge in the external ear canal. Infested rabbits shake their heads and constantly scratch their ears. Lesions may spread to the face, neck and legs (Yunker, op. cit.). *P. pienaari* Fain, a species which is morphologically distinct from *P. ovis,* causes face mange in cape buffalo in Uganda (Fain 1970a). Dogs, cats and other small carnivores often are attacked by the ear mite, *Otodectes cynotis* (Hering). Ear mites give rise to intense irritation in the external auditory canal which may lead to convulsive seizures in heavily infested animals. Secondary infestations may occur on the feet and tip of the tail.

Chorioptic mange of cattle, sheep, goats, pandas, horses and llamas results from dermal infestation by psoroptid species of the genus *Chorioptes* (Baker et al. 1956, Fain and Leclerc 1975). Although not usually as serious as psoroptic mange, heavy infestations of *Chorioptes* often bring about intense irritation in affected animals (Hirst 1922). Chorioptic mange symptoms caused by *C. bovis* (Gerlach) and *C. texanus* Hirst generally are restricted to the lower portions of the legs and the root of the tail. However, ear mange has been reported in the giant panda as the result of infestation by *C. panda* Fain and Leclerc (Fain and Leclerc, op. cit.). Sweatman (1957) synonymized the several species and varieties of the *Chorioptes bovis* "complex" on the basis of his discovery of non-specificity in these morphologically identical forms. However, *C. crewei* Lavoipierre, a parasite of the African duiker, *C. panda,* and others listed by Fain and Leclerc (op. cit.) are distinct species.

Species of the psoroptid genus *Paracoroptes* parasitize monkeys and gorillas in Africa (Zumpt 1961), and *Psoralges libertus* Trouessart causes mange symptoms in its edentate host, the tamandua, in South America (Fain and Lukoschus 1970). Other species of the psoroptid subfamily Psoralginae also are found on edentates (Fonseca 1954, Fain 1964a), while listropsoralgine psoroptids infest South American marsupials (Fain 1965a, Fain and Lukoschus, op. cit.).

South American mammals serve as hosts for the four described species of the family LOBALGIDAE (Plate 129, p. 427). *Lobalges trouessarti* Fonseca parasitizes the three-toed sloth, *Bradypus tridactylus brasiliensis. Echimytricalges surinamensis* Fain and Lukoschus, one of three species in the genus, was found in the concavities of dorsal grooved hairs of the rat *Proechimys g. guyannensis* (Fain and Lukoschus 1970).

Lemurnyssus galagoensis Fain (LEMURNYSSIDAE, Plate 128, p. 426) inhabits the nasal fossae of a lemur, *Galagosenegalensis moholi* in Africa (Fain 1957). The three described lemurnyssids of the genus *Mortelmansia* invade the nasal passages of South American monkeys (Fain 1959a and 1964b). Species of *Audycoptes* and *Saimiroptes* (AUDYCOPTIDAE, Plate 129, p. 427) infest the lip tissue of Neotropical squirrel monkeys. Audycoptids feed on sebaceous materials around sinus and normal hair follicles (Lavoipierre 1964, Fain 1968a). The audycoptid *Ursicoptes americanus* Fain and Johnston was collected from the fur of a black bear in the United States (Fain and Johnston 1970).

The family RHYNCOPTIDAE (Plate 129, p. 427) is a distinctive monogeneric group of four psoroptoid species which parasitize porcupines in South Africa (Lawrence 1956), and monkeys in Africa and South America (Fain 1965c). Rhyncoptids wedge or afix themselves on their hosts with the aid of calcarate scales and spurs on legs I-II. The chelicerae are likewise hooked so that the mite may better anchor itself into the skin. While the specific location of

Rhyncoptes recurvidens Lawrence on its porcupine host has not been determined, the species of *Rhyncoptes* from monkeys are known to embed themselves at the bases of hair follicles. Rhyncoptids may produce mange, but the etiology is not clear.

In contrast to other psoroptoid families, the PYROGLYPHIDAE (Plates 129-130)[1] is an essentially free-living group with representatives occurring as nidicoles, contaminants of stored products, and inhabitants of dwellings (van Bronswijk and Sinha 1971, Wharton 1976). Nidicolous species such as *Pyroglyphus morlani* Cunliffe, *Bontiella bouilloni* Fain, *Sturnophagoides bakeri* (Fain), *Hirstia chelidonis* Hull and *Dermatophagoides farinae* Hughes probably are saprophages in the nest material. *S. bakeri, Euroglyphus longior* (Trouessart), and species of *Dermatophagoides* are nest inhabitants which also may be found on the skin or feathers of their mammal or bird associates (Gaud 1968, van Bronswijk and Sinha, op. cit.).

At least 10 species of PYROGLYPHIDAE occur commonly in the dust of floors, furniture and bedding of human habitations throughout the world. In 1964, Voorhorst et al. implicated one of these species, *Dermatophagoides pteronyssinus* (Trouessart) as a producer of house dust allergies. At least two other pyroglyphids (*D. farinae* Hughes and *Euroglyphus maynei* (Cooreman)) have since been demonstrated to produce allergic reactions in humans (Miyamoto 1972). House-infesting pyroglyphids apparently feed on sloughed human skin, spilled foodstuffs, fungi and pollen. It is of interest to note that the average human adult sheds skin scales at a rate of 0.7-1.4 gm./day, and that as little as 180 mg. of this material is sufficient to produce and maintain mass cultures of *D. pteronyssinus* for some months (van Bronswijk and Sinha 1971). *D. pteronyssinus* and *D. farinae* have both been collected from human skin (Fisher et al. 1951, Fain 1966, 1967). Traver (1951) implicated *Dermatophagoides* sp. as the causative agent of severe scalp and facial dermatitis in man. The occurrence of *Dermatophagoides* in association with human dermatitis was further explored by Dubinin et al. (1956). A comprehensive review of pyroglyphid mites occurring in house dust is presented by Wharton (1976).

Guanolichus gabonensis Fain (GUANOLICHIDAE, Plate 130, p. 428) is a cavernicolous species found in bat guano in Africa (Fain 1968c). Its unique gnathosomal morphology is strongly reminiscent of that seen in the anoetoid family ANOETIDAE, although various setal characters suggest an affiliation with the Psoroptoidea. Fain surmises that *G. gabonensis* may be a predator of insect eggs, or possibly an arthropod parasite. It is placed in the Psoroptoidea only provisionally.

The Feather Mite Superfamilies[2]
(Analgoidea, Pterolichoidea and Freyanoidea)

DIAGNOSIS: Soft-bodied but often with weakly sclerotized podonotal, opisthonotal and lateral shields, sejugal furrow present or absent; chelicerae weakly developed, generally chelate, fixed digit may be reduced; palpi simple. Tarsi each with a stalked or sessile ambulacral disc arising terminally or subterminally, occasionally absent from a distally bifurcate pretarsal stalk. Oviporous an inverted "U", "V" or "Y" shape, or transverse, often with two pairs of greatly reduced acetabula; males with or without anal suckers, often with legs III or IV greatly

---

[1]The placement of the PYROGLYPHIDAE in the Psoroptoidea follows the conclusions of Fain (1965e), although it would be equally appropriate to include the family with the Analgoidea (page 390).

[2]Analgoidea *s. lat.* Gaud and Atyeo (1977) classify the feather mites in three superfamilies, based on certain key tarsal and pretarsal characters. Diagnoses of Pterolichoidea and Freyanoidea are presented on pages 392 and 393.

enlarged, two modified sucker-like setae may be present dorsally on tarsus IV, often with special setal or appendicular adaptations for securing themselves to their feather substrate.

The feather mite superfamilies include a large number of species which inhabit the feathers, skin or—exceptionally—the respiratory system of birds throughout the world. The majority live on the feather surface or in the feather calmus.

The great proliferation of feather mite taxa (no fewer than 19 recognized families) reflects the degree to which these forms have adapted to the many microhabitats offered by the feather substrate. Dubinin (1949, 1951) observed that there is a range of both protected and exposed feather niches in which mites may be found. In addition Dogiel (1949) and Dubinin (1951) noted that particular feathers or parts of feathers are subject to specific stresses, and that an individual feather or feather area may provide one or more unique mite microhabitats. Thus, while the primary flight feathers of the hooded crow support *Gabucinia delibata* (Robin) (Pterolichoidea, GABUCINIIDAE), the secondary flight feathers are occupied by overlapping populations of three other feather mite species (*Proctophyllodes corvorum* Vitzthum (Analgoidea, PROCTOPHYLLODIDAE), *Analges corvinus* Mégnin (Analgoidea, ANALGIDAE) and *Trouessartia corvina* (C.L. Koch) (Analgoidea, TROUESSARTIIDAE)). Each of the latter colonizes particular secondary feather regions (Fig. 41). Individual feathers also may be partitioned by feather mites. For example, *Freyana largifola* (Dubinin) and *F. anatina* (C.L. Koch) (Freyanoidea, FREYANIDAE) occupy different parts of primary feathers of their duck host (Evans et al. 1961).

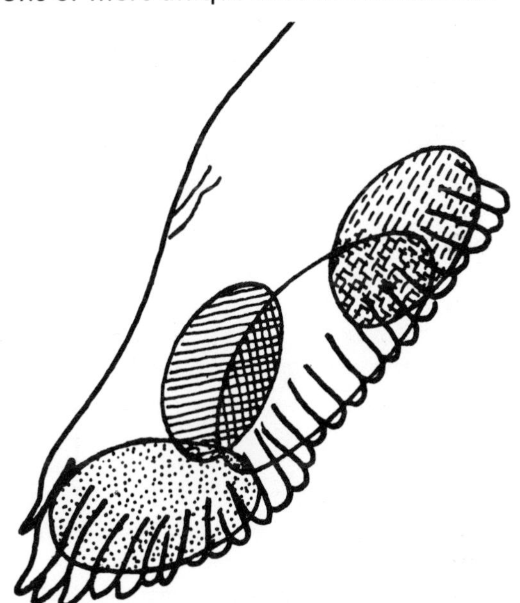

Fig. 41. Overlapping distribution patterns of *Gabucinia delibata* (Robin) (stippled area), *Analges corvinus* Mégnin (lined area), *Proctophyllodes corvorum* Vitzthum (clear area), and *Trouessartia corvina* (C.L. Koch) (broken line area) on the hooded crow (after Dubinin 1951 and Evans et al. 1961).

Adaptation to exposed niches such as those found on the primary flight feathers has led to selective modifications in body shape and/or in legs, leg epimera and leg musculature—modifications which better adapt the mites for maintaining their position in a precarious microhabitat. Exposed feather mites often have extensive dorsal sclerotization, reduced dorsal setae to eliminate wind resistance, and enlarged terminal setae to increase the effective area of contact between the mite and the feather surface (Atyeo and Gaud 1971). Examples of feather mite species which have adapted to exposed sites may be found in the pterolichoid families PTEROLICHIDAE (Plate 142, p. 440), GABUCINIIDAE (Plate 139, p. 437), and KRAMERELLIDAE (Plate 141, p. 439), and in the freyanoid family FREYANIDAE (Plate 136, p. 434). Adaptation of feather mites to protected niches such as the coverts, secondaries, contour feathers and the feather calmus has led to the acquisition of less dramatic but often distinctive modifications in body shape and pretarsal structures. Feather mites typical of protected microhabitats occur commonly in the Analgoidea (the PROCTOPHYLLODIDAE (Plate 131, p. 429), XOLALGIDAE (Plate 132, p. 430) and DERMOGLYPHIDAE (Plate 134, p. 432) are examples). Similar modifications may be seen in the pterolichoid family SYRINGOBIIDAE (Plate 138, p. 436).

Feather mites are only rarely considered to be of economic importance (Yunker 1973) although they may occur on domestic fowl and avian pets in considerable numbers. However, feather-inhabiting feather mites appear to be saprophagous rather than parasitic, probably feeding on feather fragments, desquamated skin scales, and oily secretions on the host feathers (Evans et al. 1961). Fungal spores and diatoms were recovered from the gut contents of 26 species of feather mites examined by Dubinin (1951). Diatoms predominated in species of *Freyana* occurring on aquatic birds.

The feather mite superfamilies may be distinguished from one another primarily by differences in pretarsal morphology and tarsal setation. Terminology relating to these features is presented in Figs. 38 (p. 373) and 43. General host data for feather mite families may be found in the key which follows this section (p. 404).

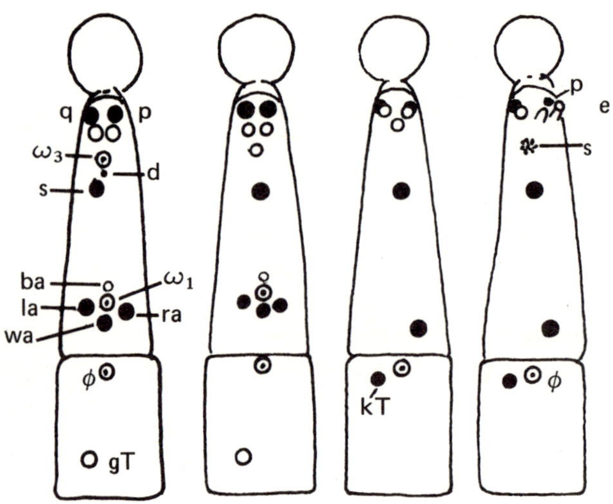

Fig. 42. Chaetotaxy of tibia-tarsi I-IV (left to right) of feather mites (after Atyeo and Gaud 1966). Solenidia are indicated by dotted circles, and tactile setae by black (ventral) and clear (dorsal) circles. Seta s (tarsus IV) is absent in the Analgoidea. Seta p and q are present only in the Pterolichoidea.

Superfamily Analgoidea
(Plates 131-136, pp. 429-434)

DIAGNOSIS: *With general feather mite characteristics (see page 388). Ambulacra variously developed, with or without distinct sclerites, sometimes small; condylophore guide usually present, often reduced or absent where ambulacral sclerites are reduced; condylophores long and tendinous in lateral aspect, flexible. Tarsi lacking apical setae p and q; tarsus IV with one or two ventral setae.*

Nine families are accommodated in the Analgoidea. They occupy a wide variety of feather microhabitats, and also may be encountered under the skin or in the respiratory passages of their avian hosts. Some are associates of dipteran or mallophagan parasites of birds.

The ANALGIDAE (Plate 135, p. 433) includes over 20 genera of mites which occur in many feather niches of both wild and domestic birds. *Megninia cubitalis* (Mégnin) and *M. ginglymura* (Mégnin), along with other species of the genus, infest the feathers of domestic fowl, parakeets and pigeons. Although *M. cubitalis* commonly infests chickens in Europe and

Africa (Zumpt 1961), it is not considered an economic pest. The PROCTOPHYLLODIDAE (Plate 131, p. 429), and TROUESSARTIIDAE (Plate 132, p. 430) are analgoid families often found on passeriform birds. These elongate mites live in the grooves between the closely spaced feather barbules, either remaining motionless or moving slowly within the confines of their narrow niche. Disturbed mites are capable of crossing rapidly between grooves in a crab-like fashion.

Species of the families DERMOGLYPHIDAE (Plate 134, p. 432), GAUDOGLYPHIDAE, and APIONACARIDAE[1] (Plate 131, p. 429) inhabit the feather calmus of their various avian hosts. Dermoglyphids of the genera *Neumannella* and *Tinamoglyphus* occur only in the quills of tinamous. Males are distinctive in having a large spur on tibia IV which opposes tarsus IV, forming an uncate grasping organ. *Dermoglyphus elongatus* (Mégnin) (DERMOGLYPHIDAE) and *Gaudoglyphus minor* (Norner) (GAUDOGLYPHIDAE) invade the quills of domestic fowl and canaries (Baker et al. 1956, Evans et al. 1961, Bruce and Johnston 1976).

Unlike other analgoids, members of the family EPIDERMOPTIDAE (Plate 133, p. 431) are normally encountered on the skin of birds (Zumpt 1961, Fain 1965b and c), or on their hippoboscid or mallophagan ectoparasites. Some are parasitic on domestic fowl and may cause a variety of symptoms on their hosts. *Epidermoptes bilobatus* Rivolta, a common skin parasite of galliform birds, has been found under the skin pellicule (Fain and Evans 1963) and was reported to cause pityriasis in chickens (Baker et al. 1956). Similarly, an infestation of *E. odontophori* Fain and Evans on an African bird was accompanied by a generalized superficial mange on the body of the host. *Microlichus avus* (Trouessart) and *M. americanus* Fain produce crateriform skin lesions and invade the feather bulbs, causing severe mange in many of their numerous hosts. Species of *Myialges* embed themselves in the superficial skin layers and give rise to pityriasis or true mange. The mange symptoms of *M. (Promyialges) macdonaldi* Evans, Fain and Bafort may be quite severe, often resulting in loss of feathers (Fain 1965b).

Gravid females of the genera *Microlichus* and *Myialges* often leave their avian hosts and attach to hippoboscid flies (nine genera and approximately 30 species) or Mallophaga (three genera and seven species) which share the same hosts (Fain 1965b, Hill et al. 1967). Females of *Myialges (Promyialges) macdonaldi* are hyperparasitic on *Ornithomyia,* attaching to the fly's abdomen and feeding on its blood (Hill et al., op. cit.). *M. (P.) macdonaldi* and *Microlichus uncus* Vitzthum both have been observed to oviposit on their fly hosts, either on the abdomen or in wing axillaries. Dispersal of hatching larvae to new hosts may then occur with dispersal of the louse or fly carriers (Evans, Fain and Bafort 1963). Females of *Microlichus avus* (Trouessart) and of two species of *Strelkoviacarus* also are found on hippoboscids, but feeding and oviposition have not been verified. Lice of the families Menoponidae and Laemobothriidae serve as hosts for females of three species of *Myialges* (Fain, op. cit.).

Members of the epidermoptid subfamily Dermationinae lack a claw-like process on tarsus I, and thus are unable to break through the superficial skin layers of their hosts. This may account for the fact that dermationines are generally less destructive to host tissue than are other epidermoptids (Fain 1965b). Nevertheless, heavy infestations of the dermationine *Rivoltasia bifurcata* (Rivolta) on chickens may lead to intense itching and pityriasis, particularly around the head. Species of the dermationine genera *Dermation, Psittophagoides* and *Passeroptes* also may produce severe lesions on their hosts (Fain. op. cit.).

---

[1]Family OVACARIDAE of Gaud and Atyeo (1975), renamed by Atyeo and Gaud (1977). The generic name *Ovacarus* is preoccupied (Stannard and Vaishampayan 1971).

Mites of the family TURBINOPTIDAE (Plate 135, p. 433) live in the nasal fossae of birds in both the Old and New Worlds (Boyd 1949, Fain 1957, 1970b, Pence 1975). Turbinoptids often occur in large numbers in infested birds, but are not considered to be of economic importance.

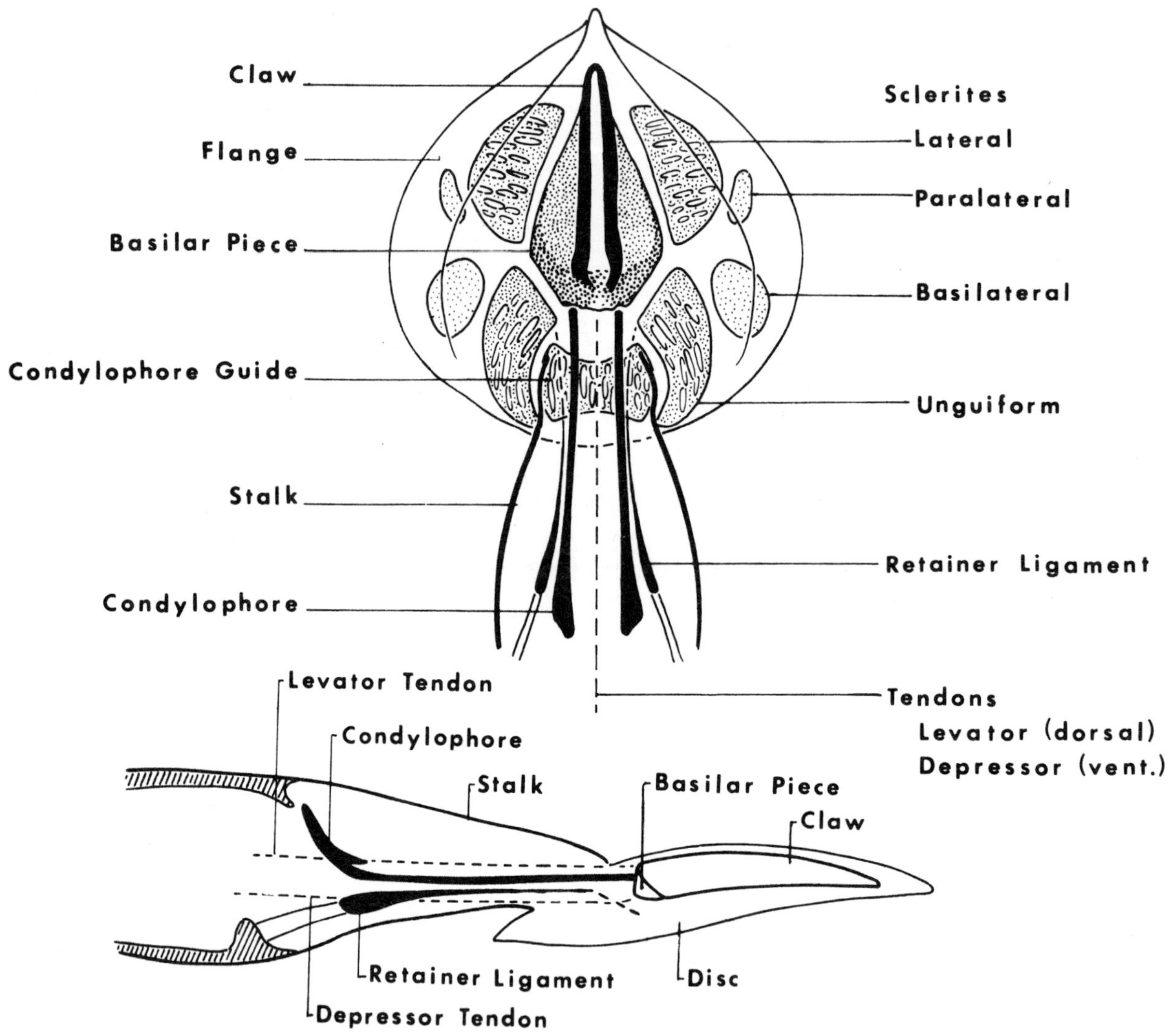

Fig. 43. Hypothetical feather mite pretarsus in dorsal (top) and lateral aspects. Dotted lines indicate uncertain structures (from Gaud and Atyeo 1977).

Superfamily Pterolichoidea
(Plates 137-142, pp. 435-440)

DIAGNOSIS: *With general feather mite characteristics (see page 388). Tarsal ambulacra with distinct sclerites, condylophore guide absent; condylophores "L"-shaped in lateral aspect, depression of condylophores results in partial retraction of ambulacra. Tarsi with apical setae p and q present, inserted ventrolaterally; tarsus IV with three ventral setae.*

The Pterolichoidea includes ten families, nine of which are external feather inhabitants of non-passeriform birds.[1] The SYRINGOBIIDAE (Plate 139, p. 437) live mainly in the quills of primary flight feathers. Syringobiids of the subfamily Ascouracarinae are remarkable in that they are extremely large (± 1000 μ), broad, well sclerotized mites with massive chelicerae (Gaud and Atyeo 1976). Known ascouracarine larvae, however, are narrow and elongate, more closely approaching the preconception of a "typical" quill mite.

The CRYPTUROPTIDAE (Plate 140, p. 438) comprises eight described genera which are confined to tinamiform hosts in the Neotropical realm (Gaud et al. 1972). Similar host group preference is noted in the elongate species of the family RECTIJANUIDAE (Plate 137, p. 435), which are known only from ducks (Atyeo and Peterson 1976). Rectijanuids appear to favor the feathers of the spinal tract rather than the wings, although *Rectijanua oxyurae* Atyeo and Peterson has been collected in considerable numbers from the flight feathers of its anatid host.

Most of the species included in the 13 presently recognized genera of GABUCINIIDAE (Plate 139, p. 437) are Neotropical, Oriental and Ethiopian forms which live on the feathers of assorted avian hosts. Most are of moderate size (350-450 μ), but one species, *Aetacarus phylloproctus* Mégnin and Trouessart, may exceed 700 μ in length (Gaud and Atyeo 1974). Members of the PTEROLICHIDAE (Plate 141, p. 439) are especially prevalent on galliform birds. *Pterolichus obtusus* Robin, a common associate of fowl in temperate realms, is little modified for a rigorous existence on feather surfaces. *Opisthocomacarus umbellifer* (Trouessart) and *Stakyonemus hystrix* (T.), on the other hand, appear to be morphologically adapted for survival on exposed feather sites subjected to great aerodynamic or hydrodynamic stress (Atyeo and Gaud 1971a). Yet their host, the hoatzin, is a sedentary gregarious galliform bird. It is assumed that the hoatzin has evolved to its present state from a highly active progenitor on which *O. umbellifer* and *S. hystrix* first acquired their remarkable setal characteristics.

Superfamily Freyanoidea
(Plates 136-137, pp. 434-435)

DIAGNOSIS: With general feather mite characteristics (see page 388). Tarsal ambulacra large, with distinct sclerites, condylophores rectangular in lateral aspect. Tarsi lacking apical setae p and q; tarsus IV with three ventral setae.

Two families are included in the Freyanoidea—the VEXILLARIIDAE and the FREYANIDAE.[2] Members of the FREYANIDAE are broad, well sclerotized species which favor exposed habitats on the primaries of large birds. The cormorant, ibis and spoonbill are typical freyanid hosts. Males of the genus *Freyanella* often illustrate bizarre asymmetry of terminal and humeral setae (Dubinin 1951, Atyeo et al. 1972) which allows them to wedge tightly between adjacent feather barbs. Vexillariids are encountered only on bucerotid birds.

[1] The OCHROLICHIDAE, a family of pterolichoid feather mites from passeriform birds, has been erected by Gaud and Atyeo (1977). It includes the genera *Ochrolichus* and *Corydolichus*.

[2] Gaud and Atyeo (1977) have erected a third family, the CAUDIFERIDAE, to accommodate the genera *Caudifera* and *Compressalges*.

Superfamily Listrophoroidea
(Plates 142-143, pp. 440-441)

DIAGNOSIS: *Soft-bodied but often with weakly sclerotized, striated or finely punctate dorsal shields, sejugal furrow usually absent; mouthparts, legs I-II, III-IV or coxal fields I-II modified for grasping or adhering to mammal hairs (males may have modified legs III, and normal legs IV without terminal suckers); normally developed legs each with a terminal stalked disc which occasionally has one or two tiny sclerotized crotchets. Oviporus an inverted "V", rarely transverse, with or without two pairs of greatly reduced acetabula; males with anal suckers.*

Listrophoroid mites are parasites of mammals, and are distinctive in that each family has a special morphological adaptation for securing itself to its host. Two of these families utilize ganthosomal structures as their primary means of attachment. Members of the LISTROPHORIDAE, for example, grasp hairs of their hosts between enlarged coxal extensions and the gnathosoma proper. The palpi of CHIRORHYNCHOBIIDAE are unsegmented and massive, and help anchor the harpoon-like chelicerae into the wing membranes of bats on which these mites are parasitic. Other listrophoroids rely primarily on leg modifications for purchase on their mammal hosts, although the chelicerae may also be adapted for anchoring in the host tissue.

The LISTROPHORIDAE (Plate 142, p. 440) is one of the larger listrophoroid assemblages and includes numerous Old and New World associates of rodents, lagomorphs, carnivores and insectivores. Fain and Hyland (1974) list seven genera and 23 species of listrophorids from North America alone. *Listrophorus* is considered the most highly evolved listrophorid genus occurring on rodents, and is common in Europe and North America. *Afrolistrophorus* and *Prolistrophorus* are widespread on rodents in sub-Saharan Africa and South America respectively (Fain 1971, 1973, Fain and Hyland, op. cit.). Lynxes and domestic cats may be infested by species of the listrophorid genus *Lynxacarus* (McDaniel 1968a, Tenorio 1974). Cats infested by *L. radovskyi* Tenorio in Hawaii developed mangy patches and a scruffy appearance. Fain and Hyland (op. cit.) also record the mink and chipmunk as hosts for *Lynxacarus* species.

The relict mountain beaver, *Aplodontia rufa rufa* is the sole host for the listrophorid *Aplodontochirus borealis* Fain and Hyland (Fain and Hyland 1972). The peculiar morphology of *A. borealis* prompted erection of a separate subfamily, the Aplodontochirinae, to contain it. Fain and Hyland consider *A. borealis* to be intermediate between more generalized listrophorids and the related fur mite family ATOPOMELIDAE (Plate 142, p. 440).

Atopomelids are flattened species which secure themselves to their hosts with modified tarsi, or tarsi and tibiae, I-II (McDaniel 1968a). The femora of legs I-II and the fused tibiotarsus of legs IV may be elongate and flattened so that the host hair may be grasped to the venter. Some species have corrugated coxal fields I-II to aid in maintaining position on the host. *Chirodiscoides caviae* Hirst, a common parasite of guinea pigs, is an example of the latter type. Members of the atopomelid genus *Listrophoroides* are more or less restricted to rodents (Fain 1970c) while *Austrochirus* species are found primarily on Australian marsupials (Domrow 1958). An interesting case of multiple speciation in the ATOPOMELIDAE was reported by Fain and Domrow (1974) in which 13 distinct species of *Cytostethum* were found on a single host, the potaroo *(Potorous tridactylus)*, in Australia. This phenomenon may be the result of microisolation due to skin or fur variations in the host.

Members of the CHIRODISCIDAE (Plate 142, p. 440) are distinctive in having strongly compressed, unsegmented legs I-II which grasp hairs in much the same fashion as does the palpal analogue in the LISTROPHORIDAE. Legs III-IV often are elongate and may terminate in strong hooklike claws. A retrorse dorsal capitular spur is present in many chirodiscids, presumably serving as an aid in preventing dislodgment from the pelage. Most chirodiscids are associated with bats (Pinichpongse 1963, Fain 1970d, 1973, Domrow and Moorhouse 1975). *Schizocoptes conjugatus* Lawrence is an exception, however, having been found on an African mole. *Chirodiscus amplexans* (Trouessart and Neumann) was collected from a bird, although McDaniel (1968b) and Domrow (1970) suggest that the avian association was accidental.

The MYOCOPTIDAE (Plates 142-143) are cosmopolitan fur mites which usually are found on rodents (Fain et al. 1970). *Myocoptes (M.) ictonyx* Fain is atypical, occurring in the fur of an African mustelid (Fain 1970e). The myocoptic mange mite, *Myocoptes musculinus* (Koch), produces a mangy condition in laboratory mice which can cause problems in animals with low resistance (Baker et al. 1956, Yunker 1973). Similar lesions have been observed on *Microtus agrestis* parasitized by *Trichoecius tenax* (Michael) (Böhm and Supperer 1958), and on squirrels infested by *Sciurocoptes sciurinus* (Hennemann) (Hennemann 1910). *T. romboutsi* (van Eyndhoven) may occur with *Myocoptes musculinus* on laboratory mice, but its pathologic effect on the host is unknown (Yunker, op. cit.). *Gliricoptes glirinus* (Canestrini), a parasite of the European dormouse, is unusual in having a pair of strongly sclerotized crotchets on the leading edge of the pretarsal suckers of legs I-II (Plate 142-7, p. 440).

Chirorhynchobia urodermae Fain and *C. matsoni* Yunker are the sole representatives of the listrophoroid family CHIRORHYNCHOBIIDAE (Plate 143, p. 441), parasites of South American bats (Fain 1967b, 1968a, Yunker 1970). As mentioned earlier, chirorhynchobiids attach to the wing membranes of their bat hosts with the aid of highly modified palpi and chelicerae. Few specimens of this bizarre group have been collected.

Superfamily Sarcoptoidea
(Plates 143-145, pp. 441-443)

DIAGNOSIS: Usually soft-bodied, commonly circular or sac-like, with or without a weakly defined prodorsal shield, paired longitudinal prodorsal sclerites sometimes present, sejugal furrow absent; chelicerae chelate, reduced, palpi simple. Legs telescoped, tarsi I-II generally terminating in stalked discs or spines, tarsi III and/or IV may terminate in long hairs, legs IV occasionally absent. Oviporus a transverse slit, acetabula absent; males without anal suckers.

Skin parasites of both mammals and birds are included in the Sarcoptoidea. Two of the three recognized families are of economic importance.

The SARCOPTIDAE (Plate 145, p. 443) live on or in the skin of many mammals, including man. *Sarcoptes scabiei* (L.) invades dermal tissues and causes a condition known as sarcoptic mange or scabies (Yunker 1973). Human scabies is characterized by erythematous patches at the sites of infestation, and by accompanying severe itching. Symptoms intensify with the continued burrowing and feeding of the mites beneath the skin, and are aggravated by the scratching reaction of the afflicted individual. Finally a weeping lesion forms over the affected area, hardening into a scab which may persist for some time. Transmission of human scabies appears to occur primarily through direct body contact rather than through contact with contaminated clothing or bed linens (Samšiňák et al. 1974). It may be highly contagious in crowded, unsanitary situations.

S. scabiei is a highly variable species which has long been considered a varietal complex. Based on a comprehensive study of morphological characters noted in specimens of *S. scabiei* from diverse hosts and regions, Fain (1968b) has identified five definitive forms of *S. scabiei* collected from 1) humans, 2) camels, 3) swine, 4) diverse African mammals (apes, goats, kudu and hartebeest), and 5) horses in South Africa and the United States. Intermediate and unclassifiable specimens were recovered from many other animals including cows, llamas, springboks, foxes, wild dogs, rabbits, coatimundi and tapirs. Sarcoptic mange occurs in many of these animals (Baker et al. 1956, Yunker, op. cit.), but the symptoms vary in severity. Heavy infestations in cattle, sheep or dogs occasionally result in death of the host animal.

Species of the sarcoptid genus *Prosarcoptes* are found in the skin of monkeys. *P. scanloni* Smiley is believed to cause depilation in its host (Smiley 1965). *Notoedres cati* (Hering), a common mange mite of domestic cats, burrows into the epidermis of the ears, neck and face of its host, and may attack the leg, genital and abdominal skin as well (Yunker 1973). *N. cati* produces a mangy condition similar to that caused by *Sarcoptes scabiei,* but symptoms are mainly confined to the head region of the host *(notoedric mange). N. cati* also infests rabbits, lynxes and coatimundi (Fain 1968b). *N. muris* (Mégnin), the ear mange mite of rats, causes mange in laboratory rodents. Other species of *Notoedres* also occur on rodents and one species, *N. galogoensis* Fain, parasitizes the primate *Galago demidovi pusillus.* Many *Notoedres* species are found on bats, as are representatives of most other sarcoptid genera. Bat-infesting SARCOPTIDAE are primarily tropical and are collected mainly from the buccal mucosa, ears and wing membranes of their hosts (Fain 1959b and c, 1965d, 1968b). Some may generate severe mange symptoms (Lavoipierre 1968). Sarcoptids of the genus *Diabolicoptes* are associated with marsupials in Tasmania (Fain and Domrow 1973). *D. sarcophilus* F. and D. was extracted from feces of the Tasmanian devil *(Sarcophilus ursinus).*

The KNEMIDOCOPTIDAE (Plates 143-144, pp. 441-442) are skin parasites of birds which produce characteristic and often serious disorders. *Knemidocoptes mutans* (Robin and Lanquetin), the scaly leg mite, burrows beneath the epidermal scales of the legs and feet of galliform birds, giving rise to inflammation and swelling which may finally cripple the host. Chickens often are attacked by *K. mutans,* and economic losses to producers may be significant. The depluming mite, *Mesoknemidocoptes laevis gallinae* (Railliet) (Dodd 1972), burrows into the skin at the base of the feathers of chickens, pheasants and geese, producing intense irritation and pruritis. Feathers in affected areas fall out, break off, or are picked out by the host (Yunker 1973). Pigeons are attacked in similar fashion by *M. laevis laevis* (Railliet) (Fain 1967c).

Various species of *Knemidocoptes* attack specific tissues of passeriform birds and often produce abnormalities in their hosts. *K. jamaicensis* Turk burrows into the legs of its host, *Turdus aurantiacus* (Turk 1950) while *K. fossor* (Ehlers), a closely related species, tunnels at the base of the bill. *K. pilae* Lavoipierre and Griffiths parasitizes psittaciform birds and may precipitate a scaly leg or scaly face condition (Lavoipierre and Griffiths 1951, Yunker 1973). *Evansacarus lari* Fain, an elongate hexapod knemidocoptid, was recovered from beneath the skin of the head of a gull, *Larus canus,* in Israel (Fain 1962a, 1967c).

The TEINOCOPTIDAE (Plate 144, p. 442) is a tropical group of bat parasites which often show considerable reduction in leg development and number (Fain 1962b). *Teinocoptes* and *Chirobia* species are found on or in the wing membrane skin of Megachiroptera, while *Bakerocoptes cynopteri* Fain, the type and only species of the genus, is found in subcutaneous pouches (in all stages) of the Malaysian bat *Cynopterus brachyotis angulatus.* The 16 species of *Teinocoptes* described by Fain (1967d) are confined to the bat family Pteropidae in the Old World tropics.

Superfamily Cytoditoidea
(Plates 145-146, pp. 443-444)

DIAGNOS *Soft-bodied but occasionally with weakly sclerotized podonotal, opisthonotal and ventral shields, with few body setae, rounded or elongate forms without a sejugal furrow; chelicerae chelate or greatly reduced, palpi simple. Legs normally developed, tarsi terminating in rounded or elongated membranous discs or claw-like spurs or spines. Oviporus a narrow inverted "V" or transverse, acetabula absent; males with or without anal suckers.*

The four families of Cytoditoidea are subdermal, respiratory or visceral parasites of birds, bats and rodents. Some are considered to be of veterinary importance.

Cytodites nudus (Vizioli), a representative of the family CYTODITIDAE (Plate 146, p. 444), invades the peritoneum of chickens, canaries and other birds (Baker et al. 1956), with primary infestations occurring in the air sacs. Severe respiratory complications result from heavy infestations, and death may occur (Yunker 1973). Other species of *Cytodites,* along with members of the genus *Cytonyssus,* have been taken from the nasal fossae, lungs, air sacs and body cavity of various birds (Fain 1960, Fain and Bafort 1964, Hyland 1969, Pence 1975). The species described as *Cytodites banksi* by Wellman and Wherry (1910) from ground squirrels is thought by Fain (op. cit.) to be a representative of *Pneumocoptes* (PNEUMOCOPTIDAE), a genus of rodent endoparasites. Two cytoditid specimens recovered from a peritoneal fold in an autopsied human in Uganda (Castellani 1907) were referred to Mégnin's *Cytoleichus sarcoptoides,* a name which has since been synonymized with *Cytodites nudus.*

The fowl cyst mite, *Laminosioptes cysticola* (Vizioli) (LAMINOSIOPTIDAE, Plate 145, p. 443) is a subcutaneous parasite in many birds, invading the skin of the neck, breast, flank, vent and thighs. Small cysts form around the mites, and often are visible as lumps under the skin of the infested bird (Lindquist and Belding 1949). The cysts calcify following death of the mites. Heavy infestations of *L. cysticola* may be fatal (Baker et al. 1956). A second species of *Laminosioptes, L. hymenopterus* Jones and Gaud, infests the skin and feathers of the eastern crow (Jones and Gaud 1962).

Mites of the family PNEUMOCOPTIDAE (Plate 145, p. 443) invade the respiratory tract of rodents and may occasionally be associated with an abnormal condition in host animals. For example, a prairie dog which had died of acute bronchopneumonia was found to be infested by *Pneumocoptes penrosei* (Weidman) (Weidman 1917). On the other hand, *P. jellisoni* Baker occupies the lung tissue of a species of *Peromyscus* (Baker 1951), but no damage to lung tissue has ever been noted. The GASTRONYSSIDAE (Plate 146, p. 444) includes eight genera of mites which infest the stomach, intestines, nasal cavities or eyes of bats and rodents. *Gastronyssus bakeri* Fain is a visceral parasite of African fruit bats (Fain 1955), while *Rodhainyssus* species invade the nasal fossae of Microchiroptera (Fain 1956, 1967e). *Opsonyssus* species are found in the nasal cavities and eyes of Old World and Asian Micro- and Megachiroptera (Pteropidae). The monotypic genera *Mycteronyssus* and *Eidolonyssus* are confined to African pteropids.

Yunkeracarus muris Fain and *Y. faini* Hyland and Clark are gastronyssids which parasitize rodents in Africa and in North America (Fain 1957a, Hyland and Clark 1959). A subspecies, *Y. faini apodemi,* has been collected from *Apodemus* in western Europe and in Korea (Fain et al. 1967). Another gastronyssid, *Sciuracarus paraxeri* Fain, was taken from the nasal fossae of a sciurid rodent in South Africa (Fain 1964c).

Useful References

Atyeo, W.T. (1966). A new genus and six new species of feather mites primarily from Tyranni (Acarina: Proctophyllodidae). J. Kansas Ent. Soc. **39**(3):481-492.

Atyeo, W.T. (1967). Two feather mite genera with polymorphic males (Analgoidea:Proctophyllodidae). J. Kansas Ent. Soc. **40**(4):465-471.

Atyeo, W.T. (1974). *Dogielacarus uncitibia* Dubinin, 1949, redescribed and reassigned (Acarina:Analgoidea). J. Kansas Ent. Soc. **47**(4):478-482. [XOLALGIDAE]

Atyeo, W.T. and N.L. Braasch (1966). The feather mite genus *Proctophyllodes* (Sarcoptiformes:Proctophyllodidae). Bull. Univ. Nebraska State Mus. **5**:1-354.

Atyeo, W.T. and J. Gaud (1966). The chaetotaxy of sarcoptiform feather mites (Acarina:Analgoidea). J. Kansas Ent. Soc. **39**(2):337-346.

Atyeo, W.T. and J. Gaud (1971a). Feather mites (Analgoidea:Pterolichidae) of the hoatzin (Aves:Galliformes). Amer. Midland Natur. **86**(1):152-159.

Atyeo, W.T. and J. Gaud (1971b). Comments on nomenclatural systems for idiosomal chaetotaxy of sarcoptiform mites. J. Kansas Ent. Soc. **44**(3):414-419.

Atyeo, W.T., J. Gaud and W.J. Humphreys (1972). The feather mite genus *Freyanella* Dubinin 1953 (Analgoidea:Pterolichidae). Acarologia **13**(2):383-409. [FREYANIDAE]

Atyeo, W.T. and P.C. Peterson (1967). Astigmata (Sarcoptiformes): Proctophyllodidae, Avenzoariidae (feather mites). Antarctic Res. Ser. **10**:97-103. [ALLOPTIDAE, AVENZOARIIDAE]

Atyeo, W.T. and P.C. Peterson (1972). The feather mite family Alloptidae Gaud, new status. I. The subfamilies Trouessartiinae Gaud and Thysanocercinae, new subfamily (Analgoidea). Zool. Anz. **188**(1/2): 56-60. [ALLOPTIDAE, TROUESSARTIIDAE]

Atyeo, W.T. and P.C. Peterson (1976). The species of the feather mite family Rectijanuidae (Acarina:Analgoidea). J. Georgia Ent. Soc. **11**(4):349-366.

Baker, E.W. (1951). *Pneumocoptes,* a new genus of lung-inhabiting mite from rodents (Acarina:Epidermoptidae). J. Parasitol. **37**(6):583-586. [PENUMOCOPTIDAE]

Baker, E.W., T.M. Evans, D.J. Gould, W.B. Hull and H.L. Keegan (1956). A Manual of Parasitic Mites of Medical or Economic Importance. Natl. Pest Control Assoc. Tech. Publ.: 170 pp.

Baker, E.W. and G.W. Wharton (1952). The Suborder Sarcoptiformes Reuter, 1909. **In** An Introduction to Acarology. Macmillan Co., New York: 320-386.

Böhm, L.K. and R. Supperer (1958). Über eine eigenartige Dermatose bei der Erdmaus *Microtus agrestis* L. durch *Myocoptes tenax* Michael 1889 und über den verschollenen *Myocoptes sciurinus* Hennemann 1910. Zeitschr. Parasitenk. **18**:223-229. [MYOCOPTIDAE]

Boyd, E.M. (1949). A new genus and species of mite from the nasal cavity of the ring-billed gull, (Acarina, Epidermoptidae). J. Parasit. **35**(3):295-300. [TURBINOPTIDAE]

Bronswijk, J.E.M.H. van and R.N. Sinha (1971). Pyroglyphid mites (Acari) and house dust allergy. J. Allergy **47**(1):31-52.

Bruce, W.A. and D.E. Johnston (1976). *Gaudoglyphus* n. gen., based on *Analges minor* Norner (Acari:Gaudoglyphidae n. fam.). Int. J. Acarology **2**(1):29-33.

Castellani, A. (1907). Note on an acarid-like parasite found in the omentum of a negro. Centralbl. Bakter. I. **43**(4):372. [CYTODITIDAE]

Dodd, K. (1972). The identity of *Knemidokoptes laevis* (Railliet, 1885) (Acari:Knemidokoptidae). Acarologia **14**(4):675-680.

Dogiel, V.A. (1949). The appearance of "concomitant species" in parasites and the evolutionary significance of these appearances. Akad. Nauk Kazakh. SSR 74 (parazitol.) **7**:3-15.

Domrow, R. (1958). A summary of the Atopomelinae (Acarina, Listrophoridae). Proc. Linn. Soc. N.S.W. **83**:40-54. [ATOPOMELIDAE]

Domrow, R. (1970). *Chirodiscus amplexans* Trouessart and Neumann redescribed (Acari:Listrophoridae). Acarologia **12**(2):415-420. [CHIRODISCIDAE]

Domrow, R. and D.E. Moorhouse (1975). Labidocarpine mites from bats in the Australian region (Acari: Chirodiscidae). J. Austral. Ent. Soc. **14**:107-112.

Dubinin, V.B. (1949). Features of the structure and biology of feather mites which inhabit the quills of feathers of terns. Zool. Zh. **28**:421-430.

Dubinin, V.B. (1951). Feather mites (Analgesoidea). Part I. Introduction to their study. Fauna U.S.S.R. **6**(5):1-363.

Dubinin, W.B. (1953). Feather mites (Analgesoidea). Part II. Epidermoptidae and Freyanidae. Fauna U.S.S.R. **6**(6):1-411.

Dubinin, W.B. (1956). Feather mites (Analgesoidea). Part III. Pterolichidae. Fauna U.S.S.R. **6**(7):1-814.

Evans, G.O., A. Fain and J. Bafort (1963). Découverte du cycle evolutif du genre *Myialges* avec description d'une éspece nouvelle (Myialgidae:Sarcoptiformes). Bull. Ann. roy. Ent. Belg. **99**:486-500. [EPIDERMOPTIDAE]

Evans, G.O., J.G. Sheals and D. Macfarlane (1961). Mites associated with vertebrates. **In** The Terrestrial Acari of the British Isles. Volume I. British Museum (Natural History), London: 132-165.

Fain, A. (1955). Un acarien remarquable vivant dans l'estomac d'une chauve-souris: *Gastronyssus bakeri* n.g., n. sp. Ann. Soc. Belge Med. Trop. **35**(6):681-688.

Fain, A. (1956). Une nouvelle famille d'acariens endoparasites des chauves-souris: Gastronyssidae fam. nov. Ann. Soc. Belge Med. Trop. **36**(1):87-98.

Fain, A. (1957a). Notes sur l'acariase des voies respiratoires chez l'homme et les animaux. Description de deux nouveaux Acariens chez un lemurien et des rongeurs. Ann. Soc. Belge Med. Trop. **37**(4):469-482. [LEMURNYSSIDAE, GASTRONYSSIDAE]

Fain, A. (1957b). Les Acariens des familles Epidermoptidae et Rhinonyssidae parasites des fosses nasales d'oiseaux au Ruanda-Urundi et au Congo belge. Ann. Mus. roy. Congo belge 8, **60**:176 pp. + xi. [EPIDERMOPTIDAE, TURBINOPTIDAE]

Fain, A. (1959a). Deux nouveaux acariens nasicoles chez un singe platyrhinien *Saimiri sciurea* (L.). Bull. Soc. roy. Zool. Anvers **12**:3-12. [LEMURNYSSIDAE]

Fain, A. (1959b). Les Acariens psoriques parasites des chauve-souris: II. *Chirnyssus myoticola* n.g., n. sp. parasite du murin *Myotis myotis* (Borkh) en Belgique. Acarologia **1**(1):119-123. [SARCOPTIDAE]

Fain, A. (1959c). Les Acariens psoriques parasites des chauves-souris: VI. Le genre *Prosopodectes* Canestrini 1897 est composite et doit tomber en synonymie de *Notoedres* Railliet 1893. Acarologia **1**(3): 324-353. [SARCOPTIDAE]

Fain, A. (1960). Revision du genre *Cytodites* (Megnin) et description de deux espèces et un genre nouveaux dans la famille Cytoditidae Oudemans. Acarologia **2**(2):238-249.

Fain, A. (1962a). Un acarien remarquable combinant les caracteres de plusieurs familles: *Evansacarus lari* n.g., n. sp. (Evansacaridae nov. fam.: Sarcoptiformes). Bull. Ann. Soc. Roy. Ent. Belg. **98**(9):125-145. [KNEMIDOCOPTIDAE]

Fain, A. (1962b). Les Acariens psoriques parasites des chauves-souris XXIII.—Un nouveau genre hexapode a tous les stades du développement (Teinocoptidae:Sarcoptiformes). Bull. Ann. Soc. Roy. Ent. Belg. **98**(28):404-412.

Fain, A. (1964a). *Edentalges choloepi* sp. n., acarien parasite cuticole du paresseux didactyle *Choloepus didactylus* (L.) (Psoroptidae:Sarcoptiformes). Z. Parasitenk. **25**:103-107.

Fain, A. (1964b). Les Lemurnyssidae parasites nasicoles des Lorisidae africains et des Cebidae sudaméricains. Description d'une espèce nouvelle (Acarina:Sarcoptiformes). Ann. Soc. Belge Med. Trop. **44**(3):453-458.

Fain, A. (1964c). Chaetotaxie et classification des Gastronyssidae avec description d'un nouveau genre parasite nasicole d'un Ecureuil sudafricain (Acarina:Sarcoptiformes). Rev. Zool. Bot. Afr. **70**(1-2):40-52.

Fain, A. (1964d). Le développement postembryonnaire chez les Acaridiae parasites cutanés des mammifères et des oiseaux (Acarina:Sarcoptiformes). Bull. Classe Sci., Acad. roy. Belg. 5, **50**:19-34.

Fain, A. (1965a). Les acariens producteurs de gale chez les edentés et les marsupiaux (Psoroptidae et Lobalgidae Sarcoptiformes). Bull. Inst. roy. Sci. nat. Belg. **41**(17):1-42.

Fain, A. (1965b). A review of the family Epidermoptidae Trouessart parasitic on the skin of birds. Parts I-II. Kon. Acad. Wetensch., Lett. Schone Kunsten Belg. (Wetensch.) **27**(84):1-176 + ix; 5-144.

Fain, A. (1965c). A review of the family Rhyncoptidae Lawrence parasitic on porcupines and monkeys. Advances in Acarology **2**:135-159.

Fain, A. (1965d). Notes sur le genre *Notoedres* Railliet, 1893 (Sarcoptidae:Sarcoptiformes). Acarologia **7**(2):321-342.

Fain, A. (1965e). Les Acariens nidicoles et détriticoles de la famille Pyroglyphidae Cunliffe (Sarcoptiformes). Rev. Zool. Bot. Afr. **72**(3-4):257-288.

Fain, A. (1966). Nouvelle description of *Dermatophagoides pteronyssinus* (Trouessart 1897). Importance de cet acarien en pathologie humaine. Acarologia **8**(2):302-327. [PYROGLYPHIDAE]

Fain, A. (1967a). Le genre *Dermatophagoides* Bogdanov 1864. Son importance dans les allergies respiratoires et cutanées chez l'homme (Psoroptidae:Sarcoptiformes). Acarologia **9**(1):179-225. [PYROGLYPHIDAE]

Fain, A. (1967b). Diagnoses d'Acariens Sarcoptiformes nouveaux. Rev. Zool. Bot. Afr. **75**(3-4):378-382. [CHIRORHYNCHOBIIDAE]

Fain, A. (1967c). Les acariens de la famille Knemidokoptidae producteurs de gale chez les oiseaux (Sarcoptiformes). Acta Zool. Path. Antverp. 45:3-145.

Fain, A. (1967d). Les acariens psoriques parasites des chauves-souris XXVIII. *Teinocoptes ituriensis* sp. n., avec une clé et une liste des espèces du genre *Teinocoptes* (Teinocoptidae:Sarcoptiformes). Rev. Zool. Bot. Afr. **75**(3-4):363-368.

Fain, A. (1967e). Observations sur les Rodhainyssinae acariens parasites des voies respiratoires des chauves-souris (Gastronyssidae:Sarcoptiformes). Acta Zool. Path. Antverp. 44:3-35.

Fain, A. (1968a). Notes sur trois acariens remarquables (Sarcoptiformes). Acarologia **10**(2):276-291. [AUDYCOPTIDAE, CHIRORHYNCHOBIIDAE]

Fain, A. (1968b). Etude de la variabilite de *Sarcoptes scabiei* avec une revision des Sarcoptidae. Acta Zool. Path. Antverp. 47:1-196.

Fain, A. (1968c). Deux nouveaux acariens cavernicoles du Gabon (Sarcoptiformes). Biol. Gabon. **4**(2):195-205. [GUANOLICHIDAE]

Fain, A. (1969a). Adaptation to parasitism in mites. Acarologia **11**(3):429-449.

Fain, A. (1969b). Some aspects of mange in man and domestic animals. Ent. Ber. VI, 29(1):103.

Fain, A. (1970a). Sur une nouvelle espèce du genre *Psoroptes* produisant la gale chez le buffle africain (Acarina:Sarcoptiformes). Rev. Zool. Bot. Afr. **81**(1-2):95-100. [PSOROPTIDAE]

Fain, A. (1970b). Nouveaux acariens nasicoles de la famille Turbinoptidae (Sarcoptiformes). Bull. Ann. Soc. Roy. Ent. Belg. **106**:28-36.

Fain, A. (1970c). Diagnoses de nouveaux listrophorides de la famille Atopomelidae (Acarina:Sarcoptiformes). Bull. Ann. Roy. Ent. Belg. **106**:275-306.

Fain, A. (1970d). Notes sur quelques nouveaux taxa de la famille Chirodiscidae (Acarina:Sarcoptiformes). Rev. Zool. Bot. Afr. **82**(3-4):280-284.

Fain, A. (1970e). Les Myocoptidae en Afrique au sud du Sahara (Acarina:Sarcoptiformes). Ann. Mus. Roy. Afr. Centr. 8(179):67 pp.

Fain, A. (1971). Les listrophorides en Afrique au sud du Sahara (Acarina:Sarcoptiformes). II. Familles Listrophoridae et Chirodiscidae. Acta Zool. Path. Antverp. 54:1-231.

Fain, A. (1973). Les listrophoridés d'Amerique Neotropicale (Acarina:Sarcoptiformes). I. Familles Listrophoridae et Chirodiscidae. Bull. Inst. roy. Sci. natur. Belg. Ent. 49:1-149.

Fain, A. and J. Bafort (1964). Les Acariens de la famille Cytoditidae (Sarcoptiformes). Description de sept espèces nouvelles. Acarologia **6**(3):504-528.

Fain, A. and R. Domrow (1973). Two new parasitic mites (Acari:Sarcoptidae and Atopomelidae) from Tasmanian marsupials. Proc. Linn. Soc. N.S.W. **98**(3):122-130.

Fain, A. and R. Domrow (1974). The subgenus *Metacytostethum* Fain (Acari:Atopomelidae): parasites of macropodid marsupials. Acarologia **16**(4):719-738.

Fain, A. and G.O. Evans (1963). On three mites of the genus *Epidermoptes* Rivolta (Acari). Ann. Mag. Nat. Hist. 13, **6**:595-608.

Fain, A. and K. Hyland (1972). Description of new parasitic mites from North American mammals (Acarina: Sarcoptiformes). Rev. Zool. Bot. Afr. **85**(1-2):174-176. [LISTROPHORIDAE]

Fain, A. and K. Hyland (1974). The listrophoroid mites of North America II. The family Listrophoridae. Bull. Inst. roy. Sci. natur. Belg. Ent. **50**(1):1-69.

Fain, A. and D.E. Johnston (1970). Un nouvel Acarien de la famille Audycoptidae chez l'ours noir *Ursus americanus* (Sarcoptiformes). Acta Zool. Path. Antverp. 50:179-181.

Fain, A. and M. Leclerc (1975). Sur un cas de gale chez le panda géant produit par une nouvelle espèce du genre *Chorioptes* (Acarina:Psoroptidae). Acarologia **17**(1):177-182.

Fain, A. and F. Lukoschus (1970). Parasitic mites of Suriname II. Skin and fur mites of the families Psoroptidae and Lobalgidae. Acta Zool. Path. Antverp. 51:49-60.

Fain, A., F. Lukoschus, J.M. Jadin and H.S. Ah (1967). Note sur un acarien du genre *Yunkeracarus* Fain, 1957 (Gastronyssidae:Sarcoptiformes). Acta Zool. Path. Antverp. 43:79-83.

Fain, A., A.J. Munting and F. Lukoschus (1970). Les Myocoptidae parasites des rongeurs en Hollande et en Belgique. Acta Zool. Path. Antverp. 50:67-172.

Fischer, A.A., A.G. Franks, M. Wolf and M. Leider (1951). Concurrent infestation with a rare mite and infection with a common dermatophyte. Arch. Dermatol. Syphilol. **63**(3):336-342. [PYROGLYPHIDAE]

da Fonseca, F. (1954). Notas de Acarologia XXXIX. Sistematica e filogenese de Psoralgidae Oudemans, Sarcoptiformes parafagistas de mamiferos com morphologia de Acari plumicolas. Mem. Inst. Butantan 26:92-167. [PSOROPTIDAE, LOBALGIDAE]

Gaud, J. (1968). Acariens de la sous-famille des Dermatophagoidinae (Psoroptidae) récoltés dans les plumages d'oiseaux. Acarologia **10**(2):292-312. [PYROGLYPHIDAE]

Gaud, J. and W.T. Atyeo (1974). Gabuciniidae, famille nouvelle de sarcoptiformes plumicoles. Acarologia **16**(3):522-561.

Gaud, J. and W.T. Atyeo (1975). Ovacaridae, une famille nouvelle de sarcoptiformes plumicoles. Acarologia **17**(1):169-176. [APIONACARIDAE]

Gaud, J. and W.T. Atyeo (1976). Ascouracarinae, n. sub-fam. des Syringobiidae, sarcoptiformes plumicoles. Acarologia **18**(1):143-162.

Gaud, J. and W.T. Atyeo (1977). A new name for *Ovarcarus* and Ovacaridae (Acarina:Analgoidea). Acarologia **18**(3):568-569. [APIONACARIDAE]

Gaud, J. and W.T. Atyeo (1977). Nouvelles superfamilles pour les acariens astigmates parasites d'oiseaux. Acarologia **19**(4):678-685.

Gaud, J., W.T. Atyeo and H.F. Berla (1972). Acariens sarcoptiformes plumicoles parasites des tinamou. Acarologia **14**(3):393-453. [ANALGIDAE, CRYPTUROPTIDAE, DERMOGLYPHIDAE]

Gaud, J. and J. Mouchet (1957). Acariens plumicoles (Analgesoidea) des oiseaux du Cameroun. I. Proctophyllodidae. Ann. Parasit. Hum. Comp. **32**(5-6):491-546.

Griffiths, D.A. and A.M. Cunnington (1971). *Dermatophagoides microceras* sp. n.: a description and comparison with its sibling species, *D. farinae* Hughes, 1961. J. Stored Prod. Res. **7**:1-14. [PYROGLYPHIDAE]

Hennemann, J.H. (1910). Uber eine noch nicht beschriebene Myocoptesräude. Oesterr. Mschr. Tierheilk. **34**:337-354. [MYOCOPTIDAE]

Hill, D.S., N. Wilson and G.B. Corbet (1967). Mites associated with British species of *Ornithomyia* (Diptera: Hippoboscidae). J. Med. Ent. **4**(2):102-122. [EPIDERMOPTIDAE]

Hirst, S. (1922). Mites injurious to domestic animals. Brit. Mus. (Nat. Hist.) Econ. Ser. 13:107 pp.

Hyland, K.E. (1969). *Cytodites therae* n. sp. from the respiratory passages of the black-billed cuckoo in North America (Sarcoptiformes:Cytoditidae). Acarologia **11**(3):597-601.

Hyland, K.E., Jr. and D.T. Clark (1959). The occurrence of the genus *Yunkeracarus* in North America (Acarina:Epidermoptidae). Acarologia **1**(3):365-369. [GASTRONYSSIDAE]

Jones, J., Jr. and J. Gaud (1962). The description of *Laminosioptes hymenopterus* n. sp. (Sarcoptiformes) from the American crow. Acarologia **4**(3):391-395. [LAMINOSIOPTIDAE]

Keh, B. (1963). The common house dust mites of the genus *Dermatophagoides* (Acarina:Pyroglyphidae). California Vector Views **20**(5):37-45.

Kurosa, K. (1976). Notes on Tarsonemini and Acaridei (Acari) from Pasoh Forest Reserve and its vicinities, Malaya. Nature and Life S.E. Asia **7**:61-81. [PLATYGLYPHIDAE]

Land, J.D., L.D. Charlet and M.S. Mulla (1976). Bibliography (1864 to 1974) of house-dust mites *Dermatophagoides* species (Acarina:Pyroglyphidae) and human allergy. J. Sci. Biol. **2**(2):62-83.

Lavoipierre, M.M.J. (1959). A description of the male and female of *Chorioptes crewei* Lavoipierre 1958 (Acarina:Psoroptidae), together with some remarks on the family Psoroptidae and a key to the genera contained in the family. Acarologia **1**(3):354-364.

Lavoipierre, M.M.J. (1964). A new family of Acarines belonging to the Suborder Sarcoptiformes parasitic in the hair follicles of Primates. Ann. Natal. Mus. **16**:191-208. [AUDYCOPTIDAE]

Lavoipierre, M.M.J. (1968). Notes on mange mites of the genus *Notoedres* (Acarina:Sarcoptidae) from mammals of eastern Asia, with a description of four new species. J. Med. Ent. **5**(3):313-319.

Lavoipierre, M.M.J. and R.B. Griffiths (1951). A preliminary note on a new species of *Cnemidocoptes* (Acarina) causing scaly-leg in a budgerigar *(Melopsittacus undulatus)* in Great Britain. Ann. Trop. Med. Parasit. **45**(3-4):253-254. [KNEMIDOCOPTIDAE]

Lawrence, R.F. (1956). Studies on South African fur mites (Trombidiformes and Sarcoptiformes). Ann. Natal Mus. **13**:337-375. [RHYNCOPTIDAE]

Lawrence, R.F. (1958). Studies on the listrophorid fur-mites of Madagascar lemurs. Mem. Inst. Sci. Madagascar **12A**:113-125. [ATOPOMELIDAE]

Lindquist, W.D. and R.C. Belding (1949). A report on the subcutaneous or flesh mite of chickens. Michigan State Coll. Vet. **10**:20-21. [LAMINOSIOPTIDAE]

Lukoschus, F.S. and J.G.J.H. Rouwet (1968). *Mycoptes ondatrae* spec. nov., ein neuer parasit von *Ondatra zibethica* L. (Listrophoridae:Sarcoptiformes). Acarologia **10**(3):483-492. [MYOCOPTIDAE]

McDaniel, B. (1968a). The superfamily Listrophoroidea and the establishment of some new families (Listrophoroidea:Acarina). Acarologia **10**(3):477-482. [LISTROPHORIDAE, MYOCOPTIDAE, RHYNCOPTIDAE]

McDaniel, B. (1968b). The genus *Chirodiscus* Trouessart and Neumann with lectotype designation of *C. amplexans* Trouessart and Neumann (Listrophoroidea:Atopomelidae). Acarologia **10**(4):653-656. [CHIRODISCIDAE]

Miyamoto, T. (1972). House dust and house dust mites. **In** Allergic Diseases, Diagnosis and Treatment (R. Patterson, ed.). J.B. Lippincott Co., Philadelphia: 114-123.

Park, C.K. and W.T. Atyeo (1974a). The pterodectine feather mites of hummingbirds: the genus *Toxerodectes* Park and Atyeo (the *lecroyae* and *gladiger* groups). J. Georgia Ent. Soc. **9**(1):18-32. [PROCTOPHYLLODIDAE]

Park, C.K. and W.T. Atyeo (1974b). The pterodectine feather mites of hummingbirds: the genus *Trochilodectes* Park and Atyeo. J. Georgia Ent. Soc. **9**(3):156-173. [PROCTOPHYLLODIDAE]

Pence, D.B. (1975). Keys, species and host list, and bibliography for nasal mites of North American birds (Acarina:Rhinonyssinae, Turbinoptinae, Speleognathinae, and Cytoditidae). Texas Tech. Univ. Spec. Publ. Mus. 8:148 pp.

Peterson, P.C. (1971). A revision of the feather mite genus *Brephosceles* (Proctophyllodidae:Alloptinae). Bull. Univ. Nebraska State Mus. **9**(4):89-172. [ALLOPTIDAE]

Pinichpongse, S. (1963). A review of the Chirodiscinae with descriptions of new taxa (Acarina:Listrophoridae). Acarologia **5**(1):81-91, **5**(2):266-278, **5**(3):397-404, **5**(4):620-627. [CHIRODISCIDAE]

Poppe, E. (1967). Die Begattung bei den Vogelmilben *Pterodectes* Robin (Analgesoidea, Acari). Z. Morph. Ökol. Tiere **59**:1-32. [PROCTOPHYLLODIDAE]

Radford, C.D. (1958). The host-parasite relationships of the feather mites (Acarina:Analgesoidea). Rev. Brasil. Ent. **8**:107-170.

Roberts, I.H. and W.P. Meleney (1970). Variations among strains of *Psoroptes ovis* (Acarina:Psoroptidae) on sheep and cattle. Ann. Ent. Soc. Amer. **64**(1):109-116.

Samšiňák, K.. E. Vorbrázkova, L. Mališ, P. Palička and K. Zítek (1974). To the possible spread of scabies through bed linen. Fol. Parasitol. **21**:89-91. [SARCOPTIDAE]

Santana, F.J. (1976). A review of the genus *Trouessartia* (Analgoidea:Alloptidae). J. Med. Ent. Suppl. 1:1-128. [TROUESSARTIIDAE]

Smiley, R.L. (1965). A new mite, *Prosarcoptes scanloni,* from monkey. Proc. Ent. Soc. Wash. **67**:166-167. [SARCOPTIDAE]

Southcott, R.V. (1976). Arachnidism and allied syndromes in the Australian region. Rec. Adelaide Children's Hospital **1**(1):97-186.

Spieksma, F.Th.M. and M.I.A. Spieksma-Boezeman (1967). The mite fauna of house dust with particular reference to the house-dust mite *Dermatophagoides pteronyssinus* (Trouessart, 1897) (Psoroptidae: Sarcoptiformes). Acarologia **9**(1):226-241. [PYROGLYPHIDAE]

Stannard, L.J. and S.M. Vaishampayan (1971). *Ovacarus clivinae,* new genus and species (Acarina:Podapolipidae), and endoparasite of the slender seedcorn beetle. Ann. Ent. Soc. Amer. **64**:887-890.

Sweatman, G.K. (1957). Life history, non-specificity and revision of the genus *Chorioptes,* a parasitic mite of herbivores. Can. J. Zool. **35**:641-689. [PSOROPTIDAE]

Sweatman, G.K. (1958). On the life history and validity of the species in *Psoroptes,* a genus of mange mites. Can. J. Zool. **36**:905-929. [PSOROPTIDAE]

Tenorio, J.M. (1974). A new species of *Lynxacarus* (Acarina:Astigmata:Listrophoridae) from *Felis catus* in the Hawaiian Islands. J. Med. Ent. **11**(5):599-604.

Traver, J. (1951). Unusual scalp dermatitis in humans caused by the mite *Dermatophagoides.* Proc. Ent. Soc. Wash. **53**(1):1-25. [PYROGLYPHIDAE]

Turk, F.A. (1950). A new species of parasitic mite, *Cnemidocoptes jamaicensis,* a causative agent of scaly-leg in *Turdus aurantiacus.* Parasit. **40**:60-62. [KNEMIDOCOPTIDAE]

Voorhorst, R., M.I.A. Spieksma-Boezeman and F.Th.M. Spieksma (1964). Is a mite (*Dermatophagoides* sp.) the producer of the house dust allergen? Allerg. Asthma **10**:329-334.

Watson, D.P. (1960). On the adult and immature stages of *Myocoptes musculinus* (Koch) with notes on its biology and classification. Acarologia **2**(3):335-344. [MYOCOPTIDAE]

Weidman, F.D. (1917). *Cytoleichus penrosei,* a new arachnoid parasite found in the diseased lungs of a prairie dog. *Cynomys ludovicianus.* J. Parasit. **3**:82-89. [CYTODITIDAE]

Wellman, C. and W.B. Wherry (1910). Some new internal parasites of the California Ground Squirrel *(Otospermophilus beecheyi).* Parasit. **3**:417-422. [CYTODITIDAE]

Wharton, G.W. (1976). House dust mites. J. Med. Ent. **12**(6):577-621. [PYROGLYPHIDAE]

Whitaker, J.O., Jr. and N. Wilson (1974). Host and distribution lists of mites (Acari), parasitic and phoretic, in the hair of wild mammals of North America, north of Mexico. Amer. Midl. Natur. **91**(1):1-67.

Yunker, C.E. (1970) A second species of the unique family Chirorhynchobiidae Fain, 1967 (Acarina:Sarcoptiformes). J. Parasitol. **56**(1):151-153.

Yunker, C.E. (1973). Mites. **In** Parasites of Laboratory Animals. R.J. Flynn, ed. Iowa State Univ. Press, Ames: 425-492.

Zumpt, F. (1961). The Arthropod Parasites of Vertebrates in Africa South of the Sahara (Ethiopian Region). Vol. (Chelicerata). Publ. S. Afr. Inst. Med. Res. **9**(1):1-457.

SUBORDER ACARIDIDA
(Plates 116-146, pp. 414-444)

KEY TO THE FAMILIES[1]

1. Tibia I with two setae or none,[2] dorsal solenidion (ϕ) generally distinct (Fig. 38a, p. 373); ambulacral pulvillus and empodial claw present on at least some of the tarsi (Plate 116-7, p. 414), pulvilli occasionally reduced but rarely absent. With two pairs of well developed genital acetabula[3] or scleritic ring-like structures (Plates 116-1, 126-1). Deutonymph may be heteromorphic, occasionally adventitious (hypopus, Plate 115c, p. 372), on vertebrates or invertebrates; other stages free-living or nidicolous, sometimes associated with other animals. . . . . .Supercohort ACARIDIDES ... 2

— Tibia I with only one seta plus the dorsal solenidion ϕ (Fig. 38b, p. 373); tarsi with membranous ambulacral discs or with claw-like setae or spines (if well defined empodial claws are present, then they are sessile and without pulvilli). Genital acetabula absent or greatly reduced (Plate 129-1, p. 427). Without a heteromorphic hypopodial stage. Parasitic or paraphagic on mammals, birds, crustaceans or (rarely) on insects, occasionally nidicolous or free-living . . . . .Supercohort PSOROPTIDES ... 16

2. Female oviporus a transverse slit bordered laterally by a pair of large sclerotized rounded or ellipsoid ring-like structures, with a second pair in the region of coxal fields IV or oviporus. Palpi each with terminal segment oriented laterally, membranous and usually with distal flagella. Pretarsi sessile, reduced. Free-living. . . . . Superfamily ANOETOIDEA (Plates 125-126, pp. 423-424) Family ANOETIDAE

— Oviporus longitudinal, with two pairs of acetabula bordering the aperture. Palpi various, but not as above. Pretarsi generally stalked, occasionally sessile and reduced or represented by sessile empodial claws 3

3. Tibia I with a dorsal solenidion (ϕ), setae absent; pulvilli sucker-like, although a weak claw may be present. Associated with insects. . . . . Superfamily CANESTRINIOIDEA ... 5

— Tibia I with two ventral setae and dorsal solenidion ϕ; pulvilli rarely sucker-like, sometimes absent or reduced, empodial claws generally distinct. Free-living, associates of vertebrates or invertebrates, or nidicolous 4

4. Tarsi without pulvilli; with strong sessile empodial claws or knob-like projections on tarsi I-II, and greatly enlarged claws III-IV which oppose ventral tibiotarsal projections to form uncate processes. Associates of pagurid crabs . . . . Superfamily EWINGIOIDEA (Plate 126, p. 424) Family EWINGIIDAE

---

[1]Hypopodes have been treated in a separate key (page 412).

[2]One seta occurs on tibia I of certain species of ROSENSTEINIIDAE, paraphages of bats (Plate 125, p. 423). Genital acetabula may be greatly reduced or absent in some rosensteiniids, further suggesting that the family may more appropriately be included with the Psoroptides (couplet 1b). The occurrence of a single seta on tibia I of the HEMISARCOPTIDAE (Plate 127, p. 425), however, is considered exceptional since the family otherwise exhibits strong affinities with the Acaridides.

[3]Reduction in genital acetabular development occurs not only in the ROSENSTEINIIDAE (see footnote 2), but also in females of the EWINGIIDAE (Plate 126, p. 424) and in adults of the HYADESIIDAE (Plate 125, p. 423).

— Empodial claws usually present and often strong, always with pulvilli (may be sucker-like in certain GLYCYPHAGIDAE); tarsi III-IV never uncate as above. Free-living, nidicolous, or associated with vertebrates or invertebrates . . . . Superfamily ACAROIDEA[1] ... 6

5. Tarsi with narrow stalked sucker-like pulvilli, claws absent. Female oviporus and anal opening contiguous, terminal; male without anal suckers. Predators of scale insects . . . . (Plate 127, p. 425) Family HEMISARCOPTIDAE

— Tarsi with large flattened pulvilli on short stalks, empodial claws present but often small. Found on insects . . . . . (Plates 126-127, pp. 424-425) Family CANESTRINIIDAE

6. With paired membranous pseudorutellar lobes or lobe remnants associated with the hypostome or with the palpal bases. Typically with three solenidia on tarsus I, two of which are inserted distally; pulvilli sucker-like. Dorsum typically rugose, tuberculate or strongly striate; with at least some dorsal setae pectinate, scale-like or otherwise ornamented. Associated with bats . . . . . (Plate 125, p. 423) Family ROSENSTEINIIDAE

— Pseudorutellar processes absent; generally with one or no distal solenidia on tarsus I, pulvilli rarely appearing sucker-like. Dorsum and dorsal setae variable 7

7. Oviporus covered by a pair of lateral paragynial shields which form a crescent, abutted posteriorly by an epigynial flap which is fused to the venter at its posterior margin. Idiosoma entirely sclerotized, weakly tanned; dorsal setae simple, minute. Free-living . . . . (Plate 123, p. 421) Family CHORTOGLYPHIDAE

— Oviporus with or without discrete shields but not forming a crescentic complex. Idiosomal sclerotization and setae variable 8

8. Each empodial claw articulates basally with a pair of short, broad, sclerotized pretarsal condylophores (Plate 116-7, p. 414) and is surrounded by a well developed fleshy pulvillus. Prodorsal shield present but sometimes indistinct, with a pair of posterolateral opisthonotal glands. Males with both anal and tarsal suckers. Mostly opaque, broad, slow moving mites in a variety of free-living habitats 9

— Condylophores attenuate, often indistinct, forming elongate connectives between empodial elements and tarsal extremities; other of the aforementioned characters may be present or absent 10

9. Idiosoma without a sejugal furrow; genital acetabula three-segmented. Small (± 300 μ), circular flattened species with strongly abbreviated opisthosoma. In bee nest. Monotypic . . . . (Plate 124, p. 422) Family PLATYGLYPHIDAE

— Idiosoma with a distinct sejugal furrow, genital acetabula each with no more than two segments. Small to large species, typically ovoid and with developed opisthosoma . . . . (Plates 116-117, pp. 414-415) Family ACARIDAE

10. Tarsi I-II pointed terminally, claw-like, pretarsi I-II each consisting of a large empodial claw and expanded pulvillus generally on a strong pretarsal stalk, joined to tarsi subterminally. Marine intertidal and freshwater algivores. . . . . (Plate 125, p. 423) Family HYADESIIDAE

---

[1]*Heterocoptes tarsii* (Fain), the only representative of the family HETEROCOPTIDAE (Plate 124, p. 422), is known only from a single male specimen (Fain 1967d). It has tentatively been assigned to the Acaroidea (page 375), but is not included in this key.

— Tarsi not as above, ambulacra I-II variable but typically arising from the distal extremities of the tarsi. Terrestrial species, occasionally in moist habitats 11

11. Idiosoma fusiform, smooth, completely sclerotized and strongly tanned; dorsal setae minute (setae *sc i* are exceptional in being elongate). In nests of rodents. . . . . (Plates 122-123, pp. 420-421) Family FUSACARIDAE

— Idiosoma not as above, often with only a prodorsal shield (if completely sclerotized, then not fusiform and with more than one pair of macrosetae dorsally) 12

12. With a sejugal furrow; opisthonotal glands generally present, often distinctive. Small (250-400 μ), weakly sclerotized mites associated with insects or inhabiting a variety of free-living habitats. . . . . (Plate 122, p. 420) Family SAPROGLYPHIDAE

— Sejugal furrow absent; opisthonotal glands variously developed, sometimes absent. Small or larger species in various free-living or nidicolous habitats 13

13. Small ($\pm$ 200 μ), ovoid; tarsi with stalked, discoid ambulacra, empodial claws absent, tarsus I without solenidion ω_2. Opisthonotal glands absent. Associated with stingless bees. Monotypic. . . . . (Plate 124, p. 422) Family GAUDIELLIDAE

— Larger species; tarsi with empodial claws and pulvilli (if ambulacra are discoid, then the body length exceeds 500 μ, and solenidion ω_2 is present on tarsus I). 14

14. Claws distinctly larger than pulvilli, often half the length of the tarsi. Associated with Hymenoptera . . . . (Plates 120-121, pp. 418-419) Family CHAETODACTYLIDAE

— Claws small or well developed, but no longer than associated pulvillus in either case . . 15

15. Coxal epimera I-II thickened and fused medially to an epigynium which borders the anterior margin of the oviporal field; male genital opening between or behind coxae IV. Tarsi I-II no more than twice the length of the adjacent tibiae. Often found in processed foods. . . . .(Plate 123, p. 421) Family CARPOGLYPHIDAE

— Coxal epimera I-II variously developed but not as above (if an epimeral-epigynial sclerite complex is present, then it extends posteriorly beyond coxal epimera II, and/or the legs are ornamented with distinctive longitudinal ribs). Tarsi I-II often three or more times the length of the adjacent tibiae. A large, diverse group found in litter, stored grain and grain products, in the nests of mammals and birds, and as insect paraphages. . . . .(Plates 117-120, pp. 415-418) Family GLYCYPHAGIDAE

16. Females (and most males) rounded or sac-like, rarely elongate; oviporus a simple transverse slit without epigynium, anterior to coxal epimera III; acetabula absent. Legs generally short, telescoped, tarsi I-II terminate in stalked ambulacral discs or spines, tarsi III and/or IV may terminate in long hairs, legs IV occasionally absent. Skin parasites of vertebrates . . . . Superfamily SARCOPTOIDEA ... 49

— Typically elongate or ovate; oviporus various, occasionally transverse, diminuitive genital acetabula often present; epigynium usually present, sometimes reduced. Legs occasionally telescoped, legs I-II may be compressed as grasping structures; with terminal or subterminal discs on those legs which are normally developed 17

17. Legs I-II, III-IV, or mouthparts modified for clasping hairs or tissue of mammal hosts. . . . . .Superfamily LISTROPHOROIDEA ... 45

— Without modifications as above. 18

18. Oviporus a slightly expanded inverted "V";[1] without indications of genital acetabula. Subdermal, visceral and respiratory parasites of mammals and birds . . . . .Superfamily CYTODITOIDEA ... 51

— Oviporus an inverted "U", "Y" or "V" shape, occasionally transverse or forming a longitudinal slit (if aperture is longitudinal, then it is bordered anteriorly and laterally by a strongly sclerotized arched epigynium), genital acetabula reduced but generally visible 19

19. Parasitic on or in mammals, hyperparasites of hippoboscid or mallophagan mammal parasites, occasionally free-living or nidicolous (species of nidicolous PYROGLYPHIDAE may be found on the feathers of birds). . . . . Superfamily PSOROPTOIDEA ... 20

— Parasitic or paraphagic on birds, ambulacra usually with a series of distinctive scleritic fragments (Fig. 43, p. 392). Feather mites[2,3] 26

20. Ambulacra I-II sucker-like, stalked; legs III-IV telescoped, tarsi III-IV terminating in setae or spines, with one long seta; legs I-II with calcarate retrorse excrescenses; gnathosoma narrow and elongate. Elongate endofollicular species 21

— Ambulacra I-IV similar in morphology (occasionally absent), or tarsi III-IV with more than one long terminal seta; legs I-II generally without calcarate excrescenses 22

21. Gnathosoma extending anteriorly to level of leg tibiae I, tarsi III-IV spur-like terminally; oviporus sometimes transverse. Palpi without conspicuous retrorse teeth. On primates and ursines . . . . (Plate 129, p. 427) Family AUDYCOPTIDAE

— Gnathosoma elongate but not extending to level of leg tibiae I, tarsi III-IV not spur-like terminally; oviporus an inverted "V". Palpi armed with conspicuous retrorse teeth. On primates and porcupines. . . . . . . (Plate 129, p. 427) Family RHYNCOPTIDAE

22. Tarsi I-IV without ambulacral discs, each tarsus terminating instead in a series of short, strong spines and a single elongate hair. Palpal tarsus minute, free, other palpal elements fused; chelicerae compressed laterally, both digits with several strongly recurved teeth. Oviporus broadly "U"-shaped, with well developed epigynium. In bat guano . . . . (Plate 130, p. 428) Family GUANOLICHIDAE

— At least some of the tarsi with terminal ambulacral discs. Palpi and chelicerae not as above. 23

23. Chelicerae stylettiform, palpal elements fused but articulated with subcapitulum. In respiratory passages of lemurs . . . . (Plate 128, p. 426) Family LEMURNYSSIDAE

— Chelicerae chelate, palpi without podomeric fusion as above 24

24. Coxal epimera I fused posteriorly in both sexes, forming a median uni- or biramous keel; opisthosoma elongate, bilobed posteriorly, setae d_5 and l_5 greatly elongated in both sexes. Skin parasites of South American edentates and rodents . . . . (Plate 129, p. 427) Family LOBALGIDAE

---

[1]*Turbinoptes strandtmanni* Boyd (TURBINOPTIDAE, Plate 135-2, p. 433) is exceptional in having a transverse genital aperture (Boyd 1949).

[2]The portion of this key which deals with feather mites (couplets 26-44) is derived, in large part, from a key by J. Gaud and W.T. Atyeo (unpublished).

[3]The family GAUDOGLYPHIDAE (Bruce and Johnston 1976) is not keyed.

— Coxal apodemes separate in both sexes or, if touching, not forming a median keel; opisthosoma of female rarely bilobed, male with or without bilobed opisthosoma, setae d_5 and l_5 variously developed 25

25. Legs III-IV of female inserted ventrally, normally developed and without long terminal setae; legs IV of male often reduced, male with or without anal suckers. Vertical setae absent in both sexes. Nidicoles of mammals or birds, or free-living . . . . (Plates 129-130, pp. 427-428) Family PYROGLYPHIDAE

— Legs III-IV of both sexes inserted laterally or ventrally, legs III usually with long terminal setae, often lacking empodial claws; legs IV normal or reduced, usually with empodial claws, legs with or without calcarate processes; male with anal suckers. Vertical setae present or absent. Skin parasites of mammals. . . . . (Plate 128, p. 426) Family PSOROPTIDAE

26. Apices of tarsi I-IV lacking setae *p* and *q*, ambulacra variously developed, condylophore guides present or absent (if ambulacra are small, then guides are reduced); condylophores rectangular or long and flexible when viewed laterally 27

— Tarsal apices with ventrolateral setae *p* and *q* (Fig. 42, p. 390), both setae typically fan-like; ambulacra with well-defined sclerites but lacking condylophore guides, condylophores "L"-shaped in lateral aspect (Plate 139-4). Primarily from non-passeriform birds. . . . . Superfamily PTEROLICHOIDEA ... 28

27. Condylophores long and flexible, tendinous in lateral view (Plate 135-5, p. 433); ambulacral sclerites well defined or reduced (if well defined, then condylophore guides are distinct; if reduced, then guides may be absent) . . . . Superfamily ANALGOIDEA ... 35

— Condylophores basically rectangular in lateral view (Plate 136-3, p. 434); ambulacral sclerites large, condylophore guides absent . . . . Superfamily FREYANOIDEA ... 44

28. Trochanters without setae; solenidion σ_1 present on genu III. Found on Apodiformes (Apodidae and Hemiprocnidae). . . . . (Plates 137-138, pp. 435-436) Family EUSTATHIIDAE

— Trochanters I-III each with one seta, rarely lacking setae on one or more trochanters (when setae are lost, then solenidion σ_1 also lost). 29

29. Oviporus a longitudinal slit surmounted by a wide "U"-shaped epigynium. Found on Anseriformes (Anatidae) . . . . (Plate 137, p. 435) Family RECTIJANUIDAE

— Oviporus an inverted "U" or "Y" shape, or transverse; epigynium variously developed or absent 30

30. Subhumeral setae (Plate 138-3, p. 436) simple, setiform; legs IV arising submarginally, epimera I fused into a median sternum. Males with sclerotized coxal fields. Usually within feather quills of Charadriiformes, Apterygiformes, and other orders . . . . (Plates 138-139, pp. 436-437) Family SYRINGOBIIDAE

— Without the above combination of characters 31

31. Tibia IV without ventral seta *kT;* genital acetabular remnants of female posterior to setae *gp.* Found on Coraciiformes, Piciformes, Falconiformes and Passeriformes (Corvidae) . . . . (Plate 139, p. 437) Family GABUCINIIDAE

— Seta *k*T usually present on tibia IV (Plate 140-2, p. 438); genital acetabula of female anterior to setae *gp* 32

32. Female with long, thin terminal external sperm duct extending at least to the level of mid-tarsus IV; setae d_3 and l_3 absent in both sexes. Found on Tinamiformes (Tinamidae) . . . . (Plate 140, p. 438) Family CRYPTUROPTIDAE

— Female with an internal sperm duct or, if external, reduced to a short tube or a simple bursa; setae l_3, or l_3 and d_3, present 33

33. Tarsus III with only one ventral seta. Found on Ciconiiformes, Strigiformes, Pelecaniformes and Falconiformes. . . . . (Plate 141, p. 439) Family KRAMERELLIDAE

— Tarsus III with two or three ventral setae 34

34. Female without epigynium, oviporus an inverted "U"-shape or transverse. Subhumeral setae in both sexes long, rarely dilated basally. Found on Columbiformes and Psittaciformes. . . . . (Plate 141, p. 439) Family FALCULIFERIDAE

— Female with an epigynium, sometimes reduced (if absent, then the subhumeral setae are dagger-like and the oviporus is shaped like an inverted "Y"). Found on many avian groups . . . . (Plates 141-142, pp. 439-440) Family PTEROLICHIDAE

35. Setae d_1-d_5 absent, with only one pair of oviporal setae; males may lack legs IV. Sac-like mites in feather quills of Charadriiformes (Charadriidae), Scolopacidae and Passeriformes (Parulidae) . . . . (Plate 131, p. 429) Family APIONACARIDAE

— With some or all of the dorsal *d* series present (Plate 132-3, p. 430), with more than one pair of oviporal setae 36

36. Seta *kT* absent on tibia IV. 37

— Seta *kT* present on tibia IV[1] 40

37. Solenidion σ_1 absent on genu II. Found primarily on Passeriformes, as well as on Coraciiformes, Trogoniformes, some Apodiformes (i.e., Trochilidae) and one species of Charadriiformes (Scolopacidae) . (Plate 131, p. 429) Family PROCTOPHYLLODIDAE

— Solenidion σ_1 present on genu II. 38

38. Gnathosoma short, quadrate; subhumeral setae usually anterior to humerals, vertical setae absent. On many avian groups. . . . . (Plates 131-132, pp. 429-430) Family XOLALIGIDAE

— Gnathosoma elongate, with divergent palpi; subhumeral setae usually posterior to humerals, vertical setae present. 39

39. Femora and genua of legs fused so that legs each comprise four functional segments. From shore, wading, and sea birds, along with certain Apodiformes (Apodidae). . . . . (Plates 132-133, pp. 430-431) Family ALLOPTIDAE

— Femora and genua of legs normally articulated. Primarily found on Passeriformes, Piciformes and Coraciiformes . . . . (Plate 132, p. 430) Family TROUESSARTIIDAE

40. Solenidion σ_1 absent on genu III. Skin parasites, or hyperparasites of hippoboscid or mallophagan parasites, of many avian orders. . . . . (Plate 133, p. 431) Family EPIDERMOPTIDAE

— Solenidion σ_1 present on genu III 41

41. With a strong claw-like dorsal or ventral apical seta or spine on some or all of the tarsi, tarsi often greatly reduced. Parasites in nasal fossae of birds . . . . (Plate 135, p. 433) Family TURBINOPTIDAE

[1] May be absent in some TURBINOPTIDAE.

— Tarsal morphology various, not as above. Feather or quill mites 42

42. Ambulacral disc of each leg spade-shaped; ambulacral sclerites well defined, central sclerite broad. Found on many avian orders. . . . . (Plate 134, p. 432) Family AVENZOARIIDAE

— Ambulacral discs rounded apically; ambulacral sclerites usually reduced, central sclerite often a thin medial band 43

43. Female with epigynium,[1] legs III-IV marginally inserted.[2] On feather surfaces of many avian groups. . . . . (Plates 135-136, pp. 433-434) Family ANALGIDAE

— Female without epigynium, legs III-IV arise submarginally. In feather quills of many avian groups . . . . (Plate 134, p. 432) Family DERMOGLYPHIDAE

44. Legs III-IV arise marginally; tarsus III with only one ventral seta, tarsus I with two setae. Ambulacral discs of tarsi I-II directed in same axes as tarsi. On Bucerotidae . . . . (Plate 136, p. 434) Family VEXILLARIIDAE

— Legs III-IV arise submarginally; tarsus III with three ventral setae, tarsus I with more than two setae. Ambulacral discs of tarsi I-II angled upwards about 45° from axes of tarsi. On shore, wading, and sea birds. . . . . (Plates 136-137, pp. 434-435) Family FREYANIDAE

45. Gnathosoma may be modified for clasping hair or tissue; legs normally developed, but often with scale-like extensions between coxae I (Plate 142-2, p. 440) 46

— Gnathosoma without modifications for clasping hair; with one or more pairs of legs modified for hair clasping, without intercoxal extensions as above 47

46. With ribbed, scale-like extensions between coxae I. Fur mites of rodents, lagomorphs, carnivores and insectivores . . . . (Plate 142, p. 440) Family LISTROPHORIDAE

— Gnathosoma greatly enlarged, strongly sclerotized; palpi pointed, without podomeric segmentation, forming a holdfast apparatus in concert with the dentate chelicerae. Parasites of South American bats . (Plate 143, p. 441) Family CHIRORHYNCHOBIIDAE

47. Legs I-II modified for clasping hair 48

— Legs III and/or IV modified for hair clasping. Fur mites of rodents. . . . . (Plates 142-143, pp. 440-441) Family MYOCOPTIDAE

48. Legs I-II compressed, without distinct segmentation; legs III-IV terminate in strongly produced claws or spines, or sucker-like discs. Primarily parasites of bats . . . . (Plate 142, p. 440) Family CHIRODISCIDAE

— Legs I-II not compressed, with terminal ambulacral discs, various segments flattened or otherwise modified for clasping hair; coxal fields I-II may be corrugated to aid in maintaining position on host. Parasites of rodents and marsupials . . . . (Plate 142, p. 440) Family ATOPOMELIDAE

49. Prodorsal shield flanked by a pair of strong longitudinal apodemes, occasionally joined posteriorly by a transverse sclerotized bridge. Sac-like or elongate species, with legs IV occasionally absent. Parasites of birds. . . . . (Plates 143-144, pp. 441-442) Family KNEMIDOCOPTIDAE

---

[1] An epigynium is absent in *Anomalges* (ex *Erolia,* Charadriidae) and reduced in some species of *Diplaegidia* (from Columbidae), but the leg insertions are distinctly marginal.

[2] The genus *Typhalges* is atypical in having submarginally inserted legs, but the female has a distinct epigynium.

— Prodorsal shield without lateral apodemes, sometimes with diverging dorsal sclerites anterior to shield (Plate 145-2). Parasites of mammals 50

50. Legs IV greatly reduced or absent, legs I-III illustrate extreme reduction where legs IV are lost (e.g., *Chirobia*). Vertical internal setae typically absent. Parasites of bats . . . . (Plate 144, p. 442) Family TEINOCOPTIDAE

— Legs IV present, generally telescoped but well developed, as are other legs. Vertical internal setae present. Parasites of mammals .(Plate 145, p. 443) Family SARCOPTIDAE

51. Legs IV inserted on posterior 1/4 of idiosoma, opisthosoma greatly reduced 52

— Legs IV inserted anterior to the position described above, opisthosoma not fore-shortened, often elongate 53

52. Coxal fields of legs form large coalesced anterior and posterior sclerites. Tarsi each with a long distidorsal seta. Parasitic in lungs of rodents . . . . (Plate 145, p. 443) Family PNEUMOCOPTIDAE

— Coxal fields may or may not be lightly sclerotized, not forming distinctive anterior and posterior sclerites. Without long tarsal setae. Respiratory and visceral parasites of birds . . . . (Plate 146, p. 444) Family CYTODITIDAE

53. Femora and genua I-II coalesced dorsally, tarsi with elongate membranous pretarsal elements; female without an epigynium. Internal parasites of domestic fowl . . . . (Plate 145, p. 443) Family LAMINOSIOPTIDAE

— Femora and genua I-II not coalesced dorsally; tarsi with or without pretarsal suckers, often terminating in a series of spurs or spur-like setae; female with an epigynium. Parasites in the nasal cavities, eyes or viscera of bats or rodents . . . . (Plate 146, p. 444) Family GASTRONYSSIDAE

GENERALIZED KEY TO HYPOPODES OF ACARIDIDES[1]

1. Venter of opisthosoma with two pairs or ridged claspers which may be partially covered by lateral flaps. Pilicolous species .... (Plate 120, p. 418) Family GLYCYPHAGIDAE, Subfamily Labidophorinae

— Venter of opisthosoma without claspers, with or without a sucker plate. 2

2. With a well developed ventral opisthosomal sucker plate (Plate 125-5, p. 423). Gnathosomal primordia *(palposoma)* usually distinctive. Phoretic primarily on insects 3

— Sucker plate reduced or absent, although primordial genital acetabula may be greatly enlarged and sucker-like. Palposoma absent or nearly so 8

3. Tarsus I with only one solenidion (ω_1); tibia I with two solenidia (ϕ_1 and ϕ_2). Legs III-IV slender, often directed forward in preserved material .... (Plate 125, p. 423) Family ANOETIDAE

— Tarsus I with two or more solenidia, tibia I with only one solenidion (ϕ_1) 4

4. Dorsum strongly ornamented with reticulations, striations or pits, usually with separated and distinct propodosomal and hysterosomal shields; tarsi I-III often with fleshy pretarsi and greatly enlarged hook-like empodial claws. Ventral epimera robust, generally forming complex coxal fields. Phoretic on Hymenoptera .... (Plate 121, p. 419) Family CHAETODACTYLIDAE

— Dorsum generally weakly sclerotized; pretarsi may or may not be elongate, but never with enlarged hook-like claws. Ventral epimera usually distinct but not as above 5

5. Empodial claws on stalked pulvillar elements; eyes usually present anteromedially on propodosoma (*Vidia* and *Psylloglyphus* (SAPROGLYPHIDAE) are exceptions); with two or three solenidia on tarsus I 6

— Empodial claws sessile, eyes generally absent; usually with two solenidia on tarsus I (*Michaelopus* (ACARIDAE) is exceptional in having three) 7

6. Solenidion α of palposoma well developed, occasionally vestigial or absent. Eyes, when present, located anteromedially, near vertical setae .... (Plate 122, p. 420) Family SAPROGLYPHIDAE

— Solenidion α absent, palposoma vestigial. Eyes present, located anterolaterally and well behind vertical setae. .... Family CARPOGLYPHIDAE

7. Palposoma poorly developed. Subelytral phoretics of *Chilocorus* species (Coccinellidae) .... Family HEMISARCOPTIDAE

— Palposoma well developed, terminal solenidia (α) greatly elongate, usually exceeding length of palposoma; sternal setae often modified into discoid structures. Primarily surface or subelytral phoretics on a wide range of insects .... (Fig. 19b, p. 52) Family ACARIDAE

8. Inert, rounded forms; sucker plate and palposoma reduced or absent. Often found in grain storages (Fig. 40, p. 377) .... Genus *Glycyphagus* (GLYCYPHAGIDAE) and *Acarus* pars (ACARIDAE)

---

[1]Hypopodes of the family FUSACARIDAE are not included.

— Ovoid, active nidicoles in mammal or bird nests, or elongate forms in connective tissue or skin of mammals or birds; sucker plate and palposoma absent 9

9. Elongate tissue inhabitants 10

— Ovoid nest inhabitants. 11

10. In tissues of mammals . . . .(Plate 120-3, p. 418) Family GLYCYPHAGIDAE, subfamilies Ctenoglyphinae, Lophuromyopinae, Echimyopinae, Muridectinae, and others

— In tissues of birds . . . . (Plate 120-2) Family GLYCYPHAGIDAE, subfamily Hypodectinae

11. In nests of birds. . . . . Family GLYCYPHAGIDAE, subfamily Hypodectinae

— In nests of mammals . . . . . . Family GLYCYPHAGIDAE, several subfamilies (Fain 1969b)

PLATE 116

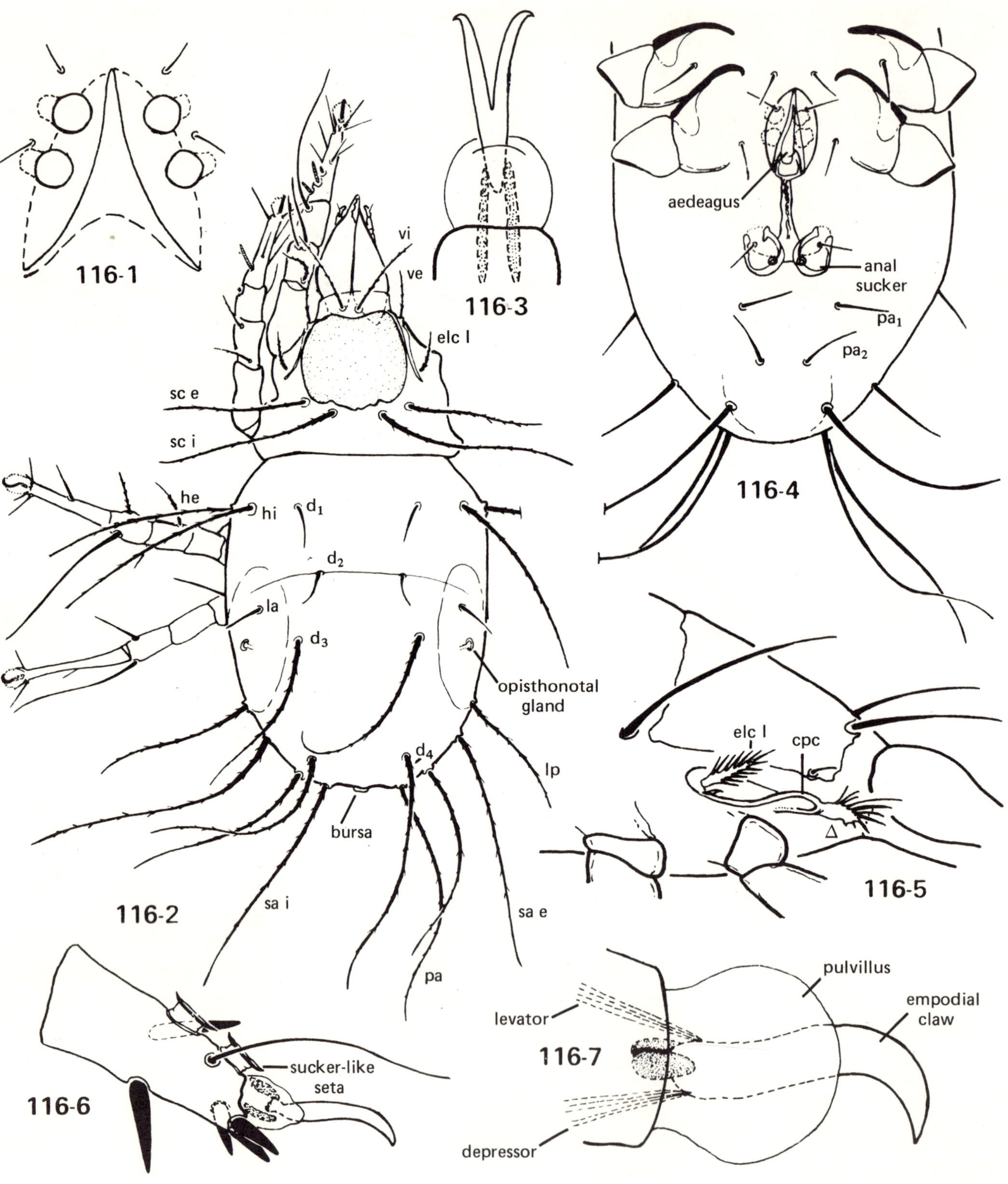

116-1 to 116-7; family ACARIDAE. **116-1**; oviporal field of an acarid mite: **116-2**; *Tyrophagus* sp. (Oregon, USA), dorsum of female: **116-3**; *Lardoglyphus* sp., pretarsus I of female: **116-4**; *Sancassania* sp. (Oregon, USA), venter of male: **116-5**; anterolateral aspect of an acarid mite (elc_1 = supracoxal seta, cpc = podocephalic canal, Δ = Grandjean's organ): **116-6**; *Rhizoglyphus echinopus* (Fumouze and Robin), tarsus IV of male: **116-7**; typical acarid pretarsus

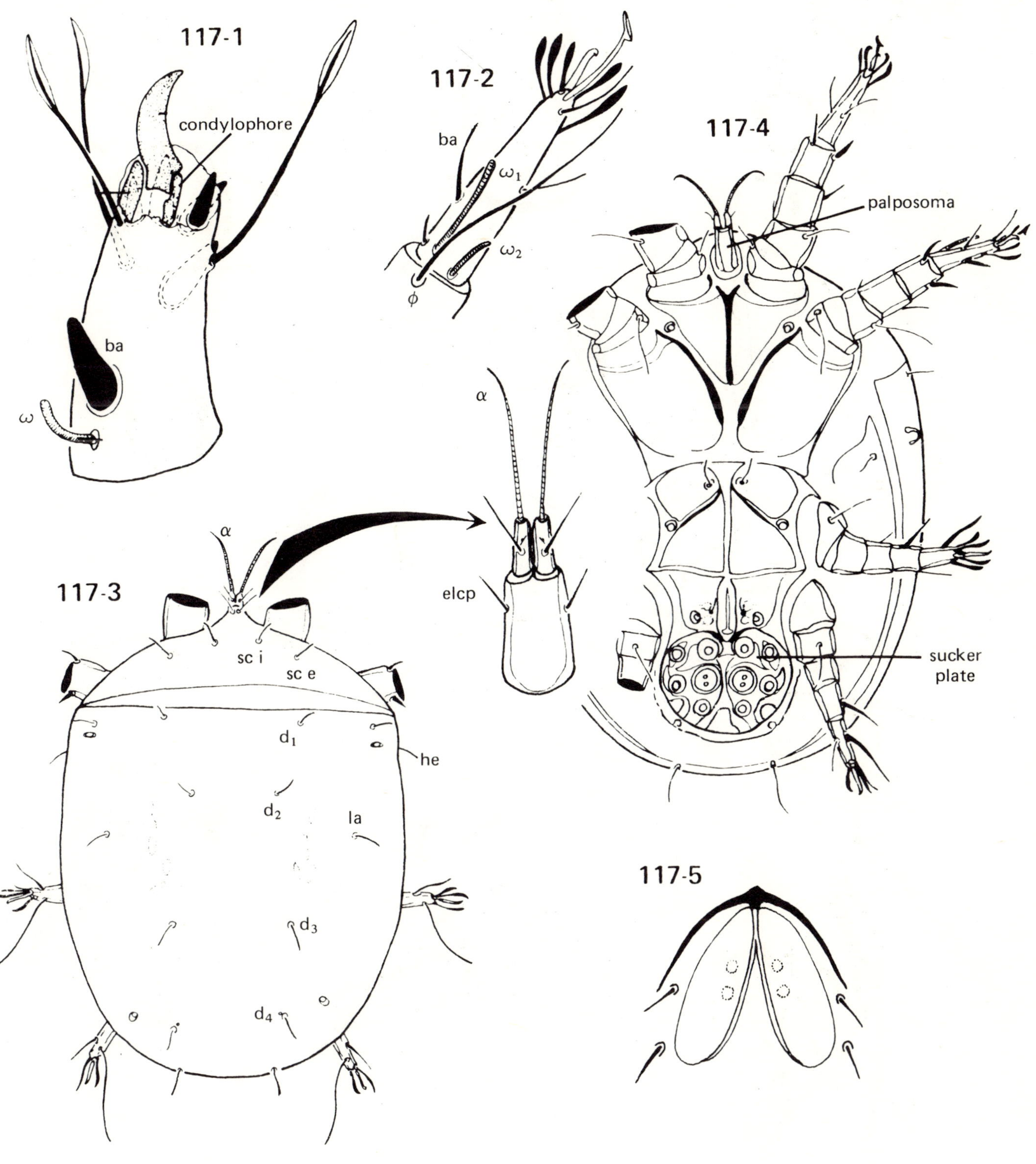

117-1 to 117-4; family ACARIDAE. **117-1**; *Rhizoglyphus* sp., tarsus II of female: **117-2**; tarsus I of hypopus (dorsal): **117-3**; dorsum of hypopus, with detail of palposoma: **117-4**; venter of hypopus

117-5; family GLYCYPHAGIDAE, *Hericia* sp., oviporal field of female

PLATE 118

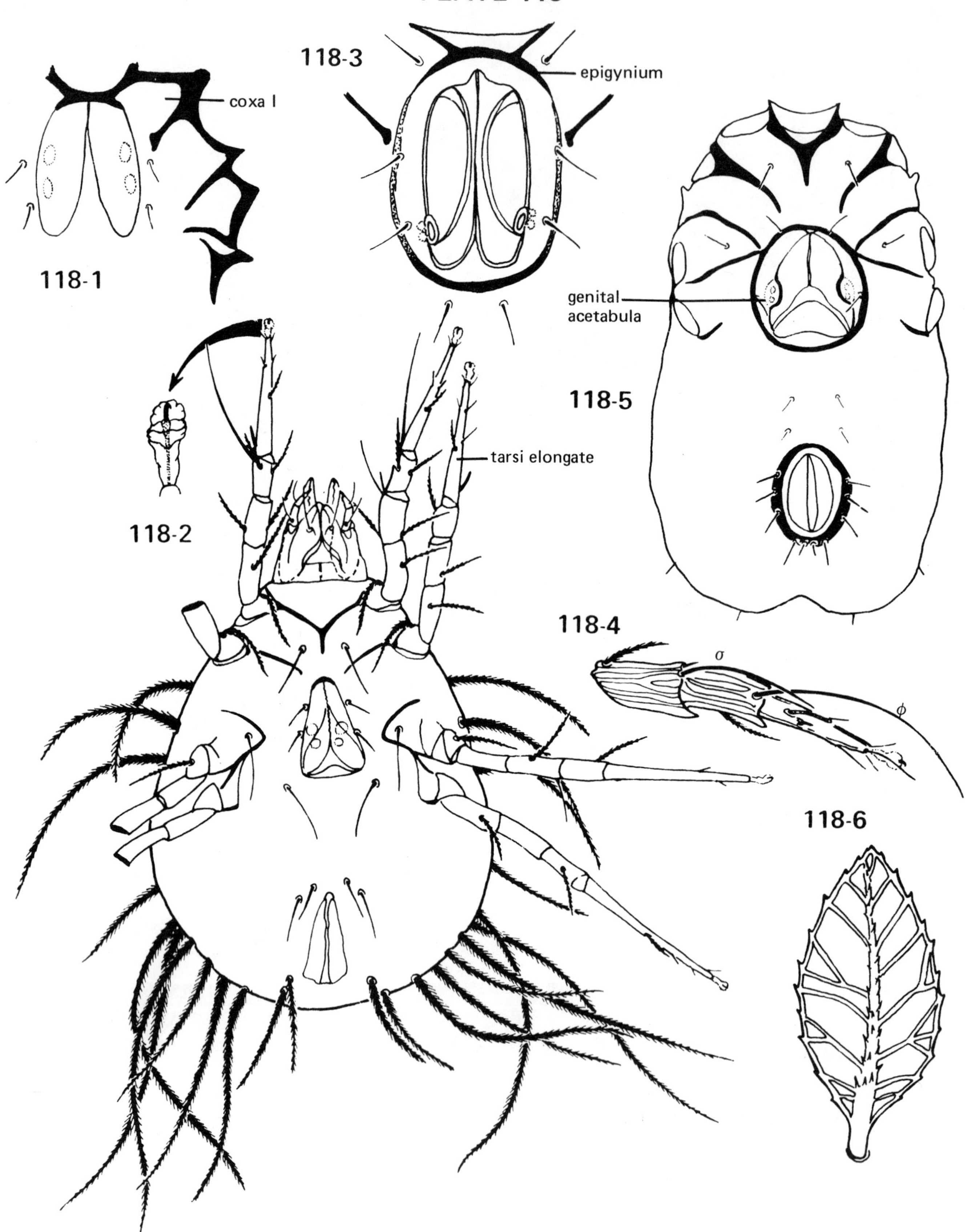

118-1 to 118-6; family GLYCYPHAGIDAE. **118-1**; *Labidophorus* sp. (Oregon, USA), coxal and oviporal fields of female: **118-2**; *Glycyphagus domesticus* (DeGeer) (Oregon, USA), venter of female with detail of pretarsus I: **118-3**; *Gohieria fusca* (Oudemans) (Oregon, USA), oviporal field of female: **118-4**; *G. fusca,* genu, tibia and tarsus I of female: **118-5**; *Xenoryctes* sp., venter of female: **118-6**; *Ctenoglyphus palmifer* (Fumouze and Robin), dorsal seta (after Hughes 1961)

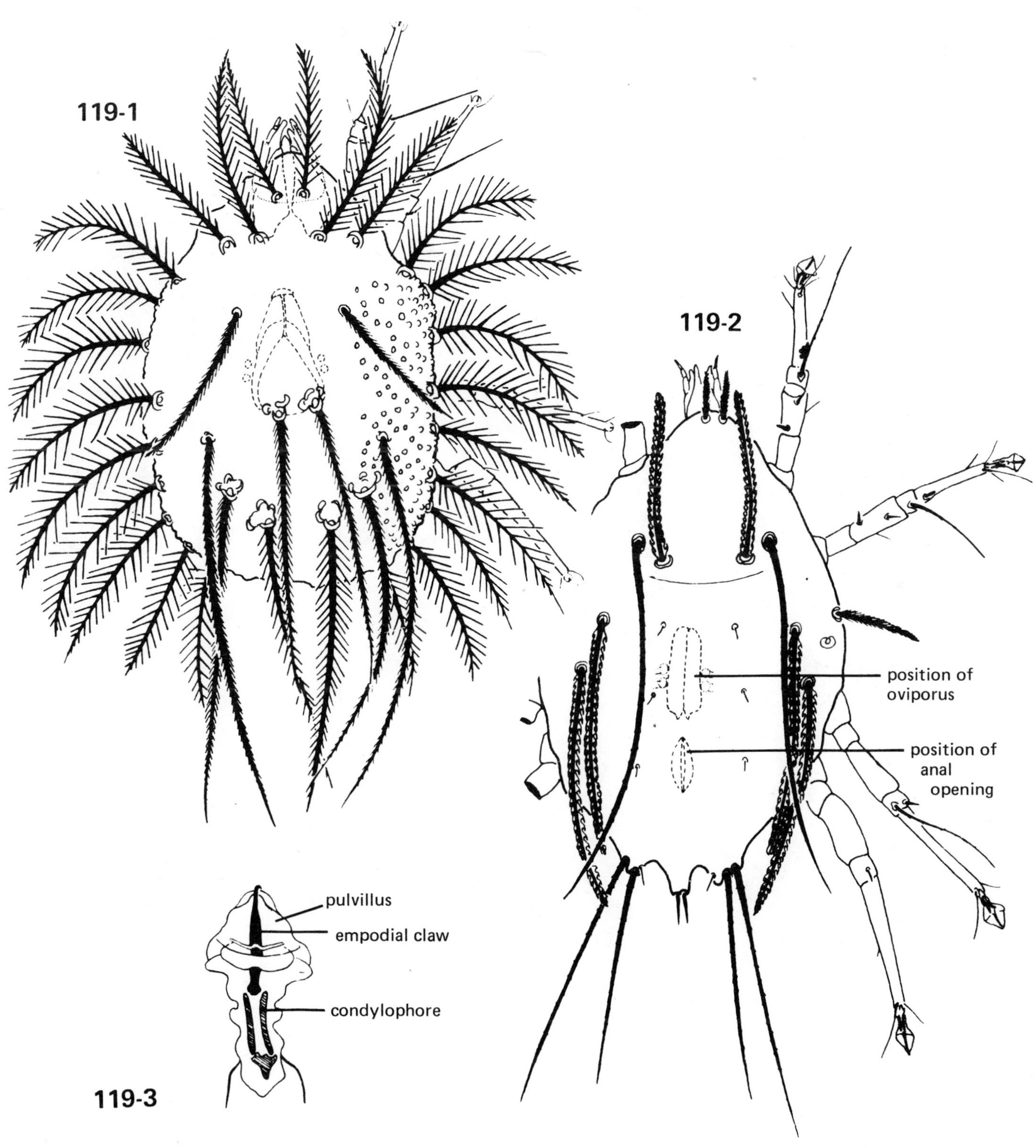

119-1 to 119-3; family GLYCYPHAGIDAE. **119-1**; *Ctenoglyphus plumiger* (Koch)(Oregon, USA), dorsum of female: **119-2**; *Melisia* sp. (India), dorsum of female: **119-3**; *Melisia* sp., pretarsus II

PLATE 120

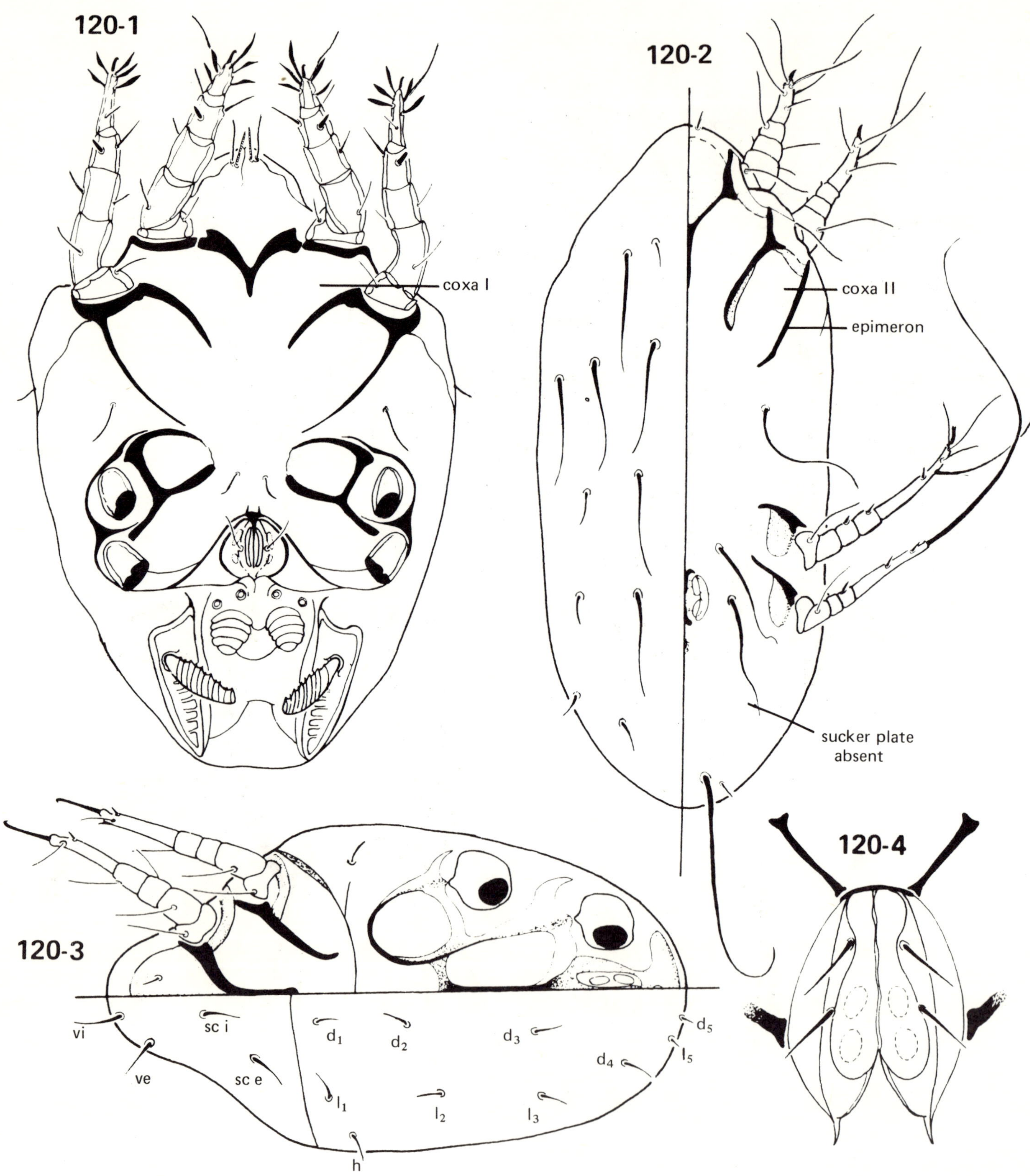

120-1 to 120-3; family GLYCYPHAGIDAE. **120-1;** *Dermacarus* sp. (Oregon, USA), venter of hypopus: **120-2;** *Neottialges (Pelecanectes) bassani* (Montagu), composite dorsum and venter of tissue hypopus (adapted from Fain 1967): **120-3;** *Rodentopus sciuri* Fain, composite dorsum and venter of tissue hypopus (adapted from Fain 1965).

120-4; family CHAETODACTYLIDAE, *Sennertia* sp. (Zaire), oviporal field of female

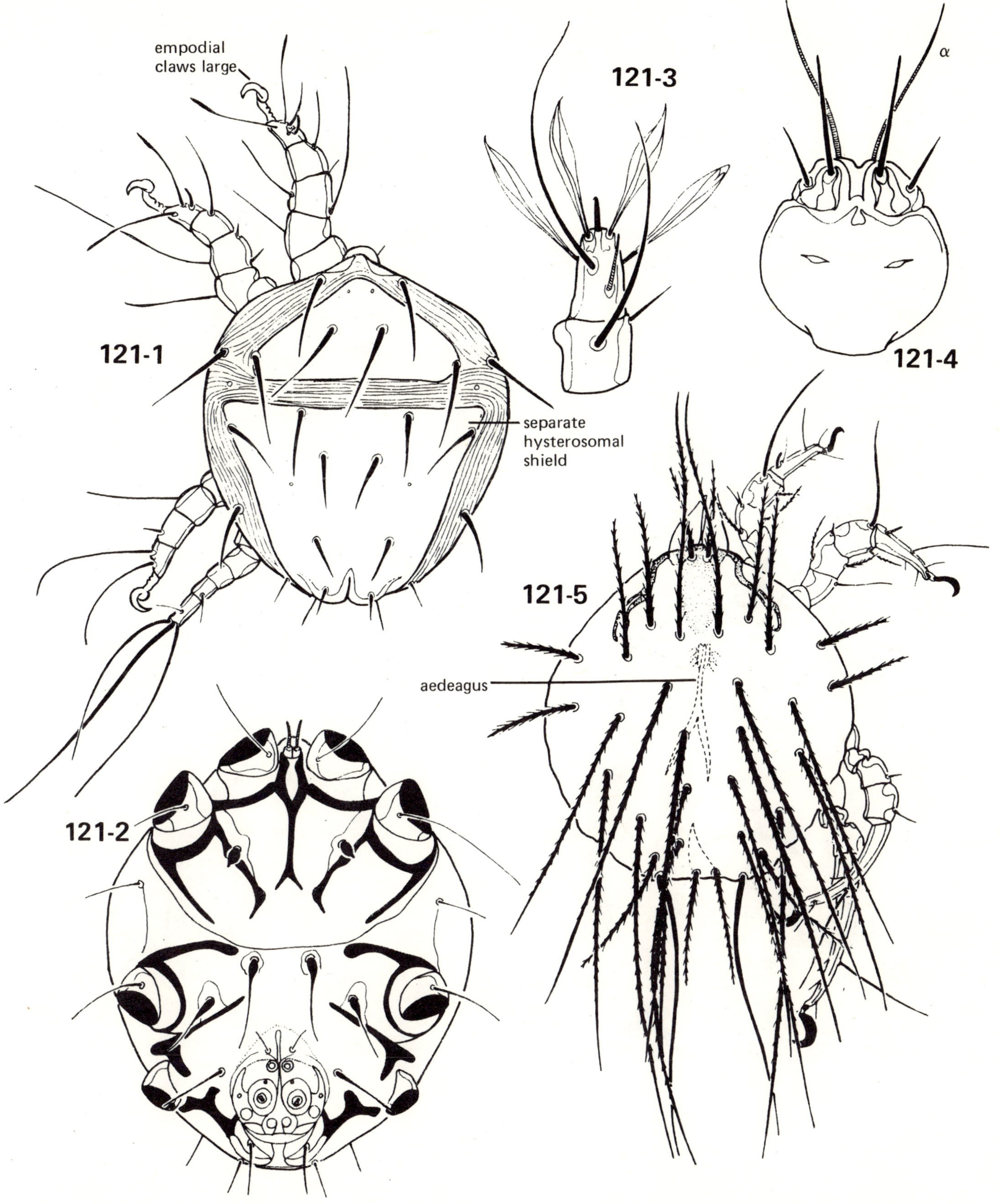

121-1 to 121-5; family CHAETODACTYLIDAE. **121-1**; *Chaetodactylus osmiae* (Dujardin) (Oregon, USA), dorsum of hypopus: **121-2**; *C. osmiae,* venter of hypopus: **121-3**; *Horstia* sp. (Mexico), tibia and tarsus I of hypopus: **121-4**; *Horstia* sp., palposoma of hypopus (dorsal): **121-5**; *Sennertia* sp. (Zaire), dorsum of male

PLATE 122

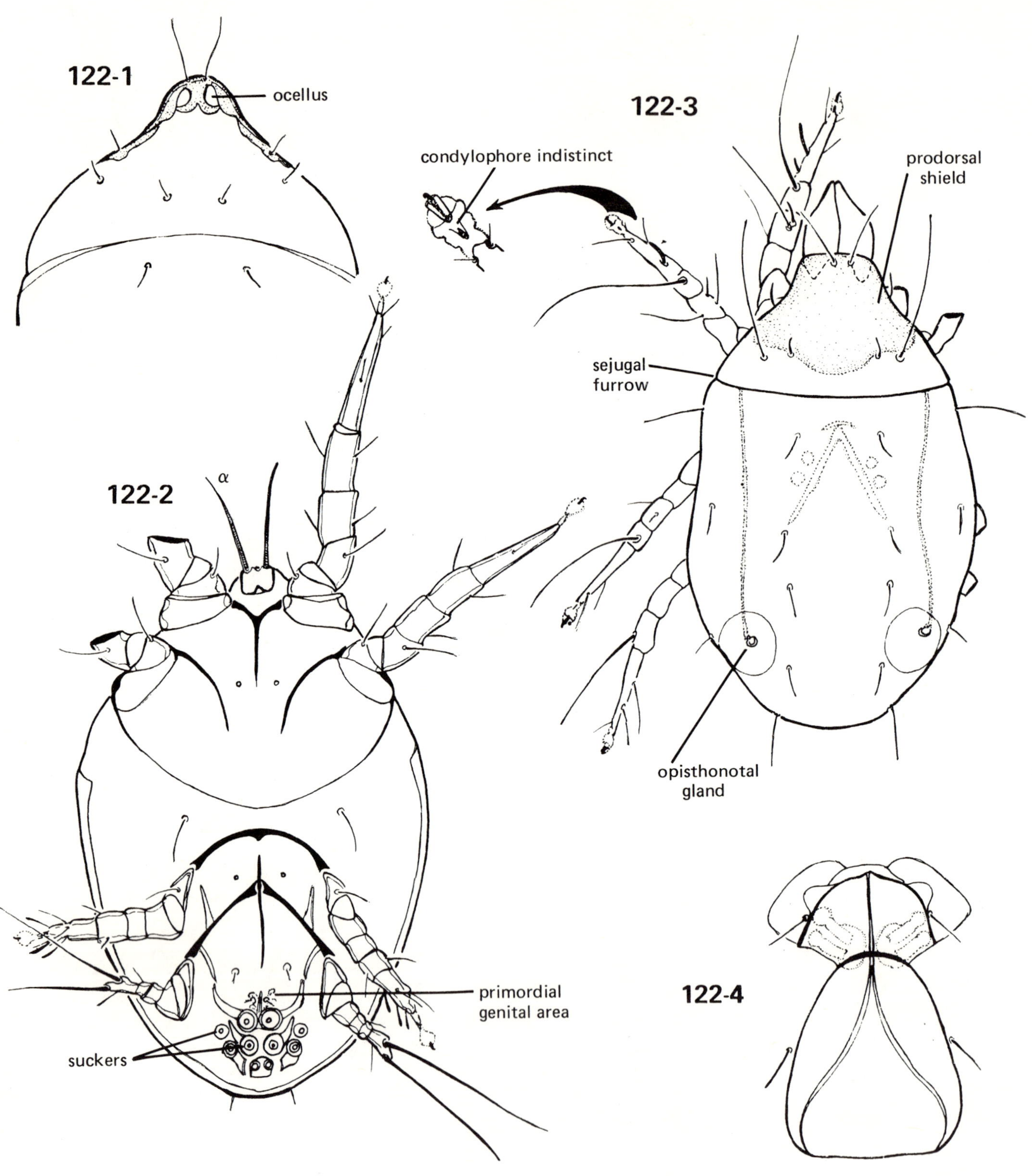

122-1 to 122-3; family SAPROGLYPHIDAE. **122-1 and 122-2**; *Calvolia* sp. (Oregon, USA), hypopus. **122-1**; propodosoma: **122-2**; venter: **122-3**; *Calvolia lordi* (Nesbitt) (Oregon, USA), dorsum of female

122-4; family FUSACARIDAE, *Fusacarus laminipes* Michael (Oregon, USA), oviporal field of female

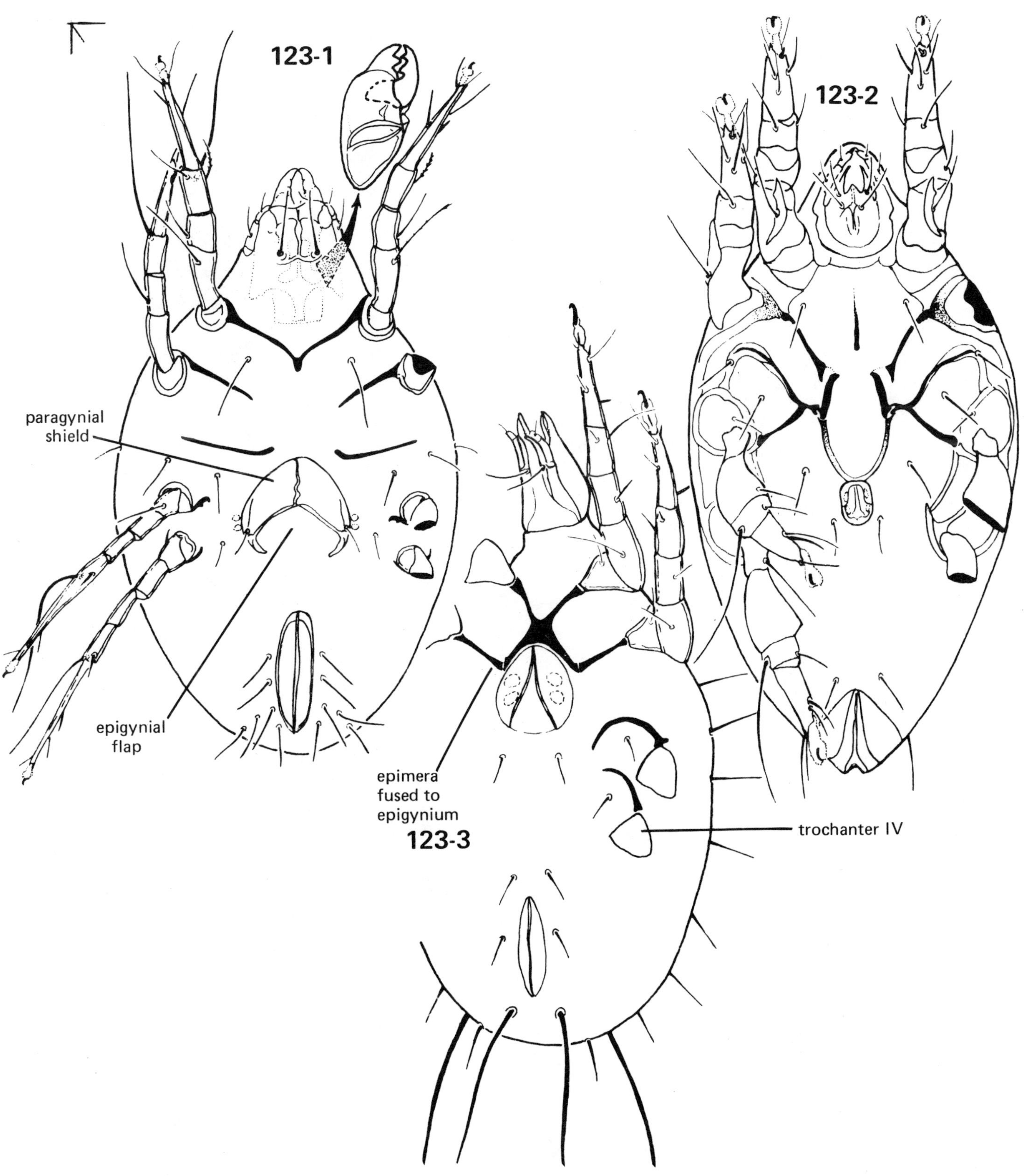

123-1; family CHORTOGLYPHIDAE, *Chortoglyphus arcuatus* (Troupeau) (Oregon, USA), venter of female with detail of chelicera
123-2; family FUSACARIDAE, *Fusacarus laminipes* Michael (Oregon, USA), venter of male
123-3; family CARPOGLYPHIDAE, *Carpoglyphus lactis* (L.), (Oregon, USA), venter of female

PLATE 124

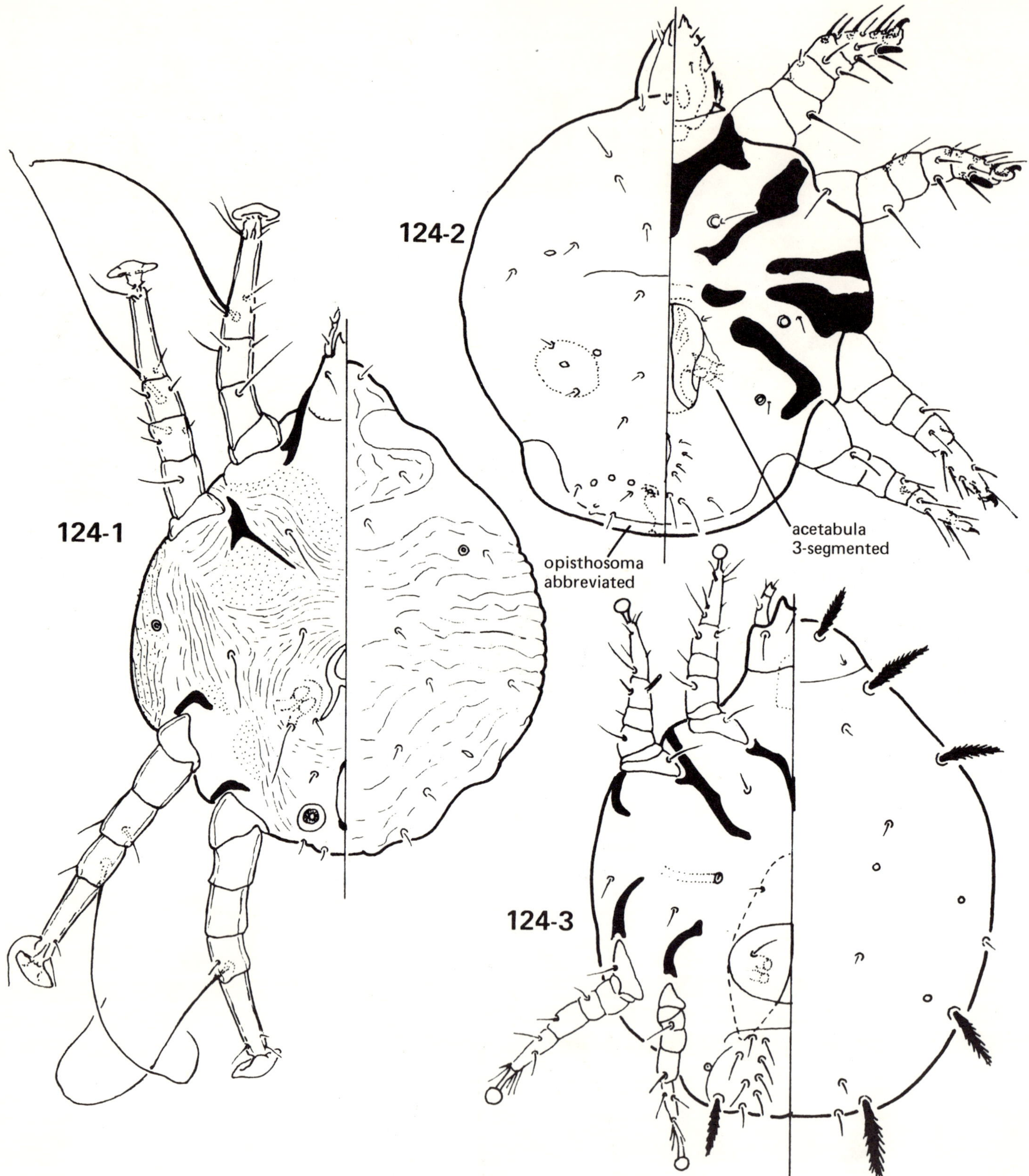

124-1; family HETEROCOPTIDAE, *Heterocoptes tarsii* Fain (Sarawak), composite dorsum and venter of male (adapted from Fain 1967)

124-2; family PLATYGLYPHIDAE, *Platyglyphus malayanus* Kurosa (Malaysia), composite dorsum and venter of female (adapted from Kurosa 1976)

124-3; family GAUDIELLIDAE, *Gaudiella minuta* Atyeo, Baker and Delfinado (Brazil), composite dorsum and venter of female (adapted from Atyeo, Baker and Delfinado 1974)

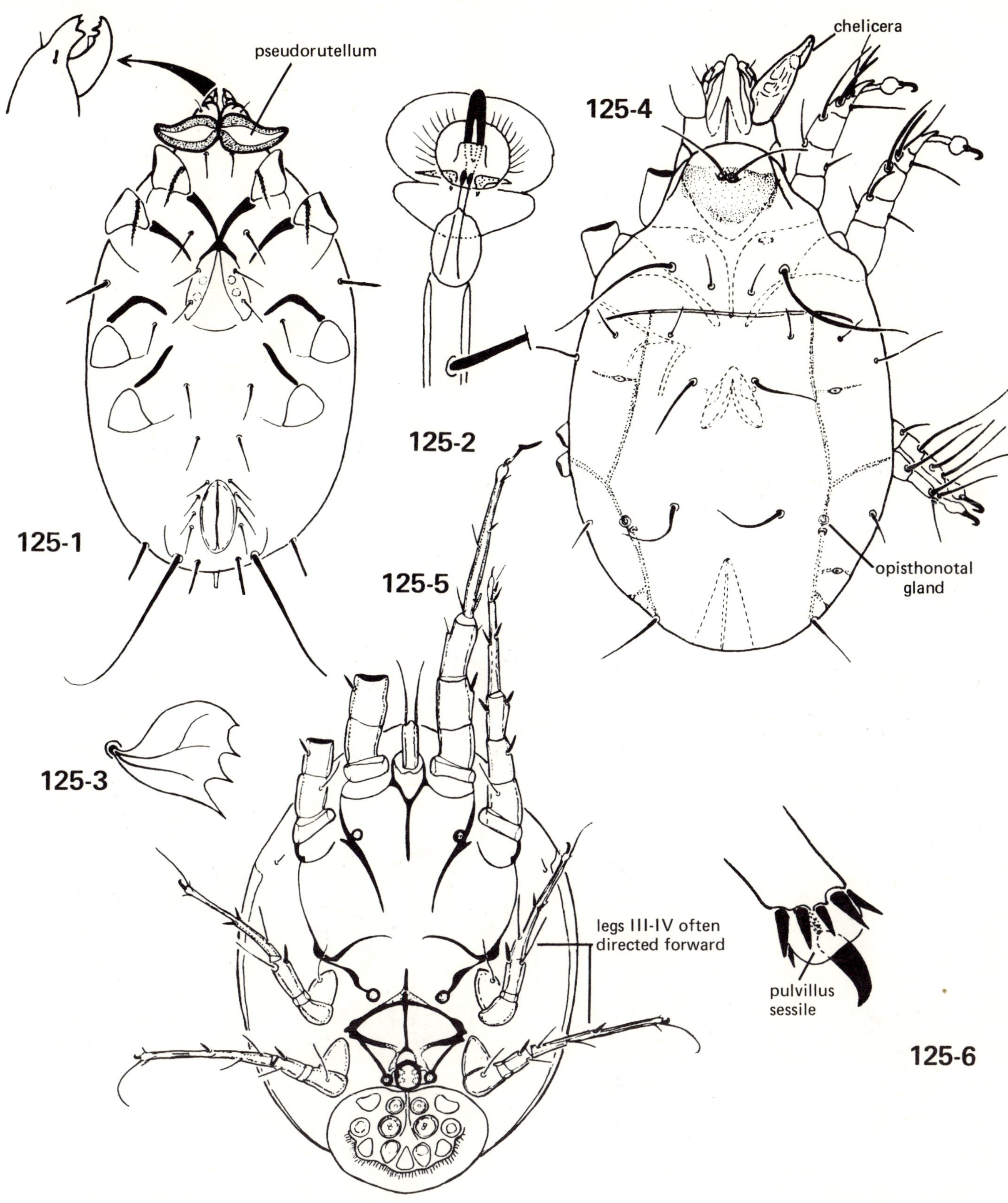

125-1 to 125-3; family ROSENSTEINIIDAE. **125-1**; *Nycteriglyphus vespertilio* Ah and Hunter (South Korea), venter of female with detail of chelicera (after Ah and Hunter 1968): **125-2**; *N. bifolum* Strandtmann, pretarsus (from Atyeo, in press): **125-3**; *N. pterophorus* Berlese, body seta (after Zakhvatkin 1941)

125-4; family HYADESIIDAE, *Hyadesia* sp. (Oregon, USA), dorsum of female

125-5 and 125-6; family ANOETIDAE, *Histiostoma* sp. (Oregon, USA). **125-5**; venter of hypopus: **125-6**; pretarsus III of female

PLATE 126

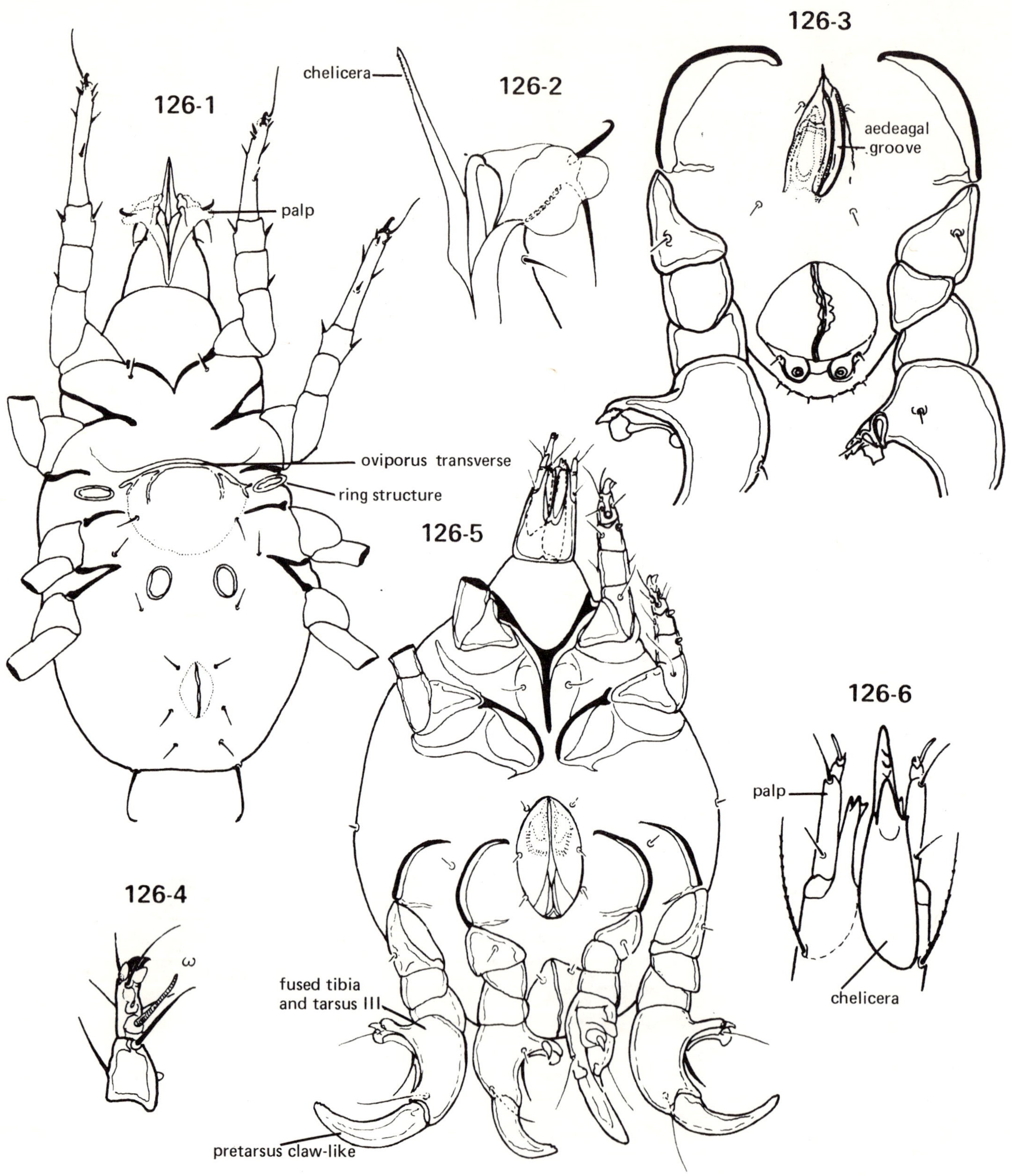

126-1 and 126-2; family ANOETIDAE, *Histiostoma* sp. (Oregon, USA). **126-1**; venter of female: **126-2**; modified palp and portion of chelicera

126-3 to 126-5; family EWINGIIDAE, *Ewingia coenobitae* Pearse (Florida, USA). **126-3**; posteroventral region of male: **126-4**; tibia and tarsus I of female: **126-5**; venter of female

126-6; family CANESTRINIIDAE, gnathosoma (dorsal) of a canestriniid (Thailand), with one chelicera removed

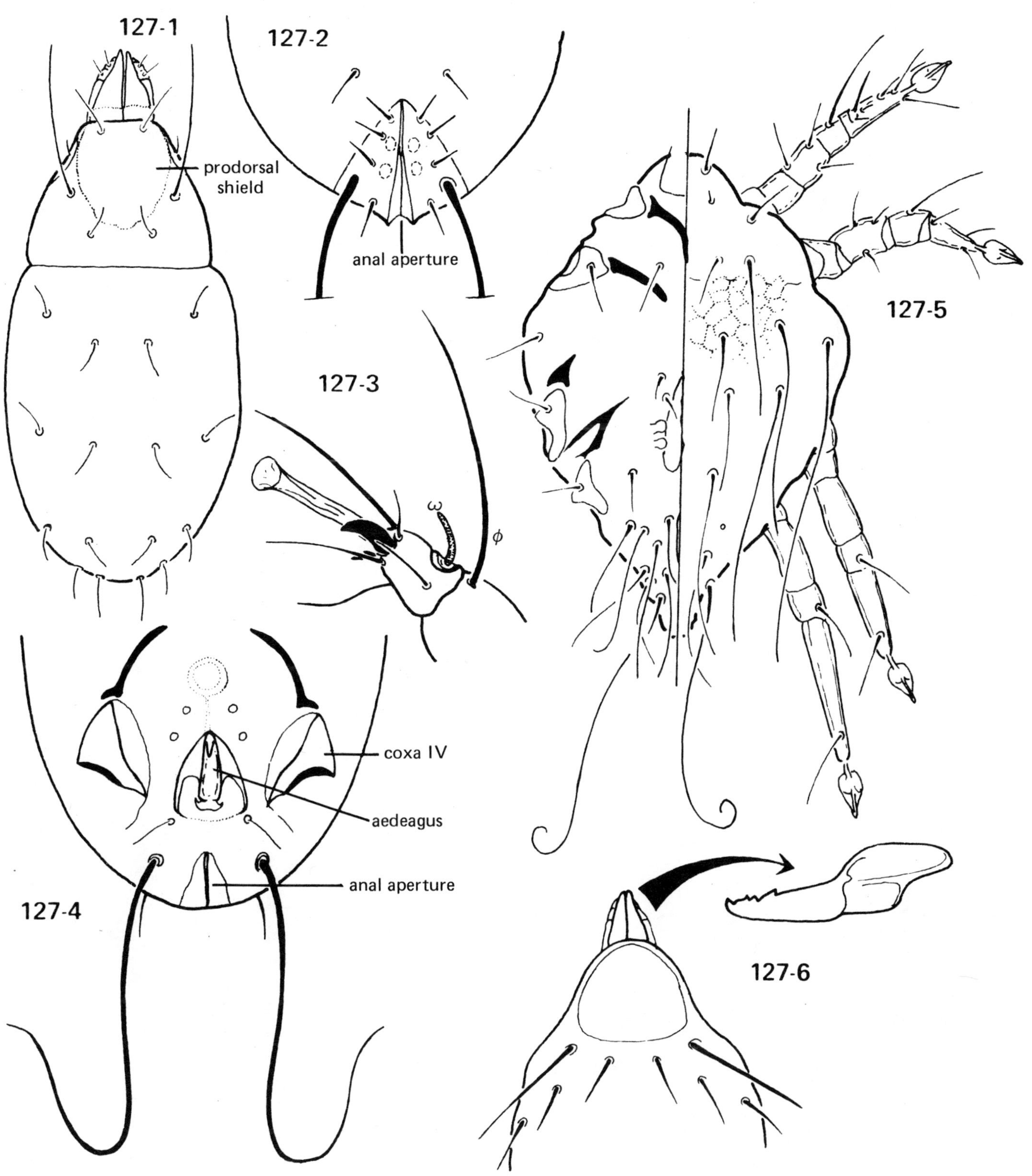

127-1 to 127-4; family HEMISARCOPTIDAE, *Hemisarcoptes malus* (Shimer) (Oregon, USA). **127-1**; dorsum of male: **127-2**; oviporal-anal area of female: **127-3**; tarsus II of male: **127-4**; posteroventral aspect of male

127-5 and 127-6; family CANESTRINIIDAE. **127-5**; *Mesophotia penicillata similis* Samšiňák (Spain), composite dorsum and venter of female (adapted from Samšiňák 1971): **127-6**; *Linobia* sp., anterodorsal aspect with detail of chelicera (after Baker et al. 1958)

PLATE 128

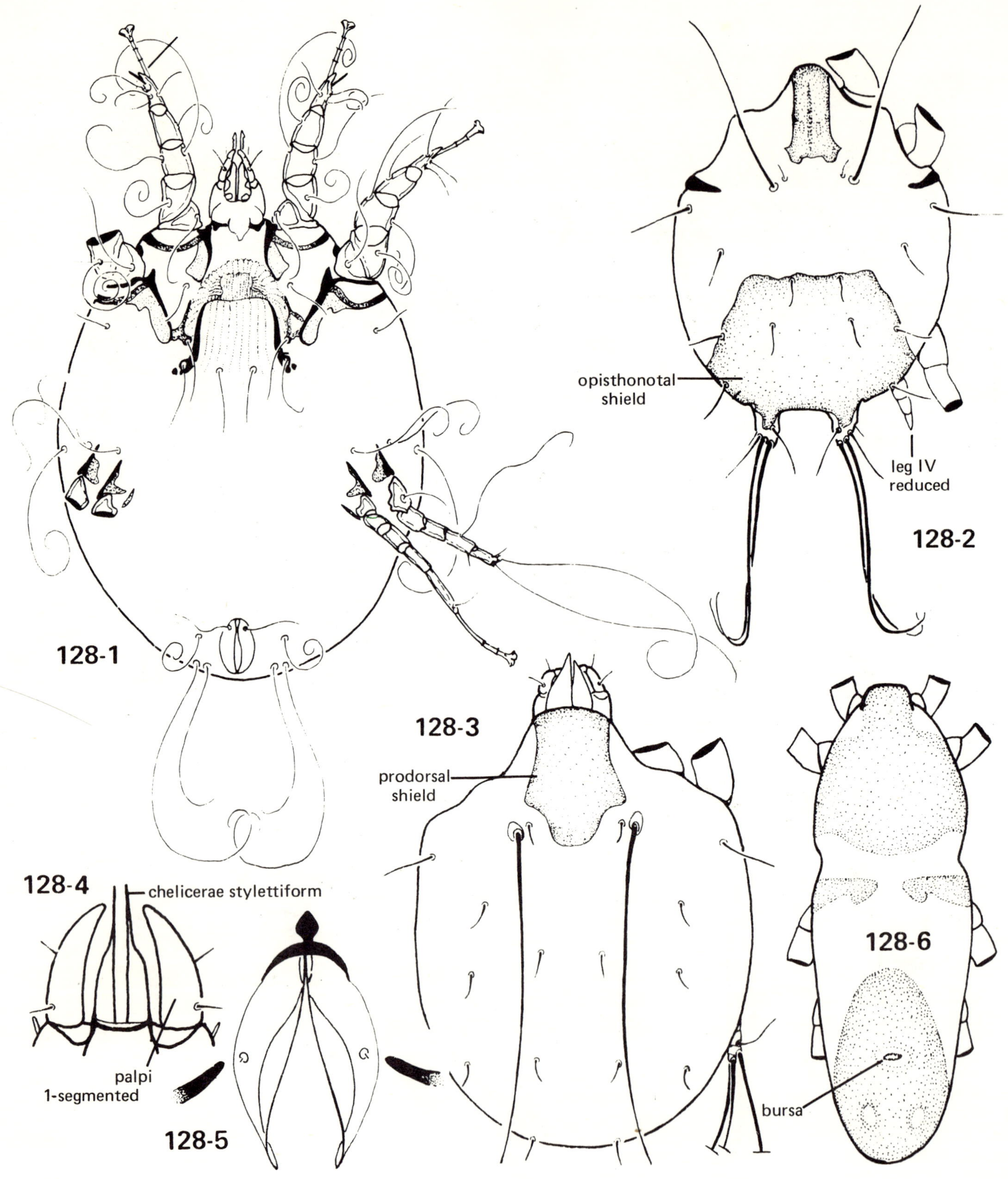

128-1 to **128-3**; family PSOROPTIDAE. **128-1**; *Psoroptes cuniculi* (Delafond) (Oregon, USA), venter of female: **128-2**; *P. cuniculi* (Colorado, USA), dorsum of male: **128-3**; *Caparinia tripilis* (Michael) (New Zealand), dorsum of female

128-4 to 128-6; family LEMURNYSSIDAE, *Lemurnyssus galagoensis* Fain (Zaire) (after Fain 1957). **128-4**; gnathosoma of female: **128-5**; oviporal field of female: **128-6**; dorsum of female

PLATE 129

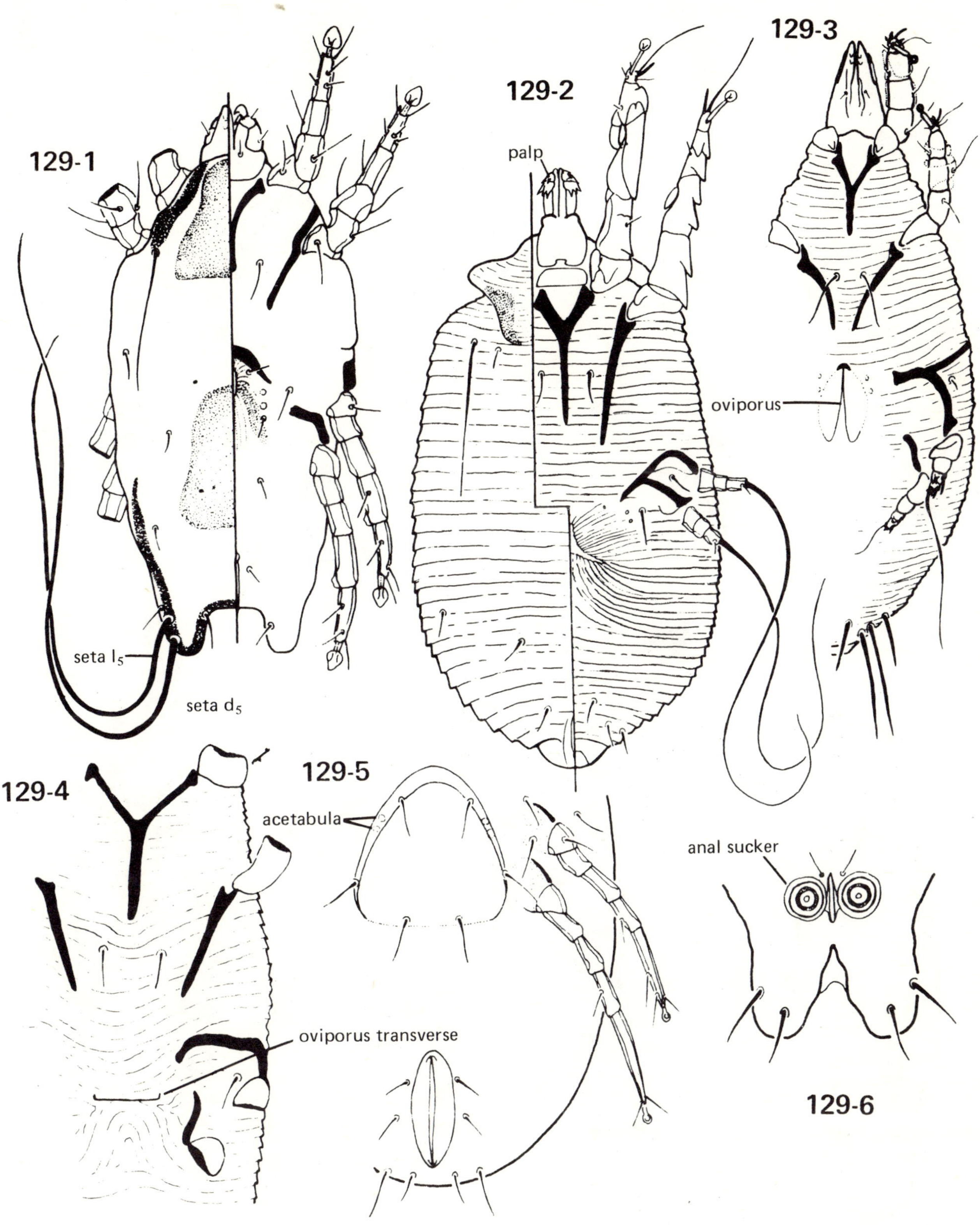

129-1 and 129-6; family LOBALGIDAE, *Lobalges trouessarti* Fonseca (Brazil) (after Fain 1965). 129-1; composite dorsum and venter of female: 129-6; posteroventral aspect of male

129-2; family RHYNCOPTIDAE, *Rhyncoptes anastosi* Fain (South America), composite dorsum and venter of female (adapted from Fain 1965)

129-3 and 129-4; family AUDYCOPTIDAE. 129-3; *Saimiroptes paradoxus* Fain (South America), venter of female (after Fain 1968). 129-4; *?Ursicoptes* sp. (Pennsylvania, USA), medioventral aspect of female

129-5; family PYROGLYPHIDAE, *Euroglyphus longior* (Trouessart) (England), posteroventral aspect of female

PLATE 130

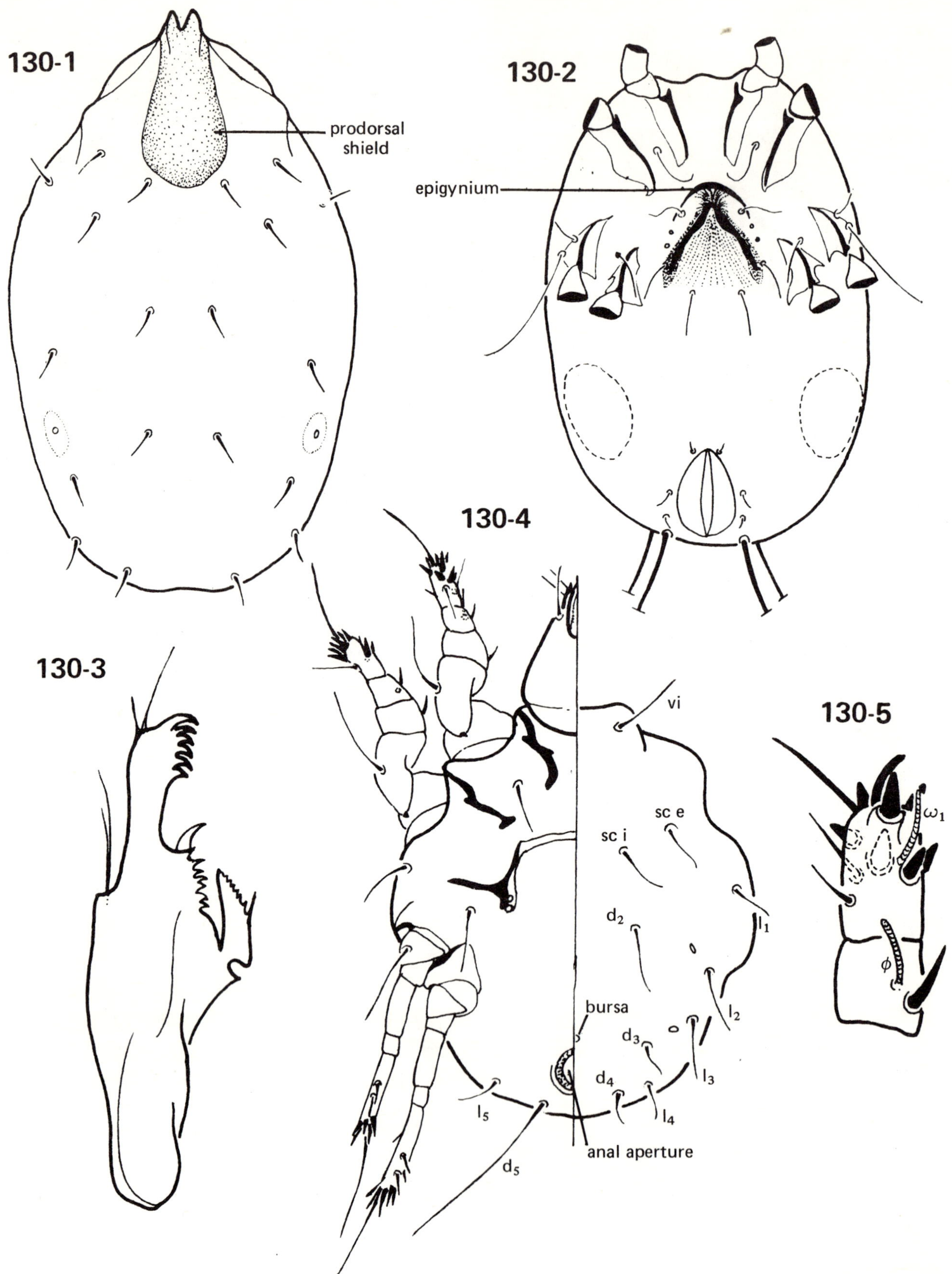

130-1 and 130-2; family PYROGLYPHIDAE. **130-1**; *Euroglyphus longior* (Trouessart) (England), dorsum of female: **130-2**; *Dermatophagoides pteronyssinus* (Trouessart) (Oregon, USA), venter of female

130-3 to 130-5; family GUANOLICHIDAE, *Guanolichus gabonensis* Fain (Gabon) (after Fain 1968). **130-3**; lateral aspect of chelicera of female: **130-4**; composite dorsum and venter of female: **130-5**; tibia and tarsus I of male

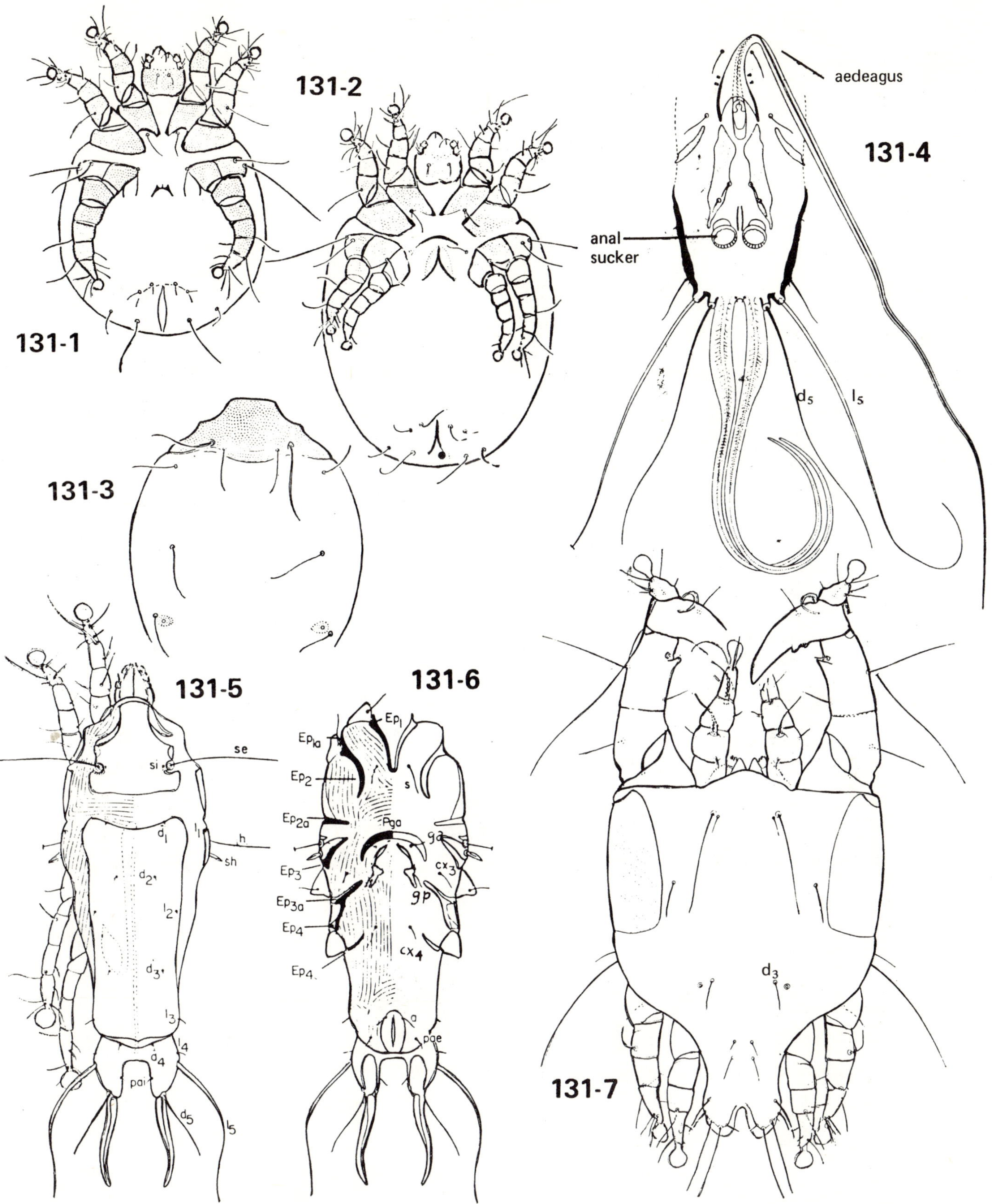

131-1 to 131-3; family APIONACARIDAE, *Atelespoda crena* Gaud and Atyeo (from Gaud and Atyeo 1975). **131-1;** venter of male: **131-2;** venter of female: **131-3;** dorsum of male

131-4 to 131-6; family PROCTOPHYLLODIDAE (from Atyeo and Braasch 1966). **131-4;** *Protophyllodes longiphyllus* Atyeo and Braasch (Texas, USA), posteroventral aspect of male: **131-5 and 131-6;** *P. glandarinus* (Koch), dorsum and venter of female, with designations for setae, epimera and epigynium (Pga)

131-7; family XOLALGIDAE; *Dogielacarus uncitibia* Atyeo (USSR) dorsum of male (from Atyeo 1974)

PLATE 132

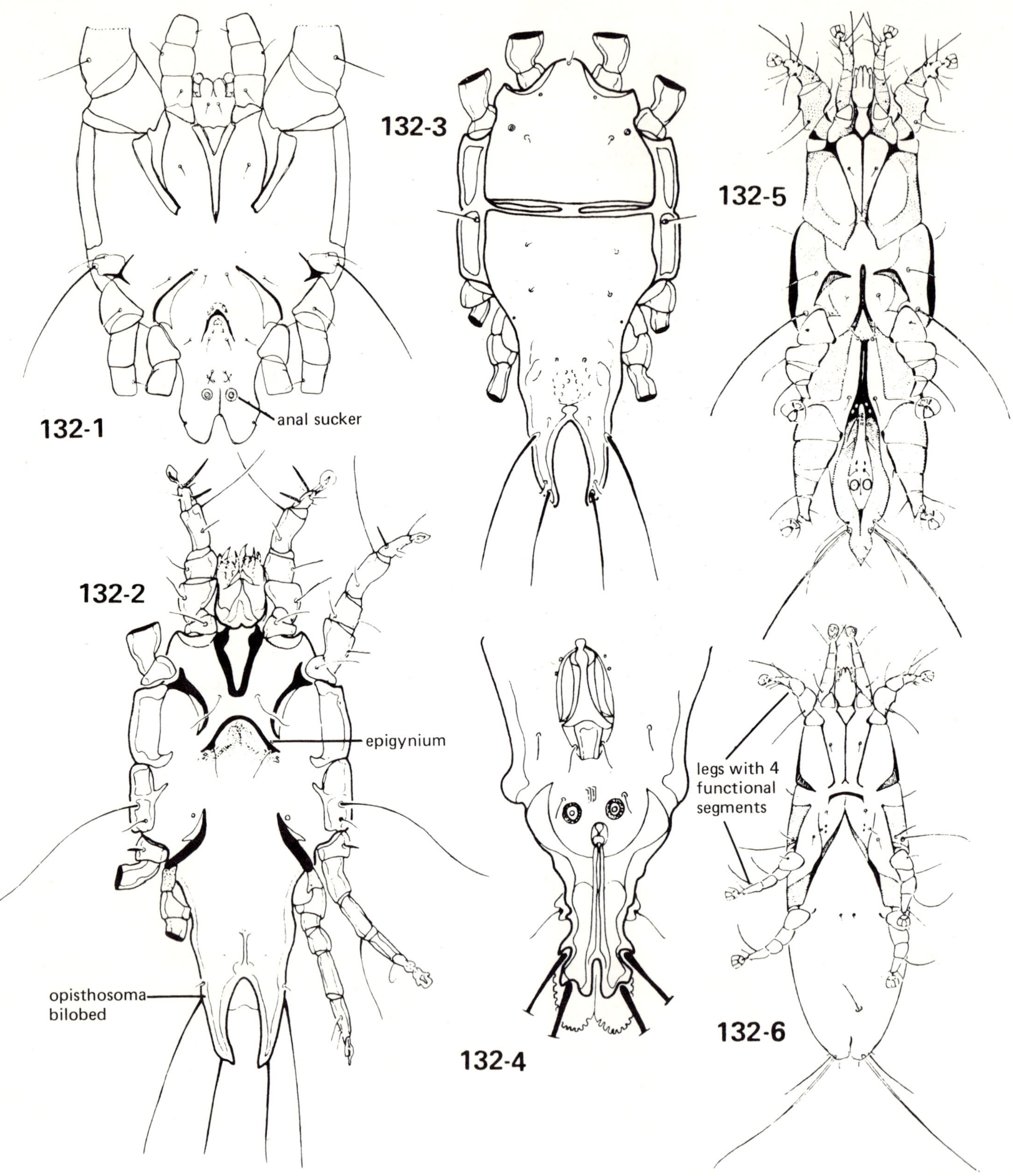

132-1; family XOLALGIDAE, *Dogielacarus uncitibia* Atyeo, venter of male (from Atyeo 1974)

132-2 to 132-4; family TROUESSARTIIDAE. **132-2 and 132-3**; *Trouessartia* sp. (Nigeria). **132-2**; venter of female: **132-3**; dorsum of female: **132-4**; *Trouessartia* sp. (England), posteroventral aspect of male

132-5 and 132-6; family ALLOPTIDAE, *Oxyalges cardiurus* Gaud and Atyeo (Falkland Islands) (from Gaud and Atyeo 1966). **132-5**; venter of male: **132-6**; venter of female

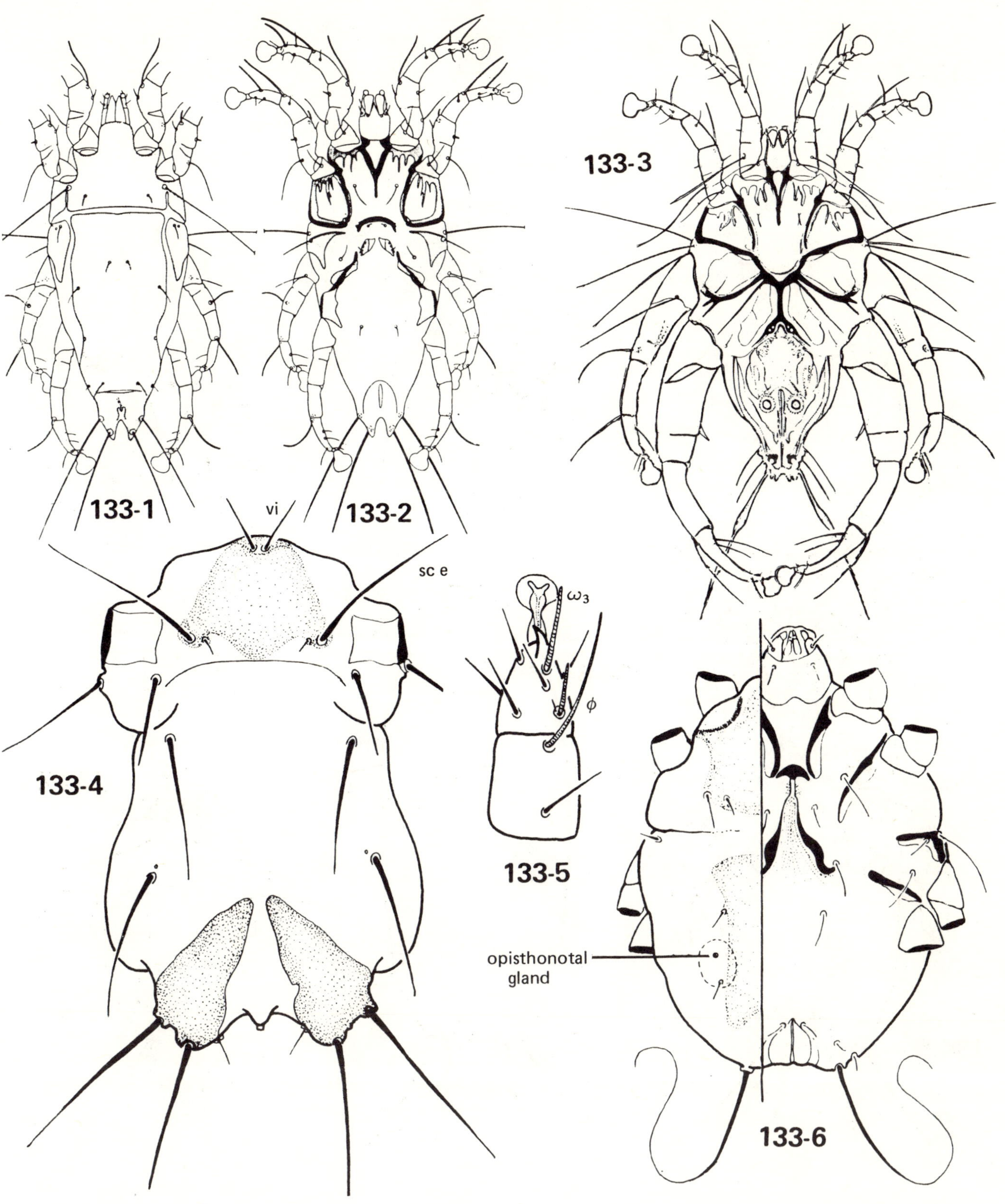

133-1 to 133-3; family ALLOPTIDAE, *Laminalloptes phaetontis* (Fabr.) (from Atyeo and Peterson 1967). **133-1**; dorsum of female: **133-2**; venter of female: **133-3**; venter of male

133-4 to 133-6; family EPIDERMOPTIDAE. **133-4**; *Strelkoviacarus* sp. (Bechuanaland), dorsum of female: **133-5**; *Epidermoptes bilobatus* Rivolta (England), tarsus I (after Fain and Evans 1963): **133-6**; *E. bilobatus,* composite dorsum and venter of female (adapted from Fain and Evans 1963)

PLATE 134

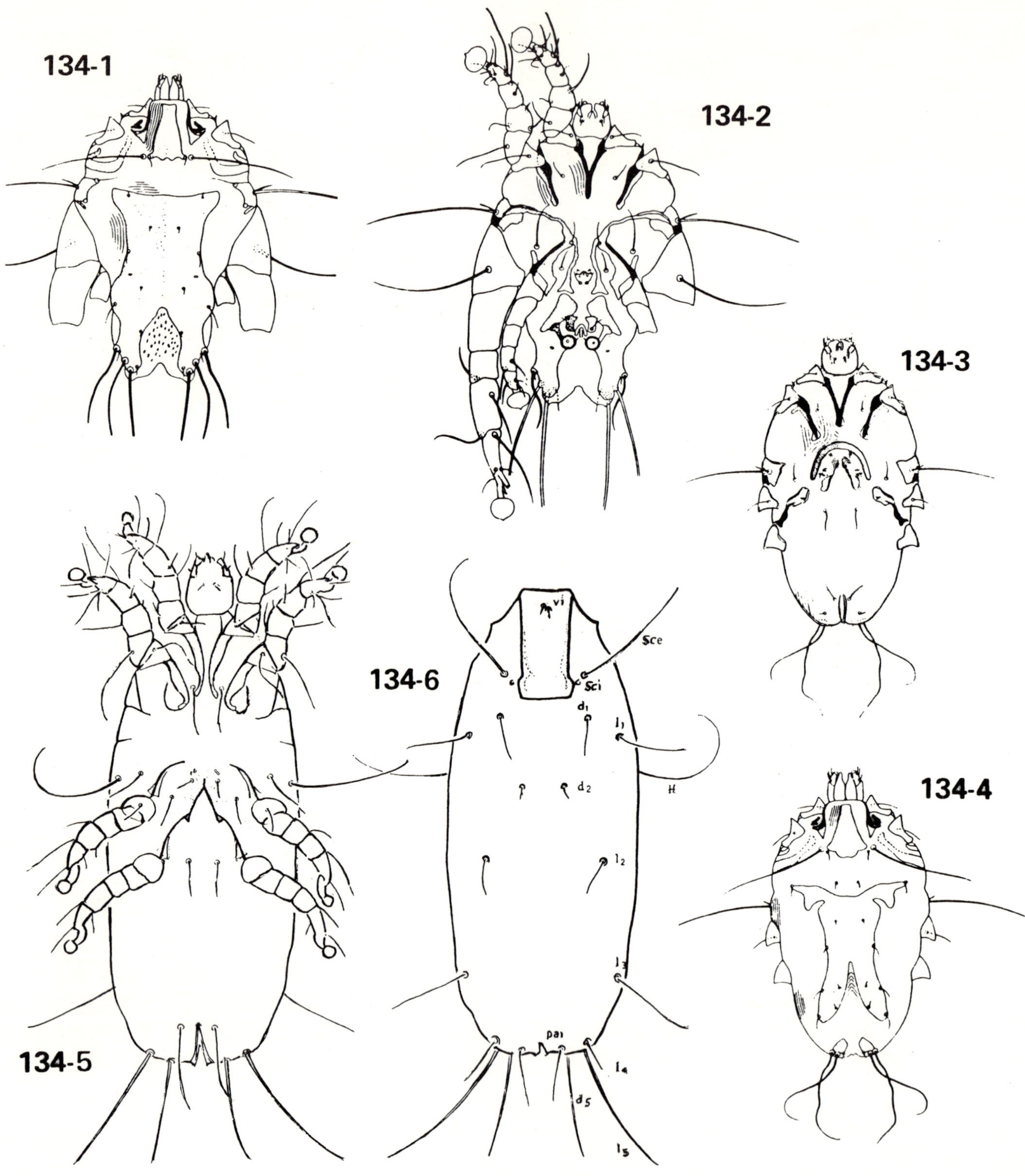

134-1 to 134-4; family AVENZOARIIDAE, *Scutomegninia phalacrocoracis* (Dubinin and Dubinina) (Maipo Island, Antarctica) (from Atyeo and Peterson 1967). **134-1;** dorsum of male: **134-2;** venter of male: **134-3;** venter of female: **134-4;** dorsum of female

134-5 and 134-6; family DERMOGLYPHIDAE, *Paralges microtrichus* Gaud (Morocco) (from Gaud 1973). **134-5;** venter of female: **134-6;** dorsum of female

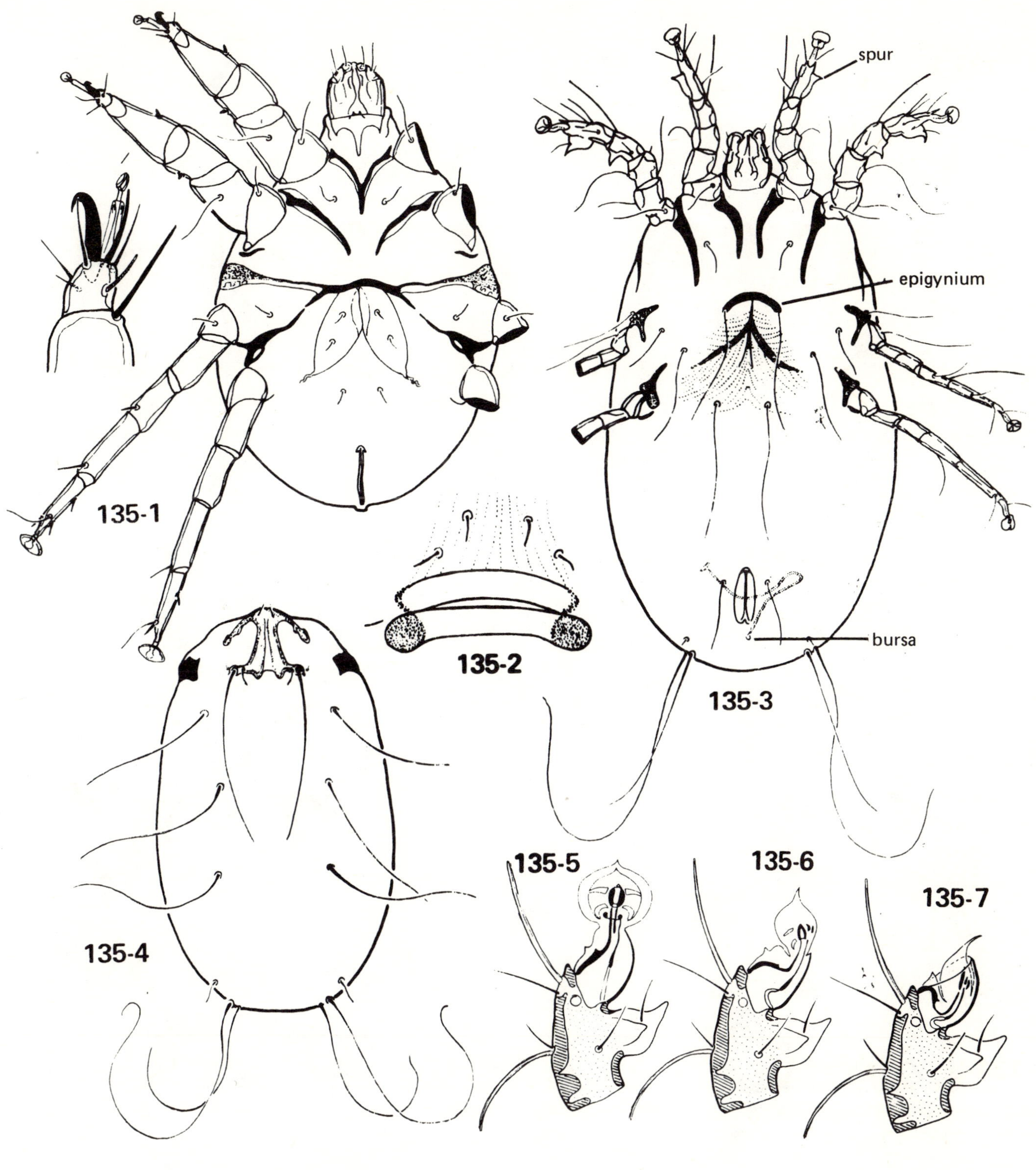

135-1 and 135-2; family TURBINOPTIDAE. 135-1; *Schoutedenocoptes* sp. (Kansas, USA), venter of female with detail of tarsus I: 135-2; *Turbinoptes strandtmanni* Boyd (Texas, USA), oviporal field of female (after Boyd 1949)

135-3 to 135-7; family ANALGIDAE. 135-3 and 135-4; *Analges* sp. (Oregon, USA). 135-3; venter of female: 135-4; dorsum of female: **135-5 to 135-7; *Anhemialges*** sp., retraction sequence of pretarsus (from Gaud and Atyeo 1977)

PLATE 136

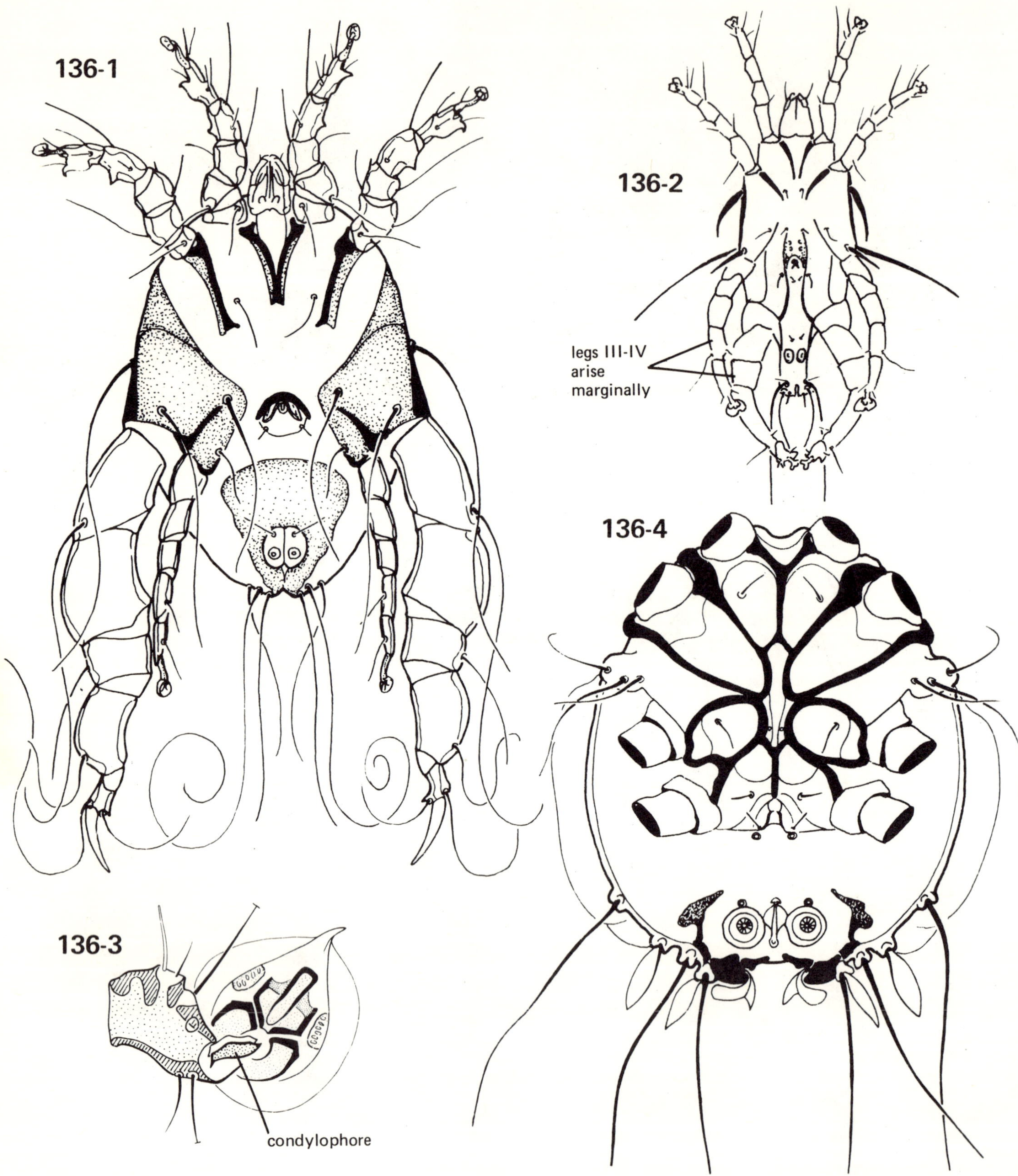

136-1; family ANALGIDAE, *Analges* sp. (Oregon, USA), venter of male

136-2; family VEXILLARIIDAE, *Pterocolurus aphyllus* Gaud and Mouchet, venter of male (after Zumpt 1961)

136-3 and 136-4; family FREYANIDAE. **136-3**; *Freyana* sp., tarsus and pretarsus of protonymph (from Gaud and Atyeo 1977): **136-4**; *Freyana* sp. (Texas, USA), venter of male

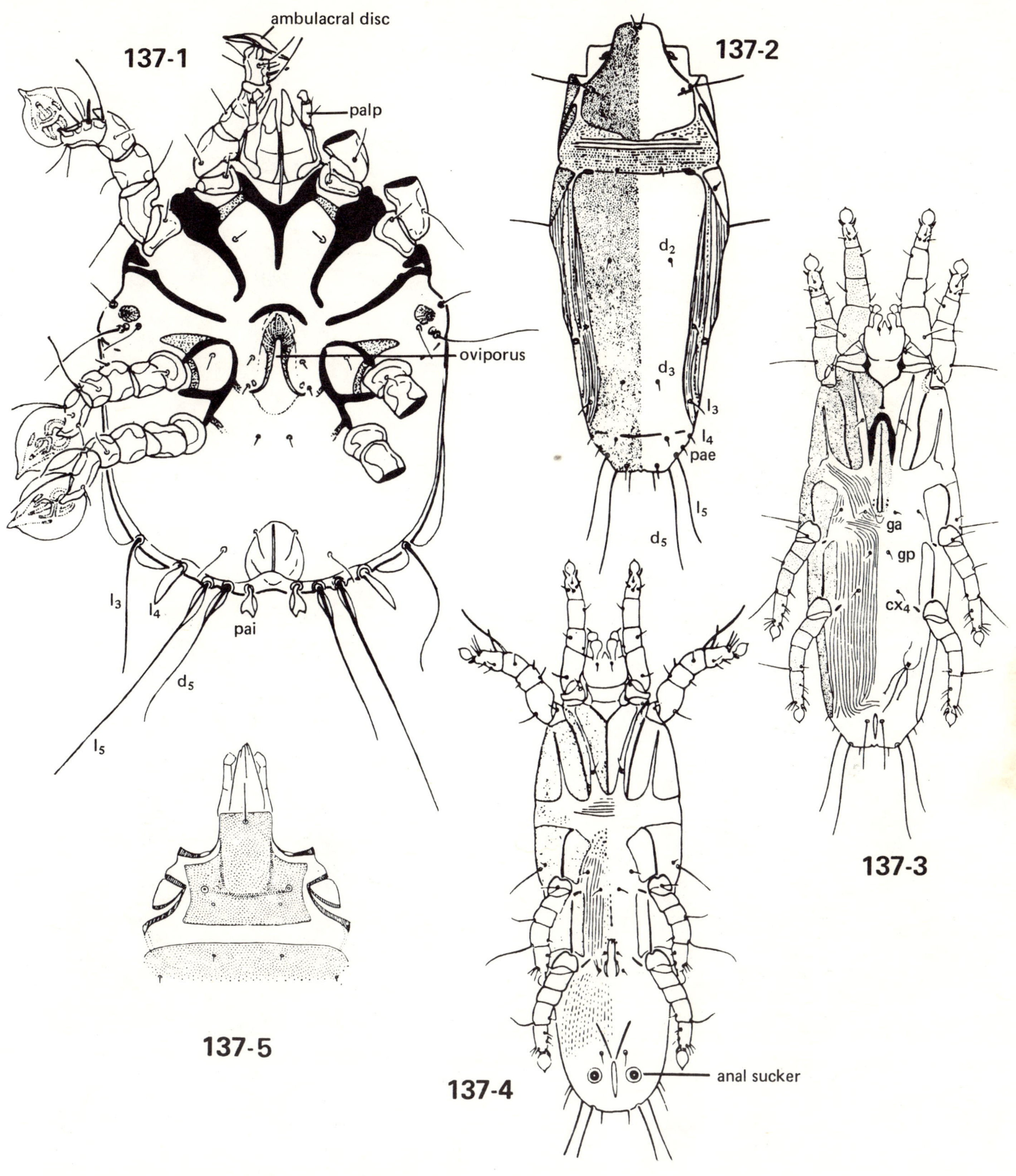

137-1; family FREYANIDAE, *Freyana* sp. (Texas, USA), venter of female

137-2 to 137-4; family RECTIJANUIDAE, *Rectijanua striata* Atyeo and Peterson (Thailand and Malaysia) (from Atyeo and Peterson 1976). **137-2**; dorsum of female: **137-3**; venter of female: **137-4**; venter of male

137-5; family EUSTATHIIDAE, *Alleusthatia ungulata* Gaud and Atyeo (Rennell Island), propdosoma of female (from Gaud and Atyeo 1967)

PLATE 138

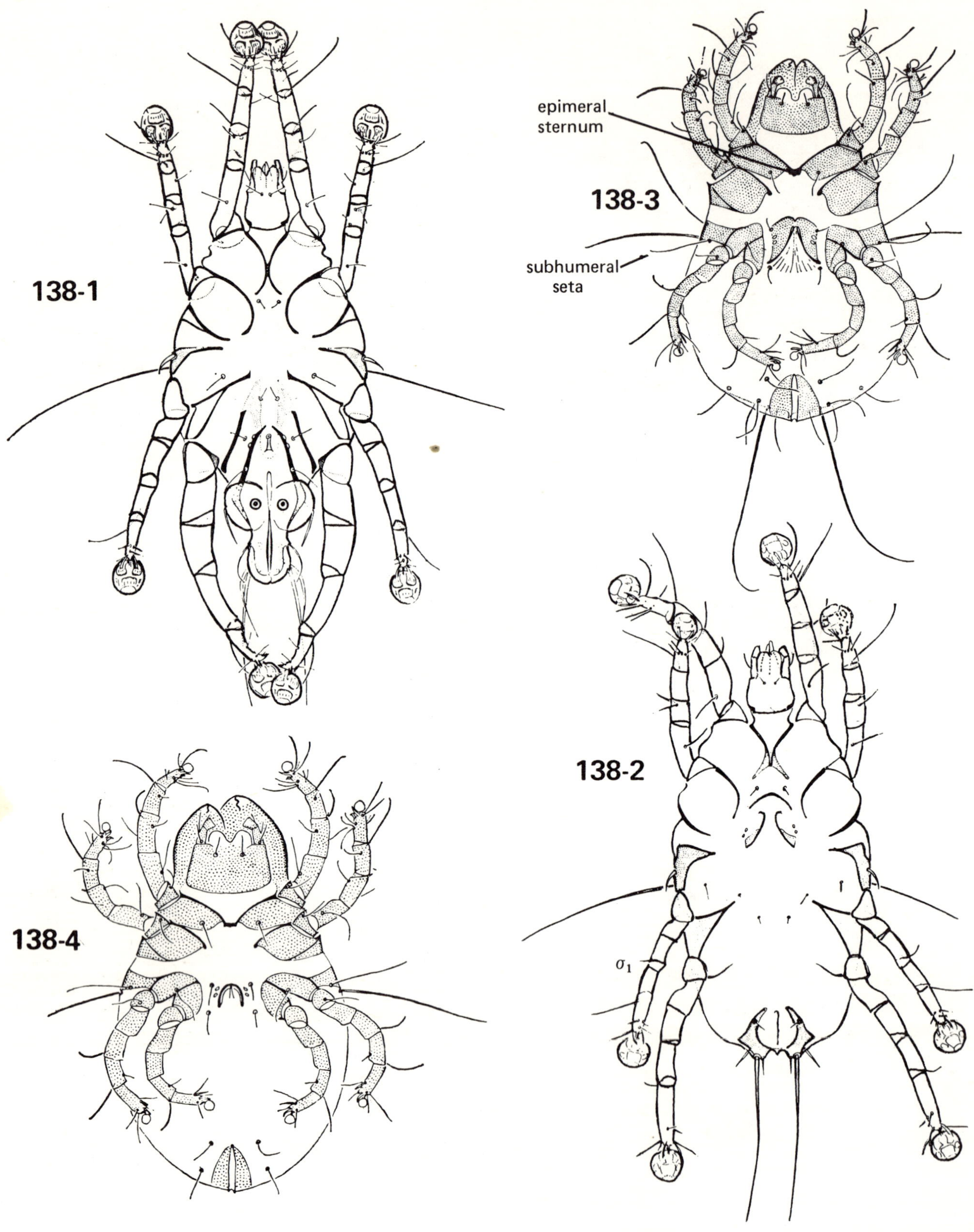

138-1 and 138-2; family EUSTATHIIDAE, *Alleustathia ungulata* Gaud and Atyeo (from Gaud and Atyeo 1967). **138-1**; venter of male: **138-2**; venter of female

138-3 and 138-4; family SYRINGOBIIDAE, *Ascogastra monstrosa* (Trouessart) (from Gaud and Atyeo 1976). **138-3**; venter of female: **138-4**; venter of male

PLATE 139

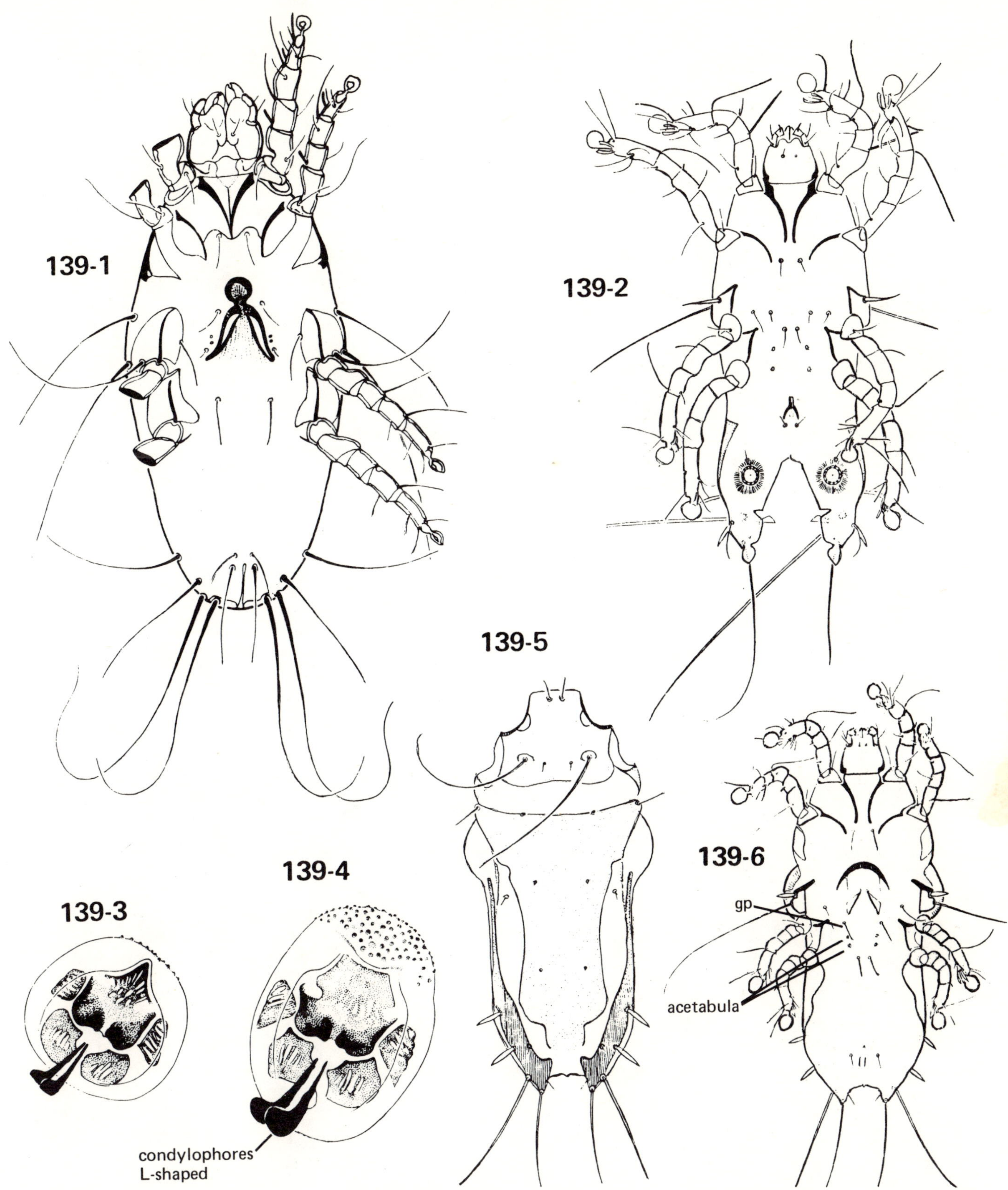

139-1; family SYRINGOBIIDAE, *Syringobia* sp. (Nebraska, USA), venter of female

139-2 to 139-6; family GABUCINIIDAE. **139-2**; *Aetacarus phylloproctus* (Mégnin and Trouessart), venter of male (from Gaud and Atyeo 1974); **139-3 and 139-4**; pretarsi of *Capitolichus* sp. (from Gaud and Atyeo 1977). **139-3**; pretarsus II of male: **139-4**; pretarsus I of female: **139-5 and 139-6**; *C. tetroplourus* Gaud and Atyeo (Colombia) (from Gaud and Atyeo 1974): **139-5**; dorsum of female: **139-6**; venter of female

PLATE 140

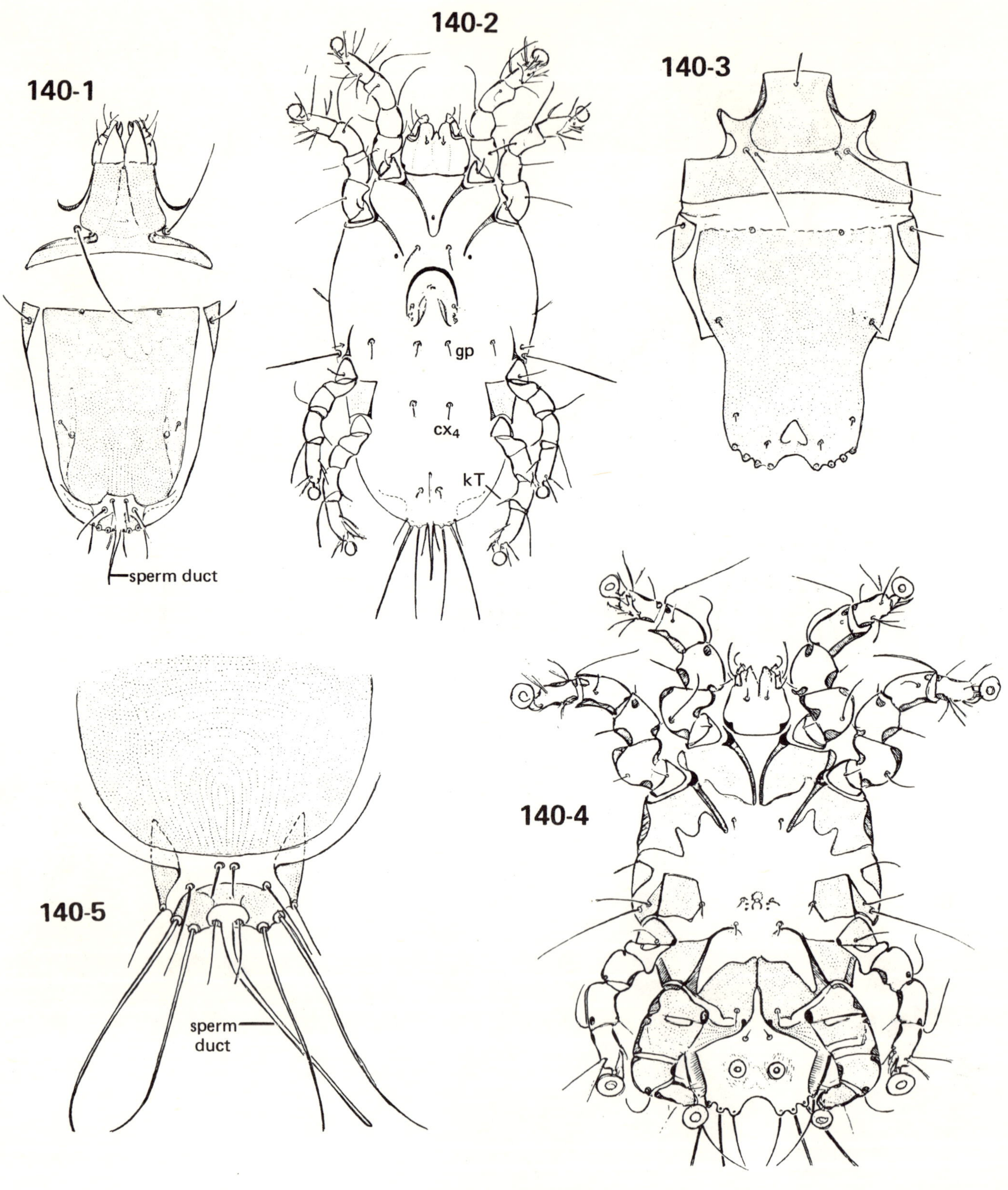

140-1 to 140-5; family CRYPTUROPTIDAE (from Gaud, Atyeo and Berla 1972). **140-1;** *Allosathes anepiandrius* Gaud, Atyeo and Berla (Brazil), dorsum of female: **140-2;** *A. anepiandrius,* venter of female: **140-3;** *Mesosathes meniscurus* Gaud, Atyeo and Berla (Mexico), dorsum of male: **140-4;** *M. meniscurus,* venter of male: **140-5;** *M. tetrasetosus* (Novaes), posterodorsal aspect of female, showing terminal sperm duct

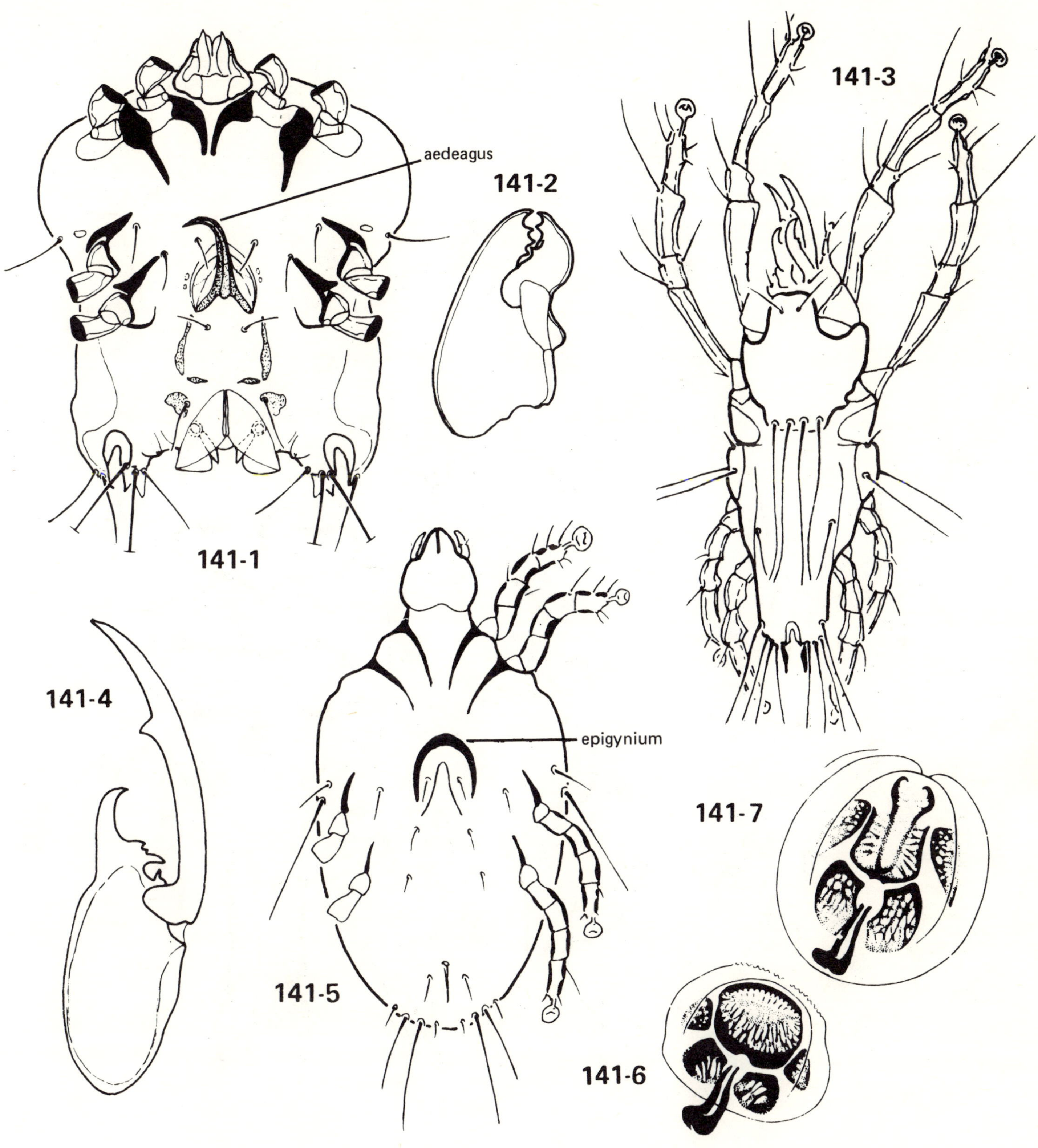

141-1; family KRAMERELLIDAE, *Kramerella* (Texas, USA), venter of male

141-2 to 141-4; family FALCULIFERIDAE. **141-2;** *Falculifer* sp., chelicera of female: **141-3;** *F. invocans* Gaud and Mouchet, dorsum of male (after Zumpt 1961): **141-4;** *Falculifer* sp., chelicera of male

141-5 to 141-7; family PTEROLICHIDAE. **141-5;** *Pterolichus francolini* Gaud and Mouchet, venter of female (after Zumpt 1961): **141-6;** *Montchadskiana* sp., pretarsus (from Gaud and Atyeo 1977): **141-7;** *Aralichus* sp., pretarsus (from Gaud and Atyeo 1977)

PLATE 142

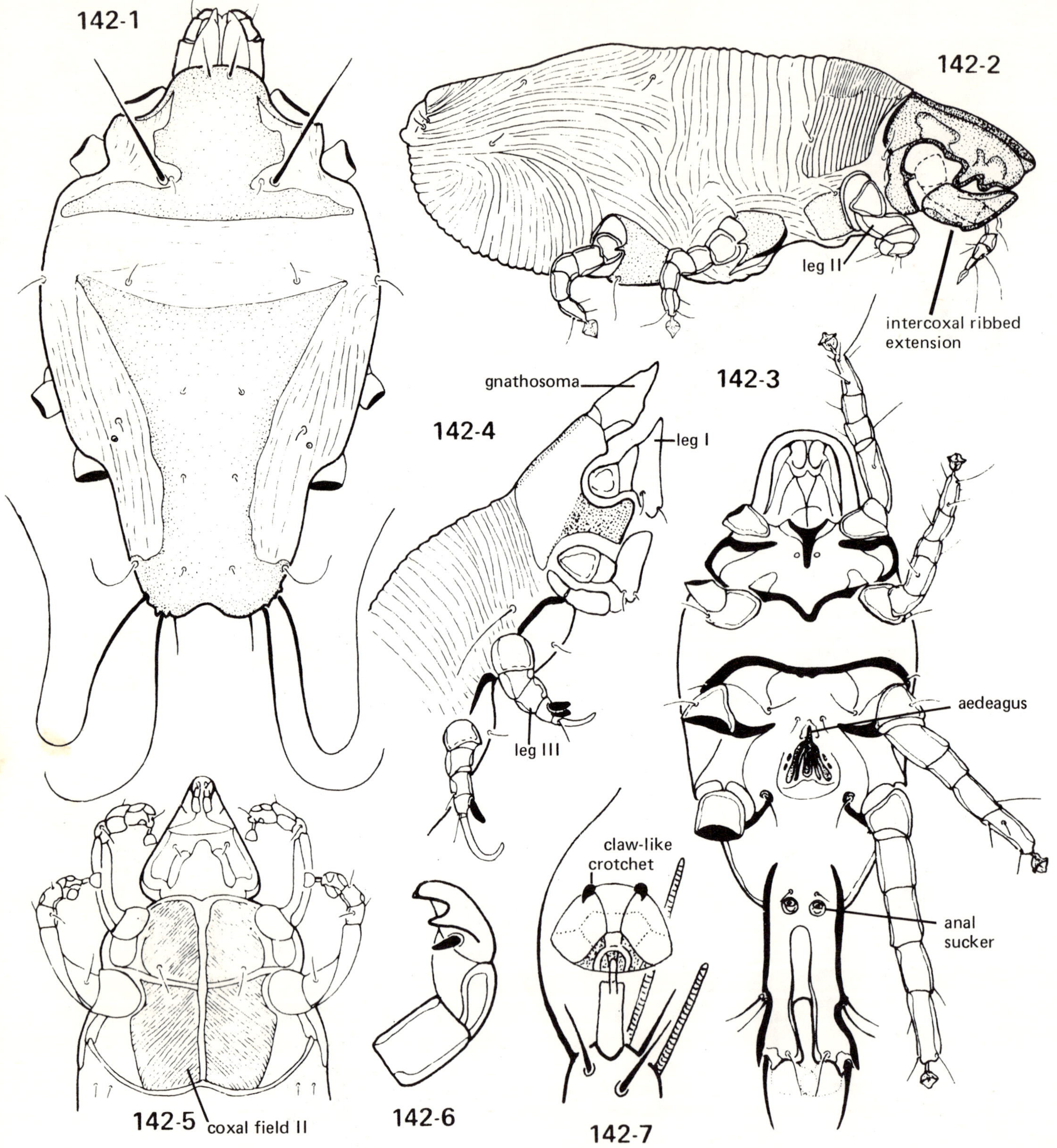

142-1; family PTEROLICHIDAE, *Pterolichus* sp. (Mexico), dorsum of male

142-2 and 142-3; family LISTROPHORIDAE, *Listrophorus* sp. (Oregon, USA). **142-2**; lateral aspect of female: **142-3**; ventral aspect of male

142-4; family CHIRODISCIDAE, *Alabidocarpus recurvus* (Womersley) (Australia), anterolateral aspect of female (after Pinichpongse 1963)

142-5; family ATOPOMELIDAE, *Chirodiscoides caviae* Hirst, anteroventral aspect of female (after Hirst 1922)

142-6 and 142-7; family MYOCOPTIDAE. **142-6**; *Trichoecius* sp., terminal portion of leg III: **142-7**; *Gliricoptes glirinus* (Canestrini), pretarsus I (ventral) (after Fain, Munting and Lukoschus 1970)

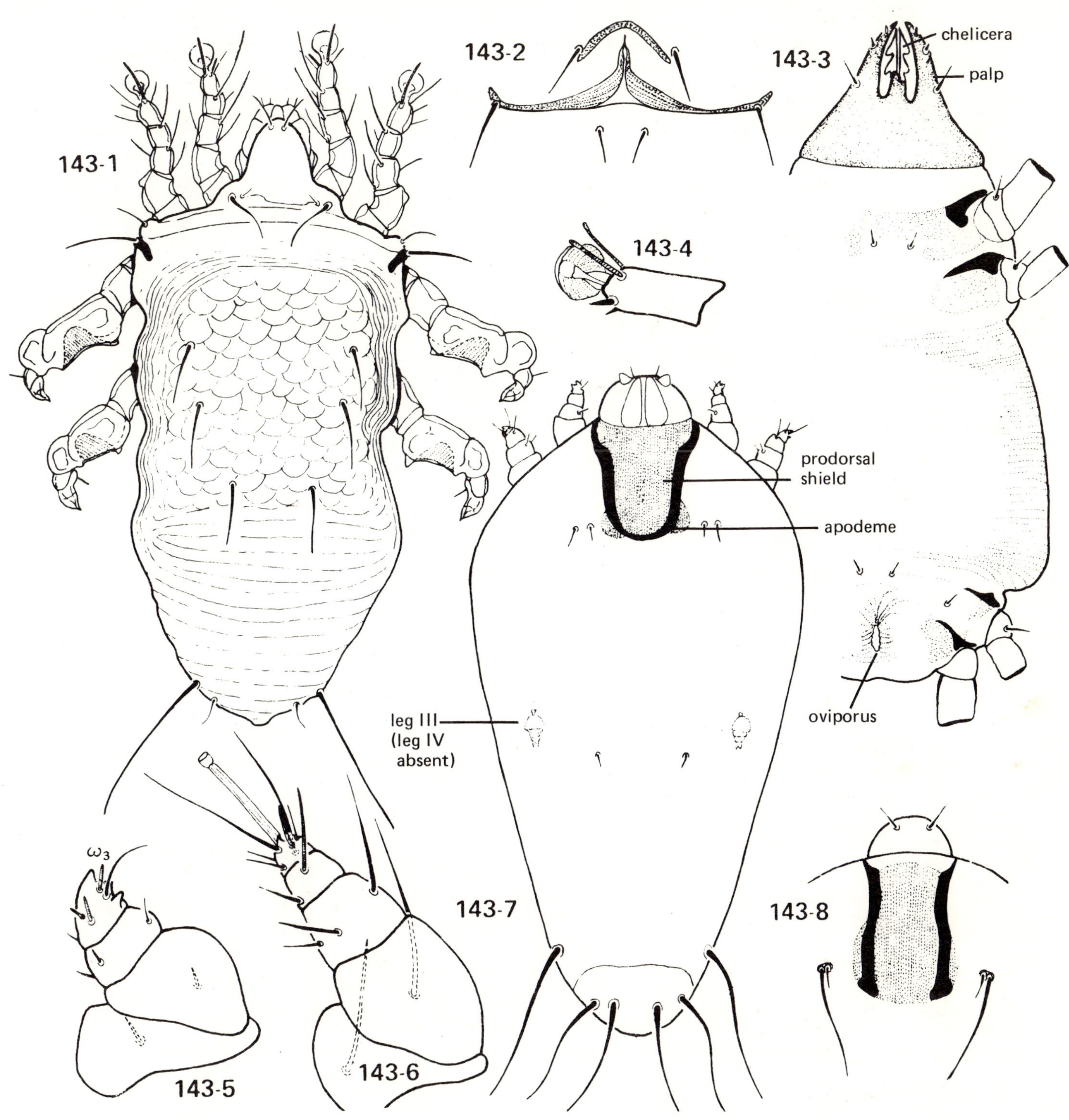

143-1 and 143-2; family MYOCOPTIDAE, *Myocoptes musculinus* (Koch) (Oregon, USA). **143-1**; dorsum of female: **143-2**; oviporal field of female

143-3 and 143-4; family CHIRORHYNCHOBIIDAE, *Chirorhynchobia urodermae* Fain (after Fain 1968). **143-3**; venter of female: **143-4**; tarsus I of female

143-5 to 143-8; family KNEMIDOCOPTIDAE. **143-5**; *Knemidocoptes mutans* (Robin & Lanquetin), leg I of female (dorsal) (after Fain and Elsen 1967): **143-6**; *K. mutans,* leg I of male (dorsal) (after Fain and Elsen 1967): **143-7**; *Evansacarus lari* Fain (Israel), dorsum of female (after Fain 1962): **143-8**; *E. lari,* propodosoma of male (after Fain 1962)

PLATE 144

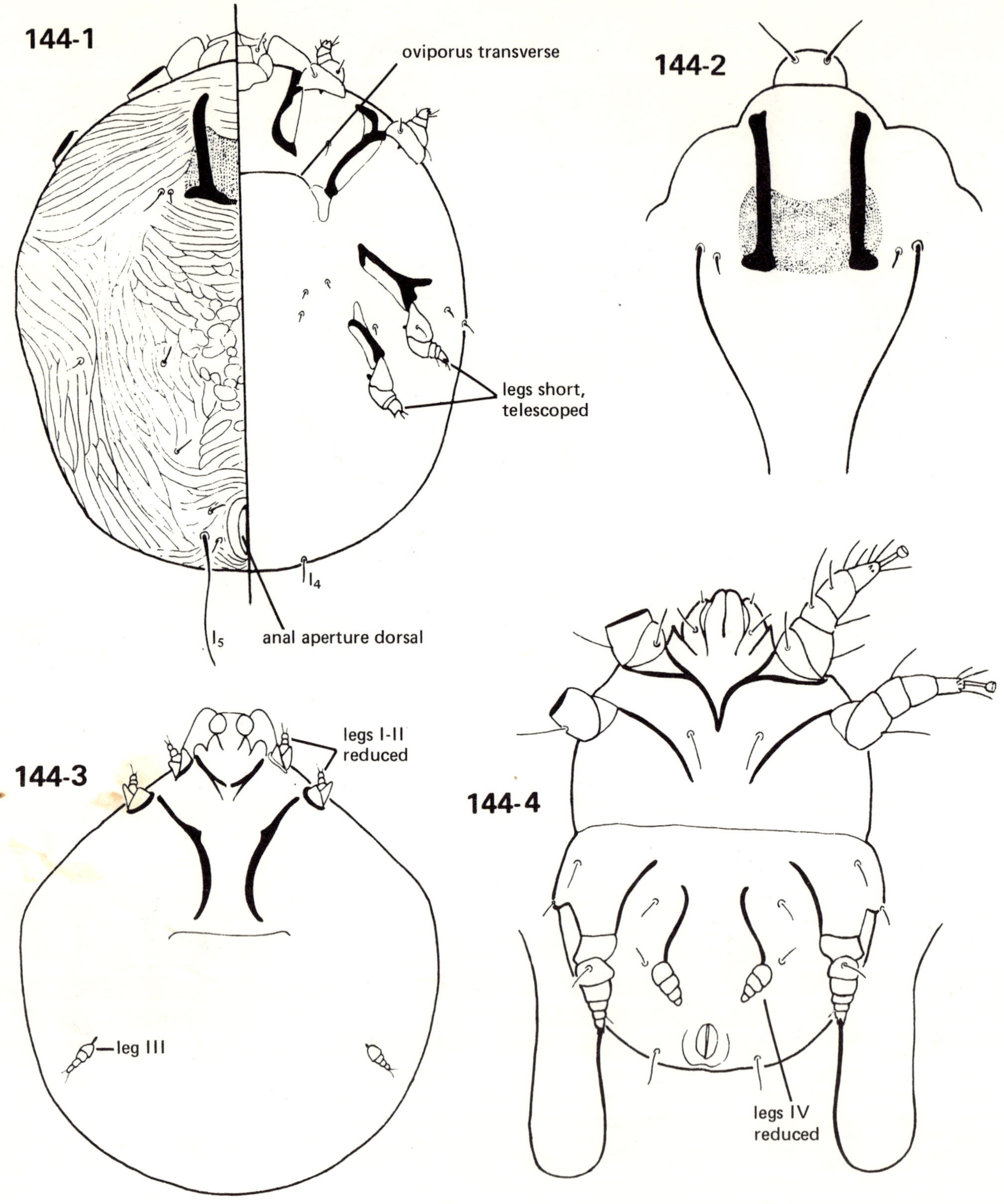

144-1 and 144-2; family KNEMIDOCOPTIDAE, *Knemidocoptes mutans* (Robin and Lanquetin). **144-1;** composite dorsum and venter of female: **144-2;** anterodorsal aspect of male

144-3 and 144-4; family TEINOCOPTIDAE. **144-3;** *Chirobia congolensis* Fain (Zaire), venter of female (after Fain 1959): **144-4;** *Teinocoptes epomophori* Rodhain (Zaire), venter of male (after Rodhain 1923)

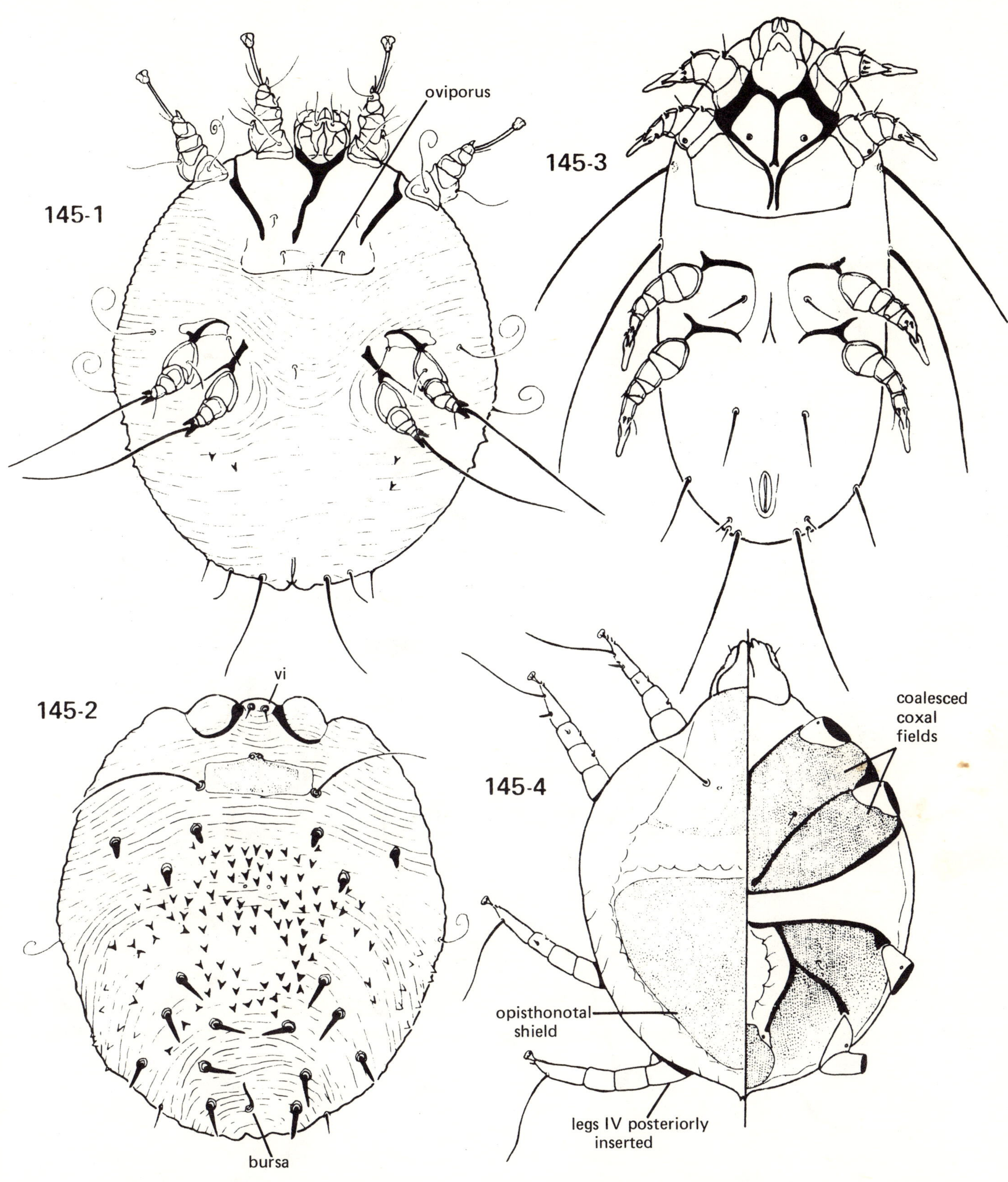

145-1 and 145-2; family SARCOPTIDAE, *Sarcoptes scabiei* (L.) (Oregon, USA). **145-1;** venter of female: **145-2;** dorsum of female
145-3; family LAMINOSIOPTIDAE, *Laminosioptes cysticola* (Vizioli), venter of female (after Hirst 1922)
145-4; family PNEUMOCOPTIDAE, *Pneumocoptes jellisoni* Baker (Idaho, USA), composite dorsum and venter of female (adapted from Baker 1951)

PLATE 146

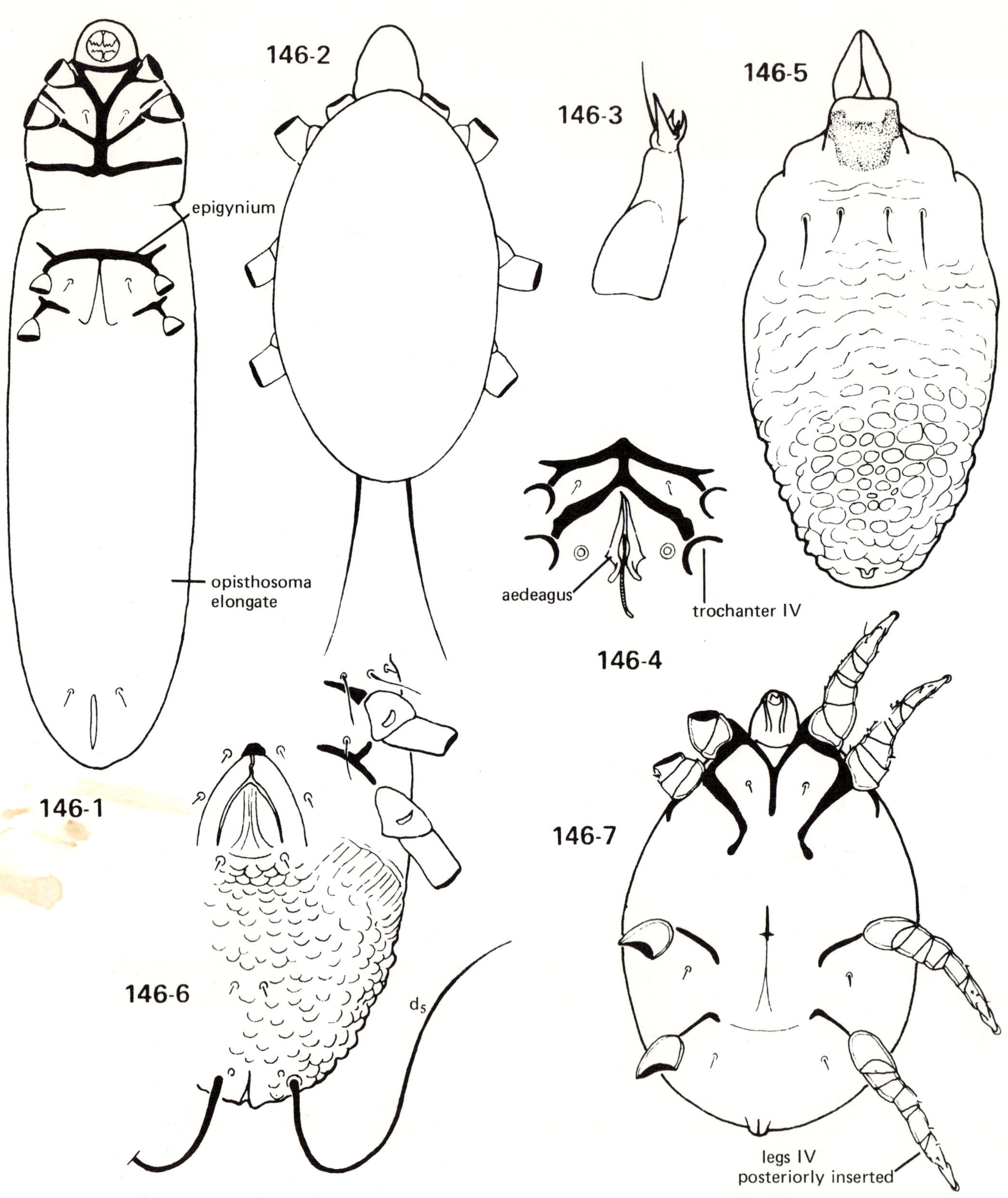

146-1 to 146-6; family GASTRONYSSIDAE. **146-1**; *Gastronyssus bakeri* Fain (Rwanda), venter of female (after Fain 1959): **146-2**; *Rodhainyssus yunkeri* Fain, dorsal profile (after Fain 1959): **146-3**; *Gastronyssus* sp., tarsus I: **146-4**; *G. bakeri,* posterior coxal and genital fields of male: **146-5**; *Sciuracarus paraxeri* Fain, dorsum of female: **146-6**; *S. paraxeri,* posteroventral aspect of female (after Fain 1964)

146-7; family CYTODITIDAE, *Cytodites nudus* (Vizioli), venter of female (after Hirst 1922)

SUBORDER ORIBATIDA

The Oribatida is a cosmopolitan group of more than 5000 species relegated to approximately 700 genera (Balogh 1972). Most oribatid species are slow-moving, strongly ornamented forms ranging in size from 200-1300 μ. Many have a laterally developed dorsal hysterosomal shield *(notogaster)* which provides pleural protection for the idiosoma. Adults often possess a discrete tracheal system consisting either of a series of ducts which open laterally between coxae II-III (Fig. 10c, p. 23), or of *brachytracheae* which open either via the acetabular cavities of the legs or through the propodosomal sensillary bases *(bothridia).* Tracheal openings are poorly defined and difficult to observe except in cleared or specially prepared material. Brachytracheal branching occurs at the acetabular origins (Woodring 1973) but tracheae are otherwise undivided. Members of the cohort Pterogasterina are considered by Wallwork (1969) to respire through a series of distinctive dorsal hysterosomal pores, the *sacculi* and the *areae porosae* (Fig. 44).

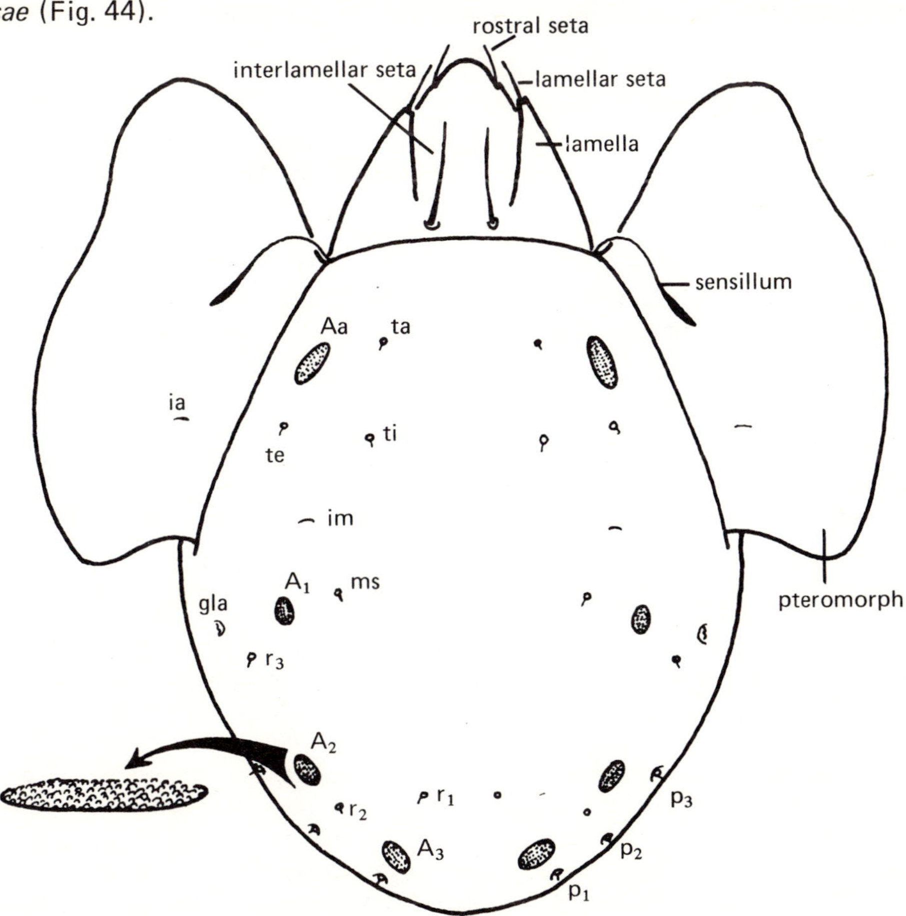

Fig. 44. Dorsum of a pterogasterine oribatid (GALUMNOIDEA), showing position of setal groups (t, ms, r, and p series), notogastral glands (gla), pori (ia, im), and the areae porosae (Aa, A_{1-3}).

The chelicerae of Oribatida are normally chelate-dentate (Plate 150-3, p. 475) but may occasionally be elongate and attenuate (Plate 158-1, p. 483). The palpi are simple sensory appendages composed of 3-5 segments. Pretarsi of the legs may consist of paired claws and a median claw-like empodium (tridactyly), or only the empodial claw may be present (monodactyly). Rarely, the pretarsi comprise two claws (bidactyly), one of which appears to be empodial.

Sexual dimorphism in the Oribatida is obscure, and determination of males and females often requires identification of internal reproductive structures. However, females are sometimes recognized by the presence of an extrusible ovipositor which may be quite large in some groups (Fig. 17c, p. 48). The ovipositor in weakly sclerotized or cleared specimens may be seen through the body wall. Males possess a short aedeagal organ which is visible as a sclerotized cone beneath the genital valves in cleared or weakly sclerotized material (Woodring 1970). The organ is adapted for extrusion and deposition of spermatophores and appended sperm packets rather than as an intromittent device. Males and females generally possess three pairs of genital acetabula (Plate 149, p. 474).

Ontogenesis in the Oribatida (Plate 147, p. 447) is either paurometabolous or holometabolous. Development in the Macropylides or "lower" Oribatida (page 453) is gradual and involves few major morphological changes (Shereef 1976). However, larvae and nymphs of the Brachypylides or "higher" Oribatida (page 458) often differ considerably from the adults in cuticular condition and pattern of idiosomal sclerotization (Grandjean 1954, 1965, Wallwork 1969).

Additional features of importance are:

1. Coxae of legs fused to the venter, often indistinguishable as separate sclerites. Legs without trichobothria.

2. A podocephalic canal present on the anterolateral aspect of the podosoma and associated with distinctive coxal glands composed of a sacculus and a convoluted labyrinth (Fig. 46, p. 453). The canal is always at least partially external.

3. A pair of well developed subcapitular rutella (Plate 150-3, p. 475). Eyes rarely present.

4. A pair of propodosomal bothridial organs or *sensilla* (pseudostigmatic organs) generally present, arising from special sensory pits or bothridia (Fig. 13, p. 28). Sensilla may be hair-like trichobothria, or variously expanded or ornamented.

5. A pair of lateral opisthonotal glands often present.

The Oribatida are primarily fungivorous or saprophagous but also consume algae, bacteria, yeasts, and higher plants (Luxton 1972). They are particularly common inhabitants of forest humus, litter, and surface soil strata in both temperate and tropical realms.

The plethora of unique and obvious morphological characters in the Oribatida usually makes specific diagnoses relatively simple. At the same time, however, there has been a tendency amongst oribatid specialists to "upgrade" the classification of the suborder, creating many monotypic genera which often are further elevated to familial rank. While such an arrangement allows for easy incorporation of new and unusual forms, it also encourages excessive proliferation of higher taxa. Such expansion is reflected in the number of monofamilial superfamilies recognized in this treatment (15 of 44), and by the number of familial categories (137) listed in the accompanying chart (pp. 450-452).

The Oribatida are grouped in two supercohorts, the largest and most complex of which is the Brachypylides. Subcohortal division of the Brachypylides is complicated by the fact that some superfamilies may partially or completely fit into either subcohort (the AMERONOTHROIDEA and the OPPIOIDEA, for example). Also, certain familial and superfamilial taxa presently recognized by specialists in the Oribatida are submerged by contiguity or overlap of basic character states. Thus it is not practical to attempt a key separation of the AMERONOTHROIDEA and CYMBAEREMAEOIDEA, nor may the CARABODOIDEA be clearly

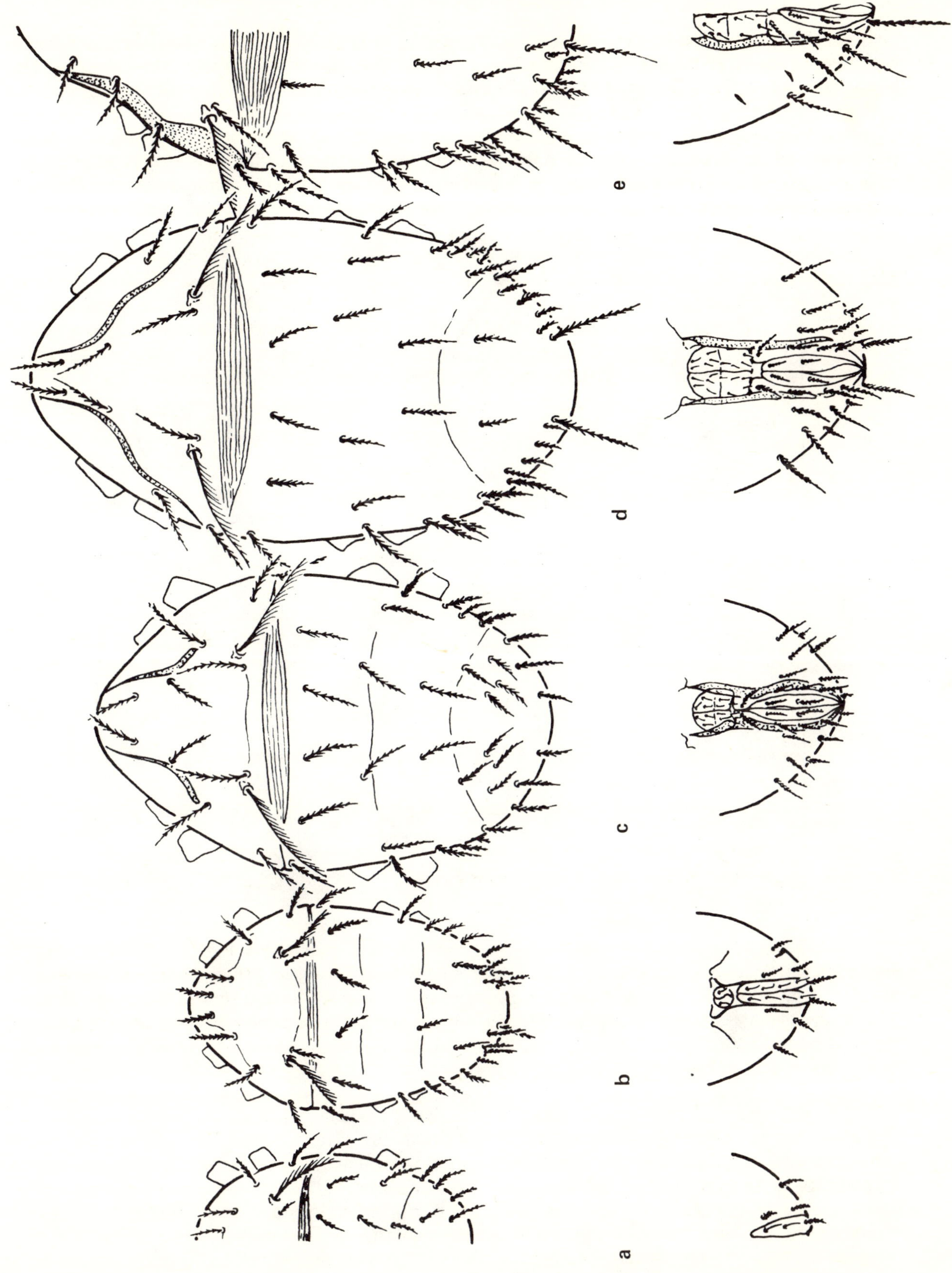

147; ontogenetic development of *Papillacarus aciculatus* Kunst (LOHMANNIIDAE) showing dorsal (top) and genital-anal aspects of each instar: a, larva; b, protonymph; c, deutonymph; d, tritonymph; e, adult (after Shereef 1976).

distinguished from the CEPHEOIDEA. Similarly, the oppioid families Oppiidae and Autognetidae are separable only on the basis of integumental characters which often are obscure. Many families of the superfamily ORIBATULOIDEA are equally difficult to define.

Higher classification of the Oribatida based on dorsal chaetotaxy has proven thus far to be of limited value. Inasmuch as homologies between setal patterns of lower and higher Oribatida have not been definitely established, it is customary to use different setal notations for each supercohort (Fig. 45).

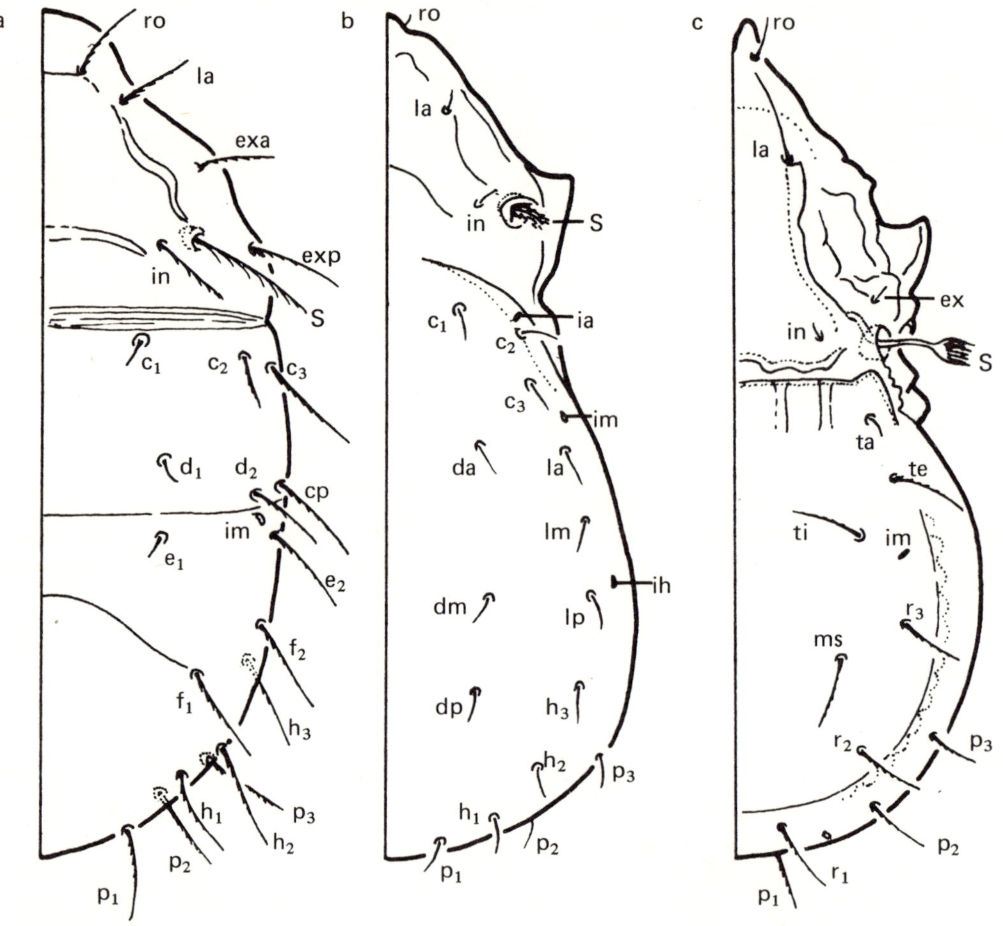

Fig. 45. Setal signatures in the Oribatida: **a**; *Torpacarus omittens* Grandjean (Macropylides, LOHMANNIOIDEA), a lower oribatid with 16 pairs of notogastral seta: **b**; *Schusteria littorea* Grandjean (Brachypylides, AMERONOTHROIDEA), a higher oribatid with 15 pairs of notogastral setae: **c**; *Autogneta pencillium* Grandjean (Brachypylides, OPPIOIDEA), a higher oribatid with 10 pairs of notogastral setae (after Balogh 1972).

The importance of immature stages in establishing a valid phylogenetic system within the Oribatida has been stressed by Travé (1964) and others. Grandjean (1954) devised a classificatory scheme for the Brachypylides based on comparative studies of immature stages, but its complexity would limit its usefulness in an introductory text. There is little doubt that analysis of cuticle condition, notogastral setal number and leg chaetotaxy in immatures will reveal relationships between families and superfamilies which existing adult-oriented classifications cannot illustrate. Analyses of known immatures have been presented by Grandjean (op. cit.) and Wallwork (1969), but relatively few families are represented in these treatments.

The superfamilial and familial classification presented below follows that of Balogh (1972). In the interests of brevity, the key to the Oribatida (page 468) is designed to be used at the superfamilial rather than the familial level. Keys to most of the presently recognized oribatid families and genera appear in three works by Balogh (1961, 1965, 1972), all of which have been used in preparing the keys and illustrations to follow. Balogh (1965) stresses that his classification is based solely on adult characters, and that it should be considered artificial for that reason. The classificatory works of Grandjean should, therefore, be consulted for an ontogenetic—and possibly phylogenetic—approach to oribatid systematics.

Useful References

Balogh, J. (1961). Identification keys of world oribatid (Acari) families and genera. Acta Zool. Acad. Sci. Hung. **7**(3-4):243-344.

Balogh, J. (1965). A synopsis of the world oribatid (Acari) genera. Acta Zool. Acad. Sci. Hung. **11**(1-2):5-99.

Balogh, J. (1972). The Oribatid Genera of the World. Akadémiai Kiadó, Budapest: 188 pp. + 71 plates.

Bulanova-Zakhvatkina, E.M., B.A. Vainshtein, V.I. Volgin, M.S. Gilyarov, L.D. Golosova, D.A. Krivolutskii, A.B. Lange, V.D. Sevast'yanov, L.G. Sitnikova and E.S. Shaldybina (1975). Group Oribatei Dugès. **In** A Key to the Soil-inhabiting Mites (Sarcoptiformes). "Nauka", Moscow: 381 pp.

Engelmann, H.-D. (1968-1976). Bibliographia Oribatologica, Numbers 1-9. Abh. Ber. Naturkundemuseums Görlitz, Leipzig: **44**(1):1-24 (1968); **44**(1):1-12 (1969); **45**(1):1-15 (1970); **46**(1):1-22 (1971); **47**(1): 1-15 (1972); **48**(1):1-19 (1973); **48**(1a):1-20 (1973); **49**(1):1-18 (1975); **50**(1):1-19 (1976).

Grandjean, F. (1951). Les relations chronologiques entre ontogenèses et phylogenèses d'apres les petits charactères discontinus des Acariens. Bull. Biol. Fr. Belg. **85**(3):269-292.

Grandjean, F. (1954). Essai de classification des Oribates (Acariens). Bull. Soc. zool. France **78**:421-446.

Grandjean, F. (1965). Complément à mon travail de 1953 sur la classification des Oribates. Acarologia **7**(4):713-734.

Grandjean, F. (1969). Considerations sur le classement des Oribates leur division en 6 groupes majeurs. 1. Les affinités de *Collohmannia gigantea* Selln. 1922. Acarologia **11**(1):127-153.

Knülle, W. (1957). Morphologische und Entwicklungsgeschictliche untersuchungen zum phylogenetischen System der Acari: Acariformes Zachv. 1. Oribatei: Malaconothridae. Mitt. Zool. Mus. Berlin **33**(1):97-213.

Shereef, G.M. (1976). Biological studies and description of stages of two species: *Papillacarus aciculatus* Kunst and *Lohmannia egypticus* Elbadry and Nasr (Oribatei-Lohmanniidae) in Egypt. Acarologia **18**(2):351-359.

Travé, J. (1964). Importance des stases immatures des Oribates en systematique et en écologie. Acarologia **6**(h.s.):47-53.

Wallwork, J.A. (1969). Some basic principles underlying the classification and identification of cryptostigmatid mites. **In** The Soil Ecosystem. J.G. Sheals, ed. Syst. Assoc. Publ. 8:155-168.

Woodring, J.P. (1970). Comparative morphology, homologies and functions of the male system in oribatid mites (Arachnida:Acari). J. Morphol. **132**(4):425-451.

Woodring, J.P. (1973). Comparative morphology, functions and homologies of the coxal glands in oribatid mites (Arachnida:Acari). J. Morphol. **139**(4):407-429.

Woolley, T.A. (1958). A preliminary account of the phylogeny of the Oribatei (Acarina:Sarcoptiformes). Proc. 10th Int. Cong. Ent. **1**:867-873.

Woolley, T.A. and E.W. Baker (1958). A key to the superfamilies and principal families of the Oribatei (Sarcoptiformes:Acarina). Ent. News **69**:85-92.

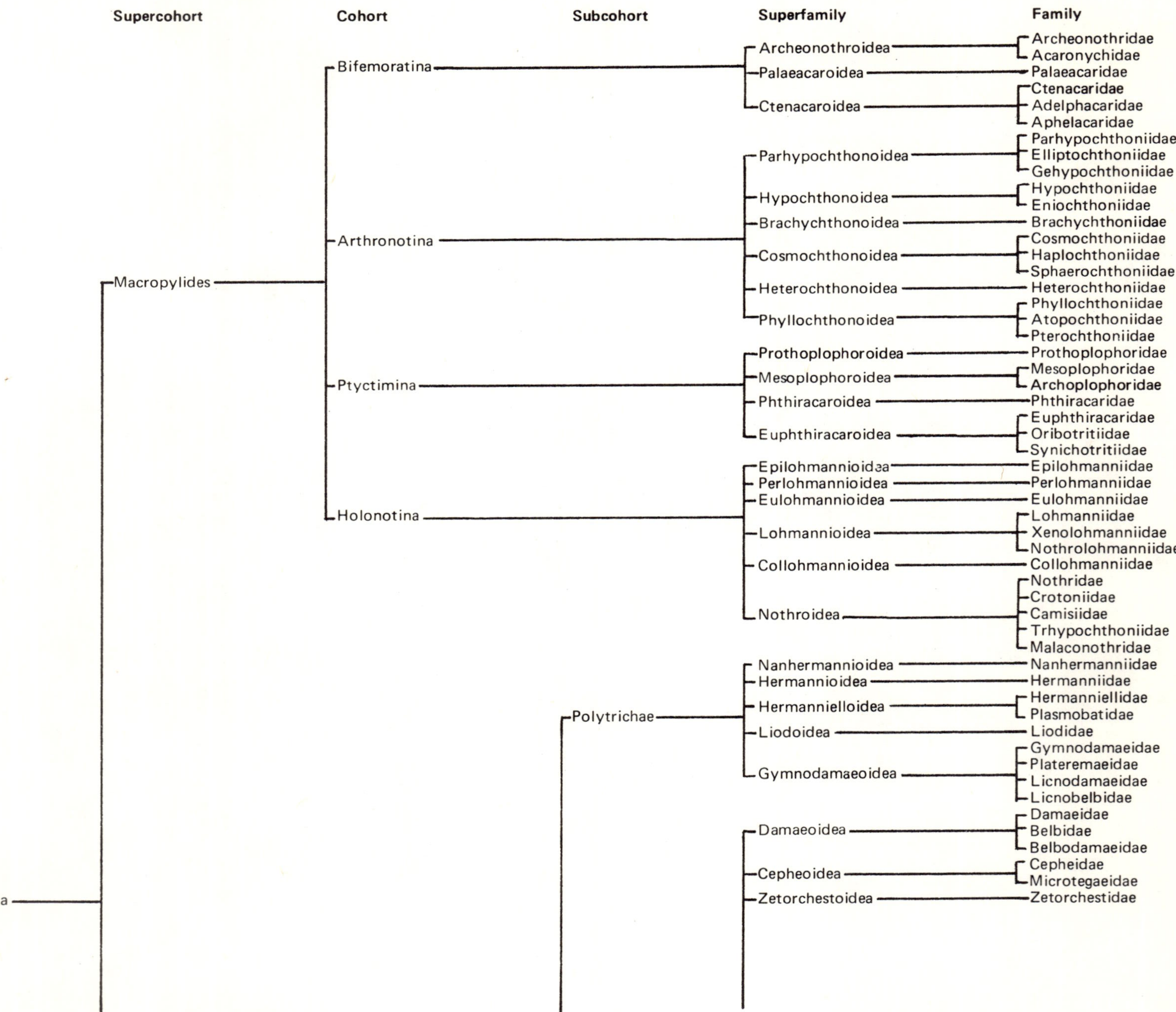
Supercohort
Cohort
Subcohort
Superfamily
Family
Archeonothroidea
Archeonothridae
Acaronychidae
Bifemoratina
Palaeacaroidea
Palaeacaridae
Ctenacaridae
Ctenacaroidea
Adelphacaridae
Aphelacaridae
Parhypochthoniidae
Parhypochthonoidea
Elliptochthoniidae
Gehypochthoniidae
Hypochthoniidae
Hypochthonoidea
Eniochthoniidae
Brachychthonoidea
Brachychthoniidae
Arthronotina
Cosmochthoniidae
Cosmochthonoidea
Haplochthoniidae
Macropylides
Sphaerochthoniidae
Heterochthonoidea
Heterochthoniidae
Phyllochthoniidae
Phyllochthonoidea
Atopochthoniidae
Pterochthoniidae
Prothoplophoroidea
Prothoplophoridae
Mesoplophoridae
Mesoplophoroidea
Ptyctimina
Archoplophoridae
Phthiracaroidea
Phthiracaridae
Euphthiracaridae
Euphthiracaroidea
Oribotritiidae
Synichotritiidae
Epilohmannioidea
Epilohmanniidae
Perlohmannioidea
Perlohmanniidae
Eulohmannioidea
Eulohmanniidae
Holonotina
Lohmanniidae
Lohmannioidea
Xenolohmanniidae
Nothrolohmanniidae
Collohmannioidea
Collohmanniidae
Nothridae
Crotoniidae
Nothroidea
Camisiidae
Trhypochthoniidae
Malaconothridae
Nanhermannioidea
Nanhermanniidae
Hermannioidea
Hermanniidae
Hermanniellidae
Hermannielloidea
Polytrichae
Plasmobatidae
Liodoidea
Liodidae
Gymnodamaeidae
Plateremaeidae
Gymnodamaeoidea
Licnodamaeidae
Licnobelbidae
Damaeidae
Damaeoidea
Belbidae
Belbodamaeidae
Cepheidae
Cepheoidea
Microtegaeidae
Oribatida
Zetorchestoidea
Zetorchestidae

Apterogasterina

Oligotrichae

Brachypylides

Eremaeoidea
Eremaeidae
Megeremaeidae

Eremuloidea
Amerobelbidae
Eremulidae
Staurobatidae
Damaeolidae
Basilobelbidae
Heterobelbidae
Ameridae
Ctenobelbidae

Liacaroidea
Gustaviidae
Metrioppiidae
Multoribulidae
Liacaridae
Xenillidae
Astegistidae
Tenuialidae

Polypterozetoidea
Charassobatidae
Eutegaeidae
Tumerozetidae
Polypterozetidae
Eremaeozetidae
Nodocepheidae
Podopterotegaeidae

Carabodoidea
Niphocepheidae
Carabodidae
Nippobodidae
Tectocepheidae

Oppioidea
Tuparezetidae
Sternoppiidae
Oxyameridae
Oppiidae
Autognetidae
Thyrisomidae
Suctobelbidae
Rhynchoribatidae
Caleremaeidae
Eremellidae
Arceremaeidae
Machadobelbidae
Spinozetidae
Anderemaeidae

Otocepheoidea
Dampfiellidae
Otocepheidae
Tokuncepheidae

Hydrozetoidea
Hydrozetidae
Limnozetidae

Cymbaeremaeoidea
Cymbaeremaeidae
Kodiakellidae
Micreremaeidae

Ameronothroidea
Ameronothridae
Podacaridae
Selenoribatidae
Fortuyniidae

(continued)

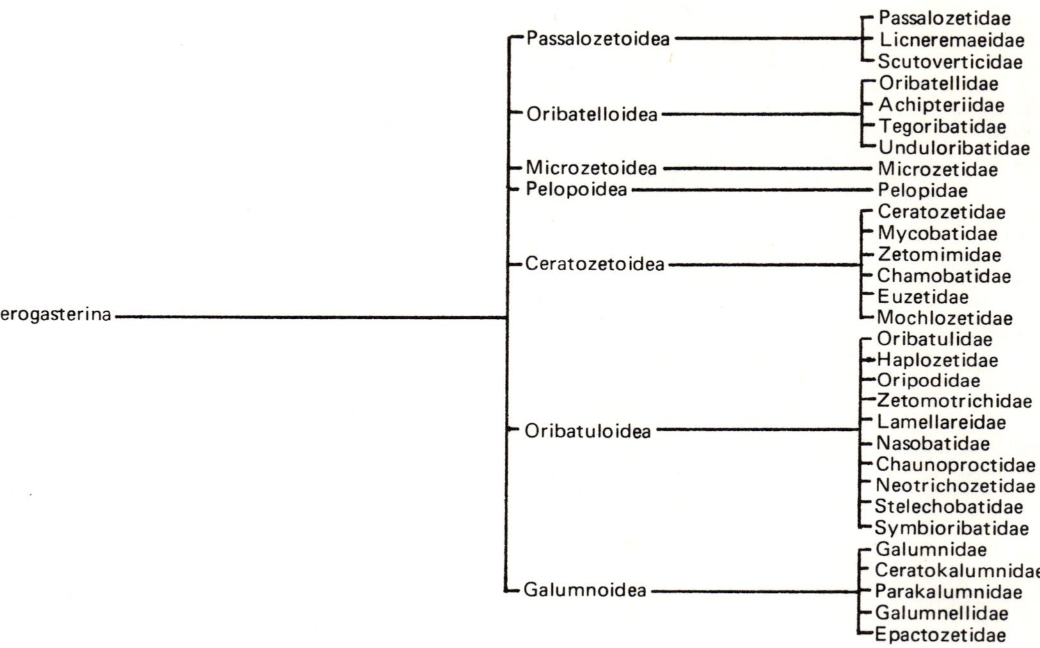

Fig. 46. Dendrogram illustrating possible relationships within the suborder Oribatida of the order Acariformes.

Supercohort Macropylides

The Macropylides, or lower Oribatida, includes 38 families grouped in 19 superfamilies which are relegated to four cohorts—the Bifemoratina, Arthronotina, Ptyctimina and Holonotina. Macropylides may be broadly characterized as species with contiguous genital and anal shields occupying the entire length of the genitoanal field, and with leg tibiae and genua of uniform length and shape. The labyrinths of the macropylide coxal glands each have three 180° turns, forming a clockwise or counter-clockwise spiral (Fig. 47a) (Woodring 1973). Members of the Pcyctimina are structured in such a way that the proterosoma (gnathosoma + propodosoma) may be folded back into the hysterosoma as a protective strategy (*ptychoidy*, Plate 153-1, p. 478). Other macropylide cohorts are normally articulated, or *aptychoid.*

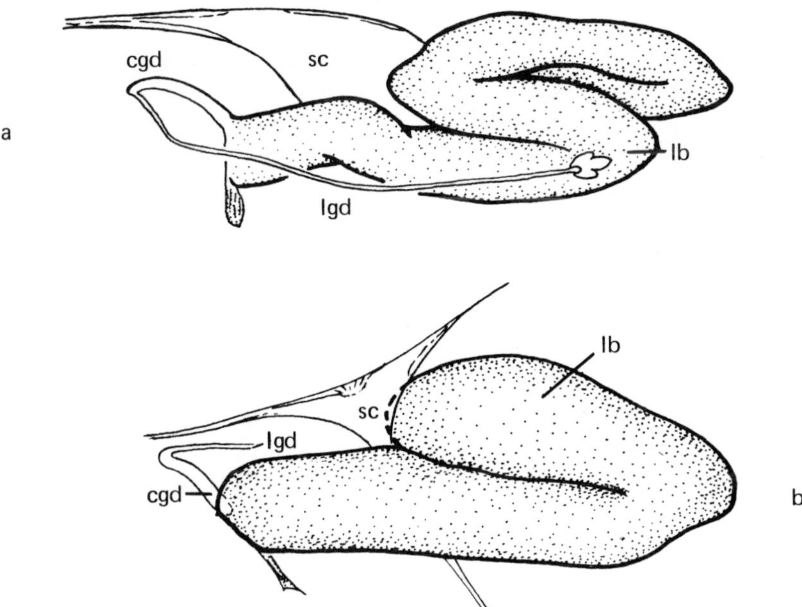

Fig. 47. Major coxal gland types in the Oribatida: **a**; *Collohmannia gigantea* Sellnick (Macropylides, COLLOHMANNIOIDEA): **b**; *Gustavia microcephala* Nicolet) (Brachypylides, LIACAROIDEA) (cgd = coxal gland duct, lb = labyrinth, lgd = lateral gland duct, sc = sacculus) (after Woodring 1973).

Cohort Bifemoratina

DIAGNOSIS: Soft-bodied, aptychoid, with or without transverse hysterosomal sutures; mostly unpigmented and whitish, often with long black setae; with a pair of piliform, fusiform or claviform prodorsal sensilla; gnathosoma visible from above. Femora of legs divided, tarsus I with four solenidia; tarsi generally monodactylous, but bi- or tridactylous at least in some immature stages. Genital valves weakly armored, with three pairs of acetabula.

The Bifemoratina are small (2-500 μ) weakly sclerotized species which may superficially resemble some of the free-living Acaridida rather than the armored higher Oribatida. Bifemoratines are widely distributed both in temperate and tropical regions (Balogh 1972), and are especially prevalent in forest litter and humus. Trägårdh (1932) suggested that bifemoratines are phytophagous or saprophagous, feeding on mosses, lichens or decayed leaves. Some appear to be very widely distributed. *Acaronychus trägårdhi* Grandjean

(ARCHEONOTHROIDEA, Plate 150, p. 475), for example, has been taken in France, Algeria, Morocco and North Carolina, USA (Grandjean 1954a). It has also been collected from moss in Colorado (Woolley 1960). The archeonothroid *Andacarus ligamentifer* Hammer is an arboreal form which lives in moist lichens and moss on tree bark (Hammer 1972). *Palaeacarus appalachicus* Jacot and *P. hystricinus* Trägårdh (PALAEACAROIDEA, Plate 151, p. 476) occur in Europe and in North America (Johnston 1967), and apparently feed on fungal hyphae in forest litter. *P. araneola* Grandjean inhabits humus under palms in North Africa and in Venezuela (Grandjean 1932a). *Aphelacarus acarinus* (Berlese) (CTENACARIODEA, Plate 150, p. 475) has been collected in various habitats in Europe and North Africa, including stored grain. The same species was taken from a grain storage and from a nest of subterranean termites in Oregon.

Cohort Arthronotina

DIAGNOSIS: Soft-bodied, aptychoid, with 1-3 transverse hysterosomal sutures; generally weakly pigmented and without long black setae, sometimes with paired opisthonotal glands, gnathosoma not completely visible from above; prodorsal sensilla usually present, variously developed. Femora of legs undivided, tarsus I with fewer than four solenidia, tarsi of immatures invariably monodactylous, some adults bidactylous.

Six superfamilies and 13 families are included in the Arthronotina. Members of the PARHYPOCHTHONOIDEA (Plate 152, p. 477) and HYPOCHTHONOIDEA (Plate 151, p. 476) apparently feed on leaf litter, soil and humus (Wallwork 1958, 1959). The hypochthonoid *Hypochthoniella minutissima* (Berlese) has been collected from moss in Chile (Hammer 1962) and from various litter habitats in Europe (Pérez-Iñigo 1968). *Eniochthonius* species may feed on fungal spores (Woolley 1960), while members of the genus *Hypochthonius* are thought to consume decaying plant and animal tissues (Bulanova-Zakhvatkina et al. 1975). Parhypochthonoids are primarily holarctic in distribution. A species of *Parhypochthonius* was recovered from an ant nest in a treehole in British Columbia, Canada.

The cosmopolitan, diminutive BRACHYCHTHONOIDEA (Plate 151, p. 476) are litter and moss inhabitants, as are the often highly ornamented COSMOCHTHONOIDEA (Plate 152, p. 477) and HETEROCHTHONOIDEA (Plate 152, p. 477). Grandjean (1950a) considers the dilated leaf-like dorsal setae of *Pterochthonius angelus* (Berlese) (PHYLLOCHTHONOIDEA, Plate 152, p. 477) to be protective devices for this fragile litter-inhabiting species. Other phyllochthonoids are similarly ornamented.

Cohort Ptyctimina

DIAGNOSIS: Well sclerotized, capable of ptychoidy, gnathosoma generally not visible from above, hysterosoma often compressed laterally; prodorsal sensilla present, occasionally minute. Femora of legs undivided, tarsus I with fewer than four solenidia; tarsi of immatures monodactylous, adult tarsi mono-, bi-, or tridactylous.

Ptyctimines are widely distributed in soil and litter, but are especially abundant in woody plant tissues, bark, and decaying tree stumps. Most ptyctimines appear to be macrophytophages (Luxton 1972), feeding on decaying parts of higher plants. They are considered to be important decomposers, especially in coniferous forests where the conifer needles are largely ignored by other saprophagous mesofauna (Bulanova-Zakhvatkina et al. 1975).

Phthiracarus borealis Trägårdh (PHTHIRACAROIDEA, Plate 153, p. 478), is a ptyctimine species which consumes deciduous leaves as well as hemlock needles (Wallwork 1958). Adults and immatures of some *Phthiracarus* and *Steganacarus* species are xylophagous, feeding in twigs. Others burrow in the woody vascular tissue of leaves (Hartenstein 1962a, Luxton 1972). An undescribed species of *Steganacarus* has been taken from holly root crowns in Oregon where it may have been browsing on root tissue. *S. balearicus* Pérez-Íñigo occurs in dry beach habitats in southern Europe (Pérez-Íñigo 1968). Its feeding habits are unknown. Members of the genus *Mesoplophora* (MESOPLOPHOROIDEA, Plate 153, p. 478) appear to consume woody tissue, and *Euphthiracarus* species (EUPHTHIRACAROIDEA, Plate 153, p. 478) tunnel and feed in twigs of deciduous and coniferous trees (Wallwork, op. cit.). A species of the euphthiracaroid genus *Oribotritia* is found on fir and cedar needles in northwestern Oregon. An undescribed mesoplophoroid was collected in substantial numbers from an ant nest in a Brazilian rain forest.

Representatives of the monofamilial superfamily PROTHOPLOPHOROIDEA (Plate 153, p. 478) are small, weakly tanned species with distinct traces of notogastral segmentation. Grandjean (1954b) notes that prothoplophoroids are mentioned only occasionally in the literature and appear to be quite rare. Large numbers of *Cryptoplophora abscondita* Grandjean, a species recorded from holarctic and Neotropical habitats, were found in damp palm litter in a garden in Algeria. *Aedoplophora glomerata* Gr. was taken from decaying root tissue in Venezuela (Grandjean 1932b). Other prothoplophoroid species have been collected in Java and South America (Grandjean, op. cit., Balogh 1972). *Prothoplophora palpalis* Berlese occurs in central Asia (Bulanova-Zakhvatkina et al. 1975).

Cohort Holonotina

DIAGNOSIS: Well sclerotized, aptychoid, gnathosoma not completely visible from above, idiosoma cylindrical or dorsoventrally compressed; prodorsal sensilla variously developed, occasionally absent. Femora of legs undivided, tarsus I with fewer than four solenidia; tarsi of immatures monodactylous, adult tarsi mono- or tridactylous.

The six superfamilies of Holonotina are represented by twelve families of soil and litter forms, many of which are widely distributed. The EULOHMANNIOIDEA (Plate 149, p. 474) and COLLOHMANNIOIDEA (Plate 154, p. 479) are monotypic holonotine taxa represented by *Eulohmannia ribagai* Berlese and *Collohmannia gigantea* Sellnick respectively. *E. ribagai* is a relict holarctic species which occurs commonly in moist forest litter. Members of the monogeneric superfamily PERLOHMANNIOIDEA (Plate 154, p. 479) are litter inhabitants but some species have been taken from pasture soils. *Perlohmannia dissimilis* (Hewitt) is a macrophytophage which damages the root systems of potato, strawberry and tulip plants (Evans et al. 1961). The LOHMANNIOIDEA (Plate 147, p. 447) are generally abundant in the tropics although some genera are known from temperate realms *(Papillacarus, Lohmannia, Cryptacarus). Papillacarus aciculatus* Kunst and *Lohmannia egypticus* Elbadry and Nasr burrow in decaying plant roots (Shereef 1976) and probably feed in these sites. *L. hispaniola* Pérez-Íñigo inhabits the root systems of xerophytic shrubs in Spain (Pérez-Íñigo 1968). *Meristacarus porcula* Grandjean lives in the bark and rotting wood of tree stumps in Panama (Grandjean 1934). Other lohmannioids have been collected from coastal bogs and moss, and one species from Zaire was found "avec insecte et arachnide".

Representatives of the holonotine superfamily NOTHROIDEA (Plate 150, p. 480) occur in moist soil and litter habitats throughout the world. Many are large (> 1000 μ) strongly armored forms with thick integuments, factors which may help explain their presence as fossils in Jurassic and Cretaceous deposits (Bulanova-Zakhvatkina 1975). *Platynothrus peltifer* (Koch) is a common and typical nothroid which feeds primarily on fungi in forest soils and litter (Hartenstein 1962b). Grandjean (1950c) recovered *Camisia segnis* (Hermann), a widely distributed arboricolous nothroid, from lichen thalli on a maple tree in France. *C. segnis* females were observed to oviposit beneath the gummy lichen surface. Additional habitats of *C. segnis* are listed by Grandjean (1936). *Nothrus palustris* C.L. Koch is a panphytophage, feeding on leaves, fungi and yeasts (Luxton 1972). Other nothroid species are reported from moss, wet grass, dune grass roots, and ferns.

Useful References

Balogh, J. (1972). The Oribatid Genera of the World. Akadémiai Kiadó, Budapest. 188 pp. + 71 plates.

Berlese, A. (1923). Centuria sesta di Acari nouvi. Redia **15**:237-262. [PERLOHMANNIOIDEA]

Bulanova-Zakhvatkina, E.M., B.A. Vainshtein, V.I. Volgin, M.S. Gilyarov, L.D. Golosova, D.A. Krivolutskii, A.B. Lange, V.D. Sevast'yanov, L.G. Sitnikova and E.S. Shaldybina (1975). Group Oribatei Dugès. **In** A Key to the Soil-inhabiting Mites (Sarcoptiformes). "Nauka", Moscow: 381 pp.

Evans, G.O., J.G. Sheals and D. Macfarlane (1961). The mites associated with plants. **In** The Terrestrial Acari of the British Isles. Volume I. Introduction and Biology. British Museum (Natural History), London: 107-131.

Ewing, H.E. (1917). A synopsis of the genera of beetle mites with special reference to the North American fauna. Ann. Ent. Soc. Amer. **10**:116-132. [MESOPLOPHOROIDEA, PHTHIRACAROIDEA, EUPHTHIRACAROIDEA]

Grandjean, F. (1932a). Au sujet des Palaeacariformes Trägårdh. Bull. Mus. nat. Hist. natur. Paris 2, **4**(4):411-426. [ARCHEONOTHROIDEA, PALAEACAROIDEA, CTENACAROIDEA, PARHYPOCHTHONOIDEA]

Grandjean, F. (1932b). La famille des Protoplophoridae (Acariens). Bull. Soc. zool. France **57**:10-36. [PROTHOPLOPHOROIDEA]

Grandjean, F. (1933). Oribates de l'Afrique du Nord (1re série). Bull. Soc. Hist. natur. Afr. Nord **24**:308-323. [MESOPLOPHOROIDEA]

Grandjean, F. (1934). Observations sur les Oribates (Arach., Acar.) (7e série). Bull. Mus. nat. Hist. natur. Paris 2, **6**:423-431. [PARHYPOCHTHONOIDEA]

Grandjean, F. (1936). Les Oribates de Jean Frédéric Hermann et de son pére (Arachn. Acar.). Ann. Soc. ent. France **105**:27-110. [NOTHROIDEA]

Grandjean, F. (1939). Observations sur les Oribates (12e série). Bull. Mus. nat. Hist. natur. Paris 2, **11**(3):300-307. [EULOHMANNIOIDEA, NOTHROIDEA]

Grandjean, F. (1947). Les Enarthronota (Acariens). Première série. Ann. Sci. natur. Zool. 11, **8**:213-248. [COSMOCHTHONOIDEA]

Grandjean, F. (1948). Les Enarthronota (Acariens). (2e série). Ann. Sci. natur. Zool. 11, **10**:29-58. [COSMOCHTHONOIDEA, PHYLLOCHTHONOIDEA]

Grandjean, F. (1950a). Les Enarthronota (Acariens). (3re série). Ann. Sci. natur. Zool. 11, **12**:85-107. [PHYLLOCHTHONOIDEA]

Grandjean, F. (1950b). Étude sur les Lohmanniidae (Oribates, Acariens). Arch. Zool. exp. gén. **87**(2):95-162. [LOHMANNIOIDEA]

Grandjean, F. (1950c). Observations éthologiques sur *Camisia segnis* (Herm.) et *Platynothrus peltifer* (Koch) (Acariens). Bull. Mus. nat. Hist. natur. Paris 2, **22**(2):224-231. [NOTHROIDEA]

Grandjean, F. (1954a). Essai de classification des Oribates (Acariens). Bull. Soc. zool. France **78**:421-446.

Grandjean, F. (1954b). Les Enarthronota (Acariens). (4e série). Ann. Sci. natur. Zool. 11, **16**:311-335. [PROTHOPLOPHOROIDEA]

Grandjean, F. (1958). *Perlohmannia dissimilis* (Hewitt) (Acarien, Oribate). Mém. Mus. nat. Hist. natur. Paris (n.s.) A, Zool. **16**(3):57-119. [PERLOHMANNIOIDEA]

Grandjean, F. (1966). *Collohmannia gigantea* Selln. (Oribate). Première partie. Acarologia **8**(2):328-357. [COLLOHMANNIOIDEA]

Grandjean, F. (1969). Considérations sur le classement des Oribates leur division en 6 groupes majeurs. I. Les affinités de *Collohmannia gigantea* Selln. 1922. Acarologia **11**(1):127-153.

Hammen, L. van der (1952). The Oribatei (Acari) of the Netherlands. Zool. Verhand. Leiden 17:139 pp. [BRACHYCHTHONOIDEA and others]

Hammen, L. van der (1959). Berlese's primitive oribatid mites. Zool. Verhand. Leiden 40:93 pp. [MESOPLOPHOROIDEA, PARHYPOCHTHONOIDEA, COSMOCHTHONOIDEA, BRACHYCHTHONOIDEA and others]

Hammer, M. (1962). Investigations on the oribatid fauna of the Andes mountains. III. Chile. Biol. Skr. Danske Vid. Selsk. **13**(2):1-96. [HYPOCHTHONOIDEA, BRACHYCHTHONOIDEA]

Hammer, M. (1971). On some oribatids from Viti Levu, the Fiji Islands. Biol. Skr. Danske Vid. Selsk. **16**(6): 1-60. [HYPOCHTHONOIDEA, PHTHIRACAROIDEA, EPILOHMANNIOIDEA, LOHMANNIOIDEA, NOTHROIDEA]

Hammer, M. (1972). Investigation on the oribatid fauna of Tahiti, and on some oribatids found on the atoll Rangiroa. Biol. Skr. Danske Vid. Selsk. **19**(3):1-65. [HYPOCHTHONOIDEA, PHTHIRACAROIDEA, EUPHTHIRACAROIDEA, LOHMANNIOIDEA]

Hartenstein, R. (1961). Soil Oribatei. I. Feeding specificity among forest soil Oribatei (Acarina). Ann. Ent. Soc. Amer. **55**(2):202-206.

Hartenstein, R. (1962a). Soil Oribatei VII. Decomposition of conifer needles and deciduous leaf petioles by *Steganacarus diaphanum* (Acarina:Phthiracaridae). Ann. Ent. Soc. Amer. **55**(6):713-716.

Hartenstein, R. (1962b). Soil Oribatei V. Investigations on *Platynothrus peltifer* (Acarina:Camisiidae). Ann. Ent. Soc. Amer. **55**(6):709-713. [NOTHROIDEA]

Jacot, P. (1930). Oribatid mites of the subfamily Phthiracarinae of the northeastern United States. Proc. Boston Soc. Nat. Hist. **39**:209-261.

Johnston, D.E. (1967). On the occurrence of two species of *Palaeacraus* in the eastern United States (Acari: Acariformes). Proc. Ent. Soc. Wash. **69**(4):301-302. [PALAEACAROIDEA]

Luxton, M. (1972). Studies on the oribatid mites of a Danish beechwood soil. I. Nutritional biology. Pedobiol. **12**:434-463.

Michael, A.D. (1888). British Oribatidae. Ray Society, London **2**:337-657 + xi + plates.

Pérez-Íñigo, C. (1968). Ácaros oribátidos de suelos de España peninsular e Islas Baleares (I.ª parte) (Acari, Oribatei). "Graellsia", Rev. Ent. Ibéricos **24**:143-238. [HYPOCHTHONOIDEA, BRACHYCHTHONOIDEA, COSMOCHTHONOIDEA, PHTHIRACAROIDEA, EPILOHMANNIOIDEA, LOHMANNIOIDAE, NOTHROIDEA]

Sellnick, M. (1928). Formenkreis: Hornmilben, Oribatei. Tierw. Mitteleur. **3**(9):1-42.

Sellnick, M. and K.-H. Forsslund (1955). Die Camisiidae Schwedens (Acar. Oribat.). Ark. Zool. **8**(2):473-530. [NOTHROIDEA]

Shereef, G.M. (1976). Biological studies and description of stages of two species: *Papillacarus aciculatus* Kunst and *Lohmannia egypticus* Elbadry and Nasr (Oribatei-Lohmanniidae) in Egypt. Acarologia **18**(2):351-359. [LOHMANNIOIDEA]

Trägårdh, I. (1932). Palaeacariformes, a new suborder of Acari. Ark. Zool. **24B**(2):1-6.

Wallwork, J.A. (1958). Notes on the feeding behavior of some forest soil Acarina. Oikos **9**(2):260-271.

Wallwork, J.A. (1959). The distribution and dynamics of some forest soil mites. Ecol. **40**(4):557-563.

Wallwork, J.A. (1962). Some Oribatei from Ghana. X. The family Lohmanniidae. Acarologia **4**(3):457-487. [LOHMANNIOIDEA]

Willmann, C. (1931). Moosmilben oder Oribatiden (Oribatei). Tierw. Deutschl. **22**:79-200.

Woolley, T.A. (1960). Some interesting aspects of oribatid ecology (Acarina). Ann. Ent. Soc. Amer. **53**(2): 251-253.

Woodring, J.P. (1973). Comparative morphology, functions, and homologies of the coxal glands in oribatid mites (Arachnida:Acari). J. Morphol. **139**(4):407-429.

Supercohort Brachypylides

The great majority of described Oribatida are included in the Brachypylides, or higher Oribatida. The 100 presently recognized brachypylide families are grouped in 25 superfamilies which are considered to comprise two cohorts–The Apterogasterina and the Pterogasterina. Cohortal separation is based on presence or absence of dorsal pores and of dorsolateral notogastral "wings", or *pteromorphae* (Fig. 44, p. 445). The supercohort may be broadly characterized as aptychoid species with rounded, generally well separated genital and anal fields on a distinct ventral shield, and with leg tibiae distinctly longer and of a different shape than the adjacent genua (Plate 159-2, p. 484). The labyrinths of the brachypylide coxal glands are doubled in a hairpin-like fashion, with the "tine" adjacent to the sacculus generally shorter than the opposing portion (Fig. 47b, p. 453) and occasionally absent (Woodring 1973).

Cohort Apterogasterina

DIAGNOSIS: Well sclerotized, without notogastral pores or pteromorphae.

The Apterogasterina includes two subcohorts which are characterized by differences in setal number on the genital valves; i.e., the Polytrichae have seven or more pairs of genital setae (Plate 155-4, p. 480), while the Oligotrichae have six or fewer pairs.

Subcohort Polytrichae

The five superfamilies grouped in the Polytrichae include nine families which are represented by genera from litter and from plants. Nymphs of *Hermannia* sp. (HERMANNIOIDEA, Plate 154, p. 479) feed in lenticels of yellow birch bark (Wallwork 1958). Related species occur with regularity in forest litter, often at high altitudes. *Hermannia subglabra* Berlese has been collected from a variety of littoral habits including the intertidal *Salicornia* zone in Spain (Pérez-Íñigo 1968) and a salt marsh in Wales. *H. convexa* (C.L. Koch), a large widely distributed form, was the subject of an intensive morphological study by van der Hammen (1968).

Nanhermannia elegantula Berlese (NANHERMANNIOIDEA, Plate 149, p. 474) is a European species which has been collected in littoral substrates (Woolley and Higgins 1958). Other members of the superfamily occur in moss, litter and treeholes. Members of the HERMANNIELLOIDEA (Plate 155, p. 480) are broadly oval, often delicately ornamented species with pronounced protruding lateral opisthonotal gland openings. Hermannielloids are primarily litter inhabitants which occur with regularity in damp forest and coastal habitats. *Plasmobates pagoda* Grandjean is arboricolous, living in epiphytic plants on tree trunks in Martinique (Grandjean 1929). *Hermanniella granulata* (Nicolet), a widespread European species, is considered a macrophytophage by Luxton (1972). While hermannielloid adults generally carry only the tritonymphal exuvium on their notogaster *(opsiopheredermy)*, the large (1-2000 μ) heavily sclerotized LIODOIDEA (Plate 155, p. 480) transport a concentric series of several nymphal exuvia *(eupheredermy)* (Wallwork 1969). Many liodoids inhabit soil and leaf litter under vegetation (Grandjean 1931, Pérez-Íñigo 1970) and appear to favor dry niches. *Licnodamaeus undulatus* (Paoli), *L. pulcherrimus* (P.) and *L. costula* Grandjean have been found in moss in several European localities (Grandjean, op. cit.), while *Licnobelba alestensis* Grandjean was recovered from a hollow stump. *Liodes theleproctus* Hermann, a common liodoid in southern Europe and northern Africa, is often arboricolous (Grandjean

1936a). A species of *Platyliodes* taken in stream drift (Woolley 1960) may have originated in the overhanging tree cover. The smaller, darkly tanned members of the GYMNODAMAEOIDEA (Plate 158, p. 481) are eupheredermous and, like the liodoids, are prevalent in dry habitats. Luxton (op. cit.) notes that *Gymnodamaeus bicostatus* (C.L. Koch)is a microphytophage.

Subcohort Oligotrichae

The Oligotrichae is extremely rich in species, and includes approximately 40% of all described oribatid genera (Balogh 1972). The subcohort presently comprises 13 superfamilies and 61 families found in virtually all soil, humus and litter habitats. Unlike other oligotrichs, many species of DAMAEOIDEA (Plate 156, p. 481) are euphederms. Some also carry a mass of debris on the notogaster (Michael 1884), a strategy which may offer a degree of protective camouflage. Damaeoids are litter, humus, and moss forms (Schweizer 1957, Pérez-Iñigo 1970) which appear to be wholly microphytophagous (Luxton 1972), browsing on a wide range of fungi, yeasts and algae. *Belba corynopus* (Hermann) also feeds readily on bacteria, and is noteworthy in being the only known oribatid species which will consume *Penicillium.* The damaeoid *Damaeus clavipes* (Hermann) is arboreal and often is encountered in dead wood (Grandjean 1936a).

Like the damaeoids, the OPPIOIDEA (Plate 159, p. 484) are considered to be exclusively microphytophagous, although one species of *Oppia* appears to be an obligate coprophage on wood borer feces (Wallwork 1958). While oppioids are a primarily tropical group, a great number of holarctic forms are known amongst the 14 described families. Oppioids are common inhabitants of moss, humus, litter and pasture sod both in moist and dry situations. Most oppioid species are accommodated in the families Oppiidae and Suctobelbidae, both of which are well represented in temperate realms. Some oppioid species are virtually worldwide in distribution (*Oppia minus* Paoli and *Oppiella nova* (Oudemans) are examples). Evans et al. (1961) described an interesting case of habitat partition by oppioid species in forest floor habitats based on species size. *Oppia ornata* (Oudemans), a relatively large (± 260 μ) species, was found to be especially abundant in the surface litter layer while the diminutive *Oppia minus* (180-200 μ) favored the deeper humus layer. *Oppiella nova,* a species of intermediate size, occupied the intermediate fermentation layer. Similar correlations have been noted by Luxton (1972) for other oribatid groups.

Feeding studies involving oppioids indicate that fungi comprise the bulk of their diet. Masses of fungal spores have been identified from the gut of *Oppia neerlandica* (Oudemans) and of *O. nitens* C.L. Koch (Luxton 1972). Shereef (1976) noted fungal species preference in reared populations of *O. sticta* Poppe and *Multioppia wilsoni* Aoki, with *Aspergillus flavus* being especially favorable for population growth of *O. sticta.*

Species of the genus *Cepheus* (CEPHEOIDEA, Plate 156, p. 481) are common in moss and litter, but may also be found in punky wood in association with other arthropods. *C. latus* (Nicolet) is a true wood borer in yellow birch twigs (Wallwork 1958). *Conoppia microptera* (Berlese), a holarctic cepheoid, was collected in vegetation at an altitude of 3000 meters in Spain (Pérez-Iñigo 1970). *Cepheus cepheiformis* has been found to be an intermediate host of the tapeworms *Cittotaenia ctenoides* and *Citto. denticulata* in Germany (Stunkard 1941).

The monofamilial superfamily ZETORCHESTOIDEA (Plate 156, p. 481) is a primarily Palearctic assemblage of oligotrichs with saltatorial legs IV, an adaptation which makes it possible for the mites to jump relatively great distances in times of danger. Zetorchestoids are encountered in plant detritus, moss, lichens and occasionally among rocks or on tree trunks or foliage (Grandjean 1951b). *Zetorchestes micronychus* (Berlese), a cosmopolitan species, is considered a panphytophage (Schuster 1956). Grandjean (op. cit.) noted the presence of fungal mycelia in the digestive systems of *Saxicolestes auratus* Gr., *Litholestes altitudinus* Gr. and *Zetorchestes flabrarius* Gr., but also observed significant amounts of pollen. All three species live on exposed rocks where wind-blown pollen from surrounding trees may easily collect.

There is great similarity in the general morphology of the superfamilies EREMAEOIDEA (Plate 158, p. 483) and EREMULOIDEA (Plate 157, p. 482), with the most obvious separating character being the condition of the pretarsi (tridactyly in the former group and monodactyly in the latter). The holarctic eremaeoid genus *Eremaeus* is well represented in both wet and dry situations in organic litter, meadow sod, and moss. *Megeremaeus* species are unusually large eremaeoids from mountain habitats in western North America (Woolley and Higgins 1968). Members of the widely distributed EREMULOIDEA are mostly litter forms. The eremuloid *Eremobelba hermosa* Hartenstein is microphytophagous (Hartenstein 1962c), as is the common European eremaeoid *Eremaeus hepaticus* C.L. Koch (Schuster 1956).

The LIACAROIDEA (Plates 157-158, pp. 482-483) comprises a heterogenous cosmopolitan group of seven oliogotrich families which are especially common in the litter-fermentation layer interface of deciduous and coniferous forest floors. *Gustavia microcephala* (Nicolet) is a microphytophage which feeds almost exclusively on bacteria (Luxton 1972). *Gustavia* species have elongate serrate chelicerae (Plate 158-1) which may be an adaptation for scraping bacterial film from damp substrates. Other liacaroids (*Adoristes ovatus* (C.L. Koch), *Xenillus anasillus* Woolley, *X. tegeocranus* (Hermann), *Liacarus coracinus* C.L. Koch), *L. tremellae* (Linnaeus), and *L. xylariae* (Schrank)) are listed by Luxton as panphytophages. Species of *Liacarus, Xenillus* and *Ceratoppia* have been implicated as intermediate hosts of tapeworms (Allred 1954).

The notogaster of the strongly sclerotized POLYPTEROZETOIDEA (Plate 158, p. 483) often is obscured by a covering of adherent dirt over a thin cerotegument. Polypterozetoids are confined for the most part to the neotropics and southwest Pacific (Hammer 1966, Balogh 1972, Piffl 1972) where they inhabit litter. The broad prodorsal lamellar appendages which cover a large part of the prodorsum of polypterozetoids are similar to the lamellae seen in many members of the superfamily CARABODOIDEA (Plate 158, p. 483). Many carabodoids also have a heavily sculptured notogaster, although a cerotegument is not present. Carabodoids often are encountered in extremely dry habitats (Pérez-Íñigo 1971) but are equally at home in the wet detritus of stream banks and damp forest floors. Most of the approximately 30 carabodoid genera are exclusively tropical groups (Balogh 1972), with notable exceptions occurring in the genera *Carabodes, Austrocarabodes* and *Odontocepheus* (Carabodidae), the genus *Niphocepheus* (Niphocepheidae), and most of the carabodoid family Tectocepheidae. *Tectocepheus velatus* (Michael) is a widespread inhabitant of forest litter and sandy habitats in temperate regions (Woolley 1960, Evans et al. 1961). Hammer (1972, 1973) also records it from dry moss in Western Samoa and from Tahiti. *T. sarekensis* Trägårdh, a closely related species, occurs in Europe, Hawaii and New Zealand (Pérez-Íñigo 1971). *Carabodes labyrinthicus* (Michael) is encountered in European forest soils but has also been collected in cave debris in Spain. *C. marginatus* (Michael), *C. minisculus* Berlese and *Odontocepheus elongatus* (Michael) have been identified from open heathland soils (Evans et al., op. cit.). *O. elongatus* is considered a panphytophage by Luxton (1972), as are various species of *Carabodes. Tectocepheus velatus,* however, is a microphytophage. A species of carabodid is a known intermediate host of the tapeworm *Anoplocephala perfoliata* in Russia (Baker and Wharton 1952).

The OTOCEPHEOIDEA (Plate 159, p. 484) is an essentially tropical assemblage of elongate species with characteristic attenuated lamellae. Otocepheoids are especially common in moist litter and in moss. Moist habitats also are favored by many of the flattened, ornate members of the CYMBAEREMAEOIDEA (Plate 160, p. 485). *Scapheremaeus arcuatus* Hammer lives in wet *Polytrichum* along a stream in Fiji (Hammer 1971), and *S. petrophagus* (Banks) occurs on water-splashed rocks in California where it apparently feeds on lichens. Other species of *Scapheremaeus* are intertidal (Hughes 1959) or litter inhabitants. *Cymbaeremaeus* species have been collected in dry sandy pine duff at high altitudes (Woolley 1960) and on the surfaces of scaly pine and spruce twigs. Both *Cymbaeremaeus* and the cosmopolitan genus *Micreremaeus* appear to favor litter substrates.

The AMERONOTHROIDEA (Plate 160, p. 485) and HYDROZETOIDEA (Plate 159, p. 484) are largely halophilous and hygrophilous species. Prodorsal sensilla are reduced or absent in both groups, suggesting that sensillary repression is related to acquisition of an aquatic habitat. Grandjean (1951a), however, notes that the sensilla are well developed in other aquatic Oribatida. Several species of *Ameronothrus* and *Hygroribates* occur in intertidal algae and lichens and in salt marsh vegetation on temperate coastlines (Grandjean 1947, Evans et al. 1961, Pérez-Íñigo 1971, Schubart 1975). The ameronothroids *Fortuynia yunkeri* and *F. marina* van der Hammen are intertidal in the southwest Pacific (van der Hammen 1963), while *Alaskozetes, Antarcticola* and *Podacarus* species inhabit algae and other substrates in the Antarctica (Grandjean 1955, Wallwork 1967). *Adhaesozetes barbarae* Hammer is well removed from intertidal influences, occurring in moist grasses on the island of Tonga (Hammer 1966).

Ameronothroids apparently utilize their well developed tracheal system *(système tracheen normal* of Grandjean 1934) for storage of air during periods of submersion in intertidal situations. A different strategy has been developed by the HYDROZETOIDEA which retain an external film of air, or plastron, on roughened portions on the notogastral and pleural margins and on the venter. These coalesce in a sejugal fossa located between coxae II-III which opens internally to the tracheal system. Hydrozetoids are found on aquatic plants, wet moss, and floating sphagnum (Grandjean 1949, Schweizer 1956, Sellnick 1960), often crawling on the undersides of leaves. Immatures and adults of *Hydrozetes* sp. (possibly *H. lemnae* Coggi) have been observed on damaged or deteriorating leaves of the aquatic plant *Salvinia rotundifolia* in Florida. It is not clear whether the mites contribute to primary leaf injury, or simply feed on previously damaged tissues or on related primary bacterial or fungal invaders. Since *S. rotundifolia* has become an important pest plant in African lakes (F.M. Young, personal communication), the role of *Hydrozetes* sp. may be of some economic interest.

Cohort Pterogasterina

DIAGNOSIS: Well sclerotized, with pteromorphae and/or notogastral pores.

Seven superfamilies and 30 families of pterogasterine Oribatida are recognized by Balogh (1972). With the exception of the PASSALOZETOIDEA, members of the cohort possess a pair of lateral pteromorphal extensions. It is generally supposed that the primary function of the pteromorphae is to provide protection for the legs (Ramsay and Wallwork 1972). The large hinged notogastral extensions seen in the PELOPOIDEA (Plate 161-1, p. 486) and GALUMNOIDEA (Plate 163-2, p. 488) are certainly well designed to function as protective devices. However, it seems unlikely that the small fixed pteromorphal shelves found in many CERATOZETOIDEA and ORIBATULOIDEA can provide more than minimal cover (Plate

162-2, p. 487). Observations on galumnoid behavior by Wallwork (1960) suggest that the pteromorphae may also serve another purpose. When exposed to above-optimal temperatures, galumnids move the pteromorphae up and down in a beating motion which may increase ventilation of the leg brachytracheae and promote evaporative cooling. Under normal conditions, the pteromorphae may restrict air flow to the brachytracheae, reducing vapor pressure across the brachytracheal apertures and controlling water loss.

The strongly sclerotized, often tuberculate or nodulate PASSALOZETIODEA (Plate 162, p. 487) usually inhabit terrestrial moss and detritus (Grandjean 1931, Sellnick 1960, Pérez-Íñigo 1971). Some, however, are arboreal (*Licneremaeus exornatus* Grandjean, for example). *Licneremaeus, Passalozetes* and *Scutovertex* are cosmopolitan passalozetoids, but most of the remaining genera are confined to the tropics (Balogh 1972). *Scutovertex minutus* (C.L. Koch) has been experimentally infected with eggs of the tapeworm *Bertiella studeri* (Stunkard 1939). *Pelops tardus* Koch, *P. planicornis* (Schrank) and *P. acromios* (Hermann) (PELOPOIDEA, Plate 161, p. 486) are known carriers of the rabbit tapeworm *Cittotaenia ctenoides,* although the widespread *P. acromios* is considered an arboricolous species (Grandjean 1936a). Most pelopoids inhabit moss and litter. Species of at least one family of the ORIBATELLOIDEA (Plate 160, p. 485) also are naturally infected by cestodes. Wardle and McLeod (1952) recovered an anoplocephaline tapeworm from a species of the oribatelloid genus *Achipteria* (family Achipteriidae). *A. coleoptrata* (L.) is a panphytophage, feeding on a variety of microbial flora in the litter layer (Luxton 1972). Another *Achipteria* species feeds on leaf epidermis (Wallwork 1958). Other oribatelloids generally are encountered in litter, moss or soil.

The small (± 300 μ) broad species of the monofamilial superfamily MICROZETOIDEA (Plate 161, p. 486) occur in temperate regions but reach their greatest development in the tropics (Grandjean 1936b). Over 30 genera are known from soil and overlying organic layers in South America, Africa, and the southwest Pacific (Balogh 1972). Microzetoids are unique in that the elongate proterosoma often equals the hysterosoma in length. In addition, a dorsal digitiform outgrowth ornaments the fixed cheliceral digit of all known microzetoid species.

The cosmopolitan CERATOZETOIDEA (Plate 162, p. 487) are found in a broad range of habitats including forest litter, humus, soil, and moss. Evans et al. (1961) found that two ceratozetoid species, *Minuthozetes semirufus* (C.L. Koch) and *Punctoribates punctum* (C.L.K.) (Mycobatidae), are the most abundant Oribatida in mineral soils in Great Britain. Forest soils support a number of species including *Ceratozetes gracilis* (Michael), *Humerobates rostrolamellatus* Grandjean, *Sphaerozetes orbicularis* (C.L. Koch) and *Trichoribates trimaculatus* (C.L.K.) (Ceratozetidae). *Euzetes globula* (Nicolet) (Euzetidae) frequents grassland soils. Ceratozetoids also occur commonly in the littoral zone (*Punctoribates quadrivertex* Halbert, for example), and are known from arctic and subarctic situations as well (*Melanozetes meridianus* Sellnick, *Edwardzetes elongatus* Wallwork and *E. dentifer* Hammer (Ceratozetidae) are examples) (Wallwork 1967). Several species are arboreal, living for the most part in moss growing on tree trunks (*Humerobates fungorum* (L.) and *Punctizetes penicillifer* Hammer are examples). Grandjean (1936a) recognized *Humerobates rostrolamellatus* as an arboreal species although it is also found in soil substrates. Large populations of *H. rostrolamellatus* often are encountered on the bark of fruit trees, presumably feeding on algae (Jeppson et al. 1975). *Diapterobates humeralis* (Hermann) and members of the tropical ceratozetoid genus *Mochlozetes* also are arboricoles (Grandjean 1930). *Trichoribates setosa* (Koch and Michael) lives in flowers where it subsists exclusively on pollen (Grandjean 1946). An unidentified species of *Trichoribates* recently was recovered from seed capsules of *Kalmiopsis leachiana,* a rare member of the Ericaceae, in southwestern Oregon. Its feeding habits are not clear. Species of *Trichoribates* and of the ceratozetid genus *Fuscozetes* are known intermediate hosts of tapeworms (Allred 1954).

The feeding habits of ceratozetoids are as diverse as are their habitats. Luxton (1972) lists species of *Chamobates, Euzetes, Ceratozetes* and *Fuscozetes* as panphytophages. *F. fuscipes* (C.L. Koch) was observed by Wallwork (1958) to feed on vascular tissue of hemlock needles, fungal hyphae, and moist decaying birch leaves. Nymphs feed on dead mites and other arthropods. *Ceratozetes gracilis* (Michael) ranges throughout the litter, fermentation, and humus layers of beech forest detritus, consuming several species of fungi, yeast and bacteria.

Like the ceratozetoids, the ORIBATULOIDEA (Plates 162-163, pp. 487-488) occur in many habitats throughout the world. They are especially prevalent in decaying organic matter on forest soils. MacFadyen (1952) found *Liebstadia similis* (Michael), *Oribatula tibialis* (Nicolet), and three species of *Scheloribates* in the arthropod community of a *Molinia* marsh in England. Species of *Scheloribates* and *Oribatula* also are common in pasture sod (Woolley 1960) and in the intertidal zone (Grandjean 1935, Evans et al. 1961). Some oribatuloids are arboreal, living in moss and lichens on tree trunks and branches (*Fijibates rostratus* Hammer and the oripodids *Cryptoribatula euaensis* and *Exoribatula marginata* Hammer are examples) (Hammer 1973). *Drymobates silvicola* Grandjean is an arboricolous oribatulid found on epiphytic plants in Martinique (Grandjean 1930).

Several oribatuloid species are important vectors of tapeworms. *Scheloribates laevigatus* (C.L. Koch) carries no fewer than eight cestode species (Allred 1954). Tapeworm cysticercoids also have been recovered from each of three other *Scheloribates* species. *Liebstadia similis* (Michael) is an intermediate host for the rabbit tapeworm, and *Oribatula minuta* (Banks) has been found to carry *Monoecocestus sigmodontis* and the sheep tapeworm, *Moniezia expansa.*

Feeding habits of oribatuloids are diverse, according to available data (Luxton 1972). *Hemileius initialis* (Berlese), a relatively weakly sclerotized oribatuloid, feeds on a variety of microbial fauna in the deeper organic layers of the forest floor. Luxton has observed the same species to be necrophagous on collembolans. *Scheloribates parabilis* Woodring and *Peloribates* sp. may also feed on dead animal tissue under certain conditions (Wallwork 1958, Woodring 1965). *S. laevigatus* was noted by Vitzthum (1943) to feed on pupae of a parasitic hymenopteran. *Rostrozetes flavus* was identified as a macrophytophage, feeding on decomposing outer sheaths of roots (Woodring, op. cit.).

The GALUMNOIDEA (Plate 163, p. 488) includes five families of relatively large mites (often $> 750\ \mu$) with a broad hysterosoma and large, wing-like pteromorphae which may fold down over the legs in times of stress (Ramsay and Wallwork 1972, Bulanova-Zakhvatkina et al. 1975). Galumnoids are recorded from many terrestrial habitats including moss, pasture sod, forest litter and rotting wood. They appear to prefer moist niches but occasionally are found in exposed situations such as dry moss on tree bark (Grandjean 1936a, Hammer 1973), or in nest material (Evans et al. 1961). *Orthogalumna terebrantis* Wallwork feeds on water hyacinth *(Eichhornia crassipes),* an important aquatic pest species in subtropical and tropical waterways (Wallwork 1965). *O. terebrantis* not only consumes previously damaged plant tissue but apparently is capable of initiating damage by penetrating the tissues with its well developed rutella (del Fosse et al. 1975). Galumnoids are listed as panphytophages by Luxton (1972), although *O. terebrantis* would appear to be a macrophytophage. *Pergalumna omniphagous* Rockett and Woodring was observed to prey on live nematodes in laboratory cultures (Rockett and Woodring 1966). *Galumna formicarius* (Berlese) may be an obligate coprophage on fecal material in the galleries of wood-boring insects (Wallwork 1958).

Probably no single genus of brachypylide Oribatida is as important in the transmission of cestode parasites as the galumnoid genus *Galumna. G. virginiensis* Jacot is a major carrier of the sheep tapeworm, *Moniezia expansa,* with *G. emarginata* (Banks) and *G. nigra* (Ewing) also serving as important intermediate hosts (Allred 1954). Other species of *Galumna* transmit several tapeworms to vertebrate hosts in Europe, Asia and North America.

Useful References

Allred, D.M. (1954). Mites as intermediate hosts of tapeworms. Proc. Utah Acad. Arts and Letters 31:44-51.

Aoki, J. (1967). A preliminary revision of the family Otocepheidae (Acari, Cryptostigmata). II. Subfamily Tetracondylinae. Bull. National Sci. Mus. Tokyo **10**:257-359. [OTOCEPHEOIDEA]

Aoki, J. (1969). Taxonomic investigations on free-living mites in the subalpine forest on Shiga Heights IBP Area. III. Cryptostigmata. Bull. National Sci. Mus. Tokyo **12**:117-141.

Arlian, L.G. and T.A. Woolley (1970). Observations on the biology of *Liacarus cidarus.* J. Kansas Ent. Soc. **43**(3):297-301. [LIACAROIDEA]

Baker, E.W. and G.W. Wharton (1952). Oribatei Dugès, 1833. **In** An Introduction to Acarology. MacMillan Co., New York: 387-438.

Balogh, J. (1961). An outline of the family Lohmanniidae Berlese, 1916 (Acari, Oribatei). Acta Zool. Hung. **7**:19-44. [LOHMANNIOIDEA]

Balogh, J. (1972). The Oribatid Genera of the World. Akadémiai Kiadó, Budapest: 188 pp. + 71 plates.

Block, W.C. (1965). The life histories of *Platynothrus peltifer* (Koch 1839) and *Damaeus clavipes* (Hermann 1804) (Acarina:Cryptostigmata) in soils of Pennine moorland. Acarologia **7**(4):735-743. [NOTHROIDEA, DAMAEOIDEA]

Bulanova-Zakhvatkina, E.M., B. A. Vainshtein, V.I. Volgin, M.S. Gilyarov, L.D. Golosova, D.A. Krivolutskii, A.B. Lange, V.D. Sevast'yanov, L.G. Sitnikova and E.S. Shaldybina (1975). Group Oribatei Dugès. **In** A Key to the Soil-inhabiting mites (Sarcoptiformes). "Nauka", Moscow: 381 pp.

Covarrubias, R. (1968a). Some observations on Antarctic Oribatei (Acarina) *Liochthonius australis* sp. n., and two *Oppia* ssp. n. Acarologia **10**(2):313-356. [OPPIOIDEA]

Covarrubias, R. (1968b). Observations sur le genre *Pheroliodes.* I—*Pheroliodes roblensis* n. sp. (Acarina, Oribatei). Acarologia **10**(4):657-695. [GYMNODAMAEOIDEA]

Engelbrecht, C.M. (1972). Galumnids from South Africa (Galumnidae, Oribatei). Acarologia **14**(1):109-149. [GALUMNOIDEA]

Evans, G.O., J.G. Sheals and D. Macfarlane (1961). Free-living Acari. **In** The Terrestrial Acari of the British Isles. Volume I. Introduction and Biology. British Museum (Natural History), London: 89-106.

del Fosse, E.S., H.L. Cromroy and D.H. Habeck (1975). Determination of the feeding mechanism of the waterhyacinth mite. Hyacinth Control J. **13**:53-55. [GALUMNOIDEA]

Grandjean, F. (1929). Quelques nouveaux genres d'Oribatei du Venezuela et de la Martinique. Bull. Soc. zool. France **54**:400-423. [HERMANNIELLOIDEA]

Grandjean, F. (1930) Oribates nouveaux de la région Caraïbe. Bull. Soc. zool. France **55**:262-284. [CERATOZETOIDEA, ORIBATULOIDEA, GALUMNOIDEA]

Grandjean, F. (1931). Le genre *Licneremaeus* Paoli (Acariens). Bull. Soc. zool. France **56**:221-250. [LIODOIDEA, PASSALOZETOIDEA]

Grandjean, F. (1933). Étude sur le développement des Oribates. Bull. Soc. zool. France **58**:30-61.

Grandjean, F. (1934). Les organes respiratoires secondaires des Oribates (Acariens). Ann. Soc. ent. France **103**:109-146.

Grandjean, F. (1935). Observations sur les Oribates (9e série). Bull. Mus. nat. Hist. natur. Paris 2, **7**(5):280-287. [ORIBATULOIDEA, ORIBATELLOIDEA]

Grandjean, F. (1936a). Les Oribates de Jean Frédéric Hermann et son pére (Arachn. Acar.). Ann. Soc. ent. France **105**:27-110. [LIODOIDEA, DAMAEOIDEA, PELOPOIDEA, CERATOZETOIDEA, GALUMNOIDEA]

Grandjean, F. (1936b). Les Microzetidae n. fam. (Oribates). Bull. Soc. zool. France **61**:60-93. [MICROZETOIDEA]

Grandjean, F. (1936c). Observations sur les Oribates (10e série). Bull. Mus. nat. Hist. natur. Paris 2, **8**:246-253. [ORIBATULOIDEA]

Grandjean, F. (1946). Observations sur les Acariens (9e série). Bull. Mus. nat. Hist. natur. Paris 2, **18**:337-344. [CERATOZETOIDEA]

Grandjean, F. (1947). Observations sur les Oribates (17e série). Bull. Mus. nat. Hist. natur. Paris 2, **19**(2):165-172. [AMERONOTHROIDEA]

Grandjean, F. (1949). Sur le genre *Hydrozetes* Berl. (Acariens). Bull. Mus. nat. Hist. natur. Paris 2, **21**(2): 224-231. [HYDROZETOIDEA]

Grandjean, F. (1951a). Comparaison du genre *Limnozetes* au genre *Hydrozetes* (Oribates). Bull. Mus. nat. Hist. natur. Paris 2, **23**(2):200-207. [HYDROZETOIDEA]

Grandjean, F. (1951b). Étude sur les Zetorchestidae (Acariens, Oribates). Mém. Mus. nat. Hist. natur. Paris (n.s.) A, Zool. **4**(4):1-50. [ZETORCHESTOIDEA]

Grandjean, F. (1952). Au sujet de l'ectosquelette du podosoma chez les Oribates supérieurs et de sa terminologie. Bull. Soc. zool. France **77**(1):13-36.

Grandjean, F. (1953). Sur les genres *"Hemileius"* Berl. et *"Siculobata"* n.g. (Acariens, Oribates). Mém. Mus. nat. Hist. natur. Paris (n.s.) A, Zool. **6**(2):117-137. [ORIBATULOIDEA]

Grandjean, F. (1954a). Essai de classification des Oribates (Acariens). Bull. Soc. zool. France **78**:421-446.

Grandjean, F. (1954b). *Posthermannia nematophora* n. g., n. sp. (Acarien, Oribate). Rev. Fr. Ent. **21**:298-311. [NANHERMANNIOIDEA]

Grandjean, F. (1955). Sur un acarien des Iles Kerguelen, *Podacarus auberti* (Oribate). Mém. Mus. nat. Hist. natur. Paris (n.s.) A, Zool. **8**(3):109-150. [AMERONOTHROIDEA]

Grandjean, F. (1956a). Observations sur les Galumnidae. 1re série. Rev. Fr. Ent. **23**:137-146. [GALUMNOIDEA]

Grandjean, F. (1956b). Sur deux espèces nouvelles d'Oribates (Acariens) apparentees a *Oripoda elongata* Banks 1904. Arch. Zool. exp. gén. **93**(2):185-218. [ORIBATULOIDEA]

Grandjean, F. (1957). L'infracapitulum et la manducation chez les Oribates et d'autres Acariens. Ann. Sci. Nat. Zool. **11**:233-281.

Grandjean, F. (1959). *Polypterozetes cherubin* Berl. 1916 (Oribate). Acarologia **1**(1):147-180. [POLYPTEROZETOIDEA]

Grandjean, F. (1960a). Les Mochlozetidae n. fam. (Oribates). Acarologia **2**(1):101-148. [CERATOZETOIDEA]

Grandjean, F. (1960b). Les Autognetidae n. fam. (Oribates). Acarologia **2**(4):575-609. [OPPIOIDEA]

Grandjean, F. (1963). Les Autognetidae (Oribates). Deuxième partie. Acarologia **5**(4):653-689. [OPPIOIDEA]

Grandjean, F. (1966a). Les Staurobatidae n. fam. (Oribates). Acarologia **8**(4):696-727. [EREMULOIDEA]

Grandjean, F. (1966b). *Selenoribates mediterraneus* n. sp. et les Selenoribatidae (Oribates). Acarologia **8**(1):129-154. [AMERONOTHROIDEA]

Grandjean, F. (1967). Nouvelles observations sur les Oribates (5e série). Acarologia **9**(1):242-272. [EREMAEOIDEA]

Grandjean, F. (1970). Nouvelles observations sur les Oribates (7e série). Acarologia **12**(2):432-460.

Hammen, L. van der (1952). The Oribatei (Acari) of the Netherlands. Zool. Verhand. Leiden 17:1-139.

Hammen, L. van der (1959). Berlese's primitive oribatid mites. Zool. Verhand. Leiden 40:1-93.

Hammen, L. van der (1963). Description of *Fortuynia yunkeri* nov. spec., and notes on the Fortuyniidae nov. fam. (Acarida, Oribatei). Acarologia **5**(1):152-167. [AMERONOTHROIDEA]

Hammen, L. van der (1968). The gnathosoma of *Hermannia convexa* (C.L. Koch) (Acarida:Oribatina) and comparative remarks on its morphlogy in other mites. Zool. Verhand. Leiden 94:1-45.

Hammer, M. (1944). Studies on the oribatids and collemboles of Greenland. Medd. Gronland **141**(3):1-210.

Hammer, M. (1958). Investigations on the oribatid fauna of the Andes Mountains. I. The Argentine & Bolivia. Biol. Skr. Dansk Vid. Selsk. **10**(1):1-129.

Hammer, M. (1962). Investigations on the oribatid fauna of the Andes Mountains. III. Chile. Biol. Skr. Dansk Vid. Selsk. **13**(2):1-65. [CYMBAEREMAEOIDEA, OPPIOIDEA, GYMNODAMAEOIDEA, EREMULOIDEA, ORIBATULOIDEA, CERATOZETOIDEA, GALUMNOIDEA]

Hammer, M. (1966). Investigations on the oribatid fauna of New Zealand. Part 1. Biol. Skr. Dansk Vid. Selsk. **15**:1-108. [POLYPTEROZETOIDEA, AMERONOTHROIDEA and others]

Hammer, M. (1971). On some oribatids from Viti Levu, the Fiji Islands. Biol. Skr. Dansk Vid. Selsk. **16**(6): 1-60. [NANHERMANNIOIDEA, HERMANNIELLOIDEA, LIODOIDEA, PASSALOZETOIDEA, MICROZETOIDEA, ORIBATELLOIDEA and others]

Hammer, M. (1972). Investigation on the oribatid fauna of Tahiti, and on some oribatids found on the atoll Rangiroa. Biol. Skr. Dansk Vid. Selsk. **19**(3):1-65. [POLYPTEROZETOIDEA, PELOPOIDEA and others]

Hammer, M. (1973). Oribatids from Tongatapu and Eua, the Tonga Islands and from Upolu, Western Samoa. Biol. Skr. Dansk Vid. Selsk. **20**(3):1-70. [OTOCEPHEOIDEA, CARABODOIDEA and others]

Hartenstein, R. (1962a). Soil Oribatei I. Feeding specificity among forest soil Oribatei. Ann. Ent. Soc. Amer. **55**(2):202-206.

Hartenstein, R. (1962b). Soil Oribatei II. *Belba kingi,* new species, and a study of its life history. Ann. Ent. Soc. Amer. **55**(4):357-361. [DAMAEOIDEA]

Hartenstein, R. (1962c). Soil Oribatei III. Studies on the developmental biology and ecology of *Metabelba montana* (Kulcz.) and *Eremobelba nervosa* n. sp. Ann. Ent. Soc. Amer. **55**(4):361-367. [DAMAEOIDEA, EREMULOIDEA]

Hartenstein, R. (1962d). Soil Oribatei IV. Observations on *Ceratozetes gracilis.* Ann. Ent. Soc. Amer. **55**(6): 709-713. [CERATOZETOIDEA]

Hughes, T.E. (1959). The free-living Acari. **In** Mites, or the Acari. Athlone Press, London: 1-22.

Jeppson, L.R., H.H. Keifer and E.W. Baker (1975). Tydeidae, Tuckerellidae, Pyemotidae, Penthaleidae, Astigmata, and Cryptostigmata. **In** Mites Injurious to Economic Plants. Univ. California Press, Berkeley: 307-326.

Kates, K.C. and C.E. Runkel (1948). Observations on oribatid mite vectors of *Moniezia expansa* on pastures, with a report of several new vectors from the United States. Proc. Helminth. Soc. Wash. **15**(1):18-33.

Lions, J.-C. (1970). La chaetotaxie gastronotique chex un Pelopsidae (Oribate). Acarologia **12**(3):612-622. [PELOPOIDEA]

Lebrun, P. (1965). Contribution a l'étude écologiques des Oribates de la litiere dans une forêt de Moyenne-Belgique. Mém. Inst. roy. Sci. nat. Belg. 153:96 pp.

Luxton, M. (1972). Studies on the oribatid mites of a Danish beech wood soil. I. Nutritional biology. Pedobiol. **12**:434-463.

MacFadyen, A. (1952). The small arthropods of a *Molinia* fen at Cothill. J. Anim. Ecol. **21**:87-117.

Michael, A.D. (1884). British Oribatidae. Ray Society, London **1**:336 + xi + plates.

Michael, A.D. (1888). British Oribatidae. Ray Society, London **2**:337-657 + xi + plates.

Oudemans, A.C. (1917). Notizen über Acari. 26 Reihe (Oribatoidea). Arch. Naturg. **83**:1-84.

Pérez-Íñigo, C. (1968). Ácaros oribátidos de suelos de España peninsular e Islas Baleares (I.ª parte) (Acari, Oribatei). "Graellsia", Rev. Ent. Ibéricos **24**:143-238. [HERMANNIOIDEA]

Pérez-Íñigo, C. (1970). Ácaros oribátidos de suelos de España peninsular e Islas Baleares (Acari, Oribatei). Parte II. "Eos", Rev. Española Ent. **45**:241-317. [HERMANNIELLOIDEA, LIODOIDEA, GYMNODAMAEOIDEA, DAMAEOIDEA, CEPHEOIDEA, MICROZETOIDEA, ZETORCHESTOIDEA, LIACAROIDEA, EREMAEOIDEA, EREMULOIDEA]

Pérez-Íñigo, C. (1971). Ácaros oribátidos de suelos de España peninsular e Islas Baleares (Acari, Oribatei). Parte III. "Eos", Rev. Española Ent. **46**:263-350. [LIACAROIDEA, CARABODOIDEA, OPPIOIDEA, AMERONOTHROIDEA, CYMBAEREMAEOIDEA, PASSALOZETOIDEA]

Piffl, E. (1972). Zur systematik der Oribatiden (Acari). (Neue Oribatiden aus Nepal, Costa Rica und Brasilien ergeben eine neue Familie der Unduloribatidae und erweitern die Polypterozetidae um die Gattungen *Podopterotegaeus, Nodocepheus, Eremaeozetes* und *Tumerozetes*). Khumbu Himal **4**(2):269-314. [ORIBATELLOIDEA, POLYPTEROZETOIDEA]

Ramsay, G.W. and J.A. Wallwork (1972). Some observations on the pteromorphs of oribatid mites (Acari: Cryptostigmata). Acarologia **13**(4):669-674.

Riha, G. (1951). Ökologie der Oribatiden im Kalksteinboden. Zool. Jahrb. **80**:407-450.

Rockett, C.L. and J.P. Woodring (1966). Biological investigations on a new species of *Ceratozetes* and of *Pergalumna* (Acarina:Cryptostigmata). Acarologia **8**(3):511-520. [CERATOZETOIDEA, GALUMNOIDEA]

Schubart, H. (1975). Morphologische Grundlagen für die Klärung der Verwandschaftsbeziehungen innerhalb der Milbenfamilie Ameronothridae (Acari, Oribatei). Zoologica 123:24-91. [AMERONOTHROIDEA]

Schuster, R. (1956) Die Anteil der Oribatiden anden Zerzetzungsvorgängen im Boden. Z. Morph. Ökol. Tiere **45**:1-33.

Schweizer, J. (1956). Die Landmilben des Schweizerischen Nationalparkes. 3. Sarcoptiformes Reuter 1909. Soc. Helvet. Sci. natur. Parc. nat. **5**(N.F.) (34):215-377.

Schweizer, J. (1957). Die Landmilben des Schweizerischen Nationalparkes. 4. Ihr Lebensraum, ihre Vergesellschaftung unter sich und ihre Lebensweise. Soc. Helvet. Sci. natur. Parc. nat. **6**(N.F.) (37):11-107.

Sellnick, M. (1960). Formenkreis: Hornmilben, Oribatei. Tierw. Mitteleur. **3**:1-42. [NANHERMANNIOIDEA, HERMANNIOIDEA, DAMAEOIDEA, ORIBATULOIDEA, CERATOZETOIDEA and others]

Shereef, G.M. (1976). Biological studies and description of stages of two species: *Papillacarus aciculatus* Kunst and *Lohmannia egypticus* Elbadry and Nasr (Oribatei-Lohmanniidae) in Egypt. Acarologia **18**(2): 351-359.

Stunkard, H.W. (1939). The role of oribatid mites as transmitting agents and intermediate hosts of ovine cestodes. Int. Kong. Ent. Berlin vii. **3**:1671-1674.

Stunkard, H.W. (1941). Studies on the life history of the anoplocephaline cestodes of hares and rabbits. J. Parasitol. **27**:299-325.

Stunkard, H.W. (1944). Studies on the life history of the oribatid mite, *Galumna* sp., intermediate host of *Moniezia expansa.* Anat. Rec. **89**(4):550.

Travé, J. (1959). Sur le genre *Niphocepheus* Balogh 1943 les Niphocepheidae, famille nouvelle. (Acariens, Oribates). Acarologia **1**(4):475-498. [CARABODOIDEA]

Travé, J. (1963). Oribates des Pyrenees-Orientales, 2e série, Zetorchestidae (1re partie): *Saxicolestes pollinivorus* n. sp. Vie et Milieu **14**(2):449-455. [ZETORCHESTOIDEA]

Vitzthum, H.G. (1943). Acarina. **In** Bronn's Klassen und Ordnungen des Tierreiches. Leipzig 5, **5**:1-1011.

Wallwork, J.A. (1958) Notes on the feeding behaviour of some forest soil Acarina. Oikos **9**(2):260-271.

Wallwork, J.A. (1959). The distribution and dynamics of some forest soil mites. Ecol. **40**(4):557-563.

Wallwork, J.A. (1961a). Some Oribatei from Ghana. IV. The genus *Basilobelba* Balogh. Acarologia **3**(1): 344-362. [EREMULOIDEA]

Wallwork, J.A. (1961b). Some Oribatei from Ghana. VII. Members of the "family" Eremaeidae Willmann (2nd series). The genus *Oppia* Koch. Acarologia **3**(4):637-658. [OPPIOIDEA]

Wallwork, J.A. (1962). Some Oribatei from Ghana IX. The genus *Tetracondyla* Newell 1956 (2nd series). Acarologia **4**(3):440-456. [OTOCEPHEOIDEA]

Wallwork, J.A. (1965). A leaf-boring galumnoid mite (Acari: Cryptostigmata) from Uruguay. Acarologia **7**(4):758-764.

Wallwork, J.A. (1967). Cryptostigmata (orbatid mites). Antarctic Res. Ser. **10**:105-122. [AMERONOTHROIDEA]

Wallwork, J.A. (1969). Some basic principles underlying the classification and identification of cryptostigmatid mites. **In** The Soil Ecosystem, J.G. Sheals, ed. Syst. Assoc. Publ. 8:155-168.

Wardle, R.A. and J.A. McLeod (1952). Zoology of the Tapeworms. Univ. Minnesota Press, Minneapolis: 780 pp.

Willmann, C. (1931). Moosmilben oder Oribatiden (Cryptostigmata). Tierwelt Deutschl. **22**(5):79-200.

Woodring, J.P. (1962). Oribatid (Acari) pteromorphs, pterogasterine phylogeny, and evolution of wings. Ann. Ent. Soc. Amer. **55**:394-403.

Woodring, J.P. (1965). The biology of five new species of oribatids from Louisiana. Acarologia **7**(3):564-576. [GALUMNOIDEA, ORIBATULOIDEA]

Woodring, J.P. (1973). Comparative morphology, functions, and homologies of the coxal glands in oribatid mites (Arachnida, Acari). J. Morphol. **139**(4):407-429.

Woodring, J.P. and E.F. Cook (1962). The biology of *Ceratozetes cisalpinus* Berlese, *Scheloribates laevigatus* Koch, and *Oppia neerlandica* Oudemans (Oribatei) with a description of all stages. Acarologia **4**(1):101-137. [OPPIOIDEA, ORIBATULOIDEA, CERATOZETOIDEA]

Woolley, T.A. (1960). Some interesting aspects of oribatid ecology (Acarina). Ann. Ent. Soc. Amer. **53**(1): 251-253.

Woolley, T.A. (1965). Eutegaeidae, a new family of oribatid mites, with a description of a new species from New Zealand (Acarina:Oribatei). Acarologia **7**(2):382-388. [CARABODOIDEA]

Woolley, T.A. (1972). The systematics of the Liacaroidea (Acari:Cryptostigmata). Acarologia **14**(2):250-257.

Woolley, T.A. and H.G. Higgins (1958). A revision of the family Nanhermanniidae (Acari:Oribatei). Proc. 10th Int. Congr. Ent. **1**:913-923. [NANHERMANNIOIDEA]

Woolley, T.A. and H.G. Higgins (1968). Megeremaeidae, a new family of oribatid mites (Acari:Cryptostigmata). Great Basin Nat. **28**(4):172-175. [EREMAEOIDEA]

SUBORDER ORIBATIDA
(Plates 148-163, pp. 473-488)

KEY TO THE SUPERFAMILIES

1. With at least one of the following characters well discernible: proterosoma *ptychoid,* or capable of being "jack-knifed" into hysterosoma (Plate 153-1, p. 478); tibia and genu of about uniform length and shape;[1] genital and anal shields contiguous, occupying entire length of ventral shield which may be divided by a horizontal, semicircular or parabolical transverse suture (Plate 149, p. 474); anal and adanal shields often separated . . . . Supercohort MACROPYLIDES ... 2

— Without the above characteristics; i.e., proterosoma *aptychoid* (not constructed so that is may be "jack-knifed" into hysterosoma); tibia longer and of a different shape than genu; genital and anal shields rounded, usually well separated and not covering entire length of ventral shield, without transverse ventral shield suture; adanal shields absent . . . . .Supercohort BRACHYPYLIDES ... 20

2. Femora of legs divided. Gnathosoma visible from above (*astegasimy,* Plate 151-5, p. 476). Tarsi may not be monodactylous in all immature stages. Small, whitish forms, with or without long black notogastral setae, and with four solenidia on tarsus I of adult. . . . . Cohort BIFEMORATINA ... 3

— Femora of legs not divided. Gnathosoma generally not visible from above (*stegasimy,* Plate 154-1, p. 479). Tarsi always monodactylous in larval and nymphal stages. Generally pigmented forms without long black body setae, and with fewer than four solenidia on tarsus I of adult 5

3. Setae c_1 (Plate 150-4, p. 475) equal in length to, or shorter than, setae c_2 4

— Setae c_1 considerably longer than setae c_2 . . . .(Plates 149-150, pp. 474-475) Superfamily ARCHEONOTHROIDEA

4. Setae c_1 equal in length to setae c_2; without broad separation between prodorsal and mesonotal shields. Notogaster with or without long black setae. . . . .(Plate 150, p. 475) Superfamily CTENACAROIDEA

— Setae c_1 shorter than setae c_2; with a broad separation between prodorsal and mesonotal shields . . . . (Plate 151, p. 476) Superfamily PALAEACAROIDEA

5. Proterosoma ptychoid (Plate 153-1, p. 478); body generally compressed laterally . . . . .Cohort PTYCTIMINA ... 6

— Proterosoma aptychoid; body cylindrical or dorsoventrally flattened 9

6. Notogaster with one or more transverse sutures posteriorly . . . .(Plate 153, p. 478) Superfamily PROTHOPLOPHOROIDEA

— Notogaster without posterior transverse sutures 7

7. With a separate ventral shield; genital and anal shields often separated, anal shield rounded. . . . . (Plate 153, p. 478) Superfamily MESOPLOPHOROIDEA

— Ventral shield absent; genital and anal shields generally contiguous, anal shields not rounded. 8

[1] Although the NANHERMANNIOIDEA (couplet 23) illustrate tibial-genual similarity, Balogh (1972) has included the superfamily in the Brachypylides.

8. Genital-anal shield complex wide, often more than 1/2 as broad as long. . . . .(Plate 153, p. 478) Superfamily PHTHIRACAROIDEA

— Genital-anal shield complex narrow, seldom more than 1/2 as broad as long; idiosoma strongly compressed laterally . (Plate 153, p. 478) Superfamily EUPHTHIRACAROIDEA

9. Notogaster with 1-3 transverse sutures . . . . Cohort ARTHRONOTINA ... 10

— Notogaster without transverse sutures. . . . . Cohort HOLONOTINA ... 15

10. Notogaster with 1 suture, sometimes continuing laterally and ventrally 11

— Notogaster with 2 or 3 sutures 12

11. Opisthonotal glands present (Plate 152, p. 477) Superfamily PARHYPOCHTHONOIDEA

— Opisthonotal glands absent . . . . . (Plate 151, p. 476) Superfamily HYPOCHTHONOIDEA

12. Notogaster with 2 sutures; prodorsal sensilla fusiform, not pectinate. . . . . (Plate 151, p. 476) Superfamily BRACHYCHTHONOIDEA

— Notogaster with 3 sutures (if 2 sutures are present, then the prodorsal sensilla are pectinate, or the notogastral setae are leaf-like or broadly expanded) 13

13. All notogastral setae broadly expanded, often covering notogaster . . . .(Plate 152, p. 477) Superfamily PHYLLOCHTHONOIDEA

— Notogastral setae simple, plumose or pectinate, occasionally dendritic but not all broad as above. 14

14. With long, smooth notogastral setae, some of which are as long as, or longer than, the notogaster; with 3 prodorsal "eyes" . . . . (Plate 152, p. 477) Superfamily HETEROCHTHONOIDEA

— Notogastral setae various, but never as above; without prodorsal "eyes". . . . .(Plate 152, p. 477) Superfamily COSMOCHTHONOIDEA

15. With a narrow preanal shield between genital and anal shields; with 10 pairs of genital setae, arranged in two longitudinal rows on each genital valve. Found in tropical and warm temperate regions . . . .(Plate 149, p. 474) Superfamily LOHMANNIOIDEA

— Preanal shield present or absent; with fewer than 10 pairs of genital setae (if more than 10 pairs, they are arranged in single longitudinal rows on each valve, as in Plate 149-6, p. 474) Cosmopolitan 16

16. Propodosoma and hysterosoma movably connected 17

— Propodosoma and hysterosoma immovably fused . . . . (Plate 155, p. 480) Superfamily NOTHROIDEA

17. With a transverse or parabolic ventral suture separating genital and anal regions 18

— Without a ventral suture as above 19

18. Ventral suture describing a concave parabola; with 9 pairs of setae on genital shield . . . . (Plate 149, p. 474) Superfamily EULOHMANNIOIDEA

— Ventral suture transverse; with 6 pairs of setae on genital shield . . . .(Plate 149, p. 474) Superfamily EPILOHMANNIOIDEA

19. Genital valves divided transversely; body flattened dorsoventrally .... (Plate 154, p. 479) Superfamily PERLOHMANNIOIDEA

— Genital valves entire; body laterally compressed .... (Plate 154, p. 479) Superfamily COLLOHMANNIOIDEA

20. Notogaster without pteromorphae, areae porosae, sacculi or pori................. .... Cohort APTEROGASTERINA ... 21

— Notogaster with pteromorphae and/or with area porosae, sacculi and pori (Fig. 44, p. 445) .... Cohort PTEROGASTERINA ... 36

21. With 7 or more pairs of genital setae.............. Subcohort POLYTRICHAE ... 22

— With 6 or less pairs of genital setae[1] .... Subcohort OLIGOTRICHAE ... 26

22. Notogaster with 16 pairs of setae, never carrying cast exuvial skins 23

— Notogaster with 6 or less pairs of setae, inserted posteromarginally; usually with exuvia dorsally 24

23. Ventral shield with a pair of lateral incisions curving medioventrally and terminating posterolaterad of the genital shield *(diagastry).* Genital and anal shields widely separated.................. (Plate 149, p. 474) Superfamily NANHERMANNIOIDEA

— Without lateral incisions on the ventral shield. Genital and anal shields contiguous, or nearly so .... (Plates 149, 154, pp. 474, 479) Superfamily HERMANNIOIDEA

24. Notogaster with a pair of protruding lateral opisthonotal glands.................. .... (Plate 155, p. 480) Superfamily HERMANNIELLOIDEA

— Notogaster without protruding lateral glands 25

25. Genital valves divided by a transverse suture...................................... .... (Plate 155, p. 480) Superfamily LIODOIDEA

— Genital valves entire .... (Plate 156, p. 481) Superfamily GYMNODAMAEOIDEA

26. Body covered with cerotegument and adherent dirt. Lamellae wide, apically rounded, protruding well beyond rostrum .... (Plate 158, p. 483) Superfamily POLYPTEROZETOIDEA

— Body not as above. Lamellae various, sometimes absent....................... 27

27. With at least one of the following characters well discernible: chelicerae long, attenuate, without movable digits, serrate distally; leg IV modified for jumping, with a thick spine; rostral setae inserted near one another, plumose, flabellate or bifurcate. . . .. (Plate 156, p. 481) Superfamily ZETORCHESTOIDEA

— Without the above characteristics 28

[1] Balogh (1972) mentions the following genera as exceptional in having more than 6 pairs of genital setae: *Trichocarabodes* Balogh, *Niphocepheus* Balogh, *Kodiakella* Hammer, and *Rhynchoribates* Grandjean.

28. Notogaster with rugose sculpturing; coarsely wrinkled, tuberculate or reticulate, or at least with thick dark chitinization. Lamellae distinct .... (Plates 156, 158, pp. 481, 483) Superfamilies CEPHEOIDEA and CARABODOIDEA[1]

— Notogaster smooth, occasionally sculptured, often darkly tanned but not thickened as above. Lamellae present or absent 29

29. Without true lamellae, at most with linear costula (Plate 159-5, p. 484) 30

— Lamellae present, variously developed. 35

30. Ventral shield with more than 4 pairs of setae. 31

— Ventral shield bearing only 4 pairs of setae (with few exceptions; e.g., *Tripiloppia* Hammer, an oppioid, has 6 pairs) 32

31. Pretarsi tridactylous .... (Plate 158, p. 483) Superfamily EREMAEOIDEA

— Pretarsi monodactylous (*Heterobelba* Berlese has tridactylous pretarsi on tarsus IV) .. .. (Plate 157, p. 482) Superfamily EREMULOIDEA

32. Notogaster hemispheric, with 8 pairs of setae arranged in two longitudinal parallel lines; legs generally long, often with swollen segments .... (Plate 156, p. 481) Superfamily DAMAEOIDEA

— Without the above combination of characters 33

33. Genital and anal shields small, widely separated; 9-10 (exceptionally 13-14) pairs of notogastral setae. Pretarsi monodactylous ....(Plate 159, p. 484) Superfamily OPPIOIDEA

— Genital and anal shields large, usually approaching one another so that the distance between them is less than length of genital shield; with 9-17 pairs of notogastral setae. Pretarsi mono- or tridactylous 34

34. Pretarsi monodactylous, with 16-17 pairs of notogastral setae .... (Plate 159, p. 484) Superfamily HYDROZETOIDEA[2]

— Pretarsi mono- or tridactylous; with fewer than 16 pairs of notogastral setae .... (Plate 160, p. 485) Superfamilies AMERONOTHROIDEA and CYMBAEREMAEOIDEA[3]

35. Body elongate; pretarsi monodactylous .... (Plate 159, p. 484) Superfamily OTOCEPHEOIDEA

---

[1]Bulanova-Zakhvatkina et al. (1975) distinguish cepheoids from carabodoids on the basis of developmental differences in notogastral humeral processes and on arrangement of notogastral setae. However, the numerous exceptions which exist prompt me to follow Balogh (1972), who keys the two superfamilies as a single entity.

[2]*Limnozetes* Hull, considered a hydrozetoid genus by Balogh (1972), has tridactylous pretarsi and possesses small pteromorphae.

[3]Bulanova-Zakhvatkina et al. (1975) distinguish ameronothroids from cymbaeremaeoids by the presence, in the former group, of an anterodorsal notogastral lenticulus and small bothridia. However, the lenticulus is absent or obscure in many ameronothroids (Balogh 1972), and the cymbaeremaeoid bothridia may be extremely small (e.g., *Glanderemaeus*). For these reasons I have chosen to follow Balogh, who keys the two superfamilies as a single entity.

— Body fusiform to ovoid; pretarsi tridactylous . . . . (Plates 157-158, pp. 482-483) Superfamily LIACAROIDEA[1]

36. Pteromorphae absent, but with either areae porosae, sacculi, or pori on the notogaster . . . . (Plate 162, p. 487) Superfamily PASSALOZETOIDEA

— Pteromorphae present, movable or immovable 37

37. Pteromorphae umbellate, movable, broad, extending both anteriorly and posteriorly. . . . .(Plate 163, p. 488) Superfamily GALUMNOIDEA

— Pteromorphae variously produced, but not as above 38

38. Chelicerae narrow, elongate *(pelopsiform)*; often with thick cerotegument and a number of fusiform notogastral setae. . . . . (Plate 161, p. 486) Superfamily PELOPOIDEA

— Chelicerae not pelopsiform; notogaster may be strongly tanned but generally is smooth and thin, without cerotegument 39

39. Chelicerae with bacilliform lateral appendages. Generally minute forms (± 250 μ) with broad lamellae. Proterosoma often equal to, or longer than, hysterosoma . . . . (Plate 161, p. 486) Superfamily MICROZETOIDEA

— Chelicerae without bacilliform appendages. Larger forms, with proterosoma distinctly shorter than hysterosoma. Lamellae narrow or wide. 40

40. Lamellae extremely wide, often meeting or fusing medially and covering major portion of prodorsum; without translamellar bridge . . . . (Plate 160, p. 485) Superfamily ORIBATELLOIDEA

— Lamellae variable; if broad, then with a translamellar bridge. 41

41. With at least one of the following characteristics well discernible; 1-5 pairs of genital setae; lamellae attenuate anteriorly, rarely with translamella or cusps; pteromorphae inconspicuous, protruding laterally from body margin . . . . (Plates 162-163, pp. 487-488) Superfamily ORIBATULOIDEA

— With 6 pairs of genital setae; lamellae usually not conspicuously attenuated anteriorly; translamella and cusps generally present; pteromorphae curved ventrad around body margin . . . . (Plate 162, p. 487) Superfamily CERATOZETOIDEA

---

[1] Balogh (1972) includes the liacaroid family TENUIALIDAE (Plate 157-5, p. 482) in the Pterogasterina on the basis of their knife-like, anteriorly projecting pteromorphae, and the presence of areae porosae, sacculi, or pori on the notogaster. It may be more appropriate to erect a separate superfamilial category for the TENUIALIDAE, relagating them to a position near the ORIBATELLOIDEA.

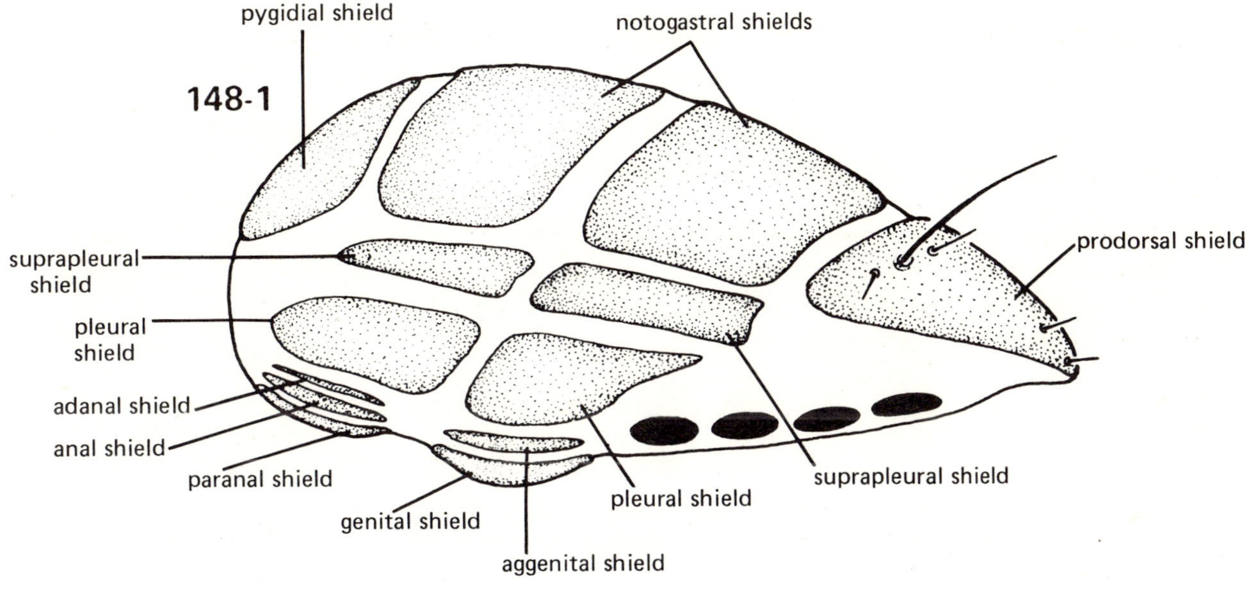

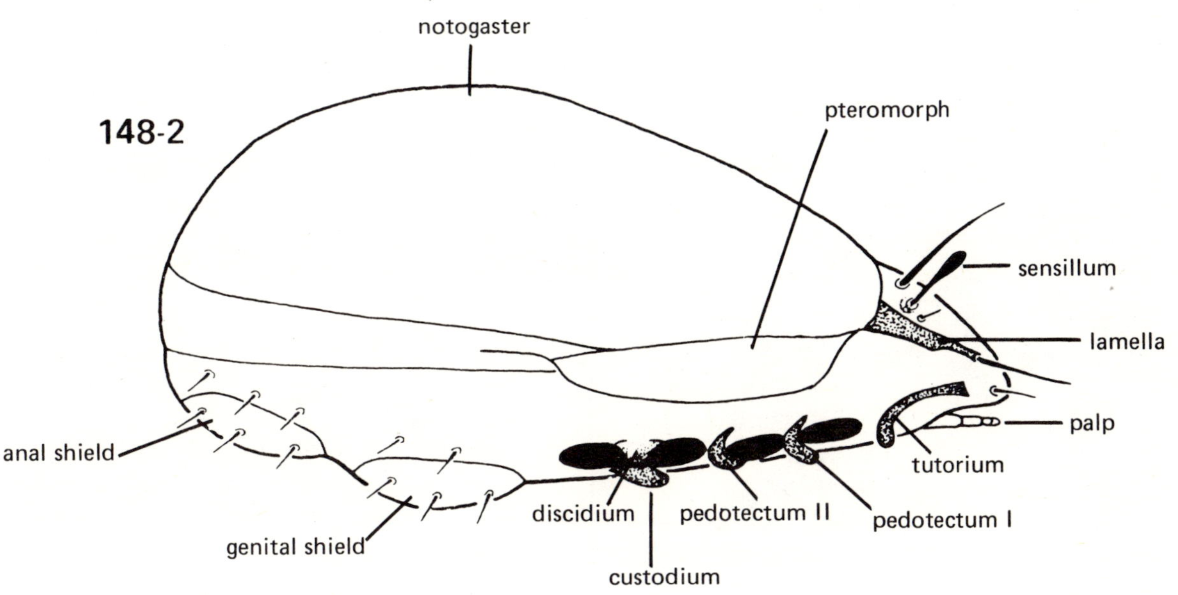

148; hypothetical Oribatida illustrating primitive (148-1) and advanced (148-2) shield states. Macropylide superfamilies such as the BRACHYCHTHONOIDEA may have a full complement of shields as shown in 148-1 (after Knülle 1957)

PLATE 149

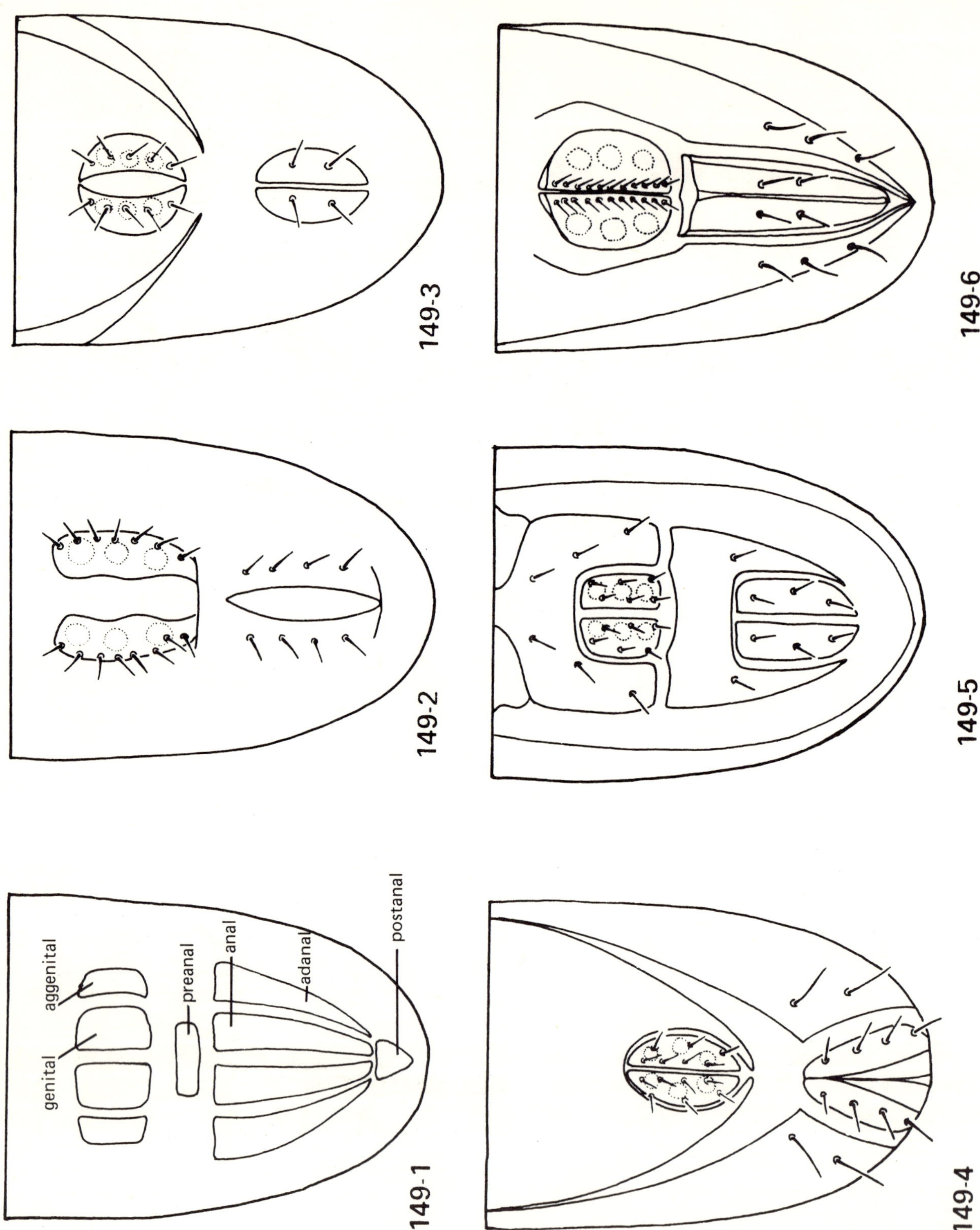

149-1 to 149-6; genital-anal shield states in lower and primitive higher Oribatida. **149-1**; hypothetical oribatid genital-anal shield state: **149-2**; *Acaronychus* sp. (Macropylides, ARCHEONOTHROIDEA), with weak shield development: **149-3**; *Nanhermannia* sp. (Brachypylides, NANHERMANNIOIDEA), illustrating *diagastry:* **149-4**; *Eulohmannia* sp. (Macropylides, EULOHMANNIOIDEA), illustrating *pseudodiagastry:* **149-5**; *Epilohmannia* sp. (Macropylides, EPILOHMANNIOIDEA), illustrating *schizogastry:* **149-6**; *Hermannia* sp. (Brachypylides, HERMANNIOIDEA), illustrating simple genital-anal condition with aggenital and adanal shields fused to genital and anal elements

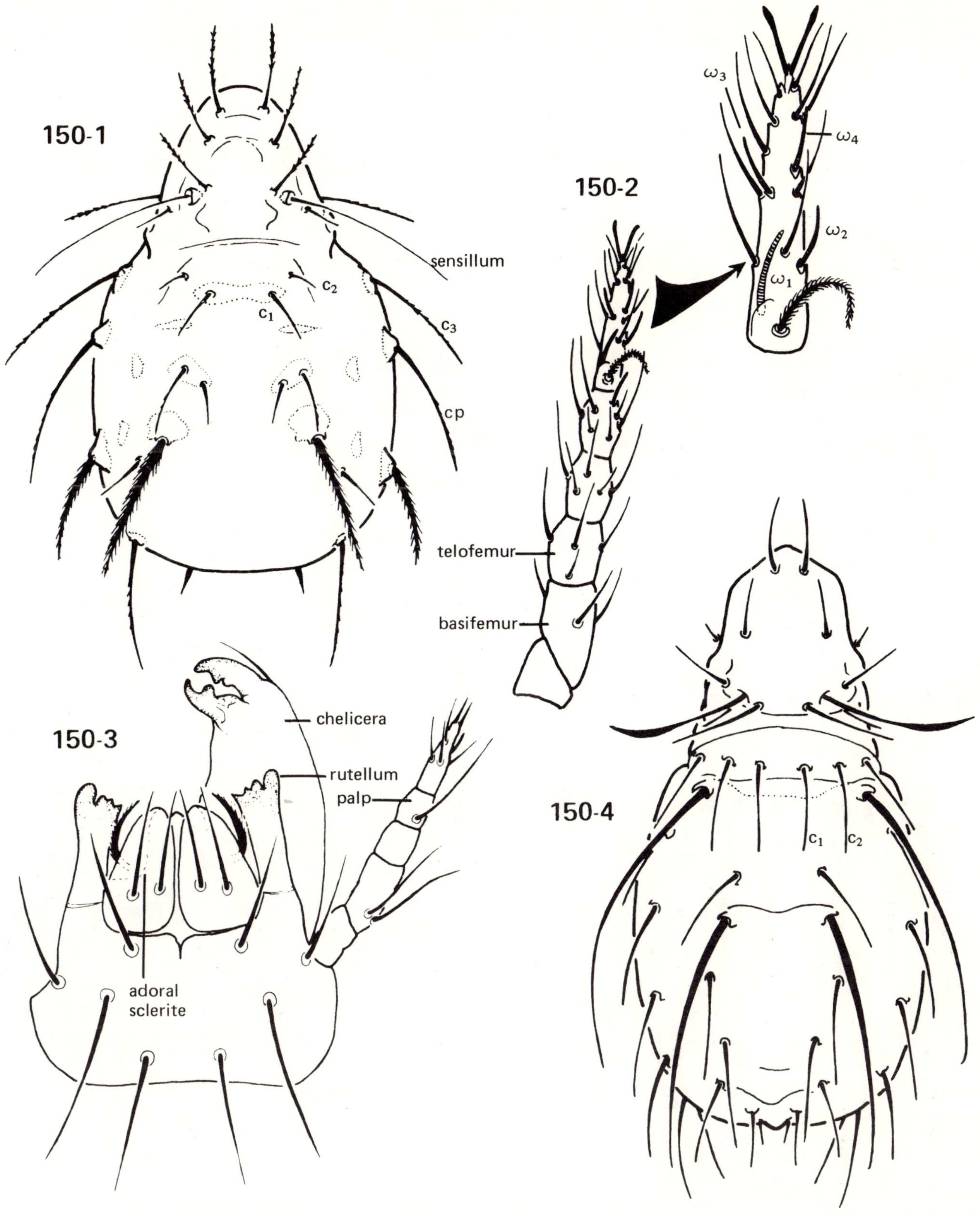

150-1 and 150-2; superfamily ARCHEONOTHROIDEA. **150-1**; *Acaronychus tragardhi* Grandjean, dorsum (after Balogh 1972): **150-2**; *Acaronychus* sp. (Oregon, USA), leg I with detail of tarsus

150-3 and 150-4; superfamily CTENACAROIDEA. **150-3**; *Ctenacarus* sp. (Oregon, USA), venter of gnathosoma: **150-4**; *C. araneola* (Grandjean), dorsum (after Balogh 1972)

PLATE 151

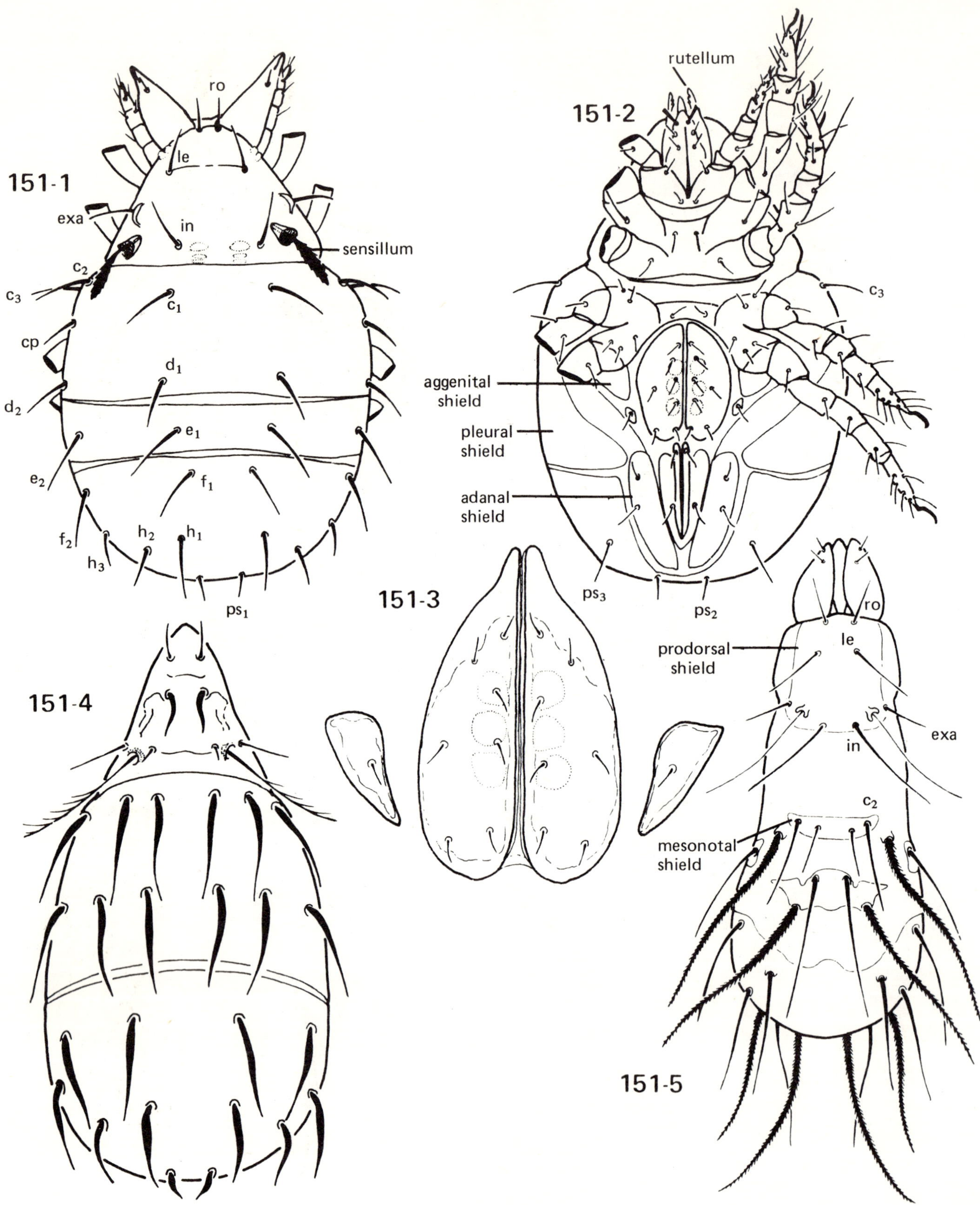

151-1 to 151-3; superfamily BRACHYCHTHONOIDEA. **151-1 and 151-2;** *Eobrachychthonius* sp. (Oregon, USA), dorsal and ventral aspects: **151-3;** *Brachychthonius* sp. (Oregon, USA), genital field

151-4; superfamily HYPOCHTHONOIDEA, *Eohypochthonius gracilis* Jacot, dorsum (after Balogh 1972)

151-5; superfamily PALAEACAROIDEA, *Palaeacarus hystricinus* Trägårdh, dorsum (after Balogh 1961)

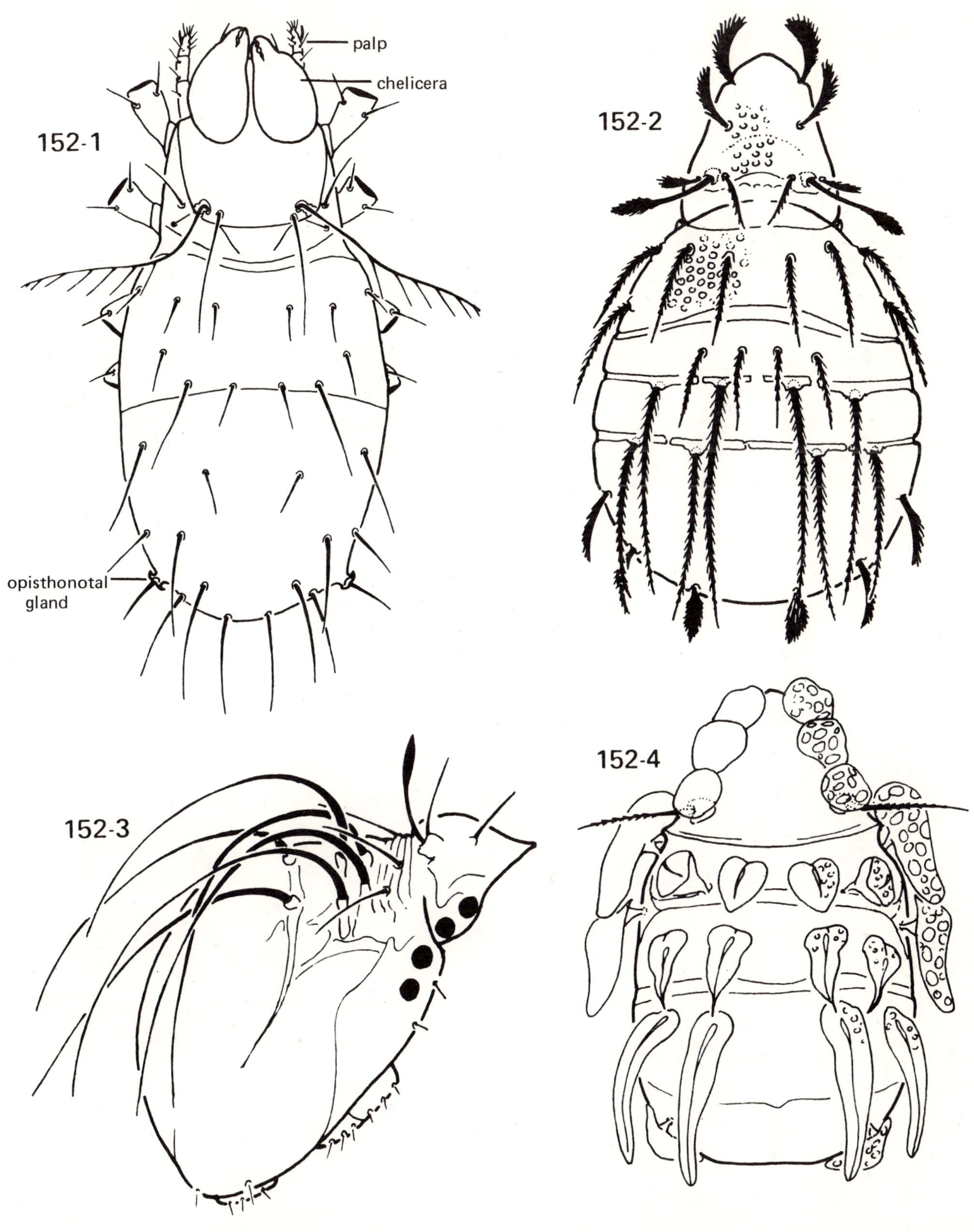

152-1; superfamily PARHYPOCHTHONOIDEA, *Parhypochthonius* sp. (British Columbia), dorsum
152-2; superfamily COSMOCHTHONOIDEA, *Cosmochthonius lanatus foveolatus* Beck, dorsum (after Balogh 1972)
152-3; superfamily HETEROCHTHONOIDEA, *Heterochthonius gibbus* Berlese, lateral aspect (after Balogh 1972)
152-4; superfamily PHYLLOCHTHONOIDEA, *Atopochthonius artiodactylus* Grandjean, dorsum (after Balogh 1972)

PLATE 153

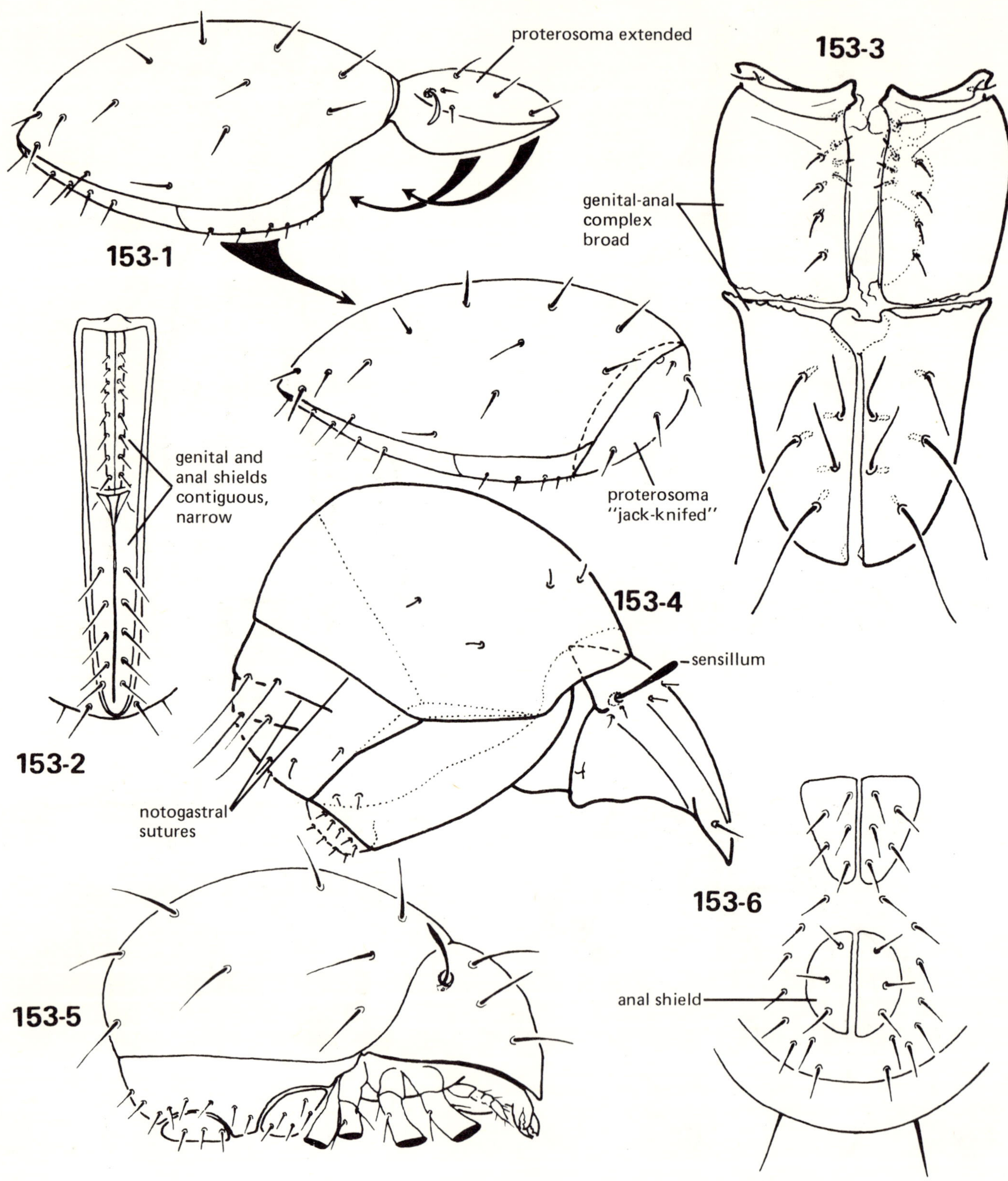

153-1 and 153-2; superfamily EUPHTHIRACAROIDEA, *Euphthiracarus* sp. (Oregon, USA). **153-1;** lateral aspect illustrating "jack-knifing" or ptychoidy: **153-2;** genital-anal field

153-3; superfamily PHTHIRACAROIDEA, *Phthiracarus murphyi* Harding, genital-anal field (after Harding 1976)

153-4; superfamily PROTHOPLOPHOROIDEA, *Aedoplophora glomerata* Grandjean, lateral aspect (after Balogh 1972)

153-5 and 153-6; superfamily MESOPLOPHOROIDEA, *Mesoplophora* sp. **153-5;** lateral aspect of female: **153-6;** genital-anal field (after Baker et al. 1958)

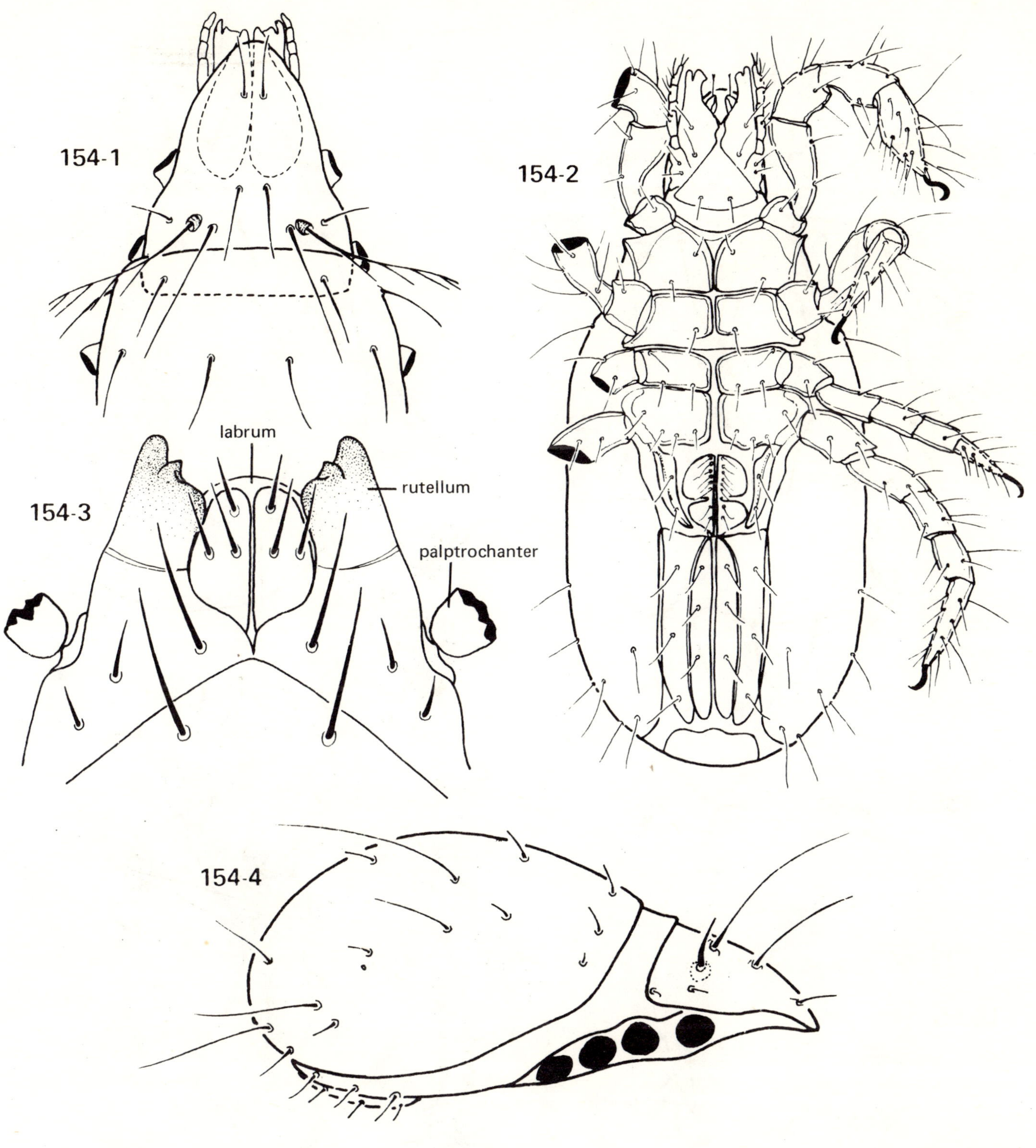

154-1 and 154-2; superfamily PERLOHMANNIOIDEA, *Perlohmannia* sp. (Oregon, USA). **154-1;** anterodorsal aspect: **154-2;** venter

154-3; superfamily HERMANNIOIDEA, *Hermannia* sp. (Oregon, USA), venter of gnathosoma

154-4; superfamily COLLOHMANNIOIDEA, *Collohmannia gigantea* Sellnick, lateral aspect (after Balogh 1972)

PLATE 155

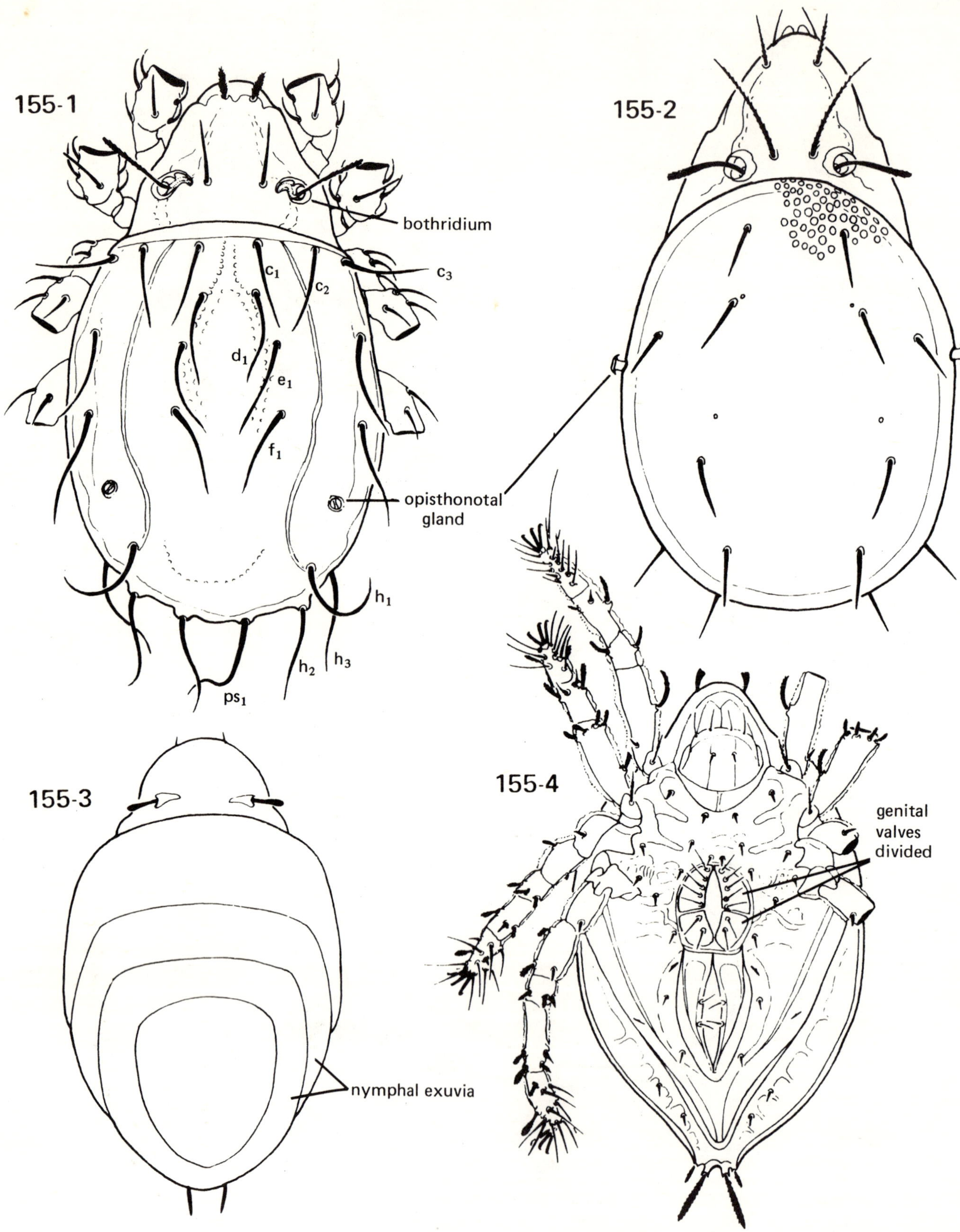

155-1; superfamily NOTHROIDEA, *Platynothrus* sp., dorsum

155-2; superfamily HERMANNIELLOIDEA, *Hermanniella* sp. (Oregon, USA), dorsum

155-3 and 155-4; superfamily LIODOIDEA. **155-3;** *Liodes theleproctus* (Hermann), dorsum (after Balogh 1965): **155-4;** *Platyliodes* sp. (Oregon, USA), venter

PLATE 156

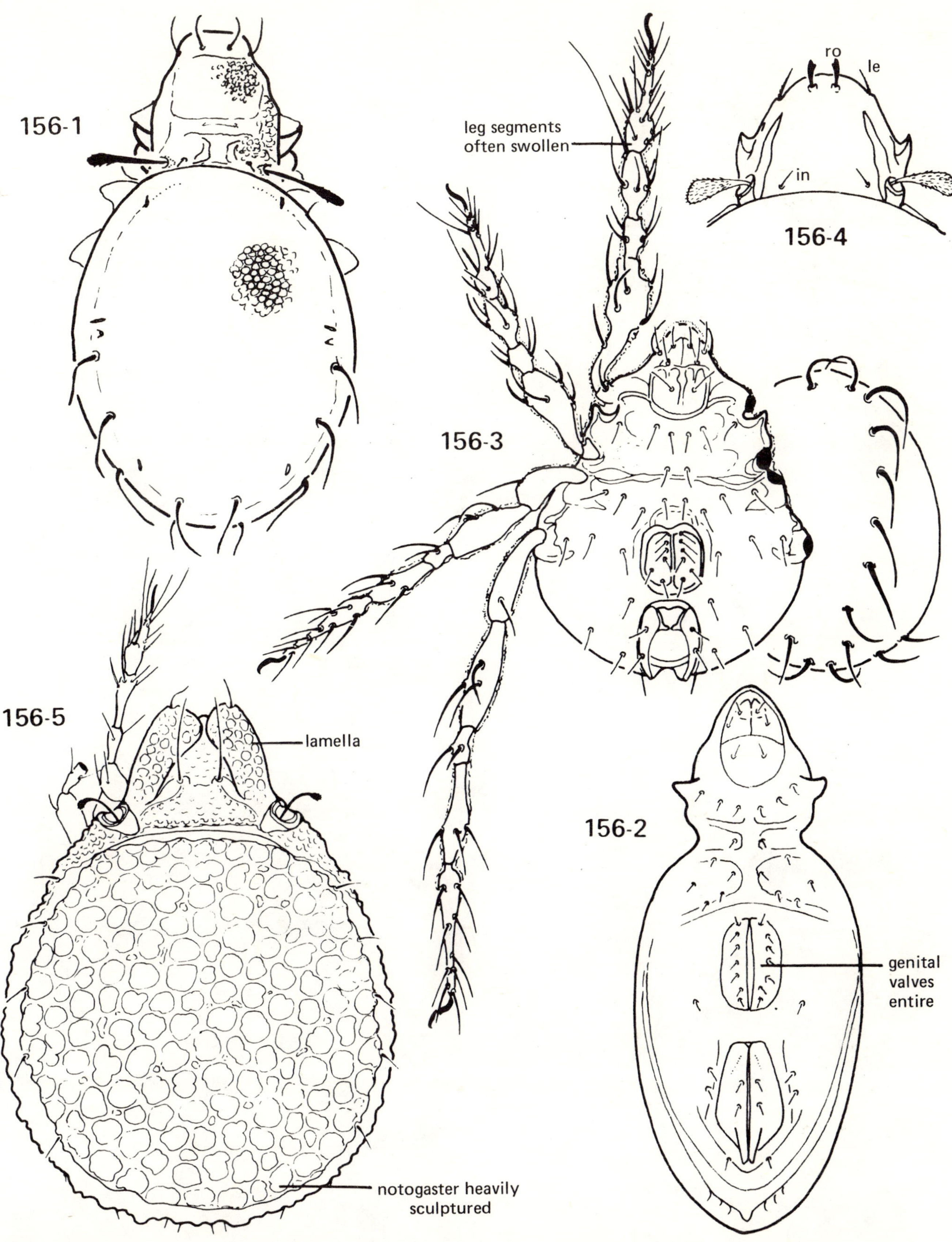

156-1 and 156-2; superfamily GYMNODAMAEOIDEA. **156-1**; *Pedrocortesella pulchra* Hammer, dorsum (after Hammer 1961): **156-2**; *Licnobelba alestensis* Grandjean, venter (after Balogh 1972)

156-3; superfamily DAMAEOIDEA, *Damaeus* sp. (Oregon, USA), venter and dorsum

156-4; superfamily ZETORCHESTOIDEA, *Zetorchestes flabrarius* Grandjean, dorsum of propodosoma (after Balogh 1965)

156-5; superfamily CEPHEOIDEA, *Cepheus* sp. (Oregon, USA), dorsum

PLATE 157

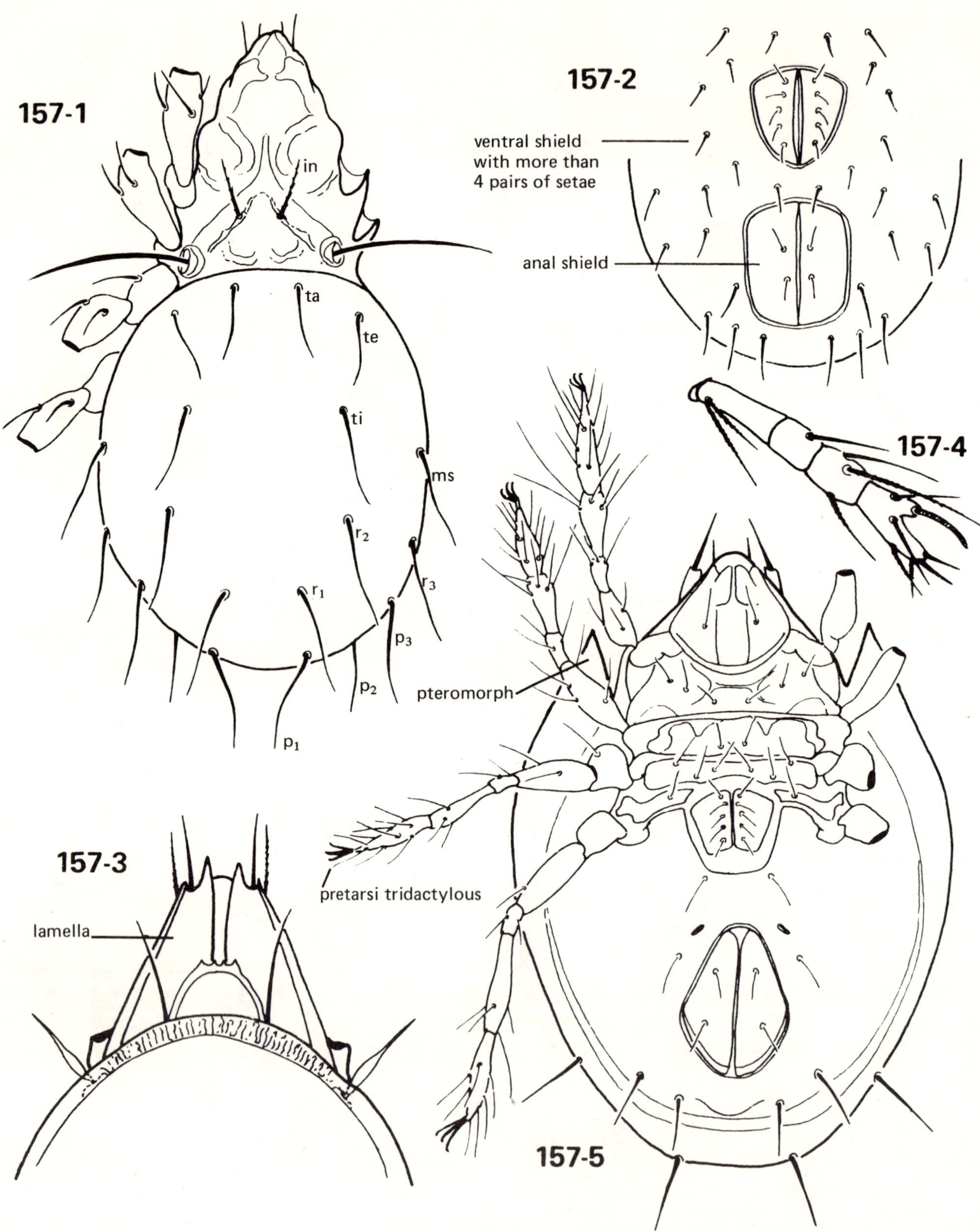

157-1 and 157-2; superfamily EREMULOIDEA, *Eremobelba flaggelaris* Jacot (Connecticut, USA). **157-1**; dorsum: **157-2**; genital-anal region

157-3 to 157-5; superfamily LIACAROIDEA. **157-3**; *Liacarus latus* (Oregon, USA), anterodorsal aspect: **157-4**; *Cultoribula* sp. (Oregon, USA), palp: **157-5**; *Hafenrefferia* sp. (Oregon, USA), venter

PLATE 158

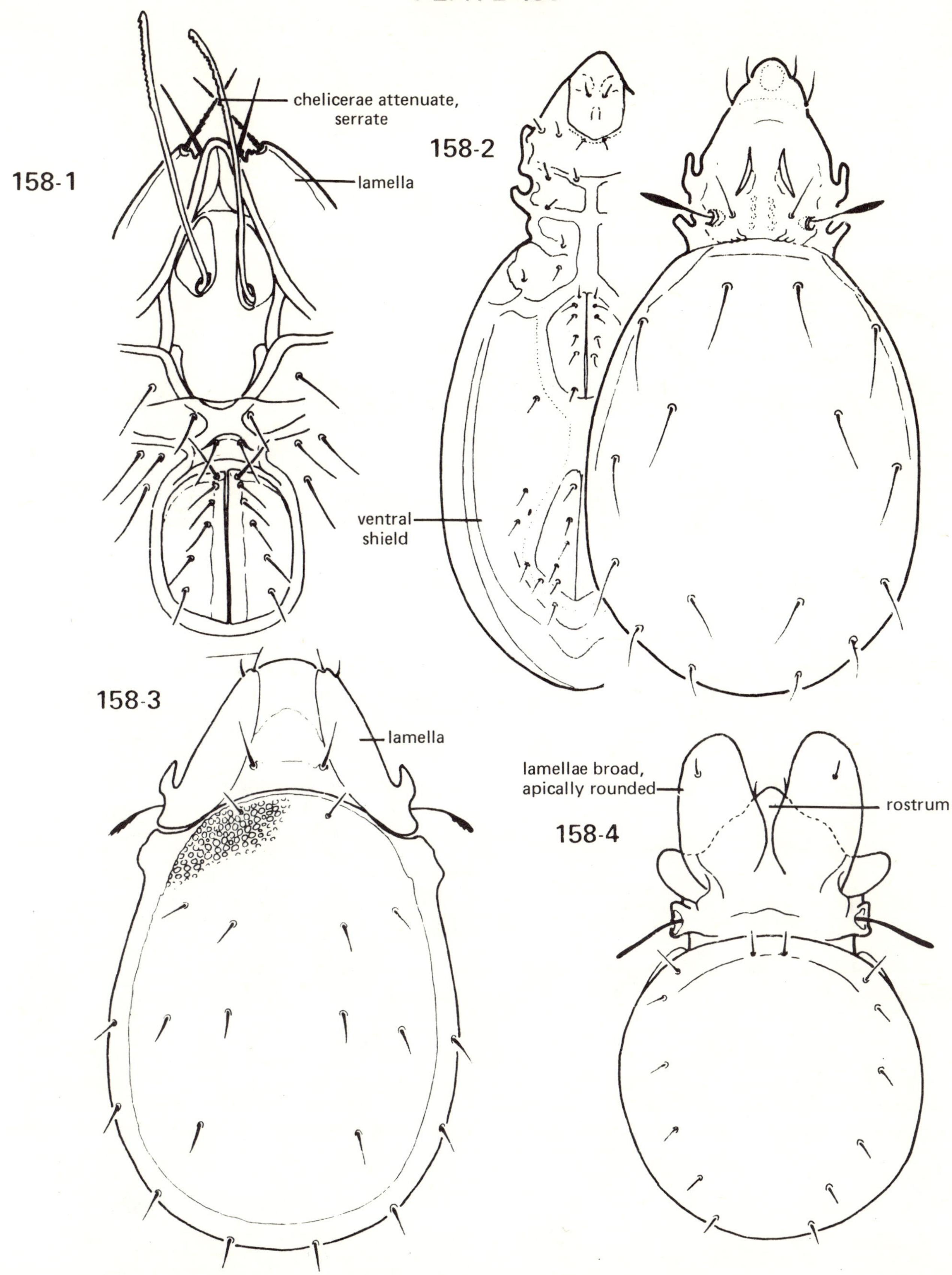

158-1; superfamily LIACAROIDEA, *Gustavia* sp. (Oregon, USA), anteroventral aspect

158-2; superfamily EREMAEOIDEA, *Eremaeus volkanovi* Kunst, ventral and dorsal aspects (after Balogh 1972)

158-3; superfamily CARABODOIDEA, *Carabodes marginatus* (Michael), dorsum (after Balough 1965)

158-4; superfamily POLYPTEROZETOIDEA, *Polypterozetes cherubin* Berlese, dorsum (after Balough 1965)

PLATE 159

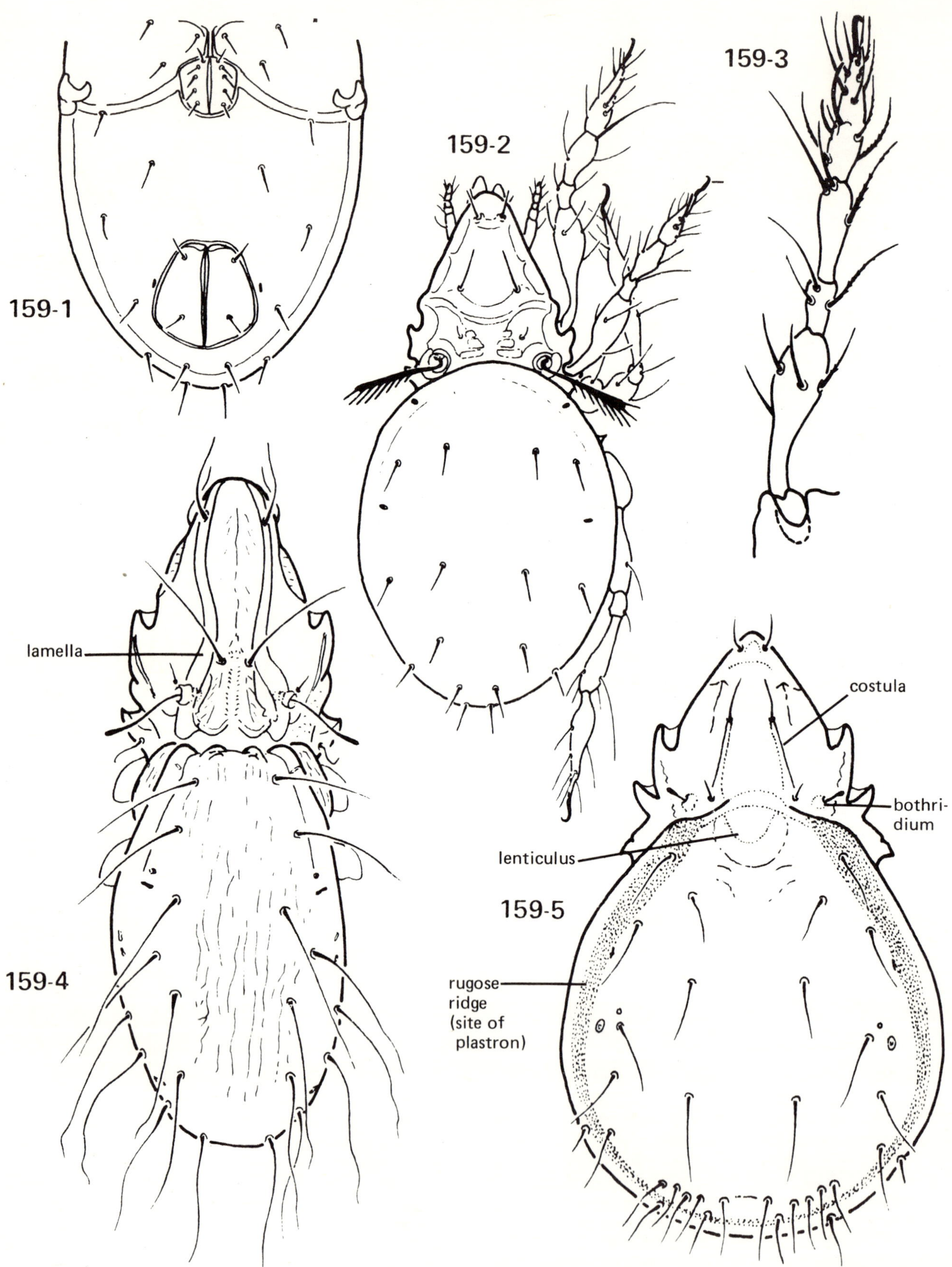

159-1 to 159-3; superfamily OPPIOIDEA. **159-1 and 159-2**; *Multioppia* sp. (Oregon, USA). **159-1**; posteroventral region: **159-2**; dorsum: **159-3**; *Suctobelba obtusa* **Jacot (Connecticut, USA), leg I**

159-4; superfamily OTOCEPHEOIDEA, *Neotrichocepheus tongaensis* Hammer (Tonga Islands), dorsum (after Hammer 1973)

159-5; superfamily HYDROZETOIDEA, *Hydrozetes parisiensis* Grandjean, dorsum (after Balogh 1972)

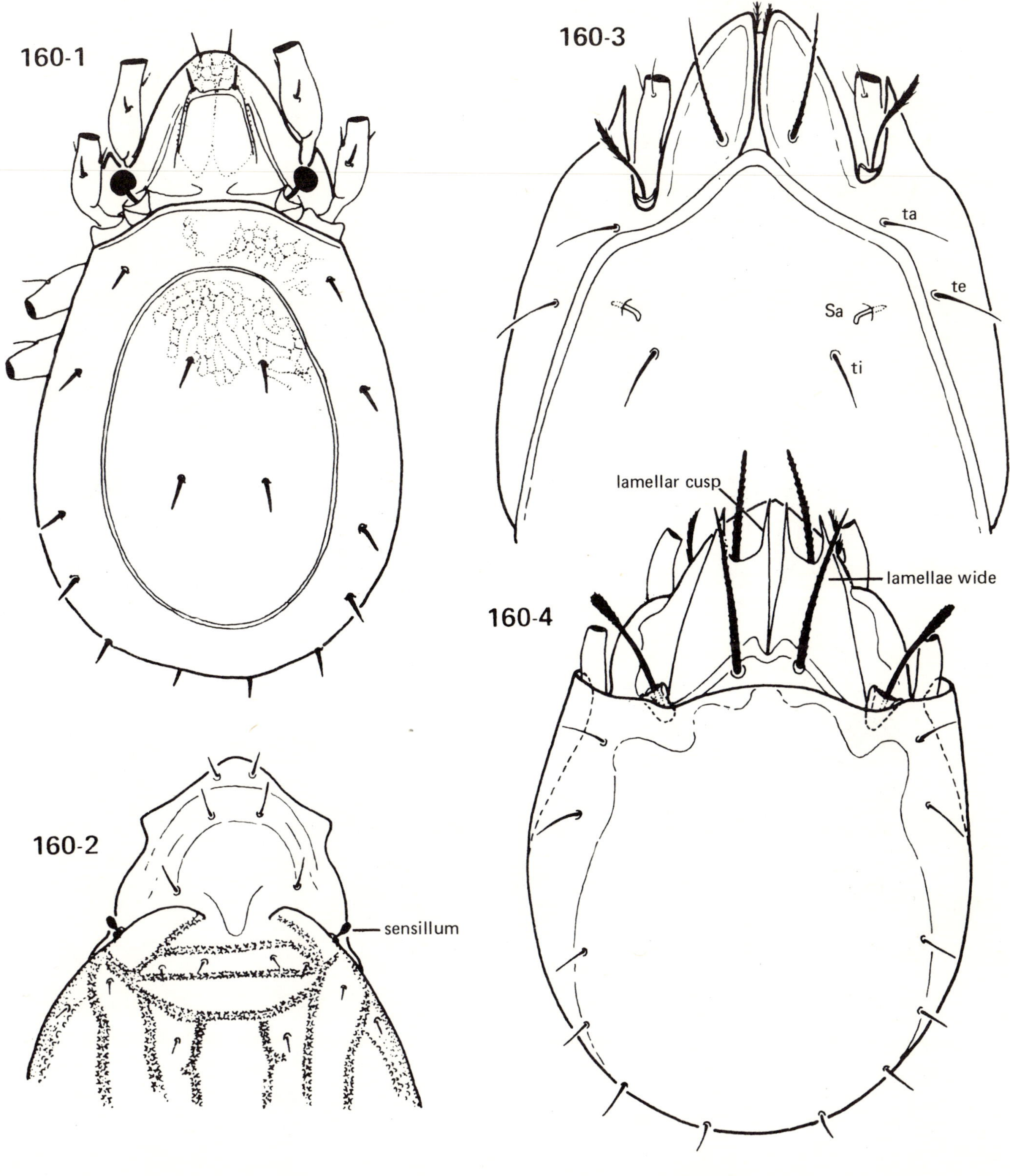

160-1; superfamily CYMBAEREMAEOIDEA, *Scapheremaeus palustris* Sellnick (Oregon, USA), dorsum

160-2; superfamily AMERONOTHROIDEA, *Ameronothrus lineatus* (Thorell), anterodorsal region

160-3 and 160-4; superfamily ORIBATELLOIDEA. **160-3**; *? Achipteria* (Oregon, USA), anterodorsal region: **160-4**; *Oribatella* sp. (Oregon, USA), dorsum

PLATE 161

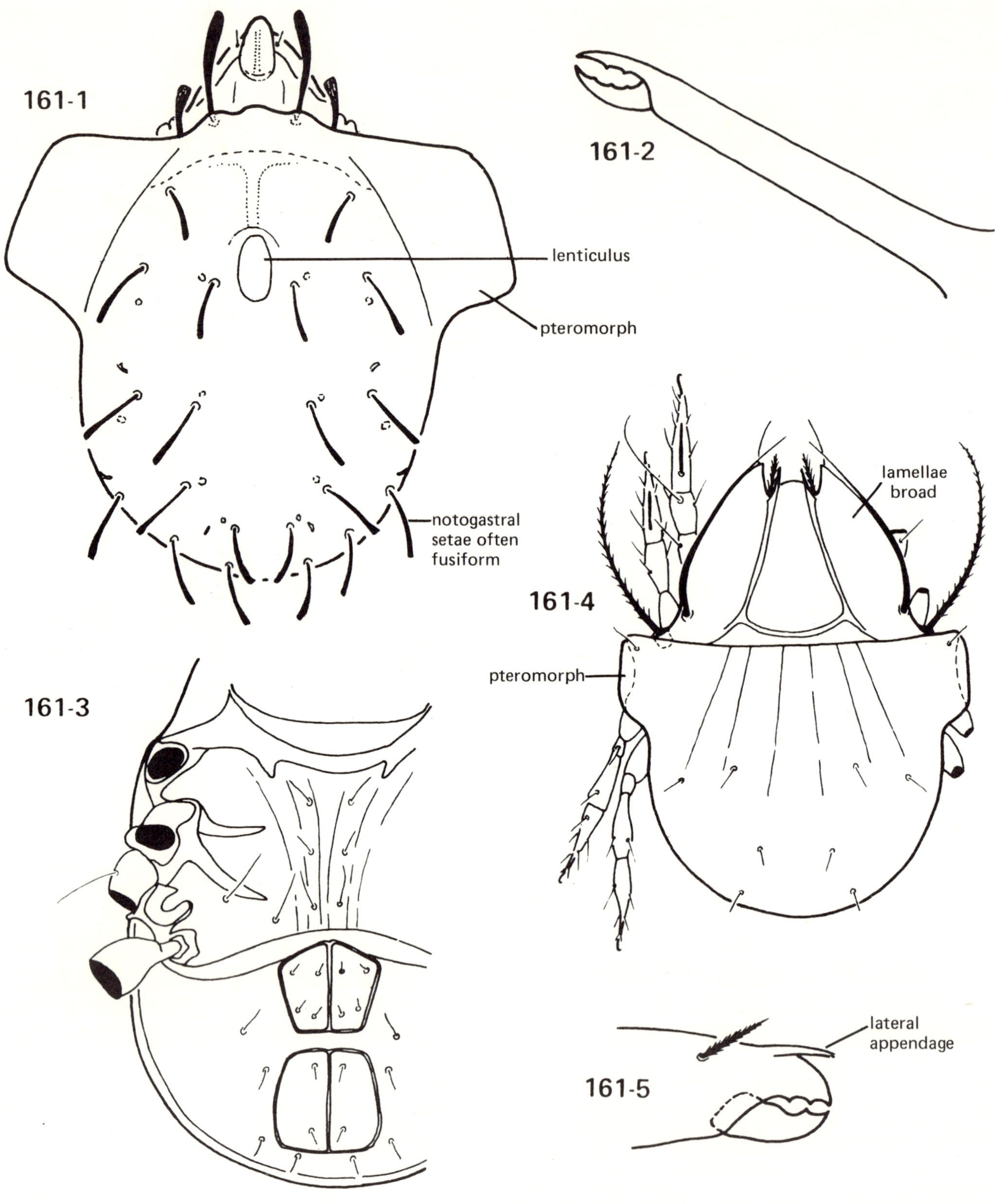

161-1 and 161-2; superfamily PELOPOIDEA. **161-1;** *Eupelops acromios* (Hermann), dorsum (after Balogh 1972): **161-2;** *Eupelops* sp., chelicera (after Baker et al. 1958)

161-3 to 161-5; superfamily MICROZETOIDEA, *Microzetes* sp. (after Baker et al. 1958). **161-3;** venter: **161-4;** dorsum: **161-5;** chelicera

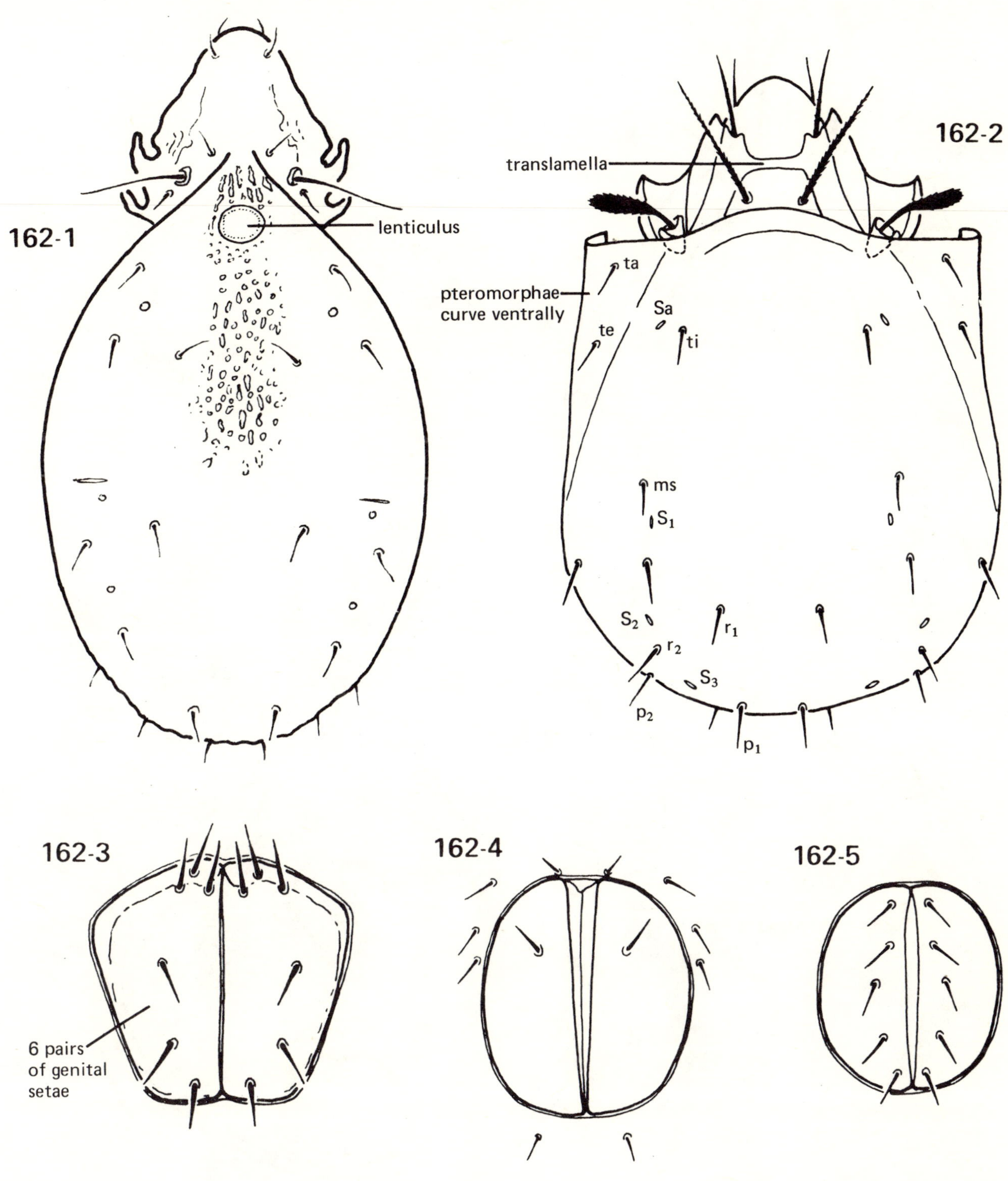

162-1; superfamily PASSALOZETOIDEA, *Passalozetes perforatus* Strenzke, dorsum (after Balogh 1972)

162-2 and 162-3; superfamily CERATOZETOIDEA. **162-2;** *Trichoribates* sp., dorsum: **162-3;** ceratozetoid genital shields

162-4 and 162-5; superfamily ORIBATULOIDEA. **162-4;** *Pirnodus detectidens* Grandjean, genital shields (after Balogh 1961): **162-5;** *Pilobates pilosellus* (Balogh), genital shields (after Balogh 1965)

PLATE 163

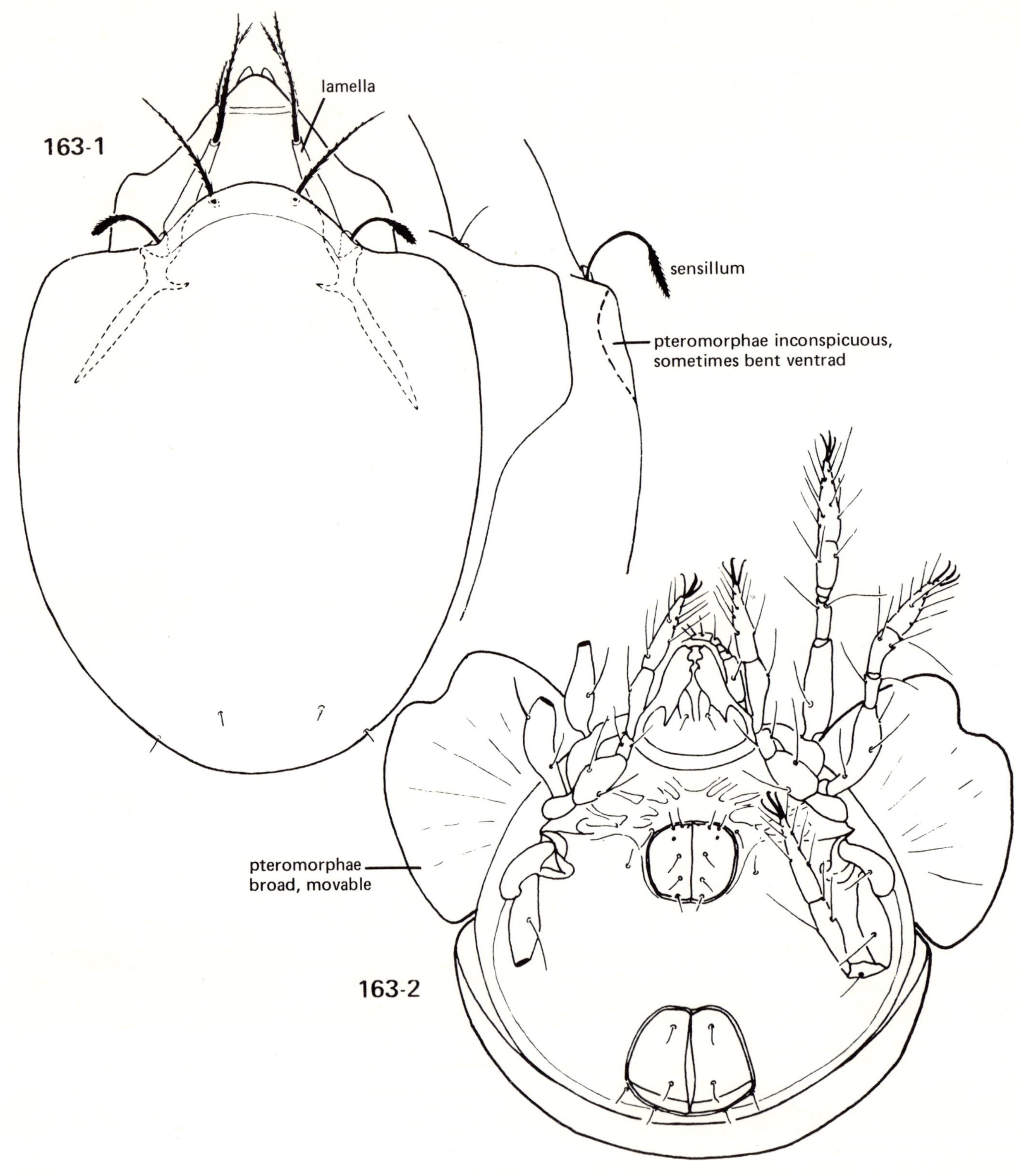

163-1; superfamily ORIBATULOIDEA, *Euscheloribates* sp. (Colorado, USA), dorsum, with pteromorphae of (from left) *Cantharozetes lucens* Hammer (after Balogh 1965) and *Xylobates* sp. illustrated for comparison

163-2; superfamily GALUMNOIDEA, *Galumna* sp. (Oregon, USA), venter

INDEX
(Numbers in bold face refer to illustrations)

A

H

I

J

K

L

M

N

O

P

R

S

T

U

V

W-X-Y-Z